fundamentals
of physics

fundamentals of physics

fifth edition

henry semat
professor emeritus
the city college of the
city university of new york

philip baumel
the city college of the
city university of new york

holt, rinehart and winston, inc.

new york   chicago   san francisco   atlanta   dallas
montreal   toronto   london   sydney

Library of Congress Cataloging in Publication Data

Semat, Henry, 1900—
    Fundamentals of physics.

    1. Physics. I. Baumel, Philip, joint author.
II. Title.
QC23.S45   1974   530   73–19985
ISBN 0–03–084747–8

Printed in the United States of America
            4 5 6 7   0 7 1   9 8 7 6 5 4 3 2

to my daughters edith and barbara
henry semat

to my wife sylvia
philip baumel

preface

The general aim of this book is to present a thorough one-year introductory course in physics, without the use of the calculus. This fifth edition has modernized its point of view by the inclusion at an early stage, of the fundamental forces (Chapter 5) and of special relativity (Chapters 11 and 12). The ideas of quantum mechanics are mentioned at appropriate places early in the text and are treated more thoroughly in the last section of the book. General relativity which until recently was on the periphery of physics but is now an integral part of the subject is presented in an elementary way in Chapter 35. Magnetism is introduced as an interaction between moving charges, and statistical physics is given a much more extensive treatment in this edition.

The size of this edition has been reduced from that of the fourth edition partly by the elimination of some topics. This has provided ample space for the newer material mentioned above. Some engineering applications have been removed and a few biological topics have been introduced. The number of problems has been increased. The problems have been graded in difficulty in numerical order at the end of each chapter. Answers to odd-numbered problems are listed in the Appendix.

A list of questions suitable for class discussion or for possible student projects appears at the end of each chapter. Practically all of the figures have been redrawn for this edition and new ones added.

We have been favored and helped by comments, criticisms, and suggestions from many teachers, many of whom have used the earlier editions and have also reviewed the original plans for this edition. We greatly appreciate their help and wish to thank them for it. Parts of the manuscript were read by Virgil E. Barnes, II, Curtis Callan, David K. Campbell, Gary R. Gruber, and Joseph H. Pifer. We are also indebted to our colleagues, Michael Arons for his help with the particle physics chapter, and Daniel Greenberger for his help with the general relativity chapter. Some of the improvements in this edition have their origin in many illuminating conversations with our colleagues Martin Tiersten and Harry Soodak. One of us, (P.B.), is also indebted to his brother, Abraham Baumel, for many discussions of physics and teaching.

Books of this type would probably never be completed without the continuous encouragement, help, and intellectual contributions of our respective wives. Ray K. Semat did all the typing for the first four editions and Sylvia Baumel took over this task for the fifth edition. Many thanks to both of them.

Henry Semat
Philip Baumel
February 1974

contents

part 1   Mechanics

1	Fundamental concepts	3
2	Motion of a particle	12
3	Force & motion	35
4	Statics	51
5	The fundamental forces	68
6	Work & energy	80
7	Circular motion	97
8	Impulse & momentum	112
9	Rotational motion	120
10	Relativity of length, time, & velocity	136
11	Relativistic dynamics	151
12	Harmonic motion	159
13	Fluids	173
14	Elasticity & surface behavior	195

part 2   Heat

15	Temperature	213
16	The first law of thermodynamics	224
17	Change of phase	237
18	The second law of thermodynamics	251
19	Statistical physics	264
20	Transfer of heat	280

part 3   Wave motion & sound

21	Wave motion	291
22	Vibrations & sound	307

part 4   Electricity & magnetism

23	Electrostatics	321
24	Current electricity	344
25	Magnetism	360
26	Electromagnetism	376
27	Alternating currents	388

28 Electronics   403

part 5   Light

29 Geometric optics   413
30 Electromagnetic waves   445
31 Interference & diffraction of
    electromagnetic waves   457

part 6   Modern physics

32 Atomic physics & quantum mechanics   477
33 Nuclear physics   507
34 Particle physics   531
35 The general theory of relativity   543

appendices

A   Tables   557
B   The greek alphabet   571
C   Answers to odd-numbered problems   573

    INDEX   581

mechanics

part 1

fundamental concepts

chapter 1

1-1
Scope of physics

Physics is a fundamental science dealing with matter and energy. Although the history of physics dates back to antiquity, the modern science of physics may be said to begin with the work of Galileo (1564–1642), who developed the method of dealing with physical problems which is essentially that used today. A considerable body of knowledge had been accumulated before Galileo's time, particularly in mechanics and to a lesser extent in optics, but very little was known about the subjects of heat, sound, magnetism, and electricity. The method introduced into physics by Galileo consists principally of three steps. The first is to use previous knowledge and experience to delimit the new problem at hand—that is, to exclude from consideration all but the few relevant factors. The second is to perform controlled quantitative experiments on these few relevant factors. Thus in his classic researches on the laws of falling bodies, Galileo paid no attention at all to the size, or shape, or color, or weight of the bodies upon which he experimented. But he did perform many careful and quantitative tests on the relationships among those factors that he considered relevant—namely, the time of fall, the vertical height of fall, and the speed of fall. The third step is the most difficult of all and is also the most difficult one to describe; it is one in which imagination plays an important role and in modern physics is very often done by a person other than the experimenter. This step may consist of an attempt to put the results of the experiments in such a form that they will fall within the compass of some known generalization or physical law. Or, if this is not possible, the scientist may attempt to present some hypothesis or some generalization, preferably of a quantitative nature, as an outgrowth of these experiments. Such an hypothesis or generalization will usually suggest some new experiments to be tried. The results of these new experiments may be found to fit in with the new ideas, they may suggest slight modifications of these ideas, or they may even show that the hypothesis or generalization is incorrect for any one of a number of reasons such as insufficient data, lack of good quantitative data, or the neglect of factors that should not have been ignored.

The results of such experiments may be presented in different ways; for example, in the form of a *table* listing the results of the measurements, or in the form of a *graph* showing the relationship between two or more measured quantities, or in the form of an *equation* relating these quantities. Considerations of utility and elegance determine which method or methods are to be used in any particular case. All of these methods will be illustrated at appropriate places throughout this book.

For convenience in studying it, the science of physics is subdivided into several branches; the traditional subdivisions are mechanics, heat, sound, magnetism, electricity, and light. To these traditional subdivisions, we must now add those on atomic and nuclear physics. In each of these subdivisions, a different aspect of the behavior of matter is studied. For example, the behavior of bodies that are acted upon by external forces is studied in the subdivision of mechanics. We shall find that under some conditions a body subject to external forces remains at rest, whereas under different conditions the body may be set into motion, or if it is in motion, its motion may be changed by the action of external forces. We shall also find that bodies possess *energy* by virtue of their motion, or position, or state. In the subdivision of heat, we shall be concerned mainly with the relationship between the physical properties of a substance and its temperature, and those changes that occur when energy in the form of heat is added to or removed from the substance. Each of the subdivisions of physics requires a special technique of experiment for measuring those quantities that are needed for expressing the behavior of matter in precise and exact forms.

Although this division of physics into several branches is useful in studying the subject, we shall find that these subdivisions are closely related to

one another in the sense that principles or laws developed in one are generally carried over for use in the other branches. Furthermore, there is one unifying principle connecting all of the subdivisions of physics; this principle, known as the *principle of conservation of energy*, has stood the test of experience for more than a century.

1-2
Relationship of physics to other sciences

The methods and results of physics have had wide applications to the other sciences, particularly chemistry, biology, geology, psychology, and medicine. The engineering sciences are frequently considered as special fields of applied physics. The development of physics has played an important part in the progress of the other sciences, but it has not been entirely a one-sided relationship. The advances in the other sciences have proved of great value and importance to physics in applying new ideas, in developing new tools for investigation, and in opening new fields of study. Some of these fields are so closely related to physics that they are given joint names such as physical chemistry, biophysics, chemical physics, geophysics, and astrophysics.

Until recently the study of the science of physics has been confined to a small group of men and women. The increasing recognition of its fundamental position in relation to the other sciences and to technology, as well as the satisfaction of one's intellectual curiosity concerning nature, has attracted an increasing number of persons to the study of physics. It is the aim of this book to enable such persons to obtain a good foundation in the fundamentals of physics.

1-3
Fundamental concepts

A large number of concepts are in use in physics and many more will be introduced as new areas of the science are explored. In physics, as in all subjects that attempt to be logical, all concepts must be defined. Usually a concept is defined by describing it in terms of other concepts, and in physics the description is usually in the form of an equation. For example the concept of *average speed* is defined as the *distance traveled* divided by the *time elapsed*. If all concepts were defined in terms of other concepts, it is obvious that sooner or later the process would become cyclic. This cyclic definition problem is one most of us have run into in using a dictionary. For example, if we look up the word "net," as in "tennis net" in a good dictionary we almost always find that the word "mesh" is used in the definition. The word "mesh" is almost always defined using the word "net." There are two ways out of this cyclic dilemma: either leave some concepts undefined (the method invoked by mathematics), where such concepts are called *primitive indefinables*, or define some concepts in another way. If dictionaries attempted to be logical rather than descriptive, they might present a page or two of undefined words or they might attempt to define some words by means of pictures. In physics a small number of concepts is defined, not in terms of other concepts but by means of an *operational definition*. An operational definition of a concept consists of a description of a procedure by which the concept may be *measured*. Thus in order to define length operationally one must give a set of rules that describe how to measure the length of something. This kind of definition does not say what length *is*; it does not say what you mean by length; it merely tells you how to measure it. These operationally defined concepts are usually called the *fundamental concepts* of physics, and all the other concepts are called the derived concepts.

Until fairly recently most physicists thought that it was necessary to have five operationally defined concepts. To some extent the choice of which particular five is arbitrary but usually *length* and *time* were chosen to describe space, time, and motion; *mass* to describe inertia and the source of the gravitational force; *electric charge* to describe the source of the electric and magnetic forces; and *temperature* to describe the properties of matter in bulk. With the recent discovery of the nuclear forces it is clear that two or three more fundamental quantities will have to be defined operationally, but since these forces are so poorly understood no agreement has been reached on which quantities will be so defined.

The process of operational definition involves two major parts: the description of a *standard*, such as 1 meter or 1 kilogram, and the description of a process for comparing the unknown to the standard. In order for this process to be operable for any unknown it is necessary to describe a procedure for setting up multiples and fractions of the standard. These will be discussed for each of the fundamental concepts at the appropriate place in this book. In addition to the conventional length, mass, time, charge, and temperature, this book will also treat *force* as a fundamental quantity, primarily for clarity of presentation, but also in part because some physicists believe that force should be thought of as fundamental rather than derived.

1-4
Concept of length, standard and units

Most of us are familiar, in a general way, with the operations involved in measuring the length of an object, or in measuring the distance between two points. Such a measurement always involves the use of some measuring rod or tape whose length is known in terms of some *standard* of length. The legal standard of length in the United States is the *meter*. From 1889 until 1960 this international standard of length was defined as the distance between the centers of two lines marked on a special platinum-iridium bar, known as the *standard meter*, when the bar is at 0°C. This standard meter is deposited at the International Bureau of Weights and Measures at Sèvres, France. Accurate copies of this standard meter, known as prototype meters, were distributed to various national physical laboratories such as the National Bureau of Standards in Washington. Additional working standards were constructed for use in the normal operations of the National Bureau of Standards. It is to be expected that their lengths may change with the passage of time, so it is essential to recalibrate them periodically against the prototype meter, and on rare occasions, against the international standard meter in France. It has long been felt desirable to have an easily reproducible standard of length that would not change with time. By the end of the nineteenth century, instruments had been developed with which it was possible to measure the wavelength of light with very great accuracy (see Chapter 31). One of these instruments was the interferometer developed by Albert A. Michelson. He was the first one to make a direct determination of the wavelength of a particular spectral line (the red line of cadmium) in terms of the standard meter. With further improvements in measuring instruments, in sources of light, and in techniques of measurement, it has become possible to define the standard meter in terms of the wavelength of a particular spectral line. The new international standard of length, adopted October 14, 1960, defines the meter as equal to the length of

1,650,763.73 wavelengths

of the orange-red line of the isotope of krypton of mass number 86.

It is now possible for any well-equipped physics laboratory to have its own standard of length for calibrating any linear measurement in terms of the meter.

The meter not only is the standard of length but is coming into increasing use as a *unit* of length in science and engineering. Another *unit of length* used in scientific work is the hundredth part of the meter and is called the *centimeter* (cm); that is,

$$1 \text{ centimeter} = \frac{1}{100} \text{ meter}$$

or

$$1 \text{ meter} = 100 \text{ centimeters}$$

The subdivision of the meter into 100 equal divisions can be made with a dividing engine, which is essentially an accurately made screw that advances through known equal distances for each rotation of the screw.

In English-speaking countries, other units of length are used in everyday life and in many engineering applications. In the United States, the *yard* is defined legally in terms of the standard meter by the relation

$$1 \text{ yard} = 0.9144 \text{ meter}$$

In engineering work, the *unit* of length is the *foot*, which is one third of a yard.

Other subdivisions and multiples of the meter and yard are in common use, and the English-speaking scientist or engineer frequently has to convert measurements and data from one set of units to another. Table 1-1 gives some of the more common units of length used, and Table 1-2 gives the conversion factors for these units. Figure 1-1 shows the relationship between the centimeter and the inch.

In actual measurements many different instruments are used, the type depending upon the magnitude of the length to be measured and the degree of accuracy required. Familiarity with such instruments is essential to a proper appreciation of the problems involved in the measurement of length. This can be gained only by work in the laboratory and in the field.

Illustrative example

An important method for measuring distances is known as the *triangulation* method which makes use of the known properties of a plane triangle. For example, the corner C of a city lot is sighted from two points A and B, as in Figure 1-2, by means of a transit. This instrument is essentially a telescope with vertical and horizontal scales for determining its line of sight. A and B are 100 ft apart, the line of sight AC makes an angle of 90° with AB, and the line of sight BC makes an angle of 75° with AB. Determine the distance AC.

One method of solving this problem is to draw the figure to some convenient scale, say 100 ft = 1 in., using ruler and protractor, and then to measure the distance AC. When so measured, the line AC will be found to be about $3\frac{3}{4}$ in., so the distance $AC = 375$ ft.

A second and more convenient method is to make use of some known trigonometric relationships. In any right triangle such as ABC, if A is the right angle, then the following relationships hold for the angle B: the *sine of B*, written sin B, is the ratio of the opposite side to the hypotenuse, or

$$\sin B = \frac{AC}{BC} \tag{1-1}$$

the *cosine of B*, written cos B, is the ratio of the adjacent side to the hypotenuse, or

$$\cos B = \frac{AB}{BC} \tag{1-2}$$

and the *tangent of B*, written tan B, is the ratio of the opposite side to the adjacent side, or

$$\tan B = \frac{AC}{AB} \tag{1-3}$$

Tables of values for these trigonometric functions are given in Appendix A.

In this problem, since the distance AC is required, the distance AB is known, and the angle B

Table 1-1 Units of length

English		metric	
3 feet	= 1 yard	100 centimeters	= 1 meter
36 inches	= 1 yard	10 millimeters	= 1 centimeter
5280 feet	= 1 mile	1000 meters	= 1 kilometer

Table 1-2 Conversion factors

1 kilometer =	0.6214 mile
1 meter	= 39.37 inches
1 yard	= 0.9144 meter
1 foot	= 30.48 centimeters
1 inch	= 2.540 centimeters

is given, we can use Equation 1-3 for determining AC; that is,

$$\tan 75° = \frac{AC}{AB}$$

from which

$$AC = AB \tan 75°$$

From the table in Appendix A, we find that

$$\tan 75° = 3.732$$

and, since

$$AB = 100 \text{ ft}$$

we get

$$AC = 100 \text{ ft} \times 3.732$$

so that

$$AC = 373.2 \text{ ft}$$

1-5
Concept of time, standard and units

All of us have some notions regarding the concept of time from our experiences with annual events, seasonal occurrences, and daily routine. The uses of watches, clocks, and radio time signals have become part of our everyday life. The accurate measurement of time has always been the special province of astronomers. Their measurement and recording of astronomical events form the basis for the determination of time. Among these events is the rotation of the earth on its axis measured with respect to the sun; the period of rotation is *one solar day*. The solar day is the time elapsed between two successive noons; noon is the instant at which the sun is observed to pass the meridian

FIGURE 1-2 Measuring distance by triangulation.

at the place of observation. The length of the solar day varies slightly throughout a year and an average over a year is a *mean solar day*. The *year* is the time it takes the earth to complete one revolution around the sun. Until recently, the mean solar day was the standard time interval. In 1960 the standard of time adopted by the eleventh General Conference on Weights and Measures was the tropical year 1900. The unit of time was then defined as 1 second = 1/31,565,925.9747 of the tropical year of 1900.

This astronomical standard of time was replaced in 1967 with a standard based on the oscillations of a particular atom. The second is now defined so that a particular frequency emitted or absorbed by cesium-133 oscillates 9,192,631,770 times in 1 second.

The instrument for measuring time intervals is the clock. The timing mechanism of a clock is some type of periodic phenomenon such as the motion of a pendulum in the gravitational field of the earth (Chapter 12), or the motion of a balance wheel of a watch or clock, or the spinning of a rotor of a synchronous motor, or the oscillations of a quartz crystal when placed in an appropriate electric field, or the electromagnetic vibration emitted by atoms when undergoing energy changes from one state to another. A pendulum clock had long been in use as an astronomical clock; within recent years, several types of atomic clocks have been developed that have much greater accuracy.

The present master clock, which determines Standard Time for the United States, utilizes one particular vibration emitted by cesium atoms, when properly excited, to control the rate of the clock system. This master clock is in the Simon New-

FIGURE 1-1 Comparison of the inch and centimeter.

FIGURE 1-3 The U.S. Naval Observatory Master Clock System. (Official U.S. Navy Photograph.)

comb Laboratory of the United States Naval Observatory in Washington, D.C. A photograph of the master clock is shown in Figure 1-3. The clock shown is, in effect, the display panel of a complex computer-controlled system that continuously records data from 25 to 30 separate oscillators located at the Naval Observatory. The computer selects the 16 most reliable oscillators and from the average of these derives the master clock time scale. Additional data are provided to the system by 30 timekeeping and monitoring stations at remote locations and by three astronomical systems. The resulting time scale is constant to better than 10^{-7} sec/day.

1-6
Concept of mass, standard and units

Most of us have some idea of the meaning of the term *mass of a body*, but unfortunately this is generally confused with the term *weight*. Since a real understanding of the concept of mass can come only after a study of the motion of bodies, for the present we shall confine our discussion to a statement of the standard of mass, the method of measuring mass, and the units commonly used.

The standard of mass is the kilogram, which is the mass of a certain piece of platinum kept at the International Bureau of Weights and Measures at Sèvres, France. Two accurate copies of the standard kilogram are deposited at the National Bureau

of Standards. The instrument used in *comparing* masses is the delicate balance, which is an equal-arm balance (see Figure 1-4) although most laboratories use single pan balances to *measure* masses. The two bodies whose masses are to be compared are placed in the pans of this instrument and, if a balance is obtained, are said to have equal masses.

The kilogram not only is the standard of mass but is coming into increasing use as a *unit* of mass in science and engineering. The system of units based upon the meter as the unit of length, the kilogram as the unit of mass, and the second as the unit of time is called the mks system of units.

Another unit of mass used in scientific work is the *gram*; the gram is 1/1000 of a kilogram. The system of units based upon the centimeter as the unit of length, the gram as the unit of mass, and the second as the unit of time is called the cgs system of units. Since the equality of masses can be determined by means of the delicate balance, it is a comparatively simple matter to construct multiples and subdivisions of the standard kilogram.

The unit of mass often used in English-speaking countries is the pound mass. In the United States the *pound mass* is defined in terms of the standard kilogram as given below

$$1 \text{ pound mass} = 0.4535924277 \text{ kg}$$

To a good approximation, a mass of one pound is equal to 453.6 grams. Stated another way:

$$1 \text{ kg} = 2.2046 \text{ pounds mass}$$

FIGURE 1-4 A delicate balance. (Courtesy of Central Scientific Company.)

The system of units based upon the foot as the unit of length, the pound as the unit of mass, and the second as the unit of time, is called the fps system of units.

1-7
Notation in powers of ten

In expressing the results of a measurement, scientists generally use a number of digits indicative of the accuracy of the measurement. For example, the distance between the two points A and B shown in Figure 1-5 may be given as 12.43 cm when measured with a meter stick whose smallest ruled intervals are millimeters. The last digit in the above measurement is the result of a guess or an estimate of the additional distance that the point B is beyond the fourth millimeter mark in the region between 12 and 13 cm. This result may be interpreted by saying that the distance AB is certainly greater

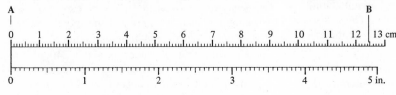

FIGURE 1-5 Estimate of last digit in measurement of distance AB is 12.43 cm.

than 12.4 cm and less than 12.5 cm; an informed guess is that it is 12.43 cm. It is possible to design instruments to improve the accuracy of the measurement, but with the instrument actually used, the last digit of the number 12.43 cm is in doubt. It is possible to improve the accuracy of the result by making several measurements using different parts of the meter stick and then averaging the results.

It may be desirable for some reason to express the distance AB in meters to the same degree of accuracy as the above measurement, in which case it would be written as 0.1243 m, or if one preferred, as 12.43×10^{-2} m, using the notation of powers of ten.

The value of the notation in powers of ten becomes more impressive when it is necessary to present the results of measurements of very small quantities or of very large quantities. For example, the distance between two neighboring atoms in a crystal of salt is

$$0.00000002814 \text{ cm}$$

This distance, when expressed in powers of ten, becomes

$$2.814 \times 10^{-8} \text{ cm}$$

Another choice available to the physicist is to change the size of the unit; this choice is frequently made. For example, when dealing with atomic dimensions, a unit of distance frequently used is the *angstrom*, which is

$$1 \text{ angstrom} = 10^{-8} \text{ cm} = 1 \text{ Å}$$

Thus the distance between two neighboring atoms in a crystal of sodium chloride can be written as 2.814 Å.

Questions

1. Consult an encyclopedia or other work of reference for the history of the origin of the standard meter.

2. Until fairly recently it was necessary for a seagoing ship to carry at least one accurate clock in order to determine the location of the ship precisely. Why? (It is interesting to note that such an instrument was not invented until well after the Age of Exploration. The history of the development of the ship's chronometer is an interesting one.)

3. Astronomers use the sidereal day for measuring time. The sidereal day is the interval be-

As an example of a large number, we may mention the Avogadro number which is the number of molecules in a gram-molecular weight of a substance; its present value is

$$6.022 \times 10^{23} \text{ molecules/gm-molecular wt}$$

The following tables of powers of ten are presented for convenience. For numbers equal to or greater than 1:

$$10^0 = 1$$
$$10^1 = 10$$
$$10^2 = 100$$
$$10^3 = 1000$$
$$10^4 = 10,000$$
$$10^5 = 100,000$$
$$10^6 = 1,000,000$$
and so on

For numbers less than 1:

$$10^{-1} = 0.1$$
$$10^{-2} = 0.01$$
$$10^{-3} = 0.001$$
$$10^{-4} = 0.0001$$
$$10^{-5} = 0.00001$$
$$10^{-6} = 0.000001$$
and so on

Some additional examples of the use of the notation in powers of ten:

The average distance between the sun and the earth is 93,000,000 mi or 93×10^6 mi.

The mass of the earth is

$$M = 5.98 \times 10^{27} \text{ gm}$$

The mass of an electron is

$$m = 9.1096 \times 10^{-28} \text{ gm}$$

tween two successive passages of a star across the meridian. Take into account the rotation of the earth and show that the sidereal day is shorter than the solar day. How many sidereal days are there in a year?

4. Give an operational definition of a scientific concept outside of physics.

5. Suggest the steps that must be taken to measure the distance between the moon and the earth by the triangulation method.

6. Knowing that the moon is about 240,000 mi away from the earth, estimate the diameter of the moon from the appearance of the moon.

7. Suggest a method based on the rules of plane geometry for subdividing a line of fixed length into 100 equal divisions.

8. Observe the clocks in your home and school. Can you determine (a) the mechanism used for obtaining periodic motion and (b) the mechanism used for running the clock?

9. How does experiment differ from mere observation? (Note that many very important experiments have consisted of observation unaccompanied by equipment.)

10. How does a scientific law differ from a tabulation of the experimental results on which it is based?

Problems

1. Determine your height in meters and your mass in kilograms.

2. Determine the number of seconds in a year.

3. English-speaking countries have used the liquid ounce as a measure of volume because that is the volume occupied by an amount of water which weighs approximately 1 ounce. A cubic centimeter of water has a mass of 1 gram. From these find the relation between 1 quart and 1 liter (10^3 cm^3).

4. Express your age in seconds.

5. Look up the following numbers and express in power-of-ten notation (a) the population of the United States, (b) the national debt, (c) the retail cost of steak in dollars per ton, (d) the diameter of a red blood cell in meters.

6. The Empire State Building is 1245 ft tall and is observed by a man 4 mi away. What is the angle between the ground and the line of sight to the top of the building?

7. A ship at sea receives radio signals from two transmitters A and B, which are 180 mi apart, one due south of the other. The direction finder shows that transmitter A is 30° south of east, while the other is due east. How far is the ship from each transmitter?

8. A mountain peak is sighted from a point 15 km away. The line of sight makes an angle of 10° with the horizontal. Determine the height of the mountain peak in meters.

9. A radio antenna is sighted from a distance of 600 ft. The top of the antenna makes an angle of 20° with the ground. (a) How high is the antenna? (b) If the distance measurement can be in error by as much as 1 ft but no more than that, and the angle measurements are very accurate, what is the maximum error in the antenna's height?

10. The width of a river is measured by marking off a distance of 250 ft parallel to the river on one of its banks. A point C is sighted on the opposite shore directly across from A and along a line from B which makes an angle of 50° with AB. Determine the width of the river.

11. Two tracking stations A and B, 1500 mi apart, sight an artificial satellite. Station A reports the position of the satellite as 37° above the line joining A and B, and B reports it to be at an angle of 53° above this line at the same time. (a) Determine the height of the satellite above the surface of the earth at this time. (b) What additional information is needed to determine its position with respect to the center of the earth?

12. The speed of light, c, appears very often in the equations of physics as does the mass of an electron, m_e. Another such quantity is a number called Planck's constant, h, which has the dimensions of energy times time. The units of energy are mass times the square of speed, divided by the square of time. Many equations are written most conveniently by treating the quantities c, m_e, and h as standards. Show that any unit that can be expressed as a combination of mass, length, and time can be expressed as a combination of c, m_e, and h.

motion of a particle

chapter 2

2-1
Motion is relative

The motions of objects had been observed and studied for centuries, but it was not until the time of Galileo (1564–1642) and Newton (1642–1727) that the laws of motion were put on a firm scientific basis. In this chapter we shall limit our discussion to the motion of particles without reference to any forces that may be involved. This subject is called the kinematics of a particle. The motions of more complex systems as well as the effects of forces that may be acting on them are discussed in subsequent chapters.

In all discussions of motion there are at least two objects involved: the body whose motion is being studied, and a body with respect to which the motion is measured. When we say that a car is moving at 40 mi/hr, we imply that the motion is relative to the road or some other object fixed on the earth. A boat sailing on a river moves with respect to the banks of the river, but it also moves with respect to the water in the river, which in turn is moving with respect to the banks. An airplane moves with respect to the ground, and it also moves with respect to the air. A rocket ship moves with respect to the air, the ground, the moon, and the stars. Thus, a complete description of motion requires that the object with respect to which the motion is described be specified. This object is usually called the *frame of reference*. The frame of reference for most of the motions commonly discussed is the earth, or some feature on it such as a road or railroad tracks. In such cases the frame of reference is usually not stated explicitly, but is implied. Unless otherwise stated, in this chapter the frame of reference will be taken to be the earth.

2-2
Position, displacement, and distance

In a mathematical description of motion, the basic step is a description of the *position* of an object at a particular instant of *time*. This is generally done by giving the coordinates of the object; but before the coordinates can be given it is necessary to choose a coordinate system and a time axis. That is, we must choose a particular location to be the origin of coordinates and, in the case of Cartesian coordinates, we must choose particular directions to be the directions of the positive x, y, and z axes, as in Figure 2-1. Further, we must choose a particular instant when the time t equals zero.

When an object moves from position A to position B it is said to have been *displaced*. If we follow the motion of the object along the path that it traverses from A to B, we can measure the *distance* it traveled. We can, however, ignore the actual path and fix our attention on the *change in position*, which is called the *displacement* from A to B, as in Figure 2-2. That is, displacement is the net change in position from the initial position A to the final position B; it is represented by an arrow running from A to B. The displacement from A to B will thus be the same no matter what path is actually traversed by the particle.

The most significant difference between distance and displacement is that distance is expressed as a number in certain units while displacement is a number in those units plus a direction. Consider, for example an automobile trip from New York to Philadelphia. The distance for this trip is a certain number of miles, a number that could be obtained from the odometer of the automobile. To describe the displacement one must look at the

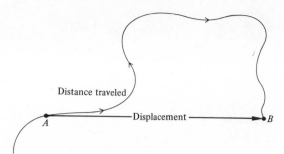

FIGURE 2-2 Distance and displacement.

initial point (New York) and the end point (Philadelphia) and describe the difference between these positions. The simplest way to do this is to draw a straight line from New York to Philadelphia on a map. This line is characterized by its length and its direction. Thus the distance from New York to Philadelphia is some number of miles, which depends on the particular roads chosen for the trip, while the displacement is a specific number of miles in a particular direction.

Consider now, two successive motions of an object. Suppose it moves from A to B and then moves from B to C (Figure 2-3). The distance traveled from A to C is the sum of the distances traveled from A to B and from B to C. The length of the displacement from A to C is not necessarily the sum of the length of the displacement from A to B and the length of the displacement from B to C. For example, consider a trip from Yale University in New Haven, Connecticut, to New York City, followed by a trip from New York to the Brookhaven National Laboratory on Long Island (see Figure 2-4). The first trip is about 75 miles and the second is about 100 miles, while a straight line distance between New Haven and Brookhaven is about 30 miles. However, from the point of view of displacement, the trip from Yale to New York to Brookhaven can be thought of as a single trip with a displacement from Yale to Brookhaven. Thus, in one sense, the sum of the 75-mile displacement and the 100-mile displacement is a 30-mile dis-

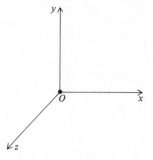

FIGURE 2-1 Cartesian coordinates.

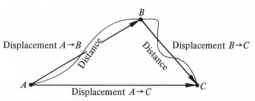

FIGURE 2-3 Distance and displacement in two successive motions.

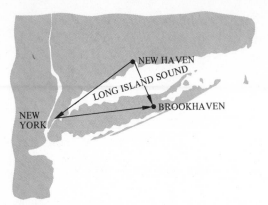

FIGURE 2-4 Displacement and distance on a map.

placement, although the distance traveled is 175 miles. It is thus necessay to develop a type of algebra that will enable us to deal with quantities having both distance and direction. This type of algebra is called *vector algebra*.

2-3
Vectors

Some quantities such as mass, time, and temperature do not have a direction associated with them, and are completely specified by a number accompanied by appropriate units. For such quantities, the ordinary rules of arithmetic are applicable. For example, a single object of mass 2 kg is equivalent to two objects, each of mass 1 kg, as far as the property of mass is concerned. It is true that one object (of mass 2 kg) is not identical to two objects (each of mass 1 kg), but if all we are interested in is the property of mass, then we can always replace one object by two of the same total mass. In that sense, we can write the simple equation:

$$2 \text{ kg} = 1 \text{ kg} + 1 \text{ kg}$$

Quantities that are specified by a number with units and for which the ordinary rules of arithmetic are valid are called *scalar* quantities. Quantities specified by pure numbers, without units, are also scalars.

As we have seen, a displacement is a quantity that has direction and is specified by a number and units. Further, as we saw in the last section, displacements are governed by peculiar rules of algebra. Two successive displacements are equivalent to a single displacement whose length may be very different from the sum of the lengths of the

original displacements. We can describe the Yale to Brookhaven trip by the equation

$$30 \text{ miles } (Y \to B) = 75 \text{ miles } (Y \to N)$$
$$+ 100 \text{ miles } (N \to B)$$

Any quantity that has direction, specified by a number accompanied by its units and is governed by the same rules of algebra that govern displacements, is called a *vector* quantity. The number (with units) is called the *magnitude* of the vector. These rules are the rules of vector algebra. In the following sections vector addition and subtraction will be described. A discussion of multiplication will be reserved for later chapters, when multiplication will be used in solving problems.

A vector is represented on a diagram by drawing an arrow. The length of the arrow is made proportional to the *magnitude* of the vector by choosing an appropriate scale. The direction of the vector is given by the direction of the arrow. In some circumstances it is useful to consider the *line of action* of the vector, which is the line along which the vector is drawn. In those cases the direction of the vector along the line of action is called the *sense* of the vector. That is, if the line of action of the vector is vertical, then the sense is either up or down. In print, symbols representing vector quantities are usually printed in boldface type, while the magnitude of vectors as well as scalar quantities are represented in italic. In handwriting, the distinction between italic and boldface is impossible for most people, so vectors are usually represented by drawing a short, horizontal arrow over the symbol. Thus the symbol printed as **A** is handwritten as $\vec{A}$.

2-4
Graphical methods of addition of vectors

Since we have defined vectors in terms of the rules of algebra of displacements, we will first consider the addition of two displacements.

In Figure 2-5, we have two displacements, **A** and **B**. If a body starts at a point, called O, and undergoes displacement **A** first, followed by displacement **B**, its motion is described by Figure 2-5b. Here the object will proceed from point O to point P, and then go to point Q, where the line OP represents the displacement **A** and the line PQ represents the displacement **B**. The total displacement, from the initial point O to the final point Q,

FIGURE 2-5 The addition of displacements by the parallelogram method. (a) The displacements which are to be added. (b) Displacement **A** followed by dispalcement **B** results in displacement **C**. (c) Displacement **B** followed by **A** results in displacement **C**'. (d) The parallelogram formed by the two triangles. The sum of the displacements **A** and **B** is the diagonal **C**.

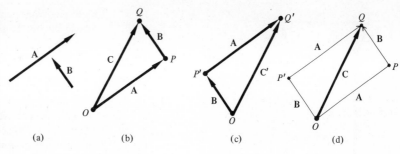

(a)   (b)   (c)   (d)

is a vector we can call **C**. Thus, **C** is equivalent to **A** followed by **B**, or

$$C = A + B$$

If the object undergoes the displacement **B** first, and then **A**, the motion is described by Figure 2-5c. Here the object goes from O to P' to Q'. The total displacement is the vector **C**'. That is, the displacement **C**' is equivalent to the displacement **B** followed by the displacement **A**, or

$$C' = B + A$$

In Figure 2-5d a parallelogram has been drawn, where OP and $P'Q$ are each equal to the displacement **A** and the sides OP' and PQ are equal to the displacements **B**. It is clear that this parallelogram is composed of the two triangles that are in Figure 2-5c and Figure 2-5d. Therefore, the displacements **C** and **C**' are equal, and

$$A + B = B + A$$

This means that the sum of two displacements, or vectors, is independent of the order in which they are taken. Further, we have two rules for adding vectors, either of which may be used.

1. To add two vectors, first draw either of the two vectors. Then, starting at the head of the first, draw the second vector. The sum of the two vectors is the vector drawn from the tail of the first vector to the head of the second. This is shown in Figure 2-6b.

2. To add two vectors, first draw either of the vectors. Then, starting at the tail of the first, draw the other. With these two lines as two of the sides, complete the parallelogram. The sum of the vectors is the vector formed by the diagonal from the point where the tails of the vectors meet. This is shown in Figure 2-6c. The sum of two vectors (or, by extension, the sum of any number of vectors) is called the *resultant*.

These two rules for the addition of two vectors can be extended to the process of addition of three or more vectors. The first rule, which is the basis of the polygon method, is applied to more than two vectors in the following way: Draw the first vector, and from its head draw the second. From the head of the second, draw the third, and from its head, draw the fourth. Continue the process until all the vectors have been drawn. The sum is the vector drawn from the tail of the first vector to the head of

FIGURE 2-6 (a) Start with two vectors. To add:
(b) Either 1. Draw one (**A**).
2. Starting at the head of the first, draw the second (**B**).
3. Complete the triangle from the tail of the first to the head of the second, forming the sum (**C**).
(c) Or 1. Draw both tail to tail.
2. Complete the parallelogram.
3. Draw the diagonal (**C**).

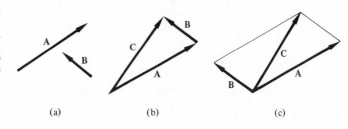

(a)   (b)   (c)

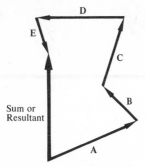

FIGURE 2-7 The sum of several vectors by the polygon method.

the last (see Figure 2-7). The second rule, also called the parallelogram method, can be applied to more than two vectors by first getting the sum of any two of the vectors and then adding this sum to the third vector. This sum can be added to a fourth vector, and so on.

We can extend the process of addition to include *subtraction* by defining the negative of a vector, and then asserting that subtraction is the same as the addition of the negative of a vector. In order to define the negative of a vector we first remind ourselves of the definition of the negative of a scalar quantity. The negative of a quantity a is that quantity which when added to a yields zero. That is, the negative of a, $-a$, is the solution of the equation

$$x + a = 0$$
$$x = -a$$

In an analogous way, we can define the negative of a vector **a** as that vector which when added to **a** gives a vector of zero magnitude. The only way two vectors can be added to yield a vector of zero magnitude is for both to have the same magnitude, but opposite directions. Thus, we can define the negative of a vector **a** to be a vector, $-$**a**, which is equal in magnitude, but opposite in direction to **a**. With this meaning of the negative of a vector, we can define the process of subtraction as the process of adding the negative of a vector. That is,

$$\mathbf{a} - \mathbf{b} = \mathbf{a} + (-\mathbf{b})$$

2-5
Resolution of vectors and analytic method of addition

Often in the solution of problems it turns out to be useful to replace a given vector **A** by two or more other vectors **B**, **C**, . . . whose sum is **A**. The vec-

tors **B**, **C**, . . . are called the *components* of **A**. This process of replacing **A** by two or more other vectors whose sum is **A** is called the *resolution* of **A** into its components. Most often the process is carried out in such a way that the components of the vector are mutually perpendicular; in that case the components are called *rectangular* components. In three-dimensional problems a vector can be replaced by three rectangular components that are mutually perpendicular. In many problems only two mutually perpendicular components are needed.

The method of resolution into rectangular components is illustrated in Figure 2-8 for the two-dimensional case, and extension to the three-dimensional case is clear. Here it is assumed that the vector **A** is given, and that the coordinate system was drawn for convenience in solving some problem so that neither axis turns out to be parallel to the edges of the paper. A perpendicular dropped from the head of **A** to the x axis produces the quantity A_x which is the x component of the vector **A**. A perpendicular dropped from the head of **A** to the y axis produces the quantity A_y, which is the y component of the vector **A**. If the angle between the vector **A** and the x axis is θ, then we have:

$$A_x = A \cos \theta \qquad (2\text{-}1)$$
$$A_y = A \sin \theta \qquad (2\text{-}2)$$

Since the angle θ is determined by the direction of the vector and the particular choice of coordinate system, it is possible for θ to turn out to be greater than 90°. In that case, it is possible for A_x or A_y or both to turn out to be negative, as in Figure 2-9.

With this method of resolving vectors into rectangular components, we can describe an analytic method of addition of vectors. Figure 2-10a shows

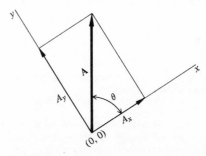

FIGURE 2-8 The resolution of a vector in rectangular components. Perpendiculars from the head of **A** to the axes produce the components A_x and A_y.

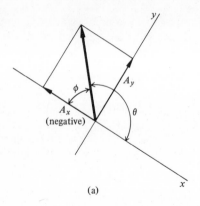

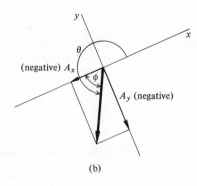

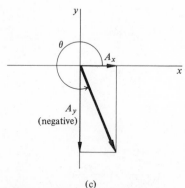

FIGURE 2-9 Situations in which one (or both) components of a vector are negative. The angle θ is the angle between the vector and the positive x axis.

two vectors, **A** and **B**, which are to be added. First, each vector is resolved into rectangular components, as in Figure 2-10b, using Equations 2-1 and 2-2. This yields four quantities, A_x, A_y, B_x, and B_y. Then the *algebraic* sum of A_x and B_x is formed as in Figure 2-10c; their sum, C_x, is the x component of the sum of the vectors **A** and **B**. Similarly, the algebraic sum of A_y and B_y yields C_y for the y component of the sum. It must be emphasized that the

components are added with appropriate signs. Then the sum of the vectors **A** and **B** is the sum of the mutually perpendicular vectors C_x and C_y; that is,

$$\mathbf{C} = \mathbf{A} + \mathbf{B} \qquad (2\text{-}3)$$

and

$$C_x = A_x + B_x \qquad (2\text{-}4)$$
$$C_y = A_y + B_y \qquad (2\text{-}5)$$

and

$$\mathbf{C} = \vec{C}_x + \vec{C}_y \qquad (2\text{-}6)$$

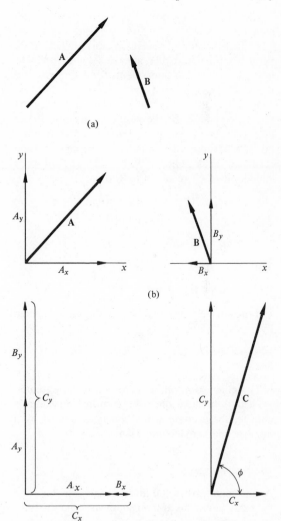

FIGURE 2-10 The addition of the two vectors. (a) The vectors which are to be added. (b) The resolution of each into components. (c) The addition of the components.

For this last equation we can introduce the notation $\mathbf{C} = (C_x, C_y)$. Then, from Figure 2-10d, it can be seen that the magnitude of $\mathbf{C}$ is given by

$$C = \sqrt{C_x^2 + C_y^2} \qquad (2\text{-}7)$$
$$C = \sqrt{(A_x + B_x)^2 + (A_y + B_y)^2} \qquad (2\text{-}8)$$

and that the angle θ between the vector $\mathbf{C}$ and the x axis is given by

$$\tan \theta = \frac{C_y}{C_x} \qquad (2\text{-}9)$$
$$= \frac{A_y + B_y}{A_x + B_x} \qquad (2\text{-}10)$$

The extension to more than two vectors may be stated as follows:

To find the sum of several vectors, select a convenient set of rectangular axes, resolve each vector into its rectangular components for each axis, and then add the respective components algebraically. The sum of the x components of these vectors is the x component of the resultant or sum of the original vectors. Similarly, the sum of the y components of these vectors is the y component of the resultant vector. Thus, if $\mathbf{A}, \mathbf{B}, \mathbf{C}, \mathbf{D}, \ldots$ are to be added, and their resultant is $\mathbf{R}$, then

$$R_x = A_x + B_x + C_x + D_x + \cdots$$

and

$$R_y = A_y + B_y + C_y + D_y + \cdots$$

with

$$R^2 = R_x^2 + R_y^2$$

and the angle θ that $\mathbf{R}$ makes with the x axis is given by

$$\tan \theta = \frac{R_y}{R_x}$$

The components of a vector are usually treated as scalars because once the coordinate system is chosen, these components can be specified by a number with an algebraic sign.

Illustrative example

A salesman travels 400 mi northeast one day. The next day he travels 100 mi southeast, and the third day he travels 150 mi south. What was his displacement over the three-day trip?

Figure 2-11a shows the polygon method of solution of this problem. In Figure 2-11b the vectors are drawn on the usual coordinate system. Then,

(a)
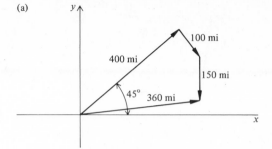

(b)
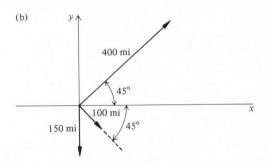

(c)
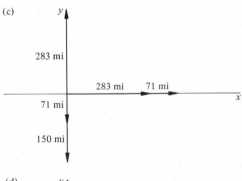

(d)
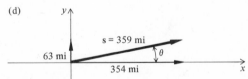

FIGURE 2-11

calling the total displacement s, we get for its components

$$s_x = 400 \text{ mi} \cos 45° + 100 \text{ mi} \cos 45° +$$
$$\qquad\qquad\qquad 150 \text{ mi} \cos (-90°)$$
$$= 283 \text{ mi} + 71 \text{ mi} + 0$$
$$= 354 \text{ mi}$$
$$s_y = 400 \text{ mi} \sin 45° - 100 \text{ mi} \sin 45° -$$
$$\qquad\qquad\qquad 150 \text{ mi} \sin (-90°)$$
$$= (283 - 71 - 150) \text{ mi}$$
$$= 63 \text{ mi}$$

Thus

$$s = \sqrt{(354)^2 + (63)^2} \text{ mi}$$
$$= 359 \text{ mi}$$
$$\tan \theta = \frac{63}{354}$$
$$\theta = 10°$$

Figure 2-11c shows the components of the original displacements and 2-11d shows the resultant **s** and its components. The displacement **s** is 359 mi, directed 10° north of east.

Having established the rules for adding vectors, it is easy to give meaning to the process of multiplication of a vector by a scalar. We define the process of multiplying a vector **v** by a scalar a to be the process of adding the vector **v** to itself a times; thus if the vector **c** is given by

$$\mathbf{c} = a\mathbf{v} \qquad (2\text{-}11)$$

then the direction of **c** is the direction of **v** and the magnitude of **c** is given by the product of a and the magnitude of v, or

$$c = av \qquad (2\text{-}12)$$

Similarly, we can give meaning to the process of dividing a vector by a scalar: if a vector **v** is divided by a scalar a, then the quotient is a vector whose direction is that of **v**, and whose magnitude is the magnitude of **v** divided by a. In symbols, this is:

$$\mathbf{c} = \frac{\mathbf{v}}{a} \qquad (2\text{-}13)$$

Then

$$c = \frac{v}{a} \qquad (2\text{-}14)$$

and the direction of **c** is the direction of **v**.

2-6
Uniform motion in a straight line

The simplest type of motion is that in which a body traverses equal distances along a straight line in equal time intervals; this type of motion is called *uniform motion in a straight line* (see Figure 2-12).

The speed v of such a body is defined as the *distance s traversed divided by the time elapsed t*; in symbols this is written

$$v = \frac{s}{t} \qquad (2\text{-}15)$$

In order to specify both the speed and the direction of motion we define the concept of *velocity*. The velocity **v** of a body in uniform motion in a straight line is defined as the *displacement **s** divided by the time t during which the displacement occurred*, or in symbols,

$$\mathbf{v} = \frac{\mathbf{s}}{t} \qquad (2\text{-}16)$$

For this simple case of uniform motion, the magnitude of the displacement is the same as the distance traversed in a given time interval. Therefore, the magnitude of the velocity is the same as the speed. In the case of uniform motion, the only difference between speed and velocity is that velocity includes both a description of the rate of motion and the direction, whereas speed includes only a description of the rate of motion. Thus velocity and displacement are vector quantities, whereas speed and distance are scalar quantities. For example, the statement that an airplane is traveling at 746 mi/hr north-northwest is a description of the velocity of the airplane, but the statement that it is traveling at 746 mi/hr is a statement of its speed.

Some aspects of the motion of a body can be represented in an informative way with the aid of graphs. In the case of uniform motion in a straight line, if we plot the speed of the body as ordinate (along the vertical axis) and the time of motion as abscissa (along the horizontal axis), the graph is a horizontal line, as shown in Figure 2-13. Since the height of this line above the abscissa is v and the length of the line is t, the product vt is the area of the rectangle formed by dropping perpendiculars from the ends of the line to the time axis. From Equation 2-14, s, the distance traversed in time t, is this same product, vt. Thus, the area under the speed-time line is the distance traversed in the time t. In the figure, the body is traveling at 40

FIGURE 2-12 Equal distances are traversed in equal intervals of time when a body moves with uniform speed in a straight line. The vector **s** is the displacement from A to B.

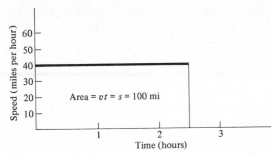

FIGURE 2-13 A graph of speed vs. time for uniform motion in a straight line.

mi/hr for 2.5 hr. The area of the rectangle is therefore 40 mi/hr × 2.5 hr = 100 mi.

If the distance is plotted as ordinate and the time elapsed as abscissa, then the resulting graph is the straight line of Figure 2-14. This straight line passes through the origin, and its slope is the speed of the motion. The slope can be found by taking any two points on the line and dividing the vertical distance between them by the horizontal distance between them. These values are not the lengths as measured by a ruler placed on the graph paper, but instead are determined from (a) the distances as read on the scale of the ordinate and (b) the times as read from the abscissa, with appropriate units for each.

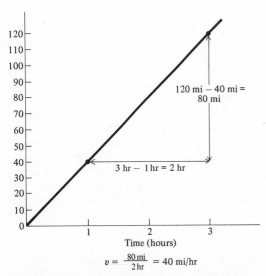

FIGURE 2-14 A graph of distance traveled vs. time for uniform motion in a straight line.

2-7
Relative velocity

Very often, the motion of a body is measured or described with respect to two different frames of reference, where the frames of reference are moving with respect to each other. For example, when a space vehicle is fired to Mars, the tracking stations are interested in both the velocity of the vehicle with respect to the earth and the velocity of the vehicle with respect to Mars. A slightly different example occurs in the case of small aircraft navigation, where it is relatively easy to measure the velocity of the plane with respect to the air, although the pilot is usually interested in his velocity with respect to the ground.

We can see how these velocities are related by considering a motorboat in a river. Figure 2-15a shows the boat with its engine turned off and the water in the river moving with a velocity relative to the earth of $\mathbf{v}_{re}$. In this case the boat will move with respect to the earth with the same velocity as the water, whereas its velocity relative to the water is zero. If the engines are started, the boat will acquire a velocity relative to the water of $\mathbf{v}_{br}$. The boat will move along with the water and in addition will move with respect to the water. Thus, the velocity of the boat with respect to the earth, $\mathbf{v}_{be}$, will be the vector sum of the velocity of the boat with respect to the water, $\mathbf{v}_{br}$, and the velocity of the water with respect to the earth, $\mathbf{v}_{re}$. In symbols,

$$\mathbf{v}_{be} = \mathbf{v}_{br} + \mathbf{v}_{re} \qquad (2\text{-}17)$$

This result can be stated in very general form: if a moving object P has a velocity $\mathbf{u}$ relative to one frame of reference A and it has a velocity $\mathbf{v}$ relative to a second frame of reference B while the first frame of reference, A, has a velocity $\mathbf{w}$ relative to the second, B, then

$$\mathbf{v} = \mathbf{u} + \mathbf{w} \qquad (2\text{-}18)$$

In using Equation 2-18 it is important to remember that if an object A has a velocity $\mathbf{u}$ with respect to object B, then object B has a velocity $-\mathbf{u}$ with respect to object A. For example, if object A is moving north and object B is standing still, then to an observer on object A it will appear that object B is moving south.

Illustrative example

A bus is traveling east to west at 60 mi/hr and approaching a crossroads. At the same time a car is

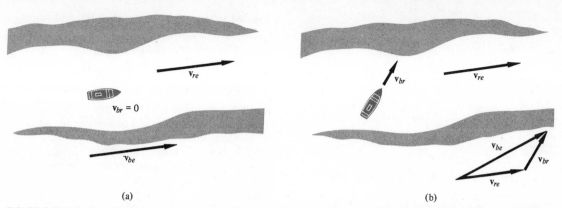

FIGURE 2-15 (a) A boat which is not moving relative to the water has a velocity relative to the earth (v_{be}), which is equal to the velocity of the water relative to the earth (v_{re}). (b) The velocity of the boat relative to the earth is equal to the vector sum of the velocity of the boat relative to the water and the velocity of the water relative to the earth.

traveling from south to north at 50 mi/hr approaching the same intersection. What is the velocity of the bus with respect to the car?

In using Equation 2-18 we must first identify the object and the two reference frames. Here we can identify the bus as the object, the road as the first reference frame, and the car as the second reference frame. The velocity of the bus with respect to the car (the second reference frame) is the unknown v. The velocity of the bus with respect to the road (the first reference frame) is u and is a westward vector of magnitude 60 mi/hr. The w of Equation 2-18 is the velocity of the first reference frame (the road) relative to the second reference frame (the car) and is therefore opposite to the given velocity of the car (with respect to the road). The vector w is then a southward vector of magnitude 50 mi/hr. The vector addition is shown in Figure 2-16. The magnitude of the velocity v—that

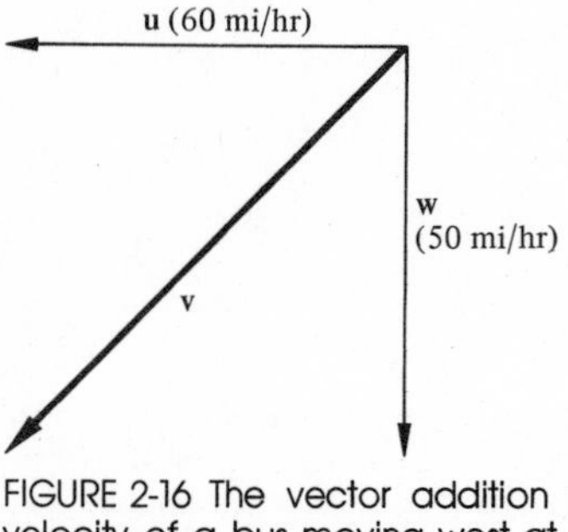

is, the speed v of the bus relative to the car—is 78 mi/hr in the direction shown in the figure.

It is easy to see that if the bus reaches the intersection first and then makes a right turn so that it is traveling northward at 60 mi/hr while the car continues northward at 50 mi/hr, the speed of the bus relative to the car is reduced from 78 mi/hr to 10 mi/hr. This sharp change in relative velocity is readily apparent to drivers and is usually interpreted as a slowing down even though both vehicles maintain constant speeds.

2-8
Instantaneous speed and velocity

Up to this point we have discussed only uniform motion in straight lines. In the most general case, objects move along paths that are not straight, and move at rates that vary during the motion. Therefore we must reconsider the definitions of speed and velocity given above.

Consider an object moving along a path such as shown in Figure 2-17a. At one instant of time the object is at some position A, and at a later time t, the object is at position B. During the time interval t, the object has traveled along the path some distance s. Then, even though the rate of motion may have varied, we can define the *average speed during the time interval* by

$$\bar{v} = \frac{s}{t} \qquad (2\text{-}19)$$

Here we have introduced the notation that a bar

FIGURE 2-16 The vector addition to find the relative velocity of a bus moving west at 60 mi/hr and a car moving north at 50 mi/hr.

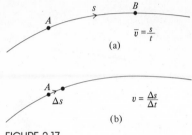

FIGURE 2-17

over a symbol denotes that the average value of the quantity is indicated.

Usually, we are interested in the speed at one instant of time, such as the speed at point A. See Figure 2-17b. We can define this quantity by considering a process such as the following. First find the average speed over the interval from A to B as above. Then consider taking the average over a time interval that starts when the objects is at A, but is shorter than the first. Then repeat the averaging successively over shorter and shorter time intervals. Ultimately, this will produce the average speed over a time interval that can be thought of as an instant of time. We can think of this process in either of two ways. In the first way, we can think of the process as a mathematical one, carried out for an infinite number of steps. In the second, we can think of carrying out the process until the time interval used in the averaging is smaller than the errors involved in a particular *measurement*.

It is a common and convenient notation to represent a small change or small increment in a quantity by writing the Greek letter Δ (delta) as a prefix for the symbol of the quantity. Thus, Δs represents a small distance and Δt represents a small time interval. In this notation, we can *define the instantaneous speed of an object, v, with*

$$v = \frac{\Delta s}{\Delta t} \qquad (2\text{-}20)$$

as Δt is taken exceedingly small. The phrase, "as is taken exceedingly small," is an integral part of Equation 2-20, and is an abbreviated way of describing the process of taking averages over shorter and shorter time intervals.

When an object is moving with uniform speed in a straight line, its instantaneous speed at any point is the same as its average speed over any interval.

We can extend the definitions of velocity in a similar way. Consider an object moving from A to B as in Figure 2-18a. In this motion, the net displacement is the vector s. Note that the distance

traveled is along the path and is different in magnitude from the displacement. Nevertheless, we can define the *average velocity* as

$$\bar{\mathbf{v}} = \frac{\mathbf{s}}{t} \qquad (2\text{-}21)$$

where t is the time it took for the object to travel from A to B. The direction of the average velocity is the same as the direction of the displacement $\mathbf{s}$.

Now, we can take the average velocity over shorter and shorter time intervals, as in Figure 2-18b. Then we can define the *instantaneous velocity* as

$$\mathbf{v} = \frac{\Delta \mathbf{s}}{\Delta t} \qquad (2\text{-}22)$$

as Δt is taken exceedingly small.

An examination of the figure shows that as Δt gets smaller and smaller, the direction of the displacement approaches the direction of the tangent to the path. Thus the *direction of the instantaneous velocity at any point is the direction of the tangent to the path at that point* as shown in Figure 2-18c. Further, as Δt is made small, the magnitude of the displacement gets closer and closer to the length of

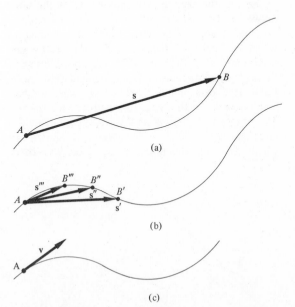

FIGURE 2-18 (a) The displacement from A to B. (b) The displacement as B is taken successively closer to A. The direction of the displacement approaches the direction of the tangent to the path. (c) The instantaneous velocity of A, which is in the direction of the tangent to the path.

the path, so the *magnitude of the instantaneous velocity is equal to the instantaneous speed.*

2-9
Acceleration

In order to continue the discussion of motion with varying velocity, we will introduce a new derived quantity, the *acceleration*. We can do this in a formal way by analogy with the definitions of speed and velocity given above.

Consider an object, as in Figure 2-19a, that moves along some path with a velocity v_A at point A, and at time t later, is at point B with some velocity v_B. Then we can define the *average acceleration* as the change in velocity divided by the time interval. In symbols, this is

$$\bar{a} = \frac{v_B - v_A}{t} \qquad (2\text{-}23)$$

We then can define the instantaneous acceleration by considering the average acceleration over shorter and shorter time intervals, or

$$a = \frac{\Delta v}{\Delta t} \qquad (2\text{-}24)$$

as Δt is taken *exceedingly small.* The units of acceleration in mks units are, from the definition,

meters per second, per second. This is usually written m/sec².

This formal definition leads to some surprising conclusions. In Figure 2-19b, we have drawn the vector diagram from which we calculate the numerator of the right-hand side of Equation 2-23. The interesting thing to note is that Δv exists (is not zero) if v_B differs from v_A in either magnitude or direction or both. Thus, if a moving object changes its *velocity* by changing its speed *or* its direction, or both, then it has a nonzero acceleration. In addition, the direction of Δv is in general parallel to neither v_A nor v_B. Only when both velocities are parallel to each other is the acceleration parallel to the velocity, and this can only occur when the motion takes place in a straight line. Thus an object will have an acceleration if its velocity changes in either magnitude or direction, or both and the direction of the acceleration depends on the details of how the velocity is changing.

In the remainder of this chapter we shall restrict our discussion to motion with constant acceleration. Since acceleration is a vector, this means we shall discuss motion in which the acceleration is constant in both magnitude and direction. Cases of varying acceleration will be discussed in later chapters.

2-10
Straight-line motion under constant acceleration

The simplest special case we can deal with is that in which the acceleration is constant in magnitude and direction and, in addition, the motion occurs in a straight line. As we shall see below, motion with constant acceleration can also occur in curved paths.

The restriction of motion to a straight line permits us to introduce the simplification of treating all the vectors involved as scalars. For example, if the line of motion is horizontal, then we can treat displacements to the right as positive numbers and displacements to the left as negative numbers. Then the vector sum of two displacements is simply the algebraic sum of these scalars. Once we have adopted the convention of calling displacements to the right positive, we are forced, by Equations 2-20 and 2-23, to refer to velocities and accelerations to the right as positive, and velocities and accelerations to the left as negative. Thus, for motion along a straight line we can treat displacement, velocity, and acceleration as scalars, with

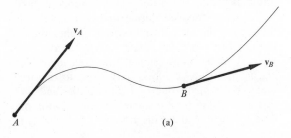

(a)

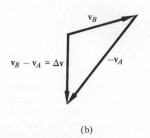

(b)

FIGURE 2-19 (a) The velocity of an object at each of two points, *A* and *B*, on a curved path. (b) The change in velocity from *A* to *B* is given by the vector difference between the two velocities.

sign, where the sign indicates the direction of the quantity. It must be emphasized that we can, if we wish, choose displacements to the left to be positive, but then to be consistent, velocities and accelerations to the left must also be taken as positive.

Since the acceleration is a constant, the average acceleration is equal to the instantaneous acceleration, and Equation 2-23, for acceleration a, can be written

$$a = \frac{v_B - v_A}{t} \qquad (2\text{-}25)$$

The convention of using signs to denote direction is very powerful when used carefully. Consider, for example, an object that is moving toward the left with a speed of 5 m/sec. If we have chosen the positive direction to be toward the right, then this object has a velocity of −5 m/sec. If one second later, this same object has speeded up, and is now moving toward the left with a speed of 7 m/sec ($v = -7$ m/sec), then from Equation 2-25 it has an acceleration given by

$$a = \frac{-7 \; \dfrac{m}{sec} - (-5) \; \dfrac{m}{sec}}{1 \; sec}$$

$$= -2 \; \frac{m}{sec^2}$$

Thus, this object, which has increased its speed, has a negative acceleration. The sign of the acceleration indicates that the change in velocity is directed toward the left, not that the speed has increased or decreased. A careful examination of Equation 2-25 should convince the reader that when the velocity and acceleration have the same sign at any given point in its path, the object is speeding up; and when the velocity and accelera-

tion have opposite signs, the object is slowing down.

2-11
Equations of motion for constant acceleration in a straight line

We can develop the equations of motion for an object moving with constant acceleration along a straight line with the aid of a graph. Consider an object which at time $t = 0$ has a velocity of magnitude v_0 and a constant acceleration a. Then at time t, the object has a velocity of magnitude v, and from Equation 2-25 we can write

$$v = v_0 + at \qquad (2\text{-}26)$$

If we assume that v_0 and a are positive, and plot the speed as ordinate and time as abscissa, we get the straight line of Figure 2-20a. Earlier (Section 2-6) we saw that in the case of constant speed the area under the speed-time graph was the distance traveled. Here, the speed is not constant. However, we can mentally divide the time t into very short intervals of time, and assume that in each interval the speed is constant and that at the end of each interval the speed instantaneously increases and then remains constant for the next time interval. This will yield the speed-time graph of Figure 2-20b. Then, during each of the small time intervals Δt, the speed-time graph *is* horizontal and the distance traveled during that time interval is the rectangular area of width Δt and height v (under the graph). It is clear that if we choose the time interval to be short enough, the area under the zigzag of Figure 2-20b is very close to the area under the straight line of Figure 2-20a, and that this area is the total distance traveled during the time t. The area under the graph of Figure 2-20a is

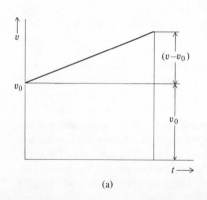

(a)

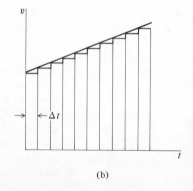

(b)

FIGURE 2-20 (a) Graph of speed vs. time for straight line motion with constant acceleration. (b) Graph of speed vs. time, assuming that the speed is constant during the time interval Δt and that the speed increases instantaneously at the end of the interval.

the sum of the areas of a rectangle of base t and height v_0 and the area of a triangle of base t and height $(v - v_0)$. Thus, the distance s traveled in the time t is

$$s = v_0 t + \tfrac{1}{2}(v - v_0)t$$
$$= \tfrac{1}{2}v_0 t + \tfrac{1}{2}vt$$
$$s = \frac{v_0 + v}{2}t \qquad (2\text{-}27)$$

From the definition of the average speed given by Equation 2-19,

$$\bar{v} = \frac{s}{t} \qquad (2\text{-}19)$$

we get for the distance traversed

$$s = \bar{v}t \qquad (2\text{-}19a)$$

By comparing Equations 2-27 and 2-19a we get

$$\bar{v} = \frac{v_0 + v}{2} \qquad (2\text{-}19b)$$

for the average speed in straight-line motion under constant acceleration. That is, in this special case, the average speed is equal to the average of the initial and final speeds.

Equation 2-27 was derived for the special case in which the initial velocity v_0 and the acceleration a were both positive. In this case, v and s are also positive, and the distance traveled is equal to the displacement. In the general case, either v_0 or a may be negative; that is, either the intercept v_0 or the slope of the line a in Figure 2-20 may be negative. In that case, if attention is paid to the signs, Equation 2-27 will give the *displacement* during time t, rather than the distance traveled.

Equations 2-26 and 2-27 are sufficient to describe completely straight-line motion with constant acceleration. For the solution of problems it is convenient to have two other equations available, one expressing the displacement in terms of the elapsed time and the acceleration, and the other expressing the final velocity in terms of the displacement and the acceleration. We can derive the first by substituting v from Equation 2-26 in Equation 2-27:

$$s = \frac{v_0 + v}{2}t$$
$$= \frac{v_0}{2}t + \frac{v}{2}t$$
$$= \frac{v_0 t}{2} + \left(\frac{v_0 + at}{2}\right)t$$

$$s = v_0 t + \frac{at^2}{2} \qquad (2\text{-}28)$$

The second equation can be derived by solving Equation 2-26 for t

$$t = \frac{v - v_0}{a}$$

and then substituting in Equation 2-27

$$s = \frac{v_0 + v}{2}t$$
$$= \left(\frac{v_0 + v}{2}\right)\left(\frac{v - v_0}{a}\right)$$
$$= \frac{v^2 - v_0{}^2}{2a}$$

or finally,

$$v^2 = v_0{}^2 + 2as \qquad (2\text{-}29)$$

Thus we have four equations:

$$v = v_0 + at \qquad (2\text{-}26)$$
$$s = \left(\frac{v + v_0}{2}\right)t \qquad (2\text{-}27)$$
$$s = v_0 t + \tfrac{1}{2}at^2 \qquad (2\text{-}28)$$
$$v^2 = v_0{}^2 + 2as \qquad (2\text{-}29)$$

where s is the displacement *from the initial position* during the time t, v_0 is the initial velocity, a is the constant acceleration, and v is the velocity at the time t. Note that since two of these equations were derived by combining the other two, only two of the four equations are independent. Therefore, these equations can be used to solve for at most two unknowns. Since there are five symbols used in the equations, this means that at least three quantities must be known. In using this set of equations, particular attention must be paid to the consistency of the units used. For example, if acceleration is expressed in feet per second per second, then velocity must be expressed in feet per second.

Illustrative example

In order to escape from the earth's gravitational field, a certain rocket must be traveling with a speed of 36,000 ft/sec. The maximum acceleration to which a human being can be subjected without losing consciousness is about 200 ft/sec², if the acceleration lasts for only a few minutes and the human being is lying down. Assuming that the rocket starts from rest, that the acceleration is constant, and that it is necessary for the pilot to be awake during the whole lift-off, what is the minimum time it takes for the rocket to reach escape speed? How far does the rocket travel while it is accelerating?

The first step in the solution of the problem is to check that in fact, three known quantities are given. It is clear that the minimum time requires the maximum acceleration, so

$$a = 200 \frac{\text{ft}}{\text{sec}^2}$$

and further, that the final speed is 36,000 ft/sec; or

$$v = 36,000 \frac{\text{ft}}{\text{sec}}$$

The phrase, "the rocket starts from rest," means that

$$v_0 = 0$$

We can then solve for the time, using Equation 2-26:

$$v = v_0 + at$$

$$t = \frac{v - v_0}{a}$$

$$= \frac{36,000 \dfrac{\text{ft}}{\text{sec}} - 0}{200 \dfrac{\text{ft}}{\text{sec}^2}}$$

$$= 180 \text{ sec}$$

We then can solve for the distance by using either Equation 2-28 or Equation 2-29. For instance,

$$s = v_0 t + \tfrac{1}{2} a t^2$$

$$= 0 + \tfrac{1}{2} \left(200 \frac{\text{ft}}{\text{sec}^2} \right) (180 \text{ sec})^2$$

$$= 3.24 \times 10^6 \text{ ft}$$

Note that it is helpful to treat the dimensions as if they were algebraic symbols. For example, in calculating the time t, we arrived at

$$t = \frac{36,000 \dfrac{\text{ft}}{\text{sec}}}{200 \dfrac{\text{ft}}{\text{sec}^2}}$$

If we think of the units as one fraction divided by another, then we invert and multiply, which yields

$$t = 180 \left(\frac{\text{ft}}{\text{sec}} \right) \left(\frac{\text{sec}^2}{\text{ft}} \right)$$

Then the "ft" cancel and one of the "sec" in the numerator is canceled by the "sec" in the denominator, leaving

$$t = 180 \text{ sec}$$

Illustrative example

The driver of an automobile that is traveling at 90 ft/sec suddenly applies the brakes. If the braking acceleration of the car is 15 ft/sec², how long does it take for the car to stop?

Here, the initial velocity of the car is 90 ft/sec and the final velocity is zero. Thus, we have chosen the positive direction to be that of the initial velocity. Since the car is slowing down, the change in velocity is directed opposite to the initial velocity, and so the acceleration is negative. Therefore

$$a = -15 \frac{\text{ft}}{\text{sec}^2}$$

$$v_0 = 90 \frac{\text{ft}}{\text{sec}}$$

$$v = 0$$

We can solve Equation 2-26 for the time t.

$$t = \frac{v - v_0}{a}$$

$$t = \frac{0 - 90 \dfrac{\text{ft}}{\text{sec}}}{-15 \dfrac{\text{ft}}{\text{sec}^2}}$$

$$t = 6 \text{ sec}$$

Illustrative example

An electron is traveling at a speed of 9×10^7 cm/sec, when an electric field is switched on that slows the electron down with an acceleration of 2×10^{14} cm/sec². (a) What is the speed and location of the electron 1 microsecond (10^{-6} sec) after the field is turned on? (b) Discuss the motion of the electron.

(a) We are given

$$v_0 = 9 \times 10^7 \frac{\text{cm}}{\text{sec}}$$

$$a = -2 \times 10^{14} \frac{\text{cm}}{\text{sec}^2}$$

$$t = 10^{-6} \text{ sec}$$

Thus

$$v = v_0 + at$$

$$v = 9 \times 10^7 \frac{cm}{sec} - 2 \times 10^{14} \frac{cm}{sec^2} \times 10^{-6} \ sec$$

$$= (9 \times 10^7 - 20) \times 10^7 \frac{cm}{sec}$$

$$= -11 \times 10^7 \frac{cm}{sec}$$

Therefore the electron is moving opposite to its initial direction. Substituting in Equation 2-28,

$$s = v_0 t + \tfrac{1}{2}at^2$$

$$= 9 \times 10^7 \frac{cm}{sec} \times 10^{-6} \ sec$$

$$\qquad - \tfrac{1}{2} \times 2 \times 10^{14} \frac{cm}{sec^2} \times 10^{-12} \ sec^2$$

$$= 90 \ cm - 10^2 \ cm$$

$$= -10 \ cm$$

Thus the *displacement* is −10 cm, or in other words, the electron is 10 cm behind its initial location. Note that this is *not* the distance the electron has traveled.

(b) Discussion of the motion: It is obvious that since the velocity after 1 microsecond (μsec) is negative, the electron must have come to rest and reversed its direction. The time t_1 that it traveled before coming to rest can be found by setting the final velocity equal to zero at this time; thus

$$0 = v_0 + at_1$$

and therefore

$$t_1 = \frac{v_0}{-a} = \frac{9 \times 10^7 \dfrac{cm}{sec}}{2 \times 10^{14} \dfrac{cm}{sec^2}}$$

$$= 4.5 \times 10^{-7} \ sec$$

or

$$t_1 = 0.45 \times 10^{-6} \ sec$$

Its displacement during this time is

$$s_1 = v_0 t_1 + \tfrac{1}{2} \, at_1{}^2$$

$$= 9 \times 10^7 \frac{cm}{sec} \times 0.45 \times 10^{-6} \ sec$$

$$\qquad - \tfrac{1}{2} \times 2 \times 10^{14} \frac{cm}{sec^2} \times (4.5)^2 \times 10^{-14} \ sec^2$$

$$= 40.50 \ cm - 20.25 \ cm = 20.25 \ cm$$

The distance s_2 that it travels in the opposite direction in the remaining time of 0.55×10^{-6} sec is then

$$s_2 = \tfrac{1}{2} \times \left(-2 \times 10^{14} \frac{cm}{sec^2}\right) \times (5.5 \times 10^{-7} \ sec)^2$$

$$= -30.25 \ cm$$

and its displacement from its initial position is

$$s = s_1 + s_2 = -30.25 \ cm + 20.25 \ cm$$

$$= -10 \ cm$$

2-12
Freely falling bodies

One of the most common examples of motion with constant acceleration is that of a body that is falling close to the surface of the earth. This is the type of motion that was first carefully studied by Galileo (Figure 2-21). The motion of a falling body is complicated, and depends on the geographic location of the object, the altitude, the size and shape of the body, and the properties of the air through which the body is falling. However, careful measurement of the motion of many bodies leads

FIGURE 2-21 Galileo Galilei (1564–1642), founder of modern scientific method, discovered the laws of motion of freely falling bodies and of bodies moving along inclined planes. He constructed a telescope with which he observed the surface features of the moon, and discovered four of the moons of Jupiter. His observations helped establish the validity of the heliocentric theory of the universe. (Courtesy of *Scripto Mathematics*.)

to the conclusion that if there were no air, all bodies would fall with the same acceleration at the same point in space, and that this acceleration varies very little with changes in location and altitude. Further, for very many bodies the effect of the air is very small. Therefore, many falling bodies may be treated as if they fell with constant acceleration. Such bodies are called freely falling bodies. The acceleration that bodies experience in the absence of air resistance is usually called the *acceleration due to gravity*. The acceleration due to gravity is represented by the symbol g and may be taken to have the value 32 ft/sec² or 980 cm/sec² or 9.8 m/sec². The direction of the acceleration due to gravity is always downward.

Illustrative example

A ball is dropped from a height of 6 m. How long does it take to hit the ground? How fast is it moving when it hits the ground?

Since the displacement, velocity, and acceleration of the object are all downward, for this problem it is convenient to choose the positive direction for vectors to be the downward direction. With this choice, the displacement of the ball is $s = 6$ m and the acceleration is $a = 9.8$ m/sec². Since the ball is "dropped," we may take the initial velocity to be zero. Then, substituting in Equation 2-28,

$$s = v_0 t + \tfrac{1}{2} a t^2$$
$$6 \text{ m} = \tfrac{1}{2}(9.8)\,\frac{\text{m}}{\text{sec}}\,t^2$$
$$t = \sqrt{\frac{12}{9.8}}$$
$$= 1.1 \text{ sec}$$

and substituting in Equation 2-26,

$$v = v_0 + at$$
$$v = (9.8)(1.1)$$
$$= 10.8\,\frac{\text{m}}{\text{sec}}$$

Illustrative example

A rock is thrown downward with an initial speed of 10 ft/sec from the edge of a 60-ft cliff. How long does it take for the rock to reach the ground, assuming it clears the cliff? Here we have

$$v_0 = 10\,\frac{\text{ft}}{\text{sec}}$$

$$a = 32\,\frac{\text{ft}}{\text{sec}^2}$$
$$s = 60 \text{ ft}$$

From Equation 2-28,

$$s = v_0 t + \tfrac{1}{2} a t^2$$
$$60 \text{ ft} = 10\,\frac{\text{ft}}{\text{sec}} \times t + \tfrac{1}{2} \times 32\,\frac{\text{ft}}{\text{sec}^2} \times t^2$$

or

$$16t^2 + 10t - 60 = 0$$
$$t = \frac{-10}{32} \pm \frac{\sqrt{100 + 3840}}{32}$$
$$= \left(\frac{-10}{32} \pm \frac{63}{32}\right) \text{sec}$$

Ignoring the nonphysical negative solution, we have

$$t = \frac{53}{32} \text{ sec}$$
$$= 1.65 \text{ sec}$$

The negative solution can be interpreted as the solution to the problem: At what time would the rock have been thrown up from the ground so that it would be at the top of the cliff at time $t = 0$ with a downward speed of 10 ft/sec?

Illustrative example

Suppose that the rock is thrown upward with a speed of 10 ft/sec so that on the way down it just misses the edge of the cliff and falls to the ground below. How high does it rise and how long does it take to reach the ground?

We can calculate the height to which the rock rises by recognizing that on the way up, the rock is slowing down until it stops; thereafter it falls. Thus for the upward part of the motion, if we continue to choose the downward direction to be the positive direction, as in Figure 2-22a,

$$v_0 = -10\,\frac{\text{ft}}{\text{sec}}$$

$$a = 32\,\frac{\text{ft}}{\text{sec}^2}$$

$$v = 0$$

Then, using Equation 2-29,

$$v^2 = v_0^2 + 2as$$
$$0 = 100 + 64s$$

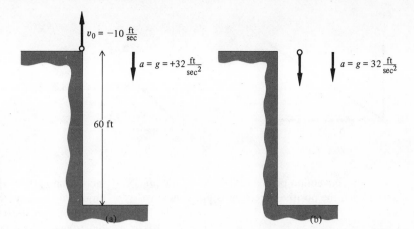

FIGURE 2-22 (a) A rock is thrown upward with initial speed of 10 ft/sec from the top of a cliff which is 60 ft high. (b) On the way down it just misses the edge of the cliff.

$$s = -\frac{100}{64} \text{ ft}$$
$$= -1.55 \text{ ft}$$

The negative sign indicates that the rock is 1.55 ft *above* its starting point when it stops. In order to find the time it takes to reach the ground, we could find the time it takes to reach this height, and then find the time it takes to fall from rest through 61.55 ft. On the other hand, we can use Equation 2-28, with

$$v_0 = -10 \frac{\text{ft}}{\text{sec}^2}$$
$$s = 60 \text{ ft}$$
$$a = 32 \frac{\text{ft}}{\text{sec}^2}$$

Thus

$$s = v_0 t + \tfrac{1}{2}at^2$$
$$60 = -10t + 16t^2$$
$$t = \frac{10}{32} \pm \frac{\sqrt{100 + 3840}}{32}$$
$$= \left(\frac{10}{32} \pm \frac{63}{32}\right) \text{ sec}$$
$$= \frac{73}{32} \text{ sec}$$

or

$$t = 2.3 \text{ sec}$$

2-13
Projectile motion

In the preceding section we dealt with the motion of an object whose initial velocity is parallel to the acceleration due to gravity, and which therefore moves in a straight line. Now we can deal with the motion of an object whose initial velocity is in an arbitrary direction. We can always resolve the initial velocity into two components, one vertical and the other horizontal. This permits us to treat the subsequent motion as two simultaneous motions, one horizontal and the other vertical. Since gravity acts only in the downward direction, the horizontal motion is motion with constant velocity. The vertical motion is that of a freely falling body.

For example, consider the motion of a skydiver who steps out of an airplane that is flying horizontally with constant velocity. The initial velocity of the skydiver is the same as that of the airplane, so he continues to move forward at the same rate as the plane. In addition, he accelerates downward. Thus, from the pilot's point of view, the skydiver remains directly below the plane, but is falling away from it. (Note that in the case of real skydivers, air resistance is not negligible, and the skydiver falls behind the plane.)

We can make this discussion analytic by considering an object whose initial velocity, $\mathbf{v}_0$, makes an angle θ with the horizontal, as in Figure 2-23a. Then the initial vertical velocity is v_{0y} and the initial horizontal velocity is v_{0x}, where

$$v_{0x} = v_0 \cos \theta \qquad (2\text{-}30)$$
$$v_{0y} = v_0 \sin \theta \qquad (2\text{-}31)$$

We also have

$$a_x = 0 \qquad (2\text{-}32)$$
$$a_y = -g \qquad (2\text{-}33)$$

where the minus sign appears in Equation 2-33 be-

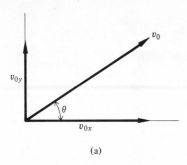

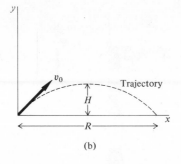

(a)

(b)

FIGURE 2-23 (a) Resolution of the initial velocity into rectangular components. (b) The trajectory of the projectile.

cause the convention of Figure 2-23 and Equation 2-31 is that the positive direction is up. If now we set

$$x = 0 \quad \text{and} \quad y = 0$$

at

$$t = 0$$

then the horizontal displacement in time t is x and the vertical displacement in time t is y. We can find x by recognizing that the velocity in the x direction is constant,

$$v_x = v_{0x} \tag{2-34}$$

so

$$x = v_x t$$
$$x = v_{0x} t$$

and, from Equation 2-30,

$$x = v_0 t \cos \theta \tag{2-35}$$

For the motion in the y direction we have, from Equations 2-26, 2-31, and 2-33,

$$v_y = v_0 \sin \theta - gt \tag{2-36}$$

From Equation 2-28 we get

$$y = v_0 t \sin \theta - \tfrac{1}{2} g t^2 \tag{2-37}$$

Equations 2-35 and 2-37 can be combined to yield the equation of the path. The result is the general equation for a parabola sketched in Figure 2-23b.

Consider the motion of an object, such as a football, that is thrown upward at some angle θ from a point essentially at ground level. The object moves forward at constant speed and rises with decreasing upward velocity. At some point it reaches a maximum height H, where it has zero vertical velocity (although it still has the constant forward velocity and the constant downward acceleration). Then the object falls, while moving forward, with

increasing vertical speed. Ultimately it reaches the ground at a point that is a horizontal distance R away from its starting point. In order to determine the *range*, R, we recognize that when the object reaches the ground

$$x = R$$

and

$$y = 0$$

Setting the left side of Equation 2-37 equal to zero, we have

$$v_0 t \sin \theta - \tfrac{1}{2} g t^2 = 0$$

Thus, in addition to the obvious solution that

$$t = 0$$

we have

$$t = \frac{2 v_0 \sin \theta}{g} \tag{2-38}$$

Then, substituting in Equation 2-35 for x,

$$x = v_0 \cos \theta \left(\frac{2 v_0 \sin \theta}{g} \right)$$
$$x = 2 v_0^2 \frac{\cos \theta \sin \theta}{g}$$

or

$$R = 2 v_0^2 \frac{\cos \theta \sin \theta}{g}$$

This can be put in somewhat simpler form by remembering that

$$2 \sin \theta \cos \theta = \sin 2\theta$$

Therefore

$$R = v_0^2 \frac{\sin 2\theta}{g} \tag{2-39}$$

Thus, the maximum range occurs when $\sin 2\theta$ has

its maximum value of 1; that is, the maximum range occurs when θ equals 45°. Similarly, we can see that the range is equal to zero for θ equals both zero and 90°. Any range less than the maximum is achieved for *two* values of θ, one less than 45° and the other larger than 45°.

We can find H by recognizing that at the maximum altitude the vertical velocity is zero; thus, from Equation 2-36,

$$v_0 \sin \theta = gt$$

or

$$t = v_0 \frac{\sin \theta}{g} \qquad (2\text{-}40)$$

Then, substituting in Equation 2-37,

$$H = (v_0 \sin \theta) v_0 \frac{\sin \theta}{g} - \tfrac{1}{2} g v_0^2 \frac{\sin^2 \theta}{g^2}$$

$$H = \tfrac{1}{2} v_0^2 \frac{\sin^2 \theta}{g} \qquad (2\text{-}41)$$

One interesting observation that can be made is that the time it takes for the object to make its whole flight from Equation 2-38 is twice the time it takes to reach its maximum altitude (Equation 2-40). That is, it takes the same time for the object to rise to its maximum height as it does for the object to fall. This is a specific illustration of the symmetry of this motion, a symmetry that follows from the fact that the path is a parabola and a parabola is a symmetric curve.

Questions

1. The circumference of the earth at the equator is about 24,000 mi and the earth turns once in 24 hr, and so a point on the equator moves with a speed of about 1000 mi/hr with respect to some frame of reference. Describe this frame.

2. Describe some examples of uniform straight-line motion that you have recently observed. Specify the frame of reference in which the motion is uniform.

3. An airplane travels due east at a constant elevation with constant speed. Is it moving with uniform velocity?

4. If there is no wind, raindrops fall vertically with uniform speed. A man driving a car on a rainy day observes that the tracks left by the raindrops on the side windows are all inclined at the same angle. What conclusions can you draw about the motion of the car? Show how the speed of the raindrops can be determined from the inclination of the tracks and the readings of the speedometer.

5. Does the speedometer of an automobile measure a scalar or a vector quantity?

6. What is the direction of a vector of zero magnitude?

7. A boy seated on one side of a train throws a ball to another boy seated directly opposite him. If the train is moving with uniform velocity, what is the path of the ball (a) with reference to the ground and (b) with reference to the train? What is the difference in each of these paths if the train moves with uniformly accelerated motion?

8. A motorboat is sailing due west on a day when there is no wind. In what direction will a flag on the ship be blown? What will be the new direction of the flag if a wind should blow from the north?

9. A train is moving with uniform speed along a level road. A man on the observation platform at the rear of the train drops an object. What is the path of the object as observed by (a) the man on the train and (b) another person standing a short distance from the track?

10. Give an example of an object that has an instantaneous velocity equal to zero but an instantaneous acceleration that is not equal to zero.

11. Give an example of a motion in which the average velocity is equal to zero but the instantaneous velocity is not equal to zero.

12. A ball is thrown up with an initial speed u. With what speed will it return to its starting point?

13. A man standing at a height H above the ground throws one object upward with an initial speed u and another downward with the same initial speed u. Compare the final velocities of these two objects when they reach the ground.

14. One object is thrown forward from the top of a building at the same instant that another object is dropped from the same point at the top of the building. Which will reach the ground first?

15. With a large enough initial velocity a rocket ship may travel in a straight line infinitely far

from the earth. Describe the direction of the acceleration and the velocity of the ship during the trip.

16. In a laboratory experiment an air rifle is clamped in position and aimed by sighting along the barrel. The target is released just as the projectile leaves the muzzle of the gun.

Show that the projectile will always hit the target.

17. Show that the speeds of a projectile are the same at any two points in its path that are at the same elevation. Compare the velocities at these two points.

Problems

1. Vector **A** is a displacement of 10 mi due north and vector **B** is a displacement of 12 mi northeast. Find **(a)** the vector 5**A**, **(b)** the vector −**B**, **(c)** the sum of **A** and **B**, **(d)** the vector **B** − **A**.

2. Find the sum of each of the pairs of vectors shown in Figure 2-24.

3. Find the difference for each of the pairs of vectors in Figure 2-24.

4. Find the vector sum of the vectors shown in Figure 2-25.

5. A man makes a trip of 100 mi in a direction 30° east of north and then a trip of 60 mi in a direction 37° south of east. What is the shortest trip that will leave him due east of the initial position? How far from his initial position will he be?

6. A driver travels 400 mi in 7 hr and 10 min. What is his average speed **(a)** in miles per hour and **(b)** in feet per second?

7. An airplane is traveling with a speed of 620 mi/hr. How far does it travel in 30 sec?

8. A runner crosses the finish line at a speed of $\frac{1}{4}$ mi/min and then comes to rest in 15 sec. Compute the average acceleration in feet per second per second.

9. The displacement-time graph of an object moving in a straight line is shown in Figure 2-26. Estimate the instantaneous velocity at $t = 3$ sec and at $t = 10$ sec.

10. For the motion described in Figure 2-26, find

the average speed during the first 3 sec and the average speed during the first 8 sec.

11. A rocket acquires a speed of 2.8 mi/sec in 60 sec. Determine the average acceleration. Compare this with the acceleration due to gravity.

12. A parade float is moving along the road at 5 mi/hr. A person is marching back and forth on the float, parallel to the direction of motion of the float, at a speed of 2 mi/hr. What are the two velocities with which the person is moving relative to the ground?

13. What would the two velocities be if the person in Problem 12 were moving from side to side on the float?

14. An automobile starting from rest acquires a speed of 60 mi/hr in 12 sec. Determine **(a)** its average acceleration and **(b)** the distance covered in this time.

15. An electron, emitted by a heated filament in a vacuum tube, acquires a speed of 6.0×10^8 cm/sec in a distance of 0.3 cm. Determine **(a)** its acceleration and **(b)** the time required to travel this distance.

16. A car approaching a turn in the road has its speed decreased from 60 mi/hr to 40 mi/hr in a distance of 150 ft. **(a)** What is the acceleration? **(b)** How long did it take to traverse this distance?

17. A stone is dropped from a bridge 30 m above the water. **(a)** With what speed does the stone

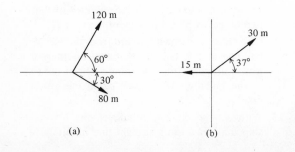

(a)   (b)

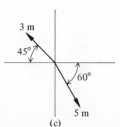

(c)

FIGURE 2-24

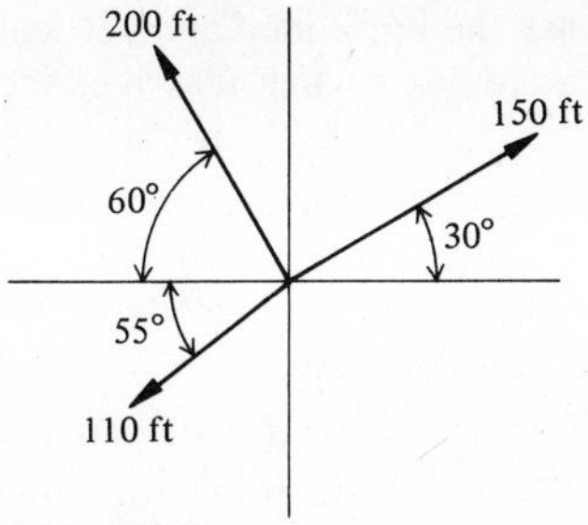

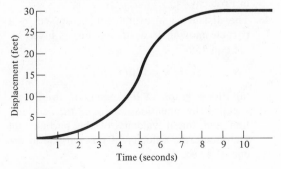

FIGURE 2-25

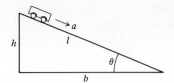

FIGURE 2-27

reach the water? **(b)** With what speed would the stone have reached the water if it had been thrown downward with a speed of 5 m/sec?

18. A boy throws a ball upward and catches it 2.6 sec later. **(a)** How high up did the ball go? **(b)** With what speed was it thrown upward?

19. A stone is dropped into a well that is 144 ft deep. If the speed of sound is 1100 ft/sec, how long after the stone is dropped is the sound of the splash heard?

20. A boy drops a stone from the edge of a cliff and 2 sec later drops another. How far apart are the two stones 3 sec after the first stone is dropped?

21. A balloon is rising at a constant rate of 8 ft/sec when an object is dropped from the balloon. How far apart are the balloon and object 3 sec later?

22. A boy throws an object straight upward to a friend who is at a window 16 ft above the ground. With what velocity should he throw the object in order that the friend catch the object at the peak of its trajectory? How much time is the object in flight?

23. With what velocity must an object be thrown upward in order to rise to a height of 30 m? How long will the object remain in the air?

24. A boy at the top of a cliff 40 m high throws a ball upward with such a velocity that it rises 15 m above the top of the cliff. At the same time his friend throws a ball downward with the same speed. The first ball misses the edge of the cliff on the way down and strikes the ground. How much time elapses between the two impacts at the ground?

25. An automobile traveling at 50 mi/hr hits a stone wall. From what height would it have had to fall in order to make the same impact with the ground?

26. A small car rolling down a frictionless inclined plane has an acceleration given by

$$a = g \sin \theta$$

If the car starts from rest at the top of the inclined plane, show that the velocity with which it reaches the bottom of the incline is given by

$$v^2 = 2gh$$

where h is the height of the raised end of the plane (see Figure 2-27).

27. A frictionless inclined plane 2.4 m long is inclined at an angle of 30° to the horizontal. A small block is projected down the plane with a velocity of 3.6 m/sec. Determine **(a)** the velocity with which it reaches the bottom of the plane and **(b)** the time it takes the block to reach the bottom of the plane.

28. The distance between two stop signs in a town is 520 ft. If the acceleration of a car, both positive and negative, is kept to a maximum value of 8 ft/sec² and if the speed limit on this road is 30 mi/hr, what is the minimum time required to traverse this distance? How much less time would have been required if the stop signs were not there?

29. A piece of paper is blown about by a gusty wind. It is subjected to two successive displacements, such that the resultant displacement is equal in magnitude to the first and at right angles to the first displacement. Find the

FIGURE 2-26

magnitude and direction of the second displacement relative to the first.

30. A pilot sets the course of an airplane due north. The plane flies at a speed of 600 mi/hr relative to the air, but there is a wind heading due east at 80 mi/hr. In what direction does the airplane fly?

31. In what direction must the pilot of the plane in Problem 30 head the plane in order to fly due north?

32. An airplane is in a holding pattern (on a windless day) flying due west at 500 mi/hr. Another airplane flying parallel to the first at the same speed is diving toward the ground at an angle of 30° to the horizontal. What is the motion of the second plane as seen by the passengers in the first?

33. A boat is drifting helplessly downstream with the current at a speed of 10 mi/hr. The boat is in midstream, $\frac{1}{4}$ mi from shore. Fortunately the skipper has a gun that can fire a rescue line with a horizontal velocity of 100 mi/hr. He sights a man on a dock downstream and decides to fire the line perpendicular to the direction of motion of the boat. Where should he fire the gun? (Assume that the man can catch the line without being hurt.)

34. A car is traveling at a constant velocity of 50 mi/hr on a level road on a rainy day. The tracks of the raindrops on the side windows make an angle of 25° with the vertical. Assuming negligible wind velocity, determine the velocity of the raindrops.

35. A ball is thrown horizontally from a height of 4 ft with a velocity of 50 ft/sec. Neglecting air resistance, determine **(a)** the time it will take for the ball to reach the ground and **(b)** the velocity it will have at the instant that it reaches the ground.

36. A sailor at the top of a 50-ft mast drops his binoculars. The ship is moving at 12 mi/hr. Where do the binoculars hit the deck?

37. A novice rifleman aims his gun directly at a target that is 300 yards away. The bullet has a muzzle velocity of 1600 ft/sec. If the gun were aimed directly at the target, by how much would the bullet miss the target?

38. A ball rolls off the top of a desk, 30 in. above the floor, and lands 2 ft away from the desk. How fast was it moving at the instant it left the desk top? What are the horizontal and vertical components of its velocity when it strikes the floor?

39. A golf ball is driven horizontally from a tee with a speed of 30 m/sec. It strikes the ground 2 sec later. How high was the tee? How far did the ball travel horizontally?

40. Two boys 100 ft apart are playing ball. When one boy throws the ball to the other it spends 2 sec in the air. How high did the ball go? With what initial velocity did the boy throw the ball?

41. At the instant a traffic light turns green an automobile accelerates at the rate of 8 ft/sec². At that same instant the automobile is passed by a truck moving with a constant speed of 25 mi/hr. How long will it take for the automobile to catch up to the truck?

42. A passenger train is traveling at 75 mi/hr when suddenly the engineer sees a freight train 1000 ft ahead moving in the same direction at 30 mi/hr. The engineer applies the brakes, causing a deceleration of 6 ft/sec². Will there be a collision?

43. A stone that has been dropped from the roof of a building is seen to pass a window 180 cm high in 0.3 sec. Determine **(a)** the average speed of the stone, **(b)** the speed with which the stone passes the lower level of the window, and **(c)** the distance from the roof to the lower level.

44. A rifle fires a bullet with a speed of 500 m/sec. If the elevation of the gun is 30° with the horizontal, determine **(a)** the range of the bullet on horizontal ground and **(b)** the velocity of the bullet when it reaches the ground.

45. **(a)** Using Equation 2-39 show that the maximum range on horizontal ground occurs when the projectile is fired at an angle of 45°. **(b)** Determine the maximum range of the projectile in Problem 44.

46. The distance s in centimeters traversed by a particle moving in a straight line is given by the equation

$$s = 180t + 50t^2$$

(a) Plot a graph of this equation. **(b)** By inspecting this equation determine the acceleration and initial velocity of the particle. **(c)** Determine the velocity of this particle at the end of 4 sec.

force & motion

chapter 3

3-1
The cause of motion

The cause of motion is a question that has occupied the minds of men for centuries. The experience of almost everybody is consistent with the idea that motion is caused by force—moving a piano once is usually enough to make that idea self-evident. But the problem of science is to make the notion precise, to state carefully the conditions and the limitations.

The medieval Scholastics had developed a set of hypotheses based on the ideas of Aristotle and Plato, which in essence asserted that there were two kinds of motion: natural and forced. Natural motions were characteristic of the things that moved and needed no force (other than the forces involved in the creation of things). From this point of view, celestial objects move in the way that they do because it is in the nature of celestial objects to move that way. Similarly, rocks fall because it is in the nature of rocks to be at the surface of the earth, and when they are displaced, they return to their natural place. This—coupled with the idea that there exist, surrounding the earth, spheres of air, water, and fire—explained many motions, such as smoke rising and rivers flowing. In each case all or part of the moving object "belonged" naturally in one of the spheres and moved naturally to that sphere. Other motions, the forced motions, involved moving an object to an unnatural place, and so involved the action of a force from some external agent. To the medieval philosophers it was obvious that this external agent must always be in contact with the moving object, and so they did not invoke gravity as a force. Although there were some serious problems, some of which were of the "How many angels can dance on the head of a pin?" variety, this set of ideas generally explained the motions that were observed. With the coming of the Renaissance the whole structure of ideas supporting the notion of natural motions crumbled and, more important, it was discovered that some motions, once started, appeared to continue *without* the presence of an external force. The subsequent reexamination of the problem experimentally and mathematically by Galileo culminated in the work of Newton, which will be discussed in detail in subsequent chapters and sections. But the central result of the Newtonian analysis of motion is that force serves to change the state of motion; forces make objects start or stop, change speed, or change direction. This result really contains several ideas, which are discussed next.

The first idea, that forces are necessary to *start* motion, is one of those abstractions from experience that is so fundamental, and learned so early in life, that it is called obvious. It is also obvious that forces do *stop* moving objects, but it is somewhat less obvious that the second idea is true; that stopping an object *always* requires the action of a force. That is, it is sometimes necessary and always possible to discover (or invent) a force that accounts for the stopping of a particular motion. For example, a billiard ball rolling on a table always slows down and stops, sooner or later. If the ball stays on the table, it is usual to ascribe the change in its motion to the existence of a force called the friction force. In fact, the friction force is so common that it is almost impossible to observe a motion in which *no* force acts; thus most of us have no experience with motions where no force acts. So the third idea, that in the absence of forces an object will move with no change—that it will move with constant speed in a fixed direction—is not obvious at all. This idea, like most of the important ideas of physics, has to be abstracted from many observations and analyses. Nevertheless, most of us have seen cases of motion in which the forces were so small that we can easily accept the idea. The billiard ball (in good condition) rolling on a billiard table (also in good condition) is subject to such small forces that it *almost* does not slow down. It is easy to extrapolate to the statement that the ball would continue rolling forever if there were no friction.

Perhaps the most difficult idea to accept is the fourth, the connection between change in direction of motion and force. Everyone will agree that if you push sideways on a moving object it will be deflected sideways; but it is not so clear to the layman that deflections always require forces. Again, it is always possible to discover a force that accounts for the change in direction of a moving object. When a car goes around a curve it changes its direction; the force that causes this change is a friction force acting sideways on the wheels of the car. This can be inferred from the fact that should the friction force be reduced by a layer of oil on the road, the sideways force would become too small, and the car would skid; that is, it would tend to continue moving straight ahead and not change its direction.

Newton's description of the causes of motion is summarized in two statements, called Newton's first law of motion and Newton's second law of motion. These are stated in the next section, along with Newton's third law, which is a statement about the nature of forces between bodies. Detailed discussion of the laws will follow in succeeding chapters and sections.

3-2
Newton's laws of motion

Newton's three laws can be stated as follows:

1. A body at rest will remain at rest, and a body in motion will continue in motion with constant speed in a straight line, as long as no unbalanced force acts upon it.
2. If an unbalanced force acts on a body, the body will be accelerated; the magnitude of the acceleration is proportional to the magnitude of the unbalanced force, and the direction of the acceleration is in the direction of the unbalanced force.
3. Whenever one body exerts a force on another, the second body exerts a force equal in magnitude and opposite in direction on the first body.

In these statements the word "body" technically should be limited to a point object, an object with no dimensions. This would compel us to treat a real body as a collection of points, which would, in turn, require that we use the methods of differential and integral calculus, and consider the forces acting on each particle. However, it can be shown that when applying Newton's laws we can use the word "body" to mean any object that has a constant size and shape and does not rotate. With this meaning, the statement will lead to no essential error, and we can understand a broad range of phenomena.

The term "unbalanced force" and sometimes the term "net force" are used to mean a nonzero vector sum of all the forces exerted on a body by all external agents. Thus it may be the force exerted by a single body, or it may be the *vector sum* of the forces exerted by two or more bodies. It must be emphasized that in all cases the forces involved come from *external* agencies.

3-3
Force

Stated simply, a force is a push or a pull; it is a vector quantity; the direction in which it acts must always be taken into consideration. For most of us this represents an adequate statement of what is meant by force.

As was said in Chapter 1, physical quantities may be defined in terms of other quantities, or may be operationally defined. Operationally defined quantities are those that are defined by describing procedures for measuring them. Force may be taken as an operationally defined quantity. That is, we will define force by describing a procedure for measurement rather then by a statement of the form, "force equals . . ." or "force is . . . ." This will not prevent us from remembering the first sentence of this section as a way of saying what we mean by force, but we will formally define force operationally.

In order to measure force we will use a helical spring as a measuring instrument. If a spring is attached at one end to a wall or ceiling, and a force is applied to the other end, the spring will stretch, as in Figure 3-1. The change in length, ΔL, can be measured and used as the reading of the instrument. Then, if the spring is calibrated in terms of a *standard force*, we can measure any other force by allowing it to act on the spring and measuring the stretch of the spring.

In order to calibrate the spring we must have a standard force. A useful force is that exerted by a heavy object on its support. Thus we can choose a particular object, label it "one pound," and hang it from a particular spring. We can measure the stretch of the spring, and then whenever the spring is stretched that same amount, we assert that a 1-lb force acted on the spring. In order to make the instrument useful we must calibrate it for a large *range* of forces. We can do this if we first describe

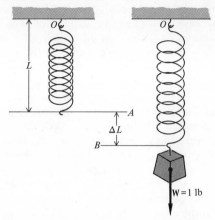

FIGURE 3-1 Use of a spring for measuring force.

a way of applying two, three, four . . . standard forces, and then a way of applying one half, one third, one fourth . . . standard forces. If we can do that, then we can make an object that applies, for example, a force of 4 lb, and another object that applies a force of $\frac{1}{2}$ lb, and then we can calibrate the spring for $4\frac{1}{2}$ lb. We can calibrate for 2 lb by making a duplicate of the standard 1-lb object (or at least make an object that stretches the spring the same amount) and then have both of the 1-lb objects supported by the spring at the same time. Note that the resulting stretch of the spring may not be twice the stretch due to 1 lb, but we can label the resulting stretch "2 lb." The extension to 3 lb, 4 lb, and so on, is obvious. We can then calibrate the spring for $\frac{1}{2}$ lb by constructing a second spring *identical* to the first, checking its behavior for 1, 2, . . . lb, and then supporting a 1-lb object by both springs simultaneously in such a way that both stretch the same amount. The extension to $\frac{1}{3}$ lb, $\frac{1}{4}$ lb, and so on, is again obvious. Thus we can end up with a standard spring calibrated to measure any unknown force exerted on it. (To be very precise, the spring is limited to a specific range of forces, to a particular geographic location, and to specific external conditions such as temperature and barometric pressure.)

At this point we have a procedure for measuring force and so we can assert that what we mean by the symbol F in an equation is the result of such measurement. In Chapter 5 we will discuss in some detail the properties of the forces that are known to exist, but in the course of discussing Newton's laws it will be useful to know something about some forces.

Most people, in the course of lifting and pushing things, have had extensive experience with the gravitational forces mentioned earlier. Of course, that experience has been confined to the surface of the earth (or close to the surface for most people), and is rarely quantitative. We all know that a gravitational force called *weight* acts on objects on or near the surface of the earth. We know that this force points down (in fact, the direction of the weight force is what we usually mean by the word "down") and that for small, simply shaped objects, the line of action goes through the object. Most of us believe that the weight is a force exerted by the earth, and generally acts as if it were exerted by a point at the center of the earth, and further we believe that if the object stays close to the surface of the earth, the magnitude of the weight stays fairly constant. Thus in a very large number of situations we know that every object we deal with will have a force acting on it; the magnitude of the force is the weight of the object and the direction of the force is downward.

We also know that in a large number of cases objects push or pull on one another when they are in contact with each other. Housewives push shopping carts; little boys pull red wagons; hammers hit nails; wind and rain erode mountains. In each case two objects in contact are involved, and at least one force acts. We can rarely say anything more definite, but for many problems some further analysis of this simple idea is useful.

Whenever two bodies are in contact we can always assume that each exerts a contact force on the other at the surface of contact. (It may turn out that the force of contact has a magnitude of zero, but since zero is a perfectly good number, it usually does not hurt to assume the existence of a force.) In general we do not know the direction of the force, but we can always resolve the force into *components*, one perpendicular to the surface of contact and the other tangent to the surface of contact (see Figure 3-2). The perpendicular component is usually called the *normal* force, and the tangential component of the contact force is what we usually mean by the term *friction force*. In many cases the force of contact between two objects is nearly perpendicular to the surface of contact; the friction force is very small. In these cases we can assume that the contact force is solely a normal force and describe the situation as frictionless.

Then every object has acting on it:
1. The force of attraction by the earth—that is, what we commonly call its weight.

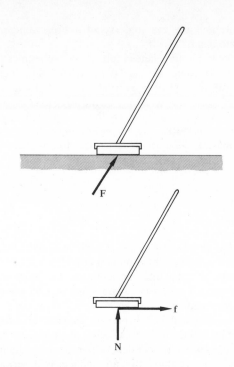

FIGURE 3-2 (a) Contact force between two objects. (b) Resolution of the contact force into two components, **N** and **f**.

2. A contact force from each body in contact with it, or one normal force and one friction force from each body in contact.

There are other (noncontact) forces, such as magnetism, but for the purpose of discussing Newton's laws, this list of forces is adequate because it allows us to analyze a large number of situations.

3-4
Newton's first law

Newton's first law states that a body at rest will remain at rest and a body in motion will continue in motion with constant speed in a straight line, as long as no unbalanced force acts on it.

The most striking thing about this statement is that bodies at rest and bodies moving at constant speed in a straight line are treated as equivalent in some way. A train at rest and a train moving at a constant speed of 100 mi/hr on a straight track are both subject to zero unbalanced force, according to Newton's first law. Consider a train of cars being pulled by a locomotive (see Figure 3-3). The weights W_1, W_2, W_3, . . . of the cars act downward.

The track exerts normal forces N_1, N_2, N_3, ... on the wheels. Since under ordinary, noncatastrophic circumstances, the train does not move up and down with respect to the track, the sum of the weights must equal the sum of the N forces, so the train is acted on by no unbalanced forces in the vertical direction. Under ordinary circumstances there are no forces acting in the horizontal direction when the train is at rest, but when the train is moving at constant speed many frictional forces act. Some of these are exerted by the track on the wheels and some by the air through which the train is moving. If these were the only forces acting then the train would be subject to an unbalanced force and would slow down. In order to keep the train moving at constant speed an external force must act on it on the forward direction. If we consider only the train, excluding the locomotive, this force is obviously supplied by the locomotive. If we consider the whole train, including the locomotive, we conclude that this force must come from the track acting on the wheels of the locomotive, since the track is the only external body (except the air) that acts on the locomotive. This leads to the somewhat surprising result that the track pushes forward on the locomotive, and at the same time pushes backward on the wheels of the cars.

One useful point of view toward Newton's first law is that it provides a description of a property of bodies, the property of *inertia*. We can think of inertia as the property of a body that makes it necessary for force to be applied to it to change its state of motion. In our discussion of Newton's second law we shall see that this property of inertia is described by and measured by the *mass* of the body.

A most fruitful question arises when we consider the frame of reference or set of axes in which the first law is described. "A body at rest" is, of course, at rest only in some frames of reference. In other frames of reference it is moving. In some frames of reference it is moving with constant speed, whereas in others it is accelerating. To an observer confined to one of these moving frames of reference, an object might appear to be moving with constant speed in a straight line, but to an observer in a different frame of reference the same object might appear to be moving with accelerated motion. For example, consider a train moving with constant speed along a track as observed by a passenger in an automobile moving on a road parallel to the track. If the automobile is moving at the same speed as the train, in the same direction, the train appears to be standing still (although, of course, the landscape appears to be moving backward); if the automobile is speeding up, the train appears to be slowing down; and when the automobile is slowing down, the train appears to be speeding up. Thus the same train may appear to be moving with constant speed, standing still, or accelerating, depending on the state of motion of the observer. On the other hand, the *forces* acting on the train are completely independent of the motion of the observer. It is clear that the forces on the train are not changed when a nearby automobile is accelerated. In this case the observers in the automobiles can always tell that they and not the train are accelerating, relative to the earth, by looking at the landscape. However in the general case, each observer assumes that he is fixed in a reference frame and measures the motion of an object relative to that frame. One can then imagine an object on which we know

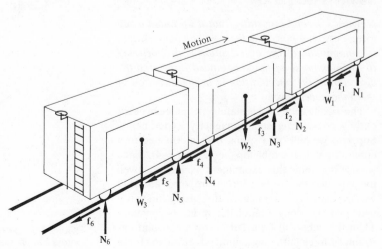

FIGURE 3-3 A train moving with constant velocity has no unbalanced force acting on it.

no net force acts, but which in some frames of reference is accelerating, in other frames of reference is moving with constant velocity, and in still other frames of reference is standing still. Thus, Newton's first law cannot be true in all frames of reference. In order for Newton's first and second laws to be valid, the motion of the object or body must be measured with respect to a "proper" coordinate system. Therefore we must describe what the proper coordinate system is.

Newton assumed that there existed one coordinate system, fixed in space, in which the first and second laws are exactly valid and that one of the jobs of physics is to discover this universal coordinate system.

Rather than discuss this grand cosmological search we will adopt the simpler approach of asserting that *any* coordinate system in which the first law is valid is one in which the second law is also valid. All frames of reference in which Newton's laws are valid are called *inertial frames of reference* and all frames of reference in which the laws are not valid are called noninertial. Thus we have reached the conclusion that Newton's first law is a test of whether a particular frame of reference is an inertial frame of reference. If in a particular frame of reference Newton's first law is obeyed, then that frame of reference is inertial and may be used as the frame of reference for Newton's second law. If on the other hand, Newton's first law is not valid in a particular frame of reference, then the frame is not inertial and Newton's second law may not be valid in that frame.

3-5
Newton's second law

Newton's second law states that if an unbalanced force acts on a body, the body will be accelerated; the magnitude of the acceleration is proportional to the magnitude of the unbalanced force, and the direction of the acceleration is in the direction of the unbalanced force.

The significance of this statement can be brought out by considering some imaginary experiments. Suppose we pull on a spring that is attached to a cube of metal on a tabletop (see Figure 3-4). Since we are imagining this experiment we might as well imagine that the friction forces are zero, and that we are capable of pulling on the spring in such a way that the elongation of the spring is constant. Then the cube will be acted on by a constant unbalanced force in the horizontal direction, the force

F_1 exerted by the spring. The cube will have an acceleration a_1 in the direction of the force F_1 and in a time t it will travel a distance $s_1 = \frac{1}{2}a_1 t^2$. Using a calibrated spring we can then pull with a different force F_2. The block will then have an acceleration a_2 and if we pull for the same length of time it will travel a distance $s_2 = \frac{1}{2}a_2 t^2$. Thus we can determine the ratio a_1/a_2 from the measured ratio of the distances s_1/s_2 and we can determine the forces from the measured elongations of the spring. Newton's second law says that the ratio a_1/a_2 will equal the ratio F_1/F_2, or

$$\frac{F_1}{F_2} = \frac{a_1}{a_2}$$

This can be rewritten

$$\frac{F_1}{a_1} = \frac{F_2}{a_2}$$

An interpretation of this equation is that the ratio of the unbalanced force on a body to the acceleration produced by that force is a constant, which we shall call M, and which is an intrinsic property of the particular body being accelerated. Thus,

$$\frac{F_1}{a_1} = \frac{F_2}{a_2} = M$$

where M is called the *mass* of the body. If we were to repeat this experiment with a different cube with the same forces F_1 and F_2 we would get, in general,

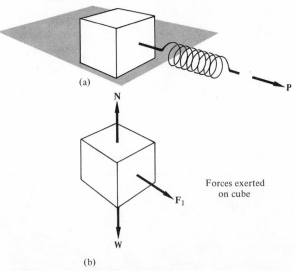

(a)

Forces exerted on cube

(b)

FIGURE 3-4 Forces on a cube which is being pulled across a frictionless tabletop.

different accelerations a_1' and a_2' and a different mass M'. Since

$$F_1 = Ma_1$$

and

$$F_1 = M'a_1'$$

we have

$$Ma_1 = M'a_1'$$

or

$$\frac{M}{M'} = \frac{a_1'}{a_1}$$

This means that if we apply the *same force* to two *different bodies*, the ratio of the accelerations is equal to the inverse ratio of the masses of the bodies. If the force is applied for the *same interval of time* to both bodies, the ratio of the accelerations will equal the ratio of the distances traveled, and the distances can be measured. Thus the ratio of the masses can be determined by the measured ratio of the distances traveled or

$$\frac{M}{M'} = \frac{s'}{s}$$

This gives the *ratio* of two masses. Now we can arbitrarily choose an object as our standard of mass (a specific piece of platinum, for example) and say it has a mass of one unit. Then we can measure the mass of any other cube (which we can call the unknown) by applying the same force to both the standard and the unknown, and measuring the distances traveled in equal times. The mass of the unknown will equal the ratio of the distance traveled by the unit mass to the distance traveled by the unknown, since

$$M = M'\left(\frac{s'}{s}\right)$$

$$M = (1)\left(\frac{s'}{s}\right)$$

With this understanding of how to measure the mass M of an object we can restate Newton's second law in symbols. In an inertial frame of reference,

$$\mathbf{F} = M\mathbf{a} \qquad (3\text{-}1)$$

where $\mathbf{a}$ is the acceleration of the body. With Newton's second law written in this form, it is easy to see that the larger the mass of the body, the smaller its acceleration when acted on by a given force. In other words, the larger the mass of a body, the less

it tends to change the state of its motion. But that is what we mean when we say an object has inertia. In that sense, mass is a measure of inertia.

Newton's second law, Equation 3-1, is among the most important scientific laws ever discovered. It says that if we know the force acting on any particle we can calculate its acceleration. Thus, if we know the acceleration at any time we can assume that the acceleration will remain constant for an instant, and so calculate the velocity and position of the particle at the end of that instant. This can be repeated for the next instant, the next, and so on. Thus, if we know the position and velocity of a particle at any time, and know what the force will be for all time, we can calculate the position and velocity of the particle at *any* future time. Therefore, if we know the laws that govern the forces, and the locations and velocities of all the particles in the universe at any given time, we should be able to *predict the positions of all the particles in the universe for any time in the future*. It is obvious that the amount of mathematics necessary to perform this job is huge, but such calculations on portions of the universe, such as the solar system, have been done. It is essentially this kind of program that permits astronomers to calculate the occurrence of eclipses and other astronomical events with such precision. The development of large digital computers has permitted such calculations to be performed very rapidly; this has been of utmost significance in the development of the space exploration program.

3-6
Systems of units

When numerical values are used with an equation involving physical quantities, such as Equation 3-1, these numerical values must be accompanied by appropriate units. There exist many units in use by scientists, technicians, and laymen, but usually in physics they must be used in consistent sets. For example, Newton's second law could have been written in the general form

$$\mathbf{F} = kM\mathbf{a}$$

where k is a constant whose value depends on the units of force, mass, and acceleration. For some choices of units, k turns out to be unity and for these units, the second law is written in the form of Equation 3-1. Such sets of consistent units are called systems of units. There exist many systems of units (as well as an enormous number of units that are not parts of complete systems of units), but this book will in general deal with only a few.

A common system among engineers and physicists is one called the mks (for meter-kilogram-second) system. In this system, the unit of mass is the kilogram, the unit of length is the meter, and the unit of time is the second. Thus the unit of acceleration is the meter per second per second. Suppose that a force acts on a 1-kg mass, and produces an acceleration of 1 m/sec². Then, from Equation 3-1 we get

$$F = 1 \text{ kg} \times 1 \frac{\text{m}}{\text{sec}^2} = 1 \frac{\text{kg m}}{\text{sec}^2}$$

This unit of force, the kilogram meter per second per second, is called a *newton*. Thus a force of 1 newton (1 nt) will give a 1-kg mass an acceleration of 1 m/sec². To illustrate, consider the force necessary to give a body of mass 50 kg an acceleration of 15 m/sec². From Equation 3-1, we have

$$F = 50 \text{ kg} \times 15 \frac{\text{m}}{\text{sec}^2} = 750 \text{ nt}$$

Another system, in use particularly among physicists, is the cgs system, where the unit of mass is the gram (1 kg = 1000 gm), the unit of length is the centimeter (1 m = 100 cm), and the unit of time is the second. In this system of units, the unit of force is a *dyne*; a force of 1 dyne will give a 1-gram object an acceleration of 1 cm/sec². It is left as an exercise to the reader to show that 1 newton = 100,000 dynes.

In the United States some engineers use systems of units based on the common units of feet, pounds, and seconds. In this book we will sometimes use one such system because most students are familiar with the units. In this system, the unit of length is the foot, the unit of time is the second, and the unit of force is the pound. The unit of mass thus becomes, from Equation 3-1,

$$m = \frac{1 \text{ lb}}{1 \frac{\text{ft}}{\text{sec}^2}}$$
$$= 1 \frac{\text{lb sec}^2}{\text{ft}}$$

In all systems of units a serious confusion often arises between the ideas of weight and mass. As we have seen, the mass of a body is a measure of its inertia. This resistance to change of motion is a characteristic of the body that is independent of its location. That is, the mass of a particular body is a property of that body, a property that is constant, unchanged by moving the body. There is no universally accepted scientific definition of weight, but it usually means the magnitude of the gravitational force exerted on a body by the earth (although it sometimes means the gravitational force exerted by the nearest astronomical body, as in discussions of the weight of an astronaut on the moon). As we will see, this force varies from point to point on the surface of the earth, and varies with altitude. That is, the weight of a body depends on the location of the body. Further, it is a property of the earth *and* the body. Thus, the two properties, mass and weight, are different and behave differently. However, as we will see in Chapter 5, the weight of a body is proportional to the mass of the body. If the body remains near the surface of the earth, the factor of proportionality is nearly constant, and is numerically equal to *g*, the acceleration due to gravity; that is,

$$W = mg \qquad (3\text{-}2)$$

Therefore, the weight of an object whose mass is 1 lb sec²/ft is 32 lb, and the weight of an object whose mass is 1 kg is 9.8 nt. It must be emphasized that this equation gives a relation between two numbers characteristic of a body. In no way does the presence of the quantity *g* imply that the body is accelerating, nor does it imply that the equation is restricted to bodies in free fall. The confusion arises partly from the fact that when one measures the weight of an object with an instrument, one can always mark the dial of the instrument with either units of force or units of mass, since they are proportional to each other. That is, we can imagine calibrating a balance (which measures the weight of objects) in pounds, but instead of marking the scale 1, 2, 3, . . . lb, we could inscribe "1 lb sec²/ft" (sometimes called a "slug") at the point where we would have written 32 lb, and similarly 2, 3, . . . lb sec²/ft at the 64-lb. 96-lb, . . . points. We would end up with a balance which measured pounds (weight), but *reads* lb sec²/ft (mass).

A common analog is the speedometer of an automobile, which *measures* typically, the rate of rotation of the drive shaft, but which *reads* the speed of the automobile. Note that if a car is up on blocks, the speedometer can read 60 mi/hr while the car is really standing still. Unfortunately, most balances read either pounds (unit of force), or kilograms (unit of mass), or grams (another mass unit), so that in nontechnical language, one speaks of an object "weighing" 50 grams, or 10 kilograms. In technical discourse one should always refer to forces and weights in units of pounds, dynes, or newtons; and one should always refer to masses in units of slugs, grams, and kilograms. To avoid confusion, phrases like "a 50-gram weight" should be avoided.

3-7
Application of Newton's second law

In the following we shall consider some typical applications of Newton's second law. The general procedure involves the following steps:

1. Choose one body to which to apply the law and which is acted upon by forces.
2. Draw a sketch of that body and draw arrows representing all of the forces that act on the body. This step is called "drawing a free-body diagram."
3. Apply Newton's second law. This may mean choosing a set of coordinate axes, resolving the force into components, and then writing Newton's second law appropriately for each axis.
4. Solve these equations. For some problems the process must be repeated for a second body (and sometimes for a third, fourth and so on) in order to obtain enough equations to solve the problem completely.

It must be noted that this procedure must be used with intelligence, skill, and sometimes with experience in order to solve problems. The procedure alone cannot be thought of as a substitute for these attributes.

In step 2 of the procedure described above, all of the forces acting on the body must be drawn and then used in step 3. For a very large class of problems, the considerations of Section 3-3 make the problem of identifying the forces acting on a body relatively simple. That is, one can usually be sure that all the forces have been included by first drawing in the gravity force (weight), and then drawing the forces exerted by the bodies that are in contact with the free body. It is important to remember that the only gravity force that acts on a body is its own weight. For example, consider one body resting on another, as in Figure 3-5. The forces acting on body B are its weight W_B and a contact force N_{AB} exerted by body A. The force exerted on body A are *its* weight W_A, a contact force N_{BA} exerted by body B, and a contact force N_{TA} exerted by the table on which A is resting. The weight of B does *not* act on A; although in a figurative sense, the

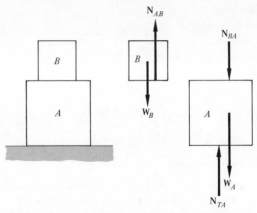

FIGURE 3-5 The forces on each of two bodies which are at rest, with one body on top of the other.

weight of B is transmitted to A by means of the contact force at the surface.

Illustrative example

A 2-kg block rests on a frictionless table. Then a horizontal force of 3 nt is applied to the block. Find the acceleration of the block.

The situation is drawn in Figure 3-6a, and the free-body diagram is in Figure 3-6b. Since the horizontal table is said to be frictionless, the contact force is a normal force N only, with no tangential component. Since all of the forces are either horizontal or vertical, the resolution into components is automatic. Then, for the vertical direction, we know that the block will neither rise spontaneously nor fall through the table, so from Equation 3-1 we have

$$N - W = ma_y$$
$$= 0$$

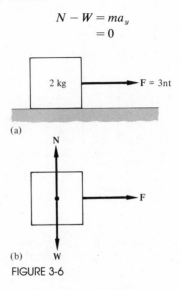

FIGURE 3-6

Therefore

$$N = W$$

Considering the horizontal direction,

$$F = ma_x$$
$$a_x = \frac{F}{m} = \frac{3 \text{ nt}}{2 \text{ kg}}$$
$$= 1.5 \frac{\text{m}}{\text{sec}^2}$$

Illustrative example

A 3000-lb car accelerates from 0 to 60 mi/hr in 8 sec. What force does the road exert on the car?

Here the body is obviously the car. A free-body diagram for the car might look like Figure 3-7a, where the forces are indicated as W, the weight of the car, and the forces exerted by the road (the only body in contact with the car except for air), as N_1, N_2, f_1, and f_2. Since cars are really three-dimensional, one should draw all four wheels and indicate the forces separately for each wheel; but

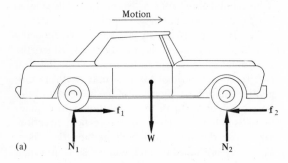

(a)

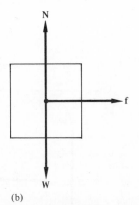

(b)

FIGURE 3-7 (a) The forces acting on an automobile. (b) The equivalent free-body diagram from which the total force exerted by the road can be calculated.

one can assume that the car is symmetric and that the forces on the left wheels are the same as those on the right. In fact, since the problem concerns itself with the motion of the car as a whole, and asks for *the force* exerted by the road, the free-body diagram might as well be Figure 3-7b, in which the car is represented by a square and the forces are the weight of the car, W, the forward component of the force exerted by the road, f, and the upward component of the force exerted by the road, N. For the horizontal direction Equation 3-1 yields

$$f = ma$$

whereas for the vertical direction

$$N - W = 0$$

The second equation gives the vertical component of the force exerted by the road N, as equal to W, the weight of the car, which is 3000 lb. The first equation gives the horizontal component as equal to ma, but in order to give a number for this force, we must do some arithmetic. Since the car weighs 3000 lb, the mass m is given by Equation 3-2:

$$m = \frac{W}{g} = \frac{3000 \text{ lb}}{32 \frac{\text{ft}}{\text{sec}^2}} = 93.7 \frac{\text{lb sec}^2}{\text{ft}}$$

Since we have assumed that the forces are constant, we may use the formulas for constant acceleration and find

$$a = \frac{v - v_0}{t}$$
$$= \frac{60 \frac{\text{mi}}{\text{hr}}}{8 \text{ sec}} = \frac{88 \frac{\text{ft}}{\text{sec}}}{8 \text{ sec}}$$
$$= 11 \frac{\text{ft}}{\text{sec}^2}$$

Therefore

$$f = 93.7 \frac{\text{lb sec}^2}{\text{ft}} \times 11 \frac{\text{ft}}{\text{sec}^2}$$
$$= 1030 \text{ lb}$$

Illustrative example

A 40-kg magician's assistant is lying on a smooth table (on stage). A very strong, thin wire attached to her passes over a smooth pulley (offstage). A 30-kg sandbag is fastened to the other end of the wire, as shown in Figure 3-8. Find the acceleration

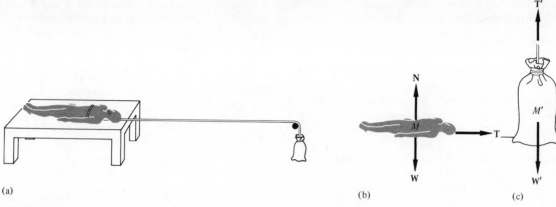

(a)

(b)   (c)

FIGURE 3-8

of the assistant and the sandbag, and the tension in the wire.

Here we obviously have two bodies, so we must draw two separate free-body diagrams, which are shown in Figure 3-8b and c. It is important to notice that the weight of the sandbag does not act on the assistant. The only forces acting on her are her weight W, the normal force N, and the pull of the wire, T. Since the table is smooth, we assume that no horizontal force (friction) is exerted by it. The force it exerts is N, normal to the table. The wire exerts a force T, in the horizontal direction. Applying Newton's second law to the assistant, we have:

$$T = Ma \qquad (3-3)$$

and

$$N - W = 0 \qquad (3-4)$$

Applying the law to the sandbag, we have

$$W' - T' = M'a' \qquad (3-5)$$

We are immediately struck by the fact that there are more unknowns than equations. Even if we use the facts that

$$W = Mg$$

and

$$W' = M'g$$

we still have T, T', a, a', and N as unknowns in Equations 3-3, 3-4, and 3-5. We still have too few equations. It is not very difficult to see that if the wire has a fixed length then the acceleration of the sandbag is equal to the acceleration of the assistant,

so we can write

$$a = a' \qquad (3-6)$$

In a later section we will show that *a massless rope or wire always exerts equal forces at its ends*, and this force is called the tension in the rope or wire. Thus we can write:

$$T = T' \qquad (3-7)$$

With these two additional equations, we can proceed:

$$W = Mg \ = 40 \text{ kg} \times 9.8 \ \frac{\text{m}}{\text{sec}^2} = 392 \text{ nt} = N$$

$$W' = M'g = 30 \text{ kg} \times 9.8 \ \frac{\text{m}}{\text{sec}^2} = 294 \text{ nt}$$

$$W' - T' \ = 294 - T = M'a = 30a$$
$$T = Ma \ = 40a$$

Combining the last two equations we get

$$294 - 40a = 30a$$

or

$$a = \frac{294}{70} = 4.2 \ \frac{\text{m}}{\text{sec}^2}$$

Then we can insert this value of a in

$$T = 40a$$
$$= 40 \times 4.2$$
$$= 168 \text{ nt}$$

It is interesting to note that this tension is less than the weight of the sandbag (294 nt). Thus if we had assumed that the weight of the sandbag acted on the magician's assistant we would have gotten the wrong number for the acceleration of the assistant.

3-8
Newton's third law

In the preceding discussion we have described the effect of a force on a body, and in Chapter 5 we shall discuss the specific characteristics of some forces. Here we will state a property of forces, a property described by *Newton's third law. Newton's third law states that whenever one body exerts a force on a second body, the second body always exerts an equal and opposite force on the first body*. This statement, which is sometimes called the law of action and reaction, says that forces come in pairs; that it is impossible to exert a force without having a second, equal, opposite force exerted back on the body that exerted the first force.

The significance of this law can be illustrated by considering several examples. If we push a trunk along the floor, the trunk exerts a force against our hands. If a ball is hit with a bat, there is not only a force exerted by the bat on the ball, but also a force exerted by the ball on the bat. The sun exerts a gravitational force on Halley's comet, and Halley's comet exerts an equal and opposite force on the sun.

For example, consider the forces involved when a rope is pulling a box upward, as shown in Figure 3-9a. Figure 3-9b shows the free-body diagram of the rope. The forces acting on the rope are: its weight, w; the force exerted by the box on the rope, T; and the pull exerted by some external agent, P. Figure 3-9c shows the free-body diagram of the box with the weight of the box, W, and the pull

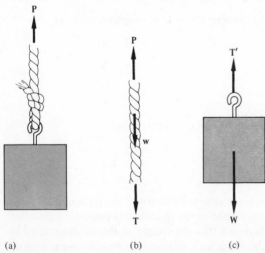

(a)   (b)   (c)

FIGURE 3-9 Forces exerted when a rope is pulling a box upward.

exerted by the rope, T'. The forces T and T' are an action-reaction pair, and so are equal, no matter what the state of the motion of the box and rope. Because the rope and the box are connected together we know that the acceleration of the rope is the same as the acceleration of the box. Applying Newton's second law to the rope only, we have

$$P - w - T = ma$$

where

$$m = \frac{w}{g}$$

Applying Newton's second law to the box only, we have

$$T - W = Ma$$

where

$$M = \frac{W}{g}$$

In the special case where $a = 0$, we have

$$T = W$$

and

$$P = w + T = w + W$$

Thus in the case in which the rope and box are at rest (or moving with constant velocity), we see that the external agent must exert a force on the end of the rope equal to the sum of the weights of the rope and box.

Although weightless ropes do not really exist, there are many situations in which the weight of the rope (or wire or cable) is so small compared with the other forces that the error introduced by neglecting the weight is small. A special case of this last example is the one in which the weight of the rope is negligible. In that case

$$P - T = ma$$

becomes

$$P - T = 0$$

or

$$P = T$$

Thus for a weightless rope the pull exerted on the upper end of the rope by the external agent, P, is equal to the pull exerted at the other end of the rope, T, no matter what the acceleration of the rope. The effect of the two external forces P and T

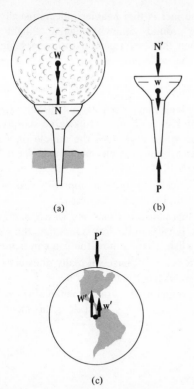

FIGURE 3-10 Forces exerted when a golf ball is on a tee.

acting on the rope is to tend to stretch it; the rope itself is said to be under *tension* and the magnitude of the tension is the value of either force P or T.

A rope can be thought of as providing a way for one object to pull on another remote object. In solving problems, it is important to remember that the tension in a rope is rarely equal to the weight of the object to which it is attached. In general it is safest to assume that the tension is an unknown or arbitrary quantity. In addition it is useful to remember that ropes can only pull, they cannot push.

As another example, consider a golf ball on a tee. Figure 3-10a shows the ball on the tee and the forces acting on the ball: the weight of the ball, W; and an upward force, N, exerted by the tee.

The forces exerted on the tee are shown in Figure 3-10b: the weight of the tee, w; an upward force exerted by the ground, P; and the downward force exerted by the ball, N'. Thus, there are five forces exerted on the golf ball and the tee: W, w, P, N, and N'.

The force N is exerted by the *tee* on the *ball* and N' is exerted by the *ball* on the *tee*. These two

forces are the action-reaction pair described by Newton's third law and so must be equal. It must be emphasized that these forces, N and N', are equal no matter what the motion of the ball or tee. They are equal if the ball is at rest or if the ball has just been dropped on the tee, or if the ball is just leaving the tee after having been struck by the club.

The force W is exerted by the earth on the ball. Therefore the reaction force is a force exerted by the ball on the earth. This force is shown as W' in Figure 3-10c. Similarly, the force w is exerted by the earth on the tee and the reaction to w is shown as a force, w', acting on the earth in Figure 3-10c. The force P is exerted by the ground on the tee and the reaction is the force P', which acts on the surface of the ground.

If we know something about the motion of the ball and the tee we can make additional statements. The simplest case is that where the ball is at rest and the tee is at rest. In that case we know that the sum of the forces on the ball is equal to zero and therefore

$$W = N$$

and that the sum of the forces on the tee is equal to zero and therefore

$$P = N' + w$$

Since N and N' are an action-reaction pair

$$N = N'$$

and so

$$P = N + w$$
$$= W + w$$

Since P and P' are an action-reaction pair,

$$P = P'$$

and so

$$P' = W + w$$

In other words, *if we know that both the ball and the tee are at rest*, then we know that the force exerted on the ground is equal to the sum of the weights of the ball and the tee. This simple, almost self-evident conclusion follows from *both* the second law and the third law. In particular, W equals N, not because they are an action-reaction pair (they act on the same body and cannot be an action-reaction pair) but because the ball is assumed to be at rest.

Questions

1. The moon does not travel in a straight line. What force accounts for this?

2. Describe the forces acting on a pole vaulter during a leap.

3. When a horse pulls a wagon, Newton's third law says that the force exerted by the horse on the wagon is equal to the force exerted by the wagon on the horse. This is erroneously interpreted to mean that the wagon cannot move. Explain how the wagon moves in spite of this equality of forces.

4. Describe the limitations on the word "body" in Newton's laws as we have stated them.

5. Define "net force."

6. Define "inertial coordinate system."

7. A man stands in a bus, facing forward, when the bus suddenly starts moving forward. What happens to the man?

8. A heavy object has a larger gravitational force exerted on it than a light object. Why would you expect them to fall at the same rate?

9. The fact is that heavier objects usually do hit the ground somewhat ahead of light objects when they are released from the same height at the same time. Why?

10. List some units of mass and some units of weight.

11. A stage magician rolls a ball across a table. The ball is observed to swerve sharply to one side when it reaches the middle of the table. Describe a possible mechanism for deflecting the ball.

12. Physically, why is it impossible to lift one's self up by one's bootstraps?

13. Fisherman often use line which breaks when the tension in the line is less than the weight of the fish. Explain how the fish can be reeled in.

14. If a "curved" baseball really does curve, what body exerts the force on it?

15. Astronauts are sometimes described as being "weightless." What does this mean?

Problems

1. What is your own mass in kilograms? What is your own weight in dynes?

2. A 3-gm table-tennis ball is partially supported by the wind, which exerts an upward force of 1000 dynes. What is the net force acting on the ball?

3. A 4000-lb elevator contains a 200-lb passenger and a 120-lb passenger. The tension in the cable is 3800 lb. What is the net force on the elevator?

4. What force is required to give a 500-gm object an acceleration of 150 cm/sec^2?

5. What constant force is required to give a body weighing 175 lb an acceleration of 8 ft/sec^2?

6. A body acquires an acceleration of 200 cm/sec^2 when subjected to a net force of 740 dynes. What is the mass of the body?

7. A net force of 6 nt is applied to a body whose mass is 3 kg. Find the resulting acceleration.

8. If the acceleration of a certain automobile exceeds 80 percent of the acceleration due to gravity, the wheels begin to spin. If the automobile weighs 3800 lb, find the maximum horizontal force that can be exerted by the ground on the wheels before the wheels spin.

9. A man is pulling a 10-lb sled on which his 40-lb child is riding. The man exerts a force of 23 lb and the acceleration of the sled is 3.4 ft/sec^2. What is the retarding force exerted by the snow on the runners of the sled?

10. A parachutist whose mass is 70 kg and whose equipment, including parachute, has a mass of 20 kg, is falling with an acceleration of 2.4 m/sec^2. What is the total resistive force?

11. A raindrop whose mass is 0.1 gm is subjected to resistive forces of 36 dynes. What is the acceleration of the drop? How large would the resistive forces have to be in order that the drop fall with constant velocity?

12. An automobile weighing 3000 lb starts from rest and acquires a speed of 45 mi/hr in 8 sec. Assuming the acceleration was constant, find the unbalanced force which acted on the car while it was accelerated. What body exerted this force?

13. A 100-gm saltshaker is sliding on a diner counter with an initial speed of 10 cm/sec. It comes to rest after having traversed 300 cm. What friction force was acting on the shaker?

14. An electron, starting from rest, acquires a

speed of 3.5×10^8 cm/sec when an electric field acts on it for a distance of 2 cm. (a) Determine the acceleration of the electron, assuming it to be constant. (b) Find the force acting on the electron. The mass of the electron is 9×10^{-28} gm.

15. A 150-lb spectator at a baseball game rises suddenly to cheer a winning home run. Assume that his legs from the knee down remain stationary and that the rest of his body (weighing 130 lb) is given an upward acceleration of 8 ft/sec². What force is required to produce this acceleration? What force is exerted by the floor on the spectator's feet?

16. A steel cable supports an elevator weighing 3500 lb. Starting from rest, the elevator acquires an upward velocity of 700 ft/min in 2 sec. (a) What is the unbalanced force acting on the elevator? (b) What is the tension in the cable?

17. A 12,000-lb truck is moving at 30 mi/hr. The truck has a forward force of 340 lb acting on it and a resistive force of 208 lb. Assume that both forces remain constant, (a) Find the acceleration of the truck (b) Find the speed of the truck after 8 seconds. (c) How far will it have traveled in the 8 sec?

18. A balloon is descending with a constant acceleration a which is less than g, the acceleration due to gravity. The weight of the balloon, basket and contents is W. What weight w of ballast should be dropped overboard in order that the balloon rise with an acceleration equal to a. Assume that dropping the ballast does not affect the forces due to the air.

19. A rifle bullet, whose mass is 1.8 gm, is traveling with a speed of 36,000 cm/sec when it strikes a block of wood. It penetrates the wood to a depth of 10 cm. Assume that the resistive force was constant, (a) Find the time it took for the bullet to stop and (b) the magnitude of the resistive force.

20. If the bullet of Problem 19 was fired from a rifle whose barrel was 70 cm long, find the average force exerted on the bullet while it was in the rifle. Assume that the speed of the bullet was unchanged in the flight from the barrel to the block of wood.

21. A fisherman is using line which breaks if the tension exceeds 8 lb. He catches a 1.5 lb fish. (a) What is the least time in which he can reel in 100 yd of line assuming that the line can be treated as horizontal and that the horizontal resisting force is 1.2 lb? (b) With what acceleration can he lift the fish out of the water?

22. A 60,000-lb rocket has an average upward force of 90,000 lb acting on it. How long does it take for the rocket to rise 100 m? How fast is it moving when it has risen 100 m?

23. A $5\frac{1}{2}$-oz baseball reaches the bat at a speed of 80 ft/sec. The ball remains in contact with the bat for 0.002 sec and then leaves the bat with a speed of 100 ft/sec, moving directly opposite to its initial motion. How large is the average force exerted by the bat?

24. A 3800-lb automobile is traveling at 60 mi/hr when the driver applies the brakes. After traveling 180 ft, the automobile has a speed of 30 mi/hr. What was the average force acting on the car?

25. A 2000-kg tow truck is pulling a 1200-kg car with a towrope, which can be assumed to be massless. Both vehicles have a forward acceleration of 1.8 m/sec². The ground exerts a 1400-nt backward friction force on the car. (a) Find the forward force exerted on the tow truck. (b) What body exerts this forward force? (c) Find the tension in the towrope.

26. A 1000-lb horse is dragging a 500-lb sled. The tension in the connecting lead between horse and sled is 340 lb and the friction force on the sled is 125 lb. (a) Find the acceleration of the horse and sled. (b) Find the forward force exerted on the horse.

27. Two masses m_1 and m_2 are attached to the ends of a cord that passes over a frictionless pulley, as shown in Figure 3-11 (such a system is sometimes called an Atwood machine). Assume that m_1 is greater than m_2 and show that the acceleration of the masses is given by

$$a = \frac{m_1 - m_2}{m_1 + m_2} g$$

and that the tension in the cord is given by

m_1 m_2 FIGURE 3-11

$$T = \frac{2m_1 m_2}{m_1 + m_2} g$$

28. A second cord is tied to the bottom of m_2 in Figure 3-11. What tension must be exerted by this cord in order that the masses move with constant speed?

29. A box whose mass is 2500 gm is on a horizontal table. A cord tied to this box passes over a frictionless pulley at the edge of the table. A cylinder whose mass is 600 gm is hung from the free end of the cord. The friction force acting on the box can be assumed to be negligible. (a) What is the acceleration of the box? (b) What is the acceleration of the cylinder? (c) What is the tension in the cord?

30. Repeat Problem 29, assuming that the friction force on the box is equal to 0.1 times the weight of the box.

31. A man of weight W is standing on a weighing scale which is on the floor of an elevator. If the elevator has an acceleration upward of a, show that the reading of the scale is F, where

$$F = W + \frac{W}{g} a$$
$$= m(a + g)$$

32. The stomach of a particular man weighs 4 lb. The man steps into an elevator which is then accelerated downward at 12 ft/sec². What force is exerted on his stomach by the rest of his body? (The fact that this force is different from the usual force of 4 lb is detected by the body. This results in the "funny feeling" most people experience when in an accelerating elevator.)

33. A 400-lb crate is on the floor of a truck. The truck is moving at 45 mi/hr when it slows down to 20 mi/hr in 2 sec. If the crate remains at rest relative to the truck during the deceleration, what force acts on the crate?

34. A body whose mass is 5 kg is placed on a smooth inclined plane that makes an angle of 30° with the horizontal. A cord attached to this body passes over a frictionless pulley; a second body whose mass is 12 kg is attached

to the free end of the cord as shown in Figure 3-12. Determine (a) the acceleration of the system and (b) the tension in the cord.

35. Repeat Problem 34 assuming that a friction force acts parallel to the plane and has a magnitude equal to 0.1 the weight of the 5-kg object.

36. A 4-kg block is on top of a 6-kg block, which is on a smooth frictionless table. A 10-nt force is applied horizontally to the 6-kg block. (a) Find the acceleration of the system. (b) Find the force exerted by the top block on the bottom block.

37. A 3-kg object and a 5-kg object are attached to the ends of a 5-m cord which passes over a fixed frictionless pulley. The pulley is 4.5 m above the floor. The objects are held at rest with the 3-kg object on the floor and then released. (a) How much time will elapse before the 5-kg object hits the floor? (b) How much time will have elapsed when the objects pass each other?

38. Three identical rocket ships, each having a mass of 3000 kg, are tied together, one behind the other. Only the first ship's engine is turned on, exerting a forward force of 2000 nt. Find (a) the acceleration of each ship and (b) the tension in each of the two connecting cables.

39. Show that if the force due to the resistance of air varies with the square of the velocity of a falling body, the limiting velocity of fall is proportional to the square root of the weight of the body.

40. A man is lifting a 100-lb bag using a cord which passes over a pulley. If the cord breaks when the tension in it exceeds 140 lb, what is the fastest time in which the man can lift the bag through a height of 15 ft?

41. One end of a weightless rod of length r is attached to the roof of a truck by means of a frictionless pivot. A ball of mass m is attached to the free end of the rod. When the truck accelerates, the rod swings over so that it makes an angle θ with the vertical. Find θ if the acceleration of the truck is equal to a.

42. Two men, one weighing 175 lb and the other weighing 130 lb, are on ice skates. Each holds one end of a taut rope. The heavier man exerts a force of 25 lb on the rope. Assume that the friction force exerted on the skates is negligible. (a) How big a force does the lighter man exert? (b) What is the acceleration of each man?

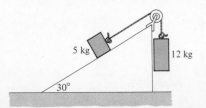

5 kg   12 kg

30°

FIGURE 3-12

statics

chapter 4

4-1
Equilibrium of a particle

Most of our primitive notions about force come from our exertion of muscular efforts on objects, often on objects at rest. We understand force because we have pushed on tables, sat on chairs, and tried to lift heavy trunks. Therefore the study of objects at rest has a basic intrinsic interest. This subject also has much practical importance. Carpenters want to build bookcases that will support heavy loads, engineers want to design bridges that will not fall, and surgeons want to pin a broken hip so that it will permit its owner to stand.

If a body is at rest and remains at rest it is said to be in *static equilibrium*. If we know that a body is in static equilibrium and therefore that it has acceleration equal to zero, then from Newton's laws we know that the net force acting on the body is equal to zero. Since the net force acting on a body is, in general, the sum of several forces, this can be written: *If a body is in static equilibrium, then the vector sum of the external forces acting on the body is zero* or, in symbols,

$$\Sigma \mathbf{F} = 0 \qquad (4\text{-}1)$$

If each of the forces is resolved into rectangular components along three mutually perpendicular axes, x, y, and z, then this vector equation can be written as three algebraic equations:

$$\Sigma F_x = 0 \qquad (4\text{-}2a)$$
$$\Sigma F_y = 0 \qquad (4\text{-}2b)$$

and

$$\Sigma F_z = 0 \qquad (4\text{-}2c)$$

Equations 4-1 and 4-2 are known to be true if a body is known to be at rest and remain at rest. It is quite a different matter to determine the *conditions* under which a body will remain at rest. In the most general case we would have to describe the conditions under which a body would maintain a fixed size and shape, not rotate and not translate (see Section 4-3). In order to simplify the discussion we will first restrict ourselves to considering objects that can be treated as points, objects we will call particles. If we do that we can ignore deformations and rotations.

If the net force on a particle is equal to zero, then from Newton's laws we know that the acceleration of the particle is equal to zero and that the particle is moving with constant velocity. If the net force on a particle is equal to zero, then the particle is said to be in *equilibrium*. If the velocity of a particle that is in equilibrium is a constant and not equal to zero, then the particle is in *dynamic* equilibrium; if the particle is at rest it is said to be in static equilibrium. Thus Equations 4-1 and 4-2 are conditions for equilibrium; if the equations are valid, then we know that the particle will remain at rest if it starts at rest and will continue moving in a straight line with constant speed if it is initially moving.

It should be emphasized that we have introduced nothing new; equilibrium is merely a special case, and Equation 4-1 is a special case of Newton's second law. Here, as in Chapter 3, we have restricted the discussion to particles, rather than extended bodies. However, even fairly large bodies may be treated as particles if the lines of action of all the external forces acting on it pass through a single point. Such a system of forces is said to be *concurrent*. If the forces acting on a rigid body are concurrent, then the state of rotation of the body will remain constant; concurrent forces do not tend to make a body rotate.

Thus, if we know that a set of forces is concurrent *and* that the sum of the forces is equal to zero, then we know that the body on which the forces act is in equilibrium.

Illustrative example

Consider the case of a body of weight **W** placed on a smooth plane inclined at an angle θ to the horizontal as in Figure 4-1. Suppose we want to find the force **F** parallel to the incline, needed to keep this body in equilibrium. As in Chapter 3, the first thing that must be done is to decide which forces act on the body. In this case the forces are: the weight **W** acting down, the force **F** acting parallel to the plane, and the push **N** of the plane. Stating that the plane is smooth implies that the force exerted by the plane on the body acts perpendicular to the plane (see Section 4-5). We can solve for **F** by drawing the vector triangle of Figure 4-1b. This vector triangle is similar to the triangle *ABC* of Figure 4-1a, and the angle between the vectors **W** and **N** is equal to the angle θ. From the similarity of the triangles we have

$$\frac{F}{W} = \frac{h}{l} \qquad (4\text{-}3)$$

and

$$\frac{N}{W} = \frac{b}{l} \qquad (4\text{-}4)$$

in which h is the height of the plane, l is the length of the plane, and b is the length of the base.

Solving Equation 4-3 for F, we have

$$F = W \frac{h}{l} \qquad (4\text{-}5a)$$
$$= W \sin \theta \qquad (4\text{-}5b)$$

Thus the force **F** necessary to keep a body in equilibrium (at rest, or moving with constant speed up or down the plane) is only a fraction of the weight **W**; this fraction is determined by the ratio of the height to the length of the plane, or the sine of the angle θ.

Similarly, solving Equation 4-4 for **N**, we get

$$N = W \frac{b}{l} \qquad (4\text{-}6a)$$
$$= W \cos \theta \qquad (4\text{-}6b)$$

Thus the force **N** that an inclined plane exerts on a body placed on the plane is only a fraction of the weight of the body; this fraction is the ratio of the base of the plane to the length of the plane, or the cosine of the angle θ.

For example, suppose that a box weighing 150 lb is placed on a smooth inclined plane 20 ft long

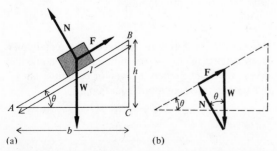

FIGURE 4-1 (a) Forces acting on a body that is in equilibrium on a smooth inclined plane. (b) Triangle of forces.

with the higher end 10 ft above the ground. The force **F** parallel to the plane necessary to keep the box in equilibrium is, from Equation 4-5a,

$$F = 150 \text{ lb} \times \frac{10 \text{ ft}}{20 \text{ ft}} = 75 \text{ lb}$$

Since the base b for this inclined plane is $10\sqrt{3}$ ft, the push **N** perpendicular to the plane is, from Equation 4-6a,

$$N = 150 \text{ lb} \times \frac{10\sqrt{3} \text{ ft}}{20 \text{ ft}} = 130 \text{ lb}$$

An alternate way to solve this problem is to resolve the force **W** into components parallel to and perpendicular to the plane. Then, from Equations 4-2a and b, we have

$$N - W_\perp = 0 \tag{4-7}$$

or

$$N - W \cos \theta = 0$$

and

$$F - W_\parallel = 0 \tag{4-8}$$

or

$$F - W \sin \theta = 0$$

Illustrative example

A rope 10 ft long and of negligible weight has one end, A, attached to a beam, while a weight of 80 lb is attached to the other end, B, as shown in Figure 4-2a. A man pushes on this weight in a horizontal direction. Determine how big a force the man exerts if the rope stays at an angle of 30° with the vertical. Also determine the tension in the rope.

There are three forces acting on the body, its weight **W** = 80 lb, the (unknown) horizontal force **F**, and the pull **S** exerted by the rope. Resolving **S** into horizontal and vertical components, we have

$$S_x = S \sin 30°$$
$$S_y = S \cos 30°$$

Then Equations 4-2a and b become

$$F - S_x = 0$$
$$F = S_x$$
$$= S \sin 30° \tag{4-9}$$

and

$$S_y - W = 0$$
$$S_y = W$$
$$S \cos 30° = W \tag{4-10}$$

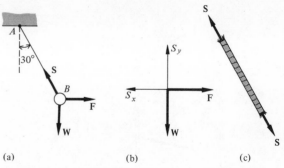

Figure 4-2 (a) The weight at B is supported by the rope which is attached to the beam at A. The horizontal force **F** is exerted by a man. (b) The forces exerted on the weight. (c) The forces exerted on the rope.

From Equation 4-10 we solve for S:

$$S = \frac{W}{\cos 30°}$$
$$= \frac{80 \text{ lb}}{\dfrac{\sqrt{3}}{2}} = 92.5 \text{ lb}$$

and from Equation 4-9

$$F = 92.5\left(\tfrac{1}{2}\right)$$
$$= 46.3 \text{ lb}$$

The rope is of course also in equilibrium. If we neglect the weight of the rope, the force diagram is that shown in Figure 4-2c and the tension in the rope is equal to S, the pull exerted by the rope. It will be noticed that the tension is greater than the weight.

4-2
Moment of force; torque

In order to extend the discussion of the preceding section to describe the conditions for equilibrium of extended bodies that are subjected to nonconcurrent forces we must introduce a way of describing the effectiveness of a force in changing the state of rotation of a body. The effectiveness of a force in changing the state of rotation of a body is called the *torque*, or *moment of force*, and depends not only on the magnitude and direction of the force but on the location of the point of application of the force.

To illustrate this, consider the process of pushing open a door, as schematically shown in Figure 4-3. It is common experience that the larger the distance r, the easier it is to open the door; that is,

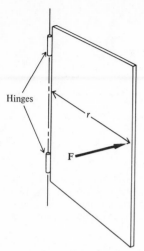

FIGURE 4-3 The force **F** is most effective at opening the door if it is perpendicular to the door and if r is as large as possible.

the greater the distance between the point of application of the force and the line of the hinges, the more effective the force is in turning the door. Further, it is almost intuitive that the force **F** should be perpendicular to the door. Any component of the force **F** that is parallel to the surface of the door is ineffective in turning the door.

In order to make this idea precise, consider a flat object (which can be treated as essentially two-dimensional) that is constrained to rotate about an axis perpendicular to the object, as shown in Figure 4-4. Then consider that a force **F**, which lies in the plane of the object, is applied at a point P, which is a distance r from the axis. The line r makes an angle θ with the force **F**. The force **F** has a moment or torque about the axis that passes through the

point O. The magnitude of the torque τ about this axis is defined as the product of the distance r and the component of the force which is perpendicular to the line OP. Since the component of **F** which is perpendicular to OP, $F_\perp$, is given by

$$F_\perp = F \sin \theta \qquad (4\text{-}11)$$

this can be written

$$\tau = rF \sin \theta \qquad (4\text{-}12)$$

Equation 4-12 can be given another interpretation by drawing a line OP' from the axis to the line of action of the force, where OP' is perpendicular to the line of action of the force. Then the length r' of the line OP' is given by

$$r' = r \sin \theta \qquad (4\text{-}13)$$

Combining Equations 4-13 and 4-12, we get

$$\tau = Fr' \qquad (4\text{-}14)$$

or, in words, the torque is equal to the product of the force and the perpendicular distance from the axis to the line of action of the force.

The force **F** will produce a rotation as shown in Figure 4-4, a rotation that can be described as counterclockwise when viewed from above. In Figure 4-5 the force F_1 will produce a counterclockwise rotation, while the force F_2 will produce a clockwise rotation. By convention, counterclockwise torques are usually called positive, while clockwise torques are usually called negative. With this convention, the net torque on the body shown in Figure 4-5 is

$$\tau = r_1 F_1 - r_2 F_2$$

The units of torque are those appropriate to the

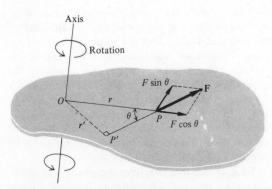

FIGURE 4-4 The magnitude of the torque about the axis is equal to $rF \sin \theta$.

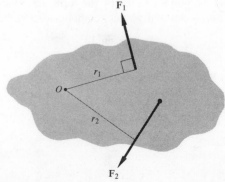

FIGURE 4-5 Torques produced by two forces, F_1 and F_2, about an axis through O.

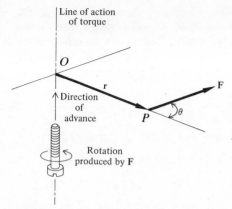

FIGURE 4-6 The right-hand screw rule for the direction of torque.

product of a force and a distance, such as a newton-meter, or pound-foot.

For a large class of elementary problems, this two-dimensional description of torque with a fixed axis is adequate, but it is interesting to extend the discussion to three dimensions.

Consider a force **F**, which acts at some point P, as shown in Figure 4-6. Then consider a vector **r**, which is drawn from a point O to the point P. We can then define a *vector*, torque, whose magnitude τ is given by

$$\tau = rF \sin \theta \qquad (4\text{-}15)$$

where θ is the angle between the vector **r** and the vector **F**. We can define the *direction* of this vector torque in two steps:

1. The *line of action* of the vector passes through O and is perpendicular to the plane formed by **r** and **F**.
2. The sense of the vector is given by a *right-hand screw rule*. To state the right-hand screw rule we imagine that an ordinary (right-hand) screw is placed along the line of action of the torque and rotated by the force **F**. The sense of the vector torque is the direction of advance of the screw when it is rotated by **F**.

We can summarize this definition of vector torque with the equation

$$\boldsymbol{\tau} = \mathbf{r} \times \mathbf{F} \qquad (4\text{-}16)$$

in which the cross symbol (×) implies both the magnitude and direction given above.

This vector cross product is a useful mathematical abbreviation that we will use several times in this book. In general, we can define the cross

product, **C**, of two vectors **A** and **B**, where

$$\mathbf{C} = \mathbf{A} \times \mathbf{B} \qquad (4\text{-}17)$$

as a vector whose magnitude C is given by

$$C = AB \sin \theta \qquad (4\text{-}18)$$

in which θ is the angle between the vector **A** and the vector **B**. The line of action of the cross product is along the line perpendicular to the plane formed by **A** and **B**, as shown in Figure 4-7. Again the sense of the cross product is determined by a right-hand screw rule. Imagine a right-hand screw placed along the line of action of **C**, and then imagine that the vector **A** is rotated through θ so that it lies along **B**, as shown in Figure 4-7b. If the screw is rotated in this same way, it will advance in the direction of the cross product. In this definition, the vector that is rotated through θ is the vector that appears *before* the cross in Equation 4-17. Thus, the vector **B** × **A** will be opposite to the vector **A** × **B**, or

$$\mathbf{A} \times \mathbf{B} = -(\mathbf{B} \times \mathbf{A})$$

The vector torque defined by Equation 4-16 is the torque about the *point O* from which the vector **r** is drawn. In most elementary problems, we are concerned with rotations about a *line* or axis, rather than rotations about a point. In such problems we deal with the component of the torque that is parallel to the axis, and call this the *torque about the axis*. In particular, in most of the problems in this chapter the forces lie in a plane and we will be interested in rotations about axes perpendicular to the plane. In that case we can use the scalar definition of torque given by Equation 4-12, with the sign convention given earlier.

4-3
Equilibrium of a rigid body

If all the points of a body move parallel to each other, the body is said to be *translating*, or in trans-

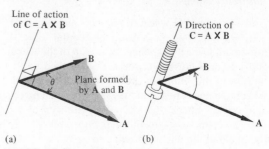

FIGURE 4-7 The vector cross product.

lation. If all the points move in circles, the body is said to be *rotating*, or in rotation. In the most general motion of bodies, both rotation and translation occur.

If a rigid body is known to be at rest, then it is neither translating nor rotating, and so two things are known about the system of forces acting on the body: (a) The net force acting on the body is equal to zero

$$\Sigma\ \mathbf{F} = 0 \qquad (4\text{-}19)$$

and (b) the net torque acting on the body is equal to zero

$$\Sigma\ \boldsymbol{\tau} = 0 \qquad (4\text{-}20)$$

Equation 4-19 is equivalent to *three* scalar equations

$$\Sigma\ F_x = 0 \qquad (4\text{-}21\text{a})$$
$$\Sigma\ F_y = 0 \qquad (4\text{-}21\text{b})$$
$$\Sigma\ F_z = 0 \qquad (4\text{-}21\text{c})$$

and Equation 4-20 is also equivalent to three equations

$$\Sigma\ \tau_x = 0 \qquad (4\text{-}22\text{a})$$
$$\Sigma\ \tau_y = 0 \qquad (4\text{-}22\text{b})$$
$$\Sigma\ \tau_z = 0 \qquad (4\text{-}22\text{c})$$

in which the subscript indicates the direction of the axis about which the torques are taken. However, the most common statics problems involve two-dimensional force systems, in which case one uses only two of Equations 4-21 and one of Equations 4-22. Thus in the most common statics problems there exist only three independent equations and therefore we can solve for at most three unknowns.

If the system of forces acting on a body satisfies Equations 4-19 and 4-20, then the body is in *equilibrium*, but not necessarily in static equilibrium. For example, it may be translating with constant speed or it may be rotating about a fixed axis.

Illustrative example

Consider a bicycle wheel of radius R and weight $\mathbf{W}$ which is mounted on a fixed horizontal axle. A horizontal force $\mathbf{F}$ is applied to the top of the wheel and an equal horizontal force $\mathbf{F}$ is applied to the bottom of the wheel as shown in Figure 4-8. The wheel remains at rest. Find the force exerted by the axle on the wheel, and show that the sum of the torques about an axis through *any* point O' (in the plane of the wheel) is equal to zero.

The free-body diagram of the wheel is shown in

Figure 4-8b. Since the wheel is symmetric, the weight of the wheel acts at its center (see Section 4-4), and we assume that the force exerted by the axle has two components, P and H. Writing the equations for the sum of the forces,

$$P - W = 0 \quad \text{or} \quad P = W$$

and

$$H - F - F = 0 \quad \text{or} \quad H = 2F$$

The sum of the torques about the axle is

$$FR - FR = 0 \quad \text{or} \quad FR = FR$$

In order to show that the sum of the torques about any axes such as that through point O' is equal to zero we will specify that the point O' has coordinates (a, b). Then the torque produced by H about an axis through O' is $+Hb$; for the other forces, the torques produced are $-F(b - R)$, $-Pa$, $+Wa$, and $-F(b + R)$. The sum of the torques is thus

$$\tau = Hb - F(b - R) - Pa + Wa - F(b + R)$$

We also have

$$P = W \quad \text{and} \quad H = 2F$$

and substituting into the equation for τ, we get

$$\tau = 2Fb - Fb + FR - Wa + Wa - Fb - FR = 0$$

Hence the algebraic sum of the torques about an axis through any point O' is zero.

In general, for two-dimensional force systems in equilibrium, the sum of the torques about an axis through *any* point in the plane of the forces is equal to zero. Thus, in solving problems we are at liberty to choose an axis about which to take torques; often this choice can minimize the labor involved in solving a problem.

Illustrative example

The horizontal bar AB in Figure 4-9 is supported at point O and has a weight $\mathbf{W}$ placed at end A. Find

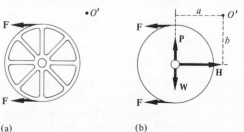

(a)   (b)

FIGURE 4-8 Forces on a bicycle wheel.

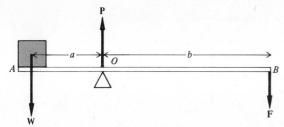

FIGURE 4-9 Forces on a long horizontal bar with a fulcrum at O and weight W placed at A.

the vertical force F that is exerted at end B and the force P exerted by the support (the *fulcrum*) at point O.

Since the force F and the weight W are vertical, the force P must also be vertical. Thus the condition for translational equilibrium becomes, simply,

$$\Sigma F_y = 0 = P - W - F$$

from which

$$P = W + F \qquad (4\text{-}23)$$

In order to apply the second condition of equilibrium, the condition for rotational equilibrium, we must choose the axis. Any point may be choosen, and in particular any of the three points A, B, or O that naturally suggest themselves may be used. However, in this problem there are two unknowns, P and F, and if the torques are taken about either O or B, the torque equation will involve only one of the two unknowns. Then the torque equation may be solved directly for that unknown. If torques are taken about point A, the resulting equation will involve both unknowns and will require a simultaneous solution with Equation 4-23. Thus, choosing point O as the axis, we have

$$\tau = 0 = Wa - Fb$$

or

$$Wa = Fb$$

and

$$W\frac{a}{b} = F \qquad (4\text{-}24)$$

Substituting Equation 4-24 in Equation 4-23, we get

$$P = W + W\frac{a}{b}$$

$$= W\left(1 + \frac{a}{b}\right)$$

A bar used in this way is called a *lever*, and the distances a and b are called the *lever arms*. From Equation 4-24 we see that the force necessary to keep the lever in equilibrium depends on the ratio of the lever arms as well as the weight of the load. Many mechanical devices, such as shovels and scissors, are basically levers of this type, where the fulcrum is between the load and the applied force. Other levers have the fulcrum located outside these forces. Many biological systems can be described as levers.

Illustrative example

A man is holding a 10-kg object with his forearm in a horizontal position as shown in Figure 4-10. The distance between the pivot in his elbow and the point of suspension of the object is 38 cm and the distance between the pivot and the point where the biceps muscle attaches to the forearm is 5 cm. If the biceps makes an angle of 90° with the forearm, how large a force does the biceps exert? (Neglect the weight of the forearm.)

The free-body diagram for the forearm is shown in Figure 4-10b. The force exerted by the upper arm on the forearm is unknown in direction and magnitude, but since there are no other horizontal forces it is immediately obvious that this force must have no horizontal component. However, the problem asks only for the force exerted by the biceps, force P, so we seek a method of finding P without bothering to find F. We can do this by writing the condition on the torques, choosing the point of application of F as the axis. With this choice, the condition on the torques for equilibrium of the body becomes

$$P(5 \text{ cm}) - W(38 \text{ cm}) = 0$$

$$P = W\left(\frac{38}{5}\right)$$

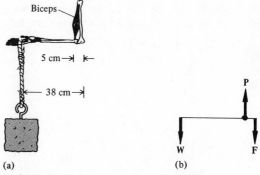

(a)   (b)

FIGURE 4-10 Forces on a human forearm.

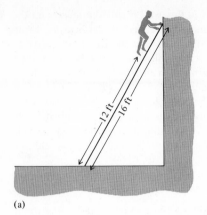

(a)

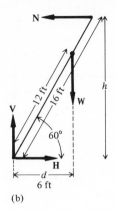

(b)

FIGURE 4-11 Forces on a weight-less ladder resting on a fric-tionless wall.

Now, W is the weight of a 10-kg body, or

$$W = 10 \text{ kg} \times 9.8 \, \frac{\text{m}}{\text{sec}^2}$$
$$= 98 \text{ nt}$$

Thus

$$P = 745 \text{ nt}$$

This force is the weight of a 76-kg object! Now that we have found P, we could find F from the first condition for equilibrium.

Illustrative example

A ladder 16 ft long rests against a frictionless wall and makes an angle of 60° with the horizontal. A 180-lb man is on the ladder, 4 ft from the top. What horizontal force is needed at the bottom of the ladder in order to prevent the ladder from slipping? Neglect the weight of the ladder.

The situation is shown in Figure 4-11. Since the wall is frictionless, it exerts no force tangential to itself, but it does exert a normal force, shown as **N** in the free-body diagram of Figure 4-11b. Similarly, the floor exerts a normal force shown as **V**. The man will exert a vertical force on the ladder, shown as **W**, and the required horizontal force is shown as **H**. From the first condition for equilibrium, we have

$$V = W$$
$$V = 180 \text{ lb}$$

and

$$N = H$$

For the second condition we have to choose an axis. If we choose an axis through the point of contact of the ladder with the floor, then

$$Nh = Wd$$

Then we can see from the geometry that

$$h = 16 \text{ ft} \times \sin 60°$$
$$= 13.8 \text{ ft}$$
$$d = 12 \text{ ft} \times \cos 60°$$
$$= 6 \text{ ft}$$

Therefore

$$N \times 13.8 \text{ ft} = 180 \text{ lb} \times 6 \text{ ft}$$
$$N = 78.3 \text{ lb} \quad \text{and} \quad H = 78.3 \text{ lb}$$

4-4
Center of gravity

Real bodies consist of large numbers of particles, each of which is attracted by the earth. If the body is near the surface of the earth, all of these forces will be parallel. The resultant of all of these parallel forces is what we usually call the weight of the body. The magnitude of this force may be mea-sured with a spring balance, equal-arm balance, or other instrument. The line of action of this re-sultant force, or weight, always passes through a single point in the body, a point called the *center of gravity* of the body. For any particular body the location of the center of gravity depends on the distribution of matter in the body. Experimentally, the center of gravity may be located by the follow-ing procedure. The body is suspended at a point near the edge of the body, as shown in Figure 4-12. If the body is allowed to come to equilibrium, then we know that the force exerted by the suspension is equal to the weight of the body, and that the sum of the torques about the point of suspension is equal to zero. Therefore the line of action of the weight, which we know goes through the center of gravity, also goes through the suspension. There-

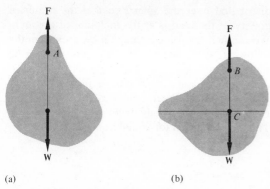

FIGURE 4-12 Method of determining the center of gravity of a body.

fore if we draw a vertical line in the body through the suspension, we know that the center of gravity is somewhere on the line. If the body is now suspended from a second point, as shown in Figure 4-12b, we will be able to draw another vertical line through this new point of suspension. The center of gravity will be at the intersection of these two lines.

In many cases of practical interest the position of the center of gravity can be calculated with the aid of a simple theorem that *the moment about any axis produced by the weight of the body acting through the center of gravity must equal the sum of the moments about the same axis produced by the weights of the individual particles of the body*. In general, application of this theorem requires the methods of the calculus. However, if the body is homogeneous—that is, made of the same material throughout—and if it has a simple geometric shape, such as a circular plate, square plate, or a cube, the results of such calculations show that the center of gravity is at the geometric center.

Illustrative example

A uniform, horizontal bar, 4 ft long and weighing 20 lb, supports five weights spaced 1 ft apart along the bar. Beginning at one end, these weights are 2 lb, 4 lb, 6 lb, 8 lb, and 10 lb, as shown in Figure 4-13a. Find the location of the center of gravity of this system.

If we treat the bar and the weights as a single system, the total weight of the system is 50 lb. The torque exerted by this total weight must equal the sum of the torques exerted by the five weights and the bar, where the torques are taken about any axis. Let us choose the axis to be through the left

end of the bar and let us assume that the center of gravity of the system is located a distance x from the left end of the bar, as shown in Figure 4-13b. Then the torque exerted by the total weight is equal to $50x$. Since the bar is uniform, the center of gravity of the unloaded bar is at the middle of the bar and so the torque exerted by the weight of the bar is equal to 20(2) lb ft. The torque exerted by each of the weights is equal to the product of the weight and its distance from the left end of the bar. Thus

$$50x = 20(2) + 2(0) + 4(1) + 6(2) + 8(3) + 10(4)$$
$$= 40 + 0 + 4 + 12 + 24 + 40$$
$$= 120 \text{ lb ft}$$
$$x = \frac{12}{5} \text{ ft}$$

Illustrative example

A uniform plank, 10 ft long and weighing 30 lb, is supported by vertical ropes at each end. A 150-lb man stands at a point 3 ft from one end of the plank, as shown in Figure 4-14a. Find the tension in each rope.

The free-body diagram for the plank is shown in Figure 4-14c, where W' is the force exerted on the plank by the man, and W is the weight of the plank. Since the plank is uniform, its center of gravity is its geometric center. From Figure 4-14b we can see that W' is equal to the weight of the man, since the man is in equilibrium. Since the plank is in equilibrium, we can write the first condition for equilibrium:

$$W + W' - T_1 - T_2 = 0 \qquad (4\text{-}25)$$

where T_1 and T_2 are the tensions in the respective ropes.

For the second condition for equilibrium we are free to choose any axis about which to find the

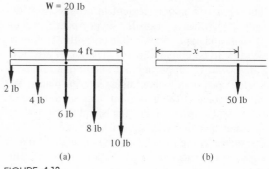

(a)   (b)

FIGURE 4-13

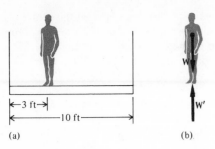

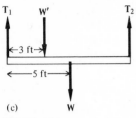

(a)

(b)

(c)   W

FIGURE 4-14

torques. Since T_1 and T_2 are both unknowns, we can simplify the calculation by choosing an axis on the line of action of one of these forces. If we choose the left edge of the plank, then the torques about that axis are

$$W'(3) + W(5) - T_2(10) = 0$$

or

$$150(3) + 30(5) - T_2(10) = 0$$

Dividing by 10, we get

$$45 + 15 - T_2 = 0$$
$$T_2 = 60 \text{ lb}$$

Then substituting in Equation 4-25, we have

$$150 + 30 - T_1 - 60 = 0$$
$$T_1 = 120 \text{ lb}$$

Illustrative example

The *arch* is a major element in architectural design. It provides a means for supporting the weight of walls and upper floors while permitting a door or window to be constructed in the wall. It also provides a way of supporting the weight of a roof over a wide open span without using interior pillars. We can see some of the important features of the arch by considering the following simplified problem. Assume that the arch shown in Figure 4-15 is a symmetric curve of height h and span s and that it supports a vertical load L, including its own weight. Further assume that the load is symmetrically

distributed over the span s so that the center of gravity of the load is at the midpoint of the arch. We can find the forces acting on the arch from the following. We assume that the ground exerts the same force on both bases of the arch (because of the symmetry), and resolve those forces into horizontal components, H, and vertical components, F, yielding the free-body diagram of Figure 4-15b. The first condition for equilibrium results in

$$L = 2F$$

or

$$F = \frac{L}{2} \tag{4-26}$$

and the fact that the horizontal forces exerted at the bases are equal and opposite. With this much symmetry, the condition on the torques gives us no additional information, which the reader may verify by taking torques about one or two points. Thus all we can conclude is that the vertical force exerted by the base on the arch is half the load, a conclusion that is almost obvious from the symmetry. We can conclude nothing about the horizontal force exerted by the ground.

We can gain some more information by something of a trick; choose half of the arch as a free body, as shown in Figure 4-15c. Then the ground exerts the same force on this new free body as it did above. From the symmetry it is reasonable to assume that only half the load acts on this free body. The location of the load on this half-arch depends on the details of the shape of the arch and the distribution of the load across the span. We can therefore assume that the load on the half-arch is a vertical force of magnitude $L/2$ located a distance b from the center of the arch. Then taking torques about an axis through the apex of the arch we have

$$\frac{L}{2} b + Hh = \frac{L}{2} \frac{s}{2}$$

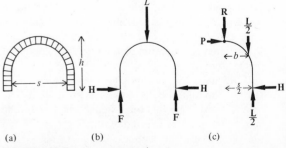

(a)   (b)   (c)

FIGURE 4-15 Forces on an arch.

FIGURE 4-16 Flying buttresses of Notre Dame of Paris (Courtesy of Professor J. Rothenberg.)

or

$$H = \frac{L}{h}\left(\frac{s}{4} - \frac{b}{2}\right)$$

and if, for example, $b = s/4$,

$$H = \frac{Ls}{8h}$$

Then from the first condition for equilibrium we have

$$P = H$$

and

$$R + \frac{L}{2} = \frac{L}{2}$$

or

$$R = 0$$

The force exerted by one half of the arch on the other half is completely horizontal. In addition, the ground must exert horizontal forces that are proportional to the load and inversely proportional to the height of the arch but which increase with increasing span. Thus the large span arches of medieval church roofs had to be supported with large horizontal forces. This is why the great cathedrals have "flying buttresses" to support the horizontal outthrust from the roof arch (see Figure 4-16).

4-5
Friction

The existence of friction may be most easily demonstrated by the experiment shown in Figure 4-17, in which a horizontal force **F** is applied to a block resting on a horizontal surface. If the force **F** is initially equal to zero and slowly increased, it is found that the block remains at rest for all values of **F** smaller than some critical value. When **F** is made larger than the critical value, the block moves. Further it is found that it is possible to have the block move in equilibrium (with constant speed) for one particular value of **F**, which may not be the same as the critical value. Since the block is in

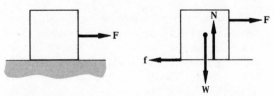

FIGURE 4-17 Experiment to show the existence of friction.

equilibrium under the action of an external force **F**, there exists an equal and opposite external force **f**, which is the friction force. Thus there are two classes of friction force, *static* and *kinetic*. The static friction force, which is exerted when there is no relative motion between the two objects involved (block and table), can take on any value between zero and a maximum value, f_{sm}. If a body is in static equilibrium, and an additional external force is applied, the friction force will attempt to adjust itself to that value which will keep the body in static equilibrium. This attempt will be successful, if the *required static* friction force is less than f_{sm}. When there is relative motion between the two bodies, the *kinetic* friction force opposes the motion and has a specific value.

Friction forces are usually described in terms of the *coefficient of friction*, μ (mu). The kinetic force f_K is given by

$$f_K = \mu_K N \qquad (4\text{-}27)$$

where **N** is the normal force exerted by one body on the other, and μ_K is the coefficient of kinetic friction. The static friction force has any value between zero and the maximum value, or

$$0 \le f_s \le f_{sm} \qquad (4\text{-}28)$$
$$f_{sm} = \mu_s N \qquad (4\text{-}29)$$

or

$$f_s \le \mu_s N \qquad (4\text{-}30)$$

where μ_s is the coefficient of static friction. The "laws of friction," usually ascribed to Coulomb, are:

1. The coefficient of static friction is a constant, independent of contact area or normal force, but depending on the nature and history of the objects in contact.
2. The coefficient of kinetic friction is a constant, independent of contact area, normal force, and relative velocity.
3. The coefficient of kinetic friction is smaller than the coefficient of static friction.

These "laws" are attempts to extrapolate simple conclusions from extraordinarily complex data. For example, the statement that the coefficients are independent of contact area assumes that the contact area is the same as the apparent surface area of one of the objects. But we know that the surface of a real object is very rough on a microscopic level; the contact surface between two ob-

jects looks like Figure 4-18. Thus the contact area is the sum of the areas of the projecting points that happen to be in contact. Clearly this area is very different from the measured, macroscopic area of contact. Further, it seems reasonable to believe that, as the normal force is increased, the tips of these points should flatten out. This will increase the area of contact. We can explain the apparent independence of the friction coefficient and area by assuming that the friction *force* is proportional to the *real area* of contact, and that this real area of contact is proportional to the normal force. With this model for the friction force, we might expect that, if the normal force is exerted for a long time, the material might flow slowly and the real area of contact would change with time. Thus, we might expect that the coefficient of friction would depend on how long the normal force has been exerted. In fact this seems to be the case. This argument says that we should treat the laws of friction given above as approximations, and should not be surprised to discover deviations from them. Further, we should not expect the laws to be obeyed in extreme cases, such as the case of very large normal forces or very large (or very small) relative forces.

The value of the coefficient of friction depends on the materials, their roughness, and to some extent, on the history of a particular sample. It turns out that the state of cleanliness of the surfaces is a most important factor, as is the surface oxidation of metals.

Illustrative example

A 20-kg block is at rest on a horizontal surface, where the coefficient of static friction is 0.25. Find the largest horizontal force which can be applied to the block without causing slipping.

The free-body diagram is shown in Figure 4-19. The conditions for equilibrium are

$$f = F$$
$$N = W$$

and from Equation 4-30, we have

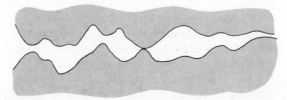

FIGURE 4-18 Magnified view of the contact surface between two objects.

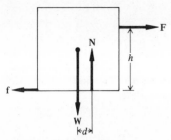

FIGURE 4-19 Forces exerted on a block.

$$f \leq \mu_s N$$

and so

$$F \leq \mu_s N$$

or

$$F \leq \mu_s W$$

Thus, the largest value of F is given by

$$
\begin{aligned}
F &= \mu_s W \\
&= 0.25 \text{ mg} \\
&= 0.25 \times 20 \text{ kg} \times 9.8 \frac{\text{m}}{\text{sec}^2} \\
&= 49 \text{ nt}
\end{aligned}
$$

As an interesting side excursion from this problem we can consider what we learn from the condition on the torques. Take as the axis the point of intersection of the friction force and the weight. The torque about this point is zero for both the friction force and the weight. Unless the force F is applied at the base of the block, the force F will have a nonzero torque about this point. In order for the body to be in equilibrium, the normal force (the only other force acting) must also produce a nonzero torque about this axis. Thus the line of action of the normal force cannot go through the center of gravity. We can see how this arises from the following argument. If the force F is exerted above the base of the block, it tends to make the block rotate clockwise in the figure. This makes the right-hand corner of the block press down harder, increasing the normal force exerted at that corner, and decreasing the normal force exerted at the left-hand corner.

If the force F is exerted on the block at a height h above the table, then the normal force will be exerted at a point a distance d in front of the center of gravity, and

$$Fh = Nd$$

or

$$d = \left(\frac{F}{N}\right)h \qquad (4\text{-}31)$$

is a consequence of the second condition for equilibrium. If the force F is applied at a point well above the base of the block, this distance d can turn out to be larger than the distance to the corner of the block. That is, it is possible to solve this equation and produce the conclusion that the normal force is *outside* the block. The significance of this absurd result is that the assumption that the block is in equilibrium is wrong; the block will rotate or tip. Equation 4-31 can then be used to determine *when* the block will tip; if d is less than the distance to the corner of the block, the block will not tip.

Illustrative example

A 10-lb block is placed in a horizontal board, and the board is slowly pivoted about one edge until the block slides down the board. The block skids when the angle θ of Figure 4-20 is equal to $30°$. What is the coefficient of static friction?

The free-body diagram is shown in Figure 4-20b. Since the block is in equilibrium,

$$
\begin{aligned}
f &= W \sin \theta \\
N &= W \cos \theta
\end{aligned}
$$

We know that

$$f \leq \mu_s N$$

or

$$W \sin \theta \leq \mu_s W \cos \theta$$

Therefore

$$\tan \theta \leq \mu_s \qquad (4\text{-}32)$$

That is, as long as the block is in equilibrium, the tangent of θ is less than or equal to the coefficient of static friction. As θ is increased from zero, this statement is true until at some angle θ, the tangent

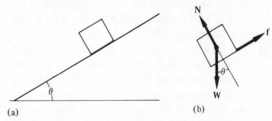

(a)   (b)

FIGURE 4-20 Forces exerted on a block which is on an inclined board.

is just equal to μ_s. For all larger angles, Equation 4-32 cannot be satisfied. Thus for all these larger angles, the block is not in equilibrium. Clearly, for this problem, this critical angle, whose tangent is equal to the coefficient, is 30°. Thus:

$$\mu_s = \tan 30°$$

$$\mu_s = \frac{1}{\sqrt{3}}$$
$$= 0.578$$

Of course, if the block is too high and narrow (see Equation 4-31), then as θ is increased, it will tip over before it begins to slide.

Questions

1. Show that if a weight is hung from the middle of a cord, it will never be possible to get the cord to be horizontal.

2. Show that when a weight is hung from the middle of a cord whose two ends are tied to points on the same horizontal level, the vertical component of the tension in each part of the cord is always equal to half the weight.

3. Using the second condition for equilibrium of a body show that when a body is in equilibrium under the action of three nonparallel forces, these forces must pass through a single point; that is, the forces must be concurrent.

4. Analyze each of the following tools as examples of levers: shovel, hammer, wrench, pliers, and scissors. Determine the position of the fulcrum in each case and point out the lever arms.

5. In each of the following a part of the body is used as a lever: rising from a chair, biting, place-kicking a football, striking a typewriter key. In each case locate the fulcrum and the lever arms.

6. There are three kinds of equilibrium, depending on what happens to the body when it is given a small displacement from its equilibrium position and then released. If the body returns to its equilibrium position, then it is in *stable* equilibrium; if it continues moving in the direction in which it was displaced, then it is in *unstable* equilibrium; and if it remains in the displaced position, it is in *neutral* equilibrium. Consider a cube that is at rest on a horizontal surface and then is tipped slightly. In which kind of equilibrium is the cube?

7. Discuss the positions of stable, unstable, and neutral equilibrium of a right circular cone.

8. How is it possible for a body not to be in equilibrium even if the resultant of all the forces acting on the body is equal to zero? Give an example.

9. A picture is hung from a nail by a single wire which is stretched across the back of the picture. Explain why the picture can hang crooked even though it has a definite mass and shape.

10. When a baseball is thrown straight upward, it is momentarily at rest at the top of the path. Is it in equilibrium? Explain.

11. On what factors does the force of friction between two objects depend? On what factors does the coefficient of friction depend?

12. How does the center of gravity of a milk container change when the container is emptied? How does it change when it is half full?

Problems

1. A 75-kg object is at rest on a plane inclined 30° to the horizontal. Find the components of the weight of the object which are parallel to and perpendicular to the surface of the plane.

2. A 10-kg object is attached to a rope which hangs from the ceiling. A horizontal force pushes the body sideways so that the rope makes an angle of 30° with the vertical. Find the tension in the rope and the magnitude of the horizontal force.

3. A 3500-lb automobile is stopped in the middle of a hill. The road rises 6 ft for each 100 ft of road. Find the friction force acting on the wheels.

4. A 2-kg block is held against a vertical frictionless wall by a force **F** which makes a 30° angle with the vertical. How large is F?

5. A man pushes a 200-lb box across a floor at constant velocity. If the coefficient of kinetic friction between the box and the floor is 0.2, how large a horizontal force does the man apply?

6. A man is attempting to push a 4000-lb automobile. He discovers that the car will not move until he applies a horizontal force of 120 lb. How large is the coefficient of static friction between the ground and the wheels?

7. In opening a 32 in.-wide door a man exerts a 2 lb force on a door knob which is 2 in. in from the free edge of the door. If the force is perpendicular to the surface of the door, how large is the torque exerted by the man? Describe the direction of the torque.

8. A student is holding a lunch tray horizontally in front of his body. Describe the direction of the torque exerted by the weight of the tray, using an appropriate diagram. If the student's arms and body are rigid, describe the origin of the torque that balances the torque of the tray.

9. In attempting to pry a large boulder out of the ground a farmer uses a 12-ft pole as a lever. He uses a rock 6 in. from the end of the pole as a fulcrum and pulls on the pole with a force of 200 lb. If the force exerted by the farmer is perpendicular to the pole, how large a force is exerted on the boulder? How large a force is exerted on the rock?

10. A 500-lb block is at rest on a frictionless incline, which makes an angle of 20° with the horizontal. What force parallel to the plane has been applied to the block to keep it in equilibrium.

11. A 10-kg block is at rest on a frictionless incline, which makes an angle of 30° with the horizontal. What horizontal force has been applied to the block to keep it in equilibrium?

12. A circus tightrope walker who weighs 140 lb is standing in the middle of a cable that is stretched between two points 60 ft apart. The man's feet are 2 ft below the points of attachment of the cable. What is the tension in the cable?

13. A 180-lb telephone lineman is supported by a belt around his waist, with his feet braced against the pole. His body makes an angle of 20° with the vertical. Assume that his center of gravity is at chest height, 4.5 ft from his feet, and that the belt is 3 ft from his feet and perpendicular to his body. Find the force exerted by the belt and the force exerted by his feet on the pole.

14. Estimate the location of the center of gravity of a human being by treating the body as homogeneous and consisting of a 10-in.-diameter cylinder that is 2.5 ft long and rests on two 5-in.-diameter cylinders each 2.5 ft long.

15. In order to pull a car out of a rut a man ties a rope to a tree and attaches the other end to the front bumper of the car. The man then pulls on the middle of the rope in a direction perpendicular to the line from the car to the tree. (a) Determine the tension in the rope if the man exerts a force of 75 lb when the angle between the two halves of the rope is 160°. (b) What force does the rope exert on the car? (c) What is the component of the force exerted on the car along the line between the car and the tree?

16. A 100-lb weight is supported by two ropes. One rope makes an angle of 30° with the vertical and the other makes an angle of 45° with the vertical. Find the tension in each rope.

17. An equal-arm balance consists of a uniform rod pivoted at its center and from whose ends are hung weights. Usually, an unknown weight W is hung from one end and known weights are hung from the other end until the rod is in equilibrium in a horizontal position. Assume now that the pivot is not in the middle of the rod. Then the known weight W_1, which balances W will not be equal to W. If the unknown weight is then suspended from the other end of the rod, it will be balanced by a different weight W_2. (a) Show that W is equal to the geometric mean of W_1 and W_2; that is,

$$W = \sqrt{W_1 W_2}$$

(b) Then verify that if W_1 and W_2 are almost the same, this is almost the same as the arithmetic mean.

$$W \approx \frac{W_1 + W_2}{2}$$

by considering the example where W_1 is equal to 50 gm and W_2 is equal to 52 gm.

18. A uniform bar 12 ft long and weighing 100 lb is horizontal and fastened to the wall at point A, as shown in Figure 4-21. The bar is supported by a 13-ft cable attached to the end B and is attached to the wall at point C, which is above A. A 500-lb weight is hung from the bar at a point 10 ft out from the wall. Find the tension in the cable and the force exerted by the wall on the bar.

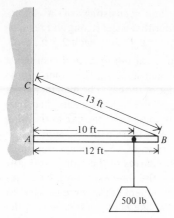

FIGURE 4-21

19. A 20-ft uniform ladder weighing 30 lb rests against a frictionless wall so that its upper end is 16 ft above the ground, as shown in Figure 4-22. A 150-lb man climbs up the ladder so that his center of gravity is 12 ft above the ground. Find the forces exerted by the ladder on the wall and ground.

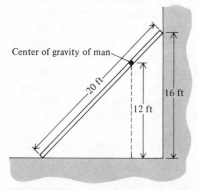

FIGURE 4-22

20. In the determination of the coefficient of sliding friction, it is found that a steel block will slide down a steel inclined plane with uniform motion when the angle of the plane is 15°. **(a)** Find the coefficient of friction between the two steel surfaces. **(b)** If the angle of the plane is increased to 30°, find the acceleration of the block.

21. A 20-kg block is at rest on a horizontal surface. The coefficient of static friction is 0.4 and the coefficient of kinetic friction is 0.3. **(a)** How large is the friction force if no body other than the horizontal surface exerts a force on the block? **(b)** How large is the friction force if a horizontal force of 30 nt is applied to the block? **(c)** How large is the friction force if a

horizontal force of 100 nt is applied to the block? **(d)** What is the minimum horizontal force needed to start the block moving? **(e)** What is the minimum horizontal force needed to keep the block moving?

22. A 100-lb block is placed on an inclined plane which makes an angle of 30° with the horizontal. A cord passes over a small frictionless pulley and a weight W is attached to the end of the cord as shown in Figure 4-23. The coefficient of static friction is 0.25 and the coefficient of kinetic friction is 0.20. **(a)** Find the maximum weight W for which the block remains at rest. **(b)** Find the minimum weight for which the block will remain at rest. **(c)** Find the weight for which the block will move up the plane at constant speed. **(d)** Find the weight for which the block will move down the plane at constant speed.

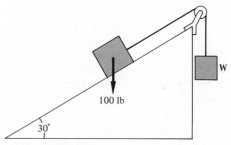

FIGURE 4-23

23. The wheel shown in Figure 4-24 weighs 30 lb and is being pulled up over the curb by the horizontal force **F**. Find the value of the force **F** which will just pivot the wheel on the edge of the curb. Note that the height of the force **F** above the curb and the horizontal distance between the curb and the line of action of the weight can be found by careful application of the Pythagorean theorem.

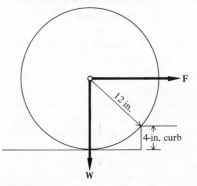

FIGURE 4-24

24. A 12-ft long flag pole is attached at one end to a vertical wall. A horizontal wire attached to the middle of the pole holds the pole at a 45° angle with the vertical. The 40-lb weight of the flag can be treated as a vertical force located 2 ft from the outer end of the pole. **(a)** Find the tension in the wire. **(b)** Find the force exerted by the wall on the pole.

25. The 20-lb chair of Figure 4-25 is being pulled at constant speed by the horizontal force **F** over a horizontal surface where the coefficient of friction is 0.2. How large is **F**? Assuming that the center of gravity of the chair is in the seat, 5 in. from the back of the chair, how large is the normal force in each leg?

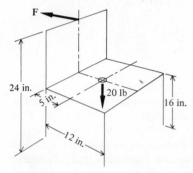

FIGURE 4-25

26. The coefficient of static friction between the block and the vertical wall of Figure 4-26 is μ_s. Show that it is impossible to start the block moving up the wall if θ is greater than the angle whose cotangent is equal to μ_s ($\theta > \cot^{-1} \mu_s$).

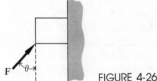

FIGURE 4-26

27. The lever system employed in the human leg is of a kind called an *end-loaded compound lever*. To illustrate some aspects of this system consider the structure shown in Figure 4-27. The lever consists of two rods of equal length l, corresponding to the thigh and the leg. Protruding from the upper rod, near the joint O (the knee) is a short stub, perpendicular to OB of length s. This corresponds to the *process* of the bone. A force **F** is exerted on this stub parallel to OB. This corresponds to the pull exerted by the thigh muscle. Assume that a

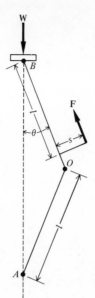

FIGURE 4-27

vertical force **W** = 200 lb is exerted at B, that $l = 25$ cm, $s = 2$ cm, and $\theta = 5°$. How large a force **F** is needed to maintain equilibrium?

28. The human back can be thought of as a uniform boom as shown in Figure 4-28, where point A is the fifth lumbar vertebra. The load $\mathbf{W}_1$ is the weight of the head, arms, and the load carried by the arms. The weight $\mathbf{W}_2$ is the weight of the rest of the upper body and is concentrated at a point halfway between A and B. The supporting force **T** is provided by muscles which are inserted at point D two thirds of the way from A to B, making an angle $\theta = 12°$. A man leans forward so $\varphi = 30°$. **(a)** Assume $W_2 = 72$ lb and $W_1 = 36$ lb. Find T and the force exerted on the fifth lumbar vertebra. **(b)** Then assume the man picks up a 50-lb weight. Find T and the force exerted on the fifth lumbar vertebra.

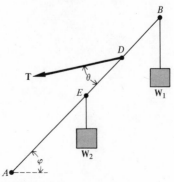

FIGURE 4-28

the fundamental forces

chapter 5

5-1
Introduction

As we have seen, one of the major aims of physics is the search for and the understanding of the nature of the fundamental forces that exist. This aim requires first, that the forces which are fundamental be identified or discovered; second, that the forces be described completely and accurately; and last, that observable phenomena be explained as arising from the action of one or more fundamental forces. As is true of so much of physics, this program has been only partially completed. We understand a great deal, but some of the most exciting and challenging current research is concerned with this particular problem.

It is now generally believed by physicists that there are only four fundamental forces: gravity, the electromagnetic force, and two forces that may be called the nuclear forces. Both gravity and electromagnetism are well understood, in the sense that there are detailed mathematical descriptions of the forces and impressively accurate experimental verifications of the predictions made from these mathematical laws. The nuclear forces, on the other hand, are poorly understood. One important reason for this is that the nuclear forces between two particles are quite large when these particles are separated by distances of the order of magnitude of the size of an atomic nucleus or less, but became practically zero when the particles are separated by distances larger than nuclear dimensions. Thus all experimental investigations of nuclear forces must be performed on collections of particles separated by no more than these extraordinarily small distances. Paradoxically, these experiments have required the use of huge accelerators with dimensions of hundreds or even thousands of feet.

The mixed success of the program of understanding the fundamental forces is paralleled by the mixed success of the program of understanding the observable phenomena in terms of combinations of fundamental forces. On the one hand, for example, we understand the motion of the planets —an effect of gravity—very well. On the other hand, we have almost no understanding of the properties of the fundamental particles, although a great deal of experimental data exist. Intermediate between these is our understanding of some common forces, such as the forces that hold solids together and the contact forces between objects. Most physicists would agree that we know what these fundamental forces are (electromagnetic), that we can accurately describe the forces and how they affect the individual particles that make up real objects, and that we largely know the nature and significant properties of the particles in many objects. Unfortunately, almost all attempts to put all this knowledge together in order to calculate a specific property of a particular material fail, chiefly because the mathematical calculations involved are overwhelming. However, if we start from some empirically known data, such as the locations of the atoms in a crystal, we can make detailed, accurate predictions of some of the other properties of materials.

In the following sections each of the forces will be described briefly. We shall discuss the gravitational force in some detail in this chapter, and touch briefly on electromagnetic forces. In subsequent chapters electromagnetism and the nuclear forces will be discussed in more detail.

5-2
Newton's law of universal gravitation

The first force that was understood in a systematic way was gravity. Sir Isaac Newton developed the theory in 1665, but for some reason he did not publish the result until 1685. One of the most striking aspects of the theory was the assertion that the gravitational force was universal; that the same force that caused apples to fall was responsible for the motion of the moon—in other words, that the mundane force of gravity was in a certain sense celestial.

Newton's law of universal gravitation states that *any two bodies in the universe attract each other with a force that is directly proportional to the product of the masses of the two bodies and inversely proportional to the square of the distance between them.*

Stated mathematically, this law becomes

$$F \propto \frac{Mm}{r^2}$$

where M is the mass of one body, m is the mass of the other body, r is the distance between them, and F is the force that one body exerts on the other. This law can be put in the form of an equation by replacing the proportionality sign by an equality sign, and a constant of proportionality; thus

$$F = \frac{GMm}{r^2} \tag{5-1}$$

where G is the constant of proportionality and is known as the *universal constant of gravitation.*

This law describes both the magnitude and the direction of the force. The direction of the force is somewhat hidden in the word "attract" in the statement of the law. The direction is made somewhat clearer (if wordier) by recognizing that "attract" means that a body exerts a force on another body that is directed back to the first body along the line joining the two bodies (see Figure 5-1).

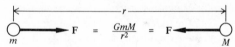

FIGURE 5-1 The gravitational force between two point objects.

A further property of gravitation is that the gravitational force obeys the *law of superposition.* That is, if there are more than two bodies, then the force exerted on any one of them is the vector sum of the separate forces exerted by each of the other bodies on that one. The separate forces are the same as they would be if there were only those two bodies present. That is, the force between any two bodies is unaffected by the presence of other bodies. This almost obvious statement becomes more significant when one recognizes that for some phenomena analogous statements are not true. For example, the force that binds two oxygen atoms together to form a molecule of gaseous oxygen is large, but the force on a third atom of oxygen is significantly reduced once a molecule has been formed. This ability of chemical forces to *saturate* is what accounts for the existence of distinct compounds. The principle of superposition states that the gravitational forces do not saturate; that a given body exerts forces on all other bodies no matter how many there are. This principle of superposition permits us to calculate (at least in principle) the forces between extended bodies.

We can treat any large object as a collection of point masses and then calculate the net force exerted by one body on another by adding the forces exerted between point masses. Figure 5-2 shows how this process might be begun for a few points. This is generally a formidable mathematical exercise for *all* the points. Newton invented the calculus in part to treat this problem. One important result of the use of calculus for this problem is that the force between two spheres (or more generally, between two spherically symmetric objects) is the same as the force between two point objects if the masses of the points is the same as the masses of the corresponding individual spheres. Another way of stating this is that *as far as external effects are concerned, a spherically symmetric homogeneous body can be considered as though its entire mass is concentrated at its geometrical center.*

It will be recalled that the units of force, mass, and distance that appear in Newton's law of gravitation have been determined in a way to be consistent with Newton's second law. Hence the constant G in the universal law of gravitation must be determined experimentally. The difficulty involved in performing such an experiment is im-

plied in the fact that the earliest reliable measurement of G was first done by Henry Cavendish in 1798, more than a hundred years after Newton published the law.

Cavendish performed this experiment by mounting two small masses on the ends of a light rod, as shown in Figure 5-3. The rod was suspended by a thin wire. When large masses M were brought near the small masses m, the gravitational forces caused the rod and the wire to twist. By comparing this twist with the twisting of this same apparatus by known forces, Cavendish was able to evaluate the very small forces between the masses. Then, since the force, masses, and distances between objects was known, he was able to solve Equation 5-1 for G.

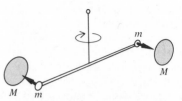

FIGURE 5-3 The Cavendish experiment. The rod is suspended by a thin wire. The gravitational forces exert torques on the rod.

The currently accepted value of G is

$$G = 6.673 \times 10^{-11} \ \frac{\text{nt m}^2}{\text{kg}^2}$$

$$= 6.673 \times 10^{-8} \ \frac{\text{dyne cm}^2}{\text{gm}^2}$$

G remains difficult to measure with high precision because the gravitational force is so weak. As an indication of the difficulty of measurement, the uncertainty in the value of G is, when expressed as a fraction of G, about 1/2000, whereas the uncertainty in most other physical constants is a few millionths.

5-3
The variation of g with attitude

One interesting example of the law of gravitation is the force exerted by the earth on objects that are near the surface of the earth. For this case we can treat the earth as an inertial frame of reference and measure the acceleration of objects with respect to the center of the earth. Further, if we assume that the earth is spherical, then we can assume that

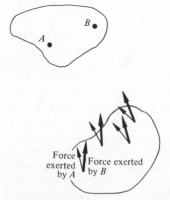

FIGURE 5-2 How one might begin to calculate the force exerted on one body by another. The point A exerts forces on each point of the other body. These forces are represented by the light arrows. The heavy arrows represent the forces exerted by B.

the earth acts as if all its mass is concentrated at its center. Then, for the gravitational force on an object of mass m, which is in free fall, we have

$$F = \frac{GM_e m}{r^2} \qquad (5\text{-}2)$$

and

$$F = ma$$

Therefore

$$a = \frac{GM_e}{r^2} \qquad (5\text{-}3)$$

where M_e is the mass of the earth and r is the distance from the object to the center of the earth.

For the special case in which the distance between the object and the center of the earth does not change much, this acceleration is a constant, at least in magnitude. Objects that are at the surface of the earth, and therefore about 20 million feet from the center of the earth, can change their altitude by several hundred feet without significantly altering the distance between them and the center of the earth. Thus for objects that remain near to the surface of the earth the acceleration due to the gravitational force exerted by the earth is essentially constant. This constant, g, is the acceleration due to gravity that we have mentioned previously.

One result that follows from this analysis of the acceleration due to gravity is that it relates g, the acceleration due to gravity, to G, the universal gravitational constant, and the size and mass of the earth. Thus since g and the radius of the earth are fairly easy to measure, once G is known, the value of the mass of the earth can be calculated. For that reason Cavendish referred to his measurement of G as an experiment to "weigh the earth." It is left as an exercise to the reader to calculate the mass of the earth from the known values of g, G, and the radius of the earth (Problem 6).

In general as we have seen, the acceleration due to gravity varies with the distance from the center of the earth. This variation can be emphasized by modifying Equation 5-3 somewhat: In that equation, a was the acceleration of an object acted on only by the gravitational force of the earth. To emphasize that let us use the symbol a_g for that acceleration. Then Equation 5-3 becomes

$$a_g = \frac{GM_e}{r^2} \qquad (5\text{-}4)$$

Now, when an object is at the surface of the earth,

r is equal to R_e, the radius of the earth; when an object is at the surface of the earth, the acceleration due to gravity is g. Therefore

$$g = \frac{GM_e}{R_e{}^2} \qquad (5\text{-}5)$$

If we multiply Equation 5-4 by $R_e{}^2/R_e{}^2$ we get

$$a_g = \frac{GM_e}{R_e{}^2}\left(\frac{R_e{}^2}{r^2}\right) \qquad (5\text{-}6)$$

In this equation we can identify $GM_e/R_e{}^2$ as g, from Equation 5-5, so we have

$$a_g = g\left(\frac{R_e}{r}\right)^2 \qquad (5\text{-}7)$$

Illustrative example

Calculate the altitude to which a rocket must be fired in order that the acceleration due to gravity is one quarter that at the surface of the earth. Here,

$$a_g = \frac{g}{4}$$

Inserting this in Equation 5-7,

$$\frac{g}{4} = g\left(\frac{R_e}{r}\right)^2$$

therefore

$$\left(\frac{R_e}{r}\right)^2 = \frac{1}{4}$$

$$\frac{R_e}{r} = \frac{1}{2}$$

Since r is the distance from the center of the earth and is equal to two earth radii, the *altitude* is one earth radius.

5-4
Inertial and gravitational mass

Thus far we have met mass in two different situations; in Newton's second law of motion mass appears as a measure of the inertia of a body, whereas in Newton's law of universal gravitation mass appears as an intrinsic property of a body, which determines the gravitational force which that body exerts. As a measure of inertia, mass is a dynamic property of a body; it influences how the body will respond to a force. In the universal law of gravitation, mass is a static property; it influences the force exerted by the body and the force exerted on the body. It might have turned out that these two types of mass were determined by different properties of a body, if nature were different. That is, one could imagine a universe in which the m that appears in $F = ma$ is different from the m that appears in the equation $F = GMm/r^2$. Nature is such that these two m's are the same in magnitude, but in a certain sense each body really has two different properties. The first property, which is measured by the m in $F = ma$, is a property called the *inertial mass*. The second property, measured by the m in $F = GMm/r^2$ is called the *gravitational mass*. The inertial mass of a body determines how that body will respond to a force, whereas the gravitational mass determines the force of interaction with another body.

The question arises, "Why are the inertial mass and gravitational mass always equal?" One legitimate answer is, "Why not?" That is, one can assume that the equality of two kinds of mass is an accident, or a coincidence. Or one can assume that this equality reveals something fundamental about nature. Albert Einstein took this second point of view in 1911. He took as an assumption the *principle of equivalence*. The principle of equivalence states that if an experiment is confined to a small region of space (such as a laboratory) then it is impossible to distinguish between the effects of acceleration and the effects of gravity. That is, if the laboratory were stationary in a strong gravitational field, then exactly the same effects would occur as if the laboratory were accelerating in the absence of a gravitational field. The principle of equivalence, which implies the equality of the two kinds of mass, is the basis of Einstein's general theory of relativity. This theory will be discussed later in the book. (See Chapter 35.)

The equality of inertial and gravitational mass has been verified experimentally to very high precision in a series of experiments carried out by Eotvos in the late nineteenth century and later refined by Dicke in the 1960s.

5-5
Electromagnetism

As we have seen, the gravitational force depends on the existence of gravitational mass. This property is possessed by every particle in the universe, including the electrons and nuclei of which matter is made. Electromagnetism depends on the existence of another property, called *electric charge*, which is possessed by some of the fundamental particles. Unlike mass, this property of charge

exists in two forms, called positive and negative, so that the force between two charged particles may be either a repulsion or an attraction, depending on the charges of the particles involved. Ordinary matter consists of equal numbers of positive and negative charges, and usually the total electromagnetic force on ordinary matter is small. Therefore in much of the subsequent discussion the word "body" will mean one of the charged fundamental particles, such as an electron or a proton. However, it *is* possible to charge macroscopic bodies and observe electromagnetic effects on such bodies.

A further complication is introduced by the fact that the electromagnetic force between charged particles depends on the motion of the charges. Historically this resulted in the discovery of electric forces and magnetic forces separately, and delayed the discovery of their connection.

The following discussion describes the electric force and briefly sketches the magnetic force and then their connection. These topics are discussed in more detail in subsequent chapters.

5-6
The electric force

Two charged particles that are at rest exert an electric force on each other, in addition to the gravitational force. For all the known particles usually described as fundamental, such as electrons or protons, the electric force is much larger than the gravitational force. The force is governed by a law, called *Coulomb's law*, which interestingly enough is algebraically the same as the law of gravitation. Figure 5-4 shows the apparatus used by Coulomb. Coulomb's law may be stated as: Any two charged particles that are at rest attract each other if they have opposite signs of charge, or repel each other if they have the same sign of charge, (as shown in Figure 5-5) with a force whose magnitude is given by

$$F = \frac{kq_1q_2}{r^2} \qquad (5\text{-}8)$$

where F is the force exerted on either charge by the other, q_1 is the charge of one of the particles, q_2 is the charge of the other, r is the distance between the charges, and k is a constant of proportionality that depends on the units chosen for force, charge, and distance.

If both charges are positive, F in Equation 5-8 is positive; if both charges are negative, F is posi-

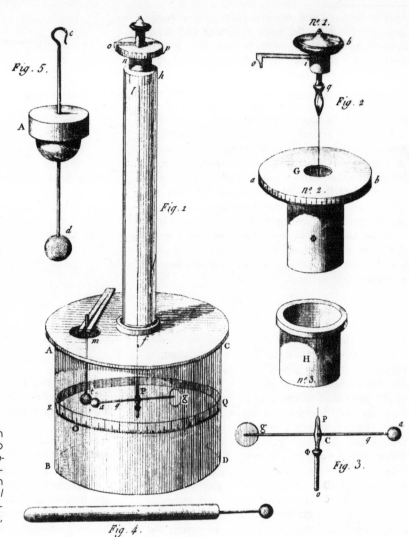

FIGURE 5-4 Coulomb's torsion balance, which was used to measure the electrical force between charged bodies. (Reproduced with permission from *Foundations of Modern Physical Science* by Holton and Roller, Addison-Wesley, Reading, Mass., 1958.)

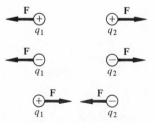

FIGURE 5-5 The direction of the force between charged objects.

tive, while if one charge is positive and the other is negative, F is negative. Since *like charges repel and unlike charges attract*, the sign convention im-

plied in Equation 5-8 is that a *positive force is a force of repulsion, while a negative force is a force of attraction.* In other words, a positive force is a force directed away from the charge exerting the force, while a negative force is a force directed toward the charge exerting the force.

The question of units for electromagnetism (which is, of course, peripheral to an understanding of the ideas) nevertheless presents a serious problem for the authors of physics texts. Most physicists professionally use one of several cgs systems of units in which the basic equations have simple forms, or in which the constants of proportionality are simple. On the other hand, most engineering applications and most practical applications are usually in mks units, because they involve equations derived from the basic equations and these derived equations are simpler in mks units. Rather than choose the alternative of using several systems of units, this text will rather arbitrarily use the mks system. The Appendix will briefly describe the conversion from this system to several of the other more common systems in use.

In the mks system, where force is measured in newtons and distance in meters, charge is measured in *coulombs*. Unfortunately, the standard operational definition of the coulomb employs a magnetic measurement that involves concepts we have not yet discussed. Therefore it is necessary to assert at this point that it *is* possible to define charge operationally without describing how this can be done. Once the coulomb has been defined this way, the constant k in Coulomb's law can be determined experimentally to be:

$$k = 8.98742 \times 10^9 \frac{\text{nt m}^2}{\text{coul}^2}$$

for which the value 9×10^9 can usually be used.

Coulomb's law, like the law of gravitation, is subject to the principle of superposition. That is, the electric force on any particle is the vector sum of the individual forces exerted by all the other particles, and the individual forces are the same as they would be if only two particles were present. One major difference between the gravitational and electric forces appears when we consider the presence of macroscopic objects or matter rather than just fundamental particles. Macroscopic objects consist of large numbers of negative and positive particles so arranged that they exert no electric forces on other objects that are reasonably far away, under ordinary circumstances. If a charged particle is placed near such a neutral collection of charges, then forces will be exerted on all the charged particles. Sometimes these forces can *rearrange* the spatial locations of the individual charges in the neutral object. This spatial rearrangement can then result in the neutral object exerting forces on nearby charged bodies and having net forces exerted on the neutral object (see Figure 5-6). Thus it is possible to have the situation where several charged particles are exerting forces on each other and then when a piece of neutral matter is brought near them, to have the force on some or all of the charged particles change. When this happens, the principle of superposition appears to fail unless it is remembered that the neutral matter really consists of charged particles.

Two charged bodies exert forces on each other

Then, when a neutral body is brought nearby, each exerts forces on the charges within the neutral body

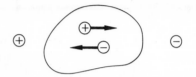

which may rearrange the charges in the neutral body

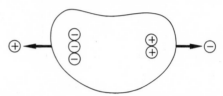

producing a net force on the neutral body and, from Newton's third law, additional forces on the original charged bodies.

FIGURE 5-6

Illustrative example

Given two electrons 1 cm apart, find the electric forces between them and compare these forces with the gravitational force each exerts on the other.

An electron is a negatively charged particle whose charge is 1.6×10^{-19} coul and whose mass is 9.1×10^{-31} kg. The electric force between two electrons is repulsive, since they both have the same sign (negative) of charge. Thus the electric force on one of the electrons is along the line connecting the two charges pointing *away* from the other charge. The force is given by

$$F = \frac{kq_1 q_2}{r^2}$$

$$= 9 \times 10^9 \frac{\text{nt m}^2}{\text{coul}^2} \times (-1.6 \times 10^{-19} \text{ coul})$$
$$\times (-1.6 \times 10^{-19} \text{ coul}) \div (10^{-2} \text{ m})^2$$
$$= +2.3 \times 10^{-24} \text{ nt}$$

The positive sign of the force is algebraic confirmation of the fact that the force is a force of repulsion.

The gravitational force between these two electrons is attractive and has a magnitude

$$F = \frac{Gm_1 m_2}{r^2}$$

$$F = \frac{6.67 \times 10^{-11} \times 9.1 \times 10^{-31} \times 9.1 \times 10^{-31}}{(10^{-2})^2}$$

$$= 5.5 \times 10^{-67} \text{ nt}$$

Thus the electric force is about 4×10^{42} times as large as the gravitational force—a huge ratio. Thus the gravitational force can be neglected in comparison with the electric force. The only reason that gravitational forces on macroscopic objects are usually bigger than electromagnetic forces is that there is only one sign of gravitational "charge" —that is, mass. There is no cancellation; the gravitational attractions of all particles add up. On the other hand, bulk matter is usually electrically neutral, so the attractive and repulsive forces virtually cancel each other.

5-7
The magnetic force

When two charged particles are moving they exert on each other: an electric force and a gravitational force, and in addition, a force that exists only when both charges are moving, called the *magnetic force*. Historically, magnetism was associated with magnets and iron, and so at first sight it is somewhat surprising to find a force between moving particles called magnetic. This connection between moving charges and magnetism was first discovered by Hans Christian Oersted, who showed experimentally in 1820 that a stream of moving charges, a current, exerted a force on a compass needle. Subsequently, others were able to arrange streams of moving charges in systems of wires in such a way that magnets exerted forces on the systems. Thus it is possible to arrange collections of moving charges so as to simulate the effects of magnets. The problem of accounting for the peculiar magnetic properties of iron is more complex but in the past 40 years it has been possible to explain the magnetic behavior of iron (and other magnetic materials) in terms of charged particles, the atomic electrons.

The magnetic force between two moving charges depends on the charges, velocities, and locations of both particles. In particular, the *direction* depends on the direction of each of the two velocities involved, and the direction of the line joining the two charges (see Figure 5-7). In addition, the *magnitude* of the force depends on the relative directions of these vectors. These details will be discussed in subsequent chapters, but for the purposes of this preliminary overview of the forces, it is

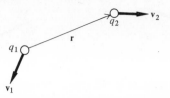

FIGURE 5-7 Vectors on which the magnetic force depends.

sufficient to say that the magnitude of the magnetic force is proportional to the product of the charges and to the product of the magnitude of the velocities, and is inversely proportional to the square of the distance between the charges. That is,

$$F \propto \frac{q_1 q_2 v_1 v_2}{r^2} \tag{5-9}$$

In order to convert this proportionality to an equation, it must be remembered that the magnitude of the force depends on the directions, so one can write

$$F = CS \frac{q_1 q_2 v_1 v_2}{r^2} \tag{5-10}$$

Here, S is a number that depends on the directions of the velocities and on the direction of the line between the charges and C is a constant that depends on the units. As will be seen in later chapters, S is a product of the sines of two angles, so it is a dimensionless number that lies between -1 and $+1$. If charge, velocity, length, and force are measured in mks units,

$$C = 10^{-7} \frac{\text{nt sec}^2}{\text{coul}}$$

One important special case of this is that of two positive charges moving along parallel paths. If the charges are moving side by side with equal velocities, v, then the force is

$$F = CS \frac{q_1 q_2 v^2}{r^2}$$

where r is the distance between the paths. For this special case $S = 1$, so

$$F = C \frac{q_1 q_2 v^2}{r^2} \tag{5-11}$$

In this case the force turns out to be attractive so, as shown in Figure 5-8, there are two attractive forces, gravitational and magnetic, and one repulsive force, the electric force. For reasonable

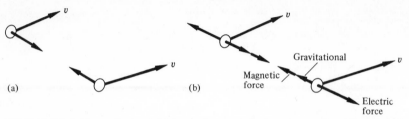

FIGURE 5-8 (a) Magnetic force between two equal charges moving with parallel, equal velocities. (b) All the forces between two equal charges moving with parallel, equal velocities.

velocities and real particles, such as protons, the magnetic force is smaller than the electric force but larger than the gravitational force.

If both charges had been negative in this example, then the force would have been exactly the same, but if one of the charges had been negative and the other positive, then the force would have had the same magnitude but would have been repulsive.

5-8
Electromagnetism—a union

As we have just seen, two moving charged particles exert three forces on one another: gravitational, electric, and magnetic. For the special case of two charges of like sign traveling along parallel paths with equal velocities, the electric force is a repulsive force and the magnetic force is an attractive force. A problem arises when we recognize that this pair of particles may be observed from many different inertial systems moving with different relative velocities. In particular it is possible to observe the particles from a coordinate system in which both particles are at rest. In this coordinate system there is no magnetic force; but the force between two particles should not be determined by the coordinate system in which they are observed. The solution to this problem requires that the electric and magnetic forces be described in a manner consistent with special relativity. (See Chapter 10.) For the purposes of this discussion, two conclusions from this may be asserted:

1. The electric and magnetic forces are really two aspects of a single interaction, the electromagnetic force.
2. For particles traveling with velocities small

compared to the speed of light (3×10^{10} cm/sec), the motions of the particles will be correctly predicted by treating the force as consisting of a magnetic force and an electric force each of which is governed by the laws given above.

5-9
The nuclear force

A reasonably accurate picture of the nucleus is that it consists of protons (positively charged particles of charge 1.6×10^{-19} coul, mass 1.67×10^{-27} kg) and neutrons (neutral particles of approximately the same mass as a proton) bound together by a force. The force between any two of these particles is much larger than the electric force and is attractive when the particles are separated by distances about the size of the nucleus (about 10^{-13} cm). For distances larger than this the force is essentially zero, and there are indications that for distances significantly smaller than 10^{-13} cm, the force becomes a repulsive force. The force does not depend on the charge or the mass of the particles, but does appear to depend to some extent on the state of motion of the particles and on some properties of the particles that may be indications of internal structure of the particles.

In the process of investigating this force and the properties of nuclei, physicists have discovered the existence of many particles that seem to be as fundamental as the electron, proton, and neutron. Some of these break up spontaneously, or decay, leaving other particles. For example, when neutrons are liberated from nuclei, they spontaneously decay, leaving protons and electrons. In this decay process a second nuclear force is involved, significantly weaker than both the first and the electromagnetic force but stronger than the gravitational force. For want of a better name, this second force is called the weak force, whereas the first nuclear force is called the strong force. The weak force, like the strong force, is independent of charge and is large only for very short distances of separation between the particles. We shall discuss these forces in Chapters 33 and 34.

5-10
The forces between objects

In some cases the force between macroscopic objects is obviously one of the fundamental forces. For example, the force exerted by the sun on the earth is clearly the gravitational force, and the force exerted on little bits of paper by a briskly rubbed rubber comb is clearly an electric force. In very many cases, however, the fundamental origin of a force is less obvious.

Consider, for example, a saucer resting on a table. The table exerts a normal force on the saucer supporting it. The distance between the nuclei of the atoms in the saucer and the nuclei of the atoms of the table is at least one atomic diameter (see Figure 5-9). Although this is a very small distance, it is at least 10,000 times as large as a nucleus. Thus on the nuclear scale of distances the nuclei of the saucer and table are very far apart. They are so far apart that the nuclear forces between saucer and table are zero. Since the gravitational force is always a force of attraction, it is hard to see how this could contribute to the normal force. This leaves the electromagnetic interaction as the primary cause of the contact force. Let us assume that the saucer consists of charged particles, some of which are moving and some of which are relatively stationary. These charged particles exert magnetic and electric forces on the charged particles in the table. Since the solids are on the whole neutral, these forces tend to cancel each other, and so the force is generally zero. When the saucer is close to the table, the details of the geometric arrangement of the charged particles becomes important. The problem of accounting for the contact force then reduces to the problem of describing a geometry that will yield the appropriate behavior. Thus, the problem of understanding the apparently simple force of contact becomes an involved geometric problem of understanding the structure of matter. The problem is further complicated by the fact that, on this scale of distance, Newton's laws are inadequate and the situa-

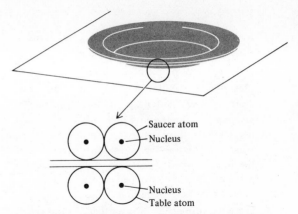

FIGURE 5-9 A saucer at rest on a table. The distance between the saucer nucleus and table nucleus is very large on the nuclear scale of distance.

tion must be described in terms of quantum mechanics.

Similarly, the force between two permanent magnets turns out to be somewhat more complex than would appear at first sight. At first it appears that the magnetic force between permanent magnets can be explained by the existence of moving charges in the magnets—the moving electrons in the atoms. Further, it is now known that electrons have intrinsic magnetic properties: stationary electrons exert magnetic forces. The difficulty with this simple point of view is that permanent magnets can be made only from iron and a few other materials; but all materials have moving electrons. Thus, in order to explain why magnets are made of iron, it is necessary to describe the structure of iron in detail. This explanation turns out to depend on the intrinsic magnetism of the electrons in the iron atoms, the specific structure of the iron atom, and the structure of iron crystals. Again the apparently simple problem of understanding the force between magnets turns out to be a problem in the structure of matter.

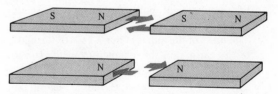

FIGURE 5-10 The forces between two bar magnets.

Questions

1. State Newton's law of universal gravitation.
2. Why is gravitation called universal?
3. What is the principle of superposition?
4. Describe a phenomenon for which the principle of superposition does not apply.
5. Imagine a universe in which the gravitational mass of aluminum is twice the inertial mass, while the gravitational mass of iron is one half the inertial mass. A piece of aluminum and a piece of iron are dropped at the same time. Which hits the ground first?
6. On what properties of a particle do the gravitational forces exerted on the particle depend?
7. On what properties of a particle do the electric forces exerted on the particle depend?
8. On what properties of a particle do the magnetic forces exerted on the particle depend?
9. The fundamental forces do not depend on the kind of matter involved, yet the magnetic force exerted by a permanent magnet does depend on the kind of matter involved. How can these be reconciled?
10. Why was a detailed law of gravitation discovered before a detailed law of electric force?
11. How does the magnetic force differ from the electrostatic force?
12. How would the acceleration due to gravity change if the radius of the earth were to double, assuming that the density of the earth remained constant?
13. What would happen to your weight if the earth were to become twice as dense, assuming no other changes?
14. Two electrons are moving parallel to each other with equal velocities. How does the force between them differ from the force exerted when they are moving with opposite velocities?
15. The force exerted on a piece of iron by a permanent magnet changes when the temperature of the magnet increases. Explain why this might happen.
16. Why should one not assume that Equations 5-1, 5-8, and 5-11 are necessarily valid for particles separated by distances small compared to the size of a nucleus?

Problems

1. What is the gravitational force between two 1-kg objects which are separated by 50 cm?
2. Two small objects have charge q_1 and q_2. When they are 1 cm apart the force between them is F. What is the force when the distance between them is (a) 2 cm, (b) 3 cm, (c) 4 cm, and (d) 5 cm? Plot a graph of the force as a function of the distance between the objects.
3. A space explorer notes that the gravitational force exerted on his ship by a star which is 1 light-year away is 10 nt. How large will the force be when he is $\frac{1}{2}$ light-year away from the star?
4. How far from the earth must an object be if the gravitational pull of the sun is exactly balanced by the gravitational pull of the earth? The mass of the sun is about 3.2×10^5 the mass of the earth, and the distance from the sun to the earth is 93×10^6 mi.
5. The mass of the moon is 0.012 that of the earth, and the radius of the moon is 0.27 that of the earth. Find the value of g on the surface of the moon.
6. From the known values of the acceleration due to gravity, the constant G in the law of universal gravitation and the radius of the earth, find the mass of the earth. From this, find the density of the earth (mass per unit volume) and compare with Newton's estimate of 5 gm/cm³.
7. At what altitude is the acceleration due to gravity 1 percent smaller than it is at sea level.
8. If the earth suddenly shrank to half its current size, without changing its mass, how much would a 70-kg man weigh?
9. The mass of the moon is 7.3×10^{25} gm, its radius is 1.7×10^8 cm; the mass of the earth is 6×10^{27} gm, its radius is 6.4×10^8 cm; and the distance between the earth and the moon is 3.8×10^{10} cm. Find the total gravitational force due to both the earth and the moon on a 75-kg astronaut (a) when he is at the surface of the earth, (b) when he is halfway between the earth and the moon, and (c) when he is on the surface of the moon. (d) Where is this total gravitational force equal to zero?
10. An electric current in a wire consists of a

stream of electrons moving parallel to each other with the same speed (on the average). Consider two such electrons moving with a speed of 0.2 cm/sec, separated by 1 mm, which is about the size of a typical wire. Find the gravitational force, the electric force, and the magnetic force acting between them.

11. A hydrogen atom consists of an electron and a proton separated a distance of 10^{-8} cm. Find (a) the gravitational force of attraction between the electron and the proton, (b) the electric force between them, and (c) the acceleration of the electron, in an inertial frame fixed on the proton.

12. An object weighed at midnight should weigh slightly more than the same object weighed at noon because of the changed distance between the object and the sun. Estimate the size of this change in weight for a 150-lb person. (The mass of the sun is 3.3×10^5 times the mass of the earth, and the distance from the earth to the sun is about 2.4×10^4 times the radius of the earth.)

13. Newton explained the existence of the tides by pointing out that the moon exerts a different gravitational force on the water that is at the surface of the earth nearest to the moon, from the force it exerts on the water on the far side of the earth. (a) Find the force exerted by the moon on 1 gm of water that is on the near side of the earth, and the force on 1 gm of water on the far side of the earth. Assume that the earth-moon separation given in Problem 9 is the distance between the centers. (b) Express the difference between these forces as a percentage of the force that would be exerted on a gram of water located at the center of the earth.

14. The sun, which has a mass of 2×10^{33} gm and a radius of 7×10^{10} cm, and is 1.5×10^{13} cm away from the earth, also exerts gravitational forces on the water on the surface of the earth. Repeat the calculations of Problem 6 for the forces exerted on the water by the sun. From the results of both problems, predict the ratio of the size of the tides due to the moon to the size of the tides due to the sun.

15. Two equal and opposite charges of magnitude 30 microcoulombs (μcoul) are 10 cm apart. A thin metal rod, 3 cm long, is placed halfway between the two charges, so that the rod lies along the line connecting the two charges. The rod remains neutral, but the internal charges are rearranged by the forces exerted by the original charges. The result is that the ends of the rod each acquire a net charge of 10 μcoul; the end of the rod that is closer to the positive charge acquires a negative charge while the end of the rod that is close to the negative charge acquires a positive charge. Find (a) the force exerted on the positive 30-μcoul charge before the rod is put in position, and (b) the force exerted on the positive charge after the rod is placed in position.

16. If a particle is located inside a spherical shell of matter, the net gravitational force exerted on the particle is equal to zero. Consider, then, a particle inside a sphere of radius R, at a distance r from the center of the sphere. The material in the shell between the radii r and R will exert no gravitational force. The material inside the radius r will exert a force that would be exerted if all the mass of the sphere of radius r were located at the center. (These statements are a consequence of the inverse square nature of the law of gravitation, and analogous statements are true for electric forces.) Show that the force on a particle inside a uniform sphere is proportional to the distance of the particle from the center of the sphere.

work & energy

chapter 6

6-1
The meaning of energy

One major dictionary defines energy in the physical sense as "the capacity to perform work." Even though most of us would be hard pressed to say exactly what is meant by "capacity" or "work," this definition adequately conveys the essential idea; a body has energy if it has the ability to alter the state or condition of another body. For example, a batted ball has energy because it is moving. If the moving baseball hits another object it can move that object or deform it.

If a baseball is held at rest at some height above the ground, it has energy because of its position; if the ball is released it will fall, accelerate, and acquire velocity; on striking another object, it can cause that object to move or it can deform that object. In this case the positional energy of the ball is converted into motional energy of the ball, but one can imagine circumstances in which the positional energy of one object is transferred to another object. In the pulley system sketched in Figure 6-1,

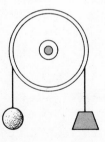

FIGURE 6-1

if the ball is slightly heavier than the weight hanging from the other end of the cord, then the ball will descend slowly and the weight will rise slowly. Some of the positional energy of the ball is being transferred to the weight. Thus we see that energy exists in at least two forms and that energy can be transferred from one form to another and from one object to another.

In order to make these ideas precise we must define energy mathematically; we have to give formulas for the motional energy of a body and for positional energy.

Because the word "capacity" in the definition above is so imprecise, no obvious formula suggests itself as the "right" one for either kind of energy. However, the idea that energy can be transformed from one form to another and transferred from one body to another suggests that we might define energy in such a way that under some circumstances the total amount of energy is constant. That is, it is possible to choose mathematical definitions of energy of motion and energy of position such that, as a consequence of Newton's laws, the sum of all of the energies for some systems remains constant. If the energy of one object in the system increases, then the energy of one or more other objects in the system decreases by the same amount. If the motional energy increases, then the positional energy decreases.

Once we have chosen these definitions of energy we can examine the behavior of the total energy of other systems. In general, if we arbitrarily pick a group of bodies and call them a system, then we will find that the energy of such a system is not constant. However, Newton's laws allow us to conclude that when the total energy of some system changes, then one can find one or more *other* systems whose energy changes by an equal and opposite amount. That is, one can either extend the system and say that the total energy of the new system is constant, or one can say that the total energy that leaves one system is transferred to one or more other systems with no loss of energy. Thus for a large number of situations *energy* is *conserved*. More important, it is possible to extend the definition so that energy is always conserved. That is, it is possible to define a small number of additional kinds of energy, such as electrical energy and nuclear energy, so that either the total energy of a system is constant, or the change in energy of the system can be accounted for by an energy transfer without loss to one or more other systems.

Energy may be transferred from one system to another. As we will see in later chapters, this energy transfer may be in the form of heat if the two systems have different temperatures. For two systems at the same temperature, the energy transfer is called *work*. Thus, if the energy of a system decreases, it must do work on another system and the energy of that other system must increase.

In the following sections, work, energy of motion, and energy of position due to gravity will be defined. The constancy of total energy, the law of conservation of energy will be illustrated for simple cases where the only forces are gravity and contact forces. Cases involving other forces or other forms of energy will be discussed in later chapters.

6-2
Work done by forces

An extremely important concept that has been developed in physics is that of the *work done* on a body by the action of some external agent that exerts a force on this body and produces motion. For example, whenever someone lifts a body, he does work by exerting a force upward on it and moving it upward. Whenever a steam locomotive pulls a train, a series of processes takes place in the steam engine of the locomotive which enables it to exert a force on the train and move it in the direction of the force. The term *work*, as used in physics, is a technical term. Whenever work is done by an external agent on a body, *the work done is the product of the force that acts on the body and the distance through which the body moves while the force is acting on it, provided that the force and the distance through which the body moves are parallel to each other.*

If a body is acted upon by a constant force **F** and is displaced through a distance s parallel to it while the force is acting on it, the work done W is

$$W = Fs \qquad (6\text{-}1)$$

For example, if a heavy trunk is pulled along the floor through a distance of 15 ft by a constant force of 50 lb acting horizontally, the work done is

$$W = 50 \text{ lb} \times 15 \text{ ft} = 750 \text{ ft lb}$$

If the constant force **F** and the displacement **s** are not parallel, then only that component of the force which is in the direction of the displacement does the work. For example, it may be more convenient to pull the trunk with a force **F** at some angle θ with respect to the floor, as shown in Figure

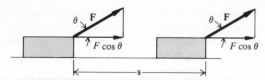

FIGURE 6-2 Work done by a force *F* acting at an angle θ to the direction of its motion is *Fs* cos θ.

6-2. The component of **F** in the direction of motion is **F** cos θ; if the trunk is moved through a distance *s*, while this force is acting on it, the work done *W* is

$$W = Fs \cos \theta \qquad (6\text{-}1a)$$

Suppose that in this case the trunk is pulled by means of a rope tied to it, and that the force acting on the trunk is 40 lb at an angle of 30°. If the trunk is moved through a distance of 15 ft, the work done by this force is

$$W = 40 \text{ lb} \times 15 \text{ ft} \times \cos 30°$$
$$= 40 \times 15 \times \frac{\sqrt{3}}{2} \text{ ft lb} = 520 \text{ ft lb}$$

Work is a scalar quantity. Both the force **F** and the displacement **s** are vector quantities. Here we have an important example of a very common occurrence in physics in which the product of two vector quantities produces a scalar quantity. We can and often do symbolize this product in some special way. We shall adopt the method of placing a center dot between the two vector quantities involved to show that we are dealing with the *scalar product* of the two vectors, thus

$$W = \mathbf{F} \cdot \mathbf{s} \qquad (6\text{-}2)$$

and shall define this scalar product of two vectors by the equation

$$\mathbf{F} \cdot \mathbf{s} = Fs \cos \theta \qquad (6\text{-}3)$$

where *F* and *s* are the respective magnitudes of the two vectors and θ is the angle between them (see Figure 6-3).

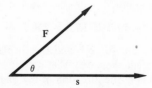

FIGURE 6-3 The work done by the force **F** in the displacement **s** is *Fs* cos θ.

If the force varies during the displacement, or if the path is curved, we can determine the small quantity of work Δ*W* done during a small displacement Δ**s** by a force **F** that is constant over this displacement and obtain

$$\Delta W = \mathbf{F} \cdot \Delta \mathbf{s} \qquad (6\text{-}4)$$

These small quantities of work can then be summed over the entire displacement.

6-3
Units for expressing work

There are several different units that are used for expressing the work done. In every case the unit must be equivalent to the product of a force by a distance. In the examples above, the unit of work contains both factors; the *foot-pound* is a unit commonly used in engineering for expressing work. In the cgs system the analogous unit would be the *dyne-centimeter*. This unit, however, is usually replaced by a single term, the *erg*: *1 erg is the work done by a force of 1 dyne acting on a body through a distance of 1 centimeter in the same direction as the force*; thus,

$$1 \text{ erg} = 1 \text{ dyne cm}$$

For example, the work done by a horizontal force of 250 dynes acting on a body for a horizontal distance of 30 cm is

$$W = Fs$$
$$= 250 \text{ dynes} \times 30 \text{ cm}$$
$$= 7500 \text{ ergs}$$

In the mks system, the unit of work is the *joule*. A joule is defined as *the work done by a force of one newton acting through a distance of one meter in the same direction as the force*; that is,

$$1 \text{ joule} = 1 \text{ nt} \times 1 \text{ m}$$

For example, if a force of 25 nt acts on a body for a distance of 12 m in the direction of the force, the work done is

$$W = 25 \text{ nt} \times 12 \text{ m}$$
$$= 300 \text{ joules}$$

The relationship between the joule and the erg can be found readily from the facts that 1 nt = 100,000 dynes, and 1 m = 100 cm, so

$$1 \text{ joule} = 10,000,000 \text{ ergs}$$
$$= 10^7 \text{ ergs}$$

6-4
Kinetic energy

The energy that a body possesses because it is in motion is called *kinetic energy*. The expression for kinetic energy can be determined by evaluating the work done in accelerating a body from some initial speed u to a final speed v; this work will produce a *change in its kinetic energy* (see Figure 6-4). Suppose that a constant force **F** acts on the body of mass m during a displacement **s** in the direction of the force; the work done is

$$W = Fs$$

Now, from Newton's second law

$$F = ma$$

and thus

$$W = mas$$

But we know that

$$v^2 = u^2 + 2as$$

and hence

$$as = \frac{v^2 - u^2}{2}$$

Substituting this value for as in the equation for the work done, we get

$$W = \tfrac{1}{2}\, mv^2 - \tfrac{1}{2}\, mu^2 \qquad (6\text{-}5)$$

The expression on the right-hand side is the *change in the kinetic energy of the body*.

The term $\tfrac{1}{2}mv^2$ is the final kinetic energy and $\tfrac{1}{2}mu^2$ is the initial kinetic energy of the body. In general, the kinetic energy $\mathcal{E}_k$ of a body is half the product of its mass and the square of its velocity; thus

$$\mathcal{E}_k = \tfrac{1}{2}mv^2 \qquad (6\text{-}6)$$

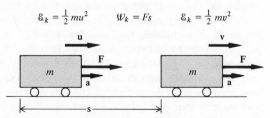

$$\mathcal{E}_k = \tfrac{1}{2}\, mu^2 \qquad W_k = Fs \qquad \mathcal{E}_k = \tfrac{1}{2}\, mv^2$$

FIGURE 6-4 The work done by a force in accelerating a body increases its kinetic energy.

Determine the kinetic energy of an automobile weighing 4000 lb moving at 60 mi/hr.

The mass of the automobile is $m = 4000/32$ lb sec²/ft and its speed

$$v = 60\ \frac{\text{mi}}{\text{hr}} = 88\ \frac{\text{ft}}{\text{sec}}$$

Putting these values into Equation 6-6, we get

$$\mathcal{E}_k = \tfrac{1}{2}\frac{4000}{32}\frac{\text{lb sec}^2}{\text{ft}} \times \left(88\ \frac{\text{ft}}{\text{sec}}\right)^2$$
$$= 484{,}000\ \text{ft lb}$$

If this car is brought to rest by an application of the brakes, the work done by the frictional forces would have to be equal to the loss in kinetic energy of the automobile, which would be 484,000 ft lb. If we assume that the average value of all the frictional forces acting on the car while it is being brought to rest is F, then the distance s that the car will move before coming to rest is given by the equation

$$Fs = \tfrac{1}{2}mv^2$$

If we compare the stopping distances of the same car for different initial speeds, assuming that the frictional forces are the same, we note that the distance s varies with the square of the speed at the time that the brakes are first applied. Thus if the brakes are applied when the car is moving at 60 mi/hr, it will travel four times as far before coming to rest as it will when the car is moving at 30 mi/hr.

It must be emphasized that energy is a scalar; even though kinetic energy is a consequence of motion it has no direction and is a scalar quantity. Since the square of the speed is always positive and the mass of an object is always positive, the kinetic energy is always positive.

Illustrative exercise

Estimate the kinetic energy of a man running at top speed. Most medical literature uses the figure of 70 kg for the mass of an average man. Since the times for the the 60-yard dash in track meets are close to 6 sec, a good estimate for "running at top speed" is about 10 m/sec. Therefore the kinetic energy is

$$\mathcal{E}_k = \tfrac{1}{2} \times 70\ \text{kg} \times 10^2\ \frac{\text{m}^2}{\text{sec}^2}$$
$$= 3500\ \text{joules}$$

6-5
Potential energy of position

In order to lift a weight W vertically through a height h, a force F must be exerted upward. If the body is moved upward with uniform velocity, it will be in equilibrium and the magnitude of F will be equal to W. The work done in lifting the body vertically is thus

$$\mathcal{W} = Wh \qquad (6\text{-}7)$$

and, since

$$W = mg$$

where m is the mass of the body, the work done can also be written as

$$\mathcal{W} = mgh \qquad (6\text{-}7a)$$

As a result of this work, the body is now in a new position in the earth's gravitational field; it is at a height h above its original position (see Figure 6-5). The *work done* has thus produced a *change* in the *gravitational potential energy of the system*, which in this case consists of the earth and the body under discussion. If we take the initial position of the body as the position of zero gravitational potential energy, the potential energy of the body at a height h above this position is equal to the work done in lifting it through this height. The potential energy $\mathcal{E}_p$ of this body can therefore be written as

$$\mathcal{E}_p = Wh \qquad (6\text{-}8)$$

or

$$\mathcal{E}_p = mgh \qquad (6\text{-}9)$$

When there are no other forms of potential energy involved, the term "gravitational potential energy" is usually called simply "potential energy."

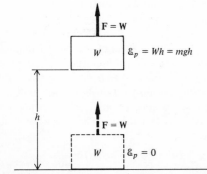

FIGURE 6-5 The work done by lifting a body increases its potential energy.

In all practical cases we are concerned only with *changes* in the potential energy of a body or system of bodies; hence the actual choice of the zero level of potential energy is arbitrary. The potential energy of a body may be either positive or negative, depending upon the choice of the zero level.

For example, an airplane flying at an altitude of 15,000 ft has a greater amount of potential energy than it had at sea level. If sea level is taken as the zero level of potential energy, then its potential energy at any other altitude h is given by Equation 6-8. If the weight of this airplane is 10 tons, its potential energy $\mathcal{E}_p$ at the altitude of 15,000 ft is

$$\begin{aligned} \mathcal{E}_p &= 10 \text{ tons} \times 15{,}000 \text{ ft} \\ &= 10 \times 2000 \text{ lb} \times 15{,}000 \text{ ft} \\ &= 3 \times 10^8 \text{ ft lb} \end{aligned}$$

Illustrative example

A body whose mass is 2.0 kg is dropped through a height of 30 m. Determine the change in its gravitational potential energy.

Let us take the original position of the body as the zero level of potential energy; its potential energy at the lower position will be less than its original energy. Its new potential energy will be

$$\mathcal{E}_p = -mgh$$

$$\mathcal{E}_p = -2.0 \text{ kg} \times 9.8 \, \frac{\text{m}}{\text{sec}^2} \times 30 \text{ m}$$

$$= -588 \text{ joules}$$

6-6
Gravitational potential energy independent of path

The expression for the potential energy of a particle in the earth's gravitational field, mgh, shows that it varies only with the height of the particle; this implies that the potential energy at any point does not depend upon the manner in which the particle arrived at that point. In other words, the change in potential energy of a particle in going from any point A to any other point B in the earth's gravitational field is *independent of the path* it traversed between A and B. This can be shown to be generally true, but we shall show it only for a simple path connecting A and B. Suppose that the path connecting A and B is a straight line such as that in Figure 6-6, and that B is at a height h above A. We may imagine that A and B are joined by a frictionless plane and that a force F moves the particle with uniform velocity along the

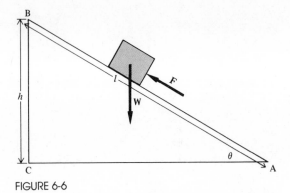

FIGURE 6-6

length l between A and B. The work done by this force is

$$\mathscr{W} = Fl$$

but we know that

$$F = W \sin \theta$$

so

$$\mathscr{W} = Wl \sin \theta$$

but

$$l \sin \theta = h$$

hence

$$\mathscr{W} = Wh$$

and therefore

$$\mathscr{E}_p = Wh$$

Thus the potential energy of the particle depends only upon the vertical distance between A and B and not upon the length of the path between them.

A field of force in which the potential energy does not depend upon the path but depends only upon the position of the particle is called a *conservative field of force*. Another conservative field that will be discussed later is the electrostatic field around an electric charge.

One consequence of the fact that the potential energy of a particle in a conservative field of force is independent of the path is that if a particle, after having traversed any path or paths whatever, returns to its original position, the potential energy returns to its original value. Thus the particle in Figure 6-6, starting at A and going to B along the path AB and returning by the paths $BC + CA$, undergoes no change in potential energy. In terms of the concept of work, work was done *against* the

force of attraction between the earth and the particle in moving it from A to B, while work was done *by* this force in moving the particle from B to C, and no work was done in moving the particle from C to A.

6-7
Conservation of mechanical energy

If an outside agent does an amount of work $\mathscr{W}$ on a particle in a conservative field of force, it may produce a change in its potential energy $\Delta\mathscr{E}_p$, or a change in its kinetic energy $\Delta\mathscr{E}_k$, or both. This may be written symbolically as

$$\mathscr{W} = \Delta\mathscr{E}_p + \Delta\mathscr{E}_k \qquad (6\text{-}10)$$

Suppose that a particle of mass m is at point A in a gravitational field where its coordinates are x_1, y_1 and its velocity is v_1, as shown in Figure 6-7. Suppose now that some external agent exerts a force on it and does an amount of work $\mathscr{W}$ in moving to point B where its coordinates are x_2, y_2 and its velocity is v_2. The change in potential energy of the particle in the gravitational field involves only the change in its y coordinate and is

$$\Delta\mathscr{E}_p = mgy_2 - mgy_1$$

and the change that occurs in its kinetic energy is simply

$$\Delta\mathscr{E}_k = \tfrac{1}{2}mv_2{}^2 - \tfrac{1}{2}mv_1{}^2$$

Hence we can write

$$\mathscr{W} = mgy_2 - mgy_1 + \tfrac{1}{2}mv_2{}^2 - \tfrac{1}{2}mv_1{}^2 \quad (6\text{-}11)$$

for the changes that occur in potential and kinetic energies of the particle as a result of the work done by the outside agent. This equation is sometimes referred to as the *work-energy equation*.

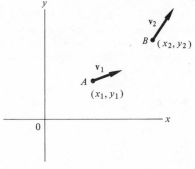

FIGURE 6-7

If the particle should move from A to B under the action of the forces of the conservative field only, then $W = 0$; in this case we can write

$$0 = mgy_2 - mgy_1 + \tfrac{1}{2}mv_2{}^2 - \tfrac{1}{2}mv_1{}^2$$

We can rearrange these terms so that they become

$$mgy_1 + \tfrac{1}{2}mv_1{}^2 = mgy_2 + \tfrac{1}{2}mv_2{}^2 \qquad (6\text{-}12)$$

The terms on the left-hand side of the equation represent the potential and kinetic energies of the particle at A and those on the right-hand side represent the potential and kinetic energies of the particle at B. This equation states that *if no external agent does work on a particle situated in a conservative field of force, its total energy remains constant*. This is a statement of the *principle of conservation of mechanical energy in a conservative field*. There may be changes from kinetic energy to potential energy and vice versa, but the total energy will remain constant.

6-8
Gravitational potential energy

In order to make the idea of positional energy more specific, we shall consider a simple situation: the lifting of an object. For example, consider an elevator that is lifted from the first to the top floor of a building. We can look at this situation from two slightly different points of view. First, we concentrate on the elevator and compare the total work done on the elevator with the change in energy of the elevator. Second, we concentrate on the work done by the lifting motor, and consider what happens to that transferred energy.

In order to make this analysis as simple as possible we shall imagine moving the elevator in such a way that we can neglect the change in kinetic energy; further, we shall neglect friction. We can see how to do this by considering the free-body diagram of Figure 6-8. When the elevator is at rest at the ground floor, the force **F** exerted by the cable, is exactly equal to the force **W** exerted by the earth. In order to lift the elevator, the force **F** must be made larger than **W**. If the change in **F** is made very small, then the acceleration of the elevator will be very small. If, in addition, **F** is allowed to remain larger than **W** for only an instant, than the elevator will accelerate for an instant, reach a small velocity, and then, if **F** becomes equal to **W**, move upward with that small constant velocity. The smaller the change in **F**, and the

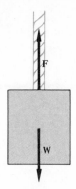

FIGURE 6-8 Free-body diagram for an elevator. The force **F** is exerted by the cable and **W** is the weight of the elevator.

shorter the time during which the change exists, the slower will be the motion of the elevator. Thus we can make the change in kinetic energy of the elevator as small as we like and, further, we can make the changes in **F** as small as we like. It is true that the smaller the change in kinetic energy, the longer the time it will take for the elevator to reach the top floor, so in order to be able to neglect the change in kinetic energy we have to be willing to wait a long time for the process to occur. When the elevator nears the top floor, we can make **F** slightly less than **W** for an instant, and the elevator will slow down and stop. Thus, if we are willing to wait long enough, we can imagine lifting the elevator through a height h, while keeping the kinetic energy and the force **F** essentially constant.

From the first point of view, where we focus our attention on the *elevator*, the change in kinetic energy of the elevator is zero, and the work is done on it by two objects, the cable and the earth. The work done by **F** is positive, since **F** and the displacement are in the same direction; the work done by **W** is negative, since **W** is opposite to the displacement. However, the work done by **F** is equal in magnitude to the work done by **W**, since the two forces remain essentially equal to each other. Thus the total work done on the elevator is zero, and the change in its kinetic energy is also zero. Thus the change in kinetic energy of the object is equal to the total work done on the object. What is most significant about this is that we can describe the process completely, using only the concepts of kinetic energy and work, without introducing positional energy. That is, as long as we consider a *single object* and the forces acting on it, we can completely describe the situation, using the work-energy theorem involving only *kinetic energy* and the *work* done by forces.

On the other hand, when we consider the *system* of objects, which consists of the earth and the elevator, and for which **W** is an *internal* force, we find that we must extend the idea of energy to include positional energy. The work done by **F** is energy transferred from an external agent, the motor, by means of the cable, to part of the system (the elevator). Since **F** is the only *external* force exerted on the system, the total work done on the system is not zero. The total change in kinetic energy of the system is, however, zero. Therefore, in order to retain the idea that the energy transferred from the motor does not disappear, we introduce the concept of positional energy. This energy, *gravitational potential energy*, is a property of the *system* rather than a property of either the elevator or the earth. Thus we can describe the lifting of the elevator, which can be viewed as a change in the separation of the parts of the system, by saying that the external agent transfers energy to the system, and that this energy serves to increase the gravitational potential energy of the system.

When the elevator is at the top floor and the force **F** is reduced to zero, the elevator will fall. If we look at the elevator as a single *object*, then we can describe the falling by saying that the change in kinetic energy comes from the work done by the gravitational force. If we look at the *system*, then we can say that the separation decreases, which means that the gravitational potential energy of the system decreases and therefore the kinetic energy of part of the system increases.

In applying Equation 6-10, with Equation 6-9 as the definition of the change in gravitational potential energy, we are dealing with a system that consists of the earth and an object whose motion is confined to be reasonably near the surface of the earth. In most such cases the motion of the earth may be ignored. Thus when we insert Equation 6-9 in Equation 6-10, we assume that the kinetic energy of the earth is constant; and so the change in kinetic energy of the earth is equal to zero. For the system then:

$$W = \Delta \mathcal{E}_k + mgh$$
$$= \left(\tfrac{1}{2}mv_f{}^2 - \tfrac{1}{2}mv_i{}^2\right) + mgh \qquad (6\text{-}13)$$

which is the same as Equation 6-11; v_f and v_i are the initial and final velocities of the object of mass m, which is lifted through a height h. Although we have discussed gravitational potential energy as a property of a system, Equation 6-13 contains no

properties of the earth. This is true because Equation 6-13 is applicable only to a special case, and in particular, to one where the motion of the earth may be neglected. Nevertheless, the equation is applicable to a large number of real situations. In applying Equation 6-13, it is quite common to refer to the *mgh* term as the change in gravitational potential energy of the object (rather than of the system). Although the language is slightly imprecise, this practice leads to no errors and is convenient, as long as we remember that it is valid only when we ignore the behavior of the earth and that the *W* of Equation 6-13 refers to the work done by all the external forces *except the gravitational force.*

6-9
Fields

The force that exists between two particles because of their masses acts no matter how far apart these masses may be. There is another way of thinking about gravitational forces and that is to imagine that in the space all around a particle of mass M there exists a *gravitational field*. Whenever any other particle of mass m finds itself in this gravitational field, it will experience a force **F** given by Equation 5-1

$$F = G_0 \frac{Mm}{r^2}$$

where r is the distance between the two masses. We can define a new term called the *intensity of the gravitational field at any point* in space as *the ratio of the force **F** that acts on a particle at this point to the mass m of the particle situated there.* Let us denote the intensity by the letter I; then

$$\mathbf{I} = \frac{\mathbf{F}}{m} \qquad (6\text{-}14)$$

and, putting in the value of F from Equation 5-1,

$$I = \frac{G_0 M}{r^2} \qquad (6\text{-}15)$$

Equation 6-15 shows that the intensity of the gravitational field varies inversely as the square of the distance from the particle of mass M. The intensity **I** is a vector quantity; its direction is that of the force **F** that acts on a particle placed anywhere in the field, and, since the force is always one of attraction, its direction is toward the mass M. We can develop a graphical method for representing the gravitational field so that it will show at a glance both the magnitude and direction of the field intensity. This is illustrated in Figure 6-9, in which radial lines are drawn converging upon the mass M; a scale can be chosen so that the number of lines passing perpendicularly through a unit area at any point, such as P, will be proportional to the intensity I at that point.

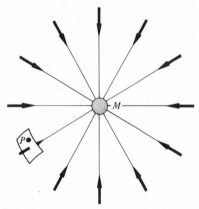

FIGURE 6-9 Radial gravitational field around a small concentrated mass M. The number of lines of force through a unit area at P is proportional to the intensity I of the gravitational field at P.

An interesting case is that of the earth's gravitational field. Newton was the first to prove that the field outside a spherical mass is identical with that of a mass concentrated at the center of the sphere. Hence, at points outside the earth's surface, the gravitational field intensity is given by Equation 6-15. However, we have been using the term "weight of a particle" to describe the force that the earth exerts on a mass m placed anywhere in its field. The intensity of the earth's gravitational field $\mathbf{I}_e$ is, therefore,

$$\mathbf{I}_e = \frac{\mathbf{W}}{m} = \mathbf{g} \qquad (6\text{-}16)$$

The term that we have been calling the acceleration of a freely falling body is identical with the intensity of the earth's gravitational field.

In analogous ways we can describe the electric field, magnetic field, and nuclear fields. In all of these cases the presence of the field is made evident by its effect on some other particle introduced into the field. But every particle has fields associated with it. The assumption that is usually made in discussing the field around any one particle is that the second particle, usually called the *test particle*, is so small that it produces a negligible

effect on the field under discussion. Thus in Equation 6-14 the mass m is assumed to be very small in comparison with the mass M of the earth.

Once we have introduced the idea of field we can think about potential energy in a slightly different way. When energy is transferred to a system of particles and changes the gravitational potential energy of the system, then the gravitational field is also changed. We can think of the change in gravitational energy as being stored in the gravitational field. That is, we can think of a system as consisting of a group of particles *and a field*. The energy of the *particles* is the *kinetic energy*, and the energy of the *field* is the *potential energy*. When energy W is transferred to a system, it can go into the particles ($\Delta\mathcal{E}_k$) or into the field ($\Delta\mathcal{E}_p$).

6-10
Potential energy of springs

As an illustration of the application of the idea of energy to a system with varying forces, we will consider the stretching or compressing of a helical spring. In order to avoid the complications of gravitational energy and friction, imagine a helical spring that is placed on a smooth horizontal table and has one end tied to a rigid support, as in Figure 6-10. Further, to avoid changes in kinetic energy, we will imagine applying forces in such a way that the motion of the spring is very slow, just as we did for the elevator.

The behavior of real springs is complicated, but for many springs the following is a good approximation. When the spring has no forces acting on it, it has length L_0. If it is stretched very slowly, the

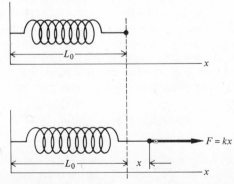

FIGURE 6-10 The magnitude of the force required to stretch a spring F is proportional to the increase in length of the spring x.

external force F acting on it is proportional to the increase in length x; that is,

$$F = kx \qquad (6\text{-}17)$$

where k is a constant, usually called the spring constant. From Newton's third law, we know that the force exerted by the spring is equal and opposite to the applied force; that is,

$$F = -kx \qquad (6\text{-}18)$$

The significance of the sign in Equation 6-18 is that the force exerted by a spring is directed opposite to the displacement of the end of the spring. For example, when the spring is compressed, the displacement is directed *toward* the support and therefore the force exerted by the spring is directed *away* from the support.

If we now imagine an external agent applying a force **F** slowly to the spring it will be doing work on the spring; but neither the kinetic energy nor the gravitational potential energy of the spring will change. Again, in order to retain the idea that the transferred energy does not disappear, we will say that the spring acquired potential energy. This kind of potential energy is sometimes called *energy of elasticity*. We can think of the spring as a system consisting of many atoms on which work is done and which acquire potential energy.

In order to write down a formula for the energy of the spring, we must calculate the work done by the variable force F. If the force F is plotted as the abscissa and the stretch x as the ordinate, the resultant graph is the straight line of Figure 6-11. As the spring is stretched from x_1 to x_2, the force varies from F_1 to F_2. When the object is stretched from x_1 to x_2, the external force does an amount of work, ΔW: the smallest value of ΔW is

$$\Delta W_{\text{smallest}} = F_1(x_2 - x_1)$$

and the largest value of ΔW is:

$$\Delta W_{\text{largest}} = F_2(x_2 - x_1)$$

If these expressions for the maximum and minimum work are compared with the figure, it can be seen that they are the areas of the two rectangles whose bases are $(x_2 - x_1)$, and whose heights are F_2 and F_1, respectively. If x_2 were closer to x_1, then the difference between F_2 and F_1 would be smaller, and the maximum and minimum work would be closer to each other. Then, for a very small change in stretch, the maximum and minimum work would be essentially the same, and the work done in stretching the spring from x_1 to a

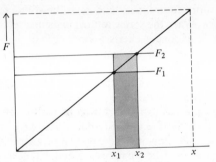

FIGURE 6-11 The work done in stretching a spring is equal to the area under the straight line.

slightly larger stretch, x_2, is the area under the straight line. Thus, in stretching the spring from zero stretch to stretch x, the work done is the area under the whole straight line. This area is the area of a triangle whose base is x, and whose height is the force F exerted on the spring when it is stretched an amount x. Thus

$$W = \tfrac{1}{2}Fx$$

and, from Equation 6-17,

$$F = kx$$

So

$$W = \tfrac{1}{2}kx^2 \qquad (6\text{-}19)$$

Therefore the elastic energy of the spring is

$$\mathcal{E}_e = \tfrac{1}{2}kx^2 \qquad (6\text{-}20)$$

This equation gives the energy stored in the spring by stretching the spring by an amount x, or the energy stored in compressing the spring by an amount x.

If a body of mass m is attached to the end of the spring as shown in Figure 6-12 and the spring is stretched by an amount x, the energy stored in the system is $\tfrac{1}{2}kx^2$. If the spring is released, this energy

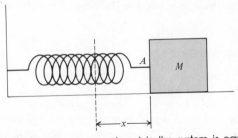

FIGURE 6-12 The energy stored in the system is equal to $\tfrac{1}{2} kx^2$.

is transformed to kinetic energy as the spring pulls the body. When the spring returns to its unstretched length, it no longer exerts a force ($x = 0$), but the body has acquired a velocity. At that point, since there is no force on the body, it will continue to move, thus compressing the spring. The compression implies that a force will be exerted by the spring opposite to the direction of motion of the body, thus slowing the body down. At some point the body will stop, and have no kinetic energy, but the spring will be compressed and have elastic potential energy. The force exerted by the compressed spring will now cause the body to accelerate back toward its initial position. Thus, in the absence of friction, or other force, the system will oscillate back and forth, indefinitely. During the motion, the sum of the kinetic energy of the body and the elastic energy of the spring is a constant; elastic energy is tranformed into kinetic energy and vice versa, but the total energy is constant. At the extremes of the motion, the body has no kinetic energy and so the total energy is, at both ends, elastic energy. The initial energy is

$$\mathcal{E}_e = \tfrac{1}{2}kx_0^2$$

where x_0 is the extreme stretch of the spring.

When the spring is unstretched, all of this energy is converted to kinetic energy, and so the velocity of the body may be obtained from

$$\tfrac{1}{2}kx_0^2 = \tfrac{1}{2}mv^2$$

from which

$$v = x_0 \sqrt{\frac{k}{m}}$$

At the other extreme of the motion, the body has no kinetic energy and all of the energy is again in the form of elastic potential energy, and so the spring is compressed by the same amount as it was originally stretched. At positions other than the extremes or the middle, the body has some kinetic energy, and the spring has some elastic energy; but the sum is equal to the original energy.

Illustrative example

A body whose mass is 200 gm is attached to the end A of a horizontal spring 40 cm long, as shown in Figure 6-12. The constant of the spring is 1500 dynes/cm. Assume that the supporting table is frictionless and that the spring has negligible mass. The body is pulled a distance of 3 cm and then

released. Determine (a) the maximum elastic energy of the system, (b) the maximum kinetic energy, and (c) the maximum speed of the body.

(a) The elastic energy of the system is

$$\tfrac{1}{2}kx^2$$

and is a maximum when x is a maximum, in this case when $x = 3$ cm. Hence the maximum elastic energy is

$$\mathcal{E} = \tfrac{1}{2} \times 1500 \frac{\text{dynes}}{\text{cm}} \times 9 \text{ cm}^2$$
$$= 6750 \text{ ergs}$$

(b) The system will have maximum kinetic energy when the body is at the equilibrium position and its value will also be 6750 ergs.

(c) The maximum velocity of the body can be obtained from the equation

$$\tfrac{1}{2}mv_0^2 = 6750 \text{ ergs}$$

or

$$\tfrac{1}{2} \times 200 \text{ gm} \times v_0^2 = 6750 \text{ ergs}$$

from which

$$v_0^2 = 67.50 \frac{\text{cm}^2}{\text{sec}^2}$$

and therefore

$$v_0 = 8.22 \frac{\text{cm}}{\text{sec}}$$

6-11
Conservation of energy

If no external agent exerts forces on a system, then the total energy of the system remains constant. This simple statement, known as the law of conservation of energy, is (as we have seen) an assumption. For some systems, the assumption has appeared to be untrue, but for all such systems, physicists have been able to discover or invent new kinds of energy, which, when properly defined, permit the retention of the idea of conservation of energy. Perhaps the simplest illustration of this is the case of a ball rolling on a horizontal surface. If the system is defined to include the ball, the surface, and the earth, then there are no external forces. Since the surface is horizontal, we may ignore changes in the motion of the earth. Thus, the only energy of significance is the kinetic energy of the ball. Thus if the ball slows down and stops, as real balls do, the energy of the system does not remain constant; it decreases to zero. This change in motion is due to friction, which is an internal force, and is not path-independent, so it would appear that energy is not conserved. We can set things right by introducing the concept of *heat*. The friction force transforms the kinetic energy into heat, which transfers the energy to another body. This subject will be treated in some detail later.

This idea that the total energy of isolated systems is constant is one of the most important contributions of physics to technology and to other sciences. It is very important in understanding such diverse systems as stars and the digestive mechanisms of animals. In physics, the idea is most useful in providing some information about complex systems. This can be illustrated by considering a relatively uncomplicated system—the simple pendulum.

As we have seen for many systems, it is convenient to define *mechanical energy*, as we did in Section 6-7, as the sum of the kinetic energies of all the parts of the system plus the potential energy of the system. If the total mechanical energy of a system is conserved, then all of the forces acting on the system are conservative, and no nonconservative forces act. Nonconservative forces are often called *dissipative* forces. The most common dissipative force is friction; in the absence of friction many systems conserve mechanical energy. The simple pendulum is such a system.

A simple pendulum consists of a small ball of weight mg attached to one end of a thin string of negligible weight and of length L. The other end of the string is attached to some fixed point O, as shown in Figure 6-13. When at rest, the string hangs vertically with the ball at its lowest position C. Let us call the energy of the pendulum zero when it is in this position. When the ball is pulled aside from its lowest position, with the string kept taut, it moves in the arc of a circle of radius L. Suppose that it is moved to position A, at which point it is at a height h above its lowest position. Because of the work done in lifting the ball through this height, its potential energy is now mgh. When the ball is released, it moves back to its lowest position and its energy, while remaining constant in amount, changes from potential to kinetic energy. At any point in its path, its total energy is the sum of the potential and kinetic energy and is equal to the original potential energy mgh.

At the lowest point C, all of its energy is kinetic, and if v is its speed at this point, then

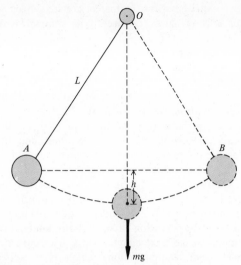

FIGURE 6-13 Motion of a simple pendulum.

$$\tfrac{1}{2}mv^2 = mgh$$

or

$$v^2 = 2gh$$

This is exactly the speed it would have acquired had it fallen through a vertical height h instead of moving in the arc of a circle. However, the direction of its motion is different. It is moving tangent to the circle so that at the lowest point its velocity is horizontal. According to Newton's first law, it tends to keep moving in this direction, but, since it is attached to a string, the ball is forced to deviate from this path and move in a circular arc. This causes it to be raised above its lowest position, thereby gaining potential energy at the expense of the kinetic energy. If no energy is lost to the outside, it will reach point B at a height h equal to that from which it started. It will then proceed back again to the lowest point C and on to the original starting position A. Then, of course, the motion will be repeated.

6-12
The role of conservation laws in physics

A conservation law is a statement that a particular quantity remains constant (or is conserved) in a particular process. Conservation of energy is only one of many such laws in physics.

Conservation laws are introduced into physics

in two different ways: as logical consequences of other laws of nature or directly as summaries of experience. For example, if we restrict our attention to simple mechanical systems in which no heat is evolved, then the principle of conservation of energy is a consequence of Newton's laws. For such systems our discussions of gravitational and elastic energies are really proofs that energy is conserved; those discussions depended on the validity of Newton's laws. On the other hand, the extension of the principle to all systems cannot be derived from Newton's laws. Our belief in conservation of energy depends on the fact that all experiments (up to this time) have been consistent with the principle. We will, in subsequent chapters, deal with conservation laws of both kinds. For example, in the next chapter we will discuss the conservation of momentum, which is a consequence of Newton's laws. Later, we will discuss the conservation of nuclear particles, which is not a derived law.

Conservation laws are introduced into physics for several different reasons. The first and simplest reason is that they are true. That is, conservation laws are stated because it is interesting to know that some things remain constant. The second reason is that they can be used to answer questions or solve problems about situations in which we have only limited information. In a very large number of real problems we simply do not know enough about the forces, the geometry, or the nature of materials to solve the problem. For most problems this ignorance forces us to resort to empirical solutions; engineering, medicine, and meteorology are full of formulas and rules that are abstracted from experiments rather than being derived from some fundamental law. For some problems we have enough information to use conservation laws even though we do not have enough to use the more fundamental laws. For example, we can calculate the velocity of a car at the bottom of a frictionless roller coaster without knowing the detailed shape of the roller coaster. In fact this problem illustrates another characteristic of the conservation laws; even if we had known the shape of the roller coaster, using the conservation of energy is an easier way to find the velocity than using Newton's second law. If we used the second law we would first find the acceleration and then calculate the velocity from the acceleration. Conservation of energy gives the velocity directly.

6-13
Gravitational potential energy at large altitudes

In deriving the gravitational potential energy earlier, we neglected the variation of gravitational force with altitude. Taking this variation into account is a problem in the calculus, so we will simply state the result.

If an object of mass m above the surface of the earth is initially at a distance r_i from the center of the earth and is then moved to a distance r_f from the center of the earth, the change in gravitational potential energy of the system can be shown to be

$$\Delta \mathcal{E}_p = -GmM_e\left(\frac{1}{r_f} - \frac{1}{r_i}\right) \qquad (6\text{-}21)$$

where M_e is the mass of the earth. It is possible to show that this general result is approximated very well by Equation 6-9 for the special case in which h is small compared with either r_f or r_i, where h is equal to the difference between r_f and r_i.

Because of the algebraic form of Equation 6-21, it is convenient to choose a location for the zero of gravitational potential energy that is independent of the specific problem. It is common to choose the zero of potential energy to be the position that is infinitely distant from the earth; that is,

$$\frac{1}{r_i} = 0$$

Then the potential energy at a distance r from the center of the earth, relative to the chosen zero reference level is

$$\mathcal{E}_p = -\frac{GmM_e}{r} \qquad (6\text{-}22)$$

Thus the gravitational energy of an object that is not infinitely far away from the earth is a negative number, in this convention.

Consider, then, an object located some finite distance from the earth, and having some kinetic energy, which is smaller than the magnitude of the gravitational energy at that location. The total energy of the system will then be negative. Then, no matter how the object moves, in the absence of external forces the total energy of the system will remain negative. Then if the object is moving away from the earth, its gravitational potential energy will get smaller in magnitude and, if it keeps going, at some point the gravitational potential energy will be equal to the total energy of the object. At that point, the kinetic energy of the object will be zero, and the object will start to return toward the earth. Thus, when the total energy of the system is negative, there is a maximum separation between the earth and the object. In order for the object to be further from the earth than this maximum, it would have to have negative kinetic energy. Another way of saying this is that a system with a negative total energy is *bound* together. In order to unbind the system, enough energy must be added to the system to make its total energy nonnegative. The minimum energy needed to unbind a system is an amount of energy equal to the magnitude of the total energy. This minimum energy, called the *binding energy* of the system, when added to the system, makes the total energy of the system zero, and permits the object to get infinitely far from the earth.

For example, consider an object that is at rest on the surface of the earth. This object, which is clearly bound to the earth, has a total energy that is just its gravitational energy. From Equation 6-22 its total energy is

$$U = -\frac{GM_e m}{R}$$

where R is the radius of the earth. If the object had a velocity V, directed away from the center of the earth, which was large enough to give the object a total energy equal to zero, then the object could *escape* from the earth. Any smaller velocity would allow the object to travel to some finite distance, and then fall back to earth. This *escape velocity* can be calculated by setting the total energy of the object, when it is at the surface of the earth, equal to zero:

$$U = \tfrac{1}{2}mV^2 - \frac{GM_e m}{R}$$
$$= 0$$

Therefore

$$\tfrac{1}{2}mV^2 = \frac{GM_e m}{R}$$

or

$$V = \sqrt{\frac{2GM_e}{R}}$$

which can be written

$$V = \sqrt{2}\,\sqrt{\frac{GM_e R}{R^2}}$$
$$= \sqrt{2R}\,\sqrt{\frac{GM_e}{R^2}}$$
$$= \sqrt{2 \times 6.4 \times 10^6}\,\sqrt{g}$$
$$= \sqrt{2 \times 6.4 \times 10^6 \times 9.8}$$

from which

$$V \approx 11 \times 10^3 \frac{m}{sec}$$

or

$$V \approx 24{,}800 \frac{mi}{hr}$$

or

$$V \approx 7 \frac{mi}{sec}$$

6-14
Power

In many cases, the time during which a given amount of work is done is of great importance. The term *power* is defined as the work done divided by the time during which this work is done, or

$$P = \frac{W}{t} \tag{6-23}$$

in which P represents the power and t the time during which the work W is done. Thus power is the rate at which work is done. Power is a scalar quantity. For example, if a hoisting engine lifts a 200-lb weight through 50 ft in 4 sec, the power supplied by the engine is

$$P = \frac{200 \text{ lb} \times 50 \text{ ft}}{4 \text{ sec}} = 2500 \frac{\text{ft lb}}{\text{sec}}$$

There are several other practical units used to express power. The horsepower (hp) is defined as

$$1 \text{ hp} = 550 \frac{\text{ft lb}}{\text{sec}} \tag{6-24}$$

In the above problem, the power delivered by the engine may be expressed as

$$P = \frac{2500}{550} \text{ hp} = 4.55 \text{ hp}$$

Another unit of power is the *watt*, which is defined

as

$$1 \text{ watt} = 1 \frac{\text{joule}}{\text{sec}} \qquad (6\text{-}25)$$

The relationship between the power expressed in watts and that expressed in horsepower is

$$746 \text{ watts} = 1 \text{ hp}$$

If in Equation 6-23 we substitute for the work W its value from Equation 6-2

$$W = \mathbf{F} \cdot \mathbf{s} \qquad (6\text{-}2)$$

where $\mathbf{F}$ is the force acting and $\mathbf{s}$ is its displacement, we get

$$P = \mathbf{F} \cdot \frac{\mathbf{s}}{t}$$

for the power delivered. Now

$$\mathbf{v} = \frac{\mathbf{s}}{t}$$

and therefore

$$P = \mathbf{F} \cdot \mathbf{v} \qquad (6\text{-}26)$$

Thus the power delivered to a body is the scalar product of the force $\mathbf{F}$ acting on it and its velocity $\mathbf{v}$. When $\mathbf{F}$ and $\mathbf{v}$ are in the same direction, Equation

6-26 becomes simply

$$P = Fv \qquad (6\text{-}27)$$

Illustrative example

An engine is delivering 1200 hp to an airplane in level flight at a uniform air speed of 300 mi/hr. Determine the total of all the resisting forces acting on the airplane.

An airplane flying with uniform velocity is in equilibrium. Hence the force supplied by the engine must be equal and opposite to all the resisting forces acting on the airplane. This force can be found by means of Equation 6-27. Now

$$1200 \text{ hp} = 1200 \times 550 \frac{\text{ft lb}}{\text{sec}}$$

and

$$300 \frac{\text{mi}}{\text{hr}} = 440 \frac{\text{ft}}{\text{sec}}$$

From Equation 6-27,

$$F = \frac{P}{v}$$

and therefore

$$F = \frac{1200 \times 550}{440} \text{ lb} = 1500 \text{ lb}$$

Questions

1. Through what height would a nickel (whose mass is about 5 gm) have to fall in order to acquire a kinetic energy of 1 erg? Through what height would it have to fall to acquire a kinetic energy of 1 joule?

2. When a moving body comes to rest its kinetic energy might have been transferred to another body or it might have been transformed into potential energy of some system or it might have been transferred in the form of heat. Give a common example of each of these processes.

3. Describe what happens to the kinetic energy of a moving tennis ball as it strikes the racket and rebounds.

4. How does the process described in Question 3 differ from the process of a baseball hitting the bat?

5. It is often said that the ultimate source of all the energy used by human beings is the sun. Illustrate this by considering the ultimate source of the kinetic energy of a person walking, the kinetic energy of an airplane, and the energy radiated by a light bulb.

6. A person winds a watch once a day, doing work on the stem. Discuss what happens to that energy.

7. Where does the energy in a waterfall come from?

8. A backpacker carries a heavy pack. How much work does he do while walking on level ground? Up a hill? Down a hill?

9. In the motion of a simple pendulum, show that the work done by the tension in the string is equal to zero.

10. Estimate the time it takes you to go up a flight of stairs. Make a calculation of the power needed to do this.

11. Derive the relationship between watts and horsepower.

12. Assuming that the total energy of the earth-sun

system is constant, explain why the kinetic energy of the earth varies during the year.

13. Since an atom, which consists of a nucleus and electrons distributed around the nucleus, is a bound system it has a binding energy. What happens to an atom if an amount of energy equal to the binding energy is added to it?

14. An iron spring is compressed and tied. It is then placed in acid and dissolved. What happened to the energy that was stored in the spring?

15. When a given number of atoms are involved in nuclear reactions, such as fission or fusion, millions of times more energy is released than when the same number of atoms is involved in chemical reactions, such as combustion. What does this tell you about binding energies?

Problems

1. A 150-lb crate is pushed horizontally across a floor a distance of 18 ft by a 35-lb force. How much work is done?

2. Calculate the work done in lifting a 3-kg object a distance of 3.4 m.

3. A woman pulls a sled by means of a cord attached to it, exerting a force of 24 lb at an angle of 30° with the horizontal. How much work is done in pulling this sled for a distance of 60 ft?

4. Determine the kinetic energy of a 40-ton airplane that is moving with a speed of 620 mi/hr.

5. Determine the kinetic energy of a 2-gm snail that moves with a speed of 2 m per day.

6. Calculate the kinetic energy of an electron that is moving with a speed of 2×10^8 cm/sec. The mass of an electron is 9×10^{-28} gm.

7. What is the change in gravitational potential energy of a 150-lb man who climbs to the top of a 180-ft tower?

8. A 20-lb sheet of wood, 4 ft × 8 ft, is lying on the ground. What is the change in its gravitational potential energy when one end is lifted so that the sheet is in a vertical position, resting on the 4-ft edge?

9. A certain automobile has a spring whose spring constant is 2×10^5 nt/m. How much energy is stored in the spring when it is compressed by 1 cm?

10. Find the gravitational potential energy of the moon, assuming a zero level at the center of the earth. The mass of the moon is 7×10^{22} kg and the moon is 3.8×10^8 m from the earth.

11. A sprinter weighing 150 lb runs 60 yd in 10 sec. What is his kinetic energy? (Assume, for simplicity, that he runs at a constant speed.)

12. A 500-gm object is dropped from a height of 12 m. What is the kinetic energy of the object just before it hits the ground?

13. A force of 1000 nt acts on an automobile so that the car accelerates from rest until it has a kinetic energy of 5×10^5 joules. Through what distance was the force exerted?

14. A body weighing 100 lb is pushed up a rough inclined plane by a force of 75 lb acting parallel to the plane. The plane is inclined at an angle of 30° with the horizontal and is 24 ft long. (a) How much work is done in moving the body to the top of the inclined plane? (b) What is its potential energy when at the top of the plane? (c) How much work was done against friction?

15. About 1 million kg of water flow over Niagara Falls per second and fall about 50 m. How much potential energy is lost per second? How many 100-watt light bulbs could this operate if all of this energy were converted to electricity?

16. An 80-lb girl slides down a hill which is 150 ft long and inclined at an angle of 30° to the horizontal. She arrives at the bottom, moving with a speed of 4 ft/sec. How much work was done against friction?

17. By "stepping on the gas" a driver speeds up a car from 15 mi/hr to 45 mi/hr in 8 sec. The weight of the car is 3600 lb. Determine (a) the change in its kinetic energy and (b) the resultant force acting on the car during this time.

18. What is your gravitational energy with respect to the center of the earth? What velocity would you have to have in order that your total energy (kinetic plus potential) be equal to zero? Express the speed in miles per hour.

19. A 1-kg book slides across a table with an initial speed of 50 cm/sec. If the constant frictional force is equal to 0.3 nt, how far does the book slide before coming to rest? (Use the energy principle directly; do not calculate the acceleration.)

20. A 10-ton railroad car is coasting freely on the tracks at a speed of 20 mi/hr when it runs into a bumper spring which has been placed on the track to prevent cars from running off the end of the track. The car is brought to rest by the spring, which in the process is compressed 1 ft. (a) Find the constant of the spring by assuming that energy is conserved and that there is no friction. (b) What happens after the car is brought to rest, assuming that no brake is applied?

21. A pendulum consists of a steel ball of mass 300 gm which is attached to a thin wire of negligible mass whose length is 1 m. The ball is pulled to one side so that the string makes an angle of 30° with the vertical and then is released. Find the velocity of the ball at its lowest point.

22. An evil scientist in a science fiction film decides to steal the moon. How much energy would it take to remove the moon from the earth's gravitational field? Use the data of Problem 10, and assume that the moon travels in a circular orbit once in 28 days.

23. What is the escape velocity from the moon? The radius of the moon is 1.7×10^8 cm.

24. An automobile is moving at 60 mi/hr on a level road and the engine is delivering 50 hp to the driving wheels. What is the total force that resists the forward motion of the car? (Most of this force is air resistance.)

25. The hammer of a pile driver weighs 1500 lb and falls through a height of 8 ft to drive a pile into the ground. (a) How much energy does the hammer have when it strikes the pile? (b) If the pile is driven into the ground a distance of 6 in. determine the average resisting force acting on the pile. Assume that energy is conserved and that no energy is converted into heat at the impact.

26. A power company uses the water descending in a waterfall to generate electric power. It uses 150,000 ft³ of water per second (a cubic foot of water weighs 62 lb) and generates 2 million kilowatts of power. Estimate the height of the waterfall.

27. A rocket is fired straight upward with a velocity that is half the escape velocity. Find the furthest distance it can go from the center of the earth.

28. A certain motor, which converts input electrical energy with an efficiency of 80 percent into mechanical energy, is used to operate a hoist. The motor is supplied with electrical energy at the rate of 5 hp. Find the velocity with which this motor can lift a 400-lb object.

29. A rocket is fired at the moon with an initial velocity of 2×10^4 m/sec. With what velocity will it strike the surface of the moon?

30. The nucleus of heavy hydrogen (deuterium) consists of a neutron and a proton bound together. Experimentally it is found that it takes about 3.5×10^{-6} erg to split the nucleus into a neutron and a proton. Assume that the neutron and the proton are each point particles and that they exert a force of attraction on each other. Assume that the attractive force has a constant magnitude F for any separation of the neutron and proton that is less than 2×10^{-13} cm and that the force becomes zero for larger separation. Find the magnitude of F.

31. Experimentally it is known that the binding energy of a hydrogen atom is 13.6 electron-volts (eV), where 1 eV is an amount of energy equal to 1.6×10^{-12} erg. It is also known that the total potential energy of the electron-proton system is equal to -27.2 eV. Assuming that the proton is at rest, find the speed of the electron. (The mass of the electron is 9.1×10^{-28} gm.)

32. A chain whose total length is l and whose total mass is m is held at rest on a frictionless table with half of its length hanging over the edge. How much work is required to pull the chain back onto the table?

33. A 5000-lb elevator is at rest on the first floor of a building when the cable snaps. Located 12 ft below the elevator is a spring, whose spring constant is 15,000 lb/ft. At the instant the cable snaps, a safety brake operates, exerting a friction force of 1000 lb, which opposes the motion of the elevator. (a) Find the speed of the elevator just before it hits the spring. (b) By how much is the spring compressed when the elevator comes to rest?

34. Two identical rocket ships are a large distance r apart and are at rest. Under the action of the mutual gravitational force they move toward each other. How fast are they moving when they strike each other? Assume that no forces other than the gravitational force act. Also assume that when they strike each other their centers are a distance D apart and that from the symmetry they must always have identical velocities.

circular motion

chapter 7

7-1
Motion along a curved path

As we saw in Chapter 2, when an object moves
along any path its velocity is directed along the
tangent to the path. If the direction of the path
changes, then the direction of the velocity changes.
Therefore, an object that moves along a nonstraight
path has an acceleration, even if the magnitude of
the velocity (the speed) is constant. If both the
speed and the direction are changing then the object has an acceleration, which is due in part to the
change in direction and in part to the change in
speed.

The total acceleration of an object consists of
two parts: the part due to change in speed is directed along the tangent to the path, whereas the
part due to change in direction is directed along
the perpendicular to the tangent. This conclusion
is reached from the following argument: Consider
an object which is moving along a curved path as
in Figure 7-1a. The figure shows the path of the
object, the positions of the object at two points,
A and B, and the velocities of the object at those
points. In order to calculate the *average* acceleration during the motion from A to B we calculate
the change in velocity, Δv, as in Figure 7-1b. Here
we have assumed that v_2 differs from v_1 in both
magnitude and direction and have drawn Δv, the
change in velocity, remembering that

$$\Delta \mathbf{v} = \mathbf{v}_2 - \mathbf{v}_1$$
$$= \mathbf{v}_2 + (-\mathbf{v}_1)$$

At this point we can draw no obvious conclusions
about the direction of the change in velocity. To do
this, let us first consider what would have happened
if the speed of the object had been constant—that

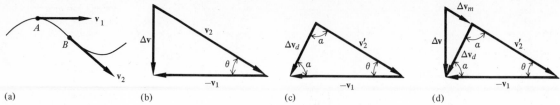

FIGURE 7-1 (a) An object moving along a curved path has velocity $\mathbf{v}_1$ at point A, and velocity $\mathbf{v}_2$ at point B. (b) The vector triangle from which the change in velocity is calculated. (c) The vector triangle in the special case that the velocities at points A and B are equal in magnitude but are in different directions. (d) The figure formed by superimposing part (c) on part (b).

is, if the magnitude of the velocity at B had been the same as the magnitude of the velocity at A. Then, instead of Figure 7-1b we would have drawn Figure 7-1c. In that figure, the velocity at point B, $\mathbf{v}_2{'}$, has the same magnitude as the velocity at point A, but a different direction; the figure is an isosceles triangle. The change in velocity, $\Delta\mathbf{v}_d$, has the subscript d to remind us that it arises from a change in direction only. If now we were to superimpose Figure 7-1c on Figure 7-1b we would have Figure 7-1d. Here we see that $\mathbf{v}_2$ is the sum of two vectors,

$$\mathbf{v}_2 = \mathbf{v}_2{'} + \Delta\mathbf{v}_m \qquad (7\text{-}1)$$

where $\Delta\mathbf{v}_m$ is that part of $\mathbf{v}_2$ which arises from a change in the magnitude of the velocity. Further, we see that the change in velocity is the sum of two vectors

$$\Delta\mathbf{v} = \Delta\mathbf{v}_m + \Delta\mathbf{v}_d \qquad (7\text{-}2)$$

Equation 7-2 separates the change in velocity into two parts, the part due to change in speed (magnitude) and the part due to change in direction. In order to find the *instantaneous* acceleration at A we must see what happens when the second point, point B, is chosen closer and closer to point A. When that is done the direction of the velocity at point B, $\mathbf{v}_2$, gets closer and closer to the direction of $\mathbf{v}_1$. Thus the direction of $\Delta\mathbf{v}_m$ gets closer and closer to the direction of $\mathbf{v}_1$. Thus, at point A, that part of the instantaneous acceleration which is due to the change in magnitude of the velocity is in the same direction as the instantaneous velocity. In order to see the direction of that part of the acceleration which is due to changing direction, we point out that the triangle of Figure 7-1c is an isosceles triangle. Then we have

$$2\alpha + \theta = 180° \qquad (7\text{-}3)$$

When we take point B closer to point A, the angle

θ gets smaller and smaller and approaches zero. Therefore, from Equation 7-3, the angle α must get closer and closer to 90°. Therefore, $\Delta\mathbf{v}_d$ gets closer to being perpendicular to the velocity. Thus we have reached the conclusion: the acceleration of an object has two components, one that is tangent to the path (parallel to the velocity) and arises from changes in the speed, and another that is perpendicular to the path and arises from changes in the direction of the motion.

Once we know this, we can draw some conclusions about the forces acting on an object. If an object is moving along a nonstraight path, then it must have a component of acceleration that is perpendicular to the path. Therefore there must be a force acting on the object that has a component perpendicluar to the path. In other words, if an object moves on a curved path there must be a sideways force acting on it.

7-2
Uniform circular motion

A common type of motion on a curved path is motion in a circle with constant speed. For this case, only the direction of the velocity is changing, so the acceleration consists only of the component perpendicular to the path. Since the tangent to a circle is perpendicular to the radius, this means that the acceleration of an object moving in a circle, with constant speed, is directed along the radius of the circle.

We can derive an expression for the magnitude of the acceleration and see the direction of the acceleration by redrawing Figure 7-1, as in Figure 7-2, for the special case of circular motion. Figure 7-2a shows the path and the velocity at two points, A and B; Figure 7-2b shows the change in velocity between points A and B.

We know from the preceding section that the

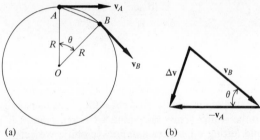

(a)   (b)

FIGURE 7-2 Calculating the change in velocity for a particle which travels on a circle with constant speed.

acceleration is directed along the radius and the only question to be resolved is whether the acceleration is directed inward along the radius or outward. From Figure 7-2b it is clear that the change in velocity, and therefore the acceleration, is directed *inward* rather than outward.

We can derive a formula for the magnitude of the acceleration from the following. First we note that in Figure 7-2a triangle OAB is an isosceles triangle with vertex angle θ. Then we note that, since the velocity is perpendicular to the radius, the angle between the velocity vectors in Figure 7-2b is the same angle θ. Since the speed is constant, the magnitudes of the two velocities are equal; therefore the triangle of Figure 7-2b is also an isosceles triangle with vertex angle θ. Thus the triangle of the velocities in Figure 7-2b is similar to the triangle of Figure 7-2a. Therefore we can write

$$\frac{\Delta v}{v} = \frac{AB}{R} \qquad (7\text{-}4)$$

where AB is the length of the chord in Figure 7-2a and v is the constant speed. When point B is taken closer and closer to point A, then the length of the chord AB gets closer and closer to the length of the arc AB. As point B is taken closer and closer to point A, Equation 7-4 becomes

$$\frac{\Delta v}{v} = \frac{\text{arc } AB}{R} \qquad (7\text{-}5)$$

and since the length of the arc is the distance traveled by the object, we have

$$\frac{\Delta v}{v} = \frac{v\,\Delta t}{R}$$

or

$$\Delta v = \frac{v^2\,\Delta t}{R}$$

from which we have

$$\frac{\Delta v}{\Delta t} = \frac{v^2}{R}$$

The acceleration is defined by

$$\mathbf{a} = \frac{\Delta \mathbf{v}}{\Delta t}$$

so we have for the magnitude of the acceleration:

$$a = \frac{v^2}{R} \qquad (7\text{-}6)$$

7-3
Centripetal and centrifugal force

We can apply Newton's second law

$$\mathbf{F} = m\mathbf{a}$$

to an object moving in a circle with constant speed and conclude from Equation 7-6 that

$$F = \frac{mv^2}{R} \qquad (7\text{-}7)$$

This means that in order for an object to move in a circle with constant speed, some *external agent* must exert a force whose magnitude is given by Equation 7-7. The direction of this force must be radially inward toward the center of the circle. This force is called the *centripetal force*. It must be emphasized that centripetal force is exerted by an external agent and that the centripetal force causes the acceleration. Equation 7-7 does not say that motion in a circle causes a centripetal force to appear.

As long as the external agent exerts the centripetal force, the object will continue to move in a circle. As soon as the centripetal force stops, the object will move in a straight line tangent to the circle. For example, consider a ball being whirled about at the end of a horizontal string, as in Figure 7-3. The centripetal force is exerted by the string on the ball. Note that strings can pull but not push. This provides empirical verification of the assertion that the centripetal force must be radially *inward*. If the string should break then the ball would have no forces (other than gravity) acting on it. It would then move with constant horizontal velocity, and since the initial velocity is tangent to the path, it would initially move tangent to the path. Subsequently it would fall under the action of gravity.

We can look at the motion of an object moving in a circle from a slightly different point of view by

FIGURE 7-3 The centripetal force exerted by the string on the ball is a pull, radially inward.

considering the energy transfer. Let us assume that we know that the only force acting on an object is perpendicular to its velocity. Then the displacement of the object during an instant of time will be parallel to the velocity, and therefore the displacement will be perpendicular to the force. Thus the work done by that force will be equal to zero and there will be no change in the energy of the moving object. Therefore, if the only force acting is perpendicular to the velocity, the kinetic energy of the object remains constant and the object moves with constant speed.

The centripetal force is exerted by some external agent on the body moving in a circle. From Newton's third law, we know that the object exerts an equal and opposite force on the external agent. This reaction force is sometimes called the *centrifugal force*. Thus, in Figure 7-3, the centrifugal force is exerted radially outward by the ball on the string.

Illustrative example

A stone weighing 0.5 lb tied to a string 2 ft long is placed on a smooth horizontal table. The other end of the string is tied to a pin at the center of the table. The stone is given a push and acquires a speed of 6 ft/sec. (a) Determine the tension in the string. (b) If the breaking strength of the string is 15 lb, determine the maximum speed with which the stone can be whirled.

(a) The mass of this stone is

$$m = \frac{0.5 \text{ lb}}{32 \text{ ft/sec}^2} = \frac{1}{64} \frac{\text{lb sec}^2}{\text{ft}}$$

The centripetal force F required to keep it moving in a horizontal circle of radius $R = 2$ ft with a speed $v = 6$ ft/sec is, from Equation 7-7,

$$F = \frac{1}{64} \frac{\text{lb sec}^2}{\text{ft}} \times \frac{36 \text{ ft}^2/\text{sec}^2}{2 \text{ ft}}$$

or

$$F = 0.28 \text{ lb}$$

(b) The breaking strength of 15 lb represents the maximum centripetal force that the string can apply to the stone. Using this value for F in Equation 7-7 and letting v be the maximum velocity, we get

$$15 \text{ lb} = \frac{1}{64} \frac{\text{lb sec}^2}{\text{ft}} \times \frac{v^2}{2 \text{ ft}}$$

from which

$$v^2 = 1920 \frac{\text{ft}^2}{\text{sec}^2}$$

and thus

$$v = 43.8 \frac{\text{ft}}{\text{sec}}$$

Illustrative example

A cylindrical spaceship 10 m in diameter is traveling from the earth to Mars. In order to keep the temperature uniform, the ship is rotated around its axis once per minute, thus exposing all surfaces uniformly to the sun. If an 80-kg spaceman were on the outside of the ship, what force would have to be exerted on him to keep him from flying off tangentially?

From Equation 7-7,

$$F = m \frac{v^2}{R}$$

The spaceman travels in circular path whose length is the circumference of the circle, in a time of 1 min. Therefore

$$v = \frac{c}{t}$$
$$= \frac{\pi D}{t}$$

where D is the diameter of the circle, which is the diameter of the spaceship. Thus

$$v = \frac{\pi \times 10 \text{ m}}{60 \text{ sec}}$$
$$= \frac{\pi}{6} \frac{\text{m}}{\text{sec}}$$
$$= 0.52 \frac{\text{m}}{\text{sec}}$$

Then we have

$$F = \frac{mv^2}{R}$$

$$= \frac{80 \,(0.27)}{5} \text{ nt} = 4.3 \text{ nt}$$

7-4
The nature of centrifugal force

In our previous discussion of circular motion, we referred the motion to a fixed frame of reference such as a stationary table or to some point on the earth. We, the observers, imagined ourselves to be on or in this fixed frame of reference. We could then easily see or imagine the *centripetal force* that acted on the body in circular motion, such as a string pulling on a stone. In each case, the *centrifugal force*, or the reaction of the body in circular motion to the object exerting the centripetal force, was also readily discernible. But if we now imagine ourselves transferred to the moving object—that is, imagine ourselves inside the moving frame of reference—the phenomena will then take on a different appearance and our description of them will be somewhat altered.

Suppose, for example, that we are inside a train that is moving at uniform speed, and that the train is leaving the straight portion of the track and entering the curved or circular portion. The passengers in the train may be seen to move, or appear to be thrown, toward that side of the car that is on the outside of the curve. If the passengers were unaware of the fact that the train was going in a circular path, they might explain their new motion or acceleration toward the outside of the circle as being due to the action of some horizontal force. Or suppose that a ball had been placed on the floor of the car while the train was moving with uniform speed along the straight portion of the track. The ball would have remained there undisturbed until the train started going around the curve. It would then roll toward the side of the car further removed from the center of the circular path. An observer inside the car might ascribe this motion to the action of a "centrifugal force" on the ball; to an observer outside the train—that is, in the stationary frame of reference—the ball would simply be moving in the same straight line along which it had been moving during the time that the train was on the straight portion of the track. This follows directly from Newton's first law of motion, since there is no unbalanced force acting on the ball.

The description of many of the phenomena that occur in circular motion may be much simpler when referred to the moving frame of reference. To avoid serious error, it is at all times necessary to know the particular frame of reference used in the discussion. The behavior of particles in a system that is moving in a circular path is sometimes compared to the behavior of particles in the earth's gravitational field. For example, we know that the effect of the earth's gravitational field is to give every particle at a particular place on the earth the same acceleration g downward. In one of the above examples we saw that a ball in a train moving in a circular path is apparently accelerated with respect to the train in a direction away from the center by the centrifugal force. This acceleration, measured inside the train, is $a = v^2/R$, independent of the mass of the ball. The same would be true for any other particle that is free to move inside the train. We may thus say that to an observer inside the moving frame of reference there exists a field very similar to a gravitational field in its effect but directed away from the center. (It should be noted that there are additional effects in a rotating reference frame that distinguish this central acceleration from a gravitational field, but for many problems the central acceleration is the most important effect.) In many practical examples, such as the effect on a person in an airplane moving in a circular path or in the description of the action of a centrifuge, this central acceleration is compared numerically to g. For example, an aviator in an airplane coming out of a power dive moves in a curved path, which we shall take as circular. It has been found from experiment that in order for the pilot to avoid blacking out in coming out of this power dive, the acceleration of the airplane in the circular motion should not exceed $6.9g$.

7-5
Banking of a curved road

A car or train rounding a curve can be considered as moving in an arc of a circle or in some cases in a series of such arcs of slightly different radii. In order to move the car in a circular path of radius R, an outside force must act on the car, and this force must be directed toward the center of the circle. In the case of an automobile rounding a curve, this force is supplied by the friction between the road and the tires. If the speed of the car is very great, the frictional force may not be sufficiently large to supply the necessary centripetal force to make the

FIGURE 7-4 Photograph of a car traveling at high speed around a banked curve. (Courtesy of General Motors Corporation.)

car move in this circular path. The car may move toward the outer part of the road and thus travel in a curve of larger radius. or it may even go off the road. This happens more often when the road is wet or icy, since the friction between the tires and the road is then much reduced.

When high speeds are normally used on curved roads, the roads are frequently banked for safer travel; that is, the outer part of the road is built at a higher level than the inner part, as shown in Figure 7-4. The proper angle θ for banking the road is one for which the normal force **N** that the road exerts on the car will have a vertical component equal and opposite to **W** and a horizontal component sufficient to produce the required centripetal force F_c for a given speed v (see Figure 7-5a). From Figure 7-5b it will be noted that

$$\tan \theta = \frac{F_c}{W}$$

and, since

$$F_c = \frac{mv^2}{R} \quad \text{and} \quad W = mg$$

we get

$$\tan \theta = \frac{v^2}{Rg} \tag{7-8}$$

If the speed of the car exceeds the proper speed

given by Equation 7-8, then there must be enough friction between the tires and the ground to supply the necessary additional centripetal force for the car to go around the curve safely.

Illustrative example

A car weighing 4000 lb goes around a curved road at 60 mi/hr. The curve is a circular arc with a 2400-ft radius. Determine (a) the centripetal force that the road must exert on the car and (b) the angle at

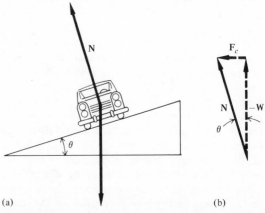

(a)   (b)

FIGURE 7-5 (a) Car moving on a curve in a road banked at angle θ to the horizontal. (b) The components of the normal force **N** are −**W** and F_c.

which the road must be banked for this to be the proper speed.

(a) The magnitude of the centripetal force is given by Equation 7-7:

$$F = m \frac{v^2}{R}$$

Now

$$v = 60 \frac{\text{mi}}{\text{hr}} = 88 \frac{\text{ft}}{\text{sec}}$$

and

$$m = \frac{W}{g} = \frac{4000 \text{ lb}}{32 \text{ ft/sec}^2}$$

hence

$$F = \frac{4000 \text{ lb}}{32 \text{ ft/sec}^2} \times \frac{(88 \text{ ft/sec})^2}{2400 \text{ ft}}$$

from which

$$F = 403 \text{ lb}$$

(b) The angle of banking, θ, is given by

$$\tan \theta = \frac{v^2}{Rg}$$

hence

$$\tan \theta = \frac{(88 \text{ ft/sec})^2}{2400 \text{ ft} \times 32 \text{ ft/sec}^2} = 0.10$$

and thus

$$\theta = 5°50'$$

7-6
Motion in a vertical circle

An interesting and important type of circular motion is that which takes place in a vertical plane— for example, when an airplane loops the loop. Occasionally a motorcycle stunt rider will ride on the inside of a vertical circular track, or a ball can be made to roll completely around the inside of a vertical circular track. The motion is not uniform, and the speed varies from point to point on the circle. Suppose we confine our attention to a particle that acquires its speed by sliding down a frictionless inclined plane (see Figure 7-6) and then starts going up on the inside of the circular track. It is obvious that the danger point is the highest point A on the track. The particle must negotiate this point with the proper speed if it is to travel safely around the track.

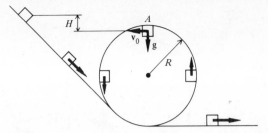

FIGURE 7-6

Suppose for a moment that the particle is at point A under the track and has zero speed. Since the force of gravity is the only force that acts on it, the particle will be simply a freely falling body with an acceleration of g downward directed through the center of the vertical circle. To negotiate the highest point safely, the particle must have a minimum speed v_0, such that its acceleration, which is v_0^2/R in a circle of radius R, is at least equal to g; that is,

$$g = \frac{v_0^2}{R}$$

from which

$$v_0^2 = Rg$$

and

$$v_0 = \sqrt{Rg} \qquad (7-9)$$

If the speed of the particle is greater than this minimum speed, its acceleration toward the center will be greater than g; this means that the track will have to supply a force toward the center to keep it moving in the circular path. If its speed is less than this minimum safe value of v_0, the particle will leave the track and follow the parabolic path of a projectile (see Figure 7-7).

One method by which a particle can acquire the necessary speed in order to travel around the vertical circle is first to slide down an inclined plane attached to this circular path; it will have to start at a sufficient height H above the highest point A of the circle to acquire the minimum safe speed v_0. If we neglect losses due to friction, the speed v that the particle will acquire in moving down a plane of height H is

$$v^2 = 2gH$$

But we know that at the top of the circle

$$v_0^2 = Rg$$

Equating these two values of the velocity, we

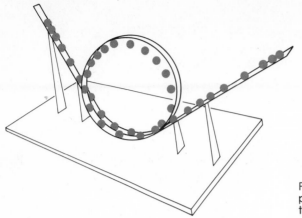

FIGURE 7-7 Path of a ball that starts on an inclined plane but does not acquire sufficient speed to loop-the-loop in a vertical curve.

obtain

$$R = 2H$$
$$H = \frac{R}{2}$$

That is, the particle must start at a height at least equal to $R/2$ above the point A in order to pass the top safely. Actually it will have to start at some point above this in order to make up for the loss in speed that is due to friction.

Illustrative example

A small toy car with ball-bearing wheels goes around a vertical circle whose radius is 2 ft. (a) Determine the minimum speed it must have at the top of the circle. (b) Determine its speed halfway down the circle. (c) Determine its speed at the bottom of the circle. If the car weighs 0.5 lb, determine the force that the track exerts on it (d) at the top and (e) at the bottom. Neglect friction effects.

(a) The minimum speed that the car must have at the top of the track to pass it safely is given by

$$v_0 = \sqrt{Rg}$$

Substituting numerical values, we get the following equation

$$v_0 = \sqrt{2 \text{ ft} \times 32 \text{ ft/sec}^2}$$

or

$$v_0 = 8 \, \frac{\text{ft}}{\text{sec}}$$

(b) There are two forces acting on the car—its weight mg and the push of the track. The force provided by the track is always at right angles to the velocity, and hence it cannot produce a change in the speed; it can only produce a change in the direction of motion. The change in speed can come only from the other force, its weight mg. It has already been shown that the speed acquired by a body moving down a smooth inclined plane does not depend upon the length of the plane but only upon its height H. This theorem can be generalized to state that the speed acquired by a body moving down any frictionless path is the same as that it would have acquired if it had fallen through the same height H. This is equivalent to the statement that mechanical energy is conserved. Hence, if the initial speed of a body is v_0, its final speed v is given by

$$v^2 = v_0^2 + 2gH$$

In this case,

$$v_0^2 = Rg \quad \text{and} \quad H = R$$

and, calling the speed halfway down $v = v_1$, we get

$$v_1^2 = Rg + 2Rg = 3Rg$$

Hence

$$v_1 = \sqrt{3Rg} = \sqrt{3 \times 2 \text{ ft} \times 32 \text{ ft/sec}^2}$$

or

$$v_1 = 13.9 \, \frac{\text{ft}}{\text{sec}}$$

(c) At the bottom of the plane, $H = 2R$; hence the speed v_2 will be

$$v_2^2 = Rg + 2g \times 2R = 5Rg$$

Hence

$$v_2 = \sqrt{5Rg} = \sqrt{5 \times 2 \text{ ft} \times 32 \text{ ft/sec}^2}$$

or

$$v_2 = 17.9 \frac{\text{ft}}{\text{sec}}$$

(d) When the car is at the top of its path, the track exerts no force on it. The only force acting on it is its weight, which, at this position, is also the centripetal force.

(e) When the car is at the bottom of its path, the force F that the track exerts is the resultant of two forces: (1) a push upward equal to the weight of the car and (2) a centripetal force mv_2^2/R to keep it moving in the circle; hence

$$F = mg + \frac{mv_2^2}{R}$$

but

$$v_2^2 = 5gR$$

$$F = mg + \frac{m \times 5gR}{R}$$

which yields

$$F = 6mg$$

so that

$$F = 6 \times 0.5 \text{ lb} = 3 \text{ lb}$$

Thus at the bottom of the path, the track must exert a force equal to six times the weight of the car, provided that the car started with the minimum safe speed at the top.

7-7
Periodicity of uniform circular motion

One of the interesting properties of uniform circular motion is that it is *periodic*; that is, a particle in uniform circular motion always takes the same time T to traverse a full circumference. This time T is called the *period* of the motion.

The relationships between the period and the other quantities used to describe the motion are readily obtainable. Referring to Figure 7-8, the distance covered by such a particle in one revolution is the length of the circumference, which is $2\pi R$, where R is the radius of the circle. From the definition of the speed of a particle as the distance traversed divided by the time, we get

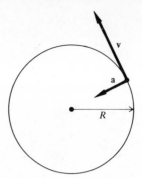

FIGURE 7-8 Directions of velocity and acceleration in uniform circular motion.

$$v = \frac{2\pi R}{T} \tag{7-10}$$

from which

$$T = \frac{2\pi R}{v} \tag{7-11}$$

The greater the speed of the particle, the shorter is the period of the motion. The period is expressed in units of time, usually in seconds.

We have already shown the acceleration of a particle in uniform circular motion to be

$$a = \frac{v^2}{R} \tag{7-6}$$

Substituting the value of v from Equation 7-10 yields

$$a = \frac{4\pi^2 R}{T^2} \tag{7-12}$$

for the relationship between the acceleration and the period for uniform circular motion.

Illustrative example

A small car moves around a circular track once every 8 sec. The radius of the track is 3 m. Determine (a) the speed of the car and (b) its centripetal acceleration. (c) If the mass of the car is 2.5 kg, determine the centripetal force exerted by the track.

(a) Substituting $T = 8$ sec and $R = 3$ m in Equation 7-10 yields

$$v = \frac{2\pi \times 3 \text{ m}}{8 \text{ sec}} = 2.35 \frac{\text{m}}{\text{sec}}$$

The acceleration can be obtained from Equation 7-12, which yields

$$a = \frac{4\pi^2 \times 3 \text{ m}}{64 \text{ sec}^2} = 1.85 \frac{\text{m}}{\text{sec}^2}$$

(c) From Newton's second law, the centripetal force F_c exerted by the track toward the center of the circle is

$$F_c = ma$$

from which

$$F_c = 2.5 \text{ kg} \times 1.85 \frac{\text{m}}{\text{sec}^2} = 4.63 \text{ nt}$$

7-8
Planetary motion

One important type of periodic motion, which has been studied and recorded for centuries, is that of the bodies constituting the solar system. Theories concerning the solar system have changed with the centuries, and to a certain extent these changes mirror man's intellectual progress. Among the early theories that held sway for many centuries was that associated with the name of Claudius Ptolemy (c. A.D. 150) and known as the *geocentric* theory. This theory was reasonably successful in explaining and predicting planetary motion with the degree of accuracy then possible of attainment. In the geocentric theory the earth was assumed to be at the center of the universe, and the sun and the planets moved around it in complicated paths.

Several centuries before Ptolemy, Aristarchus of Samos (c. 310–230 B.C.) proposed a theory in which the sun was fixed at the center of the universe and the earth revolved around the sun in a circular orbit. He also recognized that the stars appeared "fixed" in position, because their distances from the sun were tremendous in comparison with the distance of the earth from the sun. Very few of the early astronomers accepted this *heliocentric* theory of the universe; from the second century until the sixteenth century only the geocentric theory of Ptolemy was taught and used. In the latter century, Nicholas Copernicus (1472–1543) revived and extended the heliocentric theory of Aristarchus and thus started a revolution in scientific thought that was carried forward by Kepler, Galileo, and Newton. In the heliocentric theory of Copernicus (see Figure 7-9), the sun was considered at the center of the universe, the planets revolved around the sun in circular orbits, and the fixed stars were assumed to lie in a sphere surrounding the solar system.

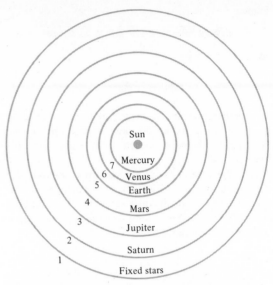

FIGURE 7-9 Orbits of the planets and the fixed stars in the heliocentric theory of the universe according to Copernicus.

The heliocentric theory was not readily accepted by scientists of that period. Tycho Brahe (1546–1601), a famous Danish astronomer, made very careful and accurate measurements of the motions of the planets and the sun. He had never become convinced of the correctness of the Copernican hypothesis, but his extensive and careful measurements, which he bequeathed to another astronomer, John Kepler (1571–1630), laid the foundations of modern astronomy. It may be noted here that Brahe's observations were made without the aid of telescopic instruments.

Kepler, from his study of the data accumulated by Tycho Brahe, deduced three laws that accurately describe the motions of the planets around the sun. Kepler's three laws are:

1. Each planet moves around the sun in an elliptic path (or orbit) with the sun at one focus of the ellipse.
2. As the planet moves in its orbit, a line drawn from the sun to the planet sweeps out equal areas in equal intervals of time (see Figure 7-10).
3. The squares of the periods of the planets are proportional to the cubes of their mean distances from the sun.

It can be seen that the simplified picture of the planetary system proposed by Copernicus is not

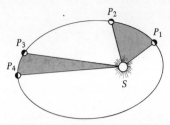

FIGURE 7-10 The path of a planet about the sun S is an ellipse. P_1, P_2, P_3, P_4, represent positions of a planet in its orbit at different times. The speed of a planet in its orbit is such that an imaginary line joining the sun and the planet would sweep out equal areas in equal intervals of time. For example, area SP_1P_2 is equal to area SP_3P_4.

sufficiently accurate; however, the elliptical orbits of the planets are not far removed from circles. In Figure 7-11 the planetary orbits are represented as circles merely to show their relative positions with respect to the sun. Kepler's third law can be put in mathematical form as follows: if r_1 is the average distance (or radius in Figure 7-10) of one planet from the sun, and r_2 is the average distance of another planet from the sun, then

$$\frac{T_1^{\,2}}{T_2^{\,2}} = \frac{r_1^{\,3}}{r_2^{\,3}} \qquad (7\text{-}13)$$

where T_1 is the period of revolution of the first planet and T_2 that of the second planet. It can be seen from the third law that the farther from the sun a planet is, the longer is its period.

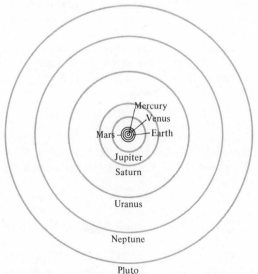

FIGURE 7-11 Relative positions of the planetary orbits with respect to the sun. (Drawn to scale.) The sun is at the center.

Kepler's second law is equivalent to the principle of conservation of angular momentum, a concept that will be discussed in Chapter 9.

7-9
Gravitation and planetary motion

Newton's law of universal gravitation (Section 5-2), when applied to planetary motion, states that the mutual force of attraction between the sun and any one of the planets is given by

$$F = G_0 \frac{Mm}{r^2} \qquad (7\text{-}14)$$

where M is the mass of the sun, m the mass of the planet, and r the distance between their centers. The sun exerts a force F on the planet and the planet exerts an equal and opposite force on the sun, in agreement with Newton's third law of motion.

Kepler's three laws of planetary motion can be derived from Newton's law of gravitation and the application of the second law, $F = ma$, to motion of the planets. The derivation of the equation for the orbit of a planet is beyond the scope of this book, but a simplified derivation of Kepler's third law can readily be given if we assume that the planets move in circular orbits around the sun. If T is the period of the planet and r the radius of its orbit, then its acceleration a towards the sun is given by

$$a = \frac{4\pi^2 r}{T^2} \qquad (7\text{-}15)$$

From Newton's second law and the law of gravitation, we have

$$a = \frac{F}{m} = G_0 \frac{M}{r^2}$$

Hence we can write

$$G_0 \frac{M}{r^2} = \frac{4\pi^2}{T^2} r$$

from which

$$T^2 = \frac{4\pi^2}{G_0 M} r^3 \qquad (7\text{-}16)$$

which is Kepler's third law that the square of its period is proportional to the cube of the average distance of a planet from the sun.

In addition to the force exerted by the sun, a

planet is acted upon by forces due to the other planets and even satellites. Because of the large distances between planets, the effects of these forces are small but not negligible. There is the interesting story of the discovery of the planet Neptune to attest to the accuracy of Newton's law of gravitation. Observations on the motion of the planet Uranus showed that its path differed slightly from that calculated when the effects of all the other planets then known were taken into consideration. It was then assumed that the difference might be due to the action of some other planet not then known. Newton's laws were used to calculate the position of this new planet; it was looked for and found at about the calculated position. This newly discovered planet was called Neptune.

7-10
Artificial satellites

It is now well known that there are many artificial satellites revolving around the earth, varying in size from very small copper needles to fairly large craft containing a variety of instruments designed to study radiations in space, to photograph clouds, or to receive television signals from stations on one continent, amplify them, and transmit them to stations on another continent. Astronauts or cosmonauts have circled the earth one or more times and returned safely; others have landed on the moon, explored regions of it and then returned to earth. Satellites have been sent around the moon, toward Venus and Mars, and to other parts of the solar system.

One of the conditions essential to get a satellite into a predetermined orbit around the earth is to give it a velocity appropriate to that orbit. Once in the orbit, the satellite can travel around the earth just as the moon does, with the force of gravity acting on it to keep it in orbit, provided that the path is outside the earth's atmosphere. Some orbits are almost circular; others approximate ellipses. It is comparatively simple to calculate the speed that a satellite has to have in order to travel in a circular path around the earth. The acceleration a of any particle traveling in a circular path of radius r and velocity v is

$$a = \frac{v^2}{r} \qquad (7\text{-}6)$$

The acceleration of a satellite moving around the earth is simply the acceleration of gravity g at the

particular position of the particle:

$$a = g = \frac{v^2}{r} \qquad (7\text{-}17)$$

for which

$$v^2 = rg$$

or

$$v = \sqrt{rg} \qquad (7\text{-}18)$$

If the satellite is not too far above the surface of the earth, we can set

$$g = 32 \ \frac{\text{ft}}{\text{sec}^2} \quad \text{and} \quad r = 4000 \ \text{mi}$$

from which

$$v = \sqrt{\frac{32 \ \text{mi}}{5280 \ \text{sec}^2} \times 4000 \ \text{mi}}$$

Therefore

$$v = 4.92 \ \frac{\text{mi}}{\text{sec}}$$

or

$$v = 5 \ \frac{\text{mi}}{\text{sec}} \ (\text{approx})$$

Since the circumference of the earth is

$$2\pi r = 2\pi \times 4000 \ \text{mi} = 25{,}000 \ \text{mi}$$

the time taken to complete one orbit will be approximately 5000 sec = 82.5 min.

The velocity of a satellite in a circular orbit of radius r around the earth varies inversely as the square root of its radius. This becomes evident if we remember that g varies inversely as r^2. Thus if we call the value of the acceleration of gravity at the surface of the earth g_e, and call R the radius of the earth, we can write

$$\frac{g}{g_e} = \frac{R^2}{r^2}$$

where g is the gravitational acceleration at a distance r from the earth's center. Hence we see that

$$g = \frac{g_e R^2}{r^2}$$

Putting this value of g into Equation 7-18 yields

$$v = \sqrt{\frac{g_e R^2}{r}} \qquad (7\text{-}19)$$

for the velocity of a satellite in a circular orbit around the earth.

Questions

1. Sometimes when a car rounds a curve on a level road at a high speed, two of its wheels leave the ground. Which two wheels remain on the ground?

2. An aviator makes a quick turn in coming out of a power dive. On the assumption that the airplane is moving in a circular path at this instant, its central acceleration has been determined as $6.9g$. Discuss the meaning of this term. How big a force is exerted on the pilot if his weight is 175 lb?

3. A boy swings a pitcher of water in a vertical circle so that the open end of the pitcher faces downward at the top of the circle. Discuss the motion of the pitcher and the water. Is there any minimum speed that the pitcher must have at the top of the circle so that the water will not spill out?

4. An aviator loops the loop in an airplane; that is, he travels in a vertical circle with the bottom of the plane always on the outside of the circle. Analyze the forces acting on the aviator when he is (a) at the top of the circle, (b) halfway down, and (c) at the bottom of the circle.

5. A ball is tied to a string and whirled in a horizontal circle on a smooth table with speed v. If the string should be cut suddenly, discuss the subsequent motion of the ball. What happens to the string after it is cut?

6. In the so-called centrifugal type of clothes dryer, the wet clothes are put into a cylinder that has holes drilled in its walls. This cylinder, with the clothes in it, is then spun rapidly. Explain the action of this dryer.

7. A passenger is sitting on the very smooth, plastic-covered bench seat of an automobile. To which side of the car does the passenger tend to slide when the car rounds a curve?

8. At what time of the year is the speed of the earth in its orbit (a) greatest? (b) smallest?

9. A glass jar is fastened on a horizontal turntable near its rim and a candle is placed inside the jar. When the candle is lit the flame will point in a vertical direction owing to the upward motion of the hot gases formed in the combustion process. If the turntable is set into rotation, will the flame be deflected toward or away from the center of the circle? Explain your answer.

10. A train is rounding a curve at high speed. A man in the train drops a ball. What will be the path of the ball as observed (a) by the man in the train and (b) by an observer who is standing outside at a distance from the train?

11. The term "weightlessness" is used to describe certain experiences of astronauts who are in satellites that are moving around the earth. Both the vehicle and the astronaut have the same acceleration g appropriate to this position in the earth's gravitational field so that the vehicle does not exert a force to support the astronaut. This produces new experiences for the astronaut. What would happen, for example, if he dropped an object?

12. A satellite in a circular orbit speeds up. What happens to the size of the orbit? Discuss what happens if an astronaut in a satellite throws a tool to another astronaut in the same satellite.

Problems

1. An airplane traveling at 350 mi/hr makes a complete circle in 2 min. Find the centripetal acceleration of the airplane.

2. Find the centripetal acceleration of a point on the surface of the earth at the equator, knowing that the earth rotates once per day and that the radius of the earth is about 4000 mi.

3. A car traveling at 45 mi/hr goes around a curve of radius 300 ft. Find the centripetal acceleration of the car.

4. An electron (mass 9.1×10^{-28} gm) is moving with a speed of 2×10^8 cm/sec in a circle of radius 5 cm. How large a force must be exerted on the electron?

5. A stone whose mass is 200 gm is attached to a cord 30 cm long and placed on a smooth horizontal table. The stone is then whirled in a circular path with a speed of 60 cm/sec. Determine the tension in the cord.

6. A car weighing 3600 lb rounds a curve of radius 900 ft at a speed of 60 mi/hr. (a) What force must the ground exert on the tires to keep this car moving in this circular path? (b) In what direction is this force?

7. A small child weighing 40 lb rides on a merry-go-round, which makes a complete circuit in 30 sec. The radius of the merry-go-round is 12 ft. How large a centripetal force is exerted on the child?

8. Assume that the moon is traveling in a circular path of radius 240,000 mi. The period of revolution is $27\frac{1}{2}$ days. (a) Find the centripetal acceleration of the moon and compare it with the value of g at the location of the moon. (b) What body exerts the centripetal force on the moon?

9. Determine the angle at which a road should be banked if the radius of the curve is 1600 ft and if it is to supply the necessary centripetal force to a car traveling at 85 mi/hr. The centripetal force is to be the horizontal component of the normal force exerted by the road on the car.

10. A 200-gm car moves on the inside of a vertical circular track as in Figure 7-6. The radius of the circle is 12 cm. (a) What is the minimum speed that the car must have at the top of the track in order to move in this circular path. (b) Assuming that it has this minimum safe speed at the top of the track, determine the speed at the bottom of the track. What is the force that the track exerts on the car (c) at the top of the track and (d) at the bottom of the track?

11. The period of Jupiter is 11.86 yr. Determine its distance from the sun with the aid of Kepler's third law.

12. Where would a planet have to be in order to have a period of 2 yr?

13. It seems reasonable that a planet must be outside the sun. The radius of the sun is about 7×10^8 m. What is the shortest possible planetary period?

14. (a) With what speed would a satellite just skim over the surface of the earth, ignoring the existence of mountains and man-made towers? (b) Compare this with the escape velocity from the earth, which is 7 mi/sec.

15. A 4000-lb car goes around an unbanked curve of radius 800 ft at 45 mi/hr. What is the least coefficient of friction that will permit the car to go around the curve without skidding?

16. A car of mass m goes around a curve of radius R. The road is unbanked and has a coefficient of friction μ. What is the maximum safe speed v?

17. An airplane is diving at a speed of 650 mi/hr; the pilot pulls it out of this dive by moving in the arc of a vertical circle. (a) Determine the minimum radius of this circle if the acceleration is not to exceed $7g$ at the lowest point. (b) Determine the force acting on the pilot, if he weighs 150 lb. (c) Determine the force on 1 gm of blood in the pilot's brain.

18. Using the results of Problem 2, determine how the weight of a body at the equator differs from the gravitational force exerted on it.

19. A 60-lb girl is swinging on a rope that is 9 ft long. At the lowest point on the path she is moving with a speed of 8 ft/sec. What is the tension in the rope at the lowest point of the path?

20. Derive a formula for the period of a satellite, in terms of the masses of the satellite and of the earth, and the distance between them.

21. The position at which a planet is closest to the sun is called the *perihelion* and the position at which it is farthest away is called *aphelion*. (a) From Kepler's second law show that the ratio of the speed at aphelion to the speed at perihelion is equal to the ratio of the planet-sun distance at perihelion to the planet-sun distance at aphelion. (b) The earth's speed at perihelion is 18.9 mi/sec and its speed at aphelion is 18.3 mi/sec. Find the ratio of aphelion distance to perihelion distance for the earth. (c) Attempt to draw the earth's orbit to scale.

22. At what distance above the earth should a satellite be placed in order that it appear to be stationary when viewed from the earth?

23. The period of the earth's orbit around the sun is 3.16×10^7 sec and the average radius of the orbit is 1.49×10^{13} cm. From these data and the value of the gravitational constant, 6.67×10^{-8} dyne cm²/gm², find the mass of the sun. Knowing the radius of the sun, 7×10^{10} cm, find the density (mass per unit volume) of the sun.

24. A common amusement park ride consists of a large cylinder that rotates about a vertical axis. The passengers stand inside the cylinder with their backs against the wall. When the cylinder is moving fast enough, the floor beneath the riders is removed. Show that the passengers will not fall if the coefficient of friction is greater than gR/v^2, where R is the radius of

the cylinder and v is the speed of the passenger.

25. A ball whose mass is 200 gm is hanging from a string attached to the roof of a car. The car is moving along a straight level road at 90 km/hr. (a) What is the resultant force on the ball? The car continues around a curve in the road at the same speed. The radius of the curve is 1500 m. (b) What is the centripetal acceleration of the car? (c) What is the acceleration of the ball? (d) Find the angle with the vertical made by the string.

26. A mass m rotates in a circle of radius r with a speed v on a smooth frictionless table. The centripetal force is supplied by a string that goes through a hole in the table at the center of the circle and is attached to a hanging mass M. Find the values of r and v necessary to keep M at rest.

27. A small object of mass m is attached to a string of length l as shown in Figure 7-12. A second string is attached to the object and pulled sideways until the first string makes an angle of 60° with the vertical and the second string is horizontal. (a) Find the tension in the first string. (b) The horizontal string is burned and the object swings down in a circular path. From the conservation-of-energy principle, find the speed of the object at the lowest point in the path. (c) Find the tension in the string when the object is at the lowest point of the path.

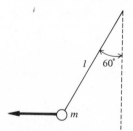

FIGURE 7-12

28. A 200-kg satellite is in an orbit 400 miles above the surface of the earth. It loses energy at the

rate of 2×10^5 joules per revolution. At the end of 1000 orbital revolutions find (a) its speed, (b) its altitude above the earth, and (c) the period of the motion.

29. If the centripetal acceleration of a man exceeds $5g$, *blackout* occurs; he loses vision because blood drains away from the retina. On the other hand, at about $3g$ excess blood can cause loss of vision because of *redout*; excess blood prevents light from reaching the retina. A man is placed in a chair at the end of a 10-m boom. The boom is rotated until either of these effects is produced. Describe the orientation of the man and the speed with which he is moving (a) when blackout occurs and (b) when redout occurs.

30. The sun appears to be moving at a speed of about 250 km/sec in a circular orbit of radius of about 25,000 light-years (one light-year is equal to 9.5×10^{17} cm). Find the total mass that is presumably responsible for this motion. Express this as a multiple of the mass of the sun (2×10^{33} gm).

31. Bode's law is a purely empirical relation which gives the average radii of the planetary orbits (it was discovered in 1772 by a J. D. Titius and publicized by Bode, but never yet has been given a complete theoretical explanation). It gives the radius R of a planet in terms of the distance from the earth to the sun (1.5×10^{11} m) as

$$R = 0.4 + 0.3(2^n)$$

where $n = 0$ gives the radius of Venus' orbit, $n = 1$ gives the orbit of earth, $n = 2$ gives the orbit of Mars. The values of 4, 5, and 6 give the orbits of Jupiter, Saturn, and Uranus, respectively. The missing value, $n = 3$, gives the approximate location of the asteroid belt. (a) Calculate the radius of the orbit of the body that fits Bode's law for $n = 3$, and (b) determine the period of the motion and (c) the average speed. The asteroids have orbits with radii ranging from this value to one about half the radius of the orbit of the earth. (d) Find the speed of the asteroid closest to the earth.

impulse & momentum

chapter 8

8-1
Impulse and momentum of a single particle

The momentum of a particle is a quantity defined as the product of the mass of the particle and the velocity of the particle:

$$\mathbf{p} = m\mathbf{v} \tag{8-1}$$

Thus momentum is a vector quantity whose direction is the same as the direction of the velocity.

If a force acts on the particle, then the momentum of the particle changes. If the force is constant, for a time t, then we have

$$\mathbf{F} = m\mathbf{a}$$

and since the force is constant, the acceleration is constant and we have

$$\mathbf{a} = \frac{\mathbf{v}_f - \mathbf{v}_i}{t}$$

where $\mathbf{v}_f$ and $\mathbf{v}_i$ are the final and initial velocities during the time interval t. Then combining these equations we have

$$\mathbf{F} = \frac{m(\mathbf{v}_f - \mathbf{v}_i)}{t}$$

or

$$\mathbf{F}t = m(\mathbf{v}_f - \mathbf{v}_i)$$
$$\mathbf{F}t = m\mathbf{v}_f - m\mathbf{v}_i$$
$$\mathbf{F}t = \mathbf{p}_f - \mathbf{p}_i \tag{8-2}$$

The product on the left side of Equation 8-2 is called the *impulse* of the force acting on the particle, and the right side is the *change in momentum of the particle*. Equation 8-2 states that the impulse acting on a particle is equal to the change in momentum of the particle.

This result can be extended to the situation where a varying force acts on the particle. We can imagine dividing the time interval into very short time intervals during each of which the force can be treated as essentially constant. For each of these short time intervals we can write Equation 8-2 and then add these equations to yield one equation. On the left side of the resulting equation is the sum of the impulses and on the right side is the total change in momentum:

$$\mathbf{F}_1 \, \Delta t_1 + \mathbf{F}_2 \, \Delta t_2 + \mathbf{F}_3 \, \Delta t_3 + \cdots = \mathbf{p}_f - \mathbf{p}_i$$

The left side of this equation can be written as the product of the *average force* $\overline{\mathbf{F}}$ and the total time interval t. Therefore for a varying force we can write

$$\overline{\mathbf{F}}t = \mathbf{p}_f - \mathbf{p}_i \qquad (8\text{-}3)$$

This equation means that the total impulse is equal to the total change in momentum of the particle, where the total impulse is the average force times the time interval during which it acts.

The major importance of Equation 8-3 is that it permits us to describe the effects of forces that act for extremely short time intervals. Consider, for example, what happens when a baseball bat hits a pitched ball. We know very little about the details of the force, but we do know that it acts for a very short time—the time of contact. During that time there is an impulse acting on the ball, an impulse that changes the momentum of the ball. This change in momentum is large enough to change the direction of the velocity. But the time of contact is so small that the displacement of the ball is very small during the time that the impulse is exerted. In general, impulsive forces (those that act for very short time intervals) change the momentum of particles, but do not change the position of the particle during the time the impulse is being exerted. After the impulse is exerted, the particle moves with constant momentum. Thus in the baseball example, the ball comes in with constant momentum, is struck by the bat, which exerts an impulse that changes the momentum without appreciable change in position, and then the ball leaves the bat with constant momentum.

8-2
Impulse and momentum for a system of many particles

If we have a system of particles, then we can write Equation 8-2 once for each particle in the system. This set of equations can usually be simplified a great deal by introducing Newton's third law. If we consider a particle A, which is part of the system, as in Figure 8-1, then some of the forces acting on A come from other particles in the system (such as B and C) and some of the forces come from particles outside the system (such as D). Since Newton's third law states that the force on A due to B is equal and opposite to the force on B due to A, the change in momentum of A due to the force from B is equal and opposite to the change in momentum of B due to the force from A. Thus, for every change in momentum of a particle in the system due to forces from other particles in the system, there is an equal and opposite change of momentum of another particle in the system. Let us define the momentum of the system, $\mathbf{P}$, as the sum of the momenta of the particles in the system:

$$\mathbf{P} = \mathbf{p}_1 + \mathbf{p}_2 + \mathbf{p}_3 + \cdots \qquad (8\text{-}4)$$

The change in this total momentum is produced by forces exerted by particles outside the system:

$$\mathbf{F}_{ext}t = \mathbf{P}_f - \mathbf{P}_i \qquad (8\text{-}5)$$

where $\mathbf{F}_{ext}$ is the force exerted by particles external to the system, $\mathbf{P}_i$ is the total initial momentum of the system, and $\mathbf{P}_f$ is its total final momentum. Equation 8-5 states that the total impulse $\mathbf{F}_{ext}t$ acting on the system is equal to the change in the total momentum of the system of particles. The

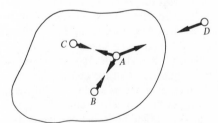

FIGURE 8-1 Particle A, which is part of the system, has forces exerted on it by particles B and C, which are part of the system, and by particle D, which is not part of the system. Each of the particles B,C, and D has a reaction force exerted on it. Since the internal forces come in equal and opposite pairs, the change in momentum of the system comes from forces exerted by particles outside the system.

most interesting application of this result is to the case in which there are no external forces. In that case, the total momentum of the system is constant, even though the individual momenta of the particles in the system may change. The law of conservation of momentum is the statement that the total momentum of a system of particles is constant if there are no external forces acting on the system. This law is among the most important and useful laws of physics.

One interpretation of Equation 8-5 is that a system of particles can be thought of as an *entity* in that the change in total momentum of the system is determined by the external forces that act on it. This point of view is made more impressive when the idea of *center of mass* is introduced. For any system of particles it is possible to find one point, called the *center of mass* (or sometimes the center of momentum) such that the total momentum of the system can be thought of as being associated with that point. That is, one can think of this point as having the total mass M of the system and as having a velocity $\mathbf{V}$, and therefore a momentum $\mathbf{P}$ that is equal to the vector sum of the momenta of all the particles of the system. Then Equation 8-5 states that the momentum of this one point is determined by the external forces. In that sense Equation 8-5, which is a description of a system, becomes identical in form to Equation 8-2, which is a description of a point particle. Thus we can replace a complex system of particles by a single particle. We can think of all the stars in a galaxy as a system and replace the entire galaxy by one point having the mass of all the stars. This point then moves in response to the external force on the galaxy, in spite of the fact that the stars move within the galaxy. It is this point of view that justifies our having treated such complex systems as a baseball or the earth as if they were single particles.

The problem of finding the center of mass is usually a problem in the calculus. It is true, however, as might be guessed, that the center of mass of rigid uniform homogeneous bodies of simple shape is the same as the geometric center. That is, the center of mass of a rigid uniform homogeneous rod is the middle of the rod and the center of mass of a sphere is the center of the sphere.

Speaking somewhat loosely, we can treat a system as if all of its mass were concentrated at its center of mass. Similarly we were able, in Chapter 4, to treat a body as if all of its weight were concentrated at its center of gravity. In a large number of cases these two points, the center of mass and the center of gravity, are identical. However, in principle, and sometimes in practice, these two points are not the same. Consider, for example, a very tall building, which can be represented by a tall box. If the box is treated as uniform in density, then the center of mass is at the geometric center. However, since the intensity of the gravitational field decreases with altitude, the gravitational force on the points near the top of the box is smaller than the gravitational force on points near the bottom. Thus, the center of gravity is *below* the geometric center.

8-3
Collisions

One interesting application of the ideas of the previous sections is to collisions between two particles. For simplicity we will deal first with the situation where there are no external forces and the motion is confined to a straight line. For example, we can imagine that the objects are sliding on a horizontal, frictionless wire. In Figure 8-2 a particle, 1, is overtaking particle 2; in (b) of the figure the two particles are colliding and in (c) particle 2 is moving away from particle 1.

If we were to attempt to analyze this situation by means of Newton's laws, we would immediately run into the difficulty that we know nothing about the forces exerted during the collision. However, since we assumed that there are no external forces on the system, we know that the total momentum of the system is constant. In particular, the momentum of the system before the collision is equal to the momentum of the system after the collision. Remembering that momentum is a vector quantity we can then write

$$m_1 v_{1i} + m_2 v_{2i} = m_1 v_{1f} + m_2 v_{2f} \qquad (8\text{-}6)$$

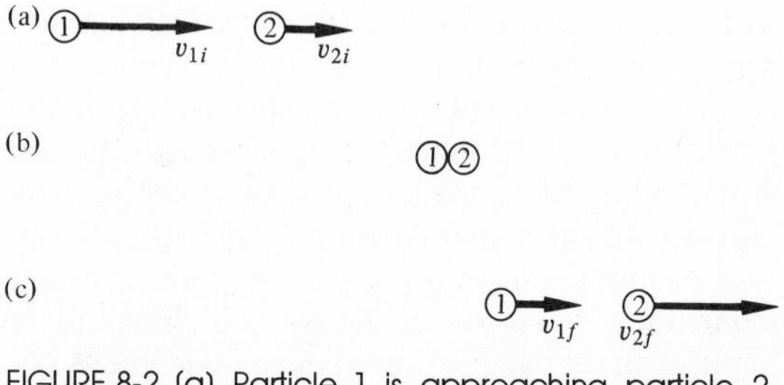

(a) ①⟶ v_{1i} ②⟶ v_{2i}

(b) ①②

(c) ①⟶ v_{1f} ②⟶ v_{2f}

FIGURE 8-2 (a) Particle 1 is approaching particle 2. (b) Particles 1 and 2 collide. (c) Particle 2 is receding from particle 1.

where the signs are chosen so that velocities to the right in Figure 8-2 are positive.

We can look at Equation 8-6 as an algebraic problem in which we assume that we know the masses and the velocities of the two objects before the collision, and are asked to find the velocities of the particles after the collision. From this point of view, there are two unknowns (v_{1f} and v_{2f}) but only one equation. Thus, in the absence of additional information we cannot solve the problem. Nevertheless, we can draw some conclusions.

For example, consider a rifle of mass M, which fires a bullet of mass m. Before the trigger is pulled, the bullet and rifle are at rest, so the initial momentum of the system is zero. After the firing, the bullet has momentum mv, so in order for the momentum of the system to be constant, the rifle must have a momentum MV in the opposite direction. Thus the rifle *recoils* with a velocity V given by

$$V = \frac{mv}{M} \qquad (8\text{-}7)$$

Similarly, we can conclude that in the nuclear process of radioactive decay, in which a nucleus emits one or more very fast particles, the residual nucleus must recoil. When the heart pumps a spurt of blood into the aorta, the entire body recoils slightly.

We can provide the additional information necessary by restricting the motion in some way. For example, we can require that the two objects stick together after the collision, in which case there is only one final velocity, and Equation 8-6 becomes

$$m_1 v_{1i} + m_2 v_{2i} = (m_1 + m_2) v_f \qquad (8\text{-}8)$$

Illustrative example

A 180-lb quarterback is standing still, ready to pass, when he is hit by an opposing tackle, who weighs 250 lb and is moving at 25 ft/sec. If the tackle holds on after he hits the quarterback, and friction with the ground can be neglected, how fast are the two of them moving just after the impact? Here

$$m_1 = \frac{W_1}{g} = \frac{180}{32}$$

$$m_2 = \frac{250}{32}$$

$$v_{1i} = 0$$

$$v_{2i} = 25 \, \frac{\text{ft}}{\text{sec}}$$

Therefore

$$\frac{180}{32}(0) + \frac{250}{32}(25) = \frac{180 + 250}{32}(v_f)$$

$$v_f = \frac{250}{430}(25) \, \frac{\text{ft}}{\text{sec}}$$

$$= 14.5 \, \frac{\text{ft}}{\text{sec}}$$

One very important kind of collision is that in which the total kinetic energy of the system is conserved. It must be emphasized that momentum is conserved if there are no external forces but that kinetic energy is conserved only for very special cases. In the collision of macroscopic objects some heat is usually generated, which means that the kinetic energy of the system is not conserved. The reason that kinetic energy conserving collisions are important is that in the collision of atoms or electrons or nuclear particles, it is often the case that kinetic energy is conserved. Furthermore, there are some macroscopic collisions, such as those among billiard balls, that almost conserve kinetic energy. A third reason for the interest in this kind of collision—which is less compelling, but nevertheless valid—is that this kind of collision is relatively easy to calculate.

Collisions in which the total kinetic energy is conserved are called *elastic collisions*. For an elastic collision between two particles whose masses are m_1 and m_2, conservation of kinetic energy yields the equation

$$\tfrac{1}{2} m_1 v_{1i}^2 + \tfrac{1}{2} m_2 v_{2i}^2 = \tfrac{1}{2} m_1 v_{1f}^2 + \tfrac{1}{2} m_2 v_{2f}^2 \qquad (8\text{-}9)$$

If the two particles are moving along the line joining their centers, the principle of conservation of momentum yields

$$m_1 v_{1i} + m_2 v_{2i} = m_1 v_{1f} + m_2 v_{2f} \qquad (8\text{-}10)$$

Equation 8-9 may be rewritten as

$$m_1 (v_{1i}^2 - v_{1f}^2) = m_2 (v_{2f}^2 - v_{2i}^2)$$

or, factoring,

$$m_1 (v_{1i} + v_{1f})(v_{1i} - v_{1f}) = m_2 (v_{2f} - v_{2i})(v_{2f} + v_{2i}) \qquad (8\text{-}11)$$

Equation 8-10 may be rewritten as

$$m_1 (v_{1i} - v_{1f}) = m_2 (v_{2f} - v_{2i}) \qquad (8\text{-}12)$$

Then, dividing Equation 8-11 by Equation 8-12, we get

$$v_{1i} + v_{1f} = v_{2f} + v_{2i} \qquad (8\text{-}13)$$

This equation can be written

$$v_{2i} - v_{1i} = -(v_{2f} - v_{1f}) \qquad (8\text{-}14)$$

The term in parentheses in Equation 8-14, $(v_2 - v_1)$, is the relative velocity of the objects, or the velocity with which object 1 approaches or recedes from object 2. The equation states that the relative velocity of the two objects in an elastic collision is the same in magnitude before and after the collision, but that the direction changes. That is, object 2 recedes from object 1 after the elastic collision with the same velocity that object 1 approached object 2 before the collision. Note that this result is independent of the masses and depends only on the conservation of momentum and the conservation of kinetic energy.

We could solve for the final velocities explicitly by simultaneous solution of Equations 8-10 and 8-14.

One interesting special case of elastic collisions is that in which both objects have the same mass. Then we have, from Equation 8-10,

$$v_{1i} + v_{2i} = v_{1f} + v_{2f} \qquad (8\text{-}15)$$

Adding this equation to Equation 8-14, we get

$$v_{2i} = v_{1f} \qquad (8\text{-}16)$$

and, subtracting Equation 8-14 from 8-15, we obtain

$$v_{1i} = v_{2f} \qquad (8\text{-}17)$$

These last two equations state that the two objects of equal mass interchange velocities in the elastic collision. Thus if an object with velocity v hits a stationary object of the same mass and the collision is elastic, then the initially stationary object will move off with a velocity v and the originally moving object will stop and remain at rest after the collision. Billiards enthusiasts will recognize this description as true, in the absence of spin.

Up to this point we have restricted the motion of the objects to a single straight line. In general, motion is three-dimensional, and the principle of conservation of momentum states that, in the absence of external forces, the total *vector momentum* of the system is constant. This means that *each component* of the momentum is conserved. One way to treat such a problem is to write Equa-

tion 8-6 once for each of the three rectangular components and proceed to solve them with such additional information as is available.

8-4
The rocket

Although real rockets are complex devices with complex motion, we can begin to analyze the motion by considering a simple idealization, represented by Figure 8-3. We assume that at some instant of time the rocket (including the unburned fuel) has a mass M and is moving straight up with a velocity V. During the next time interval Δt, some of the fuel of mass Δm is burned and leaves the rocket with a velocity v_r, *relative to the rocket*. During that time interval, the rest of the rocket acquires an additional upward velocity ΔV.

FIGURE 8-3 During an interval of time, Δt, exhaust, of mass Δm, leaves the rocket with relative velocity v_r. The rocket has a mass M and velocity **V**.

The major difficulty in the analysis of the rocket is in deciding which *system* to consider. If we consider only the rocket and treat the exhaust as external to the system, then we will be dealing with a system whose mass is not constant. None of the material we have presented up to this point is valid for such a system. Therefore we must deal with the system that consists of the rocket *and* the exhaust fuel. For that system, the initial momentum is equal to the mass of the rocket times the velocity of the rocket, MV. The final momentum consists of two parts, the momentum of the rocket and the momentum of the exhaust. The final momentum P_{fR} of the rocket is

$$P_{fR} = (M - \Delta m)(V + \Delta V) \qquad (8\text{-}18)$$

where we have taken into account the change in

mass of the rocket (due to the burning of the fuel) and the change in velocity. The final momentum of the exhaust P_{fe} is

$$P_{fe} = \Delta m(V - v_r) \qquad (8\text{-}19)$$

where $(V - v_r)$ is the velocity of the exhaust, since v_r is the velocity of the exhaust relative to the rocket. The total final momentum of the rocket is the sum of Equations 8-18 and 8-19. The change in momentum is this sum less the initial momentum, MV. This change in momentum is the external impulse on the rocket. If the rocket is close to the surface of the earth, the external impulse is $-Mg(\Delta t)$, where the minus sign indicates that the external force is down. Summarizing this in an equation:

$$-Mg(\Delta t) = (M - \Delta m)(V + \Delta V) + \Delta m(V - v_r)$$
$$- MV$$
$$= MV + M\Delta V - \Delta m V - \Delta m\ \Delta V$$
$$+ \Delta m\ V - \Delta m\ v_r - MV$$
$$= M\ \Delta V - \Delta m\ \Delta V - \Delta m\ v_r$$

If we assume that both Δm and ΔV are small, then the product $\Delta m\ \Delta V$ is much smaller than the other terms, so we can neglect that term, which leaves us with

$$-Mg\ \Delta t = M\ \Delta V - \Delta m\ v_r$$

Dividing by Δt, we get

$$M\ \frac{\Delta V}{\Delta t} = v_r\ \frac{\Delta m}{\Delta t} - Mg \qquad (8\text{-}20)$$

An exact solution of Equation 8-20 for the velocity of the rocket as a function of time is beyond the mathematical scope of this book, but the essential point of the equation is that the change in velocity of the rocket depends on the relative velocity of the exhaust and on the rate at which fuel is being burned as well as on gravity.

The left side of Equation 8-20 has the *form* of the mass of the rocket times the acceleration of the rocket. Therefore one is tempted to call the right side the external force on the rocket. If we suspend our judgment for the moment and call the right side of the equation the total force on the rocket, then the first term, $v_r(\Delta m/\Delta t)$, can be thought of as the force exerted by the exhaust on the rocket. This analysis is wrong because it assumes that Newton's law in the form $F = ma$ is correct for systems of variable mass. Newton's second law as we have discussed it is correct only for systems of fixed mass and must be modified in order to deal with varying-mass systems. Nevertheless, the description given above of Equation 8-20 is so compelling that the $v_r(\Delta m/\Delta t)$ term is usually called the *thrust of the rocket*. As long as it is remembered that this is merely a name given to a collection of symbols, this will lead to no error.

In this analysis of the rocket we have said that the change in momentum of the system is equal to the impulse of the gravitational force. In the absence of gravity (or other external forces) the increase in momentum of the rocket is equal and opposite to the change in momentum of the exhaust; that is, we know that the rocket must move up because we know that the exhaust moves down (relative to the rocket). On the other hand, if we fix our attention on the rocket itself and ask, "What pushes the rocket up?" we must look for a force that is upward and exerted by an agent external to the rocket. This upward force comes from the part of the exhaust gas that was initially moving upward. This gas strikes the upper part of the fuel chamber and pushes the rocket up.

Questions

1. A falling object strikes the ground without rebounding. What happens to the momentum of the object?

2. How does the principle of conservation of momentum apply to the motion of a simple pendulum?

3. In order for an automobile to move at constant speed, the engine must operate, and gasoline is consumed. That is, energy must be supplied to the automobile even though the momentum of the automobile is constant. Is there any conflict in this statement? Explain.

4. A man rows a boat to a dock and attempts to step out of the boat without tying the boat to the dock. Discuss the probable motion of the boat and the man.

5. A small boy runs toward his stationary red wagon and jumps into it. Discuss the motion.

6. The boy of Question 5 then jumps forward out of the wagon. Discuss the motion of the wagon and the boy.

7. An arctic explorer is trapped on an absolutely smooth, frozen lake. Ignoring the question of how he got there in the first place, describe how he can get off the lake.

8. Two explorers are on the smooth ice. One is much heavier than the other. They hold a rope between them. Discuss the motion that follows when one pulls on the rope.

9. Apply the principle of conservation of momentum to a ball bouncing off a wall (a) when the ball moves perpendicular to the wall and (b) when the ball strikes the wall at an angle less than 90°.

10. How does the principle of conservation of momentum apply to the motion of the earth around the sun?

11. Every time your heart beats, a small volume of blood is given an impulse. (a) In what system is momentum conserved? (b) Describe a procedure by which you could observe the recoil. (There are such devices, called *ballistocardiographs*.)

12. A common toy consists of a little car with a balloon attached to it, with the open end of the balloon toward the rear of the car. The balloon is blown up, the car is put down, and the air is allowed to escape. Describe the motion.

Problems

1. A 70-kg sprinter is running at the rate of 6 m/sec. Find his momentum.

2. Find the momentum of an electron (mass 9.1×10^{-28} gm), which is moving at 3×10^6 m/sec.

3. A 3600-lb automobile starting from rest acquires a speed of 60 mi/hr. Find the impulse that was exerted on the automobile.

4. What impulse is needed to have a 60-gm golf ball acquire a speed of 50 m/sec, starting from rest?

5. The acceleration of the automobile of Problem 3 is not constant but it takes 9 sec for the automobile to accelerate to 60 mi/hr. What average force was exerted on the car?

6. The golf ball of Problem 4 was struck by a club, which was in contact with the ball for 4×10^{-3} sec. What average force was exerted on the ball? What average force was exerted on the club?

7. A 2-gm bullet is fired with a muzzle velocity of 30,000 cm/sec from a 3-kg gun. What is the recoil velocity of the gun?

8. The rifle of Problem 7 is held tightly against the shoulder of a 65-kg person. Find the recoil velocity.

9. A luckless ice fisherman throws his 5-lb tackle box away with a speed of 12 ft/sec. The fisherman in his winter clothing weighs 220 lb. Assuming that the throw was horizontal and that the ice is smooth, find the recoil velocity of the man.

10. Two young ice skaters are standing in the middle of the rink. The first, whose mass is 50 kg, pushes the second and recoils with a speed of 0.8 m/sec. The second skater moves with a speed of 1 m/sec. What is the mass of the second skater?

11. A 5-ton railroad car is coasting along the track at 12 mi/hr. It runs into an identical car, which is standing still and is loaded with 20 tons of freight. The two cars couple together. Find the initial speed of this two-car train.

12. A 500-gm object falls from rest through a height of 2 m. With what momentum does it strike the ground?

13. The object of Problem 12 makes a perfectly elastic collision with the ground. (a) To what height does it rise? (b) What was the change in momentum of the object at the collision with the ground?

14. A spaceman in full equipment has a mass of 100 kg and is floating free in space. He throws a 2-kg wrench with a velocity of 2.5 m/sec to another spaceman, whose mass is 120 kg. The second spaceman catches and holds the wrench. Find the recoil velocity of each spaceman.

15. A 70-kg man jumps straight up and reaches an altitude of 3 ft. From the point of view of an observer in space, with what velocity did the earth recoil? The mass of the earth is approximately 6×10^{24} kg. (Note that Archimedes did not need a place to stand on if he wanted to move the earth; all he had to do was jump up.)

16. A 3000-lb car moving at 30 mi/hr has a head-on collision with an oncoming 20,000-lb truck moving with a speed of 25 mi/hr. (a) If the two

vehicles stick together after the collision, find the speed with which they are moving after the collision. Describe the direction of motion of the joined vehicles. **(b)** Find the total kinetic energy of the two vehicles before the collision. **(c)** Find the total kinetic energy after the collision.

17. Find the velocities of both the truck and the automobile of Problem 16 after a perfectly elastic collision.

18. A 1500-kg rocket is moving with a constant speed of 450 m/sec when it suddenly splits into two pieces. One piece, of mass 500 kg, continues moving in the original direction of the rocket with a speed of 300 m/sec. Find the direction of motion of the other piece and the magnitude of its speed.

19. A 16-lb bowling ball rolling down the alley at 5 ft/sec is struck by a 10-lb ball (thrown by an overeager child) moving in the same direction with a speed of 7 ft/sec. Assume that the child has left his bubble gum on the ball and that the two bowling balls stick together. **(a)** Find the speed of the joined bowling balls after the collision and **(b)** Find the change in kinetic energy of the system during the collision.

20. Repeat Problem 19 assuming that the collision is elastic.

21. A 2-lb catcher's mitt is hanging by a small twig from a tree 8 ft above the ground. The owner throws a $\frac{1}{2}$-lb ball at the mitt very skillfully. The ball strikes the mitt moving horizontally at a speed of 4 ft/sec, is imbedded in the pocket, and the two (ball and mitt) move together. Assume that the mass of the twig is negligible and that the force needed to break the mitt loose from the twig is also negligible. **(a)** Find the speed of the mitt and ball just after the collision. **(b)** How far from the tree will the mitt and ball hit the ground?

22. A 550-lb artillery shell has a velocity of 5000 ft/sec at the top of its path. At that point the shell breaks into two pieces of equal mass, one of which falls straight down with no initial vertical velocity. Find the velocity of the other piece.

23. A cue ball is moving with a speed of 5 ft/sec when it strikes the eight ball. The eight ball moves off at an angle of 30° with the original direction of the cue ball. The eight ball (which has the same mass as the cue ball) leaves the collision with a speed of 2 ft/sec. Find the speed

and direction of the cue ball after the collision.

24. A neutron that is initially at rest decays into three particles: an electron of momentum 7×10^{-16} gm cm/sec, a neutrino of momentum 5×10^{-16} gm cm/sec, and a proton. The neutrino is moving at right angles to the direction of motion of the electron. Find **(a)** the momentum of the proton and **(b)** the direction of motion of the proton.

25. A juggler throws a ball straight up by giving the ball a certain impulse. The ball leaves his hand, rises, and falls straight down. The juggler catches the ball and gives it an impulse that brings it to rest in his hand. Assume that the ball has a weight W and that the motion of the juggler's hand can be neglected, as can air resistance. **(a)** Show that the throwing and catching impulses are equal. **(b)** Show that the total impulse exerted on the juggler's hand is equal to the weight of the ball times the time during which the ball is in the air. Therefore, if the juggler is keeping many objects in the air simultaneously, the *average* force exerted on his hand is equal to the weight of all of the objects, including the ones in the air.

26. Solve Equations 8-10 and 8-14 to find the final velocities in an elastic collision for the case in which one of the objects is initially at rest.

27. From the results of Problem 26, **(a)** show that when a very heavy object hits a very light object the heavy object is unaffected by the collision while the light object recoils with twice the initial velocity of the heavy object. **(b)** Show that when a very light object strikes a heavy object the heavy object is unaffected by the collision but that the light object rebounds with the same speed it had originally.

28. A 10,000-kg rocket is standing vertically on the launch pad. **(a)** If the velocity of the exhaust is initially 1000 m/sec, how much gas must be ejected per second in order to have the thrust equal to the weight of the rocket? **(b)** How much gas must be ejected in order for the initial acceleration of the rocket to be 12 m/sec²?

29. A jet airplane is flying at 300 ft/sec. The engine takes in 35 lb of air per second. The engine burns $\frac{1}{2}$ lb of fuel per second, mixes the burned fuel with the air and ejects the heated exhaust with a speed of 1600 ft/sec relative to the plane. Find the thrust of the engine.

rotational motion

chapter 9

9-1
Motion of a rigid body

Our previous discussion of motion was confined almost entirely to the motion of particles. In most cases we extended or illustrated the discussion by referring to the motion of rigid bodies, but we always restricted ourselves to the situation in which every point in the body moved parallel to every other point in the body. This kind of motion, in which the velocity vectors of all the points in the body are parallel to each other, is called motion of *translation*. Imagine drawing two perpendicular lines in a rigid body; if the body moves in such a way that both lines always remain parallel to their original directions, then the body has motion of translation. If either or both lines do not remain parallel to their original directions, then the body has motion of *rotation*. A point, because it has no size, can always be thought of as translating without rotation. To illustrate these terms, consider a box sliding down an inclined plane onto a level floor. While the box is sliding down, it has motion of translation. While it is moving from the incline onto the floor, it has motion of rotation. After it is completely on the level floor, it continues with motion of translation. A ball rolling down the same plane rotates during the entire motion, and the point at its center translates down the plane and across the floor.

It is possible to discuss the motion of rigid bodies by considering the motion of each point in the body separately, but this is usually a very difficult procedure. Instead, we will take advantage of the simplifications implicit in the idea of *rigid* bodies. A rigid body is one in which each point has a *fixed* position *relative to all other points* in the body.

Thus, if we know the motion of a few points in the body we can describe the motion of all the points in the body. In addition, we can use Newton's third law to simplify the discussion of the forces acting on the particles in the body. We can divide the forces acting on any particle in the body into two classes: forces exerted by other particles in the body (internal forces) and forces exerted by particles that are not part of the body (external forces). Newton's third law tells us that the internal forces come in equal and opposite pairs, so when we form the resultant of all the forces acting on the body, the sum of the internal forces will be zero. In addition, it can be shown that the internal forces will not make the body rotate.

Even the most complicated motion of a rigid body can be treated as a combination of translation and rotation. At any instant of time the body is rotating about some line, or *axis*, while the axis continues translating. For example, the motion of a wheel of a moving automobile is a combination of the rotation of the wheel about the axle and the translation of the axle. In complex motions, such as the wobbling motion of a book thrown into the air, the identity of the points that form the axis may change from instant to instant, but it is still possible to separate the motion into one of translation and one of rotation *at any instant*. In our subsequent discussions we will restrict ourselves to the simpler kinds of motion of rigid bodies, either where the axis is fixed or where the axis moves linearly, remaining parallel to itself.

Figure 9-1a shows a rigid body rotating about a fixed vertical axis. A typical point in the body moves in a circle, whose center is a point on the axis. The motion of the body may be described by

considering Figure 9-1b, which shows a typical point P moving on a circle whose center is point O. In Chapter 7 we described this circular motion in terms of the velocity of point P. Now we are interested in the motion of point P as a *typical* point, so it will be more revealing to describe the motion in terms of the angle through which the line OP moves.

The angular motion is common to all the points in the body, whereas the velocity of a point depends on its distance from the axis. In the following sections we will first discuss the motion of a point moving in a circle in terms of angular motion, and then discuss the implications for the rigid-body motion.

9-2
Angular displacement, angular velocity, and angular acceleration

Consider a particle P, which is moving on a circle of radius R, as shown in Figure 9-2. If the particle moves from point P to point P' in a time t, the line OP rotates through an angle θ during that time. This angle θ is the *angular displacement* of the point P. There are many units for measuring angle, among which are degrees, revolutions, and radians. Radian measure of angle is convenient because if the angle θ is measured in radians, the arc length s of Figure 9-2 is given by

$$s = R\theta \tag{9-1}$$

If the particle P makes one complete revolution, the arc length is $2\pi R$; hence the number of radians in one revolution is 2π. Since the number of degrees in one revolution is 360, it follows that

$$360° = 2\pi \text{ radians}$$

or

$$1 \text{ radian} = \frac{360}{2\pi} \text{ degrees}$$

$$\approx 57.3 \text{ degrees}$$

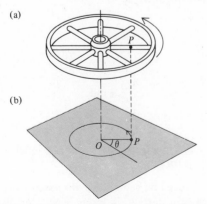

FIGURE 9-1 Rotation of a rigid body. The motion can be described in terms of the circular motion of a typical point, P.

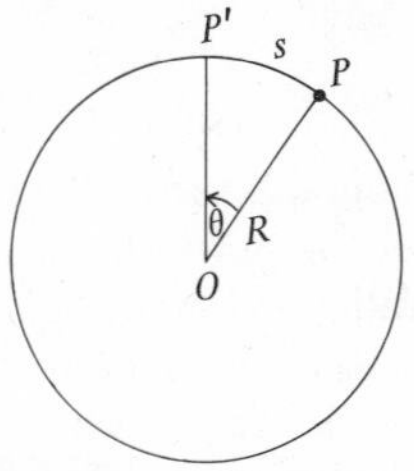

FIGURE 9-2 The distance s through which P moves is related to the angular displacement θ by $s = R\theta$, if θ is measured in radians.

The average angular speed $\bar{\omega}$ of point P is defined as the ratio of the angular displacement θ to the time interval t; that is,

$$\bar{\omega} = \frac{\theta}{t} \tag{9-2a}$$

The instantaneous angular speed ω is defined as the ratio of the small angular displacement $\Delta\theta$ to the small time interval Δt

$$\omega = \frac{\Delta\theta}{\Delta t} \tag{9-2b}$$

as Δt becomes vanishingly small.

If the initial angular speed of P is ω_i and at the end of the time interval t the particle has angular speed ω_f, then the average angular acceleration is defined by

$$\bar{\alpha} = \frac{\omega_f - \omega_i}{t} \tag{9-3}$$

The instantaneous angular acceleration is then defined as the average angular acceleration taken over vanishingly small time intervals:

$$\alpha = \frac{\omega_f - \omega_i}{\Delta t} \tag{9-4}$$

and Δt becomes vanishingly small.

These definitions are identical in form to the definitions of linear displacement, velocity, and acceleration as given in Chapter 2. Therefore the results of that chapter are applicable, with appropriate changes in symbols. In particular, Equations 2-26, 2-27, 2-28, and 2-29 can be rewritten to:

$$\omega_f = \omega_i + \alpha t \tag{9-5}$$
$$\theta = \left(\frac{\omega_f + \omega_i}{2}\right)t \tag{9-6}$$
$$\theta = \omega_i t + \tfrac{1}{2}\alpha t^2 \tag{9-7}$$
$$\omega_f^2 = \omega_i^2 + 2\alpha\theta \tag{9-8}$$

In all of the above equations the angular acceleration α is a constant.

Illustrative example

A 45-rpm phonograph turntable slows down and stops 5 sec after the power is turned off. Find the angular acceleration (assumed constant), and the number of revolutions through which the turntable moves while it is slowing down.

The initial angular speed of the turntable is 45

revolutions per minute, or

$$\omega_i = 45\ \frac{\text{rev}}{\text{min}}$$
$$= \frac{45}{60}\frac{\text{rev}}{\text{sec}}$$
$$= \frac{3}{4}\frac{\text{rev}}{\text{sec}}$$

The final angular velocity is zero, and the time is 5 sec, so from Equation 9-5 we have

$$0 = \frac{3}{4}\frac{\text{rev}}{\text{sec}} + \alpha(5\ \text{sec})$$
$$\alpha = \frac{-3}{20}\frac{\text{rev}}{\text{sec}^2}$$

The minus sign here means that the angular acceleration is opposite to the initial angular velocity; the turntable is slowing down.

From Equation 9-6 we have

$$\theta = \frac{0 + \dfrac{3}{4}\dfrac{\text{rev}}{\text{sec}}(5\ \text{sec})}{2}$$
$$= \frac{15}{8}\ \text{rev}$$
$$= 1\frac{7}{8}\ \text{rev}$$

In solving this problem we have used revolutions per second as the unit of angular speed, revolutions per second per second as the unit of angular acceleration, and we have used revolutions as the unit of angle. This is acceptable since we have used a consistent set of units. Alternatively we could have measured the angle in radians, in which case the angle through which the turntable moves while it is slowing down is

$$\theta = \frac{15}{8}\ \text{rev}$$
$$= \frac{15}{8}\ \text{rev} \times \frac{2\pi\ \text{rad}}{\text{rev}}$$
$$= \frac{15}{4}\pi\ \text{rad}$$

9-3
The relationship between linear and angular motion

In the preceding section we discussed the angular motion of a particle in terms of angular displacement, angular velocity, and angular acceleration. In a previous chapter we discussed circular motion in terms of the linear velocity and linear acceleration of the point. These two descriptions are connected by Equation 9-1. Referring to Figure 9-2, we can see that for very small time intervals, the linear displacement of an object moving on a circle is s, the length of the arc. Then, since the speed is the magnitude of the displacement divided by the time interval, we have

$$v = \frac{s}{t}$$

and, from Equation 9-1,

$$s = R\theta$$

where s is the linear displacement and θ is the angular displacement. Thus

$$v = R\,\frac{\theta}{t}$$

and, from Equation 9-2a,

$$\omega = \frac{\theta}{t}$$

so we can conclude that

$$v = R\omega \qquad (9\text{-}9)$$

or the linear speed of a particle moving in a circle of radius R is equal to the product of R and the angular speed. It should be noted that this result relates the instantaneous speed to the instantaneous angular speed when the particle is moving with varying speed.

In Chapter 7 we saw that a particle moving in a circle with constant speed has an acceleration that is directed inward along the radius and has a magnitude a given by

$$a = \frac{v^2}{R}$$

If we now refer to this acceleration as a_r (the r standing for radial) and insert Equation 9-9, we get

$$a_r = \frac{v^2}{R}$$
$$= \frac{(R\omega)^2}{R}$$
$$a_r = R\omega^2 \qquad (9\text{-}10)$$

As we saw in Chapter 7, this acceleration arises

from the change in *direction* of the velocity. Thus if a particle moves on a circle, it will have an instantaneous acceleration given by Equation 9-10, even if ω (and v) is not constant. If ω (and v) varies, then the particle will have *in addition to the radial acceleration* an acceleration that arises from the change in magnitude of v and will be directed along the tangent to the path. From Equation 9-4, we have

$$\alpha = \frac{\omega_f - \omega_i}{t}$$

Using Equation 9-9, we obtain

$$\alpha = \frac{\dfrac{v_f}{R} - \dfrac{v_i}{R}}{t}$$

This can be rearranged to give

$$\alpha = \frac{1}{R}\left(\frac{v_f - v_i}{t}\right) \qquad (9\text{-}11)$$

The quantity in parentheses in Equation 9-11 is the magnitude of the acceleration that arises from the change in speed. This is the component of the acceleration that is directed along the tangent to the circle, a_t. Thus we can rewrite Equation 9-11 to give

$$\alpha = \frac{1}{R}\,a_t$$

or

$$a_t = R\alpha \qquad (9\text{-}12)$$

In summary, if a particle moves on a circle of radius R with an instantaneous angular velocity ω and an instantaneous angular acceleration α, then it will have an instantaneous linear velocity $\mathbf{v}$, directed tangent to the circle where

$$v = R\omega$$

The particle will also have an acceleration, one component of which will be directed radially in toward the center with a magnitude

$$a_r = R\omega^2$$

The other component of the acceleration is directed along the tangent to the circle and has a magnitude given by

$$a_t = R\alpha$$

9-4
Vector angular velocity and angular acceleration

There are two possible directions for the rotation of a body about a fixed axis. If we view the rotation along the axis, the direction of rotation may be described as clockwise or counterclockwise. The wheel of Figure 9-1 is moving counterclockwise as viewed from above. For most of our purposes, this description of the motion will be adequate. One may call one direction (say counterclockwise) positive, and the other direction (clockwise) negative.

The angular velocity ω also may be represented by a vector. The usual method for doing this is to draw a vector along the axis, pointing toward the reader if the motion is counterclockwise, as illustrated in Figure 9-3, and away from the reader if the motion is clockwise. A convenient way of remembering this method of representation is by means of the *right-hand* rule: Imagine that the right hand is held so that the fingers follow the direction of rotation; the thumb will then point in the direction in which the arrow should be drawn along the axis. Figure 9-4 illustrates this. The magnitude of the angular velocity is represented by the length of the arrow drawn to some arbitrary scale.

For an object moving about a fixed axis, angular acceleration is similarly represented as a vector directed along the axis.

9-5
Torque and angular acceleration

Figure 9-5a shows a particle of mass m, which is attached to the end of a massless horizontal rod. The rod is free to pivot about the point O, so when a force is applied to the particle, the particle will move in a circle of radius R, where R is the length of the rod. In Figure 9-5b the force $\mathbf{F}$ is resolved into components F_t, tangential to the circle, and F_r, radial to the circle. The rod will exert a radial pull $\mathbf{P}$ on the particle. We can then write Newton's second law for the radial and tangential components as

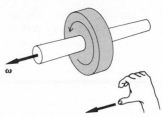

FIGURE 9-4 The right-hand rule for the direction of the angular velocity.

$$F_t = ma_t \qquad (9\text{-}13)$$

$$P - F_r = ma_r \qquad (9\text{-}14)$$

From Equation 9-12 we know that

$$a_t = R\alpha$$

which, combined with Equation 9-13, gives

$$F_t = mR\alpha \qquad (9\text{-}15)$$

Multiplying both sides of Equation 9-15 by R, the distance to the axis, we have

$$F_t R = mR^2\alpha \qquad (9\text{-}16)$$

The left side of Equation 9-16 is the magnitude of the torque τ exerted by the force $\mathbf{F}$.

Using the convention of Chapter 4 for the direction of the vector torque $\boldsymbol{\tau}$, we can write Equation 9-16 in the form

$$\boldsymbol{\tau} = mR^2\boldsymbol{\alpha}$$

and then define a new quantity by

$$I = mR^2 \qquad (9\text{-}17)$$

This gives Equation 9-16 the form

$$\boldsymbol{\tau} = I\boldsymbol{\alpha} \qquad (9\text{-}18)$$

The scalar quantity I, defined by Equation 9-17, is called the *moment of inertia* of the particle with respect to the axis through O. We can think of the moment of inertia of a body as the analog, in rotational motion, of the mass (or inertia) of a particle in linear motion. In this interpretation Equation 9-18, which describes rotation, is analogous to Newton's second law,

$$\mathbf{F} = m\mathbf{a}$$

The torque is the analog of force, the moment of inertia is the analog of mass, and the angular acceleration is the analog of linear acceleration.

FIGURE 9-3 The angular velocity vector for a rotating cylinder.

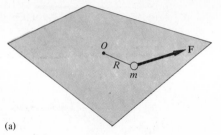
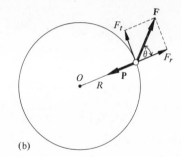

FIGURE 9-5 (a) A force **F** applied to an object of mass m, which is attached to the end of a rod of length R. The rod is free to pivot about O. (b) The force **F** is resolved into radial and tangential components F_r and F_t.   (a)

(b)

9-6
Kinetic energy of rotation

We can pursue the analogies introduced in the preceding section by considering the kinetic energy of an object that is moving on a circle. We know that

$$\mathscr{E}_k = \tfrac{1}{2}mv^2 \qquad (9\text{-}19)$$

and if we introduce

$$v = R\omega$$

we can write

$$\mathscr{E}_k = \tfrac{1}{2}m(R\omega)^2$$
$$= \tfrac{1}{2}mR^2\omega^2$$

Then introducing the definition of moment of inertia

$$I = mR^2$$

we get

$$\mathscr{E}_k = \tfrac{1}{2}I\omega^2 \qquad (9\text{-}20)$$

which is identical in form to Equation 9-19.

9-7
Angular momentum

In previous chapters we introduced the ideas of energy and momentum and with these ideas were able to describe the behavior of particles. We can see that there are some cases in which we will need at least one more idea by considering the situation represented in Figure 9-6. Here a stick is lying on a smooth horizontal table and two particles of equal mass are moving toward the stick with equal (and opposite) velocities. There are two possible cases, shown as Figure 9-6a and b: the velocities are parallel but displaced, or the velocities are along the same line. If we calculate the momentum and energy of the system before the particles hit the stick, we get identical results in the two cases. That

is, if all we knew were the energy and momentum, we would say that these two cases are indistinguishable. But we know the two cases are not the same. After the collision, the stick in Figure 9-6a will be spinning, whereas the stick in Figure 9-6b will remain at rest. In order to distinguish between these two situations, we will ascribe a new property to the moving particles, a property called the *angular momentum* of the particles.

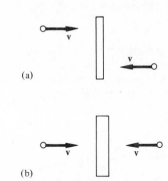

(a)

(b)

FIGURE 9-6 A rod being struck by two particles in two different ways.

The angular momentum of a particle is always taken about some axis. In Figure 9-7 we show such an axis and a particle of mass m which is at point P and has a velocity **v**. Then we draw the plane that passes through point P and is perpendicular to the axis. This plane intersects the axis at point O. We can then find v_p, the component of **v** that lies in the plane. This component is along the line AB. The distance from O to the line AB is d. The magnitude of the angular momentum of the particle, L, is defined by

$$L = mv_p d \qquad (9\text{-}21)$$

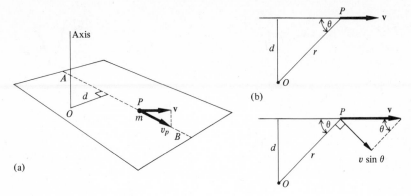

FIGURE 9-7 (a) The angular momentum of the particle about the axis is equal to $mv_p d$. (b) If the velocity is entirely in the plane perpendicular to the axis, the angular momentum is equal to $mrv \sin \theta$. (c) The component of v which is perpendicular to r is equal to $v \sin \theta$.

Angular momentum is a vector quantity, directed along the axis. The direction along the axis can be determined by a right-hand rule similar to that given above, where the fingers point along the velocity v_p, and the thumb points along the axis in the direction of the angular momentum.

Generally, the discussion of angular momentum in this book will be restricted to cases where the velocity is confined to the plane perpendicular to the axis. For those cases v is the same as v_p, and Equation 9-21 becomes

$$L = mvd \qquad (9\text{-}22)$$

or the magnitude of the angular momentum is equal to the product of the linear momentum and the distance from the axis to the line of action of the momentum. If r is the distance from O to P, then from Figure 9-7b, we have

$$d = r \sin \theta$$

and therefore

$$L = mvr \sin \theta \qquad (9\text{-}23)$$
$$= mrv \sin \theta$$

From Figure 9-7c we see that $v \sin \theta$ is the component of v that is perpendicular to r. Calling this component v_t, we find that

$$L = mv_t r \qquad (9\text{-}24)$$

or the magnitude of the angular momentum is equal to the distance from the axis to the particle, multiplied by the component of the linear momentum that is perpendicular to that distance.

The definition of angular momentum given in Equation 9-24 can be written as

$$L = mrv_t$$
$$= mr(r\omega)$$

where ω is the angular velocity of the particle about the point O. Thus

$$L = mr^2 \omega$$

and therefore

$$\mathbf{L} = I\boldsymbol{\omega} \qquad (9\text{-}25)$$

This equation is of the same form as the definition of linear momentum:

$$\mathbf{p} = m\mathbf{v}$$

With this definition of angular momentum we can return to the situation of Figure 9-6. In Figure 9-6b the particles are moving along the same line, the distance d is the same for both particles, and, since the velocities are in opposite directions, the angular momentum of one will be opposite to the angular momentum of the other about any axis perpendicular to the page. Thus, the total angular momentum of the two particles will be zero. On the other hand, the angular momenta of the two particles in Figure 9-6a will not be equal, and so the total angular momentum will not be equal to zero. The two cases are different because the total angular momentum of the system (before the collision with the stick) is different in the two cases.

9-8
Torque and angular momentum of a particle

Suppose we have a particle moving with some velocity and then a force is applied to the particle. As we have seen in Chapter 8, the force will produce a change in the momentum of the particle that will be equal to the product of the force and the time. In general, the angular momentum of the particle will also change, but as we will see, the change in angular momentum is proportional to the torque exerted by the force.

Consider the particle of Figure 9-7b, which is moving with velocity **v** and is a distance r away from the axis. From Equation 9-24, we know that the angular momentum of that particle is proportional to the component of velocity that is *perpendicular* to r. If an impulse were applied to the particle, directed along the line r (toward or away from the axis), then the only change would be in the component of the velocity *parallel* to r. Thus an impulse directed along r would not change the angular momentum of the particle. If an impulse were directed in an arbitrary direction, a *change in angular momentum* would come from the *component of the impulse directed along the perpendicular* to r. Figure 9-8 shows the particle of Figure 9-7 being acted on by a force **F**. The component of **F** perpendicular to r is $F \sin \phi$, and from Chapter 8 we have

$$(F \sin \phi)t = (mv \sin \theta)_f - (mv \sin \theta)_i$$

Then, multiplying both sides by r, we have

$$r(F \sin \phi)t = (mvr \sin \theta)_f - (mvr \sin \theta)_i$$

From the definition of torque and angular momentum this can be written

$$\tau t = \mathbf{L}_f - \mathbf{L}_i \qquad (9\text{-}26)$$

In other words, this equation states that the change in angular momentum of the particle is equal to the product of the torque acting on the particle and the time. The analogy to Equation 8-2

$$\mathbf{F}t = \mathbf{p}_f - \mathbf{p}_i \qquad (8\text{-}2)$$

is clear.

Because of the analogy between Equation 9-26 and Equation 8-2, we can immediately see that

some of the conclusions of Chapter 8 may be restated in terms of angular momentum. In particular, if we have a *system of particles* on which the external torque is equal to zero, and for which the internal torques come in equal and opposite pairs, then the total angular momentum of the system is constant. The requirement that the internal torques come in equal and opposite pairs is satisfied by Newton's third law, since the action-reaction forces are always directed along the same line. Further, it is possible to show (by the infamous "methods beyond the scope of this book") that, even when Newton's third law is not applicable, the total internal torque of a system of particles must be equal to zero. Thus, *in the absence of external torque, the angular momentum of a system of particles is constant.* Note that the system of particles need *not* be a rigid body.

If we write Equation 9-26 once for each particle in a system, and then add the equations, the sum of the internal torques will cancel and we will be left with

$$t\tau_{\text{ext}} = \Sigma(\mathbf{L}_f - \mathbf{L}_i) \qquad (9\text{-}27)$$

where τ_{ext} is the total external torque on the system and the summation sign means we are to add the change in angular momentum of all the particles in the system. This can be rewritten as

$$t\tau_{\text{ext}} = (\mathbf{L}_{\text{total}})_f - (\mathbf{L}_{\text{total}})_i \qquad (9\text{-}28)$$

or the *change in the total angular momentum of a system of particles is equal to the product of the external torque and the time during which the torque acts.*

Illustrative example

An Atwood machine, as we have seen earlier, consists of two objects of different masses attached to a cord that runs over a pulley. If we assume the pulley to have zero mass and neglect friction, we can apply the methods of this section in the following way:

If the particles start from rest, the initial angular momentum of the system is equal to zero. In the absence of friction, the external torque acting on the system about the axis of the pulley (see Figure 9-9) is given by

$$(m_2 g - m_1 g)R = \tau_{\text{ext}}$$

The angular momentum of the system is

$$(m_2 v_2 + m_1 v_1)R = L$$

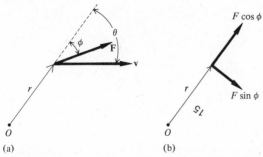

FIGURE 9-8 (a) A force **F** applied to a particle. (b) The force **F** is resolved into radial and tangential components.

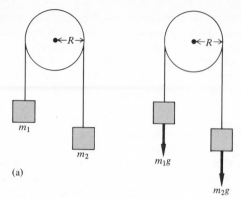

(a)

FIGURE 9-9 The Atwood machine.

and since the particles are connected by a string, which cannot stretch, they must have equal speeds. Thus, from Equation 9-27,

$$(m_2 - m_1)gRt = (m_2 + m_1)Rv$$

and so

$$v = \frac{m_2 - m_1}{m_2 + m_1} gt$$

Since this result says that the speed of the particle is proportional to the time, the objects are moving with constant acceleration; the acceleration a is given by

$$a = \frac{m_2 - m_1}{m_2 + m_1} g$$

9-9
A rigid body as a collection of particles

As we saw in Figure 9-1, if we restrict ourselves to a rigid body rotating about a fixed axis, then we can treat the rigid body as a collection of particles each of which has the same angular velocity and angular acceleration. If an external torque is applied to this body, then in general the torque will not be directed along the axis. The component of the external torque parallel to the axis will change the rate of rotation of the body, whereas the component perpendicular to the axis will act to change the direction of the axis. Thus if we assume that the axis is fixed, we are assuming that the external torques are along the axis. To illustrate this last remark through an example of a body rotating about a fixed axis, consider a pulley that rotates about a shaft, the shaft in turn being held in bearings. If an external torque is applied about an axis perpendicular to the shaft, then, as can be seen in

Figure 9-10, the shaft will attempt to twist the bearings. If the bearings are well designed, they will apply to the shaft an equal and opposite torque that is sufficient to keep the shaft in a fixed position. Frictionless bearings apply no torque along the shaft, but can apply large torques perpendicular to the shaft. As far as the pulley is concerned, these bearing torques are external torques, so the external torque on the system is zero, except when the torque is directed along the shaft. Thus, for such a system we can assume that the only relevant torques are directed along the axis of rotation.

If we now assume that we have a rigid body rotating about a fixed axis, and write Equation 9-18 once for each particle in the rigid body, and then add these equations, we will have

$$\tau_{\text{ext}} = \alpha(\Sigma I_i)$$

where the sum of the moments of inertia of each of the particles (about a fixed axis) can be written as I, *the moment of inertia of the body*,

$$I = \Sigma I_i \tag{9-29}$$

Thus for a rigid body rotating about a fixed axis we have

$$\tau_{\text{ext}} = I\alpha \tag{9-30}$$

or, the *total external torque about the fixed axis is equal to the product of the moment of inertia of the body about that axis times the angular acceleration of the body about that axis.*

Similarly, if we write Equation 9-20 once for each particle in the body, and then add all the

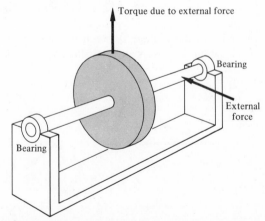

FIGURE 9-10 An external force exerts a torque that is perpendicular to the shaft. Since the shaft remains in a fixed position, the bearings must exert an equal and opposite torque.

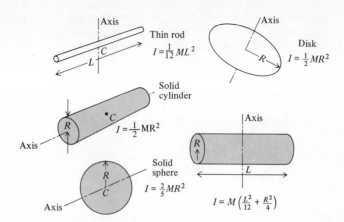

FIGURE 9-11 Moments of inertia of some rigid bodies having simple geometric shapes.

equations, we will have

$$(\mathcal{E}_k)_{\text{total}} = \tfrac{1}{2}I\omega^2 \qquad (9\text{-}31)$$

for the kinetic energy of rotation of the rigid body about the axis of rotation.

It should be noted that Equation 9-30 does not apply in general to motion where the axis is free to move. For those cases we must rely on Equation 9-28. However, Equation 9-30 does apply to cases where the axis moves parallel to itself, as in the case of a wheel rolling on a flat surface. For those cases, the fact that the axis moves parallel to itself implies that the external torque has no components perpendicular to the axis.

The moment of inertia of a body depends on the distribution of matter in the body and on the location of the axis. Therefore, it is usually very difficult to calculate the moment of inertia of a specific body, although it is possible to measure it. For uniform bodies of simple shape it is possible to calculate the moment of inertia about conveniently located axes. Figure 9-11 gives formulas for some bodies. It must be emphasized that the moments of inertia given in that figure do not necessarily apply to other axes, and that they apply only to uniform homogeneous bodies.

Illustrative example

A box of mass m is attached to a lightweight cord, which is wrapped around a pulley of mass M, radius R, and moment of inertia I. The pulley is mounted with frictionless bearings on a fixed axle. Find the acceleration of the box.

The system, which is shown in Figure 9-12, can be analyzed by means of two free-body diagrams. The first free-body diagram, for the box, is shown in Figure 9-12b. Writing Newton's second law for this body, we have

$$W - T = ma$$

or

$$mg - T = ma \qquad (9\text{-}32)$$

where T is the tension in the cord.

For the pulley we draw Figure 9-12c, where B is the upward force exerted by the shaft on the pulley, and T is the force exerted by the cord on the pulley. Writing Equation 9-30 for this body we have

$$\tau = I\alpha$$
$$TR = I\alpha \qquad (9\text{-}33)$$

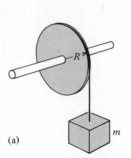

(a)

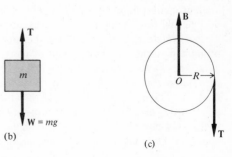

(b)

(c)

FIGURE 9-12

Since the cord does not stretch, we know that the acceleration of the box is equal to the tangential acceleration of a point on the rim of the pulley; in terms of the symbols in Equations 9-33 and 9-22 we have

$$a = R\alpha \qquad (9\text{-}34)$$

Combining Equations 9-34 and 9-33, we get

$$TR = \frac{Ia}{R}$$

$$T = \frac{Ia}{R^2}$$

and then from Equation 9-32,

$$mg - \frac{Ia}{R^2} = ma$$

or

$$a = \frac{mg}{m + (I/R^2)}$$

Illustrative example

In a common lecture demonstration a student volunteer stands on a platform that is free to rotate about a vertical axis and holds a dumbbell in each outstretched hand. The instructor rotates the platform and student and then asks the student to bring his arms close to his body. The subsequent large increase in rate of rotation can be analyzed by the following: Before the student brings his arms in, he is spinning with an angular velocity ω. Consequently the system has an angular momentum L given by

$$L = (I + 2mR^2)\omega$$

where I is the moment of inertia of the system which consists of the student and the platform, m is the mass of each dumbbell, and R is the distance between the axis of rotation and the dumbbell.

While the student is bringing his arms in he is applying a force that is internal to the system of student, platform, and dumbbells. This system has no external torque and so the angular momentum of the system is constant. After the student has brought his arms in, the moment of inertia of the dumbbells has decreased because R has decreased to r (approximately the radius of the chest of the student). The moment of inertia of the student has decreased somewhat because the arms are in closer to the axis, but this change in I is small and we can ignore it. Thus the angular momentum of the sys-

tem becomes

$$L' = (I + 2mr^2)\omega'$$

where ω' is the new angular velocity of the system. Since the angular momentum of the system is constant, we have

$$(I + 2mR^2)\omega = (I + 2mr^2)\omega'$$

or

$$\omega' = \omega \left(\frac{I + 2mR^2}{I + 2mr^2}\right)$$

It is interesting to find the kinetic energy of the system after the student has brought his arms in:

$$\mathcal{E}_k = \tfrac{1}{2}(I + 2mr^2)(\omega')^2$$
$$= \tfrac{1}{2}(I + 2mr^2)\omega^2 \frac{(I + 2mR^2)^2}{(I + 2mr^2)^2}$$
$$= \tfrac{1}{2}(I + 2mR^2)\omega^2 \left(\frac{I + 2mR^2}{I + 2mr^2}\right)$$

Before the student brought his arms in, the kinetic energy of the system was

$$\mathcal{E}_k = \tfrac{1}{2}(I + 2mR^2)\omega^2$$

Thus the kinetic energy is larger after the student brings his arms in than before. Since we have ignored external forces we must explain this increase in energy by assuming that it comes from *internal* forces. The student's arms do work on the dumbbells, work which comes from the internal body chemistry.

9-10
Precession

We have stressed the fact that the angular momentum of a rigid body about an axis remains constant unless acted upon by some external torque. In the previous sections we considered the change in the angular momentum about the *same axis* produced by the action of an external torque. The more general case, in which the torque produces an acceleration about one axis while the body is rotating about a different axis, is much beyond the scope of this book. Because of the importance of this type of motion, however, one simple case will be discussed in a qualitative manner. This is the case in which a body is spinning with a *large angular momentum* about one axis, which we shall call the *axis of spin*, and a steady torque is applied to the body to produce a *small* acceleration about an axis perpendicular to the axis of spin.

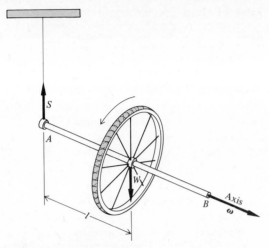

FIGURE 9-13 A gyroscope.

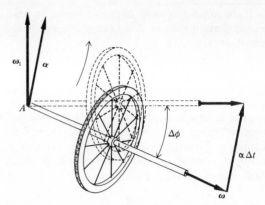

FIGURE 9-14 Precessional motion of a gyroscope with angular velocity ω about a vertical axis through A.

A good demonstration of the motion described above can be presented by mounting a bicycle wheel on an axle that protrudes beyond the hub of the wheel and then supporting one end of this axle in a loop in a string that is suspended from the ceiling, as shown in Figure 9-13. Suppose that the bicycle wheel is given a spin about its axis in the direction shown by the vector ω. (If, as we face the wheel, the arrow points toward us, the motion of the wheel about this axis is counterclockwise.) After the wheel has been given a spin, there are only two forces acting on it: its weight $\mathbf{W}$ acting through the center of gravity, and the pull $\mathbf{S}$ of the string at A. The torque produced by these two forces about a horizontal axis through A perpendicular to AB is given by Wl, where l is the distance from the center of gravity to A. This torque is clockwise, as viewed by the reader, and will produce an acceleration α in a clockwise direction about this horizontal axis.

Figure 9-14 shows the wheel spinning about a horizontal axis; its angular velocity ω is shown as a vector along this axis. The angular acceleration α is drawn as a vector at right angles to ω and also in the horizontal plane. As a result of this angular acceleration, the axis of spin will move through an angle $\Delta\phi$ in the horizontal plane with an angular velocity $\omega_1 = \alpha\,\Delta t$, where Δt is a small time interval. In this new position of the wheel, the torque produces an angular acceleration α equal in magnitude to that in the first position but now directed at right angles to the new axis of spin. This axis will turn through an angle $\Delta\phi$ in the horizontal plane in an equal time interval Δt. Thus the axis

of spin will rotate in the horizontal plane with an angular velocity ω_1, which, when viewed from above, will be in a counterclockwise direction about the string through A as an axis. This rotation of the axis is called a motion of *precession*. The rotating wheel is sometimes called a *gyroscope*, and the motion analyzed above is called *gyroscopic motion*.

The magnitude of the angular velocity of precession ω_1 is simply

$$\omega_1 = \frac{\Delta\phi}{\Delta t} \qquad (9\text{-}35)$$

Referring to Figure 9-14 we see that

$$\Delta\phi = \frac{\alpha\,\Delta t}{\omega} \qquad (9\text{-}36)$$

and thus

$$\omega_1 = \frac{\alpha}{\omega} \qquad (9\text{-}37)$$

The three vector quantities α, ω, and ω_1 are mutually perpendicular; their directions are shown in Figure 9-15.

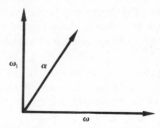

FIGURE 9-15 The three vectors α, ω, and ω_1 are mutually perpendicular.

If the angular velocity of precession ω_1 is small in comparison with ω, it is permissible to use the equation

$$\tau = \frac{\Delta(I\omega)}{\Delta t} \qquad (9\text{-}38)$$

for the relationship between the torque and the change in angular momentum, where I is the moment of inertia about the axis of spin. There is no change in the magnitude of the angular momentum about the axis of spin, only a change in its direction. Hence this equation may be written as

$$\tau = I\frac{\Delta\omega}{\Delta t} = I\alpha \qquad (9\text{-}39)$$

Putting in the value of α in terms of the angular velocities, we get

$$\tau = I\omega_1 \times \omega \qquad (9\text{-}40)$$

with the direction of τ the same as that of α.

Any rotating body can be considered a gyroscope. When a torque acts to change the direction of its axis of spin, precessional motion will occur. The motion of a top whose axis of spin is inclined to the vertical is a common example of precessional motion. The propeller of an airplane engine acts as a gyroscope when it is rotating. Whenever the airplane turns in such a way as to change the direction of the axis of spin, the gyroscope will precess, and this motion must be taken into account in making the turn.

The angular velocity of precession ω_1 can be determined from Equation 9-38. Its magnitude is

$$\omega_1 = \frac{\tau}{I\omega} \qquad (9\text{-}41)$$

where τ is the torque acting on the gyroscope and $I\omega$ is the angular momentum about the axis of spin.

Table 9-1 Linear and rotational analogs

linear		rotational
Displacement	s	θ
Velocity	$v = \Delta s/\Delta t$	$\omega = \Delta\phi/\Delta t$
Acceleration	$a = \Delta v/\Delta t$	$\alpha = \Delta\omega/\Delta t$
Mass	m	Moment of inertia I
Force	F	Torque τ
Momentum	mv	$I\omega$
Kinetic energy	$\frac{1}{2}mv^2$	$\frac{1}{2}I\omega^2$
Work	$F \cdot s$	$\tau \cdot \theta$
Power	$F \cdot v$	$\tau \cdot \omega$
	$F = ma$	$\tau = I\alpha$
	$F \cdot s = \frac{1}{2}mv_f^2 - mv_i^2$	$\tau \cdot \theta = \frac{1}{2}I\omega_f^2 - \frac{1}{2}I\omega_i^2$
	$Ft = mv_f - mv_i$	$\tau t = I\omega_f - I\omega_i$

One interesting application of this phenomenon comes from the fact that atomic nuclei, neutrons, protons, and electrons have intrinsic angular momenta; these particles act as if they were spinning solids. Each particle has a definite angular momentum characteristic of that particle. In addition, most of these particles behave as if they were small magnets; when they are placed in a magnetic field the field exerts a torque on them. Thus, in the presence of a magnetic field these particles will precess and, from Equation 9-41, the precessional velocity will depend on the torque exerted on the particle and the angular momentum. The process by which the precessional angular velocity is measured is called *nuclear magnetic resonance* (or for electrons it is called *electron spin resonance*). This process can be used to investigate the magnetic properties of nuclei or as a precise method for measuring the magnitude of a magnetic field, and as a sensitive method of assaying different chemical compounds in biological samples.

Questions

1. A solid wheel is rotating with uniform angular velocity. Compare the linear velocities of particles at different distances from the axis of rotation. Which particles have the greatest speeds?

2. Assume that the planets move in circles around the sun as a center. Describe the angular velocities of the planets in the order of their distances from the sun.

3. Assume that the planets move in circles around the sun as a center. Describe the linear veloci-

ties of the planets in the order of their distances from the sun.

4. A bicycle wheel rotating about a fixed axis with uniform angular speed is in equilibrium. Any particle in the wheel is moving with uniform circular motion about this fixed axis. (a) Is this particle in equilibrium? (b) If not, what is the force acting on this particle and in what direction does this force act?

5. Compare the linear acceleration of the center of gravity of a body rolling down a hill in-

clined at an angle θ with the linear acceleration the same body would have in sliding down a frictionless hill having the same inclination.

6. The propeller of an airplane is rotating in a clockwise direction as viewed by the pilot. Which way will the front end of the plane tend to go when this plane makes a right turn?

7. When a car is going forward, the engine and flywheel are rotating counterclockwise as viewed by the driver. In which direction will the car tend to go if the front wheels are suddenly lifted over a bump in the road?

8. Estimate the moment of inertia of the hand of a clock, an automobile wheel, and a child's toy top.

9. (a) Describe the direction of the angular momentum of the earth due to its daily rotation. (b) Attempt to estimate the magnitude of this angular momentum.

10. Why are helicopters built with two propellers? Why does this same argument not require two propellers for an airplane or even for an automobile (which has a large rotating flywheel)?

11. A diver leaves a springboard in an upright position and enters the water head first. In view of the conservation of angular momentum, how is this accomplished?

12. Explain how a diver executes a somersault.

Problems

1. The minute hand of a clock is initially pointing toward the 12 and then turns through 5π radians. Where is it pointing?

2. A wheel 3 ft in diameter is rotating with an angular speed of 2.5 rad/sec. What is the linear speed of a point on the rim?

3. The tape in a cassette tape player advances at a speed of $1\frac{7}{8}$ in./sec. At what angular speed should the hub turn if the reel of tape has a 1-in. radius?

4. The radius of the moon is 1.7×10^8 cm and the moon is 3.8×10^{10} cm away from the earth. What is the angular size of the moon as seen from the earth. Compare with the angular size of the sun which has a radius of 7×10^{10} cm and is 1.5×10^{13} cm away from the earth.

5. (a) Find the angular velocity of the earth in radians per second in its rotation about its axis. (b) Find the acceleration of a point on the equator. (c) Find the acceleration of a point whose latitude is 45°.

6. A wheel of radius 50 cm is rotating with an angular speed of 30 rev/sec and an angular acceleration of 1 rad/sec². Find (a) the linear speed of a point on the rim, (b) the radial acceleration, and (c) the tangential acceleration of a point on the rim.

7. Assume that all of the mass of a bicycle wheel is concentrated in the rim. Find the moment of inertia of a 4-lb wheel whose diameter is 26 in. about an axis that passes through the hub and is perpendicular to the plane of the wheel. (See Problem 18.)

8. The fuel supply is shut off from an engine when its angular speed is 2400 rpm. It stops rotating in 15 sec. Determine its angular acceleration, assuming it to be constant.

9. The angular speed of an automobile engine is increased from 3000 rpm to 3600 rpm in 12 sec. (a) Determine the uniform angular acceleration. (b) Determine the number of revolutions made by the engine in this time.

10. What torque will cause a sphere whose moment of inertia is 60 kg m² to reach an angular speed of 200 rev/min in 10 sec, from rest, assuming the torque to be constant?

11. A uniform disk of radius 9 cm has a mass of 30 kg, and is rotating about an axis through its center, perpendicular to the disk, with an angular speed of 3300 rpm. Find the kinetic energy of the disk.

12. A flywheel has a moment of inertia of 3×10^5 gm cm² about its axis and has an angular velocity of 3 rad/sec about this axis. Find the kinetic energy and angular momentum of the wheel.

13. A record is rotating at $33\frac{1}{3}$ rpm. A 1-gm fly is at rest on the record, 8 in. from the center. Find the angular momentum of the fly.

14. A 3600-lb car is traveling at 80 mi/hr around a circular track. The length of one lap is $\frac{1}{4}$ mi. Find the angular momentum of the car.

15. The mass of the moon is 7.34×10^{25} gm and the moon is 3.8×10^{10} cm from the earth. The moon takes $27\frac{1}{2}$ days to orbit the earth. Find the angular momentum of the moon.

16. (a) Find the angular momentum of the earth due to the daily rotation, assuming the earth to be a uniform sphere of radius 6.4×10^8 cm and mass 6×10^{27} gm. (b) Find the kinetic energy of rotation.

17. Assume that the radius of the earth is shrinking at the rate of 1 cm per day. How long would it take before the length of a day changed by 1 percent of its present value?

18. Construct an argument to show that the moment of inertia of a hoop of radius R and mass M is equal to MR^2, about an axis through the center of the hoop and perpendicular to the plane of the hoop.

19. A door has a moment of inertia of 1500 kg m² about the axis through its hinges. The door stands open at 90° to the wall, and a child pushes on the door at a point 25 cm from the axis. The child exerts a constant force of 150 nt, always exerting the force perpendicular to the plane of the door. How much time does it take to close the door?

20. A man is standing on a turntable with both arms extended, holding a 2-kg object in each hand. The man and platform are rotating with an angular speed of 8 rad/sec. The moment of inertia of the man and turntable combined is 1.5 kg m² and the distance from the axis to each 2-kg object is 1 m. Find (a) the angular momentum of this system and (b) the kinetic energy. The man pulls his hands in so that the 2-kg objects are each 10 cm from the axis. Find (c) the angular momentum of the system, (d) the angular speed of the system, (e) its kinetic energy, and (f) the gain of energy.

21. A 5-kg wheel of radius 3 m is rotating at 1800 rev/min. A brake consisting of a small leather pad is pressed against the rim of the wheel with a radial force of 70 nt. If the coefficient of friction between the wheel and brake is equal to 0.3, how long will it take for the wheel to stop? Assume that the wheel is a hoop in order to calculate its moment of inertia.

22. Find the angular momentum of a 200-kg earth satellite in an orbit that is 8000 km above the surface of the earth.

23. A hoop of radius R and mass M is spinning about an axis through the center and perpendicular to the plane of the hoop with an angular speed ω. A particle of mass m is fired with velocity V tangentially at the hoop and becomes imbedded in the hoop. The hoop comes

to rest because of this collision. Find V in terms of M, m, R, and ω.

24. Estimate the effect of the melting of the polar ice cap by assuming that the effect is to transfer mass from the polar region to the equator, thus changing the moment of inertia of the earth about the axis of rotation of the earth. How many kilograms of ice would have to melt in order that the length of a day be changed by 1 sec? How large an energy change would this involve? (Use the data of Problem 16.)

25. The moment of inertia of a rod about an axis through one end is given by $\frac{1}{3}ML^2$. A pole vaulter leaves his pole (of length L and mass M) when it is vertical and standing on end. Assume that the rod is standing still in this position and then falls, rotating about the end that remains fixed on the ground. Find the velocity of the moving end just before it strikes the ground.

26. A 50-gm ball is moving in a circle of radius 10 cm on a smooth horizontal table with an angular speed of 5 rad/sec. The ball is attached to a string, which passes through a hole in the center of the table. (a) Find the force exerted on the end of the string that is below the table. (b) The radius of the circle is decreased to 5 cm by pulling in the string. Find the change in angular momentum of the ball. (c) Find the new angular speed. (d) Find the change in energy of the ball. (e) Account for the change in energy.

27. A uniform disk of mass 4000 gm and radius 12 cm is mounted so that it can rotate about a fixed axis through the center. The axis is perpendicular to the plane of the disk and is horizontal. A cord is wrapped about the circumference of this wheel and a mass of 500 gm is attached to its free end. (a) Determine the angular acceleration of the wheel when it is released. (b) Determine the linear acceleration of the 500-gm mass. (c) Determine the tension in the cord.

28. A solid cylinder 2 ft in diameter and weighing 150 lb starts at the top of a rough plane 24 ft long and inclined at an angle of 37° with the horizontal and rolls down the plane without slipping. (a) How much energy did the cylinder have at the top of the plane (taking the zero of potential energy at the bottom of the plane)? (b) How much energy will it have at the bottom

of the plane? **(c)** Determine the angular velocity of the cylinder at the bottom of the plane. **(d)** Determine the angular acceleration of the cylinder.

29. A bicycle wheel whose radius is 30 cm is mounted on an axle 20 cm long. One end of the axle is suspended from a cord in the manner shown in Figure 9-13. The mass of the wheel and axle is 5500 gm and its moment of inertia can be calculated by assuming that the mass is located 25 cm from the axis. The wheel is made to spin with an angular speed of 18 rad/sec. Determine **(a)** the angular momentum due to the spin, **(b)** the external torque acting on the wheel, and **(c)** the angular velocity of the precession.

30. The spinning earth precesses because the earth is not a sphere but bulges somewhat at the equator. **(a)** To estimate this effect, consider a dumbbell, which consists of two objects of mass m, a distance D apart, where the center of the dumbbell is a distance R from the sun. The axis of the dumbbell is inclined at an angle of 23° to the line from the sun to the center of the dumbbell, as shown in Figure 9-16. Since the masses m are different distances from the sun, different gravitational forces are exerted. Assuming that m is about 1/300 the mass of the earth, find the torque exerted on the dumbbell. **(b)** The bulge of the earth is distributed on a hoop and the gravitational force varies during the year so the actual torque is about 1/5 of the value calculated in (a) above. Assume that this reduced torque produces a slow precession of the earth, and calculate the precessional velocity. (The moon also exerts a torque. The combined effect produces a precession with a period of 26,000 years.)

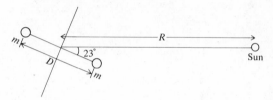

FIGURE 9-16

relativity of length, time, & velocity

chapter 10

10-1
Role of the speed of light in physics

Measurements of the speed of light and the speed of light itself have played exceedingly important roles in the development of physics over a period of more than three centuries. As we now know, the speed of light is very great—186,000 mi/sec or 300 million m/sec. It is thus not surprising that previous to the seventeenth century the general belief was that light traveled instantaneously—that is, that the speed of light was infinite. Since the measurement of speed involves the measurements of distance and time, given the timing devices available three centuries ago, the distances used had to be very large—that is, astronomical distances. Thus, the earliest measurements of the speed of light involved astronomical events; two of these will be described shortly. As the accuracy and quality of the timing devices were improved, terrestrial distances could be used. Such measurements were started in the early part of the nineteenth century and are continuing to this day, attesting to the importance that scientists attach to accurate knowledge of the speed of light.

Initially the speed of light was simply an empirical number, the speed with which light traveled through a vacuum. Later, this empirical character was expanded somewhat with the discovery that all electromagnetic radiation traveled with this same speed in vacuum. In the middle of the nineteenth century, the work of James Maxwell made the speed of light an important quantity theoretically. Maxwell was able to predict that electromagnetic radiation should exist, that this radiation should travel with a speed determined by a particular combination of electrical and magnetic constants, and that this speed was identical to the speed of light. Thus Maxwell was able to connect the speed of light with the theory of electricity and magnetism. Einstein's work was an attempt to resolve conflicts between the theory of Maxwell and the classical mechanics of Newton. As a result, as we shall see in this chapter, Einstein was able to show that the basic ideas of length and time needed revision and that in the revised theory the speed of light played an important role. Therefore the speed of light is now thought of as one of the most fundamental constants in nature.

10-2
Measurement of the speed of light

The earliest determination of a finite speed of light was made in 1675 by the Danish astronomer Olaf Roemer by observing the eclipses of the innermost moon of Jupiter. Roemer measured the time between successive disappearances of the moon behind Jupiter. Roemer observed that this time interval is not constant but depends on where the earth is in its orbit at the time the observation is made. Referring to Figure 10-1, the time between successive eclipses is approximately 42.5 hr when the earth is at points N and F (when the earth is moving neither away from nor toward Jupiter) but the time between eclipses is several seconds longer when the earth is at points such as C or D (and moving away from Jupiter) and the time is several seconds shorter when the earth is at points such as A or B (and moving toward Jupiter). In order to explain this, Roemer assumed that the light that

signals the eclipse takes time to travel from Jupiter to the earth because the speed of light is finite. Then while the earth is moving from A to B the light from each successive eclipse travels a shorter distance (because the earth is moving closer to Jupiter) and so the time interval observed between eclipses decreases. Similarly, the time interval between successive eclipses should increase while the earth travels from C to D. By appropriately adding changes in the interval as the earth moves from F to N, Roemer was able to find the time it would take light to travel from N to F. The relatively crude instruments that he used yielded a time of 22 min (a modern measurement yields a value closer to 17 min). By combining this measurement with the calculated value of the diameter of the earth's orbit, Roemer was able to estimate the speed of light.

One of the earliest terrestrial determinations of the speed of light was made in 1849 by Fizeau, who timed the passage of a beam of light a distance of 8.633 km from its source to a mirror and then back to the source. On the way from the source, the light passed between two teeth on the rim of a wheel whose speed was adjusted so that, on its return, the light failed to pass through this space but hit the adjacent tooth and was thus eclipsed. In this experiment the wheel had 720 teeth and the light was eclipsed when the speed of the wheel was 12.6 rev/sec. Hence the wheel moved only 1/1440 rev while the light traveled a distance of 17.266 km; the time required to traverse this distance was $1/1440 \times 12.6$ sec. Foucault in 1850 also measured the speed of light, using a rotating mirror to measure the time required for a beam of light to go from it to a second mirror several meters away and back to the source. Some of the best determinations of the speed of light were made by Albert A. Michelson (1852–1931). He began his experiments about 1878 and continued them for about 50 years. We shall describe one of the arrangements used in his experiments.

In Michelson's experiment, a beam of light was sent from Mt. Wilson to Mt. San Antonio and back again. The distance between the two points was measured very accurately by the United States Coast and Geodetic Survey, with a precision of about 1/5 in. in the 22-mi distance. The essential arrangement for this experiment is shown in Figure 10-2. M_1 is an octagonal mirror that is mounted on the shaft of a variable speed motor so that it rotates about an axis through its center. Light from a source S strikes mirror M_1 at an angle of 45° and

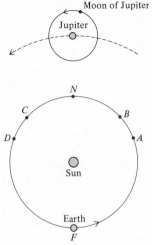

FIGURE 10-1 Roemer's method of determining the speed of light.

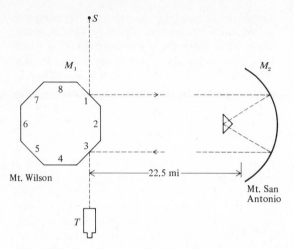

FIGURE 10-2 Michelson's method for measuring the speed of light.

is reflected from it to the distant mirror M_2. It is reflected back from this mirror to the octagonal mirror in such a way that when M_1 is stationary the reflected ray strikes section 3 at an angle of 45° and is reflected into the telescope T. When the mirror is set into rotation the reflected beam will, in general, strike section 3 at an angle other than 45° because section 3 has turned to a new position in the time required for the light to travel from M_1 to M_2 and back again. The reflected beam will not enter the telescope if it strikes a section of the mirror at an angle other than 45°. By increasing the speed of the motor, however, it is possible to bring section 2 of the octagonal mirror into the place formerly occupied by section 3 in the time required by the beam of light to go from M_1 to M_2 and back to M_1. When the motor reaches this angular speed, the light will again enter the telescope. The experiment consists of varying the speed of the motor until the reflected light comes into view again in the telescope. The speed of the motor is then measured accurately. The time taken to travel the distance of $2M_1M_2$ is one eighth of the time required for 1 rev. In these experiments the angular speed of the motor was about 530 rev/sec.

In the years since Michelson's experiments increasingly sophisticated measurements have been made of the speed of light. The current accepted value is

$$c = 299{,}792.5 \ \frac{\text{km}}{\text{sec}}$$

and recent measurements have been made with a precision of 0.02 km/sec. The very high precision of these measurements permits us to turn the principle of the Michelson and Roemer measurements around, and use the known speed of light to measure distances. By timing the flight of light we can measure the distance to the moon and make very accurate terrestrial surveys.

Unless extreme accuracy is needed we can use the approximation

$$c = 3 \times 10^8 \ \frac{\text{m}}{\text{sec}}$$

for the speed of light.

10-3
The special theory of relativity

In 1905 Albert Einstein, then a 26-year-old physicist employed as a patent examiner in Zurich, published a paper that was the basis of the theory of relativity. This theory reexamined and reformulated the most basic ideas of physics—length and time. It is based on two simple postulates, which may be stated in the following form:

1. *Fundamental physical laws should have the same mathematical forms in all inertial systems.*
2. *The speed of light in empty space is a constant and is independent of the motion of the source and observer.*

The first postulate is a statement about the laws of physics. The laws of physics are written in terms of such quantities as position, velocity, and acceleration, *relative* to some coordinate system. The first postulate says that if the laws of physics are known to be correct in one coordinate system, then they will have exactly the same mathematical form in all coordinate systems moving with constant velocity relative to the first coordinate system. This is equivalent to saying that there does not exist a single absolute reference frame and so the laws of nature must involve relative velocities rather than absolute velocities. The restriction of the coordinate systems to those moving with constant velocity is that which makes the theory the "special theory." Extending the discussion to coordinate systems moving with acceleration is the subject of the "general theory," which will be discussed later in this book.

The essential idea of the first postulate was well known in classical mechanics and follows from the work of Newton. However, the theory of electricity and magnetism implied the existence of an absolute frame of reference, called the *ether*. When measurements were made relative to the ether, the laws of electricity and magnetism had one form but the laws had another form in a coordinate system moving with constant velocity relative to the ether. This apparent difference between mechanics and electricity and magnetism disturbed Einstein, so he assumed that it was not true. He extended the classical principle of relativity, which was known to apply to mechanics, to all the laws of physics. This required a reexamination of some more fundamental ideas—those describing distance and time.

The second postulate is an extension of the idea that the ether does not exist. Einstein assumed that the velocity of light is a constant relative to all observers, rather than relative to the absolute reference frame of the ether.

The postulates were suggested by a long series of experiments in a framework of theory. Nineteenth-century physics had concluded that light was a wave that traveled in empty space. This conclusion was based on overwhelming experimental evidence, and was part of a complete, masterful theory that described all of electricity and magnetism. But it was known that every other type of wave travels in some *medium*, and that the speed of the wave is a property of the medium; the latter is assumed to be at rest. For example, sound waves travel in the air with a speed, relative to the air, that depends on the density of the air. Thus, physicists assumed the existence of a medium in which light was propagated with a speed that depended on the properties of the medium. This medium was called the *luminiferous ether*, or sometimes simply the *ether*. The assumption then, was that the ether filled all of space, and that light was propagated with a speed relative to the ether that is constant because the properties of the ether are constant. In any particular experiment, the measured speed of light would depend on the velocity

FIGURE 10-3 Albert Einstein (1879–1955). He developed the theory of relativity and revolutionized the mode of thinking about fundamental physical problems. One consequence of this theory was the extension of the concept of energy to include mass as a form of energy. Another part of his work gives us new insight into gravitational phenomena. He also developed the fundamental equation of the photoelectric effect and the theory of Brownian motion. (Official U.S. Navy Photo from Acme.)

of the emitter and observer relative to the ether. Thus it seemed reasonable that one could make a series of measurements of the speed of light under various conditions and deduce the velocity of the earth relative to the ether. All such attempts to measure the speed of the earth relative to the ether failed. Michelson and Morley performed a series of clear experiments, very carefully done, which should have given the velocity of the earth relative to the ether but always obtained the result that the earth was stationary with respect to the ether. It seemed completely unreasonable that the rotating, revolving earth, a minor satellite of a minor star, should always be at rest with respect to the universal medium for the propagation of light. The only explanation, or rather the only *simple* explanation, is in the Einstein postulates.

The postulates were *suggested* by this unsuccessful attempt to measure the velocity of the earth relative to the ether; but the postulates are *believed* because the conclusions drawn from them agree with experiment. These conclusions differ strongly from our "common sense" ideas about length and time. We can see how this violation of common sense can happen, and at the same time put some limits on the theory of relativity from the following argument. Common sense is a summary of common experience, but common experience is limited. We have no direct experience with distances as small as the size of an atom or as large as the size of a galaxy. Similarly, we have no direct experience with very short time intervals or very large intervals, and we have no direct experience with the behavior of objects moving with speeds close to that of light. Thus we can expect common-sense ideas to be valid for moderate distances or moderate times or moderate speeds. We should be prepared to discover that common-sense ideas may be wrong outside these areas. It is possible that our common-sense ideas about length or time or velocity (or other physical concepts) are wrong on the atomic scale, or the galactic scale, or for objects that are moving with speeds close to that of light. We can, however, turn this argument around and insist that common sense is at least approximately right for moderate distances, moderate time intervals, and moderate velocities. Therefore any new physical theory must give essentially the same description of this class of phenomena as does common sense. As we will see, relativity disagrees with our expectations only for objects or systems moving with speeds that are not small compared to the speed of light.

10-4
The Lorentz transformation

The special theory of relativity forces us to re-examine our understanding of the fundamental concepts of length and time. This reexamination takes the point of view that these concepts are part of a *scientific* theory. Therefore we will consider what happens when these quantities are *measured*. We will assume that the measurements are done carefully and intelligently, but that they are restricted in two ways. First we will assume that the postulates of special relativity are correct, and provide a convenient starting place to decide what measurements to make. Second, we will not assume that we can ignore the effects of motion. That is, we will assume that if a meter stick is set in motion, then we can no longer trust it to appear to be 1 m in length.

Throughout this discussion of relativity we will refer to an *observer* and his *frame of reference*. Unfortunately, these terms may bring to mind a

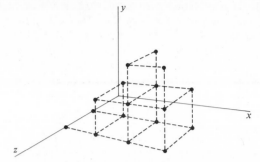

FIGURE 10-4 An observer which consists of clocks and detectors placed at the intersections of the grid.

human being looking at things—a human being subject to confusion, error, optical illusion, and psychological problems. It must be emphasized that these terms refer to intelligent scientific *measurement*, made as inanimate and free from human weakness as possible. We can for example, make a model of an observer and his reference frame of the following form. Consider a three-dimensional grid, represented schematically in Figure 10-4. At each of the intersections of the grid we can place a clock and an array of detectors. These detectors can detect the existence of objects or light, or the occurrence of events, such as explosions or chemical reactions. Then each of these clock-detector combinations can be connected to a single computer, which can record the readings of the detectors, the readings of the clocks, and the locations of the specific clock-detector. Thus, if the size of the grid is fine enough, this observer can automatically record the location of an event and the trajectory of events in space and time as accurately as necessary. The record can be retained in the computer to be examined at his leisure by a thoughtful human observer. We will assume that this system is built carefully and rationally—that all the clocks and detectors are identical, and are regularly spaced and synchronized.

Now we can assume that two such frames of reference are moving relative to each other. For simplicity, assume that the situation is as shown in Figure 10-5, where the axes are parallel to each other, and that one of the systems, called the primed system, is moving with a velocity **v** (parallel to the x axis) relative to the other system (the unprimed system). Further let us assume, at the instant the two origins coincide, that the two observers set both clocks at the origin to read zero. Then each observer can synchronize all of the rest of his clocks with the clock at the origin. (As we

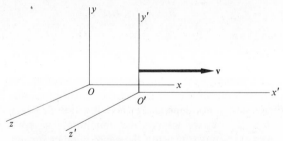

FIGURE 10-5 The primed coordinate system is moving in the x direction with velocity **v** relative to the unprimed system. The x and x' axes coincide.

will see later, this does not mean that all the clocks in the primed system are synchronized with all the clocks in the unprimed system.) It must be emphasized that since the first postulate states that the laws of physics involve relative velocities rather than absolute velocities, there is no way of telling which of the two systems is moving. The primed system is moving with a velocity **v** relative to the unprimed system and the unprimed system is moving with a velocity −**v** relative to the primed system. Relative to a third system, either or neither may be stationary.

Each system has been built carefully and measurements can be made within the system. We can now consider what happens when measurements made in one system are compared with measurements made in the other system. We can imagine the physicist in the primed system reading the output of his computer and then calling the physicist in the unprimed system on the radio to compare the results of a particular measurement.

Consider that the physicist in the primed system mounts a light bulb on his y' axis at a height above the origin that he measures to be 1 m. To the observer in the primed system, this light bulb is fixed. To the observer in the unprimed system, the light bulb is moving. The light detectors in the unprimed system will record the bulb as moving in a horizontal straight line, as in Figure 10-6. The physicist in the unprimed system can measure the height of the line and compare it (by radio) with the 1 m as measured by the primed system. Thus, if the two physicists want to, they can each adjust their measurement systems (or computers) so that both will read 1 m.

If the two physicists wish to compare measurements made on the same event, or measurements made of the same sequence of events, they should attempt to make their units of measurement identical. Figuratively, each physicist has a meter

stick, and the light-bulb experiment allows them to compare meter sticks and decide that what is 1 m to one physicist is also 1 m to the other. Within each system, each physicist can then proceed to measure the location of the detectors in his grid. He will assume that 1 m parallel to the x axis is the same as 1 m parallel to the y axis, which is also 1 m parallel to the z axis. They can check further by repeating the light-bulb experiment with a light bulb mounted on the z axis. Since there really is no difference between the y axis and the z axis except a label (both are transverse to the direction of motion), both physicists will agree on the distance between this light bulb and the x axis. We can summarize this by saying that if both physicists measure the same event they will agree on distances transverse to the direction of motion; in symbols,

$$y = y' \qquad (10\text{-}1)$$
$$z = z' \qquad (10\text{-}2)$$

They could attempt to repeat the experiment with a light bulb placed on the x' axis. They would both agree that the y and z coordinates of this light bulb are zero. To the primed observer x' would be constant, but to the unprimed observer x would be increasing at a constant rate. Thus in order to compare their measurements, they both would have to agree on *when* the unprimed observer makes his measurement. But this assumes that they can compare their clock systems and agree that they both run at the same rate. It is possible for the two observers to agree that the origin of the coordinate systems coincide, and at that point agree to have both sets of clocks read zero (see next section). After that the primed observer will measure a time t' for the occurrence of a given event and the unprimed observer will measure a time t for the same event. Although it is not obvious, we cannot assume that t and t' are the same—that is, that

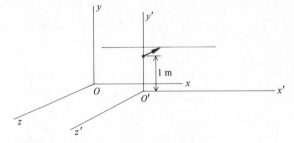

FIGURE 10-6 A light bulb fixed on the y' axis, 1 m above the origin O_1', traces a line in the unprimed coordinate system which is 1 m above the xz plane.

clocks run at the same rate in the two systems. Indeed, we shall see that the rate at which a clock runs depends upon its velocity relative to the observer.

Before 1905 and the introduction of the postulates, we would have treated the measurement of a single event in the following way. Consider an event whose location in the unprimed frame is (x, y, z, t) and in the primed frame is (x', y', z', t'), as in Figure 10-7. Then we have

$$x = x' + vt \qquad (10\text{-}3a)$$
$$y = y' \qquad (10\text{-}3b)$$
$$z = z' \qquad (10\text{-}3c)$$
$$t = t' \qquad (10\text{-}3d)$$

This set of equations is called the *Galilean* transformation. If we were to divide both sides of Equation 10-3a by t, we would obtain

$$v_x = v_x' + v \qquad (10\text{-}4)$$

This equation states that the speed measured in the unprimed system is the speed measured in the primed system plus the speed of the primed system relative to the unprimed system, and is essentially the same as Equation 2-18. Now the second postulate says that this result cannot be true if the observers are measuring the speed of light, and therefore we must alter some or all four Equations 10-3. Equations 10-1 and 10-2 are identical to Equations 10-3b and c, so the Einstein postulates require that we must change Equations 10-3a and d.

The Einstein postulates force us to find a new relation between the coordinates (in space and time) of a single event as measured in the primed coordinate system (x' and t') and the coordinates of the same event as measured in the unprimed system (x and t). In outline, we can derive the correct equation with the following argument:

Let us assume that the correct equation has the form

$$x = Ax' + Bt' \qquad (10\text{-}5)$$

and that

$$x' = Dx + Et \qquad (10\text{-}6)$$

This simple form is the most general form that is consistent with the idea that the length of an object should not depend on where it is (see Problem 27). If now we impose the condition that both observers must measure the same number for the relative velocity v of the two coordinate systems, then we find that Equations 10-5 and 10-6 become (see Problem 28)

$$x = A(x' + vt') \qquad (10\text{-}7)$$

and

$$x' = D(x - vt) \qquad (10\text{-}8)$$

The symmetry of the two frames and the requirement that both frames use identical systems of units requires that the constants A and D be identical, and so

$$x = A(x' + vt') \qquad (10\text{-}9)$$
$$x' = A(x - vt) \qquad (10\text{-}10)$$

Assume now that a flashbulb is located at the origin of coordinates and that at the instant the origins coincide the flashbulb is fired. A pulse of light will spread out from the flashbulb, and from the second postulate we know that each observer will measure the speed of the pulse to be c, the speed of light. To the primed observer, the pulse will be at a point x' on the axis at a time t', where

$$x' = ct' \qquad (10\text{-}11)$$

Similarly, the unprimed observer will measure x and t, where

$$x = ct \qquad (10\text{-}12)$$

Combining Equations 10-9, 10-10, 10-11, and 10-12, we can solve for A, obtaining

$$A = \frac{1}{\sqrt{1 - (v^2/c^2)}} \qquad (10\text{-}13)$$

It has become common to use the symbol γ (Greek letter gamma) for the constant A; thus

$$\gamma = \frac{1}{\sqrt{1 - (v^2/c^2)}} \qquad (10\text{-}14)$$

The quantity (v/c) is often represented by the symbol β, so

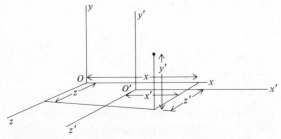

FIGURE 10-7 A single event has coordinates (x', y', z', t') and (x, y, z, t).

$$\gamma = \frac{1}{\sqrt{1-\beta^2}} \qquad (10\text{-}15)$$

Note that if v is less than c, then $\beta < 1$ and $\gamma > 1$. Thus the transformation equations become

$$x = \gamma(x' + vt') \qquad (10\text{-}16)$$

and

$$x' = \gamma(x - vt) \qquad (10\text{-}17)$$

Solving simultaneously for t and t', we have

$$t = \gamma\left(t' + \frac{\beta}{c}x'\right) \qquad (10\text{-}18)$$

$$t' = \gamma\left(t - \frac{\beta}{c}x\right) \qquad (10\text{-}19)$$

Equations 10-1, 10-2, and 10-16 through 10-19 are usually called the *Lorentz transformations*. We shall examine some of the consequences of the Lorentz transformations in the following sections, but first we can check that these equations are consistent with "common sense." If the relative velocity of the two systems v is small compared to c, the speed of light, then

$$\sqrt{1 - \frac{v^2}{c^2}} \approx 1$$

so

$$\gamma \approx 1$$

and

$$\beta \approx 0$$

so

$$\frac{\beta}{c} \approx 0$$

With these approximations, we have

$$x \approx x' + vt' \qquad \text{and} \qquad t' \approx t$$

which are the equations for the Galilean transformation. Thus, as long as we confine ourselves to systems moving slowly compared with the speed of light, we can use the results of classical (pre-relativity) physics. In the area of common experience (low speeds) our common sense is reliable.

10-5
The relativity of simultaneity

One of the most startling results of the special theory of relativity is that events that are simul-

taneous in one reference frame are not simultaneous in another. Consider two events that occur at the same time, t', in the primed reference frame. That is, we have event 1, which occurs at x_1' and t', and we have event 2, which occurs at x_2' and t'. Then we can find the times at which these two events are measured to occur in the unprimed system from Equation 10-19. Event 1 will be measured to occur at t_1, where

$$t_1 = \gamma\left(t' + \frac{\beta}{c}x_1'\right) \qquad (10\text{-}20)$$

and event 2 will be measured to occur at t_2, where

$$t_2 = \gamma\left(t' + \frac{\beta}{c}x_2'\right) \qquad (10\text{-}21)$$

Thus there will be a time difference Δt between the two events, as measured in the unprimed system, given by

$$\Delta t = t_2 - t_1 = \gamma\left[\frac{\beta}{c}(x_2' - x_1')\right] \qquad (10\text{-}22)$$

$$= \gamma\left[\frac{\beta}{c}\Delta x'\right] \qquad (10\text{-}23)$$

Therefore, if the two events are simultaneous (that is, $\Delta t' = 0$) but occur at different places in the primed system, then they will not be simultaneous in the unprimed system. The only circumstance in which two simultaneous events (as measured in one frame) are simultaneous in another frame, is when the two events are simultaneous and occur at the same place (that is, $\Delta x' = 0$). In short, simultaneity is relative.

This is very important in view of the description of an observer given in the preceding section. There we described a frame of reference as a grid of detectors and *synchronized* clocks. Figuratively, we can imagine a large number of regularly spaced clocks, all of whose hands simultaneously read the same number in the primed system. If the unprimed observer measures the locations of the hands of these clocks, he will conclude that the clocks read the same values at different times. That is, if the primed observer synchronizes his clocks, the unprimed observer will conclude that they are not synchronized. Note that this conclusion is the result of a measurement and not simply a visual effect caused by one observer looking at moving clocks. Similarly, when the unprimed observer synchronizes his clocks, the primed observer will, as a result of measurement of those clocks, conclude that they are not synchronized.

One of the major questions Einstein dealt with was this problem of how one synchronizes two clocks. In principle it is easy to synchronize two clocks that are at the same place, or right next to each other. It is somewhat more difficult to describe, unambiguously, how one can synchronize two clocks that are separated. One could imagine bringing the two clocks together, synchronizing them, and then separating them. But the act of moving them may alter their rates, and they may become unsynchronized when they are separated. In order to avoid this difficulty, one must start with the clocks separated and stationary. Then one can transmit a signal from one clock to the other, and correct for the time of transmission of the signal. That requires that the signal have a known, constant velocity. For this, the transmission of light signals is convenient because the second postulate tells that the speed of light is a constant. One simple way to synchronize two stationary clocks is by means of the following process. In the rest frames of the two clocks place a flashbulb halfway between the clocks. Then, when the bulb is fired, the light will take the same time to travel to each clock. The clocks can then be set to read the same value at the instant of receipt of the flashes.

Consider then what happens when one observer watches another observer synchronizing his clocks. Assume that the primed observer sets off a flashbulb, which is fixed in his frame of reference halfway between the two clocks, as shown in Figure

10-8a. As shown in Figure 10-8b, the pulse of light spreads out in all directions and to the primed observer reaches the two clocks at the same instant. To the unprimed observer, the two clocks are moving, as is the light. Thus, the light spreads out as shown in Figure 10-8c and d, and reaches the clock labeled A *before* it reaches clock B. Thus to the primed observer, his clocks are synchronized, and A and B read the same time at the same instant. To the unprimed observer, A reads too early, and the clocks of the primed observer are not synchronized.

10-6
Contraction of length

We can now examine the process of measuring lengths. Consider a rod that is at rest in the unprimed system, parallel to the x axis. The unprimed observer can measure the length of this object by several methods, all of which are equivalent to finding the difference of the x coordinates of the ends of the rod. Since the rod is not moving with respect to the observer, he can find these coordinates at different times. Now consider what happens when the primed observer attempts to measure the length of this same rod. The rod is moving with respect to this observer. The primed observer must measure the coordinates of the ends *and* the time at which he measures these coordinates. The simplest procedure is for the primed

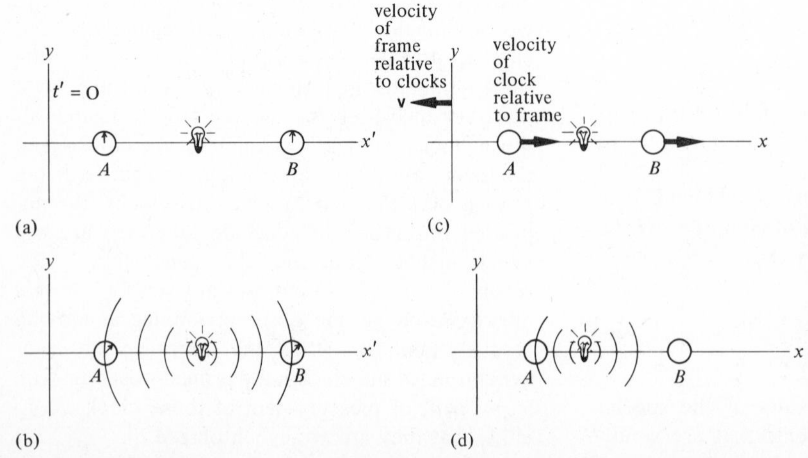

FIGURE 10-8 (a) A flashbulb halfway between two clocks (in the primed system) is fired at $t' = 0$. (b) To an observer in the primed system the light reaches both clocks at the same instant. (c) The firing of the flash bulb as seen by an observer in the unprimed system. The clocks and bulb are fixed in the primed system. (d) To an observer in the unprimed system, clock A, moving toward the light, receives the light before clock B, which is moving away from the light.

observer to measure the coordinates at both ends of the rod at the same time. That is, the primed observer measures the coordinate, x_2', of one end of the rod at time t', and at the same time, t', he measures the x coordinate of the other end, x_1'. In the primed coordinate system, the length of the rod, L', is the difference between these simultaneous readings of the locations of the ends of the rod:

$$L' = x_2' - x_1' \qquad (10\text{-}24)$$

We can then substitute in Equation 10-16 to find the coordinates of the ends of the rod in the unprimed system:

$$x_2 = \gamma(x_2' + vt') \qquad (10\text{-}25)$$
$$x_1 = \gamma(x_1' + vt') \qquad (10\text{-}26)$$

The rod is at rest in the unprimed system, so the length of the rod as measured in the frame in which it is at rest, L_0, is given by

$$L_0 = x_2 - x_1 \qquad (10\text{-}27)$$

Therefore, if we subtract Equation 10-26 from Equation 10-25 and substitute Equations 10-24 and 10-27 we get

$$L_0 = \gamma L' \qquad (10\text{-}28)$$

Thus, the length of the rod as measured by an observer moving relative to the rod, L', is shorter than the length of the rod as measured by an observer at rest with respect to the rod, L_0. This measured shortening of the rod is called the *Lorentz contraction*. Note that in this measurement process, the primed observer measured both ends of the rod at the same time (in the primed system). To the unprimed observer, the measurements will not be simultaneous. Thus, in a sense, length contraction is a consequence of the inability of the two observers to synchronize their clocks mutually.

10-7
Time dilation

Now consider two events that occur at the same place but at different times in the primed coordinate system. The first event occurs at time t_1' and position x_1', while the second event occurs at time t_2' and at the same position x_1'. To the primed observer, the time interval between these two events $\Delta t'$ is given by the difference between the times; that is,

$$\Delta t' = t_2' - t_1' \qquad (10\text{-}29)$$

The unprimed observer can measure the times at which these two events occur, t_2 and t_1, and from Equation 10-18 he will measure

$$t_2 = \gamma\left(t_2' + \frac{\beta x_1'}{c}\right) \qquad (10\text{-}30)$$

$$t_1 = \gamma\left(t_1' + \frac{\beta x_1'}{c}\right) \qquad (10\text{-}31)$$

Then the time interval between these two events, as measured in the unprimed system, is Δt, where

$$\Delta t = t_2 - t_1 \qquad (10\text{-}32)$$

Then, substituting Equations 10-30 and 10-31 in Equation 10-32 and then substituting Equation 10-29, we get

$$\Delta t = \gamma \, \Delta t' \qquad (10\text{-}33)$$

Thus the time interval between these two events, Δt, measured by an observer who is moving with a speed less than c is longer than the time interval $\Delta t'$ as measured by an observer who is at rest with respect to the two events. This result is often described with the phrase, "moving clocks run slow."

This phenomenon of *time dilation*, as described by Equation 10-33, has been verified directly by experiment. In the process of radioactive decay, a given particle "disintegrates" into two or more different particles. For any one particle, this process occurs randomly. For a large collection of radioactive particles, a specific fraction of the particles decay in a specific time. In other words, one can measure the number of radioactive particles present in a sample at a given instant of time, and then at a time t seconds later, measure the number still present (ignoring the products of the decay). Then if one repeats this pair of measurements on a different sample of the material, or on the same sample at a different time, using the same time interval t, the number of residual particles will be the same fraction of the original number. The time it takes for one half of a collection of radioactive particles to decay is called *the half-life*, and is a characteristic of the specific particles involved. Thus, the half-life of a given kind of radioactive decay is a kind of clock. One can measure the half-life of a specific decay when the decaying particles are at rest, and then measure the half-life of the same kind of particles when they are moving with a velocity v with respect to the laboratory in which the measurements are made. Equation 10-33 predicts that the moving particles should have a longer half-life than the stationary particles, in the ratio of

$$\frac{t_{\text{moving}}}{t_{\text{rest}}} = \gamma$$

This prediction has been verified many times, with several different kinds of particles.

In these experiments on the half-life of radioactive particles, the particles are observed in a frame of reference that is fixed in some laboratory. If the particles are moving with very high speeds they are observed to travel for a long time before they decay. Thus they travel a long distance, the product of their speed and the half-life, on the average. If one were to observe this process in a reference frame that is moving with the particles, they would decay in a much shorter time, their rest half-life. In that frame of reference, the laboratory is moving by the particles at a very high speed, the speed of the particles as measured in the laboratory. During the half-life of the particles, the laboratory has moved, relative to the particles, a distance equal to the product of the relative velocity and the half-life. Thus, in the laboratory frame the particles move a large distance, while in the frame of the particles the laboratory has moved a short distance. But this is just the Lorentz contraction described in the preceding section. Thus we can think of the process as either a dilation of the half-life *or* a contraction of the laboratory. Both processes are related, and both are expressions of the difficulty of synchronizing clocks and the process of measurement.

The phenomenon of time dilation can be illustrated in a most startling way by describing a famous problem, which has been the subject of much controversy and which is usually called the "twin paradox." The problem concerns two identical twins, one of whom takes a long space flight, during most of which he travels at very high speeds, approaching the speed of light. In the frame of reference of the earth, all the clocks which accompany the traveling twin run slow; that is, the elapsed time between two events as measured by the twin who remains on earth is longer than the elapsed time as measured by the space-traveling twin for the same two events. If they choose the departure and arrival as the two events, then the earthbound twin measures the duration of the trip as longer than the duration as measured by the space-traveling twin. This becomes more startling when one recognizes that chemical and biological processes proceed at a relatively fixed rate and can therefore be treated as natural clocks. Each of the twins can measure the duration of the trip by

measuring the amount of biological aging his own body has undergone. In other words, at the end of the trip, the space-traveling twin is younger than his static sibling!

This problem is called a paradox because it appears that one can argue that from the point of view of the traveling twin it is the stationary twin who is younger. If this were true, then one might say that each twin believes that he is older. This argument is wrong because it assumes symmetry between the twins; it assumes that each twin can claim that it was the other who moved away and came back. In fact only one of the twins experiences acceleration. This acceleration is accompanied by observable effects—forces and changes in the appearance of the stars. Both twins will agree on which was accelerated.

Another way of seeing this is to recognize that the time-dilation formula connects two time intervals, each of which is measured in a single reference frame, moving at constant velocity relative to the other. However, in order to describe the twin paradox one must invoke at least three such reference frames: one fixed in the earth, a second moving out away from the earth, and a third moving back toward the earth. The space traveler must switch from the second to the third reference frame in order to return to the earth.

One can apply the Lorentz transformation to this problem by assuming that one can neglect the time during which the traveler switches from the outgoing reference frame to the incoming reference frame. Then it turns out that both observers agree: the total elapsed time experienced by the traveler is shorter than the total elapsed time experienced by the stationary twin.

In October 1971 Hafele and Keating flew very accurate atomic clocks around the world on commercial jet flights and compared the elapsed time with that of identical clocks that remained stationary. In this case the moving clocks went through an infinite number of reference frames (tangent to the orbit of the airplanes around the earth), but the calculation is essentially the same as the one mentioned above. More important, a general relativistic effect (called the *gravitational red shift*) is present in this experiment due to the presence of the gravitational field. This effect predicts that the time difference between the moving and stationary clocks should depend on whether the moving clocks went eastward around the earth or westward around the earth. In the particular experiment the eastward clocks were in flight for approximately 40 hr and arrived reading approximately 60×10^{-9} sec earlier than the stationary clocks, which is in reasonable agreement with the theoretical prediction of the sum of the time dilation and gravitational red shift. The westward clocks were in flight for approximately 50 hours and returned reading 275×10^{-9} sec later than the stationary clocks, which is also in good agreement with the theory.

10-8
Relative velocity

Now that we have the transformation equations for coordinates we can construct the transformation for velocities.

Suppose a particle starts at the origin at $t' = 0$ and moves with constant speed V' parallel to the x' axis. Then at time t' it will be at point x' as measured in the primed coordinate system, whereas in the unprimed system it will be at point x at time t; these coordinates are related by

$$x = \gamma(x' + vt') \qquad (10\text{-}16)$$

$$t = \gamma\left(t' + \frac{\beta}{c}x'\right) \qquad (10\text{-}18)$$

Dividing Equation 10-16 by Equation 10-18 we obtain

$$\frac{x}{t} = \frac{x' + vt'}{t' + \dfrac{\beta}{c}x'} \qquad (10\text{-}34)$$

The left side of the equation is V, the velocity of the particle as measured in the unprimed system. If we divide numerator and denominator of the right side by t', we obtain

$$V = \frac{\dfrac{x'}{t'} + v}{1 + \dfrac{\beta x'}{ct'}} \qquad (10\text{-}35)$$

and since

$$V' = \frac{x'}{t'} \qquad (10\text{-}36)$$

we have

$$V = \frac{V' + v}{1 + \dfrac{\beta}{c}V'} \qquad (10\text{-}37)$$

or

$$V = \frac{V' + v}{1 + \dfrac{vV'}{c^2}} \qquad (10\text{-}38)$$

where V' is the velocity of the object relative to the primed system, V is the velocity of the same object relative to the unprimed system, and v is the velocity of the primed system relative to the unprimed system.

This result is to be compared with Equation 2-18, the nonrelativistic result for the same problem. Using the same symbols as were used in Equation 2-18, we write Equation 10-38 in the form

$$v = \frac{u + w}{1 + \dfrac{uw}{c^2}} \qquad (10\text{-}38a)$$

If the velocity in the primed system, u, or the relative velocity of the two coordinate systems, w, is small compared to the speed of light, c, then this result is approximately equal to the classical result

$$v = u + w \qquad (2\text{-}18)$$

If, however, both velocities are comparable to the speed of light, then the relativistic result, Equation 10-38a, is markedly different from the nonrelativistic result.

Illustrative example

A particle called a kaon is traveling with a speed which is 4/5 the speed of light, as measured in a laboratory. The kaon decays, producing, among other things, a pion which moves forward relative to the kaon with a speed 3/5 of the speed of light. What is the speed of the pion as measured in the laboratory?

We can use Equation 10-38, if we call the primed system a system that is moving along with the kaon. In that system, the pion is moving with a velocity V' where

$$V' = \tfrac{3}{5}c$$

The primed system is moving relative to the laboratory (the unprimed system) with a velocity v, where

$$v = \tfrac{4}{5}c$$

Then the velocity of the pion relative to the laboratory is V, and from Equation 10-38, we have

$$V = \frac{\tfrac{4}{5}c + \tfrac{3}{5}c}{1 + \left(\tfrac{4}{5}c\right)\left(\tfrac{3}{5}c\right)/c^2}$$

$$= \tfrac{35}{37}c$$

Thus, the pion travels at a speed which is slightly smaller than the speed of light. The nonrelativistic prediction would have been that the pion would have traveled with a speed of 1.4 times the speed of light. This kind of experiment has repeatedly verified the predictions of the theory of special relativity.

Questions

1. State the postulates of the special theory of relativity.

2. How would Einstein's postulates change if the speed of light were infinite?

3. Describe some of the consequences if nature were to change and result in the speed of light being 100 mi/hr.

4. What is meant by *time dilation*?

5. What is meant by *length contraction*?

6. If you were on a rocket ship moving with constant speed, what changes would you observe in your own rate of breathing and heartbeat? What changes would be observed by an observer fixed on the earth?

7. Explain why the effects of special relativity are usually not observed.

8. Twins are never precisely the same age. Is it ever possible for different inertial observers to disagree over which is the older (assuming that the twins are distinguishable)? Explain.

9. Why is it wrong to describe special relativity as asserting that "everything is relative"?

10. How does the concept of simultaneity enter into the measurement of length?

11. Each of two inertial observers asserts that the other's meter sticks are short. Explain this paradox by considering what each is measuring.

12. A searchlight is pointed straight up. As the earth rotates, the beam rotates also and so a point on the beam very far from the earth moves with a speed larger than c. Why is this not inconsistent with relativity?

Problems

1. A particle is moving with a speed of 2.4×10^8 m/sec in a certain frame of reference. Find β and γ for this particle.

2. An electron is moving with $\beta = 0.7$ with respect to a laboratory. (a) Find the speed of the particle in centimeters per second. (b) Find γ for the particle.

3. Two events occur at the same time in a certain inertial frame and are separated by a distance of 3 cm. An observer moves relative to that inertial reference system with a speed equal to $0.8c$. What time interval will the observer measure between the two events?

4. Two events occur at the same place in one inertial system and are separated by 3 sec in time. In a second inertial system, these two events are separated by 5 sec. What is the relative velocity of the two inertial systems?

5. A 6-ft-tall astronaut is asleep in a rocket ship that moves relative to the earth with a velocity of $0.6c$. The bed is parallel to the axis of the ship, and the direction of motion. What is the height of the astronaut as measured on earth?

6. The astronaut of Problem 5 wakes up and spends 2 min shaving. How long does an observer on earth measure the time spent shaving?

7. A space traveler is traveling toward a distant galaxy at a speed of $0.8c$ relative to the earth. Eight years after he leaves the earth his pet dog, which he has smuggled aboard, dies. How long did the dog live on the voyage, as measured by an earthbound ASPCA inspector?

8. When at rest, muons decay with a half-life of 2.2×10^{-6} sec. If a group of muons approach the earth with a speed of $0.99c$, what will be the half-life of the muons as measured by an observer on earth?

9. (a) In Problem 8, how far does a muon travel, as measured by the observer on earth, in one half-life as measured by the observer on earth? (b) How far does a muon travel in one half-life, to an observer traveling with the muons? (Both the distance and half-life are measured by the observer traveling with the muons.) (c) Calculate the velocity of the muons as measured by each of the observers in (a) and (b).

10. A cosmic-ray particle travels parallel to the plane of our galaxy with a speed of $0.99c$. The galaxy is about 10^5 light years in diameter. (a)

How long does it take the particle to cross the galaxy from the point of view of the earth? **(b)** How long does it take from the point of view of the particle? **(c)** How large does the particle think the galaxy is?

11. A particle is created and decays 10^{-21} sec later. During its lifetime it traveled through a particle detector which can only record the existence of the particle if it travels a distance of 10^{-6} cm or more. What is the minimum speed of the particle which permits its detection?

12. A rocket ship moving at speed $\frac{3}{5}c$ with respect to the earth launches a signal rocket which travels with a speed of $\frac{3}{5}c$ forward with respect to the ship. What is the speed of the signal rocket relative to the earth?

13. A pion moves to the right with speed $\frac{4}{5}c$ relative to the laboratory. The pion decays into a neutrino, which moves to the right with speed c relative to the pion. What is the speed of the neutrino relative to the laboratory?

14. The pion of Problem 13 also decays into a muon, moving to the left with a speed V relative to the pion. Find the laboratory speed of the muon in terms of V.

15. Assume that the earth is a sphere whose radius is 6500 km and that the velocity of the earth relative to the sun is 30 km/sec. How much is the earth's diameter shortened to an observer on the sun?

16. A rocket ship moves by the earth with a speed $\frac{3}{5}c$. A clock in the nose of the ship reads $t'=0$, when a clock on earth reads $t=0$. What will the clock on earth read when the rocket ship's clock reads $t' = 100$ sec?

17. A certain event occurs at $x = 150$ km, $y = 75$ km, $z = 75$ km, $t = 4$ sec. Find the coordinates of this same event in a primed coordinate system S' as described in the text, where the relative velocity of the two systems is equal to $0.95c$.

18. In one coordinate system event A occurs and then 10^{-5} sec later event B occurs, 100 m away. A second coordinate system moves by at speed $0.6c$ moving from event A to event B. Find the spatial separation and the time interval between these two events in the second system.

19. Is there a coordinate system in which the two events of Problem 18 are simultaneous? If so, describe it. If not, justify your answer.

20. A rocket ship of rest length 100 m moves past an observer on earth with a speed of $\frac{4}{5}c$. Just as the nose of the ship passes over the observer, an astronaut A in the nose of the ship sends a light pulse to an astronaut B in the tail of the ship. At that instant A and the earth observer set their clocks to read zero. **(a)** How much time does it take for the light pulse to reach B, as measured by A? **(b)** How much time does it take for the light pulse to reach B as measured by the earth observer? **(c)** At what time, as measured by A, will B pass the earth observer? **(d)** At what time, as measured by the earth observer, will B pass the earth observer?

21. A man in rocket ship A observes a second rocket ship B receding from him at speed $0.8c$ and a third rocket ship C receding in the opposite direction at the same speed. What will an observer in B measure for the speed of C?

22. Assume that a particle is moving in the primed coordinate system with a constant speed of v_y' parallel to the y' axis. By an argument similar to that of Section 10-8, show that

$$v_y = v_y'\sqrt{1 - \frac{v^2}{c^2}}$$
$$v_x = V$$

23. If a particle is moving in the primed coordinate system with a velocity whose components are v_x' and v_y', show by an argument similar to that of Section 10-8 that

$$v_x = \frac{v_x' + V}{1 + \dfrac{Vv_x'}{c^2}}$$

$$v_y = \frac{v_y'\sqrt{1 - \dfrac{V^2}{c^2}}}{1 + \dfrac{Vv_x'}{c^2}}$$

24. In a certain laboratory, light is emitted by one atom and absorbed by another atom. Show that there is no frame of reference (moving with a speed less than c) in which the emission and absorption are simultaneous.

25. A stick at rest in a frame of reference S has a length L and makes an angle θ with the x axis, so that the x component of its length is $L_x = L \cos\theta$, and the y component $L_y = L \sin\theta$. **(a)** Find the x' and y' components of the length of the stick as measured by an observer S' moving relative to S with a speed v. **(b)** Find the length of the stick as measured by S'. **(c)** Find the angle θ' that the stick makes with the x' axis.

26. The light on the nose of a rocket ship flashes on and off with a frequency f, so that the time T between flashes is given by $T = 1/f$. A second rocket ship moves by the first with a relative speed v. Show that an observer on the second ship will measure the time interval between arrivals of the flashes as T', where

$$T' = \left(\frac{1+\beta}{1-\beta}\right)^{1/2} T$$

and that therefore the second observer will measure the frequency to be

$$f' = \left(\frac{1-\beta}{1+\beta}\right)^{1/2} f$$

where

$$\beta = \frac{v}{c}$$

Since the color of light is determined by the frequency of oscillation, this implies that a moving observer will see light at a lower frequency; hence he will observe the color to be redder than the emitted light.

27. In order to show that the transformation of Equation 10-5 is the most general form consistent with the idea that the length of an object should not depend on where it is, consider the next most complicated form,

$$x = Ax' + Bt' + C(x')^2$$

The length of a rod that is fixed in the unprimed system has a value L. Show that its length, when measured in the primed system, will depend upon its coordinates in this system.

28. Starting with Equations 10-5 and 10-6 derive Equations 10-7 and 10-8. To do this assume that the primed observer measures the relative velocity by measuring the displacement, x', of the origin of the unprimed system ($x = 0$) during a time t'. Similarly the unprimed observer measures the displacement of the primed origin. Both observers agree on the magnitude of the relative velocity v. Further, one of the observers will call the relative velocity positive, while the other will call the relative velocity negative.

29. An interesting numerical example of the twin paradox was suggested by Sir Charles Darwin and can be restated in the following form: Twin A leaves the earth on New Year's Day and, traveling at $\frac{4}{5}c$, goes to Alpha Centauri, which is 4 light-years distant from the earth. He then turns around and returns to earth at the same speed. Twin B remains on earth. Each twin sends the other a radio message on New Year's Day. Ignore the time during which A turns around and show that: **(a)** A sends 6 messages while B sends 10; **(b)** A receives 1 message before reaching Alpha Centauri, and receives the remaining 9 after turning around; **(c)** B receives the first 3 messages during the first 9 yr after A leaves, and the remaining 3 during the last year of the trip.

relativistic dynamics

chapter 11

11-1
Relativistic momentum

In spite of the results of Chapter 10, where we showed that motion is properly described by Einsteinian relativity, up to this point we have assumed that dynamics (the study of forces as well as energy and momentum) is properly described by Galilean relativity. That is, we have taken Newton's second law as the basis for dynamics and, in addition, we have implicitly assumed that all inertial observers would measure the same values for mass and force. But, if all inertial observers believe that

$$\mathbf{F} = m\mathbf{a}$$

and all inertial observers agree on the values of m and $\mathbf{F}$, then all observers must agree on the value of $\mathbf{a}$. From Chapter 10 we know that this can be true for Galilean relativity, but not for Einsteinian relativity. Similarly, we have ignored special relativity in our discussion of energy and momentum.

We have been able to ignore the results of special relativity in our discussion of dynamics because special relativity gives approximately the same answers as Galilean relativity for objects moving with speeds that are small compared with the speed of light. Thus, for a huge class of phenomena (and for essentially all the phenomena known to man before the twentieth century) all of our preceding discussion of dynamics is correct in the sense that it accurately describes and predicts the results of experiments. But our preceding results (which are usually called *classical dynamics*) are incomplete because they do not correctly describe the behavior of fast-moving objects.

This leads to the idea that we must modify classical dynamics in order to formulate a *relativistic dynamics*. We must modify our ideas about force and mass, energy and momentum, so that they are consistent with relativistic notions about space and time; but these modified ideas must reduce to the classical ideas for objects moving with classical speeds. This task of modifying (or amending) classical dynamics is a large one, and mathematically very complex. Therefore, in this book we will consider only some parts of relativistic dynamics. We will discuss relativistic momentum, energy, and mass, because in some sense these are the most useful, and because they best illustrate the changes due to the special theory of relativity.

We will begin with momentum, keeping in mind two important ideas:

1. The principle of conservation of momentum has worked so well in classical physics that we will assume that momentum is conserved in relativistic dynamics. That is, we will define a quantity called momentum, which is conserved in situations analogous to those in which classical momentum is conserved.
2. Relativistic momentum will be defined in such a way that when a particle is moving with a low velocity, its momentum is approximately equal to $m\mathbf{v}$, thus ensuring that relativistic momentum

is conserved in all those situations in which we know that classical momentum is conserved.

The assumption of momentum conservation suggests that we look at a simple collision process, for which the results of Chapter 10 will permit us to calculate various quantities before and after the collision. We will be able to choose a quantity that is conserved and also becomes equal to $m\mathbf{v}$ for low velocities. We will then define this quantity to be relativistic momentum.

Let us therefore assume that there exist two observers, O and O', moving past each other with constant relative velocity, $\mathbf{V}$, in the x direction, as discussed in Chapter 10. The two observers synchronize their clocks, compare their meter sticks, and then agree to perform an experiment involving a *symmetric*, *elastic* collision. The observers are displaced in the y direction, as shown in Figure 11-1a. Observer O agrees to throw a ball, A, along the y axis toward observer O', with a speed of $-u$. The observer O' agrees to throw an identical ball, B, along the y' axis toward O with the same speed, $+u$. The observers time the throws so that the two balls collide. We assume that the collision is elastic, so each ball returns to the thrower with the same speed as it was thrown (as measured by the thrower). Each observer sees the ball that he threw move out along the y axis and return, but the ball thrown by the other observer has a component

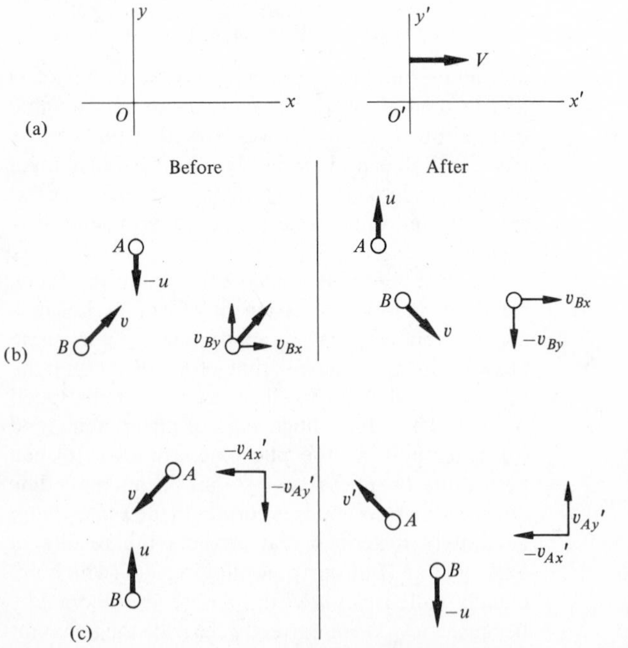

(a)

Before | After

(b)

(c)

FIGURE 11-1 (a) Two observers displaced in the y direction moving with a relative velocity V in the x direction agree to perform an experiment involving a symmetric elastic collision. (b) The collision as seen by the unprimed observer. (c) The collision as seen by the primed observer.

of velocity along the x axis. Figure 11-1b shows the collision as viewed by O: before the collision A has a velocity, $-u$, along the y axis, and B has a velocity which has components v_{Bx} and v_{By}. After the collision, O sees A as having a velocity, u, along the y axis, and B as having a velocity with components v_{Bx} and $-v_{By}$. Figure 11-1c shows the collision as seen by O': before the collision B has a velocity, u, along the y axis, and A has a velocity with components $-v_{Ax}'$ and $-v_{Ay}'$. After the collision, O' sees B as having a velocity, $-u$, along the y axis, and A as having a velocity with components $-v_{Ax}'$ and v_{Ay}'. That is, each observer sees the ball thrown by the other observer as moving with constant velocity in the x direction, and completely reversing its y component of velocity.

Now we can apply the results of Chapter 10, and express the velocities v_{Ax}, v_{Ay}, v_{Bx}', and v_{By}' in terms of the relative velocity between the observers, V, and the velocity, u. In Problem 23, Chapter 10, we showed that the relation between the components of velocity as measured in one coordinate system (v_x', v_y') and the velocity components of the same object as measured in another coordinate system (v_x, v_y) is given by

$$v_x' = \frac{v_x - V}{1 - \dfrac{Vv_x}{c^2}} \tag{11-1}$$

$$v_y' = \frac{v_y\sqrt{1 - \dfrac{V^2}{c^2}}}{1 - \dfrac{Vv_x}{c^2}} \tag{11-2}$$

To observer O, A has velocity components

$$v_{Ax} = 0$$
$$v_{Ay} = -u$$

and to observer O', the same object has velocity components $-v_{Ax}'$ and $-v_{Ay}'$. Therefore, from Equation 11-1,

$$-v_{Ax}' = \frac{0 - V}{1 - \dfrac{V(0)}{c^2}}$$

$$= -V$$

or

$$v_{Ax}' = V \tag{11-3}$$

and from Equation 11-2

$$-v_{Ay}' = \frac{-u\sqrt{1 - \dfrac{V^2}{c^2}}}{1 - 0}$$

$$= -u\sqrt{1 - \frac{V^2}{c^2}} \tag{11-4}$$

That is, observer O' sees the ball thrown by observer O moving with an x component of velocity that is just the relative velocity between the observers (which is exactly what we expect classically), but he sees the ball moving in the y direction with a velocity that is smaller than u by the factor

$$\sqrt{1 - \frac{V^2}{c^2}}$$

We can show similarly that

$$v_{Bx} = V \tag{11-5}$$
$$v_{By} = u\sqrt{1 - \frac{V^2}{c^2}} \tag{11-6}$$

Thus, each observer sees exactly what we expect *classically* for the x component of velocity, but sees a smaller y component of velocity than we would have expected classically.

We know that, classically, the quantity $m\mathbf{v}$ is conserved in this collision. Now that we know that each observer measures a *transverse* velocity that is smaller than we expect, we know that this quantity will not be conserved. The reader is invited to calculate the y component of $m\mathbf{v}$ as measured by O both before and after the collision to verify that this quantity does not remain constant during the collision.

In order to retain the principle of the conservation of linear momentum in relativistic dynamics, the momentum $\mathbf{P}$ of a particle of mass m and velocity $\mathbf{v}$ is defined as

$$\mathbf{P} = \frac{m\mathbf{v}}{\sqrt{1 - \dfrac{v^2}{c^2}}} \tag{11-7}$$

It is not very difficult, but rather cumbersome, to solve the problem of the elastic collision of two particles using the definition of linear momentum given by Equation 11-7 and show that linear momentum is conserved in this case. Other examples will be considered later.

The definition of momentum given by Equation 11-7 does become approximately equal to $m\mathbf{v}$, because for low velocities, the denominator is very close to unity.

The mass m in Equation 11-7 is assumed to be

constant, independent of the velocity of the particle. That is, we have been able to define relativistic momentum without modifying the classical quantity "mass." We will discuss this subject further in a subsequent section.

Illustrative example

A proton moving with a velocity $3c/5$ strikes a stationary nucleus and rebounds, moving back with the same speed, $3c/5$, as observed in the laboratory system. Find the momentum of the nucleus after the collision.

Taking the initial direction of motion of the proton to be positive, the initial momentum of the system, $\mathbf{P}$, is the sum of the momentum of the proton, $\mathbf{p}$, and the momentum of the nucleus. Since the nucleus is initially at rest, the initial momentum of the nucleus is equal to zero, and so,

$$\mathbf{P} = \mathbf{p}$$

hence

$$P = \frac{mV}{\sqrt{1 - \dfrac{v^2}{c^2}}}$$

$$P = \frac{\frac{3}{5}mc}{\sqrt{1 - \dfrac{9}{25}}}$$

$$= \frac{\frac{3}{5}mc}{\frac{4}{5}}$$

$$= \tfrac{3}{4}mc$$

for the initial momentum of the system. After the collision, the proton has momentum $-p$, and the nucleus has momentum R. Since the total momentum of the system is conserved, we have

$$P = -p + R$$

but

$$P = p$$

and so

$$p = -p + R$$
$$R = 2p$$
$$= \tfrac{3}{2}mc$$

for the recoil momentum of the nucleus.

11-2
Relativistic energy

In the preceding section we discussed the collision of two identical particles, where we assumed that the collision was elastic. Now we will consider an inelastic collision of two identical particles. Consider two identical particles each of mass m, which, in a reference frame O, are moving toward each other with equal and opposite speeds v as shown in Figure 11-2. After the collision, the two particles stick together, forming a composite particle. From conservation of momentum we know that, since the total initial momentum of the system was zero, the composite particle also will have momentum zero and so will be at rest in that frame of reference.

Let us now look at this same collision in a frame of reference O', which is moving to the right relative to O, with relative velocity v. In O', the particle on the right in Figure 11-2 is initially at rest, and the other particle is approaching with velocity v', whose value, from Chapter 10, is given by

$$v' = \frac{v + v}{1 + \dfrac{v^2}{c^2}} \tag{11-8}$$

The composite particle is moving with velocity $-v$ in the frame O', since it was at rest in the frame O. We could verify that momentum is conserved in the reference frame if we knew what the mass of the composite particle was. If we assume that mass is conserved in the collision, then we discover that momentum is *not* conserved in O'. In order to retain conservation of momentum we must give up mass conservation and assume that the composite particle has a mass M, which is not necessarily

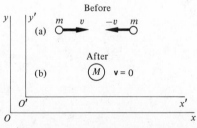

FIGURE 11-2 (a) Two identical particles approaching each other with identical speeds as seen in the unprimed reference frame. (b) The composite particle formed by the collision is at rest in the unprimed reference frame.

equal to $2m$. Then we can write that the momentum before the collision is equal to the momentum after the collision in the frame O'. Since one particle is at rest before the collision and the other has velocity v', we have

$$\frac{mv'}{\sqrt{1 - \frac{(v')^2}{c^2}}} + 0 = \frac{Mv}{\sqrt{1 - \frac{v^2}{c^2}}} \qquad \text{(11-9)}$$

Then, using Equation 11-8 which gives us the relation between v and v', we have

$$\frac{m\left(\dfrac{2v}{1 + v^2/c^2}\right)}{\sqrt{1 - \dfrac{4v^2/c^2}{\left(1 + \dfrac{v^2}{c^2}\right)^2}}} = \frac{Mv}{\sqrt{1 - \dfrac{v^2}{c^2}}}$$

which becomes

$$\frac{2mv}{1 - \dfrac{v^2}{c^2}} = \frac{Mv}{\sqrt{1 - \dfrac{v^2}{c^2}}}$$

and so

$$M = \frac{2m}{\sqrt{1 - \dfrac{v^2}{c^2}}} \qquad \text{(11-10)}$$

Thus we find that in order for momentum to be conserved, the mass of the composite particle must be greater than the sum of the masses of the particles that formed the composite.

Equation 11-10 tells us that in the theory of special relativity, mass is not a conserved quantity, but the same equation suggests that there is a new quantity, which *is* conserved. In particular, since the velocity u of the composite particle in the frame O is equal to zero, the left side of Equation 11-10 is equal to a quantity Q, which can be defined for a particle of mass M and velocity u by

$$Q = \frac{M}{\sqrt{1 - \dfrac{u^2}{c^2}}} \qquad \text{(11-11)}$$

Then Equation 11-10 can be written as

$$\frac{M}{\sqrt{1 - \dfrac{u^2}{c^2}}} = \frac{m}{\sqrt{1 - \dfrac{v^2}{c^2}}} + \frac{m}{\sqrt{1 - \dfrac{v^2}{c^2}}}$$

$$Q_M = Q_m + Q_m$$

Thus the quantity Q is conserved in this process.

Therefore the quantity Q times any constant is also constant. The constant that has been selected is c^2, where c is the speed of light. We now define a new quantity $\mathscr{E}$ by the equation

$$\mathscr{E} = Qc^2 \qquad \text{(11-12)}$$

and for a particle of mass m moving with speed v, this becomes

$$\mathscr{E} = \frac{mc^2}{\sqrt{1 - \dfrac{v^2}{c^2}}} \qquad \text{(11-13)}$$

We can attach significance to Equation 11-13 by examining its behavior for small values of the speed v.

The binomial expansion

$$(1 + x)^n = 1 + nx + \frac{n(n-1)}{2}x^2 + \cdots$$

is true for any n; hence

$$\frac{1}{(1 + x)^{1/2}} = 1 - \frac{x}{2} + \cdots$$

and

$$\frac{1}{(1 - x)^{1/2}} = 1 + \frac{x}{2} + \cdots$$

Thus if $v^2/c^2 \ll 1$, $\mathscr{E}$ is approximated by

$$\mathscr{E} \approx mc^2\left(1 + \frac{v^2}{2c^2}\right) = mc^2 + \tfrac{1}{2}mv^2 \quad \text{(11-14)}$$

So for a particle moving with low speed, the quantity $\mathscr{E}$ is equal to a constant (the mass times c^2) plus the kinetic energy. Thus, the statement that the quantity $\mathscr{E}$ is conserved is, in some sense, the statement that the sum of the mass and the kinetic energy is conserved. To be more precise, we call the quantity $\mathscr{E}$ the *total energy of the particle*, and then assert that the total energy is a conserved quantity. For a particle at rest, the total energy $\mathscr{E}$ is called the *rest mass energy*, and is given by

$$\mathscr{E} = mc^2 \qquad \text{(11-15)}$$

an equation that is by now as well known to laymen as any in physics. A particle that is moving with velocity v has mass energy given by Equation 11-15 and in addition, has energy of motion, or kinetic energy. Equation 11-13 gives us the *total* energy, so we can find the kinetic energy $\mathscr{E}_k$ as the difference between the total energy and the rest mass energy:

$$\mathcal{E}_k = \frac{mc^2}{\sqrt{1 - \dfrac{v^2}{c^2}}} - mc^2 \qquad (11\text{-}16a)$$

$$= \mathcal{E} - mc^2 \qquad (11\text{-}16b)$$

Before the development of the special theory of relativity, mass and energy were separate concepts, separately conserved. Special relativity states that mass is not conserved, but that total energy is conserved only if rest mass energy is included in the idea of energy. In that sense, the ideas of conservation of mass and conservation of energy have been merged into one idea. As an illustration of this idea, it should be pointed out that the inelastic collision we have been considering in this section is inelastic in the sense that the kinetic energy is not conserved. The total energy *is* conserved, because the lost kinetic energy appears in the form of rest mass energy of the composite particle. In all of this discussion the relativistic form for kinetic energy, Equation 11-16, must be used for all particles.

11-3
Units of energy and momentum

From Equation 11-7 and 11-13 it is easy to show that

$$\mathcal{E}^2 = p^2 c^2 + m^2 c^4 \qquad (11\text{-}17)$$

and that

$$v = c^2 \frac{p}{\mathcal{E}} \qquad (11\text{-}18)$$

Equations 11-17 and 11-18 are particularly useful in atomic and nuclear physics. For these subjects, which involve submicroscopic particles, the standard units applied to macroscopic systems become unwieldy. It has become usual to measure energy in *electron volts*, where one electron volt is the kinetic energy gained by a particle having a charge equal to that of an electron that has been accelerated through an electric potential difference of one volt. The electron volt is given by

$$1 \text{ eV} = 1.6 \times 10^{-19} \text{ joule} \qquad (11\text{-}19)$$

The mass of a particle can be expressed, by means of Equation 11-15, in terms of an equivalent amount of energy. For example, the mass of a

proton is 1.6724×10^{-27} kg. Thus, from Equation 11-15,

$$\begin{aligned}
\mathcal{E} &= mc^2 \\
&= 1.6726 \times 10^{-27} \times 9 \times 10^{16} \\
&= 15.0534 \times 10^{-11} \text{ joule} \\
&= \frac{15.0534}{1.6} \times \frac{10^{-11}}{10^{-19}} \text{ eV} \\
&= 9.4 \times 10^8 \text{ eV} \\
&= 940 \text{ MeV}
\end{aligned}$$

where

$$1 \text{ MeV} = 10^6 \text{ eV} = 1 \text{ million electron volts}$$

Similarly, the rest mass energy of an electron is 0.51 MeV.

From Equation 11-18, we have

$$\frac{v}{c} = \frac{pc}{\mathcal{E}} \qquad (11\text{-}20)$$

or

$$pc = \left(\frac{v}{c}\right) \mathcal{E} \qquad (11\text{-}21)$$

Therefore, if we know the total energy of a particle in units of electron volts, we can express the quantity pc in electron volts. This is usually abbreviated in the following way: if a particle has a momentum p such that pc is equal to, say, 1 MeV, then the particle is said to have a momentum of 1 MeV/c. The advantage of this system of units is that, if mass and energy are expressed in eV and momentum in eV/c, then Equation 11-17 can be written in the form

$$\mathcal{E}^2 = p^2 + m^2 \qquad (11\text{-}22)$$

and Equation 11-16 can be written as

$$\mathcal{E}_k = \mathcal{E} - m \qquad (11\text{-}23)$$

Thus, if an electron acquires 1 MeV of kinetic energy in a particle accelerator, then it will have a total energy

$$\begin{aligned}
\mathcal{E} &= 1.0 \text{ MeV} + 0.51 \text{ MeV} \\
&= 1.51 \text{ MeV}
\end{aligned}$$

and it will have a momentum p given by

$$\begin{aligned}
p^2 &= \mathcal{E}^2 - m^2 \\
&= 2.28 - 0.26 \\
&= 2.02
\end{aligned}$$

so that

$$p = 1.42 \, \frac{\text{MeV}}{c}$$

or, rather,

$$pc = 1.42 \text{ MeV}$$

11-4
Relativistic mass and force

In our treatment of relativistic momentum and energy we have assumed that the mass of a particle is a constant, independent of the velocity of the particle. The reader has probably seen it asserted in other places that the mass of a particle increases with the speed of the particle. It is common in other books to define a quantity called the *relativistic mass*, m_r, by means of the equation

$$m_r = \frac{m}{\sqrt{1 - \dfrac{v^2}{c^2}}} \qquad (11\text{-}24)$$

With this equation one can rewrite Equation 11-13 as

$$\mathcal{E} = m_r c^2 \qquad (11\text{-}25)$$

and Equation 11-7 as

$$\mathbf{p} = m_r \mathbf{v} \qquad (11\text{-}26)$$

Rewritten in this way these equations look simpler, and Equation 11-26 looks exactly the same as the classical definition of momentum, because all of the relativistic effects *in these two* equations are concealed in the definition of relativistic mass. As long as we deal only with Equations 11-25 and 11-26 we can think of the relativistic mass as *the* mass and then assert that the mass of a particle increases with velocity as prescribed by Equation 11-24.

The idea of relativistic mass, which varies with velocity, can lead to some confusion and error. Relativity changes our ideas about space and time and, to some extent, our ideas about dynamical concepts such as energy and momentum. It is possible to introduce these ideas without changing our ideas about the properties of particles. We can, for example, retain the idea that the mass and charge of a particle are constants. Thus, the idea of relativistic mass is an unnecessary complication. Perhaps, just as important, the idea of relativistic mass implies that we can "derive" the equations of relativity by replacing mass in classical equa-

tions by the relativistic mass. It is clear that this is wrong, when one considers Equation 11-16, which cannot be derived from the classical definition of kinetic energy by the mere substitution of relativistic mass for classical mass.

A problem related to the problem of relativistic mass involves the correct relativistic form for Newton's second law. Classically, we were able to use two expressions that are equivalent for a particle of fixed mass

$$\mathbf{F} = m\mathbf{a} \qquad (11\text{-}27)$$

and

$$\mathbf{F} = \frac{\Delta \mathbf{p}}{\Delta t} \qquad (11\text{-}28)$$

When we introduce the relativistic description of space and time, these two expressions are no longer equivalent. In particular, if we use Equation 11-28 to define force relativistically, then the force will not be parallel to the acceleration in all reference frames. The problem of choosing the correct form for the second law reduces to the problem of finding at least one force whose influence on relativistic particles is known. Knowing one such force, we can test Equations 11-27 and 11-28 and decide which is correct.

The special theory of relativity began with an investigation of the properties of light. Our understanding of the properties of light is, in turn, based on our knowledge of electricity and magnetism. Therefore, it is reasonable to examine the relativistic form of Newton's second law by means of the behavior of electric and magnetic forces. The result of such studies is that Newton's second law is written properly, at least for electric and magnetic forces, by Equation 11-28. Thus we assume that this is the correct form of Newton's second law for all forces. It must be noted that in terms of the fundamental forces, special relativity has been applied only for the electromagnetic force; gravitational forces involve the extensions of relativity called general relativity and the nuclear forces are described in terms of quantum mechanics rather than Newton's second law.

Since the relativistic form of Newton's second law is written in terms of the change of momentum rather than the product of mass and acceleration, we do not have to introduce the idea of relativistic mass. We can retain Equation 11-7 as the definition of momentum, with m representing a constant mass.

Questions

1. What is the minimum energy of a particle?

2. What is the minimum momentum of a particle?

3. Can the classical equations of physics be converted to the correct relativistic equations by substituting a mass that varies with speed for the constant mass of classical physics? Give an example.

4. Is it possible for a particle to have energy but not to have momentum? If so, give an example.

5. Is it possible for a particle to have momentum but not to have energy? If so, give an example.

6. Explain what happens to the mass, velocity, momentum, and energy of a particle if a constant force is applied to the particle for a very long time.

Problems

1. What is the momentum of a 3-gm object moving with a speed of $0.8c$?

2. An electron (mass 9.1×10^{-28} gm) is accelerated to a speed of $\frac{3}{5}c$. What is its momentum?

3. What is the total energy of the electron of Problem 2? The kinetic energy?

4. A particle has a rest mass energy of 1 erg. What is its mass?

5. What is the speed of a particle if it has a kinetic energy equal to its rest mass energy?

6. (a) Sketch a graph showing the variation of momentum with speed. (b) Sketch a graph showing the variation of total energy with speed for a particle of mass m.

7. A proton (mass 1.67×10^{-24} gm) is moving with a speed v such that its kinetic energy is 4 times the rest mass energy. Find the speed v, and the momentum (in units of GeV/c, where 1 GeV $= 10^9$ eV).

8. What is the momentum of (a) an electron having a kinetic energy of 1 GeV, (b) a pion having a kinetic energy of 1 GeV, and (c) a proton having a kinetic energy of 1 GeV?

9. Protons in a large accelerator are accelerated to a total energy of 30 GeV. Find their momentum and speed. Find the mass of the particle whose rest mass energy is equal to 30 GeV.

10. The half-life of a pion at rest is 1.8×10^{-8} sec. What is the half-life of a 400-MeV pion?

11. If electric power costs 6 cents per kilowatt-hour, how much does the energy equivalent to 1 gm of nuclear fuel cost?

12. In the fusion of four protons to form a helium nucleus in the sun, the total mass of the helium nucleus is about 5×10^{-26} gm *less* than that of the four protons. (a) How much energy is liberated in this process? (b) Two grams of hydrogen contains about 10^{24} protons. How much energy would be liberated by fusion if all of these protons formed helium?

13. Solar energy reaches the earth at the rate of 8.4×10^4 joules/m^2/min. The radius of the earth is 6.4×10^6 m and the sun is 1.5×10^{11} m away. By how much is the sun's mass decreasing per second? How much of the sun's mass reaches the earth per year? Why doesn't the earth's mass increase this much?

14. (a) Find a classical expression for the kinetic energy $\mathscr{E}_k$ of a particle in terms of the momentum p. (b) Show that the relativistic expression for $\mathscr{E}_k$ reduces to the classical expression for speeds that are small compared with the speed of light.

15. A ^{238}U nucleus decays into an alpha particle and a ^{234}Th nucleus. The alpha particle has a mass of 4.0026 atomic mass units and the ^{238}U has a mass of 238.0508 atomic mass units, where one atomic mass unit is equivalent to 931.5 MeV. The alpha particle leaves with a kinetic energy equal to 4.19 MeV. (a) Find the kinetic energy of the ^{234}Th nucleus. (b) Find the speed of the alpha particle.

harmonic motion

chapter 12

12-1
Simple harmonic motion

An extremely important type of periodic motion of a particle is one in which *the acceleration* **a** *of the particle is proportional to its displacement* **x** *from its equilibrium position and is opposite in direction to the displacement; that is,*

$$\mathbf{a} \propto \mathbf{x}$$

or

$$\mathbf{a} = -c\mathbf{x} \qquad (12\text{-}1)$$

where c is a constant of proportionality. The minus sign is used to show that the direction of the acceleration is always opposite to the direction of the displacement of the particle. The type of motion defined by Equation 12-1 is called *simple harmonic motion.*

A common example of simple harmonic motion is the motion of a body attached to one end of a helical spring that is suspended from some fixed support O (see Figure 12-1). When a body of mass m is attached to this spring and lowered slowly, it will come to rest at some point C, at which position it is in equilibrium under the action of two forces— its weight $m\mathbf{g}$ downward and the pull of the spring $\mathbf{F}_1$ upward. Suppose we now pull the mass down a distance x below C. To hold it there, we shall have to exert a force $\mathbf{F} = k\mathbf{x}$, where k is the constant of the spring (see Section 6-10). The spring will exert an equal force in the upward direction—that is, a force

$$\mathbf{F} = -k\mathbf{x} \qquad (12\text{-}2)$$

The distance **x** is called the *displacement* of the body from its equilibrium position C, and the

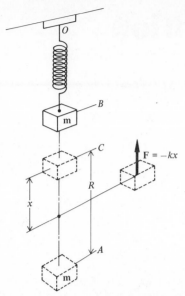

FIGURE 12-1 A body of mass m attached to a helical spring c in its equilibrium position. When the body is pulled down to A and released, it will vibrate from A to B, and back to A, repeating this motion periodically; x is the displacement of the body at a point in its path.

minus sign indicates that the force the spring exerts on it is opposite to the displacement.

If we now release the body, it will no longer be in equilibrium, and the unbalanced force acting on it will be $\mathbf{F} = -k\mathbf{x}$. As the body returns to its equilibrium position, its displacement becomes smaller and so does the unbalanced force $\mathbf{F}$. At the equilibrium position $\mathbf{F} = 0$, but the motion does not stop. From Newton's first law, we know that the body will continue to move in the same direction with the same speed unless an unbalanced force acts on it. As it moves above C, it is again displaced from its equilibrium position, and the unbalanced force $\mathbf{F} = -k\mathbf{x}$ acts on it; the direction of this force is again opposite to the displacement and decreases the speed of the body.

If originally the body attached to the spring was pulled down to position A, a distance R below C, it would move upward beyond C to a position B at a distance almost equal to R. If there were no losses due to air resistance and internal friction in the spring, the distances AC and BC would be equal. After the body has reached point B and stopped, it will return through C to A and then back again through C to B and keep up this motion indefinitely. If we time this motion, we shall find that the motion is periodic, that is, that the time to

execute a complete vibration, going from A to B and back to A, is always the same.

We can apply Newton's second law

$$\mathbf{F} = m\mathbf{a}$$

in discussing the motion of this particle. The unbalanced force acting on this particle is, from Equation 12-2,

$$\mathbf{F} = -k\mathbf{x} \qquad (12\text{-}2)$$

Equating these two expressions for the force on the particle, we get

$$m\mathbf{a} = -k\mathbf{x}$$

from which

$$\mathbf{a} = -\frac{k}{m}\mathbf{x} \qquad (12\text{-}3)$$

Equation 12-3 shows that the acceleration of this particle is directly proportional to the displacement from its equilibrium position and is always opposite in direction to that of the displacement. Equation 12-3 is identical with Equation 12-1, the defining equation of simple harmonic motion with the constant of proportionality $c = k/m$ for this specific case.

Illustrative example

A spring whose natural length, when hung from a fixed point O, is 40 cm has a 50-gm mass attached to its free end. When this mass is at the equilibrium position C, the length of the spring is 45 cm. The mass is then pulled down a distance of 6 cm and released. Determine (a) the constant of the spring, (b) its acceleration at the 6-cm point, (c) its acceleration when it has reached a point 2 cm above C, and (d) the unbalanced force that acts on it at the 2-cm point.

(a) Since the length of the spring was increased from 40 to 45 cm by hanging the 50-gm mass on it, the weight mg of this mass produced an extension of 5 cm in the length of the spring; hence

$$mg = kx$$

or

$$50 \text{ gm} \times 980\frac{\text{cm}}{\text{sec}^2} = k \times 5 \text{ cm}$$

from which

$$k = 9800\frac{\text{dynes}}{\text{cm}}$$

(b) The acceleration of the body is given by Equation 12-3

$$a = -\frac{k}{m}x \qquad (12\text{-}3)$$

At the 6-cm point, $x = 6$ cm, and thus

$$a = -\frac{9800 \text{ dynes/cm}}{50 \text{ gm}} \times 6 \text{ cm}$$

or

$$a = -1176 \frac{\text{cm}}{\text{sec}^2}$$

(c) At a distance of 2 cm above C, $x = 2$ cm; hence

$$a = -\frac{9800 \text{ dynes/cm}}{50 \text{ gm}} \times 2 \text{ cm}$$

or

$$a = -392 \frac{\text{cm}}{\text{sec}^2}$$

(d) The unbalanced force which acts on the mass is given by the equation

$$F = -kx \qquad (12\text{-}2)$$

so that

$$F = -9800 \frac{\text{dynes}}{\text{cm}} \times 2 \text{ cm}$$

from which

$$F = -19{,}600 \text{ dynes}$$

12-2
Period of simple harmonic motion

To determine the period of a particle in simple harmonic motion directly from Equation 12-3 involves the use of calculus. One method of circumventing this difficulty is to make use of a simple relationship that exists between a particle in simple harmonic motion and another particle in uniform circular motion, both having the same period. It will be shown that simple harmonic motion is the projection of uniform circular motion on a line in the plane of the circle. For convenience, we shall take this line to coincide with a diameter of the circle.

Let us consider a particle P moving with uniform circular motion, as shown in Figure 12-2. Let us call its velocity v_0 and its acceleration a_0. Its ac-

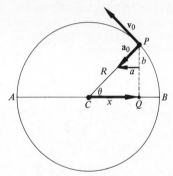

FIGURE 12-2 Projection of the motion of particle P on a diameter of the circle. a_0 is the acceleration of P; a is the acceleration of Q, the projection of P on the diameter AB.

celeration is directed toward the center C of the circle of radius R. To project this motion on a horizontal diameter AB, draw a perpendicular from the particle P onto this diameter to point Q; the particle moving with simple harmonic motion is, at this instant, situated at Q. As the particle P moves around the circular path, the particle Q moves back and forth along the diameter AB. No matter where P is, its projection on the diameter determines the position of Q. The velocity of the particle at Q is the projection of the velocity of P on AB, and the acceleration of the particle at Q is the projection of the acceleration of P on AB.

Let us first determine the acceleration a of the particle at Q. To do this, we resolve the acceleration a_0 of the particle P into two components, one of which is parallel to AB; we shall call this component a. Now we know, from Equation 7-6 of Chapter 7, that

$$a_0 = -\frac{v_0^2}{R} \qquad (12\text{-}4)$$

The minus sign is used here to indicate that a_0 is directed toward the center, while the radius R is measured outward from the center. The triangles PQC and aba_0 are similar; hence

$$\frac{a}{a_0} = \frac{x}{R}$$

where $x = QC$ and is the displacement of the particle at Q.

Substituting the value of a_0 from Equation 12-4 into the above equation yields

$$a = -\frac{v_0^2}{R^2}x \qquad (12\text{-}5)$$

From Equation 12-5 we see that since v_0 and R are constant, the acceleration of the particle at Q is directly proportional to its displacement x and opposite in direction. Hence, the particle at Q is moving with simple harmonic motion.

We have already shown in Equation 7-11 that the period T of a particle in uniform circular motion with speed v_0 is

$$T = \frac{2\pi R}{v_0} \qquad (12\text{-}6)$$

From this equation, we get

$$\frac{v_0}{R} = \frac{2\pi}{T}$$

Substituting this expression for v_0/R into Equation 12-5 yields

$$a = -\frac{4\pi^2}{T^2}x \qquad (12\text{-}7)$$

The period T is the same for the particle in simple harmonic motion as for the particle in uniform circular motion. Solving Equation 12-7 for T, we get

$$T = 2\pi\sqrt{-\frac{x}{a}} \qquad (12\text{-}8)$$

for the period of a particle in simple harmonic motion.

The period as given by Equation 12-8 is always real; this can be seen if we put in the value of x/a from Equation 12-1, which yields

$$T = \frac{2\pi}{\sqrt{c}} \qquad (12\text{-}9)$$

as a general equation for the period of a particle in simple harmonic motion.

For the case of a particle of mass m attached to a spring of constant k, we know that $c = k/m$; substituting this value into Equation 12-9 yields

$$T = 2\pi\sqrt{\frac{m}{k}} \qquad (12\text{-}10)$$

for the period of vibration of a particle attached to a spring.

Illustrative example

A particle whose mass is 30 gm is attached to a spring having a constant $k = 2400$ dynes/cm. Determine the particle's period of vibration.

The period of vibration of the particle attached to the spring is

$$T = 2\pi\sqrt{\frac{m}{k}} \qquad (12\text{-}10)$$

Substituting $m = 30$ gm and $k = 2400$ dynes/cm, we get

$$T = 2\pi\sqrt{\frac{30 \text{ gm}}{2400 \text{ dynes/cm}}}$$

from which

$$T = \frac{2\pi}{\sqrt{80}} \text{ sec} = 0.70 \text{ sec}$$

Illustrative example

A cylinder weighing 4 lb is hung from a very stiff spring whose elastic constant is 24 lb/in. Let us determine (a) the period of vibration of this cylinder and (b) its acceleration when its displacement is 3 in.

The mass of the cylinder is

$$m = \frac{4 \text{ lb}}{32 \text{ ft/sec}^2} = \frac{1}{8}\frac{\text{lb sec}^2}{\text{ft}}$$

In order to avoid having both inches and feet in the same equation, let us express k in pounds per foot:

$$k = 24 \times 12 \frac{\text{lb}}{\text{ft}} = 288 \frac{\text{lb}}{\text{ft}}$$

and

$$x = \tfrac{3}{12} \text{ ft} = 0.25 \text{ ft}$$

(a) The period can be determined by substituting the above values of m and k into Equation 12-10:

$$T = 2\pi\sqrt{\frac{m}{k}}$$

yielding

$$T = 2\pi\sqrt{\frac{\tfrac{1}{8} \text{ lb sec}^2/\text{ft}}{288 \text{ lb/ft}}}$$

or

$$T = 0.13 \text{ sec}$$

(b) The acceleration can be found with the aid of Equation 12-3; thus

$$a = -\frac{k}{m}x$$

and, substituting values for k, m, and x, we get

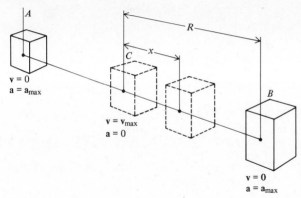

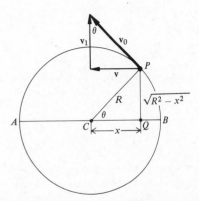

FIGURE 12-3 Path of a body in simple harmonic motion showing the values of the velocity and the acceleration at the center and at the ends of the path.

$$a = -\frac{288 \text{ lb/ft}}{\frac{1}{8} \text{ lb sec}^2/\text{ft}} \times 0.25 \text{ ft} = -576 \frac{\text{ft}}{\text{sec}^2}$$

The same result could have been obtained by using Equation 12-7.

12-3
Properties of simple harmonic motion

A term often used in discussing simple harmonic motion is the *frequency of vibration* of the particle, which is defined as *the number of vibrations per unit time*. Since T is the time for 1 vibration, the frequency of vibration f is the reciprocal of the period, or

$$f = \frac{1}{T} \tag{12-11}$$

Eliminating T between Equation 12-8 and 12-11 we get, for the frequency of vibration,

$$f = \frac{\sqrt{c}}{2\pi} \tag{12-12}$$

For the case of a particle of mass m attached to a spring of constant k, the frequency of vibration becomes

$$f = \frac{1}{2\pi} \sqrt{\frac{k}{m}} \tag{12-13}$$

Figure 12-3 shows the path of a particle in simple harmonic motion. As the particle moves in its path from A through C to B and then back through C to A to complete 1 vibration, its displacement x passes through a series of values from a maximum at A to zero at C to a maximum again at B. Let us call the value of the maximum displacement R; the maximum displacement is called the *amplitude* of the motion. It will be observed that the equations

for the period and frequency do not contain the amplitude of the motion; hence the period and frequency of a particle in simple harmonic motion are independent of the amplitude.

The velocity of a particle in simple harmonic motion can be obtained with the aid of the reference circle (see Figure 12-4). If v_0 is the velocity of the particle at P in uniform circular motion, then the velocity v of the particle at Q in simple harmonic motion is the component of v_0 parallel to the diameter AB.

Since the triangle v, v_1, v_0 is similar to the triangle PQC, we have

$$\frac{v}{v_0} = \frac{\sqrt{R^2 - x^2}}{R}$$

From the definition of the period, we know that

$$T = \frac{2\pi R}{v_0}$$

FIGURE 12-4 The velocity v of the particle at Q moving with simple harmonic motion is the horizontal component of the velocity v_0 of the particle at P moving with uniform circular motion.

hence

$$v = \frac{2\pi}{T} \sqrt{R^2 - x^2} \qquad (12\text{-}14)$$

or, in terms of the frequency $f = 1/T$, we have

$$v = 2\pi f \sqrt{R^2 - x^2} \qquad (12\text{-}15)$$

Equation 12-15 shows that the velocity is a maximum when the displacement $x = 0$, that is, at point C, and the velocity is zero when $x = R$, that is, at points A and B.

Illustrative example

A particle attached to a spring has a frequency of 4 vibrations/sec and an amplitude of 6 cm. Determine (a) the period of vibration, (b) the maximum velocity of the particle, (c) the velocity of the particle when its displacement is 2 cm, (d) the acceleration of the particle when its displacement is 2 cm, and (e) its maximum acceleration.

(a) The period of vibration is the reciprocal of the frequency; hence, from Equation 12-11,

$$T = \tfrac{1}{4} \text{ sec} = 0.25 \text{ sec}$$

(b) The velocity of the particle at any position is given by Equation 12-15:

$$v = 2\pi f \sqrt{R^2 - x^2}$$

The velocity is a maximum at the center where $x = 0$; hence, from Equation 12-15,

$$v_{max} = 2\pi f R$$

and, substituting the values

$$f = 4 \frac{\text{vibrations}}{\text{sec}} \qquad \text{and} \qquad R = 6 \text{ cm}$$

we get

$$v_{max} = 2\pi \times 4 \frac{\text{vibrations}}{\text{sec}} \times 6 \text{ cm}$$

or

$$v_{max} = 151 \frac{\text{cm}}{\text{sec}}$$

(*Note:* The term *vibrations* has no physical dimensions; hence vibrations/sec is equivalent to 1/sec.)

(c) The velocity when $x = 2$ cm is obtained by substituting this value in Equation 12-15:

$$v_2 = 2\pi \times 4 \frac{\text{vibrations}}{\text{sec}} \sqrt{(36 - 4) \text{ cm}^2}$$

or

$$v_2 = 8\pi \sqrt{32} \frac{\text{cm}}{\text{sec}} = 142 \frac{\text{cm}}{\text{sec}}$$

(d) The acceleration can be obtained from Equation 12-7:

$$a = -\frac{4\pi^2}{T^2} x$$

This equation can be rewritten in terms of the frequency f, since $f = 1/T$, so that

$$a = -4\pi^2 f^2 x$$

Substituting

$$f = 4 \frac{1}{\text{sec}} \qquad \text{and} \qquad x = 2 \text{ cm}$$

we get

$$a_2 = -4\pi^2 \times 16 \frac{1}{\text{sec}^2} \times 2 \text{ cm}$$

from which

$$a_2 = -1263 \frac{\text{cm}}{\text{sec}^2}$$

(e) The maximum acceleration will occur at $x = R = 6$ cm. Since the acceleration is directly proportional to the displacement, the maximum acceleration will be three times the acceleration at the 2-cm point or

$$a_{max} = -3 \times 1263 \frac{\text{cm}}{\text{sec}^2} = -3789 \frac{\text{cm}}{\text{sec}^2}$$

12-4
Equations of motion for simple harmonic motion

The equations of motion thus far derived for a particle in simple harmonic motion give the acceleration and the velocity in terms of the position of the particle. It is frequently desirable to know these values at every instant of the motion—that is, in terms of the time. These equations may also be found with the aid of the reference circle. Referring to Figure 12-4, the displacement x of the particle in simple harmonic motion is the projection of the radius R on to AB at any instant of the motion. If θ is the angle that R makes with AB, then

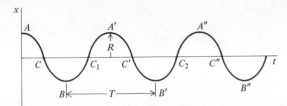

FIGURE 12-5 Graph of the displacement x against the time, t, of a particle in simple harmonic motion. ω is the period and R, the amplitude of the motion.

$$x = R \cos \theta \qquad (12\text{-}16)$$

The motion of the particle P in the reference circle takes place with uniform angular velocity ω; the relationship between ω and θ is simply

$$\omega = \frac{\theta}{t}$$

or

$$\theta = \omega t \qquad (12\text{-}17)$$

Substituting this value of θ into Equation 12-17 yields

$$x = R \cos \omega t \qquad (12\text{-}18)$$

for the displacement x of the particle in simple harmonic motion as a function of the time. Figure 12-5 is a graph of this equation with x as the ordinate and the time t as the abscissa. The maximum displacement or the amplitude of the motion is R. The period T is the time for a complete vibration—that is, starting from A, say, and going through C to B and back to A. This is indicated on the graph as the time between the two positions AA'; this is also the time required for motion from B to B' or C to C'. However, the time required to go from C to B and back to C, indicated on the graph by C_1, is only $T/2$.

Points A, A', and A'' are said to be in the same *phase*; points B, B', and B'' are also in the same phase. A and B, however, are out of phase by half a period. C and C' are in the same phase, but C and C_1 are out of phase by half a period. Thus, the period T is the time between two successive positions of the particle in the same phase of its motion.

The relationship between T and ω can be found by remembering that the linear speed v_0 of the particle moving in the reference circle is given by

$$v_0 = \omega R$$

Substituting this value of v_0 into Equation 12-6

yields

$$T = \frac{2\pi}{\omega} \qquad (12\text{-}19)$$

The equation of motion thus becomes

$$x = R \cos \frac{2\pi}{T} t \qquad (12\text{-}20)$$

This equation may also be written in terms of the frequency f; remembering that

$$f = \frac{1}{T} \qquad (12\text{-}11)$$

we find that

$$x = R \cos 2\pi f t \qquad (12\text{-}21)$$

The velocity v of the particle can also be determined with the aid of the reference circle. From Figure 12-4 we find that

$$v = -v_0 \sin \theta \qquad (12\text{-}22)$$

or

$$v = -v_0 \sin \omega t \qquad (12\text{-}22a)$$

or

$$v = -v_0 \sin \frac{2\pi}{T} \qquad (12\text{-}22b)$$

The minus sign is introduced to show that the velocity v is directed toward the left at this value of θ. A graph of this equation is shown in Figure 12-6, with the points marked to correspond to those in Figure 12-5. The amplitude of the curve is the maximum velocity v_0 that occurs at the equilibrium position C, with zero velocity at the ends A and B.

The value of the maximum velocity may be written in any of the following ways:

$$v_0 = \omega R \qquad (12\text{-}23a)$$

$$v_0 = \frac{2\pi}{T} R \qquad (12\text{-}23b)$$

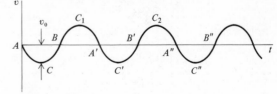

FIGURE 12-6 Graph of velocity of a particle in simple harmonic motion.

or

$$v_0 = 2\pi f R \qquad (12\text{-}23c)$$

The equation for the acceleration a of a particle in simple harmonic motion

$$a = -\frac{4\pi^2}{T^2} x \qquad (12\text{-}7)$$

may also be written as follows:

$$a = -4\pi^2 f^2 x \qquad (12\text{-}7a)$$

and

$$a = -\omega^2 x \qquad (12\text{-}7b)$$

from which

$$a = -4\pi^2 f^2 R \cos 2\pi f t \qquad (12\text{-}24)$$

Illustrative example

A small body hanging from a long spring is pulled down a distance of 10 cm from its equilibrium position and then released. It is observed to vibrate with a period of 1.5 sec. Determine (a) its maximum velocity, (b) its velocity at the end of 0.5 sec, (c) its position at this time, and (d) its acceleration at this time.

(a) Its maximum velocity v_0 can be obtained from Equation 12-23b:

$$v_0 = \frac{2\pi}{T} R$$

with $R = 10$ cm and $T = 1.5$ sec, so that

$$v_0 = \frac{2\pi \times 10 \text{ cm}}{1.5 \text{ sec}} = 41.9 \frac{\text{cm}}{\text{sec}}$$

(b) Its velocity v at the end of 0.5 sec can be obtained with the aid of Equation 12-22b as follows:

$$v = -v_0 \sin \frac{2\pi}{T} t$$

which becomes

$$v = -41.9 \frac{\text{cm}}{\text{sec}} \sin \frac{2\pi}{1.5} \times 0.5$$

from which

$$v = -41.9 \frac{\text{cm}}{\text{sec}} \sin \frac{2\pi}{3}$$

or

$$v = -41.9 \frac{\text{cm}}{\text{sec}} \sin 120°$$

hence

$$v = -41.9 \frac{\text{cm}}{\text{sec}} \times 0.866$$

so that

$$v = -36.0 \frac{\text{cm}}{\text{sec}}$$

(c) Its position x at the end of 0.5 sec can be found with the aid of Equation 12-20:

$$x = R \cos \frac{2\pi}{T} t$$

so that

$$x = 10 \text{ cm} \cos \frac{2\pi}{1.5} \times 0.5$$

or

$$x = 10 \text{ cm} \cos 120° = -5 \text{ cm}$$

(d) Its acceleration a is given by

$$a = -\frac{4\pi^2}{T^2} x \qquad (12\text{-}7)$$

hence its acceleration at the end of $t = 0.5$ sec when it is at $x = -5$ cm is

$$a = \frac{4\pi^2}{2.25 \text{ sec}^2} \times 5 \text{ cm} = 87.7 \frac{\text{cm}}{\text{sec}^2}$$

12-5
The simple pendulum

The motion of a pendulum is another interesting example of periodic motion. This property was first discovered by Galileo as a result of his observations of the periodic motions of lamps that were suspended by means of cords. Because of their periodic motion, pendulums are used in the construction of clocks. Although Galileo had designed a pendulum clock, he never actually constructed one. The first pendulum clock was constructed by the Dutch physicist Christian Huygens (1629–1695) in 1657. He also developed the mathematical theory of the pendulum. Newton also studied the motion of a pendulum and experimented with pendulums made of different materials and of different lengths.

We shall simplify the problem by confining our attention to the motion of a *simple* pendulum. A simple pendulum consists of a string of negligible weight, one end of which is attached to some fixed support O; a small mass, called a *pendulum bob*, is attached to the other end of the string (see Figure 12-7). When at rest, the bob is at C vertically below O and is in equilibrium under the action of two forces—its weight $m\mathbf{g}$ and the tension $\mathbf{S}$ in the string. When pulled aside to some position A and released, it travels in a circular arc through C to a point B on the other side. In moving the pendulum to A, it was actually lifted through a height h. If we make friction effects negligible, point B which it reaches will also be at a height h above C. It will then travel back through C to A, and the motion will be repeated. Both theory and experiment show that the period of the motion does depend upon the length of arc ACB. However, if this arc is made very small, so that it approximates a straight line, then the motion of the pendulum is simple harmonic.

To derive the expression for the period of a simple pendulum, let us consider the forces that act on the bob at the point A. These forces are its weight $m\mathbf{g}$ downward and the tension $\mathbf{S}$ in the string in the direction AO (see Figure 12-7a). The resultant $\mathbf{F}$ of these two forces is shown in Figure 12-7b. $\mathbf{F}$ is perpendicular to $\mathbf{S}$ and hence to OA. If we drop a perpendicular from A onto OC, it will intersect it at D. The triangle OAD is similar to the triangle of forces F, mg, S. Hence

$$\frac{F}{mg} = \frac{DA}{AO} = \frac{x}{L}$$

where L is the length of the pendulum and $x = DA$. When the amplitude of the motion is very small, point D will practically coincide with point C, and $DA = CA = x$, the displacement of the particle.

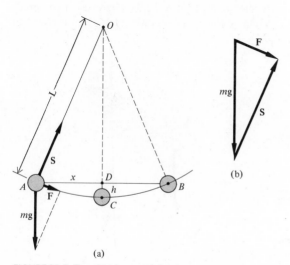

(a)

FIGURE 12-7 The simple pendulum.

Using a minus sign to indicate that the direction of F is opposite to that of x, we get

$$F = -\frac{mg}{L}x \qquad (12\text{-}25)$$

Using Newton's second law

$$F = ma$$

we get, for the acceleration of the pendulum bob,

$$a = -\frac{g}{L}x \qquad (12\text{-}26)$$

Equation 12-26 shows that the acceleration of the pendulum bob is proportional to its displacement and opposite in direction when the amplitude of vibrations is small. This motion of the pendulum is therefore simple harmonic.

The period of the pendulum can be found by substituting the value of a/x from Equation 12-26 into Equation 12-8, obtaining

$$T = 2\pi \sqrt{\frac{L}{g}} \qquad (12\text{-}27)$$

Equation 12-27 shows that the period of a simple pendulum does not depend upon the mass of the pendulum bob. This is in agreement with the results of the experiments of Newton with pendulum bobs of different masses. At any particular place on the earth's surface, the period depends only upon its length L. If a pendulum of known length is taken to different parts of the earth and its period is determined at each place, the value of g can readily be computed. However, actual pendulums used for such determinations are made of rigid bars; these are called physical or compound pendulums. It is possible to determine the equivalent simple pendulum length of a physical pendulum and then use Equation 12-27 for determining g with it; this is done in Section 12-6.

Illustrative example

A simple pendulum is 100 cm long. (a) Determine its period at a place where $g = 980$ cm/sec². (b) The measured value of its period at another place is 2.03 sec. Determine the value of g at this place.

(a) The period of the pendulum can be determined with the aid of the equation

$$T = 2\pi \sqrt{\frac{L}{g}} \qquad (12\text{-}27)$$

Substituting values for L and g in this equation, we

get

$$T = 2\pi \sqrt{\frac{100 \text{ cm}}{980 \text{ cm/sec}^2}} = 2.01 \text{ sec}$$

(b) To determine the value of g, let us solve Equation 12-27 for g, obtaining

$$g = \frac{4\pi^2 L}{T^2}$$

and, substituting $L = 100$ cm and $T = 2.03$ sec, we get

$$g = \frac{4\pi^2 \times 100 \text{ cm}}{(2.03 \text{ sec})^2} = 958 \frac{\text{cm}}{\text{sec}^2}$$

12-6
Angular simple harmonic motion

The analogy between linear and angular motion can be extended to simple harmonic motion by defining *angular simple harmonic motion* as that in which the angular acceleration α of a body is proportional to its angular displacement θ from its equilibrium position and opposite in direction, or

$$\alpha = -c\theta \qquad (12\text{-}28)$$

where c is a constant of proportionality. The motion of the body in this case will be periodic with a period given by

$$T = \frac{2\pi}{\sqrt{c}} \qquad (12\text{-}9)$$

An example of angular simple harmonic motion is the motion of the balance wheel of a watch. Such a wheel is mounted on an axle whose ends are pivoted in jeweled bearings. One end of a very fine hairspring is attached to the axle, the other end to the frame (see Figure 12-8). When the wheel rotates through an angle θ from its equilibrium position, the spring exerts a torque τ proportional to this angle to return it to its equilibrium position; thus

$$\tau = -K\theta \qquad (12\text{-}29)$$

where K is the twist or torsional constant of elasticity of the spring. Since

$$\tau = I\alpha \qquad (9\text{-}18)$$

where I is the moment of inertia about the axis, we get

$$\alpha = -\frac{K}{I}\theta \qquad (12\text{-}30)$$

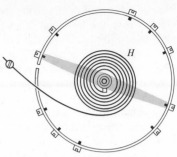

FIGURE 12-8 Balance wheel of a watch. H is the hairspring.

Comparing this with Equation 12-28, we find that

$$c = \frac{K}{I} \qquad (12\text{-}31)$$

so the period of oscillation of the balance wheel is given by

$$T = 2\pi \sqrt{\frac{I}{K}} \qquad (12\text{-}32)$$

A rigid body that is suspended from some fixed point O at a distance h from its center of gravity C and is free to rotate about an axis through O is called a *physical pendulum* (see Figure 12-9). When in equilibrium, it will hang with C vertically beneath O. If pulled aside so that the line OC makes an angle θ with the vertical, and then released, it will oscillate with periodic motion but not necessarily simple harmonic motion.

Let us specify the position of the body by the angle θ measured from the vertical to the line OC as shown in Figure 12-9. There are two forces acting on this body—its weight Mg and a force F at

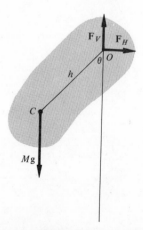

FIGURE 12-9 A physical pendulum.

the point O whose vertical component is F_V and horizontal component is F_H. Since we are interested only in the rotational motion about O, the torque about this axis is simply

$$\tau = -Mgh \sin \theta$$

The minus sign is used because the direction of the torque is opposite to the direction of the angular displacement θ. Using the equation

$$\tau = I\alpha$$

we get the equation for the angular acceleration

$$\alpha = -\frac{Mgh}{I} \sin \theta \qquad (12\text{-}33)$$

If the angular displacement is kept small, then, to a good approximation,

$$\sin \theta \approx \theta$$

and the equation for the angular acceleration becomes

$$\alpha = -\frac{Mgh}{I} \theta \qquad (12\text{-}34)$$

This now satisfies the condition for angular simple harmonic motion as given by Equation 12-28. The period of the physical pendulum under these conditions is

$$T = 2\pi \sqrt{\frac{I}{Mgh}} \qquad (12\text{-}35)$$

The use of this equation for the determination of the moment of inertia of a rigid body is fairly obvious. Once its moment of inertia has been determined about some axis, its moment of inertia about a parallel axis through the center of gravity can then be calculated.

All actual pendulums are physical pendulums. It is interesting to compare a physical pendulum with an ideal simple pendulum that has the same period. It will be recalled that the period of a simple pendulum is given by

$$T = 2\pi \sqrt{\frac{L}{g}} \qquad (12\text{-}27)$$

By comparing this with Equation 12-35 for the period of a physical pendulum we find that

$$L = \frac{I}{Mh} \qquad (12\text{-}36)$$

This value of L is called the *equivalent simple pendulum length*.

12-7
Energy in simple harmonic oscillators

As we have seen in Chapter 6, the potential energy of a stretched spring, $\mathcal{E}_p$, is given by

$$\mathcal{E}_p = \tfrac{1}{2}kx^2 \qquad (6\text{-}18)$$

Thus, if a body of mass m is attached to the end of a spring, whose elastic constant is k, the total energy of the system will be, as was shown in Chapter 6,

$$\mathcal{E} = \tfrac{1}{2}mv^2 + \tfrac{1}{2}kx^2 \qquad (12\text{-}37)$$

As the particle executes simple harmonic motion, the total energy is constant, since we have assumed the absence of friction or other dissipative forces. The potential energy is a maximum at the extremes of the motion, when the displacement is equal to the amplitude. When the displacement is less than the amplitude, the potential energy decreases and becomes equal to zero when the particle is at the equilibrium position. At the equilibrium position,

$$x = 0$$
$$\mathcal{E}_p = \tfrac{1}{2}kx^2$$
$$= 0$$

and

$$\mathcal{E}_k = \tfrac{1}{2}mv^2$$
$$= \mathcal{E}_T$$

where $\mathcal{E}_T$ is the total energy of the system.

When the displacement is equal to the amplitude, R, the particle is instantaneously at rest,

$$v = 0$$

and so

$$\mathcal{E}_k = 0$$

and

$$\mathcal{E}_p = \tfrac{1}{2}kx^2$$
$$= \tfrac{1}{2}kR^2$$
$$= \mathcal{E}_T$$

Thus

$$\mathcal{E}_T = \tfrac{1}{2}kR^2 \qquad (12\text{-}38)$$

Equation 12-38 illustrates a general property of simple harmonic motion; that the total energy is proportional to the square of the amplitude of the motion.

Combining Equations 12-38 and 12-37, we have

$$\tfrac{1}{2}kR^2 = \tfrac{1}{2}mv^2 + \tfrac{1}{2}kx^2 \qquad (12\text{-}39)$$

or

$$v^2 = \frac{k}{m}(R^2 - x^2)$$

from which we get

$$v = (R^2 - x^2)^{1/2}\sqrt{\frac{k}{m}} \qquad (12\text{-}40)$$

which is essentially the same result as Equation 12-14 or 12-15.

Questions

1. A billiard ball hits the edge of the billiard table perpendicularly, bounces back and strikes the opposite edge, bounces off it and continues back to the opposite edge, and so forth. Is the billiard ball moving with simple harmonic motion?

2. Compare the direction of the acceleration of a particle in simple harmonic motion with that of a particle in uniform circular motion.

3. Is the acceleration of a particle in simple harmonic motion equal to zero anywhere in its path?

4. Is the velocity of a particle in simple harmonic motion equal to zero anywhere in its path?

5. What is the relationship between the frequency of a particle in simple harmonic motion and its acceleration?

6. What is the effect on the period of the motion of a spring if the mass attached to it is increased so that it is four times the original mass?

7. What is the function of a pendulum in a clock?

8. Two simple pendulums have equal lengths, but the mass of one pendulum is four times that of the other. Compare their periods.

9. In an experiment for the determination of g using a simple pendulum, the student is usually advised to keep the length of arc, or the amplitude of the motion, small. Why is this precaution necessary?

10. A pendulum clock is taken into an elevator cab. What is the effect on the period when the elevator cab is accelerated (a) upward and (b) downward?

11. Geologists sometimes use precise measurements of the period of a pendulum in the search for deposits of ores or oil. Explain how this might work.

12. Give an example of simple harmonic motion other than those given in the text.

Problems

1. A 6-in.-long spring has a force constant of 50 lb/in. How long is the spring when a 20-lb weight is suspended from it?

2. A certain spring scale is marked from zero to 500 gm on a scale whose total length is 10 cm. What is the spring constant?

3. A 200-gm object is hung from the scale of Problem 2 and set in simple harmonic motion. What is the acceleration of the object when the scale reads 250 gm?

4. When a 200-lb passenger sits in a certain automobile the car settles 0.1 in. because of the compression of the springs. If the springs support 800 lb of the car as well as both the passenger and a 110 lb driver, find the period of simple harmonic motion of the springs. (Assume that the shock absorbers do not operate; usually they would quickly damp out any oscillations.)

5. A "seconds pendulum" has a period of 2.0 seconds. (Why this nomenclature?) Find the length of a simple pendulum that is a seconds pendulum at a place where $g = 980.0$ cm/sec^2.

6. A seconds pendulum that kept accurate time at a place where g was 980.0 cm/sec^2 is found to lose 2 min/day at a new location. Find g at this location.

7. A tuning fork vibrates at 256 oscillations/sec with an amplitude of 0.2 mm. What is the speed at the middle of the motion?

8. An object oscillates at the rate of 10 vib/sec with an amplitude of 8 cm. What is the speed of the object when it is 2.5 cm away from the middle of its path?

9. The balance wheel of a certain watch beats 5 times per second, and has a moment of inertia of 0.1 gm cm^2. What is the torsional constant of the spring?

10. A 20-gm object moves in simple harmonic motion with an amplitude of 80 cm and fre-

quency of 60 vibrations/sec. What is the restoring force on the object at the extremes of the motion? What is the maximum velocity of the object?

11. A body whose mass is 900 gm hangs from a vertical spring whose constant is 150,000 dynes/cm. The body is pulled down a distance of 6 cm and then released. Determine (a) the period of the motion, (b) the resultant force on the body when at the 6-cm point, and (c) the acceleration at this position.

12. In Problem 11 determine (a) the velocity of the body when its displacement is 3 cm and (b) the maximum velocity.

13. A body that has a mass of 60 gm is attached to a spring and set into vibration. The measured value of the period is 0.7 sec. Determine (a) the constant of the spring, (b) the velocity of the body at its equilibrium position if the amplitude is 6 cm, and (c) the maximum acceleration.

14. When a cylinder whose mass is 5.2 kg is hung from a spring and set into vibration the frequency is 2.4 vib/sec. When another cylinder is substituted for the first one its frequency is 3.2 vib/sec. (a) Determine the mass of the second cylinder. (b) Discuss the advantages and disadvantages of this method of measuring mass.

15. A body hanging from the end of a long spring is pulled down a distance of 15 cm and then released. It oscillates with a period of 2.4 sec. Determine (a) its maximum velocity, (b) the positions where its velocity is one-half the maximum, and (c) the time taken to reach these positions starting from the initial position.

16. A thin metal reed fixed at one end is set into vibration with a frequency of 250 vib/sec. The amplitude of vibration of the free end of the reed is 0.3 cm. Determine (a) the maximum acceleration of a point on the free end of the reed, (b) its maximum velocity, (c) its velocity when the displacement is 0.15 cm, and (d) the time required to reach this position from the initial position.

17. A simple pendulum 1.0 m long having a mass of 250 gm is displaced through an angle of 8° and released. Determine (a) the resultant force acting on it at the position of maximum displacement, (b) its maximum acceleration, and (c) its maximum velocity.

18. An avant-garde musician wants a large, low-pitched gong struck regularly, and a smaller, high-pitched gong struck at precisely three times that rate. He constructs a simple pendulum 120 cm long to strike the large gong and a pendulum 40 cm long to strike the smaller gong. The value of g is equal to 980 cm/sec². (a) What is the period of each of the pendulums he built? (b) How long should the shorter pendulum have been in order to satisfy the stated desire of the musician?

19. A 10-kg object is suspended from a spring and oscillates with a period of 0.5 sec and an amplitude of 3 cm. How much energy was transferred to the object initially?

20. The pan of a certain postal scale goes down 0.2 cm when a 25-gm object is weighed. A 300-gm object is placed on the scale and then the pan is pushed down, transferring 0.45 joule of energy to the spring. Determine (a) the period of the motion after the pan is released, (b) the amplitude of the motion, (c) the maximum velocity, and (d) the maximum acceleration.

21. An 8-kg disk of radius 10 cm is suspended from a long wire, which is attached to the center of the disk. It is found that a force of 2 nt tangential to the disk rotates the disk through 90°. The force is removed. How long does it take for the disk to execute 20 vibrations?

22. A small pendulum has a frequency of 10 vib/sec, where g is equal to 980 cm/sec². When it is in an elevator cab, which is accelerated upward, its frequency is 12 vib/sec. Determine (a) the value of g in the elevator cab and (b) the acceleration of the elevator cab relative to the earth.

23. A more accurate equation for the period of a simple pendulum when the maximum angle that the string makes with the vertical α is

$$T = T_0\left(1 + \frac{1}{4}\sin^2\frac{\alpha}{2}\right)$$

where T_0 is the period when α is very small. Determine the value of T when $\alpha = 30°$ for a pendulum whose length is 1.0 m.

24. Prove that the formula for the period of a physical pendulum gives the formula for the period of a simple pendulum as a special case.

25. You are given a meter stick, which you are told was initially uniform, but which has now been loaded with a small point mass. You are

asked to determine the location and mass of the load but you are not told the mass of the meter stick before the load was added. Show that you could determine both unknown masses and the location of the load by three operations: **(a)** weighing the stick, **(b)** finding the balance point of the stick, and **(c)** measuring the period of the stick when it is used as a physical pendulum.

26. A small particle is on a membrane, which is vibrating vertically with a frequency of 2000 vib/sec. The amplitude is increased slowly from zero. At what amplitude will the particle lose contact with the membrane?

27. A certain spring (of unstated constant) is hanging vertically. The spring is held in one hand so that it cannot stretch and a weight is attached to the end of the spring. The weight is suddenly released and descends a distance d before rising again. Find the period of the motion in terms of the distance d and constants.

28. A *conical* pendulum consists of an object of mass m which is suspended by a string of length l and which is moving in a horizontal circle with constant speed. Show that if the angle that the string makes with the vertical is small, the period of a conical pendulum is the same as that of the same pendulum when used as a simple pendulum.

fluids

chapter 13

13-1
Three phases of matter

From our everyday experience, we have become familiar with the fact that matter occurs in three different forms—*solid*, *liquid*, and *gas*. Under ordinary conditions, stone, iron, copper, and chalk, for example, are solids; water, oil, and mercury are liquids; and air, hydrogen, and carbon dioxide are gases. Each one of these forms is called a *phase*. The phase of a substance is determined by its temperature and pressure. The study of changes of phase will be considered in Chapter 17. For the present, we shall confine our discussion to the application of the principles of mechanics to bodies that remain in the same phase.

Liquids and gases are sometimes grouped together as *fluids* because they flow very readily upon the application of an external force, while solids do not. A solid has a definite size and a definite shape, and these change only very slightly when subject to external forces. Liquids, on the other hand, although they do possess a definite size or volume, change their shape very readily. Liquids at rest generally take the shape of the containing vessel. If the containing vessel is open to the atmosphere, or if its volume is greater than that of the liquid put into it, there will be a *free surface* at the top of the liquid. A mass of gas differs from a mass of liquid in that it has neither definite size nor definite shape. A mass of gas, no matter how small, will completely fill the container. The volume of the gas is the volume of the container.

13-2
Density

Different substances differ greatly in their physical properties. Information about these properties is desirable in deciding which substance is most suitable for a particular use. Such information is necessary in order to be able to predict the probable behavior of the substance under a variety of physical conditions. Some of these properties, known as the *physical constants* of the substance, have been measured and tabulated for easy reference. Among these constants is the *density* of a substance at specific temperatures and pressures.

The *density* of a substance is defined as *the ratio of the mass of a sample of the substance to its volume*; that is,

$$\rho = \frac{M}{V} \qquad (13\text{-}1)$$

where ρ (Greek letter rho) is the density of the substance, M the mass of the sample, and V its volume. The densities of solids and liquids vary slightly with changes of temperature and pressure, while the densities of gases vary greatly with changes of temperature and pressure. The temperature and pressure should always be specified when the densities of gases are given. In the cgs system, the densities of substances are expressed in grams per cubic centimeter. The density of water is 1 gm/cm³, while the density of air at 0°C and atmospheric pressure is 0.001293 gm/cm³. Table

13-1 gives the densities of some of the more common substances.

In the mks system, the density of a substance is expressed in kg/m³. Since 1 kg = 1000 gm and 1 m³ = 10⁶ cm³, the numerical value of the density in mks units will be 1000 times that in cgs units. Thus the density of water is 1000 kg/m³ and that of air is 1.293 kg/m³.

In the British engineering system, the density of a substance should be expressed in lb sec²/ft. This term, however, is seldom used. Instead, it is replaced by the term *weight density D*, where

$$D = \frac{W}{V} \qquad . \quad (13\text{-}2)$$

is the weight per unit volume of the substance. Since

$$W = Mg$$

and

$$\rho = \frac{M}{V}$$

then

$$D = \rho g \qquad (13\text{-}3)$$

that is, the weight density is the product of the mass

Table 13-1 Densities of some common substances

solids	density in gm/cm³	liquids	density in gm/cm³
Aluminum	2.70	Alcohol	0.79
Brass	8.44–8.70	Ether	0.74
Copper	8.93	Glycerin	1.26
Cork	0.22–0.26	Mercury	13.596
Glass, common	2.4–2.8	Oil, olive	0.92
Glass, flint	2.9–5.9	Oil, paraffin	0.8
Gold	19.3	Water	1.00
Ice	0.917		
Iron	7.9		
Lead	11.34		density at 0°C, 760 mm Hg in gm/cm³
Nickel	8.8	gases	
Osmium	22.5		
Platinum	21.37		
Silver	10.49	Air	0.001293
Tungsten	19.3	Ammonia	0.000771
Uranium	18.7	Carbon dioxide	0.001977
Wood, cedar	0.31–0.49	Helium	0.000178
Wood, ebony	0.98	Hydrogen	0.000090
Wood, elm	0.54–0.60	Oxygen	0.001429
Wood, white pine	0.35–0.50		
Zinc	6.9		

density ρ and the gravitational acceleration g. Thus, the mass density of water is 1.94 lb sec²/ft and its weight density is 62.4 lb/ft³. The weight density of any other substance listed in Table 13-1 can be determined by multiplying the numerical value listed there by 62.4. The weight density of aluminum, for example, is 168.5 lb/ft³.

13-3
Pressure

There is a difference in the manner in which a force is applied to a fluid and the way it is applied to a solid. A force can be applied to a single point of a solid, but it can be applied only over a surface in the case of a fluid. In a discussion of the results of the application of forces to fluids, it is therefore convenient to introduce a new term called *pressure*. If a force F is applied to the surface of a fluid and acts over an area A perpendicular to it, then the pressure P is defined as

$$P = \frac{F}{A} \qquad (13\text{-}4)$$

Pressure is a scalar quantity. The pressure may be expressed in dynes per square centimeter, in pounds per square foot, in pounds per square inch, or in any other appropriate set of units.

13-4
Pressure due to weight of a liquid

Suppose that a cylindrical jar of cross-sectional area A is filled with a liquid to a level a distance h from the bottom, as shown in Figure 13-1. If W is the weight of the liquid, then the pressure P on the bottom of the jar is, from Equation 13-4,

$$P = \frac{W}{A}$$

The volume V of the liquid can be expressed as

$$V = Ah$$

If ρ is the density of the liquid, and M its mass, then, from Equation 13-1,

$$\rho = \frac{M}{V}$$

or

$$M = V\rho = Ah\rho$$

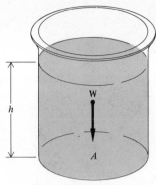

FIGURE 13-1 Pressure on the base of a cylindrical jar produced by the weight of the liquid it contains.

Now

$$W = Mg$$

and thus

$$W = Ah\rho g$$

Substituting this value of W in the equation for the pressure, we get

$$P = \frac{Ah\rho g}{A}$$

or

$$P = h\rho g \qquad (13\text{-}5)$$

We may also write this in terms of the weight density D, in which case the above equation becomes

$$P = hD \qquad (13\text{-}5a)$$

Equation 13-5 gives the pressure not only at the bottom of the liquid but at any point in the liquid. The height h is then interpreted as the height of the liquid above the point in question. To verify this let us consider a portion of the liquid in the form of a vertical circular cylinder of height h and cross-sectional area A somewhere inside the liquid as shown in Figure 13-2. Suppose that the top of this cylinder is at a distance h_1 below the free surface of the liquid; the bottom surface will be at a distance $h_1 + h$ below the free surface. Since the liquid in the container is at rest, this cylinder is in equilibrium, so the vector sum of all the forces acting on it is zero. The force $\mathbf{F}_1$ acting downward on the top surface is produced by the pressure P_1 of the liquid above it and is given by

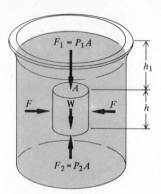

FIGURE 13-2 Forces on a portion of the liquid in the form of a cylinder of height h and cross-sectional area A. The top of the cylinder is a distance h_1 below the surface of the liquid.

$$F_1 = P_1 A$$

The force $\mathbf{F}_2$ acting upward on the bottom surface is produced by the pressure P_2 of the liquid above this surface and is given by

$$F_2 = P_2 A$$

The additional vertical force acting on this liquid is its weight $\mathbf{W}$, which may be written as

$$W = Mg = \rho A h g$$

where Ah is the volume of the liquid in the cylinder. From the conditions of equilibrium, we have

$$F_2 = F_1 + W$$

so that

$$P_2 A = P_1 A + \rho A h g$$

from which

$$P_2 = P_1 + \rho h g$$

or

$$P_2 - P_1 = \rho h g$$

Thus the difference in pressure between any two points in a liquid depends only upon the vertical distance between these points.

If the pressure P_1 on the top surface is taken as zero, in which case $h_1 = 0$ and the top surface is the free surface of the liquid, then the pressure on the bottom surface is

$$P_2 = P = \rho h g$$

This is identical with Equation 13-5.

There are also forces acting on the sides of the cylinder of liquid shown in Figure 13-2. These forces are at right angles to the cylindrical surface everywhere; at any given level below the surface, the vector sum of the forces acting horizontally must be zero, since the cylinder of liquid is in equilibrium. The magnitude of the force acting on any part of the surface will vary directly with the height of liquid above the point in question. The direction of this force will always be at right angles to the surface, real or imaginary, at this point.

The pressure at the bottom of a vessel depends upon the height of the liquid above it, the density of the liquid, and the value of g at that place; it does not depend upon the area of the base or upon the shape of the vessel. This can be demonstrated with the aid of a series of vessels that have different shapes but whose bases have the same cross-sectional area. Three such vessels are shown in Figure 13-3. Each vessel is fitted with a brass screw thread at its base so that it can be mounted on a pressure gauge. When these vessels are filled to the same level with a liquid such as water, the pressure readings will be found to be the same in all three cases.

Each of the vessels in Figure 13-3 contains a different quantity of water, but the level in each is at the same height h above the bottom, and the pressure at the bottom is the same in each one. It may appear surprising that different weights of liquids can produce the same pressure, but the

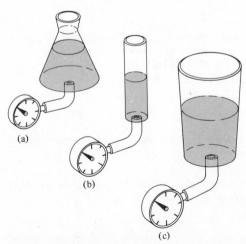

FIGURE 13-3 Pressure at the bottom of a vessel is independent of the shape of the vessel; it depends only upon the height of the liquid and its density. Here jars are filled to the same level with the same liquid; pressure gauges attached show identical readings.

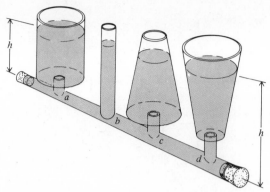

FIGURE 13-4 The level of a liquid in a series of connected vessels is the same in each vessel if the liquid is at rest.

paradox disappears if we analyze the forces that act on the bottom of each vessel. In the cylinder of Figure 13-3b, the weight of the cylindrical column of water of area A is the only force acting downward on the bottom. The liquid also exerts forces on the walls of the cylinder, and these walls in turn exert equal forces in the opposite direction on the water. But since the walls are vertical, the forces are horizontal and have no components in the vertical direction. But in Figure 13-3a the walls flare inward and exert forces that have vertical components directed downward. Although the weight of water in this vessel is less than that in the cylinder, it can be shown that the vertical component of the force exerted by the walls is just equal to the difference between these weights. Hence the force on the bottom of this vessel is equal to that on the bottom of the cylindrical vessel. When the sides of the vessel flare outward, as in Figure 13-3c, it is easy to see that they contribute a force that has a vertical component upward to support the weight of water in excess of that of a cylinder of cross-sectional area A. Hence the force on the bottom of this vessel is also equal to that of a cylindrical column of water of height h and cross-sectional area A.

Another consequence of the fact that the pressure at any point in a given liquid depends only upon the height of liquid above that point is that the level of a liquid in a series of connected vessels, such as those shown in Figure 13-4 will be the same in each one. If we consider such points as a, b, c, and d in the connecting tubes when the liquid is at rest, the force due to the liquid to the right of each point must be equal to the force due to the liquid at the left of the point; otherwise there will be a

flow of liquid in the direction of the greater force.

The pressure, as given by Equation 13-5, is usually expressed in dynes/cm², or lb/ft², or lb/in.². Frequently, however, the pressure is expressed simply as the height of a column of liquid that produces it, such as a pressure of 2 in. of water, or a pressure of 30 in. of mercury, or a pressure of 760 mm of mercury. Whenever necessary, these values can always be converted to the more appropriate ones by means of Equation 13-5.

Illustrative example

Determine the pressure at the bottom of a column of mercury 70 cm high.

The pressure can be found with the aid of Equation 13-5:

$$P = h\rho g \qquad (13\text{-}5)$$

with $h = 70$ cm, $\rho = 13.60$ gm/cm³, and $g = 980$ cm/sec². Substituting these values in Equation 13-5, we get

$$P = 70 \text{ cm} \times 13.6 \frac{\text{gm}}{\text{cm}^3} \times 980 \frac{\text{cm}}{\text{sec}^2}$$

$$= 933{,}000 \frac{\text{dynes}}{\text{cm}^2}$$

or

$$P = 9.33 \times 10^5 \frac{\text{dynes}}{\text{cm}^2}$$

Illustrative example

A dam is built to impound water in a reservoir. The level of the water is 20 ft above the bottom of the dam. Determine the pressure (a) 6 ft below the surface and (b) 20 ft below the surface. Assuming the face of the dam to be vertical and 60 ft across, determine (c) the force on the dam.

(a) The appropriate form of the equation for the pressure is

$$P = hD \qquad (13\text{-}5a)$$

with

$$h = 6 \text{ ft and } D = 62.4 \frac{\text{lb}}{\text{ft}^3}$$

Hence

$$P = 6 \text{ ft} \times 62.4 \frac{\text{lb}}{\text{ft}^3}$$

from which

$$P = 374.4 \; \frac{\text{lb}}{\text{ft}^2}$$

(b) At $h = 20$ ft,

$$P = 20 \text{ ft} \times 62.4 \; \frac{\text{lb}}{\text{ft}^3} = 1248 \; \frac{\text{lb}}{\text{ft}^2}$$

(c) Since the pressure varies uniformly from zero at the top of the water to 1248 lb/ft² at the bottom, the force against the dam will also vary uniformly from the top of the water surface to the bottom. The total force against the wall can be determined by considering the *average* pressure to act over the entire area A of the wall. In this case the average pressure is half its largest value; that is, $P_{av} = 624$ lb/ft². The area A over which this pressure acts is

$$A = 20 \text{ ft} \times 60 \text{ ft} = 1200 \text{ ft}^2$$

and the total force acting on this dam is

$$F = P_{av}A$$

so that

$$F = 624 \; \frac{\text{lb}}{\text{ft}^2} \times 1200 \text{ ft}^2$$

from which

$$F = 748{,}800 \text{ lb} = 374 \text{ tons}$$

13-5
Pressure in a confined liquid

In addition to the pressure due to its weight, a confined liquid may be subjected to an additional pressure by the application of an external force. Suppose the liquid is in a cylinder, as shown in Figure 3-5, and that a tight-fitting piston is placed on the surface of the liquid. If a force F is applied to the piston, it will remain practically in the same position since the compressibility of liquids is very small. If A is the area of the piston, this external force produces a pressure $P = F/A$ in the liquid. This additional pressure is transmitted throughout every part of the liquid and acts on all surfaces in contact with the liquid. This is sometimes known as *Pascal's principle*, and may be stated as follows:

Whenever the pressure in a confined liquid is increased or diminished at any point, this change in pressure is transmitted equally throughout the entire liquid.

There are many practical applications of this

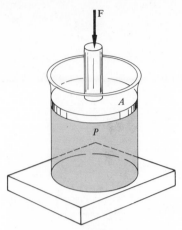

FIGURE 13-5 A pressure P is produced in a confined liquid by the application of a force F on a piston of area A in contact with the liquid.

principle. The operation of the hydraulic press is based upon this principle. The hydraulic press, which is sketched in Figure 13-6, consists essentially of two connected cylinders, one of small cross-sectional area a, the other of large cross-sectional area A, each fitted with a piston. A liquid, usually oil or water, is supplied to it from a reservoir. By exerting a force F on the small piston, the additional pressure produced is $P = F/a$. This pressure is transmitted throughout the liquid and hence acts on the larger piston of area A. The force that can be exerted by this piston is then PA. If this hydraulic press is designed to lift a weight W, then

$$W = PA = F \; \frac{A}{a}$$

or

$$\frac{W}{F} = \frac{A}{a} \tag{13-6}$$

The weight that can be lifted with the aid of a hydraulic press by the application of a force F is multiplied in the ratio of the areas of the two pistons.

There are many other practical applications of Pascal's principle. Most automobiles are now equipped with hydraulic brakes. By pushing down on the brake pedal, the driver exerts a force on the piston of a cylinder containing a light oil. This force produces an increase in pressure that is transmitted equally through hollow tubes to the brakes.

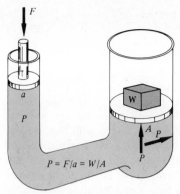

FIGURE 13-6 Hydraulic press. The pressure P is the same everywhere throughout the liquid, if the pressure due to the weight of the liquid is negligible.

13-6
Balanced columns of liquid

A convenient method for comparing the densities of two liquids is shown in Figure 13-7. Some mercury is first poured into the U-tube to prevent the two liquids from mixing. One of the liquids, say alcohol, is poured in on the right side and the other liquid, say water, is poured in on the left side and adjusted until the level of mercury is the same in both tubes. Since the pressure on each mercury surface is now the same, we can write, from Equation 13-5,

$$P = h_1\rho_1 g = h_2\rho_2 g$$

from which

$$\frac{h_1}{h_2} = \frac{\rho_2}{\rho_1} \qquad (13\text{-}7)$$

Equation 13-7 shows that the heights of the two columns are in the inverse ratio of their densities. Thus the alcohol will stand at a higher level than the water.

13-7
Archimedes' principle

The fact that some objects float in water while others sink to the bottom has been known for centuries; Archimedes (287–212 B.C.) was the first to discover the principle underlying these phenomena. To understand Archimedes' principle, it is necessary to consider the forces acting on a body totally immersed in a liquid. It will be convenient to use a glass cylinder with graduations on it indicating the volume of the liquid at different levels. Suppose that liquid is poured into the cylinder and its volume is noted. Now let us take some regularly shaped object of known volume V and immerse it in the liquid, as shown in Figure 13-8. The liquid will rise to a new level and will show that the body displaces a volume of liquid V equal to its own volume.

Let us now consider the forces acting on this body when it is immersed in the liquid. There is, of course, the weight W of the body pulling it down. There is also an additional force B acting upward on this body. This can be understood by imagining the volume now occupied by the body to be occupied instead by an equal volume of liquid. This volume of liquid would have been in equi-

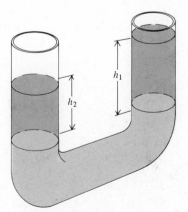

FIGURE 13-7 Balanced columns of liquid of different densities; the heights of the two columns are in the inverse ratio of the densities of the liquids.

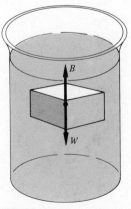

FIGURE 13-8 Bouyant force on a block immersed in a liquid.

librium, which means that its weight must have been supported by the action of the rest of the liquid. This support comes from the difference in pressure between the top and bottom of this volume. Hence, no matter what material occupies this volume, there will be a force upward on it equal to the weight of the liquid displaced. This upward force B exerted by the rest of the liquid is known as the *buoyant force*.

Archimedes' principle is a generalization of the result obtained above: it states that *any object partly or completely immersed in a fluid is buoyed up by a force equal to the weight of the fluid displaced*. This principle is applicable to both liquids and gases.

The question that still remains to be answered is whether the body that is completely immersed in the liquid will go up or down. The result depends upon the difference between the weight of the body and the buoyant force due to the liquid. If the weight W of the body is greater than that of an equal volume of liquid, the resultant force on it will be downward, and the body will sink to the bottom. If its weight W is less than the weight of liquid displaced, the resultant force will be upward; the body will be forced upward, and part of its volume will rise above the surface so as to establish equilibrium. Equilibrium will be established when the weight of liquid displaced be-

comes equal to the weight of the body; it floats with part of it below and part of it above the surface of the liquid.

A ship afloat, for example, displaces its own weight of water. The weight of a ship is frequently expressed in terms of the weight of water it displaces. Thus there are ships of 10,000 tons displacement, 15,000 tons displacement, and so forth. Of course, as the ship is loaded with fuel, freight, and passengers, it displaces a correspondingly greater amount of water, and more of it is submerged in the water. There is usually a definite waterline marked on a ship indicating the limit to which a ship may be submerged and still be safe.

A submarine is designed so that it can take water into specially built tanks to make its weight greater than the weight of the volume of water equal to its own volume. It then submerges completely and may, if necessary, rest on the bottom of the ocean. To enable the vessel to rise to the surface, water is forced out of the tanks with the aid of pumps.

Illustrative example

A cylinder of brass 6 cm high and 4 cm² in cross-sectional area is suspended in water by means of a string so that its upper surface is 7 cm below the surface of the water, as shown in Figure 13-9. Determine (a) the force acting on the top of the cylinder, (b) the force acting on the bottom of the cylinder, and (c) the buoyant force acting on this cylinder.

(a) The force F_1 acting on the top of the cylinder is that due to the pressure of the water above it and is

$$F_1 = P_1 A$$

Since

$$P_1 = h_1 \rho g$$
$$= 7 \text{ cm} \times 1 \frac{\text{gm}}{\text{cm}^3} \times 980 \frac{\text{cm}}{\text{sec}^2}$$
$$= 6860 \frac{\text{dynes}}{\text{cm}^2}$$

then

$$F_1 = 6860 \frac{\text{dynes}}{\text{cm}^2} \times 4 \text{ cm}^2$$

or

$$F_1 = 27{,}440 \text{ dynes}$$

This force pushes down on the cylinder.

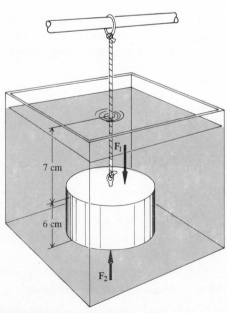

FIGURE 13-9

(b) The force F_2 acting on the bottom of the cylinder is that due to the pressure produced by the column of water above it. This column is 6 cm + 7 cm = 13 cm high.

Calling this pressure P_2, we get

$$P_2 = h_2 \rho g$$

$$= 13 \text{ cm} \times 1 \frac{\text{gm}}{\text{cm}^3} \times 980 \frac{\text{cm}}{\text{sec}^2}$$

or

$$P_2 = 12{,}740 \frac{\text{dynes}}{\text{cm}^2}$$

Since

$$F_2 = P_2 A$$

$$F_2 = 12{,}740 \frac{\text{dynes}}{\text{cm}^2} \times 4 \text{ cm}^2$$

or

$$F_2 = 50{,}960 \text{ dynes}$$

This force F_2 acts upward on the cylinder.

(c) The buoyant force B is the net force upward caused by the difference in pressures in the liquid. The forces that act on the walls of the cylinder are all directed horizontally and their resultant is zero, as can be seen from the symmetry of the figure. Therefore the buoyant force is simply the difference between the two vertical forces F_1 and F_2; thus

$$B = F_2 - F_1$$

or

$$B = 50{,}960 \text{ dynes} - 27{,}440 \text{ dynes}$$

so that

$$B = 23{,}520 \text{ dynes}$$

and acts upward.

It is interesting to compare this buoyant force with the weight of water displaced. The volume of the cylinder is $h \times A = 6 \text{ cm} \times 4 \text{ cm}^2 = 24 \text{ cm}^3$. This is also the volume of water displaced. The weight of this displaced water is

$$w = mg$$

$$= \rho V g$$

$$= 1 \frac{\text{gm}}{\text{cm}^3} \times 24 \text{ cm}^3 \times 980 \frac{\text{cm}}{\text{sec}^2}$$

or

$$w = 23{,}520 \text{ dynes}$$

As expected, this is the same as the buoyant force on the cylinder.

13-8
Specific gravity

From what has gone before, it appears that the ratio of the weight of a body to the weight of an equal volume of water will determine whether the body sinks or floats. This ratio is termed the *specific gravity* of the body. If its specific gravity is greater than 1, the body will sink; if it is less than 1, it will float. *The specific gravity is also the ratio of the density of the body to the density of water.* Since it is a ratio, the specific gravity is represented by a pure number and is the same no matter what system of units is used. Since the density of water in the cgs system is 1 gm/cm³, the specific gravity of a substance has the same numerical value as its density in the cgs system.

Illustrative example

A block of aluminum is attached to a spring balance. When the block is suspended in air, the spring balance reads 250 gm. When the aluminum block is lowered until it is completely immersed in water, the spring balance reads 160 gm. When the aluminum block is lowered until it is completely immersed in alcohol, the spring balance reads 180 gm. Determine (a) the specific gravity of aluminum and (b) the specific gravity of alcohol.

At any one place the mass of a substance is directly proportional to its weight, and since, in determining specific gravities of substances, we need only the ratios of their weights, we can use the masses of the substances in grams instead of converting them to dynes.

(a) The buoyant force provided by the water is the difference between the weight of the aluminum block in air and its weight when immersed in water; that is,

$$B = 250 \text{ gm} - 160 \text{ gm} = 90 \text{ gm}$$

The aluminum block thus displaces 90 gm of water. Its specific gravity S_G, which is the ratio of the weight of the aluminum to the weight of an equal volume of water, is thus given by

$$S_G = \frac{250}{90} = 2.78$$

(b) The amount of alcohol displaced by the aluminum block is

$$250 \text{ gm} - 180 \text{ gm} = 70 \text{ gm}$$

The volume of 70 gm of alcohol is the same as that of 90 gm of water, since each is equal to the volume of the aluminum block; hence the ratio of their weights is the specific gravity of the alcohol; that is,

$$S_G = \frac{70}{90} = 0.78$$

13-9
Atmospheric pressure

The atmosphere is a layer of air surrounding the earth; its thickness has been estimated at about 600 or 700 miles. Since air has weight, this layer of air produces a pressure. This pressure can be measured by an adaptation of the balanced-column method—that is, balancing the pressure due to the column of air above us against the pressure produced by a column of mercury. A simple way of doing this is to take a glass tube about 3 ft long, closed at one end, and fill it completely with clean mercury. The open end is closed temporarily by placing the thumb over it. The tube is then inverted, and the end is placed in an open dish of mercury. When the thumb is removed, the column of mercury will drop slightly and come to equilibrium at a height of about 30 in. above the open level in the dish, as in Figure 13-10. The atmosphere exerts a pressure P on the open surface of mercury in the dish, and this is transmitted to the liquid in the tube. This pressure is balanced by the pressure due to the mercury in the tube at a height h above the open surface of the dish. This pressure is therefore

$$P = h\rho g \qquad (13\text{-}5)$$

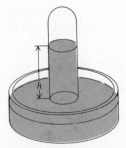

FIGURE 13-10 Mercury barometer.

where ρ is the density of mercury. The instrument used for measuring atmospheric pressure is called a *barometer*.

At any one place, atmospheric pressure varies slightly from day to day. For scientific work, standard atmospheric pressure is defined as the pressure equivalent to that produced by a column of mercury 76 cm high at 0°C. This height corresponds to 29.92 in. The density of mercury at 0°C is, from Table 13-1, 13.60 gm/cm³; substituting this value into Equation 13-5, we get

$$P = 76 \text{ cm} \times 13.60 \frac{\text{gm}}{\text{cm}^3} \times 980 \frac{\text{cm}}{\text{sec}^2}$$

from which

$$P = 1{,}013{,}000 \frac{\text{dynes}}{\text{cm}^2} \qquad (13\text{-}8)$$

This pressure can also be expressed as

$$P = 14.70 \frac{\text{lb}}{\text{in.}^2} \qquad (13\text{-}9)$$

13-10
Compressibility of gases: Boyle's law

The compressibility of gases was first studied by Robert Boyle (1627–1691). Suppose we have a mass of gas in a cylinder with a tight-fitting piston, on which a force F is exerted producing a pressure $P = F/A$, where A is the area of the piston, as shown in Figure 13-11. The gas will be subject to this pressure and will occupy a volume V determined by the distance of the piston from the end of the cylinder. By increasing the force on the piston to a new value F_1, the pressure on the gas will be increased to a new value P_1. It will be observed that the piston will descend, decreasing the volume of the gas. Unless precautions are taken, the temperature of the gas will also increase. This can be avoided by surrounding the cylinder with a water jacket, or else by waiting until the gas has cooled to room temperature. If the new volume V_1 is measured at the same temperature as the original volume V, it will be found that

$$\frac{V_1}{V} = \frac{P}{P_1} \qquad (13\text{-}10)$$

Or, stated in words, at constant temperature the volume of a mass of gas varies inversely as the pressure.

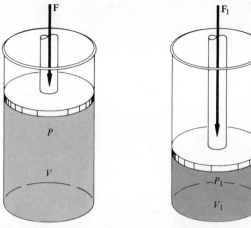

FIGURE 13-11 Gas in a cylinder is compressed by increasing the force on the piston, that is, by increasing the pressure of the gas.

A more convenient form for the equation is

$$P_1V_1 = PV = \text{constant} \qquad (13\text{-}11)$$

that is, *the product of the pressure and the volume of an enclosed gas at constant temperature remains constant.* This statement is known as *Boyle's law.* For example, 1.293 gm of air occupy a volume of 1000 cm^3 at a pressure of 1 atm at 0°C. If the pressure is increased to 2 atm and the temperature kept at 0°C, then the volume occupied by this gas will be 500 cm^3.

Since the density of a gas varies inversely with the volume, a decrease in volume means an increase in density. Hence, when the pressure in a mass of gas is increased, its density is also increased in the same ratio. Thus Boyle's law may also be written as

$$\frac{P}{P_1} = \frac{\rho}{\rho_1} \qquad (13\text{-}12)$$

in which ρ is the density of the gas at pressure P, and ρ_1 is the density of the gas at pressure P_1.

In the above example, by doubling the pressure of the air, the density was doubled.

Illustrative example

A steel tank contains 2 ft^3 of oxygen at a gauge pressure of 200 lb/in.^{2}. What volume will this gas occupy at the same temperature at atmospheric pressure?

The gauge pressure reading is the difference be-

tween the pressure of the gas in the cylinder and the pressure of the atmosphere. Hence the pressure P_1 of the oxygen in the cylinder is 214.7 lb/in.^{2}; the pressure P of the atmosphere is taken as 14.7 lb/in^2. Using Boyle's law in the form $P_1V_1 = PV$ and substituting numerical values, we get

$$214.7 \, \frac{\text{lb}}{\text{in.}^2} \times 2 \text{ ft}^3 = 14.7 \, \frac{\text{lb}}{\text{in.}^2} \times V$$

from which

$$V = 29.2 \text{ ft}^3$$

13-11
Archimedes' principle applied to gases

We can think of ourselves and all objects on the earth as situated at the bottom of an ocean of air. Every object in this ocean is buoyed up by a force equal to the weight of air it displaces. If this buoyant force is greater than the weight of the object, it will rise. Balloons are made by filling a flexible bag with a gas such as hydrogen or helium that is less dense than air. The balloon will rise if its total weight is less than the weight of the air it displaces. As it rises, the density of the atmosphere becomes smaller, and, if the volume of the balloon does not change, the buoyant force on it becomes smaller. The balloon will stop rising at the level at which the buoyant force equals the weight of the balloon.

As the balloon rises, the pressure of the gas inside the flexible bag becomes greater than the atmospheric pressure. This difference in pressure causes the bag to expand, displacing a greater quantity of air and increasing the buoyant force. For the balloon to descend, valves must be provided to allow some of the gas to escape so that the volume of the balloon is decreased. It is quite a common practice to send up a set of instruments attached to one or more small balloons. Meteorological stations send up radiosondes regularly. These radiosondes contain instruments that measure the pressure, temperature, and relative humidity of the atmosphere, and a radio transmitter that sends signals to a receiving station on the ground. They rise until the difference between the pressures inside and outside the balloon causes the bag to burst. When this happens, a parachute attached to the instrument box opens so that the radiosonde falls slowly. Most of the instrument boxes are later recovered.

13-12
Steady flow of liquid

When a liquid flows through a pipe in such a way that it completely fills the pipe and as much liquid enters one end of the pipe as leaves at the other end in the same time, then the liquid is said to *flow at a steady rate*. At any point in the liquid, the motion does not change with time; as a small volume of liquid leaves that point, another particle comes to it and moves with the same velocity that the preceding particle had there. When the velocity of the liquid is not too great, and the change in shape and size of the pipe is gradual, with no sharp edges or obstructions, a particle of liquid moves through the pipe along a path known as a *streamline*. We can map the flow of liquid through the pipe by drawing a series of streamlines following the paths of the particles of liquid, as shown in Figure 13-12; these lines will be close together where the liquid is moving rapidly and farther apart in regions of the pipe where the liquid is moving slowly.

Since the liquid is incompressible and there are no places in the pipe in which the liquid can be stored, the volume of liquid that flows through any cross-sectional area perpendicular to the streamlines in any interval of time must be the same everywhere in the pipe. The small volume of liquid ΔV, which flows through a small distance Δs past a point where the cross-sectional area of the pipe is A, is

$$\Delta V = A \, \Delta s$$

If this flow takes place in a short time interval Δt, then the rate of flow Q, or the quantity of liquid that flows through this area per unit time, is

$$Q = \frac{\Delta V}{\Delta t}$$

from which

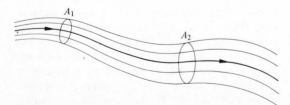

FIGURE 13-12 Streamlines of a liquid flowing through a pipe at a steady rate.

$$Q = \frac{A \, \Delta s}{\Delta t}$$

or

$$Q = Av \tag{13-13}$$

where $v = \Delta s/\Delta t$ is the velocity of the liquid at this point.

Consider two typical areas A_1 and A_2 perpendicular to the streamlines. The volume of liquid Q passing through area A_1 per unit time is simply

$$Q = A_1 v_1$$

where v_1 is the velocity of the liquid at this point. Similarly, the volume of liquid passing through A_2 per unit time is

$$Q = A_2 v_2$$

Since these two quantities must be equal for steady flow, we have

$$Q = A_1 v_1 = A_2 v_2 \tag{13-14}$$

Equation 13-14 is called the *equation of continuity*. From this equation we see that

$$\frac{v_1}{v_2} = \frac{A_2}{A_1} \tag{13-15}$$

Equation 13-15 states that the velocity of the liquid at any point in the pipe is inversely proportional to the cross-sectional area of the pipe. The liquid will be moving slowly where the area is large and will be moving rapidly where the area is small.

Illustrative example

Water flows out of a horizontal pipe at the steady rate of 2 ft³/min. Determine the velocity of the water at a point where the diameter of the pipe is (a) 1 in. and (b) $\frac{1}{2}$ in.

The area A_1 of the 1-in. portion of the pipe is

$$A_1 = \frac{\pi \times 1}{4 \times 144} \text{ ft}^2 = 0.0055 \text{ ft}^2$$

and the area of the $\frac{1}{2}$-in. pipe is

$$A_2 = 0.0014 \text{ ft}^2$$

also

$$Q = 2 \, \frac{\text{ft}^3}{\text{min}} = \frac{2}{60} \, \frac{\text{ft}^3}{\text{sec}}$$

Using Equation 13-13, we get for the velocity through the 1-in. pipe

$$\frac{2}{60} \frac{\text{ft}^3}{\text{sec}} = 0.0055 \text{ ft}^2 \times v_1$$

from which

$$v_1 = 6.10 \frac{\text{ft}}{\text{sec}}$$

From Equation 13-15, the velocity of the water through the $\frac{1}{2}$-in. pipe is four times that through the 1-in. pipe, so

$$v_2 = 24.4 \frac{\text{ft}}{\text{sec}}$$

13-13
Bernoulli's theorem

There is a very important relationship between the pressure at any point in a fluid and the velocity of the fluid at that point. This relationship can be derived most conveniently by considering the steady flow of an incompressible liquid through a frictionless horizontal pipe of varying cross section, as shown in Figure 13-13. Let us consider two regions in this pipe: region 1, where the cross-sectional area is A_1, and region 2, where the cross-sectional area is A_2. These areas are perpendicular to the streamlines in their respective regions. Imagine a quantity of liquid moving through the pipe that at this instant is contained between the areas A_1 and A_2. There is a force $F_1 = P_1 A_1$ acting toward the right on surface A_1 and a force $F_2 = P_2 A_2$ acting toward the left on surface A_2, where P_1 is the pressure in region 1 and P_2 is the pressure in region 2. In a small time interval Δt, the liquid will have moved to the right; its motion may be described by saying that the area A_1 has moved a small distance Δs_1 in this time and area A_2 has moved a distance Δs_2 in the same time, such that

$$A_1 \, \Delta s_1 = A_2 \, \Delta s_2 = V \qquad (13\text{-}16)$$

where V is the volume of liquid that has moved through the pipe in this time. This follows directly from the equation of continuity.

Now the work W done by the forces acting on

this volume of water is

$$W = F_1 \, \Delta s_1 - F_2 \, \Delta s_2$$

or

$$W = P_1 A_1 \, \Delta s_1 - P_2 A_2 \, \Delta s_2 \qquad (13\text{-}17)$$

From Equation 13-16, the expression for the work done can be written as

$$W = (P_1 - P_2)V \qquad (13\text{-}18)$$

The net effect of this work is to move a quantity of liquid of volume V from region 1 where its velocity is v_1 to region 2 where its velocity is v_2, thus changing its kinetic energy. If m is the mass of this volume of liquid, the change in its kinetic energy $\Delta \mathscr{E}_k$ is given by

$$\Delta \mathscr{E}_k = \tfrac{1}{2}mv_2{}^2 - \tfrac{1}{2}mv_1{}^2 \qquad (13\text{-}19)$$

From the principle of work and energy we get

$$(P_1 - P_2)V = \tfrac{1}{2}mv_2{}^2 - \tfrac{1}{2}mv_1{}^2 \qquad (13\text{-}20)$$

In deriving Equation 13-20 it was assumed that the pipe was horizontal and that there was no change in the potential energy of the mass m of liquid as it moved from region 1 to region 2. If the pipe is not horizontal and region 1 is at a height h_1 above some reference level and region 2 is at a height h_2 above the same level, then

$$\Delta \mathscr{E}_p = mgh_2 - mgh_1 \qquad (13\text{-}21)$$

where $\Delta \mathscr{E}_p$ is the change in potential energy.

Applying the principle of work and energy to this case we now obtain

$$(P_1 - P_2)V = \tfrac{1}{2}mv_2{}^2 - \tfrac{1}{2}mv_1{}^2 + mgh_2 - mgh_1 \qquad (13\text{-}22)$$

Dividing through by V and setting $\rho = m/V$, we get

$$P_1 - P_2 = \tfrac{1}{2}\rho v_2{}^2 - \tfrac{1}{2}\rho v_1{}^2 + \rho gh_2 - \rho gh_1$$

Rearranging terms yields

$$P_1 + \tfrac{1}{2}\rho v_1{}^2 + \rho gh_1 = P_2 + \tfrac{1}{2}\rho v_2{}^2 + \rho gh_2 \qquad (13\text{-}23)$$

Equation 13-23 is a mathematical statement of

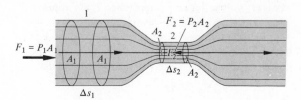

FIGURE 13-13 Steady flow of a liquid through a non-uniform horizontal pipe.

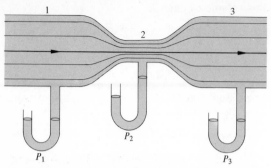

FIGURE 13-14 The pressure in the narrow portion of the pipe is less than in the wide portion.

Bernoulli's theorem, which states that at any two points along a streamline the sum of the pressure, the kinetic energy per unit volume, and the potential energy per unit volume has the same value. This may be written as

$$P + \tfrac{1}{2}\rho v^2 + \rho g h = \text{constant} \qquad (13\text{-}24)$$

along a streamline.

In the motion of a fluid through a horizontal pipe in which changes in potential energy along a streamline are negligible, Equation 13-24 becomes

$$P + \tfrac{1}{2}\rho v^2 = \text{constant} \qquad (13\text{-}25)$$

This equation shows that in the region where the velocity is small, the pressure is large, while in the region where the velocity is large, the pressure is small.

Bernoulli's theorem can be verified experimentally by attaching pressure gauges or manometers at various points along the pipe. Figure 13-14 shows three pressure gauges: at points 1 and 3, the pressures are the same since the cross-sectional areas are equal; at point 2, the pressure is less than at 1 and 3 since the area here is smaller and the kinetic energy of the water is greater.

Although Bernoulli's theorem has been derived for the steady flow of an incompressible liquid, it can be applied to the steady flow of gases. There are many practical devices whose operation is based upon Bernoulli's theorem. Among these are the aspirator or filter pump, and the atomizer or sprayer. Of the highest importance is its application to the design of the wings of an airplane so that the pressure on the top of the wing will be less than on the bottom of the wing, thus providing the force necessary to sustain the airplane in flight.

Illustrative example

The horizontal pipe sketched in Figure 13-15 is known as a *Venturi tube;* the constricted region is known as the throat of the tube. Suppose that water is flowing through a Venturi tube at the rate of 100 ft³/min. The pressure in the wide portion is 15 lb/in.² and the diameter is 6 in. Determine the pressure in the throat of the tube if its diameter is 3 in.

The area of the tube A_1 is

$$A_1 = \frac{\pi}{16} \text{ ft}^2$$

The area of the throat is

$$A_2 = \frac{\pi}{64} \text{ ft}^2$$

The velocity v_1 in the tube is, from Equation 13-13,

$$\frac{100}{60} \frac{\text{ft}}{\text{sec}} = \frac{\pi}{16} \text{ ft}^2 \times v_1$$

from which

$$v_1 = 8.5 \frac{\text{ft}}{\text{sec}}$$

and the velocity in the throat of the tube is

$$v_2 = 4 \times 8.5 \frac{\text{ft}}{\text{sec}} = 34 \frac{\text{ft}}{\text{sec}}$$

Let us write Equation 13-25 as

$$P_1 + \tfrac{1}{2}\rho v_1{}^2 = P_2 + \tfrac{1}{2}\rho v_2{}^2 \qquad (13\text{-}25a)$$

with

$$P_1 = 15 \frac{\text{lb}}{\text{in.}^2} = 15 \times 144 \frac{\text{lb}}{\text{ft}^2} = 2160 \frac{\text{lb}}{\text{ft}^2}$$

and

$$\rho = \frac{D}{g} = \frac{62.4 \text{ lb/ft}^3}{32 \text{ ft/sec}^2}$$

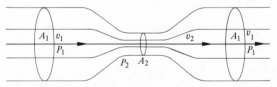

FIGURE 13-15 The Venturi tube.

Solving Equation 13-25a for P_2 and substituting numerical values we get

$$P_2 = 2160 + \frac{1}{2}\frac{62.4}{32} \times (8.5)^2 - \frac{1}{2}\frac{62.4}{32} \times (34)^2$$

or

$$P_2 = 2160 + 70 - 1130 = 1100 \frac{lb}{ft^2}$$

or

$$P_2 = \frac{1100 \text{ lb}}{144 \text{ in.}^2} = 7.64 \frac{lb}{in.^2}$$

The Venturi tube has several practical applications. If manometers or pressure gauges are placed at the throat and at the wide portion of the tube, the difference in pressures between these two regions and a knowledge of the cross-sectional areas can be used to measure the rate of flow of a fluid. Used in this way, it is called a *Venturi meter*. Wind tunnels designed for experiments with models of airplanes are also built in the shape of a Venturi tube.

13-14
Applications of Bernoulli's theorem

With a properly designed wing surface, the motion of the air will be along streamlines in the neighborhood of the surface of the wing. This is more likely to be the case when the angle between the wing and its direction of motion through the air is small; this is usually called a low *angle of attack*. The motion of the air past a wing at different angles of attack is illustrated in the photographs reproduced in Figure 13-16. Smoke was used to make the flow of air visible. The streamlines above the wing are crowded together closely because the air is moving more rapidly than below the wing. According to Bernoulli's theorem, since the air is moving more rapidly above the wing that below, the pressure above the wing is less than the pressure below it. The pressure below the wing is usually the pressure of the undisturbed atmosphere at that level, while the pressure above the wing is less than atmospheric. It is this difference in pressure that supplies the lifting force when the angle of attack is low.

At higher angles of attack, the streamlines do not follow the wing surfaces very closely, and turbulence of the air sets in, particularly near the rear of the wing. Wherever turbulence sets in, the pressure is not reduced below atmospheric pressure. At very high angles of attack, the wing acts more nearly like a sail. The lifting effect in this case is due to the force exerted on the wing as the air is deflected from it.

An idea of the magnitude of the lifting effect caused by the streamline motion of the air at low angles of attack can be obtained from the following data. The measured value of the pressure on the top of a wing was found to be 14.55 lb/in.^{2} while the pressure on the bottom of the wing was 14.70 lb/in.^{2}. The difference is 0.15 lb/in.^{2} or 21.6 lb/ft^2. If the wing had a spread of 40 ft and a width of 8 ft

FIGURE 13-16 Smoke is used in the NACA smoke tunnel at Langley Field, Virginia, to make the flow of air visible. Note the smoothness of the air flow in the lower pictures. When the angle of attack has been increased to 10 degrees, the air flow begins to separate from the upper surface of the airfoil (center view); and when the angle is increased to 30 degrees, the flow separates completely from the upper surface. Turbulence behind the trailing edge of the airfoil may be observed. (Reproduced from *Journal of Applied Physics*, August, 1943, with permission from the National Advisory Committee for Aeronautics.)

or an area of 320 ft², the lifting force would be 6900 lb.

One of the factors determining the lift of an airplane is the speed of the air relative to the plane; it is called the *air speed*. In taking off, the airplane should be headed into the wind to provide greater air speed. The angle of attack is very small while the plane is running along the ground, so that the entire lift is due to the Bernoulli effect. Once the plane is off the ground, the angle of attack is increased to provide a higher rate of climb. The lift in this case is due partly to the Bernoulli effect and partly to the force of the air against the wing as it is deflected from its lower surface.

Bernoulli's theorem can also be used to explain the curving of a baseball. When the ball is pitched it not only is thrown forward but also is given a spin. Suppose that the ball is thrown to the left and given a spin in a clockwise direction, as shown in Figure 13-17. If it were not for the spin, the air speed would be the same at the top and bottom and would be equal in magnitude and opposite in direction to the velocity of the ball. But, because of the spin of the ball, the air near its surface is forced around in the direction of spin. This means that the speed of air on top of the ball is greater than at the bottom. This will create a difference in pressure between top and bottom, with the greater pressure on the bottom. There will thus be a force upward causing the ball to deviate from the usual path. If we are looking down on the ball as it moves forward, this force will produce a deflection to the right. If the direction of spin is reversed, the ball will be deflected to the left.

Bernoulli's theorem in the form given by Equation 13-23 can be used to determine the velocity with which a liquid comes out of an orifice in a tank in which the top surface of the liquid is at a

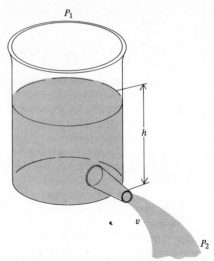

FIGURE 13-18 Velocity of efflux $v = \sqrt{2gh}$.

height h above the orifice. Referring to Figure 13-18 let us assume that a particle of liquid moves along a streamline from the top surface of the liquid to the orifice. The pressure at the top surface is simply the barometric pressure, and so is the pressure at the orifice; hence

$$P_1 = P_2$$

Equation 13-23 now becomes

$$\tfrac{1}{2}\rho v_1{}^2 + \rho g h_1 = \tfrac{1}{2}\rho v_2{}^2 + \rho g h_2 \qquad (13\text{-}26)$$

The velocity of the particle at the top surface is very small and if the tank is large enough that the level does not change appreciably in a short time interval, we can set $v_1 = 0$. The height h of the liquid above the orifice is simply

$$h = h_1 - h_2$$

Putting these values into Equation 13-26 yields

$$\rho g h = \tfrac{1}{2}\rho v_2{}^2$$

from which

$$v_2{}^2 = 2gh$$

Let us set

$$v_2 = v$$

so we can write

$$v^2 = 2gh$$

or

$$v = \sqrt{2gh} \qquad (13\text{-}27)$$

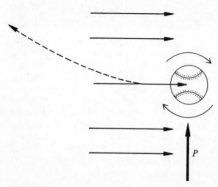

FIGURE 13-17 Curving of the path of a baseball is produced by giving it a spin when pitching it.

Equation 13-27, sometimes called *Torricelli's theorem*, states that the speed of the liquid coming out of a tank filled to a height h above the opening is exactly the same as if the liquid had fallen through the same height. If the pipe were provided with an outlet directed vertically upward, the liquid would rise to a height h equal to the level of the liquid in the tank (see Figure 13-19). In this discussion we have neglected the effect of friction in the pipe. If friction, or rather viscosity, is taken into consideration, the velocity with which the liquid comes out of the tank will be less than that given by Equation 13-27.

13-15
Fluid friction; viscosity

When a fluid, either a liquid or a gas, is set into motion, different parts of the fluid move with different velocities. For example, if a jar of water is tilted so that the water starts flowing out, the top layer of the water moves over the lower portion of water. Just as there is friction when one surface of a solid slides over another, so there is friction when one layer of a fluid slides over another. This friction in fluids is called *viscosity*. When a fluid flows through a cylindrical pipe, for example, the part of the fluid that is in contact with the pipe adheres to it and remains at rest. We may think of the rest of the fluid as divided into concentric cylindrical layers. The cylindrical layer next to the stationary layer of fluid moves slowly past it. The next inner cylindrical layer moves with greater velocity, and the velocity of each succeeding inner layer increases as we go toward the center. A difference in pressure between the two ends of the pipe is needed to maintain a steady flow through it and

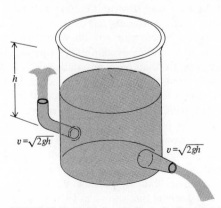

FIGURE 13-19 Velocity of efflux of a liquid depends only on the height of the liquid above the orifice.

oppose the force due to the viscosity of the fluid. One method for measuring the viscosity of the fluid is to determine the volume of fluid flowing in unit time through a pipe of given diameter under a known difference of pressure. The greater the viscosity, the smaller is the volume of fluid flowing through the pipe in unit time. For example, glycerin at 20°C has a viscosity 850 times that of water at 20°C, and the viscosity of water is about 55 times that of air at 20°C. There is, however, one important distinction between the viscosity of a liquid and that of a gas: the viscosity of a liquid decreases with a rise in temperature, whereas the viscosity of a gas increases with a rise in temperature.

The resistance experienced by a solid in moving through a fluid is due essentially to the viscosity of the fluid. A certain amount of the fluid adheres to the surface of the solid and moves with it. There is relative motion between this layer of fluid and the layer adjacent to it. The latter, in turn, moves relative to the next outer layer, and so on until we come to a layer of fluid that remains at rest. An additional cause of the resistance experienced by objects moving through fluids is the turbulence set up in the fluid. In this case the fluid can no longer be considered as made up of layers; these layers are broken up, forming eddies and sometimes waves.

13-16
The siphon

A siphon is a bent tube used for transferring a liquid from one vessel to another one at a lower level. Figure 13-20 shows a siphon in operation. Liquid is flowing through the bent tube from A to B in the direction of the arrows and will continue to flow as long as there is a difference in levels between A and B. To understand the operation of the siphon let us imagine that A and B are at the same level and that the bent tube is filled with liquid; in this case there will be no flow of liquid. If the level of B is now lowered so that it is at a distance H from the top of the bend while the level of A remains at a distance h below the top, liquid will flow from left to right. To prevent the flow, we would have to supply an external force on the lower surface sufficient to produce an additional pressure equal to that produced by a column of liquid of height $H-h$. But this is simply the difference in the lengths of the columns of liquids in the two arms of the siphon. When there is no additional

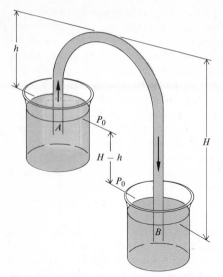

FIGURE 13-20 Operation of a siphon.

external force, liquid flows through the siphon because the column of liquid in the longer arm, being heavier than the column of liquid in the shorter arm, pulls the liquid over the bend in the tube.

Another way of considering the action of a siphon is to imagine it already full of liquid and a valve placed at the highest point to stop the flow of liquid. Consider the pressures on the two sides of this valve. The pressure P_L on the left side will be that due to the atmosphere less that due to the column of liquid of height h, or

$$P_L = P_0 - h\rho g$$

where ρ is the density of the liquid. Similarly, the pressure P_R on the right side of the valve will be

$$P_R = P_0 - H\rho g$$

The difference in pressure on the two sides of the valve will be

$$P_L - P_R = (H - h)\rho g \qquad (13\text{-}28)$$

producing a force toward the right. If the valve is opened or removed, the liquid will flow from left to right and will continue to flow as long as there is a difference in pressure.

It will be noted that the pressure of the atmosphere P_0 does not appear in Equation 13-28. Hence, once the siphon has been started, it should be possible to keep it operating even if P_0 is reduced to practically zero. The column on the right, being heavier than that on the left, can pull the

liquid over the bend, provided that the cohesive forces between the molecules of the liquid are sufficiently large and there are no bubbles of air or vapor to break the column. In ordinary operation, liquids, particularly water, are never free of air bubbles, so it would be practically impossible to operate such a siphon in a vacuum.

13-17
The centrifuge

A centrifuge is a machine designed to rotate a liquid at high speed in order to produce a more rapid separation of the particles suspended in the liquid than would occur if it were allowed to stand and be acted upon by the earth's gravitational field. In some types of centrifuge, such as a cream separator, the separated constituents flow out continuously during its operation. In other types, the rotational motion is stopped after a certain time has elapsed, and the separated constituents are then removed.

To understand the operation of a centrifuge, consider a particle of mass m and weight W placed in a liquid; this particle will experience a buoyant force B equal to the weight of liquid displaced by it. If m_0 is the mass of the displaced liquid, the buoyant force $B = m_0 g$, and, since $W = mg$, the difference between these forces is

$$W - B = (m - m_0)g \qquad (13\text{-}29)$$

If W is greater than B, the particle will move down in the direction of g. Its motion will be opposed by a resisting force owing to the viscosity of the liquid. To increase the motion downward, it will be necessary to increase the difference between W and B. An examination of Equation 13-29 shows that the only way this can be done is to put the system in a region where g has a larger value. In our discussion of circular motion we showed that a particle in a rotating system experiences an effect equivalent to that of a gravitational field directed away from the center; this effect is an acceleration that is independent of the mass of the particle. The magnitude of this acceleration is $a = v^2/R$, where v is the speed of the particle and R is its distance from the axis of rotation. By making the speed v large enough, we can produce any desired value of a equivalent in effect to cg, where c is a number greater than one. Hence, if a liquid is put into a tube such as that shown in Figure 13-21 and the tube is rotated at high speed, the particles in the liquid will separate out very rapidly. Those

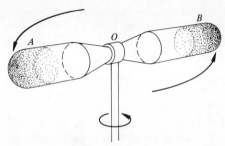

FIGURE 13-21 The centrifuge. Tubes *A* and *B* are whirled around in a circular path about an axis through *O*. Particles of greatest density go to ends of tubes farthest removed from *O*.

particles that are denser than the liquid will be found at the bottom—that is, farthest removed from the axis of rotation *O*. Those particles that are less dense than the liquid will be found near the top of the liquid—that is, closest to the axis of rotation.

Illustrative example

A liquid containing some solid particles is poured into the cup of a centrifuge. It is then rotated by means of an electric motor at a constant speed of 6000 rev/min. Determine the acceleration of a particle at a distance of 12 cm from the axis of rotation. Express this in terms of *g*.

A particle revolving at 6000 rev/min makes 100 rev/sec. The period of this motion $T = 0.01$ sec $= 10^{-2}$ sec. The acceleration of a particle moving with uniform speed in a circle is given by

$$a = \frac{4\pi^2 R}{T^2}$$

Substituting the values $R = 12$ cm and $T = 10^{-2}$ sec, we get

$$a = \frac{4\pi^2 \times 12 \text{ cm}}{10^{-4} \text{ sec}^2}$$

or

$$a = 474 \times 10^4 \ \frac{\text{cm}}{\text{sec}^2}$$

Taking

$$g = 980 \ \frac{\text{cm}}{\text{sec}^2}$$

we get

$$a = \frac{474 \times 10^4}{980} g = 4840g$$

Some modern centrifuges are being operated at speeds of about 20,000 rev/sec. Modern ultracentrifuges develop centripetal accelerations as large as $10^9 g$.

Questions

1. Using Archimedes' principle prove that the specific gravity of a solid is equal to the weight of the solid divided by its apparent "loss of weight" in water.

2. A steel battleship is sunk in the ocean where the depth is about 5 mi. How far down will the ship go?

3. Explain why a balloon can float at a definite level in the atmosphere, whereas a submarine cannot float at a definite level below the surface of the ocean.

4. Why is a radiosonde sent up with many small balloons instead of with one large balloon?

5. Estimate the total force exerted on you by the atmosphere.

6. A physician measuring a patient's blood pressure notes it as 125-88. What do these numbers mean in terms of the pressure of the blood?

7. A glass tumbler is filled to the brim with water and a thin aluminum disk is placed on top of the glass. The glass is inverted, with the disk held in place. The hand supporting the disk is removed and the disk remains in place, pressed against the glass. Explain. Try this experiment, using any light material for a cover if aluminum is not available.

8. The experiment of Question 7 is repeated with the tumbler partly full of water. Explain what will happen.

9. An empty glass tumbler is inverted and pushed down into a dish of water. Will the water level inside the tumbler be the same as outside the glass. Explain. Try this experiment.

10. Two objects of very different densities are on opposite pans of an equal-arm balance, which indicates that they have the same mass. Do they have the same mass? Explain what would happen if the balance were placed in an evacuated chamber.

11. Why is a difference of pressure usually necessary to force a liquid through a horizontal pipe of uniform cross section?

12. There is a steady flow of liquid through a horizontal pipe of nonuniform cross section. Does a typical small volume of the liquid remain in equilibrium as it flows through this tube?

13. If a ball is placed in a funnel and a stream of air is blown through the stem of the funnel against the ball, the ball will remain in the funnel.

Account for this on the basis of Bernoulli's theorem.

14. Explain how you would pitch a ball to produce (a) an outcurve and (b) a drop.

15. Explain the effect of spin on the motion of a "Frisbee," which can be thought of as an inverted pie plate.

16. Hold two sheets of paper together and blow air between them. Instead of becoming separated, the sheets will come together. Explain this.

17. Will a siphon work in a vacuum?

Problems

1. A 180-lb man stands on his own two feet, each of which has an area of 40 in.² What pressure does he exert on the ground?

2. The man of Problem 1 puts on spiked shoes, each of which has six spikes. The area of the tips of each spike is 0.1 in.² What pressure does the man exert?

3. A suspicious king gives a jeweler 1 kg of gold to be fashioned into a crown. The king tests the crown, measuring its volume by determining the amount of water the crown displaces. The jeweler has adulterated the crown by substituting copper, so that the resultant alloy is 10 percent copper. How much water does the king expect to be displaced and how much water is actually displaced? The jeweler is clever enough to present the king with a crown that weighs 950 gm and with 50 gm of pure gold dust.

4. Find the pressure at the bottom of a flask of mercury that is 12 cm deep.

5. A merchant measures the pressure at the bottom of a barrel, 1 m high, as 8×10^4 dynes/cm². Does the barrel contain pure olive oil, as labeled?

6. A block of metal weighs 6 lb in air and 5.2 lb when immersed in water. What is the density of the metal?

7. A sphere, 1 cm in diameter, is made of the lightest common glass (see Table 13-1) and dropped in a glass of water. Find the buoyant force on the sphere.

8. What is the total downward force exerted by the atmosphere on a sheet of 8.5 in. by 11-in. paper that is lying on a tabletop? In view of

this, how is it possible to lift the paper from the table as easily as experience indicates?

9. During normal breathing the gauge pressure of the air in the lungs changes from zero to about 20 mm of mercury. Assume that the mass of air does not change (which is not in general true) and that the temperature stays constant at 20°C. Find the maximum density of the air in the lungs.

10. Water flows through a horizontal pipe of varying cross section at the rate of 5 ft³/min. Determine the velocity of the water at a point where the diameter of the pipe is (a) 1.5 in. and (b) 2 in.

11. Oil flows through a 12-in. pipeline with a speed of 3 mi/hr. How many gallons of oil are delivered per day by this pipeline? (*Note:* 1 gal = 231 in.³)

12. At a place in a pipeline where the diameter is 6 in. the speed of a steady stream of water is 9 ft/sec. (a) What will be the speed of the water in that portion of the pipeline where the diameter is 5 in.? (b) At what rate in cubic feet per minute is water being delivered by this pipeline?

13. Blood flows in a particular capillary which is 2×10^{-3} cm in diameter and 0.5 mm long, at a speed of 0.5 mm/sec. The pressure difference between the ends of the capillary is 1500 dynes/cm². (a) What volume of blood flows per second? (b) How much work is done per second to maintain this flow?

14. A water storage tank is filled to a height of 20 ft. (a) With what speed will water come out of a valve at the bottom of the tank if friction is

negligible? **(b)** To what height will the water rise if the valve is directed upward?

15. Water flows steadily through a Venturi tube at the rate of 50 ft³/min. At a place where the diameter of the tube is 6 in. a pressure gauge reads 20 lb/in.². Determine the pressure in the throat of the tube, where the diameter is 2 in.

16. In a wind tunnel experiment the pressure on the upper surface of a wing was 13.02 lb/in.² while the pressure on the lower surface was 13.30 lb/in.². Determine the lifting force on the wing if the wing can be treated as a rectangle of length 60 ft and width 9.5 ft.

17. A metal sphere whose mass is 64 gm is attached to one arm of an equal-arm balance by means of a string. When the sphere is completely immersed in water a mass of 50 gm is sufficient to balance it. Determine **(a)** the volume of the sphere and **(b)** the density of the metal, assuming that the sphere is solid.

18. A glass tube 1 cm² in cross-sectional area and 50 cm long is fitted into the top of a cylindrical bottle 20 cm high and 5 cm in diameter. The bottle is filled with water. Find **(a)** the mass of water in the bottle, **(b)** the pressure on the top of the bottle, **(c)** the pressure on the bottom of the bottle, and **(d)** the force on the bottom of the bottle. Then more water is poured into the tube so that the tube is now filled. Find **(e)** the additional mass added, **(f)** the pressure on the top of the water in the bottle, **(g)** the pressure on the bottom of the bottle, and **(h)** the force on the bottom of the bottle.

19. Hoover Dam is 1180 ft long and 726 ft high. **(a)** What is the pressure at the bottom of the dam when the reservoir behind it is full? **(b)** Assuming the face of the dam to be a plane rectangle, determine the force pushing against the dam.

20. A block of wood whose mass is 100 gm has a hole drilled in it, removing 8 gm of wood. The specific gravity of the wood is 0.8. The hole in the wood is filled with lead. Will this block sink or float when placed in water?

21. A raft is made in the form of a rectangular box 9 ft by 12 ft by 5 ft deep. The raft weighs 3000 lb. **(a)** How deep will this raft go when placed in fresh water? **(b)** What load can this raft carry without sinking? **(c)** What is the total force on the bottom of the raft when so loaded?

22. A cube of iron 6 cm on an edge is placed in a dish of mercury. The density of the iron is 7.7 gm/cm³. **(a)** How much of this cube is immersed in the mercury? **(b)** If water is poured over the mercury to a depth of 8 cm, what will be the depth of the iron in the mercury?

23. The pistons of a hydraulic press have diameters of 2 in. and 9 in., respectively. A force of 28 lb is applied to the smaller piston. **(a)** What is the pressure developed by the application of this force? **(b)** What weight can be lifted by the larger piston as a result of this force?

24. Apply the principle of work and energy to the operation of a hydraulic press. Show that Pascal's principle follows from this law.

25. The pressure for the hydraulic lifts in service stations is usually supplied by an air compressor. If the maximum available pressure is 100 lb/in.² and the lifting piston has a diameter of 1 ft, find the weight of the heaviest vehicle that can be lifted.

26. Most human beings can just float in fresh water with a negligible volume in the air. Find the average cross-sectional area of a 5-ft 2-in. woman who weighs 110 lb. Compute the circumference of a circle of this area.

27. A stone of weight 180 lb has a specific gravity of 3.0. A block of wood of specific gravity 0.6 is to be fastened to the stone so that both will float in water almost completely submerged. What is the maximum weight of wood that may be used?

28. A glass tube 100 cm long and closed at one end is sunk, open end down, to the bottom of the ocean. When it is drawn up, a marker indicates that the water had risen to a point 5 cm from the top. Calculate the depth of the ocean at this point.

29. A vertical cylinder, closed at the bottom, contains air at room temperature. A piston weighing 30 lb is placed in the open end and comes to rest 10 in. from the bottom of the cylinder. The area of the piston is 100 in.². A load of 350 lb is now placed on the piston. Where does the piston come to equilibrium?

30. An airtight bag is filled with air while lying on the shore of a lake. It is then lowered under water by means of a rock tied to it with a string. At what depth will the volume be one half as much as it was before immersion?

31. A U-shaped tube containing water has one end

connected to a city gas supply outlet. The difference in levels in the two arms of the tube is 3.0 in. Determine the pressure of the gas.

32. The gauge pressure of the air in a tire is 28 lb/in.². If the volume of the tire is 1.9 ft³, what volume will the air occupy at atmospheric pressure? Take atmospheric pressure to be 15 lb/in.².

33. The gauge pressure in the tire of Problem 32 is increased to 32 lb/in.² by adding more air. Find the change in density of the air.

34. A balloon is partly inflated with 6000 ft³ of hydrogen at 0°C. (a) Determine the buoyant force on the balloon. (b) What is the total load that the balloon can lift? (c) If the balloon rises to 10,000 ft above sea level where the pressure is 10 lb/in.², determine the new volume of the hydrogen, assuming the balloon does not burst.

35. Oil of specific gravity 0.9 flows through a tube 3 cm in diameter at a pressure of 1.8×10^6 dynes/cm². At one portion, the tube narrows down to 2 cm in diameter and the pressure drops to 10^6 dynes/cm². (a) Determine the velocity of the oil in the wider portion of the tube. (b) Determine the rate at which oil flows through this tube.

36. Air, of density 1.29×10^{-3} gm/cm³, flows through a horizontal pipe 8 cm in diameter. The pipe narrows down to 2 cm in diameter and then widens again to 8 cm where the pipe is open to the atmosphere. At the constriction a narrow pipe descends to a dish of mercury. The level of the mercury in this vertical pipe is 6 cm above the level in the dish. How fast is the air moving in the wide portion of the horizontal pipe?

37. The lower end of a siphon is 12 ft below the level of the water surface in the tank. (a) Determine the speed with which the water flows out of the open end. (b) If the siphon has an inside diameter of 8 mm, determine the rate at which water is siphoned out.

38. The water behind a levee is 3 m high. A hole appears in the levee, 80 cm below the top of the water level. Find (a) the velocity with which the water escapes, (b) the distance from the hole where the water strikes the ground, and (c) the velocity of the water when it strikes the ground.

39. A beaker partially filled with water is placed on a scale pan and found to have a mass of 600 gm. A string is attached to a stone and held so that the stone is completely submerged in the water but does not touch the beaker at any point. The scale now reads 680 gm. When the string is released and the stone rests on the bottom of the beaker, the scale reads 800 gm. Determine the mass of the stone and the density of the stone.

40. A piece of wood of height h and cross-sectional area A floats in water. Show that if it is pushed down a distance x below its equilibrium position and released, the resultant force on it will be proportional to x and opposite in direction. Discuss its subsequent motion, assuming the friction of water to be negligible.

41. A barge m ft long, n ft wide, and h ft high is floating in a canal lock loaded with p lb of iron. In a fit of pique a sailor dumps the entire load of iron over the side into the water in the lock. Find the change in height of the water level in the lock. The lock is L ft long and W ft wide.

42. (a) Estimate the buoyant force of the air on you. (b) Calculate the buoyant force exerted by air on a piece of gold whose mass is the same as your mass.

43. Biological laboratories often store solutions in large jugs on high shelves. The solutions are dispensed through a tube that is attached to a spout at the base of the jug. In such a system the flow rate of the fluid would decrease as the level of the fluid in the jug fell. In order to keep the flow rate constant, each such jug is fitted with a tight, one-hole stopper, with a glass (or plastic) tube, open to the air, that reaches almost to the bottom of the jug, as shown in Figure 13-22. Show that the velocity v of the fluid as it leaves the dispensing tube depends on the height h of the bottom of the tube, rather than on the height of the top surface of the liquid.

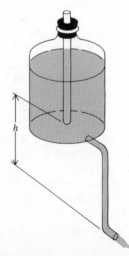

FIGURE 13-22

elasticity & surface behavior

chapter 14

14-1
Internal forces in solids

When an object or system is subjected to external forces it generally undergoes a change in size or shape, or both. We have thus far touched very lightly on such changes; for example, we have considered the change in length of an elastic spring and the change in volume of a gas. The changes produced in a system by the action of external forces depend on the physical properties of the materials of which the system is composed. The properties of the materials depend on the properties of the molecules (or atoms) of which the material is composed and on the characteristics of the forces between the molecules. In addition, the geometric arrangement of the molecules is a very important determinant of the properties of the material. A study of the properties of matter leads to information that is of practical value to both the physicist and engineer, and also gives us some information about these internal properties of materials.

When a substance is in the solid phase, the forces that exist between atoms and molecules cause them to form definite geometric patterns or crystalline structure. Sometimes these crystals grow to a fairly large size, as in the case of rock salt or quartz (see Figure 14-1). Sometimes the crystals are very small and can be seen only with the aid of a microscope.

The mechanical properties of crystals depend on the nature of the bonding forces between molecules. There are four different kinds of bonding forces. The simplest kind of bonding is called *ionic bonding*; here one atom loses an electron and becomes a positively charged ion, while another atom gains the electron and becomes a negatively charged ion. The two charged ions are then at-

FIGURE 14-1 Photograph of a large quartz crystal grown at the Bell Telephone Laboratores. (Courtesy of the Bell Telephone Laboratories.)

tracted by electric force. Common table salt, sodium chloride, consists of a three-dimensional array of sodium ions and chlorine ions bound by these ionic forces into a crystal with a basically cubic form. Ionic bonds are very strong, as evidenced by the fact that it takes very high temperatures to melt such crystals.

Another very strong bond is the *covalent* bond. In this bond, electrons are shared by two atoms. The great hardness of diamond arises from the fact that each of the carbon atoms has a covalent bond with each of its four nearest neighbors. These bonds are arranged in space so as to construct a rigid three-dimensional crystal. In graphite the four covalent bonds lie in a plane, so the graphite crystal consists of strong layers, but the layers can slip over each other. In many crystals the bonding is a mixture of ionic and covalent bonds. Perfect ionic or covalent crystals (with no impurities or dislocations) may be thought of as single large molecules.

Another kind of bonding is *molecular bonding*; here the atoms are bound together by covalent bonds into small molecules and these molecules are bound together weakly to form the crystal. Sugar is one such crystal.

The metals are a separate class of crystals. Here the electrons are not associated with any one atom or even with a local group of atoms. In *metallic bonding* the electrons are shared by the whole crystal. One or two electrons leave each atom in the crystal and form a gas of electrons. The positive ions are bound together by this electron gas.

This metallic bonding is relatively weak compared to ionic or covalent bonding. Perfect metallic crystals are relatively soft and can be deformed rather easily. Interestingly enough, impurities, dislocations, or imperfections in the crystal structure make the crystal relatively strong. Pure iron is soft, but when small amounts of carbon in the form of fine grains are added, steel is formed. The hardness of steel arises from these imperfections. Similarly, many metals such as copper are normally soft, but can be *work-hardened* by hammering or repeated flexing. These processes introduce small dislocations or imperfections in the crystal structure.

14-2
Elasticity

When an object is subjected to the action of external forces, it may undergo a change in size or shape. If, after these forces are removed, the object returns to its original size and shape, the substance of which the object is composed is said to be *elastic*. In order to make quantitative measurements of the elastic properties of a substance, we find it convenient to introduce two new terms: these are (a) *stress* and (b) *strain*.

Stress is defined as *the ratio of the internal force F, brought into play when the substance is distorted in any way, to the area A over which this force acts.* Thus

$$\text{stress} = \frac{F}{A} \qquad (14\text{-}1)$$

In the cgs system, the stress is expressed in dynes per square centimeter; in the British Engineering system, it is expressed in pounds per square foot. In most engineering practice, the stress is expressed in pounds per square inch.

Strain is defined as *the ratio of the change in size or shape to the original size or shape.* As a ratio, strain has no physical dimensions; that is, it is a numeric. Methods of expressing the strain will be shown in the various cases.

The relationship between stress and strain was first given by Robert Hooke (1635–1703) and is known as *Hooke's law*. *Hooke's law states that, for an elastic body, the ratio of the stress to the strain produced is a constant*, or

$$\frac{\text{stress}}{\text{strain}} = K \qquad (14\text{-}2)$$

where K is called the *modulus of elasticity*. The units for K are the same as those for stress, since strain is expressed as a pure number.

In this chapter, we shall limit our discussion to two types of stress: (a) that involved in a change of length and (b) that involved in a change of volume.

14-3
Tensile stress and strain

As an example of the stress set up inside a substance, let us consider the increase in the length of a rod produced by the action of two forces, each equal to F, applied at the ends of the rod, as shown in Figure 14-2a. These forces are applied by means of clamps C_1 and C_2 attached to the ends of the rod.

If L is the original length of the rod and if Δl is the increase in length produced by the application of the forces F, then the strain produced is

$$\text{strain} = \frac{\text{increase in length}}{\text{original length}} = \frac{\Delta l}{L} \qquad (14\text{-}3)$$

To determine the stress in the rod, let us take

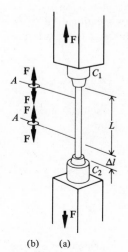

(b)   (a)

FIGURE 14-2 Tensile stress in a rod.

any cross-sectional area A through the rod and consider the forces that act on it (see Figure 14-2b). The entire rod is in equilibrium under the action of the two external forces, each of magnitude F but in opposite directions, so as to extend the rod. These forces produce a *tension* inside the rod. If we consider a cross-sectional area A near the clamp C_2, it is acted upon by the external force F downward, and, since it is in equilibrium, it must also be acted upon by a force F upward exerted by that part of the rod that is immediately above this area. If we take any other parallel cross-sectional area A, it will be acted upon by a force downward equal to F exerted by that part of the rod below it and another force upward also equal to F exerted by that part of the rod above it. The effect of these two forces is to tend to separate the rod across this section; it is opposed by the forces of attraction between the molecules on the two sides of this section. The stress in the rod is the ratio of one of these forces F to the cross-sectional area A, or

$$\text{stress} = \frac{F}{A} \qquad (14\text{-}1)$$

If the material of the rod is elastic, then we know, from Hooke's law, that

$$\frac{\text{stress}}{\text{strain}} = K \qquad (14\text{-}2)$$

Putting in the values of the stress and strain found above and replacing the letter K by Y, we get

$$\frac{F/A}{\Delta l/L} = Y \qquad (14\text{-}4)$$

Y is called *Young's modulus*, after Thomas Young (1773–1829), who devised careful experiments for the determination of ratio of stress and strain. The values of Y for several substances are listed in Table 14-1.

The extent to which a substance remains elastic as the tensile stress is increased can be determined only by experiment. Figure 14-3 shows the results of a typical experiment on a metallic rod. In this figure, the stress is plotted as ordinate and the strain as abscissa. This curve is obtained by exerting a force, measuring the strain, and then removing this force. If the rod returns to its original length, it is elastic. The straight-line portion of this curve from O to E represents the values of stress and strain for which the rod remained elastic. The curved portion from E to B represents the values

Table 14-1   Elastic constants of some solids

material	Young's modulus		bulk modulus	
	in dynes/cm² × 10¹¹	in lb/in.² × 10⁶	in dynes/cm² × 10¹¹	in lb/in.² × 10⁶
Aluminum, rolled	6.96	10.1	7	10
Brass	9.02	13.1	6.1	8.5
Copper, rolled	12.1–12.9	17.5–18.6	14	21
Duralumin	6.89	10.0		
Iron, cast	8.4–9.8	12–14	9.6	14
Glass, crown	6.5–7.8	9.5–11.3		
Lead	1.47–1.67	2.13–2.42	0.8	1.1
Nickel	20.0–21.4	29.0–31.0	26	34
Platinum	16.67	24.18		
Steel, annealed	20.0	29.0	16	23
Tin	3.92–5.39	5.69–7.82		
Tungsten, drawn	35.5	51.5	20	29

of stress and strain for which the rod is no longer elastic. For example, if a stress of value *CD* is applied to the rod, and then removed, it is found that the rod does not return to its original length. It is said to have a permanent set. The point *E* is called the *elastic limit* of the material. At point *B*, the stress was great enough to break the rod. The value of this stress is known as the *breaking strength* of the material.

A very simple and instructive experiment to demonstrate the above properties of a metal can be performed by attaching one end of a long copper wire to a support in the ceiling and hanging a weight of, say, 1 lb on the other end (No. 16 B. & S. gauge wire will do). (See Figure 14-4.) A yardstick mounted in a base is placed on the floor, and a pointer near the top of the yardstick is set opposite a reference mark on the wire. Additional weights of 1 lb are then added to the wire, and its elongation is noted. At first this elongation will be small but observable. The additional weights can be removed

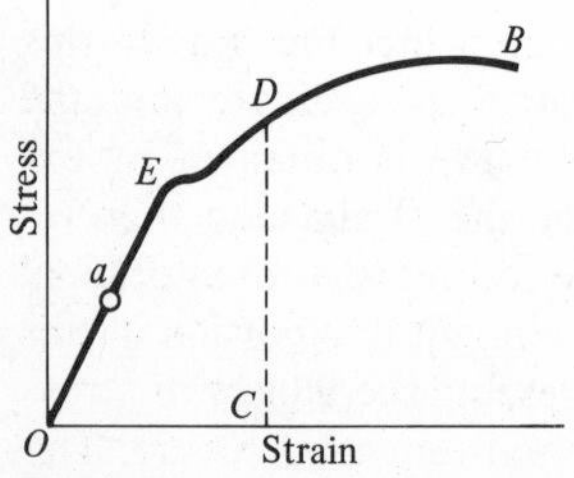

to see whether the wire returns to its original length.

After about 6 lb have been added to the wire, the increase in length for each additional pound weight will become more and more marked when the elastic limit is passed. With sufficient weights, the breaking strength will finally be reached.

Illustrative example

A piece of No. 16 B. & S. gauge copper wire 3 ft long is suspended from a rigid support and supports a load of 4 lb. Determine (a) the stress in the wire, (b) the increase in length produced by the 4-lb load, and (c) the strain produced.

(a) A wire of No. 16 B. & S. gauge has a diameter of 0.05082 in. and a cross-sectional area of 0.00203 in.². The stress in this wire is, from Equation 14-1,

$$\text{stress} = \frac{F}{A} = \frac{4 \text{ lb}}{0.00203 \text{ in.}^2} = 1970 \frac{\text{lb}}{\text{in.}^2}$$

(b) The increase in length can be determined by solving Equation 14-4 for Δl. Now

$$Y = \frac{F/A}{\Delta l/L} \qquad (14\text{-}4)$$

hence

$$\Delta l = \frac{F}{A} \times \frac{L}{Y}$$

The value of Young's modulus for copper is given in Table 14-1 and is

FIGURE 14-3 Stress-strain curve for a ductile material.

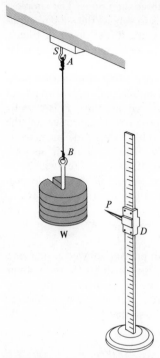

FIGURE 14-4 A copper wire AB attached to support S in ceiling. W is load on wire, D is a yardstick, and P is a movable pointer.

$$Y = 17.5 \times 10^6 \, \frac{\text{lb}}{\text{in.}^2}$$

Substituting this value for Y in the equation for the increase in length of the wire, for F/A the value determined in (a), and for L the value 36 in., we get

$$\Delta l = 1970 \, \frac{\text{lb}}{\text{in.}^2} \times \frac{36 \text{ in.}}{17.5 \times 10^6 \text{ lb/in.}^2}$$

from which

$$\Delta l = 0.0041 \text{ in.}$$

(c) The strain produced in the wire is, from Equation 14-3,

$$\text{strain} = \frac{\Delta l}{L} \qquad (14\text{-}3)$$

Hence

$$\text{strain} = \frac{0.0041 \text{ in.}}{36 \text{ in.}}$$

or

$$\text{strain} = 0.000113 = 1.13 \times 10^{-4}$$

14-4
Compressive stress and strain

If the ends of a rod of some material are subjected to the action of two forces, each of magnitude F but directed so as to diminish its length, the rod is said to be under *compression*, and the stress inside the rod is a *compressive stress* (see Figure 14-5a). If we consider the forces that act across any cross-sectional area A, that part of the rod to the right of this area exerts a force F toward the left, while that part to the left of this area exerts a force to the right (see Figure 14-5b). The compressive stress in the rod is the ratio of either one of these forces to the area over which it acts; that is,

$$\text{compressive stress} = \frac{F}{A}$$

The molecular forces that are brought into play by the action of the external forces must be forces of repulsion, and these are of an electrical nature.

The strain produced by the compressive stress is the ratio of the decrease in length Δl to the original length L; that is,

$$\text{compressive strain} = \frac{\Delta l}{L}$$

If the material of which the rod is made is elastic, then experiment shows that, within the elastic limit, the ratio of compressive stress to compressive strain, or Young's modulus for compression, is identical in value to Young's modulus for tension for the same material.

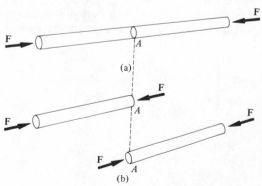

FIGURE 14-5 (a) Rod put under compression by action of two external forces each equal to F. (b) Any cross section A is acted upon by an internal force F to the left due to the section of rod on its right, and by an internal force F to the right due to the section of rod on its left.

Illustrative example

A steel rod 5 cm long and 0.12 cm² in cross-sectional area is to be compressed until its length has been decreased by 3 percent. Determine (a) the compressive stress that must be produced in the rod and (b) the force that must be applied to each end of the rod.

(a) From Equation 14-4 we can write the compressive stress as

$$\frac{F}{A} = \frac{\Delta l}{L} \cdot Y$$

$$= 0.03 \times 20 \times 10^{11} \frac{\text{dynes}}{\text{cm}^2}$$

or

$$\frac{F}{A} = 6.0 \times 10^{10} \frac{\text{dynes}}{\text{cm}^2}$$

(b) The force F that must be applied to each end to produce this stress is

$$F = 6.0 \times 10^{10} \frac{\text{dynes}}{\text{cm}^2} \times 0.12 \text{ cm}^2$$

or

$$F = 7.2 \times 10^9 \text{ dynes}$$

14-5
Volume change: bulk modulus

In both compressive and tensile stresses, the stress acts along one direction in the body and produces a change in only one dimension. The change produced in the cross-sectional area of a rod under compression or tension is practically negligible. To produce equal strains in all three dimensions of a homogeneous solid, it is necessary to have equal

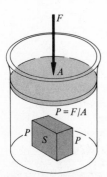

FIGURE 14-6 Hydrostatic pressure $P = F/A$ used to produce a change in volume of the solid S immersed in the liquid.

stresses along these three directions. The simplest method of doing this is to subject the solid to a uniform hydrostatic pressure P, as shown in Figure 14-6. The solid is placed in some liquid in a jar with a tight-fitting piston of area A. By applying an external force F to the piston, an increased pressure $P = F/A$ is transmitted uniformly throughout the liquid. This increased pressure P is usually called *hydrostatic pressure*, irrespective of the nature of the liquid. This hydrostatic pressure is the same on all sides of the solid and is also the stress inside the solid. If we call V the volume of the solid and ΔV the change in its volume produced by the stress P, then, from Hooke's law,

$$\frac{\text{stress}}{\text{strain}} = \frac{P}{\Delta V/V} = K$$

Since an increase in pressure always produces a decrease in volume, the modulus K for volume change will always be a negative number. To avoid having a negative number, let us define the *bulk modulus $B = -K$*, so that

$$B = -\frac{P}{\Delta V/V} \qquad (14\text{-}5)$$

Not only solids but also liquids and gases undergo volume changes when subjected to changing pressures. Equation 14-5 is also applicable to liquids and gases. Table 14-1 lists values of the bulk modulus for several solids, and Table 14-2 lists them for a few liquids.

The volume change of a gas when subjected to hydrostatic pressure can be demonstrated very simply by taking a balloon filled with air and submerging it in water contained in a glass cylinder with a tight-fitting piston. The balloon can be kept submerged by tying a piece of lead to it (see Figure 14-7a). The hydrostatic pressure can be increased by any desired amount P by applying a force F to the piston of area A. The decrease in volume of the balloon can be readily observed, and, if a measure of the change in volume is desired, it

Table 14-2　Bulk modulus of liquids

material	in dynes/cm²
Carbon disulfide	0.15×10^{11}
Ethyl alcohol	0.09×10^{11}
Glycerin	0.45×10^{11}
Mercury	$2.6 \ \times 10^{11}$
Nitric acid	0.03×10^{11}
Water	0.23×10^{11}

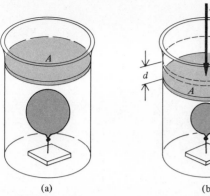

(a)　　　　　　　　　(b)

FIGURE 14-7 Demonstration of a volume change of a gas in a balloon by the application of additional hydrostatic pressure $P = F/A$. The volume change is Ad.

is merely necessary to measure the distance d through which the piston has been moved (see Figure 14-7b).

Illustrative example

A piece of copper 3 cm by 4 cm by 4 cm is placed in a steel cylinder filled with oil. The pressure of this oil is increased from 1 atm to 101 atm. Determine (a) the stress, (b) the strain, and (c) the change in volume of the copper.

(a) The stress P is the same as the increase in the hydrostatic pressure, and this is

$$P = (101 - 1) \text{ atm} = 100 \text{ atm}$$

and, since

$$1 \text{ atm} = 1.013 \times 10^6 \frac{\text{dynes}}{\text{cm}^2}$$

then

$$P = 1.013 \times 10^8 \frac{\text{dynes}}{\text{cm}^2}$$

(b) To determine the strain produced, let us solve Equation 14-5 for $\Delta V/V$, obtaining

$$\frac{\Delta V}{V} = -\frac{P}{B}$$

The value of the bulk modulus of copper, as given in Table 14-1, is

$$B = 14 \times 10^{11} \frac{\text{dynes}}{\text{cm}^2}$$

Hence

$$\text{strain} = \frac{\Delta V}{V} = -\frac{1.013 \times 10^8}{14 \times 10^{11}}$$

or

$$\frac{\Delta V}{V} = -7.24 \times 10^{-5}$$

(c) The change in volume is

$$\Delta V = -7.24 \times 10^{-5} \times V$$

Now

$$V = 3 \times 4 \times 4 \text{ cm}^3 = 48 \text{ cm}^3$$

Hence

$$\Delta V = -7.24 \times 10^{-5} \times 48 \text{ cm}^3$$

or

$$\Delta V = -3.47 \times 10^{-3} \text{ cm}^3$$

The volume of the copper is thus decreased by about 0.0035 cm^3.

Illustrative example

Show that if a gas that obeys Boyle's law undergoes small pressure changes at constant temperature, its bulk modulus is equal to the pressure of the gas.

The bulk modulus is given by

$$B = -\frac{P}{\Delta V/V} \qquad (14\text{-}5)$$

where P is the *change in pressure* of the gas and ΔV the *change in its volume*. Let us assume that the pressure of the gas is changed by a small fraction, say 1 percent, so that if P_1 was the original pressure, the change in pressure will be

$$P = 0.01P_1$$

From Boyle's law we know that if the pressure is increased by 1 percent, the volume will be decreased by 1 percent. Now $\Delta V/V$ is the fractional change in volume; hence

$$\frac{\Delta V}{V} = -0.01$$

The minus sign is used to show that there was a decrease in volume.

Substituting the above values for P and $\Delta V/V$ in Equation 14-5, we get

$$B = \frac{-0.01P_1}{-0.01} = P_1$$

Thus when the change in pressure is very small, the bulk modulus of a gas that follows Boyle's law is equal to the initial pressure.

From the above analysis, it can be seen that if the pressure of a gas is changed by any amount while the temperature is kept constant, the bulk modulus will vary and, at any stage of the process, will be equal to the pressure of the gas at that stage.

14-6
Cohesion and adhesion

The fact that molecular forces have a short range would lead us to expect some distinctive types of phenomena to be observable at the surfaces of substances. Conversely, the appearance of these surface phenomena should lead to information about these molecular forces. For example, if we take two pieces of metal, each with an accurately plane surface, and bring them together, there will be no observable force between them until the two surfaces are placed in very good contact. Once they are placed in good contact, a very great force will be required to pull them apart. This experiment shows that the forces between the molecules in the two surface layers have a very short range of effectiveness. The above experiment can be readily performed with two pieces of steel with clean plane surfaces, or with two pieces of lead with clean plane surfaces. The force of attraction acting between molecules of the same material is sometimes called *cohesion*.

If some water is poured into a glass vessel, the free surface of the water will be a level surface, that is, horizontal, except at the region of contact with the glass; at this region, the water will be seen to cling to the glass for a short distance

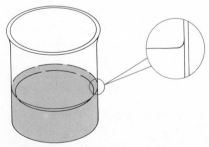

FIGURE 14-8 Free surface of water in a glass jar is level (horizontal) except near the glass.

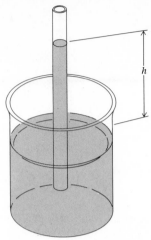

FIGURE 14-9 Level of water in a capillary tube is at a height *h* above level in the larger vessel.

above the level surface (see Figure 14-8). This phenomenon can be accentuated by immersing a glass tube with a narrow bore, a so-called "capillary tube," into the water (see Figure 14-9). The level of the water inside the capillary tube will be found to be considerably higher than the level in the larger jar. Furthermore, an examination of the surface of the water in the capillary tube shows that it is not plane but is concave upward.

The above experiment can be repeated with a variety of capillary tubes of different materials and a variety of liquids. In many cases the liquid will be found to rise in the capillary tube to a level above that of the liquid outside the tube. In other cases the level in the capillary will be lower than that outside the tube. For example, if a glass capillary tube is immersed in mercury contained in a larger dish, the level of the mercury will be lower in the capillary tube than outside the tube (see Figure 14-10). If a glass U-tube is constructed with one arm about 1 cm in diameter and the other about 0.2 cm in diameter, and mercury is poured into the tube, the level in the narrower tube will be lower than that in the wider tube (see Figure 14-11a). If water is poured into such a U-tube, the level of water will be higher in the narrower tube, as shown in Figure 14-11b.

One method of accounting for the behavior of liquids in capillary tubes is to assume that there are forces of attraction, also of short range, between the molecules of the liquid and the molecules of the solid at the surface of contact. This type of attractive force between molecules in the surface

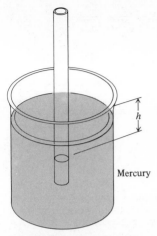

FIGURE 14-10 Level of mercury in a glass capillary tube is at a level h below that in the large vessel.

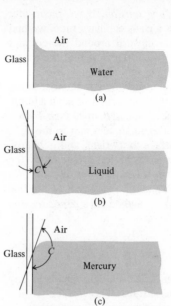

FIGURE 14-12 Angles of contact. (a) 0° between water and glass; (b) angle of contact $C < 90°$ between a liquid and glass. (c) angle of contact $C > 90°$ between mercury and glass.

of one substance for those in the surface of another substance is called *adhesion*, to distinguish it from the force of cohesion between like molecules. If the force of adhesion is greater than the force of cohesion, the liquid will cling to the solid surface; that is, the liquid will wet the solid. Water and glass, and oil and glass are both examples of liquids that wet the solids.

The angle between the liquid surface and the solid surface at the region of contact is an indication of the relative value of the forces of cohesion and adhesion. This angle is known as the *angle of contact*. For water and glass, the angle of contact is zero degrees (see Figure 14-12a). For some other liquid, the angle of contact will have some value C,

as shown in Figure 14-12b. In this case the relative value of the force of adhesion to that of cohesion is not so great as in the case of water and glass. If the force of cohesion is much greater than the force of adhesion, as in the case of mercury and glass, the angle of contact C is greater than 90°. For mercury and glass, $C = 139°$ (see Table 14-3).

14-7
Surface tension

We have seen that the liquid inside a capillary tube has a curved surface, and if the tube is circular the liquid surface may be nearly spherical. The interesting phenomena associated with liquid surfaces can be most easily explained by introducing

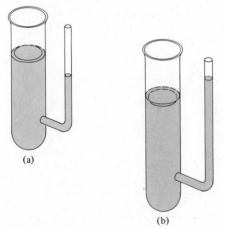

FIGURE 14-11 (a) U-tube containing mercury. (b) U-tube containing water.

Table 14-3 Contact angles

liquid	tube	angle, degrees
Alcohol	Glass	0
Ether	Glass	0
Glycerin	Glass	0
Mercury	Glass	139
Water	Glass	0
Water	Paraffin	107

the concept of *surface tension*. If we reconsider the case of water in a glass capillary tube, we find that there is a force upward around the circular region of contact between the glass and the water. The liquid surface behaves as though it is under tension, with the pull everywhere tangential to the surface. If we take a small section of such a liquid surface and imagine a line *AB* in it (see Figure 14-13), this line will experience a pull to the right produced by the surface to its right, and it will experience an equal pull to the left produced by the surface to the left of it. If *F* is the force due to either part of the surface and *L* is the length of the line *AB*, the ratio of *F* to *L* is called the *surface tension* *S*; thus

$$S = \frac{F}{L} \qquad (14\text{-}6)$$

The surface tension *S* acts at right angles to any line imagined in the surface. The surface tension is usually expressed in dynes per centimeter.

That a liquid surface behaves as though it is under tension can be demonstrated in a variety of experiments. Let us construct a rectangular wire frame having one side movable; this can be done by curving the ends of a wire *AB* so that it slides easily on two legs of the frame (see Figure 14-14a). We can pick up a film on this frame by dipping it in a soap solution. This film will have two rectangular surfaces. The film will tend to contract, and, since *AB* is movable, it will pull this wire toward *CD* with some force *F*. As the surface contracts, the thickness of the film will increase; that is, molecules will leave the surface and enter the liquid between the two surfaces.

To keep the wire *AB* in equilibrium, a force *F* to the right has to be applied to it. This force can be used to measure the surface tension. If *l* is the length of the wire, the length of the surface which exerts this force *F* is 2*l*, since there are two surfaces to this film. The surface tension *S* is therefore

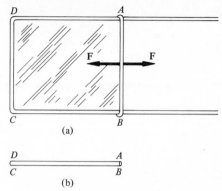

FIGURE 14-14 (a) Wire frame with movable slide *AB* used to measure the surface tension of a film in the frame *ABCD*. (b) Shows the thickness of the film.

$$S = \frac{F}{2l}$$

If the surface area is increased by moving the wire *AB* through a distance *x*, the work done is

$$\mathcal{W} = Fx$$

and, since

$$F = 2lS$$

then

$$\mathcal{W} = S \times 2lx \qquad (14\text{-}7)$$

Now 2*lx* is the increase in the surface area of the film; setting 2*lx* = *A* and solving Equation 14-17 for *S*, we get

$$S = \frac{\mathcal{W}}{A} \qquad (14\text{-}8)$$

The surface tension thus represents the work done per unit area in increasing the area of the film.

Another simple experiment to illustrate the con-

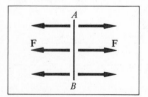

FIGURE 14-13 Any line *AB* in a surface is acted on by equal forces *F* at right angles to *AB* produced by the surfaces on either side.

FIGURE 14-15 A circular wire frame with a loop of thread loosely tied to it. (a) A soap film on the wire frame with a loop in it. (b) Film inside loop broken. Soap film on the outside of the loop pulls at right angles to it, giving it circular shape.

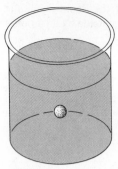

FIGURE 14-16 Spherical shape of drop of oil suspended in an alcohol solution of equal density.

tractile force of a film is to dip a metal frame, with a looped piece of thread loosely tied to it, into a soap solution. When the frame is taken out of the solution, a soap film will be stretched across it and the looped thread will be collapsed in it (see Figure 14-15a). If we now take a heated needle and puncture the film inside the loop, the contractile force of the film on the outside of the loop will stretch it into a circular shape (see Figure 14-15b). The circular shape results from the fact that the surface tension acts at right angles to every part of the looped thread.

The contractile force in the surface of a liquid is the cause of the spherical shape of a liquid drop. A simple experiment to show the formation of a spherical drop can be performed by preparing a solution of alcohol and water that has the same density as some heavy engine oil. The oil will not mix with this solution. If the oil is now introduced into the middle of this solution by letting it flow out of a glass tube placed there, the oil will assume a spherical shape and remain suspended in the

solution (see Figure 14-16). The contractile force in the surface makes it assume the smallest possible area consistent with its volume. The shape of such a surface is spherical.

The surface tension of a liquid depends upon the nature of the liquid and the nature of the substance on the outside of the liquid surface—that is, whether it is air or the vapor of the liquid itself. The values of the surface tensions of some liquids are given in Table 14-4. The surface tension depends upon the temperature of the system, decreasing as the temperature rises.

14-8
Capillarity

We have already shown that if a capillary tube is inserted into a liquid, the levels inside and outside the tube will differ by an amount h. In some cases, the liquid will be higher in the capillary tube; in other cases, it will be lower, depending upon the relative values of the forces of adhesion and cohesion.

The concept of surface tension enables us to obtain a simple relationship between the difference in levels h inside and outside the capillary tube and the radius r of this tube. Let us take the case of a liquid such as water, which rises in the tube to a height h above the water outside the tube (see Figure 14-17). The contact angle between the water and the glass is $0°$. The force F acting upward along the circle of contact between the water and the glass must be equal to the force F caused by the surface tension S in the water. The force F

Table 14-4   Surface tension

liquid in contact with air	temperature in °C	surface tension in dynes/cm
Ethyl alcohol	20	22.3
Water	0	75.6
	20	72.8
	60	66.2
	100	58.9
Mercury	25	473
Olive oil	20	32
Glycerin	20	63.1
Soap solution		26

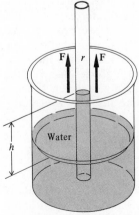

FIGURE 14-17 Force F acting on surface film in capillary tube when the contact angle is zero.

can be evaluated by considering the surface tension S as acting along the circumference of the circle of contact $2\pi r$, so that

$$F = S \times 2\pi r$$

This upward force supports the weight of the column of water of height h. If ρ is the density of the liquid, the mass m of this liquid is

$$m = \rho \times \pi r^2 h$$

since $\pi r^2 h$ is the volume of this column of liquid. Its weight W is

$$W = mg = \pi r^2 \rho g h$$

Equating the upward force F to the weight of the column of liquid, we get

$$2\pi r S = \pi r^2 \rho g h$$

from which

$$h = \frac{2S}{r\rho g} \qquad (14\text{-}9)$$

Thus the height to which water will rise in a capillary tube varies inversely as the radius of the tube. This is known as *Jurin's law*.

If the contact angle is not zero but has some value C, then the upward force is the vertical component of F and is therefore

$$F \cos C = 2\pi r S \cos C = \pi r^2 h \rho g$$

from which

$$h = \frac{2S \cos C}{r\rho g} \qquad (14\text{-}10)$$

The same analysis will hold if the surface in the capillary is depressed by an amount h. In this case the vertical component of the force F—that is, $F \cos C$—acts downward and is opposed by the force due to the difference in pressure between the liquid outside the tube and that inside the tube. When the angle of contact is greater than 90°, the value of its cosine is a negative number; hence h will be negative, indicating that the level is depressed in the capillary tube.

Illustrative example

Two glass capillary tubes, each 1 mm in radius, are put into two different liquids, one in water and the other in mercury. Compare the liquid levels in the two tubes.

Let us take the level of the liquid outside each capillary tube as the zero reference level. From

Equation 14-10, the level of the liquid inside the capillary will differ from that outside by an amount

$$h = \frac{2S \cos C}{r\rho g}$$

For the case of water, $C = 0°$, hence $\cos C = 1$; $S = 73$ dynes/cm; $r = 0.1$ cm; $\rho = 1$ gm/cm³; and $g = 980$ cm/sec². Letting $h = h_1$ for water, we get

$$h_1 = \frac{2 \times 73 \text{ dynes/cm}}{0.1 \text{ cm} \times 1 \text{ gm/cm}^3 \times 980 \text{ cm/sec}^2}$$

from which

$$h_1 = \frac{146}{98} \text{ cm} = 1.5 \text{ cm}$$

For the case of mercury, $C = 139°$, hence $\cos C = \cos 139° = -\cos 41° = -0.755$; $S = 473$ dynes/cm; $r = 0.1$ cm; $\rho = 13.6$ gm/cm³; and $g = 980$ cm/sec². Letting $h = h_2$ for mercury, we get

$$h_2 = -\frac{2 \times 473 \text{ dynes/cm} \times 0.755}{0.1 \text{ cm} \times 13.6 \text{ gm/cm}^3 \times 980 \text{ cm/sec}^2}$$

from which

$$h_2 = -0.536 \text{ cm}$$

14-9
Pressure and curved surfaces

From our study of hydrostatics, we know that if a liquid is at rest in a series of connected vessels, the pressure at any one level must be the same in all the vessels, and the pressure at a higher level is less than that at a lower level by an amount equal to $h\rho g$, where h is the difference in levels. Let us apply this rule to a liquid in a capillary tube which is connected to a much larger vessel (see Figure 14-18). The atmospheric pressure P_0 acts on the open surfaces in the two vessels. If the liquid surface in the capillary is at a height h above that in the larger vessel, then the pressure P_1 just under the curved surface must be less than the pressure P_0 outside it by an amount equal to the pressure of the column of liquid of height h; that is,

$$P_0 - P_1 = h\rho g \qquad (14\text{-}11)$$

Now the height h is related to the surface tension S and the radius of the capillary tube r by

$$h = \frac{2S \cos C}{r\rho g} \qquad (14\text{-}10)$$

From Equation 14-10 we get

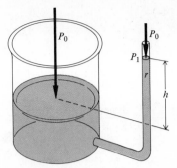

FIGURE 14-18 Pressure P_1 under convex surface is less than pressure P_0 over concave surface.

$$h\rho g = \frac{2S \cos C}{r}$$

hence

$$P_0 - P_1 = \frac{2S \cos C}{r} \qquad (14\text{-}12)$$

The pressure on the convex side of the surface is thus less than the pressure on the concave side.

In the special case where the surface is spherical, such as that shown in Figure 14-19, the pressure difference is simply related to the radius of the sphere. The relationship between the radius r of the tube and the radius R of the spherical surface can be obtained very readily with the aid of Figure 14-19. O is the center of the sphere, $OA = R$, and, since AB is tangent to the sphere, it is perpendicular to OA. Hence the angle between R and r is equal to the contact angle C, so

$$r = R \cos C \qquad (14\text{-}13)$$

Substituting this value for r in Equations 14-12 yields

$$P_0 - P_1 = \frac{2S}{R} \qquad (14\text{-}14)$$

Thus the difference in pressure on the two sides of a spherical surface due to the surface tension depends inversely upon the radius of the sphere. P_0 is the pressure on the concave side of the surface, and P_1 is the pressure on the convex side. The above result holds for a whole sphere as well as for any portion of it. Thus, if an air bubble is formed in the water, the pressure inside exceeds that outside by $2S/R$.

In the case of a soap bubble blown in air, the pressure difference is practically twice as great as that given by Equation 14-14. Since the thickness

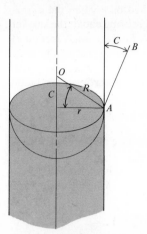

FIGURE 14-19 Relationship between radius R of spherical surface, radius r of the capillary tube, and contact angle C. $r = R \cos C$.

of the soap film is very small, the excess pressure inside is practically $4S/R$, where R is the average radius of the two surfaces of the soap bubble and S is the surface tension of the soap solution.

The smaller the soap bubble, the greater is the pressure inside it. This can be demonstrated with a very simply experiment using the apparatus sketched in Figure 14-20. A long glass tube has a three-way stopcock at its center. One end of the glass tube A is dipped into a soap solution, and a film is formed on it. The stopcock is then turned so that air blown into it goes only to the film at A, blowing it out into a soap bubble of small size. The stopcock is then turned so that side A is closed and side B is open. The above process is repeated, and a larger soap bubble is blown on B. The stopcock is now turned so as to connect A and B but is shut off from the outside. Air will flow from A to B, since the pressure inside A is greater than that inside B. The larger bubble will thus get larger, and the smaller bubble will get smaller;

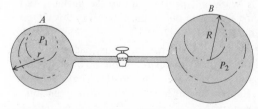

FIGURE 14-20 Two soap bubbles of different radii blown at the ends of a glass tube. Smaller soap bubbles will get still smaller and large ones will get larger.

equilibrium can be reached only when the curvatures of the two surfaces become equal; the smaller bubble will then consist of only a small section of a large sphere equal in radius to that of the large bubble. A similar phenomenon occurs in aneurysms, in which a weakened portion of a blood vessel balloons out.

Illustrative example

Determine the excess pressure inside a small raindrop 3 mm in diameter.

Since the raindrop is a sphere with a single surface, the excess pressure ΔP inside the sphere

is, from Equation 14-14,

$$\Delta P = \frac{2S}{R}$$

Using the values $S = 73$ dynes/cm and $R = 0.15$ cm, we get

$$\Delta P = \frac{2 \times 73 \text{ dynes/cm}}{0.15 \text{ cm}}$$

or

$$\Delta P = 973 \frac{\text{dynes}}{\text{cm}^2}$$

Questions

1. In common speech, rubber is said to be elastic whereas steel is not usually described as elastic. Distinguish between this meaning of the word "elastic" and that used in this chapter.

2. Distinguish between the elastic limit of a substance and its breaking strength.

3. Two wires, one twice as long as the other, are made of the same material and have the same cross-sectional area. Compare the changes in length of the two wires when they are subjected to the same stress.

4. Two wires made of the same material and of the same length are hung from a suitable beam. The diameter of one wire is twice that of the other. Equal loads are placed on each wire. Compare the strains produced in them.

5. Which is more compressible, steel or water?

6. Tinsmiths and other sheet-metal workers use large, scissorlike shears to cut sheet metal. Explain how they cut.

7. Water can rise to a height h when a capillary tube of a certain diameter is placed in it. Suppose this tube is inserted into the water so that a height $h/2$ projects above the surface. Will the water run out of the tube? Try this experiment.

8. When a camel's-hair brush is placed in water, the bristles stay apart. When the brush is taken out of the water, the bristles come together. Explain this in terms of the surface tension of water.

9. One plate of soup has a uniform layer of liquid fat covering it; another plate has small drops of liquid fat floating on top. Which plate is hotter? (This effect in part explains the difference in flavor between hot and cool soup.)

10. A thin steel needle is placed horizontally on the surface of water without breaking the surface. What forces act on this needle? Does Archimedes' principle apply?

11. A capillary tube has a narrow constriction in one part. Some water is drawn into this part of the tube. Will the water stay in the constricted part or flow into the wider part? Explain in terms of the surface tension of water.

12. A capillary tube has a narrow constriction in one part. Some mercury is drawn into this part of the tube. Will the mercury stay in the constricted part or flow into the wider part?

13. Account for the rise of ink in blotting paper.

14. Account for the rise of oil in the wick of an oil lamp.

15. A tall, stiff piece of blotting paper is placed with its surface vertical in a shallow dish containing a solution that consists of several chemicals dissolved in water. Each of the chemicals has a distinctly different color and the molecules of each have distinctly different masses. After a few minutes the paper is removed and allowed to dry. The dried paper shows horizontal bands of color. Explain. (This is the basis of *paper chromatography*, an important technique used in biological research.)

16. What is the meaning of the phrase "pouring oil on troubled waters"? (Ignore the ecological implications.)

17. Take a narrow-necked bottle and pour water into it until it is about two thirds full. With skill, this bottle can be inverted so that, even though the open end is down, the water will not flow out of it. (a) Describe the shape of the water surface at the open end of the bottle. (b) Is the pressure of the air inside the bottle greater than, less than, or equal to the atmospheric pressure? Explain.

18. Compare the tension of a soap bubble with that of a rubber balloon. Is Hooke's law applicable to the soap film? to the rubber balloon?

19. How can you demonstrate that the force acting to keep two blocks of metal together when their plane surfaces touch is not due to atmospheric pressure acting on them?

20. Waterproof canvas is treated so that the fibers are coated and the holes in the weave are small, but not completely closed. This permits air to move through the canvas but not water. Explain how the surface tension of water can produce this effect. (Note that touching the inside of a wet tent causes a leak.)

Problems

1. A steel wire 30 cm long is stretched to a length of 30.5 cm by a load. What is the strain?

2. A copper wire 2 m long and 0.02 cm² in cross-sectional area supports a load of 5 kg. What is the stress?

3. A cube, 1 ft on a side, sinks to the bottom of the sea and shrinks 0.1 in. in all directions. What is the strain?

4. A steel rod 6.0 in. long and 0.5 in. in diameter is used as a piston in a cylinder to produce a pressure of 4000 lb/in.². Determine the decrease in its length produced by this stress.

5. A beaker 4 cm in diameter contains glycerin at 0°C. Find the force exerted by one half of the surface of the glycerin on the other half.

6. How high will water at 20°C rise in a glass capillary of internal diameter 0.8 mm?

7. What size of air bubble in water at 20°C has internal pressure 1 atm greater than the pressure of the surrounding water?

8. A copper wire 200 cm long and 0.2 cm in diameter is suspended from a rigid framework. An object whose mass is 7 kg is hung at the end of the wire. Determine (a) the stress in the wire and (b) the strain produced.

9. A steel wire 2 m long and 0.04 cm in diameter supports a cylinder whose mass is 6 kg. Determine (a) the stress in the wire, (b) the strain produced, and (c) the elongation of the wire.

10. A 200-kg box is being lifted by a cable that is 3 m long. The cable is made of 20 strands of steel wire, each 0.01 mm in diameter. How much does the cable stretch? (Assume that the box is moving with negligible acceleration.)

11. A steel cable 25 m long is to be used as an elevator cable, exerting a maximum force of 15,000 nt. What is the total cross-sectional area of all the strands of the cable if the cable is permitted to stretch no more than 1 cm?

12. A metal wire 150 cm long and 0.3 mm in diameter has series of cylinders hung from it in succession. Each cylinder has a mass of 100 gm. The measured changes in length, expressed in centimeters, are 0.014, 0.029, 0.042, 0.056, and 0.070. Plot a graph with the stress as ordinate and the strain as abscissa; from the slope of this graph determine the Young's modulus for this wire.

13. A steel rod 5 in. long and 0.6 in. in diameter is to be used as a piston in a cylinder to produce a pressure of 4500 lb/in.². Determine the change in length of the rod.

14. What increase in pressure is required to decrease the volume of a cubic meter of water by 0.01 percent?

15. At what depth below the surface of a lake is the density of water 0.1 percent greater than the density of the water at the surface?

16. Glycerin is subjected to a pressure of 800 atm. Find the percentage change in its density.

17. Three capillary tubes—0.4, 0.6, and 1.2 mm in diameter, respectively—are supported in a jar of water. Determine the height to which the water will rise in each of these tubes.

18. It is sometimes asserted that water reaches the leaves of a tree by capillarity. How large would the capillary tubes have to be in order for the water to rise to the leaves at the top of a 60-ft tree? (Assume that the contact angle is equal to 0°.)

19. Three holes 1.5, 2.0, and 2.8 mm in diameter,

respectively, are bored in a block of paraffin. The paraffin is partly immersed in water. Determine the level of the water in each hole.

20. A capillary tube 1.2 mm in diameter is placed in a soap solution. The liquid rises to a height of 1.1 cm above the level of the rest of the surface. Assuming the contact angle to be zero, determine the surface tension of this solution.

21. A soap film is formed on a rectangular frame 4 cm by 10 cm, as shown in Figure 14-14. (a) Determine the force that the film exerts on the shorter wire. (b) If the wire is moved through a distance of 5 cm, determine the amount of work done. Assume that the temperature remains constant in this process.

22. Calculate the excess pressure inside a raindrop that is 6.0 mm in diameter and has a temperature of 20°C.

23. Determine the excess pressure inside a soap bubble that is 4 cm in diameter; assume the temperature to be 20°C.

24. (a) How much energy is stored in the surface of a raindrop 4 mm in diameter at 20°C? (b) How much energy is stored in the surface of a spherical drop formed by 100 of these raindrops?

25. Two rectangular glass plates are spaced 0.8 mm apart; they are partially immersed in a dish of water at 20°C, with the plates placed so that the air space between them is vertical. Determine how high the water will rise in the space above the level of the water in the dish. (Consider the forces acting on a surface film 1 cm long in contact with each plate; balance these forces against the weight of water lifted through a height h.)

26. A father makes a wooden boat out of wood of specific gravity 0.5. The boat is a rectangle 6 in. long, 3 in. wide, and 1 in. thick. When the boat is floating, the father drops a small amount of soap solution near one of the 3-in. edges, reducing the surface tension to one third its normal value. Determine (a) the net force acting on the boat and (b) the acceleration of the boat.

27. A hollow glass tube has a soap bubble of 8 cm diameter formed on one end and another bubble of 3 cm diameter on the other end. Determine the pressure difference at the ends of the tube. Explain what will happen as a result of this pressure difference.

28. Two glass plates, each having a large surface, are clamped together along one edge and separated by spacers a few millimeters thick along the opposite edge to form a wedge-shaped air film. These plates are then placed vertically in a dish of colored liquid, with the clamped edge and the open edge both vertical. Calling x the horizontal distance from the edge where the thickness of the air film is zero, show that the vertical distance y through which the liquid rises in the air space varies inversely as x.

29. Considering a vertical wire that carries a load as a spring, find the spring constant in terms of Young's modulus and the dimensions of the wire.

30. When a wire is stretched, work is done and the wire acquires elastic potential energy. From the results of Problem 29 and the known energy stored in a spring, show that the energy per unit volume is equal to one half the product of the stress and strain.

heat

part 2

temperature

chapter 15

15-1
Microphysics and macrophysics

Up to this point we have dealt with the behavior of particles, making only occasional references to the fact that matter in bulk is made of great numbers of particles. As we have said earlier, the laws of nature we have discussed could be applied to each of the particles in matter and, in principle, all the properties of matter could be derived, but the mathematical difficulties with this straightforward procedure are overwhelming.

Physics has adopted two different points of view toward the problem of attempting to gain a theoretical understanding of the behavior of matter. The first procedure, which is usually called *statistical mechanics*, starts with our knowledge of the behavior of the individual atoms, molecules, and electrons. The difficulty of dealing with large numbers of atoms is handled by introducing the mathematics of statistics. Appropriate, cleverly chosen *averages* yield information about the properties of a collection of atoms. This study, which focuses on the behavior of the microscopic constituents of matter, is sometimes called *microphysics*.

The other point of view is the other extreme, ignoring the existence of the microscopic structure. In this procedure, usually called *thermodynamics*, the macroscopic properties and behavior of matter are measured and then described by laws of nature that appear to have no connection with Newton's laws (or quantum mechanics). Thus, in *macrophysics*, such concepts as temperature and heat are described without reference to the atomic nature of matter. In spite of the apparent restrictiveness of this approach, very powerful, fundamental laws of nature have been discovered. It then remains for physics to explain the macrophysical laws in terms of microphysics. But even without such an explanation, the macrophysical laws remain correct and useful.

In this chapter we will describe a most important macrophysical concept, *temperature*. We will adopt a thermodynamic approach, ignoring the particulate nature of matter. In a later chapter we will discuss the microphysical interpretation of temperature.

15-2
Thermal equilibrium

We can begin to understand temperature in simple human terms; there are objects that feel hot and there are objects that feel cold. In these terms, temperature is a measure of hotness and coldness.

We can then observe that there are measurable properties of most things that are correlated with their hotness: the length of a metal rod usually increases as it gets hotter; the color of very hot glowing objects changes as they get hotter. Electrical, magnetic, frictional, and elastic behavior change with hotness. This permits us to decide, with less concern for the fallibility of human senses, when the temperature of a particular body is constant. When one of these properties that change with hotness, properties called *thermometric properties*, is constant, then they all are constant and the temperature is constant.

If we take two objects at random and have them touch each other, then in general, the temperatures of both will change. In such cases we always find that the hotter one will cool down and the cooler one will heat up. If both objects are isolated from the rest of the universe, then sooner or later both will reach some constant temperature. This is described by saying that the two objects will come to *thermal equilibrium* with each other. We can detect this state of equilibrium by the absence of change of the thermometric properties of both objects.

We can now imagine three objects, A, B, and C, such that A is in thermal equilibrium with B, and A is also in thermal equilibrium with C. Then we know *from experience* that B will be in thermal equilibrium with C. This idea, that two objects each separately in thermal equilibrium with a third object will be in thermal equilibrium with each other, is called the *zeroth law of thermodynamics*. This idea is so fundamental that it seems obvious, but analogous statements are not always true.

The idea of thermal equilibrium is the most fundamental thing we know about temperature. We can think of temperature as that which characterizes an equilibrium state. When two objects are in thermal equilibrium with each other, then they have the *same temperature*. When two objects are not in thermal equilibrium they have different temperatures. When two objects are put in thermal contact and they come to thermal equilibrium, their temperatures change from initially different values to the same final value.

15-3
Thermometers

In order to *define* temperature in a way that is useful in physics, we have to describe a procedure for measuring it. An instrument for measuring temperature is called a *thermometer*. Once we have completely described a thermometer, we can define temperature as the reading of the thermometer.

Basically, a thermometer is a system with an easily measurable thermometric property. When a thermometer is used to measure the temperature of a system, it is put in thermal contact with the system and allowed to come to thermal equilibrium. In general, this will change the temperature of the system whose temperature is being measured. In order to minimize this effect, thermometers are usually desiged so that they produce very small changes on the systems whose temperatures they measure, usually by making them as small as possible.

There are many different kinds of thermometers in use, based on many different thermometric properties. These include:

1. Liquid-in-glass, such as the common mercury thermometer, in which thermal expansion of the liquid causes the length of the column to vary (Figure 15-1a).
2. Bimetal thermometers, in which two different metals are welded or riveted together so that they form a single straight piece at room temperature. When the temperature is raised, this strip will bend in an arc because one of the metals will expand more than the other (Figure 15-1b).
3. Electrical-resistance thermometers, which utilize the fact that the electrical resistance of many materials changes markedly with temperature (Figure 15-1c).
4. Thermocouples, which operate on the principle that when two wires of different metals are

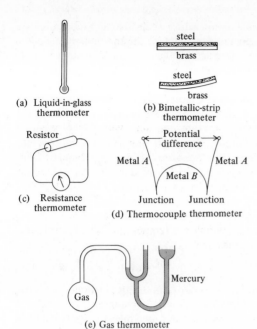

(a) Liquid-in-glass
    thermometer

steel
brass

steel
brass

(b) Bimetallic-strip
    thermometer

Resistor

(c) Resistance
    thermometer

Potential
difference

Metal *A*   Metal *A*

Metal *B*

Junction   Junction

(d) Thermocouple thermometer

Gas

Mercury

(e) Gas thermometer

FIGURE 15-1 Several common thermometers.

connected at their ends, forming two junctions, and the two junctions are kept at different temperatures, a measurable electric potential difference is developed that depends on the two temperatures (Figure 15-1d).

5. The gas thermometer, which is based on the principle that when a container of gas is kept at constant volume, the pressure of the gas changes with temperature; if the gas is kept at constant pressure, then the volume changes with temperature (Figure 15-1e).

15-4
Common temperature scales

In order to define a *scale* of temperature we have to choose one thermometer and describe a scheme for assigning numbers to particular temperatures. We will illustrate this by describing a procedure for calibrating a mercury-in-glass thermometer with the most common temperature scales, Celsius (formerly called centigrade) and Fahrenheit.

A typical mercury-in-glass thermometer consists of a glass capillary tube (a tube with thick walls and a narrow channel) with a glass bulb at one end. See Figure 15-2. Mercury fills this bulb and extends into the capillary; the rest of the capillary is essentially empty and is sealed. In order to

define either Celsius or Fahrenheit temperature scales, the position of the end of the mercury column is marked at two fixed points; the *ice point* and the *steam point*. The ice point is determined by immersing the thermometer in a mixture of pure water and melting ice; the mixture must be at thermal equilibrium with air at a pressure of one standard atmosphere. The steam point is determined by immersing the thermometer in the steam of pure water that is boiling at a pressure of one standard atmosphere. The Celsius scale is defined by assigning the value zero degrees to the ice point, and the value of 100 degrees to the steam point, and assuming that the length of the mercury column is proportional to the temperature. That is, if the mercury column is halfway between the marked positions of the ice point and the steam point then it is assumed that the temperature of the thermometer is 50° Celsius or 50°C. The Fahrenheit scale is defined by assigning the value 32°F to the ice point and the value of 212°F to the steam point, and assuming proportionality between the length of the column and the temperature.

This way of defining a temperature scale is consistent and more or less useful as long as we choose one thermometer. We could define the Celsius scale using a specific size and shape mercury-in-glass thermometer, using a specific kind of glass. Alternatively, we could have used alcohol in glass or an electrical-resistance thermometer, or a constant-volume helium gas thermometer. At first it might seem that the kind of thermometer should not matter as long as the thermometers are cali-

°F   °C

212 — — 100

32 — — 0

FIGURE 15-2 A mercury-in-glass thermometer with the ice point and the steam point marked in both the Fahrenheit and Celsius scales.

brated at the same points. However, it turns out that the kind of thermometer does matter. If one carefully builds and calibrates a Celsius mercury-in-glass thermometer, it will read 0° for the melting point of ice, and 100° for the boiling point of water. A carefully built and calibrated Celsius resistance thermometer can be calibrated to read exactly the same values for the same two points. For other temperatures, the two thermometers will yield slightly different values. That is, if both thermometers are used to measure the normal boiling point of ethyl alcohol, they will both read approximately the same temperature, but not exactly the same. A third kind of thermometer, such as a constant-volume helium thermometer will read a third value slightly different from either of the other two. Thus in order to define a temperature scale precisely, we must describe the specific thermometer as well as the calibration procedure.

The difficulty with the requirement for specifying the kind of thermometer is that each kind of thermometer is useful for a limited range of temperatures. A mercury-in-glass thermometer is not very useful for measuring temperatures above the boiling point of mercury, or above the melting point of glass. A solution to this difficulty is to define a temperature scale in a given temperature range by means of one kind of thermometer. In a different temperature range a different kind of thermometer is used. This kind of temperature scale, usually using several fixed points in each range, is frequently used in research where precise temperature measurement is necessary. However, this procedure has the conceptual difficulty that it defines temperature in a way that depends critically on the properties of specific substances (mercury, for example) and on specific structures (the capillary tube).

The search for a temperature scale that is independent of specific properties of specific substances leads to the Kelvin temperature scale, which is discussed in the next section.

15-5
The Kelvin temperature scale

If we perform the experiment mentioned in the preceding section, that of measuring the same temperature with different kinds of thermometers, we find that gas thermometers differ from each other much less than the other kinds of thermometers. This observation suggests that it might be possible to define a temperature scale that is independent of the specific substance in the thermometer. Such a temperature scale is based on the constant-volume gas thermometer.

A constant-volume gas thermometer, such as that shown in Figure 15-3, consists essentially of a flask so arranged that the volume of the gas in the flask may be kept constant over a large range of temperature. When the temperature of the gas changes, the pressure of the gas changes. The thermometric property of this thermometer is the gas pressure, which is measured by a device that can be thought of as a pressure gauge.

In order to construct a temperature scale one can assume that the temperature T is proportional to the pressure P:

$$T = aP \qquad (15\text{-}1)$$

where a is a constant of proportionality. Since Equation 15-1 has only one constant in it, this assumption requires that one measure only *one* fixed point. The fixed point is chosen to be the *triple point of water*.

The triple point of water is the temperature at which pure water is in equilibrium with both water vapor and ice. That is, one can put pure water in a sealed container, called a *triple-point cell* (see Figure 15-4), and vary the pressure and temperature in the container until ice, water, and water vapor are in equilibrium with each other simul-

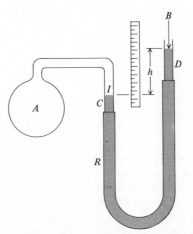

FIGURE 15-3 A constant volume gas thermometer. The height of point D is adjusted so that the mercury surface at the left is always at point I. Then the volume of gas is constant. The pressure is read by measuring h.

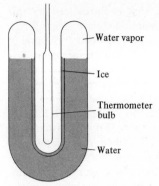

FIGURE 15-4 Schematic representation of a triple-point cell.

taneously (see Section 17-11). This will occur at only one temperature (0.01°C) and pressure. At any other temperature or pressure it will be possible to have only two of the phases in equilibrium: water and ice, water and water vapor, or ice and water vapor. Just as the ice point was arbitrarily chosen to be 0°C, the triple point of water is chosen to be 273.16 K. (As the reader might guess, this value is not as arbitrary as it might be; the choice is made for historical reasons. The Kelvin scale was originally arrived at by extrapolation from the Celsius scale, and this choice for the triple point introduces as little change as possible into the earlier work.) In order to calibrate a constant-volume gas thermometer, we immerse the thermometer in a triple-point cell, and read the pressure, P_3. From Equation 15-1, we have then

$$273.16 = aP_3 \qquad (15\text{-}2)$$

Equation 15-2 can be solved for a; when a is substituted back in Equation 15-1, we obtain

$$T = \frac{273.16}{P_3} P \qquad (15\text{-}3)$$

Then to measure any other temperature, such as the normal boiling point of ethyl alcohol, the thermometer is immersed in the boiling alcohol, and the pressure P_A is read. The temperature of the boiling alcohol is then

$$T_A = \frac{273.16}{P_3} P_A$$

If we made the thermometer with a different kind of gas or even a different amount of gas, then we would measure a slightly different temperature for the boiling point of alcohol, even though we followed this calibration procedure exactly.

The significance of the gas thermometer begins to be revealed when the following series of experiments is conducted. Imagine that we build a certain gas thermometer, and with it measure a specific temperature—the normal boiling point of alcohol, for example. In the process of making this measurement we have measured P_3, the pressure of the thermometer at the triple point of water, and also T_A, the boiling point of alcohol. We can now remove some of the gas from the thermometer. This changes the thermometer, so it will need to be recalibrated. When the thermometer is immersed in the triple-point cell, the pressure $P_3{}'$ will be different from P_3, the pressure when the thermometer, *with more gas in it* was in the triple-point cell. Recalibrated, the thermometer can then be used to measure the temperature of boiling alcohol. This temperature $T_A{}'$ will be slightly different from T_A, the temperature of boiling alcohol as measured by the thermometer with more gas in it. We can then repeat the procedure, extracting more gas, recalibrating, and measuring the temperature of boiling alcohol. At each stage, we will have two numbers, the pressure in the thermometer when it is in the triple-point cell, and the temperature of boiling alcohol. We can then plot the measured temperature against the pressure at the triple point. The striking thing is that as the amount of gas decreases, the points always fall on a straight line, as shown in Figure 15-5. This straight line can be extrapolated to a point at which P_3 is equal to zero. This point

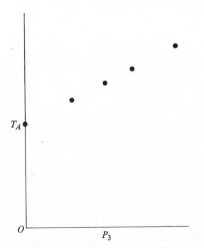

FIGURE 15-5 The temperature of boiling alcohol, T_A, as measured by a constant volume gas thermometer with different amounts of gas, and consequently with different pressures at the triple point, P_3. The graph becomes a straight line for low values of P_3.

represents the temperature that would be measured by this thermometer if it contained no gas.

On can now imagine making a different constant-volume gas thermometer with a different kind of gas in it. The entire process of measurement, gas extraction, and remeasurement can be repeated in order to measure the same temperature, that of boiling alcohol. This process will yield different points and a different straight line, but exactly the same extrapolated temperature, as shown in Figure 15-6. All constant-volume gas thermometers give the same extrapolated temperature, independent of kind of gas. We can then define this temperature to be the Kelvin temperature of boiling alcohol.

We can put this definition of the Kelvin temperature scale in the form of an equation:

$$T = 273.16 \lim_{P_3 \to 0} \left(\frac{P}{P_3} \right) \qquad (15\text{-}4)$$

where

$$\lim_{P_3 \to 0} \left(\frac{P}{P_3} \right)$$

means that one extrapolates the ratio to the value it has when P_3 is equal to zero.

The value of this definition of a temperature scale is that it gives meaning to the concept of temperature which is independent of a specific chemical substance. It is limited, in that it has meaning only for temperatures at which at least one gas exists. But gases do exist over a large range of temperatures.

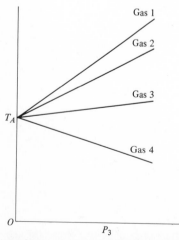

FIGURE 15-6 When different kinds of gas are used in a constant volume gas thermometer, different temperatures are measured for a specific process. However, all the gases give the same extrapolated temperature.

Another important idea is introduced by this definition of temperature, *absolute zero*. From Equation 15-1 we can see that when P is equal to zero, for all values of P_3, the temperature T is equal to zero. That is, if we cool any amount of gas down to a Kelvin temperature of zero, the pressure of the gas becomes zero. Since negative gas pressures are inconceivable, this implies that 0 K is the lowest possible temperature. This conclusion is on shaky ground, since all known gases liquefy at temperatures above 0 K, and therefore there cannot be any gas thermometers at or near that temperature. However, it is possible to show, by other means, that in fact, 0 K is the lowest temperature. (See Section 18-4.) Further, and perhaps more significant, it is impossible to cool any object to that temperature, although one can get very close. Researchers have cooled some systems to temperatures as low as a thousandth of a degree Kelvin.

With the choice of a given in Equation 15-2, the relation between a temperature t measured on the Celsius scale and the same temperature T measured on the Kelvin scale (remembering that the triple point of water is 0.01°C) is

$$t = T - 273.15 \qquad (15\text{-}5)$$

Thus, the zero of the Kelvin scale is a temperature t_0 Celsius, where

$$t_0 = -273.15°C$$

Similarly, the Kelvin temperature of the ice point is

$$T_i = 273.15 \text{ K}$$

The Kelvin temperature of the steam point T_s is

$$T_s = 373.15 \text{ K}$$

15-6
Thermal expansion of solids

When the temperature of a solid rod is increased, its length is also increased. Rods made of different materials will increase in length by different amounts, even though their temperature changes and original lengths are the same. Figure 15-7 shows one method of measuring the expansion of a metal rod. One end of the rod is fastened firmly to a support at A, while the other end slides freely through a guide at B. Since the change in length is usually very small, it can be magnified by attaching a lever to it pivoted at O and having a pointer at the other end that moves over a scale. The rod may be heated by a gas flame, or it may be sur-

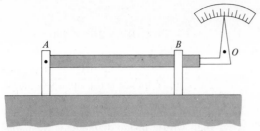

FIGURE 15-7 Method of measuring the expansion of a metallic rod.

rounded by a jacket containing a liquid whose temperature can be varied. The term *coefficient of linear expansion* is used to designate the fractional change in length of a rod per degree change of temperature. If L_i is the length of the rod at the ice point, and L its length at any other temperature t on the Celsius scale, then the average value of the coefficient of linear expansion α is given by

$$\alpha = \frac{L - L_i}{L_i t} \tag{15-6}$$

from which

$$L = L_i(1 + \alpha t) \tag{15-7}$$

Average values of the coefficients of linear expansion of some common substance are listed in Table 15-1. These values are for temperature changes on the Celsius scale. If the temperature changes are measured on the Fahrenheit scale, each of these coefficients should be multiplied by $\frac{5}{9}$. It will be noticed that the coefficients of linear expansion are generally very small, but this does not mean that they can be neglected. In the building of railroads, bridges, and other structures that are subject to wide variations of temperature owing to seasonal changes, allowances must be made for the changes in length that are likely to occur.

Table 15-1   Coefficients of linear expansion

substance	α	
Aluminum	22×10^{-6} per °C	
Brass	18	” ”
Copper	16	” ”
Glass (Pyrex)	3.2	” ”
Invar	0.7	” ”
Lead	28	” ”
Platinum	9.0	” ”
Steel	11	” ”
Tungsten	4.4	” ”

Illustrative example

The steel span of the George Washington Bridge over the Hudson River is about 1 mile long. The extremes of temperature in New York may be as great as 120°F. Determine the maximum change in length of the span between summer and winter.

Equation 15-7 will have to be modified slightly, since the length of the bridge at 32°F is not given in this example. If we call L_0 the length of the bridge at the temperature t_0, and L its length at any other temperature t, then Equation 15-7 may be rewritten as

$$L = L_0[1 + \alpha(t - t_0)] \tag{15-7a}$$

or

$$L - L_0 = \alpha L_0(t - t_0)$$

Now, letting $L_0 = 5280$ ft, $t - t_0 = 120°F$, and

$$\alpha = \frac{5}{9} \times 11 \times \frac{10^{-6}}{°F}$$

we get

$$L - L_0 = \tfrac{5}{9} \times 11 \times 10^{-6} \times 5280 \times 120 \text{ ft} = 3.87 \text{ ft}$$

If two metals, say brass and steel, are welded or riveted together so that they form a single straight piece at room temperature, then when the temperature is raised, this strip will bend in the form of an arc with the brass on the outside, as shown in Figure 15-1b. This is because brass has a greater coefficient of expansion than steel. A bimetallic strip of this kind is used very frequently as an element in thermostats and other temperature-control devices.

When the temperature of a solid is raised, it expands in three dimensions. So far we have confined our attention to the expansion in one direction only. Experiments can be performed to measure the coefficient of linear expansion of a solid for each of three directions at right angles to one another. It is found that certain crystals have different coefficients of expansion in different directions. But many solids commonly used have the same coefficient of linear expansion in every direction; such a substance is said to be *isotropic* with respect to this property. If we consider a cube of an isotropic solid of length L_i and volume V_i at the ice point, then

$$V_i = L_i{}^3 \tag{15-8}$$

If V is its volume at any other temperature t above the ice point, and L is the length of one of its edges,

then

$$V = L^3$$

but

$$L^3 = L_i^3 (1 + \alpha t)^3$$

so that

$$V = V_i(1 + \alpha t)^3$$

Now

$$(1 + \alpha t)^3 = 1 + 3\alpha t + 3\alpha^2 t^2 + \alpha^3 t^3$$

Since the value of α is very small, the value of α^2 and of α^3 will be much smaller than α, so that, in general, the last two terms of the above expansion may be neglected, which yields

$$(1 + \alpha t)^3 = 1 + 3\alpha t$$

hence

$$V = V_i(1 + 3\alpha t) \tag{15-9}$$

If we now introduce a new term called the *coefficient of volume expansion* β of a substance and defined as the fractional change in volume per degree in temperature, then we can write

$$\beta = \frac{V - V_i}{V_i t} \tag{15-10}$$

from which

$$V = V_i(1 + \beta t) \tag{15-11}$$

A comparison of Equation 15-9 and 15-11 shows that, for isotropic solids,

$$\beta = 3\alpha \tag{15-12}$$

or that the coefficient of volume expansion of an isotropic solid is three times its coefficient of linear expansion.

Illustrative example

Determine the change in volume of a copper sphere of 4 cm radius when it is heated from 0°C to 400°C.

The volume of a sphere is given by

$$V = \tfrac{4}{3}\pi r^3$$

in which r is the radius of the sphere. Now

$$V_i = \tfrac{4}{3}\pi \times (4\ \text{cm})^3 = 268\ \text{cm}^3$$

From Equation 15-10,

$$V - V_i = \beta V_i t = 3\alpha V_i t$$

therefore

$$V - V_i = 3 \times 16 \times 10^{-6} \times 268 \times 400\ \text{cm}^3$$
$$= 5.15\ \text{cm}^3$$

It may be interesting to note that the change in volume of the copper sphere is the same whether it is solid or hollow. We may think of a solid sphere as consisting of a solid central core and a hollow spherical shell whose inside diameter is the same as the diameter of the core. Before the sphere is hollowed out, the two fitted perfectly at all temperatures; hence the inside diameter of a hollow sphere should always be the same as that of the solid sphere at the same temperature. The same argument can be extended to a container of any other shape. Hence, when solid containers expand, the change in volume of the container may be calculated as though the inside were filled with a solid of the same substance.

15-7
Thermal expansion of liquids

As a general rule a liquid expands when its temperature is raised; the notable exception to this rule is water, which, in the limited region of temperature from 0°C to 4°C, contracts when its temperature is raised. Above 4°C, water expands with an increase in temperature. In this discussion of changes in volume of a liquid, it is assumed that the pressure on the liquid remains constant while the temperature is changing. The most convenient way of determining the coefficient of volume expansion of a liquid is to put a known volume of liquid into a glass bulb with a narrow tube at one

FIGURE 15-8 Expansion of a liquid in a container when heated.

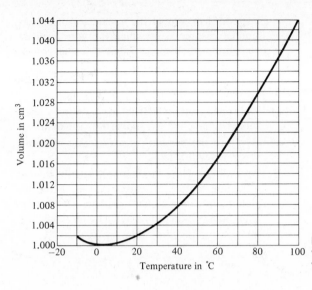

FIGURE 15-9 Curve showing the volume of a gram of water as its temperature is raised from −23°C to 100°C with its minimum volume or maximum density at 4°C.

end and to note the level of liquid in the narrow tube, as sketched in Figure 15-8. When the temperature is raised, both the glass container and the liquid expand. Since liquids generally have greater coefficients of expansion than glass, the level of the liquid will rise in the narrow tube. Hence only the *relative expansion* of the liquid with respect to the container can be measured directly by this method. Since the coefficient of volume expansion of the solid is known, the coefficient of volume expansion of the liquid can be calculated.

The behavior of water at atmospheric pressure in the range from −20°C to 100°C is shown in the graph in Figure 15-9, in which the volume of 1 gm of water is plotted as ordinate and the temperature as abscissa. It will be noted that the density of water is a maximum at 4°C. An interesting result due to this property is that in the winter the temperature of the water at the bottom of a lake may be 4°C, while the temperature of the water at the surface may be 0°C, or the water may be frozen.

Questions

1. Ice cubes are placed in a glass of water to cool the water. Explain this in terms of the zeroth law.

2. When you read a mercury-in-glass thermometer you measure a length. (a) What do you measure when you read a bimetallic thermometer, such as in an ordinary house thermostat? (b) What do you measure when you read a constant-volume gas thermometer?

3. What kind of thermometer might be used to measure the temperature of a fly?

4. Is there any reason that the temperature of the steam point is chosen to be a larger number than the temperature of the ice point?

5. There are many reasons for preferring the metric system of measurement of mass and length to the English system. What advantage (if any) is there to using the Celsius system over the Fahrenheit system?

6. Why is the value of the triple point not chosen to have a simple round number in the Kelvin scale?

7. Since the absolute temperature scale is independent of the specific gas in the thermometer, one can use other criteria in choosing which gas to use. What might some of these be?

8. Follow the procedure used in defining the coefficient of linear expansion and coefficient of volume expansion of a solid and define the coefficient of area expansion of a solid. Show that, to a good approximation, the coefficient of area expansion is twice the coefficient of linear expansion of the solid.

9. Account for the fact that the coefficients of

linear expansion given in Table 15-1 must be multiplied by $\frac{5}{9}$ if used in connection with Fahrenheit temperature changes.

10. The inside diameter of a steel ring is slightly less than the diameter of a steel shaft. It is desired to put the steel ring on the shaft. Would you suggest heating or cooling the ring to get it on the shaft?

11. When a mercury-in-glass thermometer is put into a hot liquid, the column of mercury first descends and then rises. Try this experiment and explain the results.

12. In order to supply electricity to the filament of a light bulb, wires must pass through the glass. When the wires are carrying current they are heated. How should the coefficient of expansion of the metal of which the wire is made compare with the coefficient of expansion of the glass?

13. The coefficient of expansion of rubber is a negative number. Explain what would happen if a weight were hung from piece of thin-walled rubber tubing and the tubing were heated.

14. Why are the sections of a concrete roadway separated with pitch?

15. An ordinary glass tumbler will usually shatter if boiling water is poured into it, while an aluminum cup will not. Explain. (Note that aluminum conducts heat much better than glass does.)

Problems

1. Normal body temperature is 98.6°F and fever temperatures often are 103°F. Express both of these in degrees Celsius.

2. The boiling point of sulfur is 444.6°C. Convert this to the Fahrenheit scale of temperature.

3. The melting point of mercury is −38.87°C. What is this temperature on the Fahrenheit scale?

4. The critical temperature of hydrogen is −240°C. What is the critical temperature of hydrogen on the Fahrenheit scale?

5. At what temperature will the readings of a Celsius and a Fahrenheit thermometer be equal?

6. How long must a brass rod be in order that its length will increase by 0.05 in. as a result of a change in temperature of 20°C?

7. An iron steam pipe laid underground between two buildings is 180 ft long. How much room must be provided for expansion if its temperature will change from 25°F to 212°F?

8. Find the coefficient of volume expansion of glass.

9. A steel cube is 3 cm on a side at room temperature, 25°C. Find the change in volume of the cube when it is plunged into ice water.

10. A gasoline station has 5000 gallons of gasoline delivered at night, when the air temperature is 0°F. The next morning, the temperature has warmed up to 30°F. How many gallons of gasoline does the station have available for sale, from that delivery? (Take the coefficient of volume expansion of gasoline to be 10^{-3} per °C.)

11. Suppose the ice point was chosen to be 100° and the temperature of the steam point to be 0°, precisely the reverse of the present Celsius scale. What would one measure for normal human body temperature?

12. On the Reaumur scale, which is sometimes used in continental Europe, the ice point is 0°R and the steam point is 80°R. What would 70°F read on a Reaumur thermometer?

13. The temperature of a certain star is known to be 6000 K. Express this temperature in Celsius and Fahrenheit. Explain why astronomers sometimes describe such a temperature as "about 6000°," omitting an indication of which scale they are employing.

14. The following data were taken with a constant-volume air thermometer similar to that shown in Figure 15-3: barometric pressure 754 mm, height of column C 48.4 cm, height of column D 44.7 cm at the ice point; height of column C 48.4, height of column D 71.0 cm at the steam point: height of column C 48.4 cm, height of column D 64.0 cm when the air bulb is surrounded by warm water. Find the temperature of the warm water.

15. The distance between two markers is measured with a steel tape at 25°C; the reading of the tape is 60 ft. If the calibration of the tape is correct at 0°C, determine the distance between the markers.

16. A steel bushing in the form of a ring is to have a shrink fit on a shaft; that is, it is to be heated so that it just fits on the shaft and then, as it cools down, shrink to fit tightly. The shaft is 1.752 in. in diameter and the inside diameter of the bushing is 1.748 in. at 72°F. To what temperature must the bushing be heated in order that it will just slip over the shaft?

17. A copper rivet, 5 mm in diameter, is 10^{-4} mm too large to go through a hole in an aluminum plate when both the rivet and the plate are at 0°C. To what temperature should both be heated in order that the rivet will just fit in the hole?

18. A glass flask has a volume of 1 liter at 0°C. What will be its volume at 80°C?

19. A brass rod 30 cm long and 5 mm in diameter is heated from 25°C to 40°C. What force is needed to compress the rod back to its original length?

20. A steel bar, 3 in. wide and $\frac{1}{8}$ in. thick, is bent to form the rim of a wooden wagon wheel. The wheel is 36 in. in diameter at 20°C. The steel rim is formed, and welded so that it is slightly undersized at 20°C, but when heated to 150°C it just fits on the wheel. (a) To what circumference was the rim formed at 20°C? (b) What tension exists in the rim after it cools down, assuming that the wheel stays fixed in size?

21. Water is poured into a 2-liter calibrated glass flask until a level of 100 cm³ is reached. The readings are correct at 20°C. If the system is heated from 20°C to 75°C, determine the new reading on the flask for the volume of the water.

22. A 250-cm³ aluminum cup is filled to the brim with water; both the cup and the water are at 0°C. The cup is heated to 100°C. (a) Does the water overflow or does the water level drop? (b) Find the difference in volume between the heated water and the heated cup.

23. A block of copper, 3 cm on a side, is heated from 20°C to 60°C. What pressure must be applied equally to all sides of the cube in order to keep it from expanding?

24. A steel bomb is filled with water at 20°C. If the system is heated to 100°C, determine the in-crease in pressure of the water, assuming that the expansion of the steel is negligible.

25. A clock regulated by a seconds pendulum made of brass is correct when the temperature is 70°F. (a) Find the period of the pendulum when the temperature is raised to 92°F. (b) Find the gain or loss, in seconds per day, of the clock at 92°F.

26. If the temperature of a barometer changes, while the atmospheric pressure remains constant, the height of the mercury column will increase because of the thermal expansion of the mercury (neglecting the expansion of the glass and the expansion of the measuring scale). Show that, if the temperature increases by ΔT, the height of the barometer will increase by Δh, where

$$\Delta h = \beta h\ \Delta T$$

in which β is the coefficient of volume expansion of mercury.

27. A liquid of coefficient of volume expansion β partly fills a capillary tube. The material of which the tube is made has a coefficient of linear expansion α. The length of the column of liquid is originally l; the bore of the capillary tube is uniform. (a) Show that the length of the liquid column will change by Δl, if the temperature is changed by ΔT, where

$$\Delta l = (\beta - 2\alpha)l\ \Delta T$$

(b) Find the change in measured length of the column, as indicated by graduations on the tube.

28. In the ordinary pocket watch and in most wristwatches, the rate is determined by a balance wheel and hairspring executing angular harmonic motion, as shown in Chapter 12. As the temperature increases, the torsional constant of the spring decreases and the wheel increases in size, increasing the moment of inertia. Both of these effects can be compensated for by making the wheel of two different metals bonded together, and splitting the wheel at two places. (a) Which of the two metals should be on the outside? (b) Describe the process by which this arrangement compensates for both effects.

the first law of thermodynamics

chapter 16

16-1
Flow of heat

When two objects initially at different temperatures come to equilibrium with each other, the temperature of one rises and the temperature of the other falls. It is almost impossible to think about this process without assuming that something flows from one object to the other. That which flows between a hot object and a cold object is called *heat*. (One could, and popular language sometimes does, refer to cold flowing from the low temperature to the high temperature, but this idea turns out to be wrong.)

The use of the word "flow" when applied to heat is one of the many remnants in our language of the eighteenth-century idea that heat was a fluid. In the most popular form of this idea, the fluid was called caloric, and had many properties such as:

1. The more caloric an object had, the higher its temperature.
2. Caloric is conserved.
3. Caloric repels itself, but is attracted to the atoms of matter.

With these properties we could explain many phenomena, in addition to understanding how objects come to thermal equilibrium. For example, we can understand the behavior of a pot of water placed on a stove. Since the flame has more caloric than the water or the pot, and the caloric repels itself, the caloric will flow from the flame to the pot, and then to the water. The caloric will be attracted to each atom of the pot and the water, and so each atom will end up surrounded by a coating of caloric. Because the caloric repels itself, the force of attraction between atoms will be reduced, causing the metal pot and the water to expand. Ultimately, when enough caloric has flowed into the water, the force of attraction between water molecules will decrease enough to cause the water to become a gas. The water will boil.

Among the major opponents of the caloric theory was Benjamin Thompson (1753–1814), who was born in Colonial America and, while serving in the government of Bavaria, took the title of Count Rumford. Among Rumford's duties was the supervision of the Bavarian armory, where he performed significant experiments on the heat generated in the process of boring cannon. He was able to measure the heat generated by the process of drilling the cannon barrel. This total heat extracted from the metal was significantly larger than the total amount of heat that could have been put into the metal during the manufacture of the metal and the gun. The caloric theory explained the heat generated by friction in rubbing or drilling in terms of the caloric being squeezed out of the material. But here, Count Rumford was able to show that the heat developed required more caloric to be squeezed out than was available. Rumford surmised that the heat generated by the boring machine arose from the work done on the machine by the horses that provided the motive power for the machines. He also made a crude estimate of the relationship between the work done and the amount of heat developed. This quantity has come to be known as the mechanical equivalent of heat (see Section 16-3). By the middle of the nineteenth century, the idea that heat is a form of energy became definitely established.

16-2
Measurement of heat; specific heat

When two objects, initially at different temperatures, are placed in contact, the temperatures of both bodies will change because heat flows from one to the other. That is, all the heat that leaves the hot object will enter the cold object. Thus, we could definite a unit of heat in terms of the temperature changes of objects. However, experience shows that the temperature change in the process of coming to thermal equilibrium depends on many things such as the materials used, the initial physical states of these materials, and in many cases their final states as well. It is thus necessary to define a unit of heat in terms of specific processes using a specific material. The unit of heat called a *calorie* is defined as follows:

One calorie is that amount of heat which, when added to one gram of water, at one atmosphere pressure, will raise the temperature of the water from 14.5°C to 15.5°C.

In engineering work the unit of heat is the *Btu*, an abbreviation of the words British thermal unit; *the Btu is that quantity of heat which will raise the temperature of one pound of water from 63°F to 64°F.* The pound is here used as a unit of mass. One Btu is equivalent to 252 cal.

Another unit of heat is the *large calorie*, or the *kilogram calorie* (sometimes shortened to *kilocalorie*); *the large calorie is the heat required to raise the temperature of one kilogram of water one degree Celsius.* The large calorie is equivalent to one thousand calories. The large calorie is frequently used in biology, in dietetics, and in some engineering work. In this book, unless otherwise stated, the term *calorie* will always refer to the small calorie.

With these definitions in mind, we can now investigate the thermal properties of many materials. One method used is called the *method of mixtures.* Keeping the pressure fixed at one atmosphere, we can add materials, at known temperatures, to water at some other known temperature and measure the temperature change for each substance when thermal equilibrium has been reached. From the temperature change of the water, we can find the amount of heat transferred between the object under study and the water. Suppose we have an object of mass m, into which amount of heat Q has been transferred, producing a temperature change ΔT. We can then define a quantity called the specific heat, c, by

$$c = \frac{Q}{m \, \Delta T} \qquad (16\text{-}1)$$

This quantity turns out to be characteristic of the chemical and physical identity of the material; for example, the specific heat of copper is different from the specific heat of lead.

Once we have measured the specific heat of materials, we can use these values to measure the amount of heat transferred in processes in which no water is used. In fact, it is possible to generalize the measurements to processes that take place at pressures other than 1 atm and even to processes in which the pressure varies. When that is done, the value of the specific heat will be found to depend on the process; the specific heat at constant pressure will differ from the specific heat measured with the volume of the object kept constant. For solids and liquids, most experiments are done with open containers, and thus the pressure is auto-

Table 16-1

element	specific heat in cal/gm °C	at. wt.	molar heat capacity in cal/mole K
Aluminum Al	0.21	26.98	5.5
Copper Cu	0.093	63.54	5.9
Iron Fe	0.11	55.85	6.7
Lead Pb	0.031	207.2	6.4
Silver Ag	0.056	107.88	6.2
Tin Sn	0.060	118.70	7.6
Zinc Zn	0.095	65.38	6.2

matically kept constant at 1 atm, and so the tabulated specific heats for most solids and liquid materials are for processes that take place at the constant pressure of 1 atm. Although the specific heat of solids and liquids does vary with temperature, this variation is small over modest temperature ranges. The values given in Table 16-1 are average values over such ranges of temperature about room temperature, which may be taken as 25°C.

When a certain amount of heat is added to a material we can calculate the temperature change by using Equation 16-1. Conversely, when the temperature of an object changes by a certain amount, we can use Equation 16-1 to calculate the heat that was added to the object.

Illustrative example

Suppose that there are 300 gm of water at 20°C in a copper container, usually called a *calorimeter*, of 100 gm mass. A piece of iron of 400 gm mass has been heated separately in a steam bath at atmospheric pressure, so that its temperature is 100°C. If this piece of iron is put into the water in the calorimeter, the temperature of the iron will drop, while that of the water and the copper calorimeter will rise. The water should be stirred (with an object of small heat capacity) until the temperature has reached its final equilibrium value, say t_f. The calorimeter should be placed in a well-insulated container to avoid heat exchanges with the outside. If this is properly done, then the heat that is given out by the iron in cooling from 100°C to temperature t_f will be equal to the heat added to the water and the copper calorimeter to raise their temperature from 20°C to t_f.

Now, the heat given out by the iron is, from Equation 16-1,

$$Q = 400 \text{ gm} \times 0.11 \frac{\text{cal}}{\text{gm °C}} (100° - t_f)$$

The heat added to the water and the copper calorimeter is

$$Q = 300 \text{ gm} \times 1 \frac{\text{cal}}{\text{gm °C}} (t_f - 20°)$$
$$+ 100 \text{ gm} \times 0.093 \frac{\text{cal}}{\text{gm °C}} (t_f - 20°)$$

We can find the final temperature t_f by equating these two quantities of heat, obtaining

$$400 \times 0.11 (100° - t_f)$$
$$= (300 + 100 \times 0.093) (t_f - 20°)$$

from which

$$t_f = 30°C$$

16-3
Heat as a form of energy

Through the first half of the nineteenth century the idea grew that heat is a form of energy. The idea was suggested by a physician, Julius Robert von Mayer, and put on the firm footing of careful measurement by James Prescott Joule. Joule spent a major part of his scientific career measuring the *mechanical equivalent of heat*. He constructed a large number of systems into which energy was transferred, sometimes by mechanical work and sometimes by electrical means. The systems were constructed (see Figure 16-1) so that this mechanical energy "disappeared" and the temperature of some part of the system was increased. From the measured properties of the materials in the system and the measured temperature rise, Joule was able to calculate the heat necessary to produce the same temperature rise. He then was able to calculate

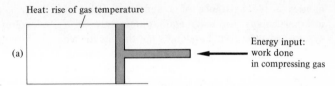

(a)

Heat: rise of gas temperature

Energy input: work done in compressing gas

Piston compressing gas in cylinder

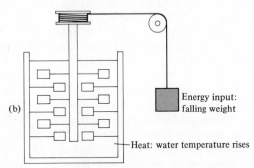

(b)

Energy input: falling weight

Heat: water temperature rises

Falling weight turning paddle wheel in water

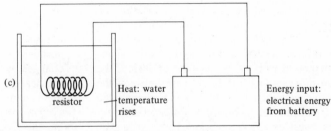

(c)

resistor

Heat: water temperature rises

Energy input: electrical energy from battery

FIGURE 16-1 Heat is a form of energy.

Electric current from battery heats coil immersed in water

the ratio of this amount of heat to the total amount of energy that entered the system. For all systems, the ratio had the same value, $J = 4.18$ joules/calorie. Thus, if W is the work done on the system to produce a given temperature change and Q is the quantity of heat necessary to produce the same temperature change in this system, then the mechanical equivalent of heat J is defined as

$$J = \frac{W}{Q} \qquad (16\text{-}2)$$

Its present value, as determined from the results of many different experiments, is

$$J = 4.185 \text{ joules/cal}$$
$$= 4.185 \times 10^7 \text{ ergs/cal}$$
$$= 778 \text{ ft lb/Btu}$$

The fact that the mechanical equivalent of heat is a constant implies two things: (a) calories and

joules are different units of measurement for the same thing; (b) for the systems measured, all of the energy is conserved, in the sense that all of the input energy (measured in joules or ergs) appeared ultimately in the form of heat (measured in calories). This second idea, that total energy, including heat, is conserved, was amplified by others in the nineteenth century and will be discussed in this and subsequent chapters in more detail.

16-4
The first law of thermodynamics

The first law of thermodynamics is a restatement of the principle of conservation of energy; it asserts that the total energy change of a system is equal to the sum of the heat transferred to the system and the work done on the system by all external forces.

The central idea of the first law of thermodynamics is the existence of a definite amount of internal energy for each equilibrium state of a system. In order to understand what this means we shall consider a system, which for the sake of definiteness we shall choose to be the gas in the cylinder of Figure 16-1a. If the gas is in equilibrium, it has constant thermometric properties; the pressure, volume, and temperature of the gas are fixed. The system is said to be in a particular *equilibrium state*. If we plot pressure, temperature, and volume as the three coordinate axes of a graph, we can represent each equilibrium state by a point as shown in Figure 16-2a. If a system is in one equilibrium state, it can be brought to a different equilibrium state, but only if heat is added (or subtracted) or work is done, or both heat and work are transferred to (or from) the system. Figure 16-2b shows two different equilibrium states, labeled A and B. There are many different ways of bringing the system from one state to the other, some of which are represented by the lines drawn from A to B. Each of these is called a *path*. While the system is changing from one state to the other, along a particular path, a certain amount of heat, Q, flows into the system, and the system does a certain amount of work W, on its surroundings. If the system moves along a different path between the same two equilibrium states, then in general the heat Q and the work W will be different. For example, consider two equilibrium states of a gas, both of which have the same temperature, as shown in Figure 16-2c. It is possible to bring the gas from one state to the other by many paths, two of which are shown in the figure. In one path the pressure of the gas is reduced, keeping the volume constant, which brings the gas from state A to state C. Then the volume is increased at constant pressure,

bringing the system to state B. In the other path the volume is increased at constant pressure first, bringing the gas from state A to state D, and then the pressure is decreased at constant volume, bringing the gas to state B. For each of these paths from A to B, both heat and work are transferred. However, the amount of heat transferred is different for the two paths. Now, for any one path, the quantity $Q - W$ is the net energy transferred into or out of the system. The first law of thermodynamics states that this net energy transfer is the same *for all paths* between the same two equilibrium states, even though Q differs for each path and W differs for each path. Thus when a system goes from equilibrium state A to equilibrium state B, we can define the change in internal energy, $U_A - U_B$ by

$$U_A - U_B = Q - W \qquad (16\text{-}3)$$

Since the change in internal energy is independent of the path, it is reasonable to think of U_A as the internal energy of equilibrium state A, and to think of U_B as the internal energy of equilibrium state B. Thus, when a system is in an equilibrium state it has a definite internal energy. When it changes to another equilibrium state its internal energy changes by a definite amount. While it is changing, the heat flow into the system will depend on how the system is changing and the work done by the system will depend on how the system is changing, but the total internal energy change depends only on the initial and final equilibrium states.

It is difficult to avoid thinking about the form in which this internal energy exists. One is tempted to make models involving moving molecules, but it is important to recognize that from a thermodynamic point of view, it is unnecessary to make such models. Macrophysically, it does not matter in

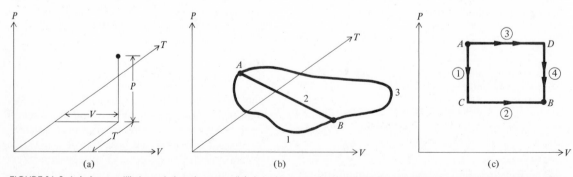

FIGURE 16-2 (a) An equilibrium state of a gas. (b) A system can be brought from state A to state B by many paths. (c) Both the work done, W and the heat transferred, Q, are different for the two paths, but the internal energy change, $U_A - U_B = Q - W$, is the same.

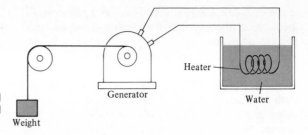

FIGURE 16-3 A system in which gravitational potential energy from the falling weight is transferred to internal energy of the water.

what form the internal energy exists. Experimentally we know that internal energy changes depend on the initial and final equilibrium states, and therefore each equilibrium state must correspond to a definite amount of internal energy. A description of the internal energy of a system in terms of the energies of its constituent particles is an important part of microphysics. We shall discuss some aspects of this topic in later chapters.

In using Equation 16-3 it is important to remember that it describes the energetics of a definite system and that Q is taken to be positive when heat flows into the system (Q is negative when heat flows out of the system), and that W is taken as positive when the system does work on the rest of the universe (W is negative when the universe does work on the system). For example, consider a complex system, as shown in Figure 16-3, that consists of a falling weight, which turns an electric generator; current from the generator flows through a heater, which is immersed in a bucket of water. For this entire system

$$Q = 0$$

because no heat flows into or out of the system; since the gravitational force does work on the system while the weight falls through a distance h,

$$W = -mgh$$

Therefore, from Equation 16-3 we have

$$U_2 - U_1 = mgh$$

or, the internal energy of the system increases by mgh. If, on the other hand, we had taken the system to be the water alone, then the work done by the objects outside the system would be equal to zero, but the heat Q would flow from the heater into the system, and so

$$U_2 - U_1 = Q$$

The essential point is that the net energy transfer into the system ($Q - W$) must appear in the

form of internal energy of the system. This internal energy may then be extracted either in the form of heat, or work, or both. (As we shall see later, the internal energy cannot be extracted totally in the form of work.)

16-5
The ideal gas

The volume of a particular sample of matter depends on the pressure that is exerted on it, and on the temperature. Usually, this dependence is very complicated, but in principle it can be written in the form of an equation. The equation relating the equilibrium values of temperature, pressure, volume, and mass is called the *equation of state*.

The simplest equations of state are those describing the equilibrium states of gases, but even in this case no one has ever been able to write down a single equation that exactly describes any one particular gas over the entire range of pressure, temperature, and volume. Very accurate *approximations* over fairly broad ranges of the variables have been written down and verified by comparison with experiment. For essentially all gases, these approximate equations of state take on the same, simple form when the density (mass per unit volume) of the gas is low. Thus almost all gases are described approximately, but with fairly high accuracy, by the same equation of state, an equation that is called the *ideal gas law*.

It is convenient then to imagine the existence of an ideal gas, which exactly obeys the ideal gas law. We can describe the behavior of these ideal gases by describing the ideal gas law, and then apply the results to real gases as long as we remember that for real gases the results are only approximate.

The ideal gas law has the simple form

$$PV = nRT \tag{16-4}$$

P is the pressure of the gas, as described in Chapter 13. The volume, V, is simply the total volume of the container. The temperature, T, is the Kelvin temperature of the gas, which is uniform throughout the gas when the gas is in equilibrium. The symbol n represents the number of moles of the gas, a quantity proportional to the mass of the gas. Every chemical element is characterized by a number, called the *atomic weight*. These numbers are tabulated in the Appendix, where it is found, for example, that the atomic weight of hydrogen is 1, the atomic weight of nitrogen is 14, and the atomic weight of oxygen is approximately 16. Every chemical compound is characterized by a quantity called the molecular weight, which is equal to the sum of the atomic weights of the constituents of the molecule. Thus, hydrogen gas, which consists of molecules containing two hydrogen atoms, has a molecular weight of 2 (one for each atom); water vapor consists of molecules containing two hydrogen atoms and one oxygen atom, and so has a molecular weight of 18; oxygen gas consists of molecules containing two atoms, so the molecular weight of oxygen is 32. The number of moles, n, of a gas of mass m (in *grams*) where the gas has a molecular weight M is given by

$$n = \frac{m}{M} \qquad (16\text{-}5)$$

Thus, 1 mole of a gas is that amount of gas that has a mass equal to the molecular weight. It must be emphasized that m must be measured in grams in Equation 16-5. Thus 1 mole of H_2 has a mass of 2 gm; 1 mole of H_2O has a mass of 18 gm.

The quantity R in Equation 16-4 is a constant whose value depends on the units used to measure the pressure and volume. If the pressure is measured in dynes per square centimeter and the volume in cubic centimeters, then R becomes

$$R = 8.31 \times 10^7 \frac{\text{dyne cm}}{\text{mole K}} \qquad (16\text{-}6)$$

which can be written

$$R = 8.31 \times 10^7 \frac{\text{erg}}{\text{mole K}} \qquad (16\text{-}7)$$

or

$$R = 8.31 \frac{\text{joules}}{\text{mole K}} \qquad (16\text{-}8)$$

and, using the mechanical equivalent of heat (4.18 joules/cal), this can be written

$$R = 1.986 \frac{\text{cal}}{\text{mole K}} \qquad (16\text{-}9)$$

$$\approx 2 \frac{\text{cal}}{\text{mole K}}$$

Illustrative example

Find the volume occupied by 1 mole of an ideal gas at a pressure of 1 atm and a temperature of 0°C.

One atmosphere is a pressure of approximately 10^6 dynes/cm², and so

$$PV = nRT$$

or

$$V = \frac{nRT}{P}$$

$$V = \frac{1 \text{ mole} \times 8.31 \times 10^7 \dfrac{\text{erg}}{\text{mole K}} \times 273.2 \text{ K}}{10^6 \dfrac{\text{dynes}}{\text{cm}^2}}$$

$$= 22.4 \times 10^3 \text{ cm}^3$$
$$= 22.4 \text{ liters}$$

Thus, 22.4 liters of hydrogen will have a mass of approximately 2 gm, while the same volume of oxygen will have a mass of approximately 32 gm, if the pressure is 1 atm and the temperature is 0°C.

If the temperature of an ideal gas is kept fixed, from the ideal gas law we have

$$PV = nRT$$
$$= \text{constant}$$

and so the pressure and volume at state 1 (P_1 and V_1) are related to the pressure and volume in state 2 (P_2 and V_2) by

$$P_1 V_1 = P_2 V_2 \qquad (16\text{-}10)$$

This equation is called *Boyle's law*, after Robert Boyle, who discovered the result experimentally in 1660.

Similarly, if the pressure is kept constant, we have

$$\frac{V_1}{T_1} = \frac{V_2}{T_2} \qquad (16\text{-}11)$$

which is sometimes called *Charles' law* or *Guy-Lussac's law*. If the volume is kept constant we have

$$\frac{P_1}{T_1} = \frac{P_2}{T_2} \qquad (16\text{-}12)$$

This last result is, in fact, the definition of the Kelvin scale. That is, the Kelvin temperature scale was defined in terms of a constant volume gas thermometer in the limit that the density of the gas becomes exceedingly small. It is precisely in the limit of low densities that real gases behave like ideal gases.

From the point of view of the first law of thermodynamics, the most important characteristic of an ideal gas is that the internal energy of an ideal gas depends only on the temperature. When the temperature of an ideal gas is constant, then the internal energy is constant.

16-6
Work done by an ideal gas

Whenever a gas expands against some external force, it does work on the external agency; conversely, whenever a gas is compressed by the action of some outside force, work is done on the gas. To calculate the work done by a gas, consider a gas enclosed in a cylinder with a tight-fitting piston. The piston may be connected by means of a piston rod to a crankshaft or other mechanical device on which it exerts some force. The force F acting on the piston due to the pressure P exerted by the gas is given by

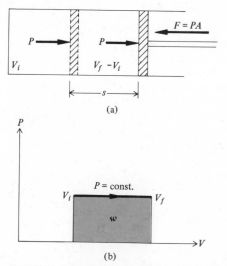

(a)

(b)

FIGURE 16-4 (a) Expansion of a gas at constant pressure. (b) Graphical representation of work done by a gas expanding at constant pressure.

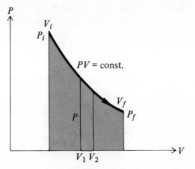

FIGURE 16-5 Graphical representation of work done during an isothermal expansion of a gas.

$$F = PA \qquad (16\text{-}13)$$

in which A is the cross-sectional area of the piston and also of the cylinder (see Figure 16-4a). Suppose that the piston is pushed out a small distance s and that the pressure of the gas remains constant. Then the work W done by the gas in moving the piston is

$$W = Fs = PAs \qquad (16\text{-}14)$$

But As is the change in volume of the gas. Hence the work done by an expanding gas at constant pressure is

$$W = P(V_f - V_i) \qquad (16\text{-}15)$$

where V_i is the initial volume of the gas and V_f is its final volume.

The work done by the gas may be found from a graph in which the pressure is plotted as ordinate and the volume as abscissa, as shown in Figure 16-4b. A constant pressure is represented by a horizontal line; the line extends from volume V_i to volume V_f. The work done by the gas is represented by the area under this line extending down to the V axis. If the gas is compressed at constant pressure, the work done will still be represented by the area under this line, but it is considered negative since the volume is decreased.

If, during the expansion, the pressure of the gas changes, the work done can be calculated by taking small volume changes during which the pressure remains practically constant, multiplying each such volume change by the appropriate pressure, and then summing all of these products. This can easily be done graphically, as illustrated in Figure 16-5. In the case chosen, the temperature of the gas is maintained constant during the expansion. The relationship between pressure and volume is given by *Boyle's law*

$$PV = \text{constant} \qquad (16\text{-}10)$$

and thus the pressure decreases as the volume increases. If we consider any small change in volume from V_1 to V_2, the work done is $P(V_2 - V_1)$ and is represented by the area of the small rectangle shown shaded in the figure. The work done by the gas in expanding from a volume V_i to a volume V_f is then the sum of such small areas and is equal to the area under the curve down to the V axis and included between the vertical lines representing the values of V_i and V_f. A process that takes place at constant temperature is known as an *isothermal* process.

Illustrative example

During one part of the operation of a diesel engine, the gas expands at a constant pressure of 800 lb/in.², while the volume changes from 12 in.³ to 16 in.³. Calculate the work done by the gas during this process.

The work done during a constant-pressure expansion is

$$
\begin{aligned}
\mathscr{W} &= P(V_f - V_i) \\
&= 800 \,\frac{\text{lb}}{\text{in.}^2}\,(16 - 12)\,\text{in.}^3 = 3200\,\text{in. lb} \\
&= 266.7\,\text{ft lb}
\end{aligned}
$$

16-7
Specific heats of a gas

When the temperature of a gas is changed by the addition of heat to it, both its pressure and its volume can be changed. We may, however, choose the conditions under which heat is added to the gas. One simple method is to keep the volume of the gas constant; the quantity of heat added to unit mass of a gas at constant volume to produce a change in temperature of one degree is called the *specific heat at constant volume* and will be denoted by c_v. Another method is to keep the pressure of the gas constant while adding heat to it. The quantity of heat that must be added at constant pressure to raise the temperature of unit mass of gas by one degree is called the *specific heat at constant pressure* and will be denoted by c_p. The specific heat at constant pressure is greater than the specific heat at constant volume. This differ-

Table 16-2 Specific heats of gases

gas	c_p in $\dfrac{cal}{gm\,°C}$	c_v in $\dfrac{cal}{gm\,°C}$	ratio c_p/c_v
Air	0.242	0.173	1.40
Ammonia	0.523	0.399	1.31
Carbon dioxide	0.200	0.154	1.30
Hydrogen	3.40	2.40	1.41
Nitrogen	0.248	0.176	1.41
Oxygen	0.218	0.156	1.40

ence can be accounted for if we note that in order to keep the pressure constant while the temperature is being raised, the volume of the gas must be allowed to expand. But when the volume increases at constant pressure, the gas does work on the surrounding atmosphere or mechanism connected to the movable part to the gas container, equal to the pressure times the change in volume. Hence c_p is greater than c_v because, in addition to raising the temperature of the gas, some of the heat that is added to it is transformed into work by the expansion of the gas at constant pressure.

The values of c_p and c_v at room temperature and atmospheric pressure, for some of the common gases, are listed in Table 16-2.

Illustrative example

A cylinder fitted with a movable piston that contains 4 gm of nitrogen at 0°C and 2 atm occupies a volume of 1.6 liters. The nitrogen is heated at this constant pressure until its temperature is 127°C. Determine (a) the heat supplied to the nitrogen, (b) the work done by the nitrogen in expanding, and (c) the change in internal energy of the nitrogen.

(a) Since the heat Q was supplied to the nitrogen at constant pressure, we can write Equation 16-1 in the form

$$Q = mc_p(t_f - t_i)$$

Now $m = 4$ gm, $c_p = 0.248$ cal/gm °C, and $t_f - t_i = 127$°C; therefore

$$
\begin{aligned}
Q &= 4\,\text{gm} \times 0.248\,\frac{\text{cal}}{\text{gm °C}} \times 127°C \\
&= 126.0\,\text{cal}
\end{aligned}
$$

(b) The work done by the nitrogen in expanding at constant pressure is given by

$$\mathscr{W} = P(V_f - V_i) \qquad (16\text{-}15)$$

$V_i = 1.6$ liters; V_f can be found with the aid of

Equation 16-11, which can be written in the form

$$V_f = \frac{V_i}{273.2} \times T$$

where T is the Kelvin temperature of the gas when its volume is V_f. In this case

$$T = 273.2 + 127 = 400.2 \text{ K}$$

hence

$$V_f = 1.6 \text{ liters} \times \frac{400.2}{273.2}$$

from which

$$V_f = 2.34 \text{ liters}$$

The work done is therefore

$$\mathscr{W} = 2 \text{ atm} \times (2.34 - 1.60) \text{ liters}$$
$$= 1.48 \text{ atm liters}$$

Now

$$1 \text{ atm} = 1.013 \times 10^6 \frac{\text{dynes}}{\text{cm}^2}$$

and

$$1 \text{ liter} = 10^3 \text{ cm}^3$$

hence

$$\mathscr{W} = 1.48 \times 1.013 \times 10^9 \text{ ergs}$$
$$= 1.50 \times 10^9 \text{ ergs}$$

(c) To determine the change in internal energy of the nitrogen, we make use of the first law of thermodynamics:

$$U_f - U_i = Q - \mathscr{W} \qquad (16\text{-}3)$$

In part (a) the value of Q was determined in calories, while in part (b) the value of $\mathscr{W}$ was determined in ergs; we must thus use the mechanical equivalent $J = 4.185 \times 10^7$ ergs/cal in the equation, so that

$$U_f - U_i = JQ - \mathscr{W}$$
$$= 4.185 \times 10^7 \frac{\text{ergs}}{\text{cal}} \times 126.0 \text{ cal}$$
$$- 1.50 \times 10^9 \text{ ergs}$$

from which

$$U_f - U_i = (5.27 \times 10^9 - 1.50 \times 10^9) \text{ ergs}$$
$$= 3.77 \times 10^9 \text{ ergs}$$

Thus the internal energy of the nitrogen was increased by 3.77×10^9 ergs in this process.

16-8
Adiabatic processes

Whenever the expansion or compression of a gas takes place without the transfer of heat to or from the gas, the process is said to be *adiabatic*. In the ideal experiment, the gas should be placed in a cylinder or other container whose walls are perfectly insulated. In practical cases, expansions or contractions that take place so rapidly that practically no heat is transferred to or from the gas may be considered adiabatic. During an adiabatic process, the pressure and volume of a gas are related by the equation

$$PV^\gamma = \text{constant} \qquad (16\text{-}16)$$

where γ (Greek letter gamma) is a number between 1.0 and 1.67 and depends upon the gas used. For air, γ has the value 1.4; it also has this value for hydrogen, oxygen, and other diatomic gases. It has the value 1.67 for the inert gases, such as helium, neon, and argon. Furthermore, it can be shown that

$$\gamma = \frac{c_p}{c_v} \qquad (16\text{-}17)$$

for each gas.

Let us apply the first law of thermodynamics to an adiabatic compression of a gas. During this process no heat is added to or removed from the gas, so that $Q = 0$. Since work is done on the gas, then, according to our sign convention, it will be called $-\mathscr{W}$. Putting these values into the mathe-

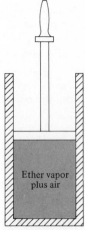

FIGURE 16-6 During adiabatic compression the temperature will rise sufficiently to ignite the ether.

matical statement of the first law as given by Equation 16-3, we get

$$0 = U_f - U_i - W$$

or

$$W = U_f - U_i \qquad (16\text{-}18)$$

that is, the work done on the gas results in an increase of the internal energy of the gas. This increase in internal energy shows itself by the increase in temperature of the gas. In a similar manner, the work done by the gas during an adiabatic expansion results in a decrease of the internal energy of the gas and shows itself by a decrease in temperature of the gas.

The increase in temperature that takes place during the adiabatic compression of a gas can be demonstrated by taking a thick-walled glass cylinder, closing one end, and fitting the tube with a piston (see Figure 16-6). The piston should be moistened by dipping it into some ether and then inserted into the cylinder. If the piston is pushed down rapidly, the temperature of the air will rise sufficiently during the compression to ignite the ether. No heat is added during this process; the work done in compressing the air is converted into additional internal energy.

Questions

1. What is thermal equilibrium?
2. Define specific heat.
3. From the point of view of the caloric theory, explain the fact that it takes 80 cal to melt 1 gm of ice.
4. What is the difference between heat and temperature?
5. Give an example in which one system increases in internal energy while a second system decreases in internal energy and the energy is transferred in the form of heat.
6. Give an example of a system that increases or decreases in internal energy without a transfer of heat.
7. In a classic demonstration two apparently identical steel balls are dropped onto a hard steel plate. One ball rebounds almost to the original height. The other ball is hollow and filled with lead shot and so hits the plate with a thud and hardly rebounds at all. Discuss this from the point of view of the first law of thermodynamics.
8. People often rub their hands together in winter to warm them up. Discuss this from the point of view of the first law.
9. Why is it easier to measure the specific heat at constant pressure of a solid than it is to measure the specific heat at constant volume?
10. The body temperature of a human being is almost constant, but it is higher than the temperature of the surroundings (except in very rare circumstances). Discuss this from the point of view of the first law.
11. When an ideal gas expands isothermally it does work on some outside agency. Does the internal energy of the gas change in this process? If not, what is the source of the energy that enables the gas to do the work?
12. Does an ideal gas do work when it expands adiabatically? If so, what is the source of the energy that enables it to do this work?
13. The rubber of an automobile tire is a very good insulator, yet the temperature of the air inside increases when an automobile is driven. Why?
14. The air temperature in a house is increased from 62°F to 72°F while the family is waking. During this process, the volume of the air in the house remains constant and the pressure remains constant. How can this be done without violating the ideal gas law?

Problems

1. If 1 cal of heat is added to 1 gm of copper, find the temperature change of this piece of copper.
2. How much heat is required to raise the temperature of 400 gm of water from 60°C to 63°C?
3. It is found that 450 cal are needed to raise the temperature of 85 gm of a substance from 20°C to 30°C. What is the specific heat of the substance?
4. A 100-gm aluminum cup is cooled by 1°C when

some water is poured into the cup. The water warmed by 1°C. How much water was added?

5. A hot water bottle containing 800 gm of water cools from 50°C to 20°C. How much heat was given off by the water?

6. How many moles of water are there in a glass that contains 250 gm of water?

7. A tank of hydrogen has a volume of 2 ft³ and contains the gas at a pressure of 250 lb/in.². How large a volume will the gas occupy at 1 atm, assuming the temperature remains constant?

8. A certain mass of gas occupies 20 liters at 300 K and 1 atm pressure. The gas is heated to 600 K and allowed to increase in pressure to 4 atm. What is the volume of the gas?

9. A tank of volume 4 m³ contains hydrogen at 20°C and 5 atm pressure. What is the mass of the hydrogen?

10. A 100-gm copper cup is initially at 20°C, and 250 gm of water, initially at 90°C, is poured into it. Assume that the specific heat of the water is constant and that all of the heat that leaves the water flows into the cup. Find the final temperature of the water and the cup.

11. A 150-gm sample of metal, initially at 99.5°C, is put in a copper cup that contains 200 gm of water at 20°C. The cup has a mass of 150 gm. The final temperature of the system is 25°C. Find the specific heat of the metal sample.

12. A lead ball strikes the ground after having fallen through a height of 200 m. Assume that half the kinetic energy of the ball is converted into internal energy in the lead. Determine the temperature rise of the lead.

13. The water in Niagara Falls drops 160 ft. If all the energy is converted into internal energy, find the rise in temperature of the water.

14. A copper cylinder whose mass is 50 gm contains 150 gm of water. The cylinder is dragged along a rough horizontal floor by a force of 80,000 dynes, and the acceleration produced is observed to be 180 cm/sec². (a) Determine the force of friction between the floor and cylinder. (b) If the cylinder is moved through a distance of 60 m, determine the rise in temperature of the water and the cylinder, assuming that all the work done against friction is converted into heat in the cylinder.

15. A horse can do about 750 joules of work per second. In a Rumford cannon-boring experiment a horse drives a lathe that turns a cannon barrel. The barrel is immersed in water. From calorimetric measurements it is known that it takes 1.2×10^6 cal to boil this mass of water. It takes $2\frac{1}{2}$ hours of operation of the lathe to boil the water. From these data, find the mechanical equivalent of heat.

16. How much work is done by the expansion of a gas in a cylinder if the pressure remains constant at 200 lb/in.² and the volume of the gas changes from 15 in.³ to 24 in.³?

17. A gas in a cylinder occupying a volume of 1000 cm³ at a pressure of 1 atm is compressed adiabatically to a volume of 25 cm³. If $\gamma = 1.5$, find the final pressure of the gas.

18. A steel tank contains 500 gm of air at 24°C. The cylinder is heated to a temperature of 90°C. Neglecting expansion of the tank, determine (a) the amount of heat added to the air and (b) the change in its internal energy.

19. In a certain process 500 cal of heat are supplied to a system and at the same time 150 joules of work are done on the system. Find the change in internal energy of the system.

20. A gas expands at a constant pressure of 1 atm while 5000 cal of heat are added to the system. Find the change in volume if the internal energy is constant during the expansion.

21. An adult man takes in about 3×10^6 cal per day in the form of food. Describe what happens to this energy. Compare this amount of energy with the amount of energy supplied to a 100-watt light bulb during one day.

22. A steel cylinder is fitted with a movable piston and contains 175 gm of air at 28°C, a pressure of 1 atm, and a volume of 1.5×10^5 cm³. The air is heated at constant pressure to 85°C. Determine (a) the amount of heat supplied to the air, (b) the work done by the air, and (c) the change in internal energy.

23. A steel tank contains 25 gm of helium at 15°C and 1 atm pressure. The helium is heated to a temperature of 80°C at constant volume. Find (a) the heat supplied to the helium and (b) the change in internal energy.

24. How much heat is required to raise the temperature of the air in a sealed room from 50°F to 72°F, assuming that the volume of this room remains constant.

25. A tank contains 28 gm of nitrogen at 5 atm

pressure and 30°C. Some time later it is discovered that the pressure of the gas is 1 atm and the temperature is 20°C. How much nitrogen leaked out of the tank?

26. A gas is initially at pressure P_0 and volume V_0. The pressure is increased to P_f, with the volume kept constant. Then the volume is increased at constant pressure to V_f. During this process 20,000 cal of heat flow into the system and 30,000 joules of work are done on the gas. The same gas is returned to its initial pressure and volume by first decreasing the pressure at constant volume and then decreasing the volume at constant pressure, during which 10,000 joules of work are done on the gas. How much heat flows into the gas during this return?

change of phase

chapter 17

17-1
Phases of a substance

A substance that has a definite chemical composition can exist in one or more *phases*, such as the vapor phase, the liquid phase, or the solid phase. When there are two or more phases of a substance in equilibrium at any given temperature and pressure, there will always be surfaces of separation between the phases. The change of phase of a substance from solid to liquid is called *melting* or *fusion*; heat must be added to the solid to melt it. The quantity of heat that must be added to melt a unit mass of the substance at a constant temperature is called the *heat of fusion*. Conversely, to freeze the substance—that is, to change its phase from liquid to solid—heat must be removed from it. The change of phase from liquid to vapor is called *vaporization* and involves the addition of heat to the substance. The quantity of heat that must be added to vaporize a unit mass of the substance at a constant temperature is called its *heat of vaporization*. Conversely, to *condense* the substance—that is, to change its phase from vapor to liquid—requires the removal of heat from the substance. A third type of phase change—that from solid to vapor—is called *sublimation*. Heat must be added to produce this change of phase; the quantity of heat required to change a unit mass of a substance from solid to vapor at a constant temperature is called the *heat of sublimation*. Conversely, when a substance is condensed directly from the vapor to the solid phase, heat must be removed from the substance.

The changes that occur when a substance in the solid state is heated to the melting point and the liquid thus formed is heated to the boiling point are illustrated graphically in Figure 17-1. Suppose we start with 100 gm of ice at −20°C and supply

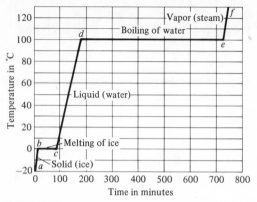

FIGURE 17-1

heat to it at the rate of 100 cal/min. We will assume that the substance is heated uniformly throughout its volume and is in a suitable container. Since the specific heat of ice is 0.5 cal/gm °C, the temperature of the ice will increase at the rate of 2°C/min until it reaches 0°C; this is shown by the straight line ab. At this temperature the ice will begin to melt at the rate of 1.25 gm/min, the temperature remaining 0°C until all of the ice is melted; this is shown by the horizontal line bc. This process will take 80 min, and during this time there will be a mixture of ice and water in equilibrium until all of it is in the liquid phase.

With heat being supplied at the same rate to the water in the liquid phase, its temperature will rise at the rate of 1°C/min, as shown by the line cd. During this process, care must be taken to prevent the water from evaporating into the atmosphere. We may imagine the surface of the water to be covered by a very light substance, so the pressure on the water will remain practically 1 atm. The temperature of the water will rise until it reaches 100°C, at which time the water will begin to boil; that is, some of it will change to vapor. The volume of the substance will now begin to increase markedly, but the temperature of the system will remain 100°C; this is the temperature of both the liquid and the vapor. With heat being supplied at the rate of 100 cal/min, it will take 540 min to convert all of the water into steam at 100°C; this is shown by the horizontal line de. If the heating is continued beyond this time, the temperature of the steam will rise at the rate of about 2°C/min, as shown by the line ef.

Many substances are known to exist in several different solid phases; ice, for example, can exist

in six different solid phases; sulfur is known to have four different solid phases. These solid phases are distinguished principally by the different groupings of the molecules to form different types of crystals. In rare cases, notably helium, two different liquid phases are known to exist. There can be only one vapor phase of a substance.

17-2
Vaporization; vapor pressure

One method for studying the process of vaporization and the properties of a vapor is illustrated in Figure 17-2. This is a modification of the simple barometer. A tube A about 1 m long is first filled with mercury and then inverted and put into a long reservoir R containing mercury. The level of the mercury in A will be at the barometric height above the level in R. The space above the mercury in A contains mercury vapor at a very low pressure; this region is sometimes called a *Torricelli* vacuum. For the purpose of the present experiment, we shall neglect the effect of this mercury vapor. Suppose

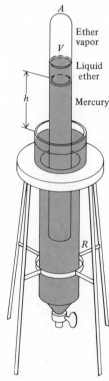

FIGURE 17-2 Method of measuring the saturated vapor pressure of ether.

we take a small quantity of liquid ether in an eye dropper and put it into the open end of tube A, always keeping it under the top of the mercury in R. Since ether is less dense than mercury, the ether will rise to the top of the mercury column and vaporize into the space above it. We shall observe that the level of the mercury column in A is now much lower than before, and there probably is no trace of the liquid ether. If we now push the tube A slowly into the reservoir R, we shall find that at some stage in this process a small layer of liquid ether will appear on top of the column of mercury in A. As the tube is pushed down still further, the thickness of this layer of liquid increases, showing that more vapor is condensing. If we reverse this procedure and now move the tube A up, the thickness of the layer of liquid ether will decrease and will finally disappear completely.

An analysis of this experiment shows that the ether vapor exerts a pressure; this pressure is equal to the difference between the actual barometric pressure B and the height h of the mercury column in tube A. Let us designate the vapor, when there is no liquid present, as an *unsaturated vapor*. The unsaturated vapor behaves as a gas. When the tube A is pushed down into R, the volume of the unsaturated vapor is decreased and its pressure is increased. If we continue to move A down, the volume of the unsaturated vapor will decrease and its pressure will increase until a certain volume is reached, at which point some of the vapor condenses into the liquid phase. If the volume of this system is decreased still further, more vapor condenses into liquid, but the pressure remains constant as long as there is vapor present and the temperature remains unchanged. At these conditions of pressure and temperature, there is equilibrium between the liquid and its vapor, and the vapor is said to be a *saturated vapor*. The pressure of a saturated vapor does not depend upon its volume. If we modify this experiment so that we can vary the temperature of the liquid ether and its saturated vapor, it will be found that the pressure of the saturated vapor increases rapidly with increasing temperature.

17-3
Boiling

Our most common experience with the process of boiling is the boiling of water in a dish that is open to the atmosphere. The water is usually heated by means of a flame or an electric heater. A ther-

Table 17-1   Normal boiling points

substance	temperature in °C
Sulfur	444.6
Mercury	356.7
Water	100.0
Alcohol (ethyl)	78.3
Ether	34.6
Oxygen	−183.0
Nitrogen	−195.8
Hydrogen	−252.8
Helium	−269

mometer placed in the water will show an increase in temperature until the boiling point is reached; the temperature will then remain constant during the process of boiling. When water is boiling, the liquid is changing to vapor. If the dish is open to the atmosphere, the pressure of the water vapor just above the liquid surface must be equal to that of the atmosphere. The *boiling point* is that temperature at which the pressure of the vapor is equal to the pressure exerted on the liquid. When the pressure on the liquid is 76 cm of mercury, the boiling point of the water is 100°C (or 212°F or 373.15 K). The boiling point of a liquid at a pressure of 76 cm of mercury is called the *normal boiling point*. The normal boiling points of some substances are given in Table 17-1.

It has long been known that water will boil at a temperature lower than 100°C if the pressure of the atmosphere is less than 76 cm of mercury—for example, at the top of a mountain—and that its boiling point will be higher than 100°C if the atmospheric pressure is greater than 76 cm of mercury. For example, if we take a flask partly filled with water, heat it, and allow the water to boil vigorously for a few minutes, the air above it will be driven off by the steam. If the flask is now closed by means of a rubber stopper, there will be only water vapor or steam above the liquid surface. The pressure on the liquid will be due to the pressure of the water vapor. If this sealed flask is maintained at a constant temperature of 100°C by being placed in a steam bath, the pressure of the vapor in the sealed flask will remain at the value of 76 cm of mercury. This is its normal boiling point. There is a state of equilibrium between water and its saturated vapor (steam) at 100°C and a pressure of 76 cm of mercury. From the molecular point of view we may think of this type of equilibrium as one in which there is a constant interchange of molecules between the liquid and the vapor phases; just as many

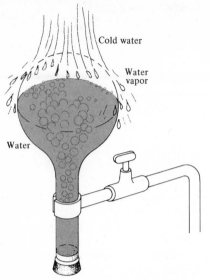

FIGURE 17-3 Water boiling under reduced pressure.

molecules go from the liquid to the vapor phase as from the vapor back to the liquid phase.

If this sealed flask is turned over and clamped in the overturned position, as shown in Figure 17-3, and cold water is poured over it, the water in the flask will start boiling vigorously. If the flask is allowed to cool in the room for some time and cold water again is poured over it, the water in the flask will start boiling vigorously again. The effect of pouring cold water over the flask is to reduce the pressure of the vapor below the saturation pressure. There is no longer equilibrium between the liquid and the vapor, and the liquid starts vaporizing once more. Thus, when the pressure on the water is less than 76 cm of mercury, it boils at a temperature less than 100°C.

The relationship between the temperature of the water and the pressure of its saturated vapor can be investigated by placing the water in an iron boiler that has a thermometer and a pressure gauge fitted into it. The water is first boiled vigorously to drive off the air above it, and the boiler valve is then closed. For low temperatures the boiler may be immersed in a cool bath and the pressure and temperature may be measured. It will be found that as the temperature of the water decreases, the vapor pressure also decreases. Some of the water vapor condenses, and equilibrium is established between the liquid and its vapor at this lower temperature and pressure.

To study the relationship between temperature and pressure at higher temperatures, the boiler can be surrounded by electric heater coils. The current through the coils can be adjusted to the value required to maintain the water at any desired temperature. It will be found that at temperatures above 100°C the pressure of the saturated vapor is greater than 76 cm of mercury. The curve *OA* in Figure 17-4 shows the relationship between the pressure and temperature of water and its saturated vapor. Any point *P* on this curve represents a definite temperature and pressure at which water is in equilibrium with its saturated vapor (steam). This curve is called the *vaporization curve* of water. Similar vaporization curves can be obtained for other liquids.

17-4
Heat of vaporization

The process of vaporization requires the addition of heat to the liquid. *The quantity of heat per unit mass required to vaporize a liquid at a constant temperature is known as its heat of vaporization.* Experiments show that the heat of vaporization of a liquid depends upon the temperature at which vaporization takes place; the higher the temperature, the smaller the heat of vaporization. For example, in the case of water, the heat of vaporization at 100°C is 540 cal/gm or 972 Btu/lb. At 20°C, however, the heat of vaporization of water is 590 cal/gm, and at 300°C it is 331 cal/gm. The heat of vaporization is also the quantity of heat per unit mass liberated when a substance condenses at a constant temperature from the vapor to the liquid phase. Thus, when steam at 100°C is condensed to water at the same temperature, 540 cal of heat are liberated for each gram of steam that is condensed.

In general, if *m* is the mass of a substance that is changed from the liquid to the vapor phase at

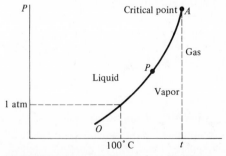

FIGURE 17-4 Vaporization curve of water.

constant temperature and if Q_V is the heat of vaporization of the substance, then the quantity of heat Q that must be added to produce this change of phase is

$$Q = mQ_V \qquad (17\text{-}1)$$

Equation 17-1 also holds when a mass m of the substance is changed from the vapor to the liquid phase. In the latter case, Q is the quantity of heat that must be removed from the vapor, or that is given out by the vapor, in this change of phase. One of the methods for measuring the heat of vaporization of water is to take steam from a boiler and add it to a known quantity of water. In this process the steam is first condensed to water and then cooled from the boiling point down to the final temperature of the mixture. At the same time, the temperature of the cool water is raised from its initial value to the final temperature of the mixture.

Illustrative example

To 200 gm of water at 25°C in a copper calorimeter whose mass is 120 gm are added 5 gm of steam at 100°C. The resulting final temperature is 39.2°C. Determine the heat of vaporization of water.

If Q_V is the heat of vaporization of water, the quantity of heat given out when the 5 gm of steam are condensed to water at 100°C is

$$Q = 5 \text{ gm} \times Q_V$$

The additional heat given out by the 5 gm of condensed steam in cooling from 100°C to 39.2°C is

$$5 \times (100 - 39.2) \text{ cal} = 304 \text{ cal}$$

Equating the heat thus liberated to the heat absorbed by the water and the copper calorimeter that raises their temperature from 25°C to 39.2°C, we get

$$5 \text{ gm} \times Q_V + 304 \text{ cal} = 200 \times 14.2 \text{ cal}$$
$$+ 120 \times 0.093 \times 14.2 \text{ cal}$$

from which

$$Q_V = 539 \frac{\text{cal}}{\text{gm}}$$

17-5
The critical point

The vaporization curve is not indefinite in extent; it has both a lower limit and an upper limit. The upper limit is known as the *critical point*; the

Table 17-2   Critical constants

substance	critical temperature in °C	critical pressure in atmospheres
Ammonia	132	112
Carbon dioxide	31	73
Ether	194	35.5
Helium	−268	2.3
Hydrogen	−240	13
Nitrogen	−147	34
Oxygen	−119	50
Water	374	218

temperature and pressure of the critical point are known as the *critical temperature* and *critical pressure*. At temperatures above the critical temperature the substance cannot exist as a liquid; that is, no matter how great the pressure, it cannot be put in the liquid phase. At the critical temperature the densities of the liquid and the vapor are equal. The heat of vaporization is zero at the critical temperature. A distinction is sometimes made between the vapor states above and below the critical temperature: above the critical temperature it is usually called a *gas*; below the critical temperature it is called a *vapor*. The critical temperature of water is 374°C and the critical pressure is 218 atm. The critical point of carbon dioxide is 31°C and 73 atm. The critical points of some of the more common substances are given in Table 17-2.

Two substances known as gases at ordinary temperatures have very low critical temperatures. They must first be cooled to these low temperatures before they can be liquefied. Helium has the lowest critical temperature, −268°C or 5 K.

17-6
Evaporation

A liquid such as water or alcohol, when left open to the atmosphere, will in time *evaporate*; that is, the liquid will change to a vapor and go into the atmosphere. If the atmosphere above the liquid surface is set into motion by means of a fan, the rate of evaporation will be increased. The heat that is required to vaporize a liquid must come either from the external surroundings or from the remaining liquid. If the liquid is placed in a fairly well-insulated container or if the process of evaporation is so rapid that the liquid cannot get sufficient heat from the surrounding bodies, the temperature of the liquid will be lowered.

If a large jar is placed over the liquid container,

some of the liquid will evaporate until equilibrium is established between the liquid and its vapor—that is, until the pressure of the vapor in the air is equal to the saturated vapor pressure at the temperature of the air. The air is then said to be *saturated* with this vapor. If the jar over the liquid container is removed and a fresh supply of unsaturated air is blown over the liquid, evaporation will start again. Evaporation will continue as long as the vapor pressure in the air is less than the saturated vapor pressure. The rate of evaporation will depend upon the difference between the saturated vapor pressure and the actual vapor pressure in the air. Ether, for example, has a relatively large vapor pressure at room temperatures; it therefore evaporates very rapidly into a moving stream of air.

There are some similarities and some differences between the processes of evaporation and boiling. In each case there is a change of phase from liquid to vapor. During the process of boiling, the pressure of the vapor is the *saturated vapor pressure*, corresponding to the temperature of the liquid and vapor and is equal to or greater than the surrounding atmospheric pressure; whereas, during evaporation, the vapor pressure is less than the saturated vapor pressure. During evaporation, the part of the liquid near the surface changes to vapor; whereas, during boiling, bubbles of vapor can form in the body of the liquid, rise to the surface, and go out above the surface of the liquid. In both cases, the heat of vaporization at a given temperature of the liquid is the same.

As mentioned above, if during evaporation sufficient heat is not supplied from the outside to vaporize the liquid at a constant temperature, the temperature of the liquid will be lowered; that is, the energy necessary to vaporize the liquid will come from its internal energy. This can be demonstrated very readily by placing a small beaker of water under a bell jar, as shown in Figure 17-5, and pumping the air and water vapor from it very rapidly. If the pressure in the bell jar is kept below the saturated vapor pressure of the liquid, the water will be converted to vapor very rapidly. Very little heat can be transferred to the water from the outside; hence the internal energy of the water will continue to decrease; it will be evidenced by the drop in its temperature. If the evaporation is sufficiently rapid, the temperature of the water may drop to the freezing point, and some of the water will begin to freeze.

17-7
Melting; fusion

Heat must be added to a substance to change it from the solid to the liquid phase; conversely, when the substance freezes, that is, changes from the liquid to the solid phase, it will give out heat. The temperature at which melting, or its converse freezing, takes place depends upon the pressure. As long as the pressure on a substance remains constant, the temperature at which it melts remains constant. We have already made use of this fact in choosing the melting point of ice at atmospheric pressure as one of the fixed points of the thermometer. For convenience in calibrating thermometers at temperatures above 100°C, use is made of the melting points of other substances, such as zinc, sulfur, and gold, at atmospheric pressure. Table 17-3 lists the melting points of some of the substances at atmospheric pressure.

The volume changes that accompany the process of melting are rather small. In most cases, the melting of a substance is accompanied by an increase in its volume. In a few exceptional cases, such as ordinary ice and type metal, there is a decrease in volume on melting. This unusual behavior of type metal is of practical value in the casting of type: the molten metal is poured into a mold, and, as it freezes, it expands and completely fills every part of the mold, yielding sharp, clear type.

The expansion that takes place when water freezes has many important consequences. When water in a lake freezes, the ice floats on top; its specific gravity is 0.92. In some lakes a solid layer of ice may be formed on the top during winter.

To vacuum pump

FIGURE 17-5 Rapid evaporation of water when the pressure above it is reduced causes it to cool very rapidly. If the evaporation is sufficiently rapid, the remaining water may freeze.

Table 17-3 Melting points at atmospheric pressure

substance	temperature in °C	substance	temperature in °C
Aluminum	660	Nickel	1455
Copper	1083	Platinum	1773.5
Gold	1063.0	Silver	960.5
Lead	327.4	Tungsten	3410
Mercury	−38.87	Zinc	419.5

Thereafter, freezing takes place more slowly, since ice is a poor conductor of heat. If such lakes are sufficiently deep, the water is never completely frozen, so aquatic life can continue throughout the winter in the water below the frozen surface. The temperature of this water may be as high as 4°C.

The quantity of heat per unit mass that is required to melt a solid at constant temperature is called the heat of fusion of the substance. If m is the mass of solid that is melted and Q_F is its heat of fusion, then the heat Q that must be added to the solid to change its phase to that of a liquid is

$$Q = mQ_F \qquad (17\text{-}2)$$

Conversely, if a liquid freezes at constant temperature, it will give out a quantity of heat Q, as given by Equation 17-2.

The heat of fusion of water is about 80 cal/gm or 144 Btu/lb at 0°C. This means that about 80 cal must be supplied to melt 1 gm of ice or 144 Btu to melt 1 lb of ice. When water freezes, 80 cal are liberated for each gram of ice formed. The heat of fusion of ice may be determined very readily by putting a known mass of ice into a sufficiently large quantity of water so that all of the ice will melt. By measuring the original temperature of the ice and the original temperature of the water and the final temperature of the system, the heat of fusion may be calculated.

Illustrative example

Twenty grams of ice originally at −10°C are put into 250 gm of water originally at 80°C contained in a glass beaker. The final temperature of the system is 67.8°C. Calculate the heat of fusion of ice, neglecting the heat given out by the glass beaker.

The heat absorbed by the ice can be considered in three separate stages.

(a) Specific heat of ice is 0.5 cal/gm °C; hence the heat required to raise its temperature from −10°C to 0°C is

$$20 \times 0.5 \times 10 \text{ cal} = 100 \text{ cal}$$

(b) The heat required to melt it at 0°C is, from Equation 17-2,

$$Q = 20 \text{ gm} \times Q_F$$

where Q_F is the heat of fusion of ice.

(c) The heat required to raise the temperature of the 20 gm of water thus formed to 67.8°C is

$$20 \times 1 \times 67.8 \text{ cal} = 1356 \text{ cal}$$

All of this heat must come from the cooling of the 250 gm of water from 80°C to 67.8°C, which will liberate 3050 cal. Equating these two quantities, we get

$$3050 \text{ cal} = 1456 \text{ cal} + 20 \text{ gm} \times Q_F$$

from which

$$Q_F = 79.7 \frac{\text{cal}}{\text{gm}}$$

Sometimes so large an amount of ice is placed in the water that not all of it can be melted by the heat liberated by the water in cooling down to 0°C. If the final state of this system is a mixture of ice and water, its temperature is 0°C. In general, it is good practice in the solution of problems to determine whether there is sufficient heat available to melt all of the ice.

Illustrative example

A block of ice whose mass is 110 gm originally at −10°C is placed in a glass beaker containing 150 gm of water at 60°C. Determine the final state of the system; neglect the heat liberated by the beaker.

The maximum amount of heat that can be liberated by the water in cooling from 60°C to 0°C is

$$150 \times 60 \text{ cal} = 9000 \text{ cal}$$

To raise the temperature of 110 gm of ice from −10°C to 0°C requires

$$110 \times 0.5 \times 10 \text{ cal} = 550 \text{ cal}$$

This leaves 8450 cal to melt the ice. Since the heat of fusion is 80 cal/gm, enough heat is available to melt 105.6 gm of ice at 0°C. The final state of the system will be a mixture containing 4.4 gm of ice and 255.6 gm of water at 0°C.

17-8
Dependence of melting point on pressure

The temperature at which a solid melts depends slightly upon the pressure. For most substances, the temperature of the melting point increases with increasing pressure. For the few substances that expand on freezing, the temperature of the melting point decreases with increasing pressure. Ice, for example, will melt at a temperature lower than 0°C if the pressure on it is greater than 1 atm. The curve OB in Figure 17-6 shows the variation of temperature with pressure for the melting points of ice. Each point on the curve represents a definite temperature and pressure at which ice and water are in equilibrium.

The fact that an increase in pressure leads to a lowering of the melting point of ice gives rise to a series of interesting results. One of these is the common experience of making snowballs by pressing the loose snowflakes tightly. The increased pressure causes some of the snow to melt even though its temperature is lower than 0°C; when the pressure is released, the melted snow refreezes, forming the snowball. This process of melting at a temperature lower than 0°C because of the increased pressure and then refreezing when the pressure is removed is known as *regelation.*

In ice skating and skiing, the ice and snow are melted even though the temperature is below 0°C, and a thin film of water is formed under the skate or the ski. This water refreezes as soon as the pressure is removed, since its temperature is below 0°C. Recent experiments show that at very low temperatures, friction between the blade of the skate and the ice plays an important part in melting the ice and forming the thin layer of water. Similarly, friction between the ski and snow aids in melting the snow. The water formed in each case refreezes when the pressure is reduced to atmospheric pressure.

17-9
Nonequilibrium states

By cooling a liquid slowly, it is possible to lower its temperature below the freezing point without solidification taking place. A substance in this nonequilibrium state is called a *supercooled liquid.* If the supercooled liquid is disturbed mechanically or if a small crystal of the substance is put into the liquid, it immediately starts to solidify, and its temperature rises to the normal freezing temperature. The heat needed to raise the temperature comes from the heat of fusion liberated by that portion of the substance which solidifies. Water has been cooled down to −20°C without changing to ice. A small crystal of ice put into this supercooled water causes solidification to start around it as a nucleus, and the temperature of the system rises to 0°C. This is of particular interest in meteorology, since the process of supercooling takes place in the formation of some clouds.

If one quickly cools a sample of very clean saturated vapor, the vapor will not immediately condense but will remain in a nonequilibrium state. The vapor is said to be *supersaturated.* Any ions present serve as nuclei around which the vapor condenses to form drops. Energetic charged particles moving through the vapor will free some of the electrons from some of the atoms in the vapor, leaving a trail of ions. Thus, the supersaturated vapor will condense along the path of the particles. This process is the basis of the cloud chamber, which was used extensively for particle detection. Cloud chambers were replaced by bubble chambers, which consist essentially of a superheated liquid. Such a liquid is in a nonequilibrium state at a temperature above its boiling point. The liquid begins to boil at points where ions serve as nuclei. Thus, in the bubble chamber, the trail of charged particles is marked by bubbles.

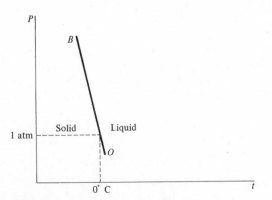

FIGURE 17-6 Fusion curve of water.

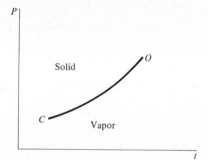

FIGURE 17-7 Sublimation curve of water.

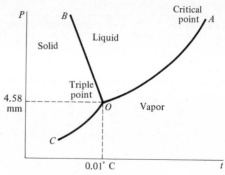

FIGURE 17-8 Triple point of water.

17-10
Sublimation

The change from the solid directly into the vapor phase, though very common, is not usually observed directly, because the more common vapors are usually colorless. A piece of solid carbon dioxide, which is white and usually called dry ice, goes directly into the vapor phase at atmospheric pressure. It does not melt, because the liquid phase does not exist at ordinary temperatures and atmospheric pressure. Another common example is the sublimation of tungsten in an ordinary tungsten lamp. When the filament is hot, some of the tungsten goes directly into the vapor phase, and when the vapor comes into contact with the cooler glass of the bulb, it condenses on it. This is the explanation for the blackening of the inside of a tungsten lamp.

To focus our attention on the process of sublimation, let us consider a flask containing some ice, and let us suppose that the air has been completely removed from this flask. If we keep this flask at a temperature of about $-10°C$, the ice will sublime, forming water vapor in the tube. This process will continue until the vapor pressure reaches a value of 1.97 mm of mercury. Thereafter the vapor pressure will remain the same as long as the temperature of the system is $-10°C$. In this state, the ice and the water vapor are in equilibrium. If the temperature is lowered, the equilibrium vapor pressure will be lowered; if the temperature is raised, the equilibrium vapor pressure will be raised. The curve OC in Figure 17-7 shows the relationship between the vapor pressure and the temperature of sublimation for ice. The equilibrium vapor pressure drops rapidly with decreasing temperature.

17-11
Triple point

The three curves for water—the vaporization curve, the fusion curve, and the sublimation curve—are plotted on a single graph in Figure 17-8. A point on any curve represents a state of equilibrium between two phases at a definite temperature and pressure. All three curves intersect at one point O, known as the *triple point*. At this point all three phases of water are in equilibrium. The temperature of the triple point is $+0.010°C$, and the pressure is 4.58 mm of mercury. As long as the temperature and pressure are maintained at these values, ice, water, and water vapor will coexist in the same flask and remain in equilibrium. If the temperature should be raised slightly, the ice will melt, and the state of the system will be represented by a point on the curve OA. If conditions are changed in any way, one of the phases will disappear. The three phases can exist in equilibrium only

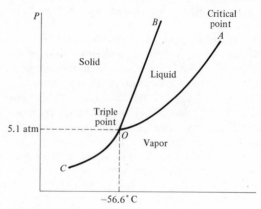

FIGURE 17-9 Triple point diagram for CO_2.

Table 17-4 Triple points of water

phases in equilibrium	temperature in °C	pressure
Ice I, liquid, vapor	+0.010	4.579 mm
Ice I, liquid, ice III	−22	2115 kg/cm²
Ice III, liquid, ice V	−17	3530
Ice V, liquid, ice VI	+0.16	6380
Ice I, ice II, ice III	−34.7	2170
Ice II, ice III, ice V	−24.3	3510
Ice VI, liquid, ice VII	+81.6	22400

at the temperature and pressure of the triple point.

The triple point of water is now the only standard fixed point on the Kelvin scale of temperature; its value is $T_3 = 273.16$ K (see Section 15-5).

The triple-point diagrams of most substances differ from that of water in that the melting-point curve OB has a positive slope; that is, the temperature of the melting point increases with pressure. Figure 17-9 is the triple-point diagram for carbon dioxide; the temperature of its triple point is −56.6°C and the pressure is 5.1 atm. It can be seen that at a pressure of 1 atm, CO_2 cannot exist as a liquid.

Substances that have more than one solid phase have several triple points. Ice is known to exist in several different solid phases. The known triple points of water are listed in Table 17-4. The common form of ice is called ice I. The other forms of ice exist only at very high pressures.

17-12
Humidity of the atmosphere

One of the most important constituents of the atmosphere is water vapor. The amount of water vapor in the air is a variable quantity. As we have seen, a mass of air is saturated when the pressure of the water vapor in that mass of air is equal to the saturated vapor pressure at the temperature of the air. Figure 17-10 is a curve showing the saturated vapor pressure as a function of the temperature of the air, while Table 17-5 gives the saturated vapor pressure at various temperatures. Usually, however, the actual vapor pressure is less than the saturated vapor pressure. The term *relative humidity* is defined as the ratio of the actual vapor pressure to the saturated vapor pressure at the temperature of the air. Thus, if r is the relative humidity, p the actual vapor pressure, and P the saturated vapor pressure at the temperature of the air, then

$$r = \frac{p}{P} \qquad (17\text{-}3)$$

and is usually expressed in percentages. For example, suppose that the actual vapor pressure is 3.0 mm of mercury when the temperature of the air is 50°F. Since the saturated vapor pressure at this temperature is 9.2 mm of mercury, the relative humidity is

$$r = \frac{3.0}{9.2} = 0.325 = 32.5 \text{ percent}$$

At any given temperature, the mass of water vapor in the air is proportional to the pressure of the water vapor. Hence the relative humidity can also be defined as the ratio of the mass of water vapor in a given volume of air to the mass of water vapor required to saturate it. Table 17-5 gives the mass of water vapor in a cubic meter of saturated air at various temperatures.

A simple instrument for measuring the relative humidity is the hair *hygrometer*. The length of the

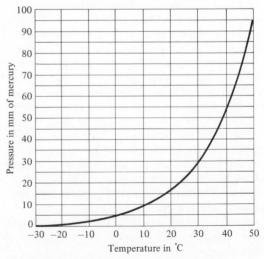

FIGURE 17-10 Saturated vapor pressure curve.

Table 17-5 Properties of saturated water vapor

temperature in °C	temperature in °F	pressure in milli-meters of mercury	grams of water vapor in a cubic meter of air
−10	14	2.15	2.16
−5	23	3.16	3.26
0	32	4.58	4.85
5	41	6.54	6.80
10	50	9.21	9.40
15	59	12.79	12.83
20	68	17.54	17.30
30	86	31.82	30.37
40	104	55.32	51.12
50	122	92.51	
60	140	149.41	
70	158	233.7	
80	176	355.1	
90	194	525.8	
100	212	760.0	
120	248	1489.1	
140	284	2710.9	
160	320	4636	
180	356	7520	
200	392	11659	

hair increases as its cells absorb moisture from the air. The hair hygrometer consists of a bundle of human hairs exposed to the atmosphere; one end of the bundle is attached to an adjustable screw and the other end to a lever that moves a pointer over a scale. The apparatus is sketched in Figure 17-11. The scale is calibrated in terms of the relative humidity.

17-13
Dew-point temperature

It is a common experience to observe moisture condensing on the outside surface of vessels containing cold beverages. This moisture is produced by the condensation of the water vapor from the air onto the cold surface. When the relative humidity is less than 100 percent, the vapor pressure is less than the saturation pressure. The temperature to which the air must be lowered in order to become saturated with the mass of water vapor in it remaining constant is called the *dew-point temperature*. When the dew-point temperature is known, its location on the saturation pressure curve of Figure 17-10 will also give the actual vapor pressure in the air. Since the saturated vapor pressure at the temperature of the air is also known from this curve, the relative humidity is easily determined.

Human comfort depends on relative humidity as well as upon temperature, and this fact must be taken into account in air-conditioning equipment. Unless the temperature and relative humidity fall within certain limits, sometimes called the comfort zone, most of us experience discomfort.

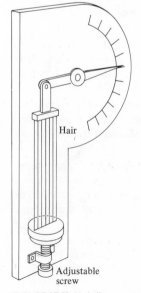

Hair

Adjustable screw

FIGURE 17-11 Hair thermometer.

A simple laboratory method for determining the dew-point temperature is to put some warm water into a polished tin can and then add a quantity of ice to it. Stir the mixture and observe the readings of a thermometer immersed in it. At some definite temperature, moisture will begin to condense on the outside of the can; note this temperature. Now warm it slowly and note the temperature at which the moisture on the outside disappears. The average value of these two temperatures is the dew-point temperature. This is the temperature of the thin layer of air very close to the walls of the can.

Illustrative example

When the temperature of the air is 86°F, a dew-point determination shows that the dew-point temperature is 50°F. Determine the relative humidity of the atmosphere.

From Table 17-5 the saturated pressure P at 86°F is 31.8 mm. At the dew-point temperature, 50°F, the saturated vapor pressure p is 9.2 mm. The relative humidity r is therefore

$$r = \frac{p}{P} = \frac{9.2}{31.8} = 0.29 = 29 \text{ percent}$$

Questions

1. Explain why water pipes sometimes burst in the wintertime.

2. Winter clothing is often stored with "mothballs" during the summer. Usually by the end of the fall the mothballs have disappeared. Explain.

3. When the pressure on ice is increased, will its melting point be higher or lower than 0°C?

4. When the pressure on a piece of copper is increased above atmospheric pressure, will its melting point be higher or lower than 1083°C?

5. What is in the bubbles that are formed when water boils?

6. Draw a horizontal line on the triple-point diagram; discuss the changes in the phase of the system as one proceeds along this line.

7. Draw a vertical line on the triple-point diagram so that it intersects curves *OC* and *OB*. Discuss the changes that take place as one proceeds upward along this line.

8. In hot climates water kept in a porous earthen jar in the shade is much cooler than the surrounding atmosphere. Account for the cooling process.

9. The temperature of a person with a high fever may be brought down by a sponge bath using a mixture of alcohol and water. Why is this mixture preferred to plain water at the same temperature?

10. What are the essential differences between the processes of boiling and evaporation?

11. A pressure cooker is set at 15 lb. What is the steam pressure inside the cooker? Why is food cooked more rapidly in a pressure cooker?

12. What is the function of the water when eggs are boiled?

13. A burn produced by steam is usually more serious than one produced by boiling water at the same temperature. Why?

14. What factors determine the pressure of the saturated water vapor in the air?

15. It is fairly common for the weather bureau to report a humidity of less than 100 percent while it is raining. Explain.

16. On cold winter mornings the inside of the windows of some homes have a coating of frost, while the windows in other homes do not. Assuming that the temperatures of the rooms are equal, account for this difference.

17. Consider the processes that take place in the following cycle: 1 gm of ocean water is evaporated into the air, is formed into part of a cloud, precipitates in the form of rain or snow, and then descends in the form of water in streams or rivers back to the ocean. Describe the energy changes that occur.

18. What is the difference between a gas and a vapor?

19. Why do the bubbles form at the bottom of the pot when water is boiled?

Problems

1. How much heat is required to vaporize 40 gm of water at 100°C?

2. How much heat will be given out by 2 lb of steam when it condenses to water at 212°F?

3. How much heat is required to change 50 gm of ice at 0°C to 50 gm of water at 25°C?

4. There are 500 gm of water at 30°C in a container whose heat capacity is negligible. How much steam at 100°C must be added to the container to raise the temperature to 75°C?

5. How much heat must be extracted from 50 kg of water at 0°C in order to freeze it?

6. The heat of vaporization of liquid oxygen at its normal boiling point (−182.97°C) is 51 cal/gm. A small electric heater in the liquid oxygen tank of a rocket ship supplies 5 watts of electric power. Assuming that the pressure is 1 atm, find the amount of oxygen that is vaporized in 1 min.

7. What is the relative humidity of the air in a room if its temperature is 25°C and the vapor pressure of water in the air is 8 mm of mercury?

8. What is the relative humidity if the temperature of the air is 68°F and it contains 10 gm/m³ of water vapor?

9. The relative humidity of the air in a room is 60 percent when the temperature is 72°F. What is the vapor pressure?

10. When the air temperature was 25°C the dew-point temperature was found to be 15°C. Find the relative humidity.

11. Find the equilibrium temperature when 20 gm of ice at 0°C are placed in an aluminum calorimeter containing 190 gm of water at 50°C. The calorimeter has a mass of 120 gm.

12. A copper calorimeter of 200 gm mass contains 250 gm of water at 15°C. If 30 gm of ice at 0°C are placed in the water, determine the final temperature and the amount of ice, if any, remaining.

13. An aluminum calorimeter of 100 gm mass contains 215 gm of water at 20°C. A block of ice at 0°C is placed in the water. The ice melts and a negligible sliver of ice remains floating on the water. What was the mass of the block of ice?

14. If 500 gm of lead shot at 100°C are placed in a dry cavity in a block of ice and then covered with another piece of ice, how much ice will be melted?

15. Five hundred grams of water at room temperature (20°C) are poured into an aluminum saucepan, also at room temperature, which has a mass of 400 gm. The pan is put on an electric hotplate, which supplies 600 watts. How much time elapses before the water begins to boil? How much time elapses before the water has boiled away completely?

16. A 100-gm block of ice at 0°C is placed in an insulated container. (a) How much steam must be added to just melt the ice? (b) How much steam must be added to have the mixture end up at 25°C? Assume the steam starts at 100°C.

17. A copper calorimeter whose mass is 250 gm contains 400 gm of water and 180 gm of ice in equilibrium. Seventy grams of steam at 100°C are added to the system. Determine the final temperature of the system.

18. An aluminum cup of 120 gm mass contains 250 gm of water and 20 gm of ice at 0°C. A piece of silver whose mass is 350 gm is heated to 140°C and dropped into this calorimeter. Find the final temperature.

19. A galvanized pail (iron with a 1 percent coating of zinc) contains 1 kg of snow. The mass of the pail is 1500 gm. Steam is passed through the snow. The final temperature of the system is 30°C. (a) How much steam was condensed in the process? (b) If the steam started at 100°C you can neglect the steam that bubbled out of the snow. Why?

20. A 10-kg hammerhead, made of iron, is dropped from a height of 1 m onto a soft surface. (a) If 50 percent of the energy is absorbed by the hammerhead as heat, what is the temperature change of the iron? (b) If the surface is a thin foil of silver whose mass is 1 gm, what is the temperature change of the silver? (Assume that the remaining 50 percent of the energy is absorbed by the silver and that the specific heat of the silver remains constant.) (c) If the hammer were raised and struck repeatedly, how many strokes would be necessary to melt

the silver, whose heat of fusion is 21 cal/gm, assuming that the process was so rapid that the silver did not have time to cool off between strokes?

21. When 1 gm of water boils at 1 atm pressure, its volume changes from 1 cm³ in the liquid phase to 1671 cm³ in the vapor phase. Apply the first law of thermodynamics to the process and calculate (a) the work done by the fluid in expanding against the external pressure and (b) the change in internal energy in this process.

22. A solar house absorbs energy from the sun and stores it by heating a substance during the day. During the night the substance cools off, warming the house. Consider designing such a system in which the substance is heated from 75°F to 120°F and cools back to 75°F; in the process the substance is required to store 10^9 cal of energy. (a) How much water would be necessary if water were used as the substance, neglecting the thermal behavior of the container of the water? (b) How much of a certain solid would be necessary if the solid has the following properties: specific heat of the solid, 0.46 cal/gm °C; specific heat of the melted solid 0.68 cal/gm °C; specific gravity 1.6, melting point 32°C, heat of fusion 58 cal/gm?

23. In order for water to boil, bubbles must form. This means that the pressure inside the bubble must exceed that outside by $2S/r$, where S is the surface tension and r is the radius of the bubble. Thus the temperature must be larger than the normal boiling point if the bubbles are reasonably small; the smaller the bubbles are, the higher the pressure inside the bubbles and the higher the temperature. From the vapor pressures in Table 17-5 estimate the size of the bubbles formed if the water pressure is 1 atm, the temperature of the water vapor in the bubble is 200°C, and the surface tension of water is assumed to be 50 dyne/cm. (Note that as heat is added, the pressure increases and r increases; the bubble is in unstable equilibrium and grows explosively.)

the second law of thermodynamics

chapter 18

18-1
Heat engines

Throughout most of history, man has been limited in his capacity to produce food and goods because the only sources of labor were the muscles of men and animals. The economy of Europe was vastly altered by medieval developments in waterwheels and the invention of the windmill; but still the source of most energy for production remained muscles. Engines were invented in the eighteenth and nineteenth centuries and permitted man to replace the labor of muscles with work done by machines; these engines generally derive energy from heat developed in the burning of chemical fuels, such as coal, oil, gasoline, and natural gas, and more recently from the transformation of nuclear energy into heat. There are devices in which energy is directly transformed into work without first being transformed into heat, but here we shall be concerned with heat engines.

Although we are primarily interested in the general principles that govern the behavior of engines, we shall first look at how two of the most common engines work. This will provide us with a vocabulary with which to discuss the operation of engines in general, and will also provide some illustrations of the general principles involved.

The four-stroke cycle gasoline engine used in most automobiles consists of several cyclinders, of the type shown in Figure 18-1. The piston in each cylinder is connected to a piston rod, and all the piston rods are connected to a single crankshaft. The piston rods are connected to the pistons and crankshaft in such a way that when the pistons are moved up and down in the cylinders, the crankshaft is made to rotate. This rotation of the crankshaft is transmitted ultimately to the wheels.

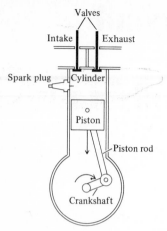

FIGURE 18-1 A four-stroke cycle engine.

In one end of the cylinder there are two holes, called the intake and exhaust ports. The intake and exhaust valves serve to seal these ports shut or to open them at various times in the operation of the engine. There is, in addition, a spark plug, which produces an electric spark to ignite the fuel.

This gasoline is broken up into tiny drops and mixed with large amounts of air in the *carburetor*. This fuel-air mixture is drawn into the cylinder while the piston is moving down. (See Figure 18-2.) At the proper time, the intake port is closed and the piston moves upward, compressing the mixture. The high-pressure gas is then ignited by an electric spark. The rapid combustion generates a large amount of heat. This raises the temperature and pressure of the residual gas, which then pushes the piston down. On its return stroke, the exhaust port is opened and the piston pushes the remaining burnt gas out.

The typical steam engine consists of a similar arrangement of crankshaft, piston rod, and piston in a cylinder which has intake and exhaust ports. The major difference is that the fuel is burned outside the cylinder. The heat is used to generate steam at a high pressure. The steam is injected into the cylinder and allowed to expand, thus pushing the piston down. On the return, the piston pushes the cooled steam out through the exhaust port.

From a thermodynamic point of view, the most important characteristic of both of these engines is that they serve to transform heat into work when the engine is connected to an external device. In order to discuss this thermodynamic process we must choose a system into which heat flows and which does work on an external agency. We can choose the *steam* in the steam engine as the system. Then we can describe the operation of the system as a *cycle*: The system (the steam) is heated in the boiler to a high temperature and pressure and then is fed into the cylinder, where it expands, causing the piston to move, and then the steam is exhausted from the cylinder. The exhausted steam then starts a new cycle by being reheated in the boiler. From this point of view the important thing about the engine is the steam, which we call the *working substance*. Heat flows into the working substance and the working substance does work through the pistons, rods, and crankshaft on the external agent.

The gasoline engine is more difficult to describe from the thermodynamic point of view because the source of heat and the working substance are com-

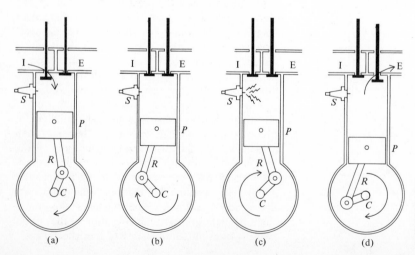

FIGURE 18-2 The strokes in the operation of an internal combustion engine: (a) intake stroke; (b) compression stroke; (c) ignition and power stroke; and (d) exhaust stroke.

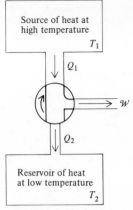

FIGURE 18-3 Schematic diagram of the operation of a heat engine.

bined; they are the gases in the cylinder. But we can construct an abstraction in which we think of the gasoline first as a fuel, or source of heat, and then the burning gas as the working substance into which the heat flows and which then does work on the pistons.

Thus we can think of these engines, and all heat engines, in terms of four systems: a *source* of heat at a high temperature, a *working substance*, a *cooler reservoir* into which waste heat is rejected, and a *mechanical system* on which the working substance does work. See Figure 18-3. The working substance serves to transform the input heat into work and is therefore the essential part of the device. Paradoxically, the pistons, cylinders, rods, and shafts, which are usually thought of as *the* engine, can be considered as forming an auxiliary device for transmitting the work to an external system.

18-2
Thermodynamic efficiency

In the steam engine and the gasoline engine, and in all other types of heat engines, only part of the heat input is converted into work. When the working substance goes through a complete cycle, it returns to its original state; hence its internal energy returns to its original value so that $U_f = U_i$. During this cycle, an amount of heat Q_1 is supplied to the working substance by the source at high temperature, an amount of work W is done by the working substance on some external agency, and an amount of heat Q_2 is delivered to the cooler reservoir as the exhaust. Applying the first law of thermodynamics to this cycle of operations we have

$$Q_1 - Q_2 = W \qquad (18\text{-}1)$$

since $U_f - U_i = 0$. All quantities are expressed in the same units.

The *thermodynamic efficiency*, e, of a heat engine cycle is defined as

$$e = \frac{\text{work done}}{\text{heat added}}$$

or

$$e = \frac{W}{Q_1} \qquad (18\text{-}2)$$

Substituting the value of W from Equation 18-1 yields

$$e = \frac{Q_1 - Q_2}{Q_1}$$

or

$$e = 1 - \frac{Q_2}{Q_1} \qquad (18\text{-}3)$$

for the thermodynamic efficiency of a heat engine cycle. *The efficiency of all known heat engines is less than one.* (Other engines also have less than 100 percent efficiency, but not for directly thermodynamic reasons, and not *in principle*.)

In ordinary language, when we describe the efficiency of an engine we compare the work done by the engine to the input energy, but we usually think of the work done by the output shaft of the machine. In Equation 18-2 we have described the efficiency of the working substance. The work done by the working substance is transmitted by mechanical devices and linkages to the output shaft of the machine. In this process of mechanical energy transfer, additional losses occur because of friction, vibration, and deformations. Thus for any real engine we can think of the efficiency of the machine as a whole as consisting of two parts: the thermodynamic efficiency of energy transformation by the working substance (as given by Equation 18-2) and the efficiency of mechanical energy transfer from the working substance to the output shaft. The overall efficiency of all practical engines is smaller than the efficiency given in Equation 18-2. Basically, the inefficiency of a heat engine lies in the fact that some heat is necessarily discarded in its operation. No one has ever been able to devise a working substance cycle that operates without discarding heat. All attempts to build such a cycle have failed. This history of failure has been translated into a law of nature, which will be discussed in the next section.

18-3
The second law of thermodynamics

Based on all man's experience with heat engines, physics has reached a generalization known as the *second law of thermodynamics*. One way of stating this law is: *It is impossible to construct an engine whose sole effect is the extraction of heat from a reservoir at a single temperature and the conversion of this heat completely into mechanical work.* In this statement, the phrase, "reservoir at a single temperature" means a source of heat that is so large that when heat is transferred from it to the engine, the temperature of the source is unchanged.

This law is now more than simply a generalization from experience. It has become firmly established because conclusions from it have always been verified. The law can be put in quantitative terms, as we shall see, and applied to fields other than heat engines. For example, extensive use has been made of the second law of thermodynamics for the description of many chemical and engineering processes.

In the form stated above, the law states that all engines must reject waste heat. In addition, it states that some processes are not *reversible*. For example, all of Joule's experiments on the mechanical equivalent of heat involved processes that converted all of the input work into heat. If the only limitation on these processes were the first law of thermodynamics, then one might imagine running these processes backward and converting all the heat into work. But the second law of thermodynamics states that this is impossible, or that these processes are irreversible.

Another way of stating the second law is: *It is impossible to construct an engine whose sole effect is the transfer of heat from one body to another at a higher temperature.*

This second way of stating the second law looks different from the first way, but it is possible to prove that they are *logically equivalent*. That is, it is possible to show that if either form is true, then the other form must be true and also to show that if either form is false, then the other form must also be false.

In this second form, the second law states that it is impossible to build a refrigerator that does not require work input from an external agent. It also states that the process of heat transfer from a hot object to a cold object is not a reversible process.

The second law of thermodynamics states that many processes are irreversible. This means, first of all, that these processes cannot be made to run backward spontaneously. For example, heat will not spontaneously flow from cold to hot objects. There is additional significance to the second law. Consider a process such as heat transfer. It is possible to construct a sequence of processes that have the reverse effect; that is, one can build refrigerators to transfer heat from a cold to a hot body. The second law states that work must be done on the refrigerator. This work must come from some external agent. Therefore, the operation of the refrigerator requires that this *external agent be changed*. Thus, although the refrigerator does do the reverse of spontaneous heat transfer, it must also do something else: it must change the system from which the work is derived. Similarly, a heat engine which converts heat into work also does something else: it rejects heat into some other system. The second law says that certain processes are irreversible: the reverse processes do not happen spontaneously and it is impossible to make *only* the reverse process occur, without altering something else in the universe.

A favorite activity of inventors for centuries has been the search for perpetual-motion machines. Most proposed devices violate the first law of thermodynamics. They purport to get something for nothing, and so it is usually easy to see why they will not work. However, there are many proposed perpetual-motion machines that do not violate the first law, but they do violate the second law. Such machines propose to extract energy from large reservoirs of energy such as the atmosphere or the ocean. Unfortunately, the second law requires that these devices must reject heat. In order to extract heat in the first place, the engine must be cooler than the reservoir. Therefore, the rejected heat cannot be exhausted to the same reservoir because the heat will not flow spontaneously. The heat must be rejected to a colder reservoir. The inventors of perpetual-motion machines neglect to describe this colder reservoir.

18-3
The Carnot cycle

The second law of thermodynamics states that it is impossible to build a heat engine that is 100 percent efficient. Accepting this law of nature leads to a search for a description of the most efficient engine possible. Such an engine, the maximally efficient engine, is called a *Carnot engine.*

Sadi Carnot was a physicist who made major contributions to thermodynamics. He published only one work, *Reflections on the Motive Power of Heat and on Proposed Machines to Develop That Power*, in 1824, at the age of 28. That work, which was unnoticed by the scientific community until it was publicized by Lord Kelvin 25 years later, contained the first statement of the second law and a description of the Carnot engine. It is interesting that Carnot discovered the *second* law before the *first* law was accepted by physics, and it is not clear whether Carnot himself believed the first law.

The Carnot engine is only a description of the cycle through which a working substance is taken, ignoring the problem of the efficiency of mechanical energy transfer. In order to describe the engine, we must first be clear about the sources of inefficiency in the process of transformation of heat into work.

During the operation of an engine, the working substance goes through a cycle of processes. If the working substance is a fluid it is compressed, or expanded, and gains heat or loses heat, but it is periodically returned to its initial state. These individual processes may be either reversible or irreversible. Since the working substance returns to its initial state, one half of the cycle must be the reverse of the other half, as far as the working substance is concerned. Therefore, if all of the processes are reversible, the cycle may be performed without changing anything in the rest of the universe; if, on the other hand, *any part* of the cycle is irreversible, the rest of the universe must be changed by the cycle of the working substance. This change will require energy—energy that could have been used for work. Thus in order to maximize the efficiency of the engine, *all processes of the working substance must be reversible.*

As a practical matter no real processes are reversible. In order for a process to be reversible, the system must always be in equilibrium. It is difficult to imagine how a system that is in equilibrium can change its state and remain in equilibrium while it is changing. One can imagine a system that is *almost* in equilibrium and changing slowly. The closer it is to equilibrium, the slower the change. Similarly we can think of systems undergoing almost reversible processes, where the less the deviation from reversibility, the slower the

process. In this way, we can conceive of reversible processes as an abstraction, although real processes can be nearly reversible. Thus, the Carnot engine is an idealization, in the same sense that the ideal gas is an idealization of the behavior of real gases.

In the operation of the engine (see Figure 18-3) we have to deal with at least three processes: heat transfer into the working substance, the performance of work by the working substance, and the rejection of heat by the working substance. We know that heat transfer between two systems is an irreversible process, but if the temperature difference between the two systems is made smaller and smaller, the process becomes less and less irreversible. Thus, in order to build an engine in which the input heat is transferred reversibly, the engine must be maintained at a temperature that is infinitesimally smaller than the temperature of the source of heat. Similarly, when the heat is rejected in the exhaust, the engine should be maintained at a temperature that is infinitesimally higher than the system into which the heat is exhausted.

Thus, we are led to think of the engine operating between two temperatures as shown in Figure 18-3. There exists a reservoir of heat at the higher temperature. Heat flows from this reservoir into the engine, while the engine is maintained at a temperature that is almost equal to the temperature of the reservoir. At some other point in the cycle, heat flows out of the engine into the lower temperature reservoir, while the engine is maintained at a temperature that is almost equal to the temperature of the cold reservoir. Thus we know that during the operation of a Carnot engine at least two processes must occur: heat transfer into the engine while the engine is kept at constant high temperature, and heat transfer out of the engine while the engine is kept at constant low temperature. A Carnot cycle must contain two *isothermal processes.*

At some point in the cycle the temperature of the engine must be changed from the high temperature to the low temperature, and at another point the temperature must be increased from the low temperature to the high temperature. During these temperature changes, the temperature of the engine is distinctly different from the temperature of either the hot reservoir or the cold reservoir. Should heat flow while the temperature is changing, the process would be irreversible. Therefore these temperature changes take place without the flow of heat—that is, *adiabatically.*

A Carnot engine, then, is an engine in which the working substance undergoes a *Carnot cycle*. A Carnot cycle consists of the following four reversible processes:

1. Heat Q_1 is transferred into the engine isothermally at temperature T_1.
2. The temperature of the engine is decreased adiabatically from T_1 to T_2.
3. Heat Q_2 is rejected from the engine isothermally at temperature T_2.
4. The temperature of the engine is raised adiabatically to temperature T_1, and the cycle starts over again.

During one complete cycle an amount of work, W, is done by the engine. Note that work may be done during the isothermal processes as well as the adiabatic processes, and that during part of the process, negative work may be done.

With this description of the Carnot cycle, Carnot was able to prove that the efficiency of a Carnot engine depends only on the temperatures T_1 and T_2, and does not depend at all on the working substance. A Carnot engine (which it must be remembered is an idealization) can be thought of as operating with any material for a working substance. All Carnot engines operating between the same temperatures will have the same efficiency. In addition, Carnot was able to prove that no engine operating between two temperatures is more efficient than a Carnot engine operating between the same two temperatures (see Problems 21 and 22). A Carnot engine is the most efficient engine possible; or rather, all *real* engines must be less efficient than Carnot engines operating between the same two temperatures.

Since we are at liberty to choose any substance for a working substance for a Carnot engine, let us choose an ideal gas, since its equation of state is so well known. For any Carnot engine, the thermodynamic efficiency is given by

$$e = \frac{W}{Q_1} \tag{18-2}$$

or

$$e = 1 - \frac{Q_2}{Q_1} \tag{18-3}$$

For a Carnot engine using an ideal gas as a working substance, we can show, from the results of Chapter 16, that

$$\frac{Q_1}{Q_2} = \frac{T_1}{T_2} \tag{18-4}$$

or

$$e = 1 - \frac{T_2}{T_1} \tag{18-5}$$

Equation 18-5 gives the efficiency of any Carnot engine operating between temperatures T_1 and T_2 and therefore gives the upper limit to the efficiency of real engines operating between those temperatures.

Illustrative example

Find the maximum efficiency of a steam engine that extracts heat from a boiler operating at 100°C and exhausts into air at room temperature, 20°C.
  Here

$$T_1 = 373 \text{ K}$$
$$T_2 = 293 \text{ K}$$

So the efficiency of a Carnot engine would be

$$e = 1 - \frac{293}{373}$$
$$= 0.22 = 22 \text{ percent}$$

Thus no such steam engine could possibly be more than 22 percent efficient. It is clear that in order to increase the theoretical efficiency, it is necessary to increase T_1 and/or decrease T_2. For that reason steam engines are usually operated with boilers under high pressure, which makes the boiling point of water high, raising T_1. Similarly, many engines use cold water for the exhaust reservoir, reducing T_2.

18-5
Absolute thermodynamic temperature scale

The efficiency of a Carnot engine is related to the Kelvin temperatures of the reservoirs by Equation 18-4. As we have seen, the Kelvin temperature scale is defined in terms of a gas thermometer and so has meaning only as long as gases exist. On the other hand, the efficiency of a Carnot engine depends only on temperature, and a Carnot engine may have any substance for a working substance. Therefore, we can use Equation 18-5 to define a temperature scale that is identical to the Kelvin scale, but also has meaning for temperatures at which no gas exists. That is, we can define a temperature T_2 in terms of a fixed point T_1, and the efficiency e of a Carnot engine operating between T_1 and T_2 as

$$T_2 = T_1(1 - e) \qquad (18\text{-}6)$$

The temperature scale defined in this way is called the *absolute thermodynamic temperature scale*.

On this temperature scale we can give meaning to

$$T = 0$$

If a Carnot engine operates between a hot reservoir at a finite temperature and a cold reservoir which is at absolute zero, the efficiency of the engine will be 1, or 100 percent. For no other temperature will the engine have 100 percent efficiency. As we have said earlier, temperatures close to absolute zero have been achieved, but absolute zero is unattainable.

18-6
Entropy

The idea of entropy, as well as the term itself, was invented by Clausius, who chose the word *entropy* partly because of its similarity to the word *energy*. We will describe entropy in a formal way by considering the reversible transfer of heat, Q, from one system to another. In order for the transfer to be reversible both systems must be at essentially the same temperature, T. We can then define the increase in entropy, ΔS, of the system into which the heat flows by

$$\Delta S = \frac{Q}{T} \qquad (18\text{-}7)$$

The entropy of the system from which the heat flows will decrease by the same amount. If we have a system in some equilibrium state we can bring it to some other equilibrium state by adding heat to it in a series of infinitesimal, reversible heat transfers, thus moving the system along a reversible path. There are many reversible paths by which a system may be brought from a particular equilibrium state to another equilibrium state. As was said earlier, the amount of heat transferred will depend on which path the system traverses. It turns out, however, that the total entropy change is the same for all reversible paths connecting two specific equilibrium states. In other words, when a system is in one equilibrium state and then is brought to another equilibrium state by any reversible path,

its entropy changes by a definite amount. That means that we can think of each equilibrium state as having a specific amount of entropy. Since all of the definitions have involved *changes* in entropy, we have no way of describing which state has zero entropy, just as we had no way of describing the state of zero potential energy. Therefore, for any particular system we can arbitrarily choose one specific equilibrium state to be the state of zero entropy and then use Equation 18-7 to find the entropy of any other state.

We can now find the total entropy change for two interesting processes. In the operation of a Carnot engine there are three systems that we have to consider: the hot reservoir, the cold reservoir, and the working substance. In one cycle the working substance returns to its original equilibrium state. Therefore the entropy of the working substance does not change in one cycle:

$$\Delta S_{WS} = 0 \qquad (18\text{-}8)$$

The hot reservoir transfers heat, Q_1, reversibly at temperature T_1, and so the entropy change of the hot reservoir ΔS_H is given by

$$\Delta S_H = -\frac{Q_1}{T_1} \qquad (18\text{-}9)$$

The cold reservoir absorbs heat Q_2 at temperature T_2 and so the entropy change of the cold reservoir, ΔS_C, is given by

$$\Delta S_C = \frac{Q_2}{T_2} \qquad (18\text{-}10)$$

The total entropy change of the universe is given by the sum of these:

$$\Delta S = \Delta S_{WS} + \Delta S_H + \Delta S_C$$
$$= \frac{Q_2}{T_2} - \frac{Q_1}{T_1} \qquad (18\text{-}11)$$

From Equation 18-4 we have

$$\frac{Q_1}{Q_2} = \frac{T_1}{T_2}$$

or

$$\frac{Q_1}{T_1} = \frac{Q_2}{T_2}$$

and inserting this in Equation 18-11 we have

$$\Delta S = 0 \qquad (18\text{-}12)$$

In the operation of the Carnot engine the total

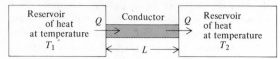

FIGURE 18-4 Conduction of heat.

entropy change of the universe is equal to zero. In fact, the total entropy change of the universe is equal to zero in any reversible process.

In the process of heat conduction, which is shown schematically in Figure 18-4, there are also three systems: the hot reservoir, the cold reservoir, and the metal bar. After the heat has been flowing for a while, each point of the bar is in equilibrium, and so the entropy of the bar is constant. Thus, the change in entropy of the bar, ΔS_B, is

$$\Delta S_B = 0 \qquad (18\text{-}13)$$

The hot reservoir loses heat Q at temperature T_1 and so

$$\Delta S_H = -\frac{Q}{T_1} \qquad (18\text{-}14)$$

while the cold reservoir gains the same amount of heat at the temperature T_2, or

$$\Delta S_C = \frac{Q}{T_2} \qquad (18\text{-}15)$$

The total entropy change of the universe is given by the sum, or

$$\begin{aligned} \Delta S &= \Delta S_B + \Delta S_H + \Delta S_C \\ &= \frac{Q}{T_2} - \frac{Q}{T_1} \\ &= Q\left(\frac{1}{T_2} - \frac{1}{T_1}\right) \end{aligned} \qquad (18\text{-}16)$$

Since the temperature T_2 is less than T_1, this total entropy change is a positive number. Thus, in the process of heat conduction the total entropy change of the universe is positive; another way of stating this is that the entropy of the universe is increased by this irreversible process. In general the entropy of the universe is increased in any irreversible process. Since all real processes are irreversible, this means that the total entropy of the universe is constantly increasing. This, combined with the statement that the entropy of the universe is unchanged by reversible processes, is another form of the second law of thermodynamics.

18-7
The drive to thermal equilibrium

One form of the second law of thermodynamics is the statement that all irreversible processes increase the total entropy of the universe, and all reversible processes keep the total entropy of the universe constant. In short,

$$\Delta S_{TOT} \geq 0 \qquad (18\text{-}17)$$

Throughout our discussion of thermodynamics we have dealt with systems in equilibrium states. We know that when two systems are brought into contact with each other they will come to equilibrium; in general, both systems will *change* to a new state. The increase in entropy that accompanies all real (irreversible) processes describes the direction of the change in each system. The total entropy of the two systems before they are brought into contact is less than the total entropy of the two systems after they have come to equilibrium with each other. We can then describe the tendency of systems to come to equilibrium by saying that systems change in such a way as to maximize the total entropy.

This tendency to reach equilibrium, or maximize entropy, can be seen in phenomena that do not obviously involve heat. Gases with different pressures tend to a single equilibrium pressure when allowed to mix. Solutions with different concentrations reach a uniform concentration when allowed to mix. Chemical reactions reach equilibrium. In each of these cases it is possible to calculate the total entropy of the systems and show that the processes proceed in the direction that maximizes entropy, when equilibrium is reached.

In each of these processes in which two systems come to equilibrium, each of the systems is described by some property that is different for each of the systems before equilibrium is reached, and is equalized at equilibrium. As examples of such properties, we can list: temperature for objects coming to thermal equilibrium, pressure for two gases that mix, and concentration for two solutions that mix. In each case we can think of the property as a generalized coordinate. The two systems initially have different values of this coordinate, and at equilibrium they have the same value. The initial difference in the values of the "coordinate" is what drives the system to equilibrium. For example, if a dilute solution is brought

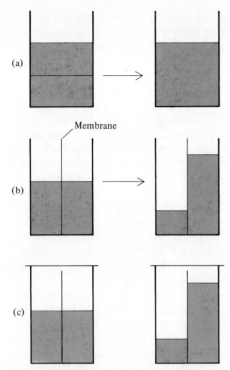

Membrane

FIGURE 18-5 Methods by which equilibrium may be achieved between a dilute and concentrated solution: (a) diffusion; (b) flow of solvent through a semipermeable membrane; and (c) evaporation and condensation of solvent.

into contact with a concentrated solution, the difference in concentration drives the systems toward uniform concentration. The dilute solution tends to become more concentrated and the concentrated solution tends to become more dilute. In general the larger this difference of coordinate is, the faster the process tends toward equilibrium. In order for a reversible process to occur, this difference in coordinates must be infinitesimally small.

We can illustrate this drive to equilibrium by considering two solutions, one dilute and the other concentrated, which are brought into contact in several ways. The heavier solution can be poured into a container and then the lighter solution carefully poured on top of the first. Initially the two solutions will be separated, as in Figure 18-5a. With a lapse of time, the two solutions will *diffuse* into each other and end up as a single solution of uniform concentration, as shown in the figure. Another way is to divide the container into two

halves by means of a *semipermeable membrane*; a semipermeable membrane is a sheet of substance that permits the solvent (usually water) to pass through it, but does not permit the dissolved substance to pass through it. In this arrangement, shown in Figure 18-5b, the difference in concentration is equalized by the flow of solvent through the membrane, from the dilute solution to the concentrated solution. Thus, with time, the level of the solution that was initially concentrated will rise and the level of the initially dilute solution will fall. If the container is closed and divided into two halves by a solid wall that does not reach to the top, as in Figure 18-5c, the equalization of concentration can be achieved eventually by evaporation and recondensation of the solvent. Again, the level will rise in the originally concentrated side, and the level will fall in the dilute side. Note that we can explain these processes without really understanding the mechanism by which a semipermeable membrane works, or how concentration affects evaporation rates. All we really know is that systems tend toward eventual equilibrium and that if any mechanism is available to achieve equilibrium, then systems will use that mechanism to reach equilibrium.

Illustrative example

A 50-gm sample of water at 40°C is mixed with 50 gm of water at 42°C. Find the entropy change.

Actually, this process is irreversible, but we can calculate the entropy change by assuming that each sample of water changed from its initial state to its final state reversibly. The final state temperature of the water is 41°C, so we calculate the entropy change when 50 gm of water is warmed reversibly from 40°C to 41°C and the entropy change when 50 gm of water is cooled from 42°C to 41°C.

When heat ΔQ flows into water at temperature T, the entropy change of the water, ΔS, is given by

$$\Delta S = \frac{\Delta Q}{T}$$

In this problem the temperature of the water is continually changing as the water warms from 40°C to 41°C. Therefore, to do this problem exactly we would have to use the methods of calculus. However, because the temperature *change* of the water is small, we can calculate the entropy change by a reasonable approximation. Assume that *all* of the heat flows into the water while it is at its average temperature. Thus, for the water that is cooled,

$$\Delta S = -\frac{\Delta Q}{T}$$

$$= -\frac{\Delta Q}{314.5 \text{ K}}$$

and, for the water that is warmed,

$$\Delta S = +\frac{\Delta Q}{313.5 \text{ K}}$$

The total entropy change, ΔS, is

$$\Delta S = \Delta Q \left(\frac{1}{313.5} - \frac{1}{314.5} \right)$$

Now since the specific heat of water is equal to c, where

$$c = 1 \frac{\text{cal}}{\text{gm K}}$$

$$= 4.2 \frac{\text{joules}}{\text{gm K}}$$

we have for the total entropy change:

$$\Delta S = mc \left(\frac{1}{313.5} - \frac{1}{314.5} \right)$$

$$= 50(4.2) \left(\frac{1}{313.5} \right) \left(\frac{1}{314.5} \right)$$

$$\approx 0.002 \frac{\text{joule}}{\text{K}}$$

18-8
The direction of time

Among the most elementary ideas that all human beings share is the knowledge that time runs in a single direction. We all know that today is inherently different from both yesterday and tomorrow. We all know that plants and animals die, that mountains erode and houses crumble. We know that time cannot run backward, because then impossible or miraculous things would happen: the dead would rise and grow young, and eroded, smooth rocks would become angular and sharp.

Imagine making a motion picture of natural processes that runs for enough time to see these normal aging effects. This movie could be shown running either forward or backward through the projector. Run forward, the movie would look normal. Run backward, the movie would look impossible, miraculous, or like a work of fiction. Knowing that the film was taken of real things, any person could tell if the film were being run through the projector backward.

On the other hand, imagine making a movie of some of the elementary processes described by Newton's second law or by relativistic mechanics —a two-body collision, for instance. For these elementary processes, a film run backward looks just as normal as the film run forward. This is because neither Newtonian mechanics nor Einsteinian mechanics is capable of distinguishing between time running forward or time running backward. If time runs forward, mechanics predicts that certain processes will occur. If time runs backward, mechanics predicts that the reverse processes will occur. But both the forward and reverse processes are possible processes, which do not seem unreasonable or miraculous.

The primitive idea that time has a fixed direction enters physics through the second law of thermodynamics. All natural processes increase the entropy of the universe. Therefore, time flows in such a direction as to increase the entropy of the universe. If we made a movie in which each frame showed a number equal to the total entropy of the universe in one corner of the screen, we could always tell if the movie were running backward or forward. Run in the forward direction, the numbers would always increase, while run in the backward direction, the numbers would decrease. It is tempting to think that aging and erosion are indications of the increase in entropy.

This connection between entropy and time is interesting because entropy is a macrophysical or thermodynamic idea. Entropy is a concept associated with systems that consist of many particles. That is, physics is able to describe or understand the flow of time only in terms of processes involving many particles.

The development of the second law of thermodynamics occupies a unique place in the history of science, but it illustrates an important characteristic of science. The second law of thermodynamics began as an abstraction about machines, which are man-made devices and the products of technology. A fuller understanding of the law leads to an understanding of some of the most fundamental ideas about nature. It is possible that human beings might have traveled some other route to reach an understanding of the drive to equilibrium and the direction of time, but the fact is that we did not. Historically, this understanding is a direct outgrowth of the previous technology. The technology was developed for many reasons, but no one thought about, or built, steam engines in order to gain an understanding of equilibrium. Nevertheless, the research that went into the mundane enterprise of improving steam engines had the unpredictable, but important, consequence of expanding man's understanding of the universe. This improved understanding has led to other ideas and to other technologies that have almost no connection with engines. This discovery of new, wide-ranging ideas from apparently narrow research appears again and again in science. Relativity began with some technical, limited questions about the speed of light and led to a deeper understanding of space and time. The investigations of an obscure monk (Gregor Mendel) on the color of peas led to an understanding of genetics.

Questions

1. Look up the following heat engines: (a) diesel, (b) Wankel, and (c) Stirling. Identify the working substance, heat source, and the general characteristics of the cycle.

2. Human beings can be characterized as heat engines in that they do work and discard heat. Identify the source of heat.

3. List some engines that are not heat engines.

4. What is a heat reservoir?

5. Describe several processes in which a given amount of work is converted completely into heat.

6. Since an automobile engine wastes heat by discarding it in the exhaust, why not just plug up the exhaust pipe?

7. The Carnot engine operates between two reservoirs at fixed temperatures. Why not use several reservoirs at different temperatures as sources of heat and several reservoirs at different temperatures as sinks?

8. Can you cool the kitchen by opening the door of the refrigerator? If not, devise a way in which the refrigerator can be used as a way of cooling the kitchen.

9. From the discussion of the drive to equilibrium, guess at the effect of dissolved salts on the vapor pressure of a solution. From this deduce the effect on the boiling point.

10. One of the most pervasive problems of modern times is thermal pollution, the dumping of large amounts of heat into the air and particularly into the waters. Discuss the implications of the second law of thermodynamics for this problem.

11. Water is poured into a cup and allowed to stand. Discuss the circumstances under which the entropy of the water increases and those under which it decreases.

12. An inventor claims to have invented a machine that draws in ocean water, circulates the water through a secret device extracting energy, and discards the chilled water into the sea. Would you as a patent examiner reject this device as violating the second law of thermodynamics?

13. The foreman of a factory, having discovered the second law of thermodynamics, proposes to increase the efficiency of the engines in the factory by attaching a refrigerator to the exhaust. Discuss this idea.

14. A drop of ink is placed in a dish of glycerin. (a) Discuss the subsequent behavior of the ink. (b) Will this take place on a time scale measured in seconds, days, or years?

15. Discuss how you can tell if each of the following items is new or old: (a) a wooden house, (b) a stone house, (c) a painting, (d) a document handwritten in Sanskrit.

16. Glass is an amorphous solid that behaves like a very thick liquid. How could you use this fact to decide whether a piece of window glass in an old house was old or a recent replacement?

17. The second law of thermodynamics is sometimes described as predicting the "heat death" of the universe. Discuss the meaning of this phrase.

Problems

1. A heat engine takes in 4000 cal per cycle and does 1800 joules of work per cycle. (a) What is the efficiency of the engine? (b) How much energy is discarded per cycle?

2. A certain steam engine does work at the rate of 75 kilowatts and is 17 percent efficient. How much coal does it have to burn per hour if the combustion of the coal yields 2.5×10^6 cal per pound of coal?

3. What is the efficiency of a Carnot engine operating between 240°C and 24°C?

4. A Carnot engine absorbs 4000 cal per cycle and rejects 1500 cal per cycle. (a) What is the efficiency? (b) If the heat is absorbed from a reservoir at 500°C, what is the temperature of the cold reservoir?

5. A Carnot engine does 2500 joules of work per cycle and rejects 1000 cal of heat to a cold reservoir. What is the efficiency of the engine?

6. A flask containing 1000 gm of water is warmed from 14°C to 15°C. Find the entropy change of the water.

7. A heat reservoir at 200°C gives off 2500 cal of heat. Calculate the entropy change of the reservoir.

8. A human being absorbs 2×10^6 cal per day from food, which is at approximately body temperature (37°C). What is the entropy change of the food?

9. What is the entropy change of 1 gm of ice when it melts at 0°C?

10. If a heat engine is run backward it operates as a refrigerator, extracting heat Q from a cold reservoir while work W is done on the refrigerator. Refrigerators are measured in terms of a quantity called the coefficient of performance, K, where K is given by

$$K = \frac{Q}{W}$$

Find the coefficient of performance of a Carnot refrigerator in terms of the temperatures of the reservoirs.

11. (a) Determine the thermal efficiency of a Carnot engine that operates between temperatures of 350°C and 0°C. If 5000 cal of heat are supplied to the engine, (b) how much work is done and (c) how much heat is rejected to the low-temperature reservoir?

12. A large steam turbine operates between the temperatures of 750°C and 18°C with an efficiency of 35 percent. (a) Find the efficiency of a Carnot engine operating between the same temperatures. (b) How much power in the form of heat must be supplied to this engine in order

that it produce useful work at the rate of 1 million kilowatts? (c) How much power would have to be supplied to produce 1 million kilowatts if the engine were a Carnot engine? (d) Upon combustion, oil yields 10 million calories per ton; one ton of oil occupies 5 barrels. How many barrels of oil does the turbine use per day? (e) How many would a Carnot engine use?

13. The coefficient of performance of a refrigerator is 5.3; this refrigerator is used to remove 8000 cal from a supply of food. (a) If electricity costs 6 cents per kilowatt-hour and the electric motor is assumed to be 90 percent efficient, how much did it cost to cool the food? (b) How much heat is supplied to the surrounding air?

14. The motor in a certain freezer has a power input of 250 watts. If the freezer is maintained at −8°C and the outside air is at 20°C, (a) what is the maximum amount of heat which can be extracted from the freezer in 10 minutes, and (b) what is the shortest time in which this freezer can convert 1000 gm of water at 0°C to ice cubes at 0°C?

15. Some houses are heated with heat pumps. These can be thought of as refrigerators that cool the ground outside the house and warm the inside of the house. If the ground outside the house is at 0°F and the inside of the house is at 70°F, what is the maximum heat delivered for each kilowatt-hour of electricity consumed?

16. In a measurement of specific heat, 100 gm of lead at 30°C is immersed in 200 gm of water at 20°C. Find the total entropy change of the system as it comes to equilibrium.

17. One end of a copper rod is in contact with a heat reservoir at 200°C and the other end with a reservoir at 20°C. Find the change in entropy of the system when 1500 cal are conducted by the rod.

18. If 50 gm of water at 30°C are mixed with 40 gm of alcohol (specific heat 0.6 cal/gm °C) at 26°C, find the total change in entropy.

19. An ideal gas initially at a pressure of 3 atm, volume 10,000 cm³, and temperature of 300 K undergoes the following cycle: It is heated at constant volume to a pressure of 4 atm; then it is heated at constant pressure to a temperature of 600 K; then it is cooled at constant volume to its original pressure; and finally it

is cooled at constant pressure to its original volume. (a) Sketch this cycle on a P-V graph, labeling the end of each of the processes with the numerical values of the pressure, volume and temperature. (b) Find the number of moles of gas. (c) Calculate the heat added during the heating processes, assuming $c_p = 29.3$ joules/mole K and $c_v = 21$ joules/mole K. (d) Calculate the heat removed during the cooling processes. (e) Find the work done. (f) Find the efficiency.

20. M_1 grams of a liquid, of specific heat c_1, initially at temperature T_1 are mixed with M_2 grams of a liquid, of specific heat c_2, initially at temperature T_2. (a) Find the final temperature T in terms of M_1, M_2, T_1, T_2, c_1, and c_2. (b) If the temperature difference $(T_2 - T_1)$ is small, show that the entropy change, ΔS, due to this mixing is given by

$$\Delta S = Q \frac{(T_2 - T_1)}{2T^2}$$

where Q is the heat transferred from one liquid to the other in the mixing.

21. Suppose we have two Carnot engines E and E', which have different working substances but which operate between the same two reservoirs. Since they are Carnot engines they can be run forward as engines producing work, or backward as refrigerators absorbing work and transferring energy out of the cold reservoir into the hot reservoir. Assume then that engine E is operating forward, producing work W, and that engine E', operating as a refrigerator, absorbs all this work W. Then assume that the efficiency of E is greater than the efficiency of E'. Prove that this means that the two engines treated together as a single engine transfer heat from the cold reservoir to the hot reservoir without requiring work. This violates the second law. Thus the efficiency of E cannot be greater than that of E'. Complete the proof to show that the efficiency of both Carnot engines must be the same.

22. By an argument similar to that of Problem 21, show that the efficiency of an irreversible engine is less than or equal to that of a reversible engine.

statistical physics

chapter 19

19-1
The kinetic theory of gases

In the preceding chapters we discussed thermal phenomena from the macrophysical, or thermodynamic, point of view, which ignores the internal structure of matter. In this chapter, we will discuss some of these same phenomena from the microphysical point of view, in which we take into account the internal structure.

The simplest illustration of the methods of microphysics is the *kinetic theory of gases*. In this theory, we construct a very simple model. The model is so simple that we know that it ignores most of the real structure of real gases. Yet the behavior of this model is the same as that of the ideal gas, and the ideal gas is a good approximation to real gases. Therefore, most of the behavior of real gases can be explained by the kinetic model. The real structure of gases influences the details, or the deviations from the ideal gas law.

The basic assumption in the model is that gases consist of *molecules*, which may be considered as *point masses*. That is, the model ignores the size of real molecules and the internal structure of the molecules. It is assumed that there are very *large numbers* of molecules in any sample and that these molecules are in *random* motion. This permits us to calculate *average* values of physical properties and treat them as truly representative. It is assumed that Newton's laws of motion govern the behavior of the molecules and that all collisions are completely *elastic*. Finally, it is assumed that there are no forces acting on the molecules except during collisions.

One striking result of these assumptions is that they predict that *no* collisions among molecules will occur. We can calculate the number of collisions by first looking at one molecule, chosen at random, and then calculating the *probability* that this molecule will experience a collision. We can then find the average probability that a collision will occur. This average probability multiplied by the number of molecules yields the number of collisions. Let us first assume that the molecules are not points but have some size. Then if we look at one molecule and follow it through the gas, it will sweep out some volume, as shown in Figure 19-1. In order for a collision to occur, some part of the target molecule must project into this swept-out volume (at the right time). The larger the cross-sectional area of the molecule, the more likely is the occurrence of a collision; the probability of a collision is proportional to the cross-sectional area of the molecule. But in the ideal gas model, we assume that the molecules have no size at all. Therefore, we have assumed that no collisions will occur among molecules. Thus, the only collisions that occur are between the molecules and the walls of the container. For simplicity we will assume that the container is a cube of edge d and volume $V = d^3$. Let us further assume that there are N molecules, each of mass m. Therefore, the total mass M of gas is given by

$$M = mN \qquad (19\text{-}1)$$

In this theory, the pressure exerted by the gas on the walls of the container is due to the impact of the molecules on the walls, and at equilibrium, this is the same as the pressure throughout the gas. To calculate this pressure, let us assume that the impact of a molecule with the wall is an elastic impact; that is, if a molecule is approaching the wall with a velocity v and momentum mv, then it will leave the wall with a velocity $-v$ and momentum $-mv$. The change in momentum of the molecule produced by this impact will thus be $-2mv$. To determine the pressure on the walls of the container, let us first calculate the force exerted by the molecules on one of the six faces of the cube, say the face $BCDE$ of Figure 19-2, and then divide by its area.

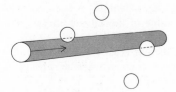

FIGURE 19-1 Volume swept out by a molecule as it moves through a gas

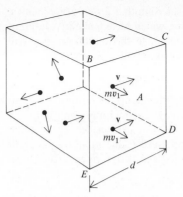

FIGURE 19-2 Molecules with equal velocity components near face *BCED* of the cube.

Let us consider those molecules that at some instant are very close to this face. Only those molecules whose velocities have components perpendicular to this face, and directed toward it, will strike it and rebound. Suppose we consider a small number of molecules that have the same value v_1 for this velocity component. The number of these molecules that will strike this face during a small time interval t will be one half of the number contained in a small volume Al, where A is equal to the area of the face of the cube and $l = v_1 t$; the other half having this velocity component v_1 are moving away from the wall. If n_1 represents the number of molecules per unit volume that have this velocity component v_1, then the number striking this face of the cube in time t will be

$$\frac{n_1}{2} A v_1 t$$

Since each such molecule will have its momentum changed by $-2mv_1$ as a result of this impact, the impulse imparted to the wall will be equal and opposite to it, or $+2mv_1$. The impulse $F_1 t$ on the wall produced by these collisions in time t will then be

$$F_1 t = \frac{n_1}{2} A v_1 t \times 2mv_1$$

from which

$$F_1 = A n_1 m v_1{}^2 \qquad (19\text{-}2)$$

The pressure on the wall produced by the impact of these molecules is

$$p_1 = \frac{F_1}{A} = n_1 m v_1{}^2 \qquad (19\text{-}3)$$

We can now consider another group of molecules, n_2 per unit volume, which have a slightly different velocity component v_2 in this direction; they will produce an additional pressure p_2, given by

$$p_2 = n_2 m v_2{}^2$$

In this way, we can break up the gas into different groups of molecules, each group contributing a similar term to the pressure on this face of the cube. The total pressure P due to all the different groups of molecules will therefore be of the form

$$P = n_1 m v_1{}^2 + n_2 m v_2{}^2 + n_3 m v_3{}^2 + \cdots \quad (19\text{-}4)$$

The symbol $(+ \cdot \cdot \cdot)$ means that we have to add terms similar to those that precede it to include all the different groups of molecules that have velocity components in a direction perpendicular to this face of the cube.

Equation 19-4 can be simplified by introducing a new term called *the average of the squares of the components of the velocities of all the molecules moving perpendicular to face A* and defined by the equation

$$\overline{v_A{}^2} = \frac{n_1 v_1{}^2 + n_2 v_2{}^2 + n_3 v_3{}^2 + \cdots}{n} \quad (19\text{-}5)$$

in which n represents the total number of molecules per unit volume. Substituting this value of $\overline{v_A{}^2}$ in Equation (19-4), we get

$$P = nm\overline{v_A{}^2} \quad (19\text{-}6)$$

There will be a similar expression for the pressure on each of the six faces of the cube, except that the factor $\overline{v_A{}^2}$ will be replaced by the appropriate average of the squares of the components of the velocities of the molecules for that particular face.

The velocity v of any one molecule may be in any direction; it can be resolved into three mutually perpendicular components v_x, v_y, v_z. The magnitude of v in terms of the magnitudes of these components is given by

$$v^2 = v_x{}^2 + v_y{}^2 + v_z{}^2 \quad (19\text{-}7)$$

There will be a similar equation for the square of the velocity of each molecule of the gas in terms of the squares of its three mutually perpendicular components. If we add the squares of the component velocities in the x direction and divide this sum by the total number of molecules, we will get the

average value of the square of this velocity component; it will be represented by $\overline{v_x{}^2}$. Similarly, $\overline{v_y{}^2}$ and $\overline{v_z{}^2}$ will represent the average squares of the velocities in the y and z directions, respectively. By adding these average squares of the three velocity components, we get

$$\overline{v^2} = \overline{v_x{}^2} + \overline{v_y{}^2} + \overline{v_z{}^2}$$

where $\overline{v^2}$ is the average of the squares of the velocities of all the molecules. Since the velocities of the molecules have all possible directions, the average value of the squares of the velocity in any one direction should be the same as in any other direction, or

$$\overline{v_x{}^2} = \overline{v_y{}^2} = \overline{v_z{}^2}$$

so that

$$\overline{v^2} = 3\overline{v_x{}^2} \quad (19\text{-}8)$$

If we take the x direction as perpendicular to the face A, we can write

$$\overline{v^2} = 3\overline{v_A{}^2} \quad (19\text{-}9)$$

so that Equation 19-6 becomes

$$P = \tfrac{1}{3}nm\overline{v^2} \quad (19\text{-}10)$$

Now

$$nm = \rho \quad (19\text{-}11)$$

where ρ is the density of the gas; hence Equation 19-10 may be written as

$$P = \tfrac{1}{3}\rho\overline{v^2} \quad (19\text{-}12)$$

The expression for the pressure of an ideal gas may be written in a more instructive way by noting that the kinetic energy of 1 molecule is $\tfrac{1}{2}mv^2$, and that $\tfrac{1}{2}nm\overline{v^2}$ can be considered as the total kinetic energy of translation of all the molecules in a unit volume under these conditions. Setting

$$\mathcal{E}_1 = \tfrac{1}{2}nm\overline{v^2} \quad (19\text{-}13)$$

the expression for the pressure can be written

$$P = \tfrac{2}{3}\mathcal{E}_1 \quad (19\text{-}14)$$

that is, the pressure of an ideal gas is equal to two thirds of the total value of the kinetic energy per unit volume of the molecules of the gas.

If the volume of the gas is V and the total number of molecules is N, then the number of molecules per unit volume is

$$n = \frac{N}{V} \qquad (19\text{-}15)$$

so the expression for the pressure of a gas can be written as

$$P = \tfrac{1}{3} \frac{N}{V} m \overline{v^2} \qquad (19\text{-}16)$$

from which

$$PV = \tfrac{2}{3} \cdot \tfrac{1}{2} N m \overline{v^2} \qquad (19\text{-}17)$$

From a comparison of Equation 19-17 with the ideal gas law as given by Equation 16-4, it follows that the average value of the kinetic energy of the molecules of an ideal gas is proportional to its absolute temperature T. If we set

$$\tfrac{1}{2} m \overline{v^2} = \tfrac{3}{2} kT \qquad (19\text{-}18)$$

where k is a constant of proportionality, then Equation 19-17 may be written as

$$PV = NkT = cT \qquad (19\text{-}19)$$

in which $c = Nk$. Thus, on the basis of the assumptions stated in this analysis, we have been able to derive the general equation of state for an ideal gas.

It is important to emphasize that Equation 19-18, which relates the temperature to the average kinetic energy, is valid only for the ideal gas. It is not a general statement, and it is not a definition.

We know that real molecules have size and structure, and that they do exert forces on each other and on the walls of the container, even when they are not in contact. Nevertheless, this model gives a good approximation of the behavior of gases. Therefore we know that these real properties of real molecules play a small role in determining the behavior of gases. This small role may, however, be very important in some circumstances. For example, we know that gases can be liquefied under the right conditions. The kinetic theory does not describe this phenomenon, because it ignores the forces between molecules. Similarly, the kinetic theory as we have described it ignores the transfer of energy to the molecules from the walls of the container, because it assumes that all collisions are elastic. Thus, it cannot describe the process of heating the gas. It is possible to extend the theory to deal with many such phenomena, but this would in general take us outside the scope of this book. We will, in subsequent sections, deal with a few

generalizations from the theory and some consequences of this simple model.

19-2
Avogadro's hypothesis

The success that we had had so far with the kinetic theory of gases suggests other extensions. Suppose that we consider two different ideal gases, each occupying the same volume V at the same temperature T and exerting the same pressure P. Let one of the gases contain N_1 molecules, each of mass m_1, and let the other gas contain N_2 molecules of mass m_2. We can then write Equation 19-16 for the first gas as

$$PV = \tfrac{1}{3} N_1 m_1 \overline{v_1^2} \qquad (19\text{-}20)$$

and for the second gas

$$PV = \tfrac{1}{3} N_2 m_2 \overline{v_2^2} \qquad (19\text{-}21)$$

Since the product PV is the same for each gas, we may write

$$N_1 m_1 \overline{v_1^2} = N_2 m_2 \overline{v_2^2} \qquad (19\text{-}22)$$

Since the temperatures of the two gases are the same, we can write

$$\tfrac{3}{2} kT = \tfrac{1}{2} m_1 \overline{v_1^2} = \tfrac{1}{2} m_2 \overline{v_2^2}$$

or

$$m_1 \overline{v_1^2} = m_2 \overline{v_2^2} \qquad (19\text{-}23)$$

Dividing Equation 19-22 by Equation 19-23 yields

$$N_1 = N_2 \qquad (19\text{-}24)$$

That is, two different gases that occupy equal volumes at the same temperature and pressure have equal numbers of molecules.

This result was first predicted by Avogadro in 1811 from the known properties of gases. *Avogadro's hypothesis stated that all gases occupying equal volumes at the same temperature and pressure contain equal numbers of molecules.* The molar volume, that is, 22.4 liters at 0°C and 76 cm mercury pressure, is the volume occupied by a gram molecular weight, or mole, of gas. Hence each gram molecular weight of a substance contains exactly the same number of molecules, which we shall denote by N_0. The accepted value of this number, determined experimentally, is

$N_0 = 6.022 \times 10^{23}$ molecules per gram
molecular weight

This number N_0 is called the *Avogadro number*.

If we are dealing with 1 mole of an ideal gas at temperature T and pressure P, then Equation 19-19 can be rewritten as

$$PV_1 = N_0 kT$$

where V_1 is the volume of this mole of gas and $N = N_0$. Equation 16-4 can be rewritten as

$$PV_1 = RT$$

since, for this case, $n = 1$. A comparison of these two equations shows that

$$N_0 k = R$$

from which

$$k = \frac{R}{N_0} \qquad (19\text{-}25)$$

Since R is the universal constant for a mole of a gas and N_0 is the number of molecules per mole, k then represents the universal constant per molecule. It is usually called the *Boltzmann constant*; its value can be obtained from the known values of R and N_0, and is

$$k = 1.38 \times 10^{-16} \frac{\text{erg}}{\text{K}}$$

19-3
Equipartition of energy

In the simple point-mass model of gases we identified the average kinetic energy of the molecules as proportional to the temperature of the gas. The kinetic energy of Equation 19-18 is only that kinetic energy due to the *random* motion of the molecules within the container. The kinetic energy that arises from the motion of the gas as a whole is *ordered*, and does not contribute to the temperature.

When we attempt to extend the kinetic theory to take into account the properties of real molecules, we run into a similar problem. Real molecules have structure and size, and therefore may have energy in addition to kinetic energy of the molecule as a whole. The molecule may spin and have energy of rotation of the form $\frac{1}{2}I\omega^2$. A molecule with structure may have internal vibrations, and therefore have internal kinetic energy and internal potential energy. We are faced with the problem of relating this kind of energy, which is internal to a molecule, to the temperature.

In order to discuss this question we must first discuss an idea called the *number of degrees of freedom* of a system.

In order to describe the position of a point in space, we must supply three numbers. These may be the Cartesian coordinates, x, y, and z, as shown in Figure 19-3a, or some other set of coordinates. In order to describe the location and orientation of a rigid body we must supply *six* numbers, as shown in Figure 19-3b. As shown in that figure, we can locate the end of a hammer handle by giving three numbers (x, y, and z) and then locate the direction of the handle by giving two more numbers, the angles ϕ and θ of the figure. Finally, we can locate the orientation of the hammerhead by giving the angle ψ. If the handle were made of rubber and could stretch, then we would require a seventh number to describe the length of the handle. The number of quantities needed to completely specify the position and orientation of an object is called the *number of degrees of freedom* of that object.

Each molecule in a gas can be described by the number of degrees of freedom of that molecule. Point molecules or rigid spheres have three degrees of freedom. A dumbbell-shaped molecule (two spheres connected by a rod) is described by five degrees of freedom. Complex, rigid molecules have six degrees of freedom and nonrigid molecules have more than six degrees of freedom.

In thermodynamics we dealt with systems that actually consist of many particles. For such a system we can find the total number of degrees of freedom of the *system* by adding the number of degrees of freedom for each particle in the system. For a gas consisting of N identical point molecules, the total number of degrees of freedom, F, is given by

$$F = 3N$$

whereas for a gas consisting of N complex rigid molecules, F is given by

$$F = 6N$$

The principle of equipartition of energy asserts that *for a system in equilibrium, each degree of freedom has the same average kinetic energy and that this average kinetic energy per degree of freedom is proportional to the temperature.* This principle clearly applies to the simple point-mass model we considered in Section 19-1. For that model, each molecule has three degrees of freedom

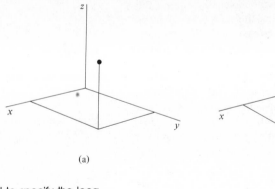

FIGURE 19-3 Coordinates needed to specify the location and orientation of a body: (a) for a point, and (b) for a complex rigid body.

and each molecule has an average energy u, given by

$$u = \tfrac{3}{2}kT \qquad (19\text{-}26)$$

and so the average energy per degree of freedom, u_f, is

$$u_f = \tfrac{1}{3}u \qquad (19\text{-}27)$$
$$= \tfrac{1}{2}kT \qquad (19\text{-}28)$$

For a dumbbell-shaped molecule, each degree of freedom would have this same average kinetic energy and so the average total kinetic energy of the molecule, u_t, would be

$$u_t = 5\left(\tfrac{1}{2}kT\right) \qquad (19\text{-}29)$$

since each molecule has five degrees of freedom. Of this total energy, some of it, u_k, would be kinetic energy of translation of the molecule. The rest would be kinetic energy of rotation of the molecule, u_r. Since there are three degrees of freedom associated with translation, we have

$$u_k = \tfrac{3}{2}kT \qquad (19\text{-}30)$$

and, since two degrees of freedom are associated with rotation,

$$u_r = \tfrac{2}{2}kT \qquad (19\text{-}31)$$

Thus, for a gas consisting of such dumbbell-shaped molecules, the total internal kinetic energy U would be

$$U = N\left(\tfrac{5}{2}kT\right) \qquad (19\text{-}32)$$

while the total internal kinetic energy of translation would be

$$U_k = N\left(\tfrac{3}{2}kT\right) \qquad (19\text{-}33)$$

and the total internal kinetic energy of rotation would be

$$U_r = N\left(\tfrac{2}{2}kT\right) \qquad (19\text{-}34)$$

In a gas that consists of a mixture of different kinds of molecules, each degree of freedom has the same average kinetic energy. All the molecules would have the same average kinetic energy of translation. However, complex molecules would have more total energy, since they would have energy associated with additional degrees of freedom.

One of the strange results of quantum mechanics is that the energy associated with internal degrees of freedom is quantized. That is, the kinetic energy of rotation of a molecule must be one of a set of discrete values. It may be $\mathcal{E}_1$, $\mathcal{E}_2$, or $\mathcal{E}_3$, but it may not be any value between $\mathcal{E}_1$ and $\mathcal{E}_2$. Internal energy of vibration is similarly restricted to one of a set of discrete values. For each of these energies corresponding to internal degrees of freedom, there is a lowest value of allowed energy. In the light of this, we can consider what happens to a gas that consists of complex molecules and is in equilibrium at a low temperature. According to the principal of equipartition, each degree of freedom has an *average* kinetic energy equal to $\tfrac{1}{2}kT$. If the temperature is low, this average energy is small compared with the lowest allowable energy of rotation and vibration. Thus, for every molecule with the lowest allowable energy of rotation, there must be many molecules that have no energy of rotation at all in order for the *average* energy per rotational degree of freedom to be $\tfrac{1}{2}kT$. For low-temperature gases, the number of molecules that have rotation or vibration is small, and the lower the temperature, the fewer molecules exhibit this internal motion. Thus, as the temperature drops, the internal motion is "frozen out." For low temperatures the molecules behave as if they had no structure, and the gas behaves more like an ideal gas.

As the temperature rises to the point where $\frac{1}{2}kT$ is not small compared to the lowest allowable energy of rotation or vibration, more and more molecules will be rotating or vibrating and the internal structure of the molecules will be more apparent.

19-4
Specific heat and internal energy

The kinetic theory of gases, in spite of its very simple assumptions, does predict some of the properties of gases. For example, we have shown that the total internal energy of a gas consisting of point-mass particles is proportional to the temperature. Experiments have been performed to determine whether this is the case for real gases. The results have not been very accurate because the experimental difficulties are great. The main conclusion from such experiments is that the dependence of the internal energy of a real gas upon other factors—pressure and volume—is very slight. It is reasonable to assume that the dependence on these other factors arises from the fact that the molecules do have nonzero size and do exert forces on each other when they are not in contact.

Another interesting point can be obtained by applying the first law of thermodynamics to an ideal gas. The statement of the first law is

$$Q = U_f - U_i + W$$

Suppose that a mole of gas is heated at constant volume; in this case,

$$W = P\,\Delta V$$
$$= 0$$

since

$$\Delta V = 0$$

Then

$$Q = C_v\,\Delta T$$

where C_v is the molar heat capacity at constant volume; that is,

$$C_v = c_v M_w \qquad (19\text{-}35)$$

where M_w is the molecular weight of the gas and c_v is its specific heat at constant volume. For this process the first law can be written as

$$C_v\,\Delta T = U_f - U_i$$
$$= \Delta U \qquad (19\text{-}36)$$

Now, for a point-mass molecule, from Equation 19-26, the total energy is

$$u = \tfrac{3}{2}kT$$

One mole of gas contains N_0 molecules, so the total energy of one mole of gas is

$$U = \tfrac{3}{2}N_0 kT \qquad (19\text{-}37)$$

and from Equation 19-25,

$$R = N_0 k$$

so

$$U = \tfrac{3}{2}RT \qquad (19\text{-}38)$$

Comparing Equation 19-36 and 19-38, we find

$$C_v = \tfrac{3}{2}R \qquad (19\text{-}39)$$

Similarly, using Equation 19-29 we find that the molar heat capacity at constant volume for a dumbbell-shaped molecule is

$$C_v = \tfrac{5}{2}R$$

Thus we would expect that the molar heat capacity at constant volume for *monatomic* molecules, would be approximately $\tfrac{3}{2}R$ or approximately 3 cal/mole K, while the molar heat capacity of diatomic molecules would be approximately 5 cal/mole K. Table 19-1 shows that this is nearly the case. Further, the table shows that the number of

Table 19-1 Molar heat capacities of selected gases

substance	chemical symbol	C_p $\left(in\ \dfrac{cal}{mole\ K}\right)$	C_v	$C_p - C_v$
Argon	A	4.97	2.98	1.99
Helium	He	4.97	2.98	1.99
Neon	Ne	4.97	2.98	1.99
Hydrogen	H_2	6.89	4.85	2.04
Nitrogen	N_2	6.96	4.97	1.99
Oxygen	O_2	7.02	5.02	2.00
Carbon dioxide	CO_2	8.83	6.80	2.03
Ammonia	NH_3	8.92	6.80	2.12

degrees of freedom of a complex gas like ammonia is approximately equal to seven.

If we now consider a constant pressure process, one in which the gas is heated and its temperature raised by an amount ΔT, we have

$$Q = C_p \, \Delta T \tag{19-40}$$

where C_p is the molar heat capacity at constant pressure. The first law of thermodynamics can then be written as

$$C_p \, \Delta T = U_f - U_i + P \, \Delta V \tag{19-41}$$

We have shown above that

$$U_f - U_i = C_v \, \Delta T$$

From the equation of state of an ideal gas for one mole, we have

$$PV = RT$$

or

$$P \, \Delta V = R \, \Delta T$$

Substituting in Equation 19-41, we have

$$C_p \, \Delta T = C_v \, \Delta T + R \, \Delta T$$

or

$$C_p - C_v = R$$
$$\approx 2 \, \frac{\text{cal}}{\text{mole K}} \tag{19-42}$$

This result is in good agreement with the data of Table 19-1.

19-5
Diffusion of gases

Everyone is familiar with the fact that if a bottle of ammonia is opened in a room, the odor of the ammonia soon pervades the entire room. The ammonia is said to *diffuse* through the air in the room. The same is true of any other gas or vapor; that is, it will diffuse through the air and ultimately fill the entire volume. To illustrate this process, fill a flask with a gas that is lighter than air, such as fuel gas, and place it over a similar flask containing air, keeping the necks of the bottles open, as shown in Figure 19-4. Even though the gas is lighter than air, it will diffuse into the air in the lower bottle; at the same time, some of the air from the lower bottle will diffuse into the gas above it. This can be

FIGURE 19-4 Diffusion of gases. After a short time, some fuel gas will be found in the lower flask and some air will be found in the upper flask.

tested by separating the bottles and applying a lighted match to each one.

Gases can also diffuse through porous surfaces. Figure 19-5a shows a porous cup C with a glass tube T sealed to its open end and dipping into a jar of water. A glass jar G is inverted over the porous cup. If a stream of helium, for example, is led into the space between the porous cup and the glass jar, the helium will diffuse into the porous cup faster than the air diffuses out of it. The pressure inside the cup is thus increased, and the gas is forced out of it through the tube T. The gas will bubble through the water and go out into the atmosphere. If the stream of helium is shut off and the glass jar G is removed so that only air surrounds the porous cup C, the helium will diffuse out of the porous cup faster than air will diffuse into it. The pressure inside the cup will decrease below atmospheric pressure, as evidenced by the rise of the water in the tube T (see Figure 19-5b).

When a mixture of gases of different densities is allowed to diffuse through a porous surface, the least dense gas will diffuse through at the greatest rate. This can be seen by an examination of Equation 19-23:

$$m_1 \overline{v_1^2} = m_2 \overline{v_2^2} \tag{19-23}$$

If we take the square roots of the average squares of the velocities, we find that their ratio is

$$\frac{v_{1s}}{v_{2s}} = \sqrt{\frac{m_2}{m_1}} \tag{19-43}$$

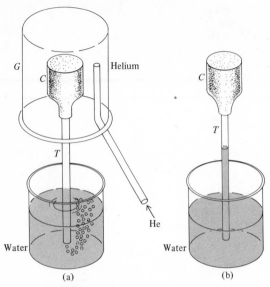

FIGURE 19-5 Diffusion of gases through a porous cup.

v_{1s} and v_{2s} are called the *root mean square* (rms) values of the velocities. These velocities vary inversely as the square roots of the masses of the molecules. This is one form of *Graham's law of diffusion*.

For example, if we have a mixture of hydrogen and helium diffusing through a porous surface, the ratio of the rms values of their velocities is

$$\frac{v_H}{v_{He}} = \sqrt{\frac{m_{He}}{m_H}}$$

The molecular weight of hydrogen is about 2, and the molecular weight of helium is about 4. Since each mole of a substance contains the same number of molecules, the ratio of the molecular weight is equal to the ratio of the masses of the respective molecules. Hence

$$\sqrt{\frac{m_{He}}{m_H}} = \sqrt{\frac{4}{2}} = \sqrt{2}$$

or

$$\frac{v_H}{v_{He}} = \sqrt{2} \approx 1.4$$

The gaseous mixture coming through the porous surface will thus be richer in hydrogen than the original mixture. By passing this new mixture through a second porous surface, the mixture can be further enriched in hydrogen.

An important application of a process of diffusion through a porous surface or a "barrier," as it is sometimes called, is to the enrichment of uranium so that it will contain a larger percentage of the lighter *isotope* than is usually found in natural uranium. Natural uranium consists of 99.3 percent of the isotope of mass 238, symbol ^{238}U, and 0.7 percent of the isotope of mass 235, ^{235}U. A gaseous compound is formed of uranium and fluorine known as uranium hexafluoride; this is a combination of 6 fluorine atoms and 1 uranium atom, UF_6. Since the atomic weight of fluorine is 19.0, the mass m_1 of the compound formed with ^{235}U is

$$m_1 = 235 + 6 \times 19 = 349$$

The mass m_2 of the compound formed with ^{238}U is

$$m_2 = 238 + 6 \times 19 = 352$$

Applying Graham's law to this process, we get

$$\frac{v_1}{v_2} = \sqrt{\frac{352}{349}} = 1.0043$$

for the ratio of the respective rms velocities of the light and heavy molecules. This ratio is sometimes called the separation factor of the diffusion process. Since this separation factor is very small, it is necessary to use a great many such diffusion processes in succession, in order to get a final product which has a large percentage of uranium-235.

19-6
Brownian motion

If very fine particles are put into a liquid and then examined with the aid of a microscope, it will be observed that the particles are executing haphazard motions. These random motions never cease. This phenomenon was first observed by the botanist Robert Brown in 1827. The motion of these particles has since become known as *Brownian motion*. The same type of motion can be observed if very fine particles are suspended in a gas. A simple way of observing Brownian motion is to blow some smoke into a glass cell that has flat sides and then examine it under a microscope, as shown in Figure 19-6.

The Brownian motion of a particle is produced by the bombardment of the particle by the molecules of the fluid in which it is suspended. The particle is not bombarded equally from all sides, so there is a resultant force on it. The particle is accelerated by this resultant force, and, as it moves through the fluid, it also experiences a force due to the viscosity of the fluid opposing the motion. The

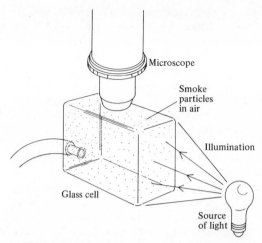

FIGURE 19-6 Method of observing Brownian motion of smoke particles in the air.

distance that a particle moves in any one direction is very small. The direction of its motion is continually changing, because the direction of the resultant force produced by the unequal bombardment by the molecules of the fluid is a matter of chance. Many observers noticed the similarity between the Brownian motion of the suspended particles and the motion that the molecules of an ideal gas are assumed to have.

Einstein, in 1905, developed a theory for the Brownian motion on the assumption that these particles behaved as though they were large-sized molecules of an ideal gas suspended in a fluid. From this theory he developed two important results which could be checked experimentally. One of these is that the particles have a vertical distribution similar to the distribution of air in the atmosphere; that is, that the density of the particles, or the number of particles per cubic centimeter, should be greatest at the bottom and least at the top. Perrin, in 1908, performed a series of experiments using resinous particles of known sizes suspended in a variety of liquids, such as water and alcohol, and counted the number of particles at different heights in the solution (see Figure 19-7). If n_1 is the number of particles per cubic centimeter at any level in the liquid and n_2 is the number of particles per cubic centimeter at a height h above the first position, then Einstein's theory leads to the equation

$$\frac{n_2}{n_1} = 1 - \frac{N_0 mg}{RT}\left(1 - \frac{\rho_0}{\rho}\right)h \qquad (19\text{-}44)$$

where N_0 is the Avogadro number, m is the mass and ρ is the density of the particle in the liquid, ρ_0 is the density of the liquid, T is the absolute temperature of the liquid, R is the universal gas constant, and g is the acceleration of gravity. Since all the quantities in Equation 19-44 except N_0 are known or measurable, Perrin was able to determine the Avogadro number from these experiments.

Another result of Einstein's theory concerned the displacement of a particle in Brownian motion in a given time interval t. Suppose the position of the particle is observed at any instant and plotted on x-y coordinates, and then after the lapse of, say, 30 sec, its position is again observed and plotted; the line joining these two positions is the displacement of the particle in this time. Perrin performed a series of experiments in which he determined the displacements of a particle in known time intervals for as long as it remained in the field of view of the microscope. A typical graph of the displacements of such a particle is shown in Figure 19-8. Einstein developed the equation

$$\overline{x^2} = \frac{2RT}{N_0 K}t \qquad (19\text{-}45)$$

where x is the x component of any displacement in time t, and $\overline{x^2}$ is the average of the squares of these displacements. R is the universal gas constant, T is the absolute temperature of the liquid containing the particles, K is a constant dependent upon the viscosity of the liquid, and N_0 is the Avogadro number. Since all the quantities are known or mea-

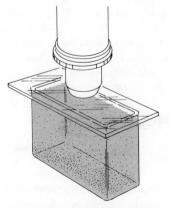

FIGURE 19-7 Method of observing the vertical distribution of particles in a fluid. As the microscope is moved vertically, different layers of the particles come into focus.

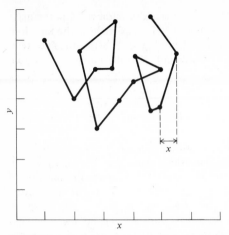

FIGURE 19-8 Trajectory of a particle in Brownian motion. Each line connecting two dots represents the displacement of a particle in a fixed time interval.

surable except N_0, the results of these experiments could then be used to determine the Avogadro number. In one set of experiments he measured 1500 displacements of particles of 0.367×10^{-4} cm radius. His early results led to a value of 6.85×10^{23} for N_0, but, with improved techniques, the values of N_0 determined from Brownian motion experiments are very close to the present accepted value of 6.022×10^{23} molecules per mole.

19-7
The distribution of energy

The kinetic theory permits us to give a mechanical interpretation of some thermodynamic ideas, such as internal energy and temperature, for ideal gases. We can think of the internal energy as the total kinetic energy of the molecules (for the point-mass model) and the temperature as a quantity that is proportional to the average kinetic energy. In fact we can extend these interpretations to other systems; we can think of the internal energy of any system as the sum of the random kinetic energy of translation of the molecules, the rotational and vibrational kinetic energy of the constituents of the molecules, and the total potential energy of the molecules and the molecular constituents. Using the principle of equipartition of energy, we can think of the temperature as a quantity which is proportional to the average energy per degree of freedom. In making this identification of temperature we have to recognize that some degrees of freedom are "frozen out." For example, in most

solids, the atoms are more or less fixed to specific locations, which means that the translational degrees of freedom are effectively frozen out. Thus, for a solid, it is somewhat misleading to think of temperature as proportional to the kinetic energy of the molecules in the same sense as for a gas. Here, the temperature is proportional to the average random energy, which is in part kinetic energy and in part potential energy.

In constructing the model and extension from it, we have dealt with averages. In effect, we have treated the molecules as if they all had the same velocity and internal motion. Actually, some of the molecules have large velocities and some have small velocities, some have large amounts of rotational kinetic energy and some have none. The various quantities of interest about the molecules have some distribution, where each quantity is described by some *distribution function*. The distribution function for velocity is a mathematical equation (or table of numbers) that tells us for each value of velocity what fraction of the molecules has that velocity.

The problem of calculating these distributions is a formidable one. For example, in order to calculate the distribution of velocities in an ideal gas, one would have to first assume some initial distribution and then apply Newton's laws to each of the molecules in order to calculate the change in the distribution. After repeating this process of calculating changes in distribution many times, one would hope that the distribution would stop changing; that the distribution would become some equilibrium distribution. Just as important, one would hope that this equilibrium distribution would be the same for all choices of initial distribution. In any case this problem is so large a mathematical job that no one has ever done it.

The problem of finding other, more tractable procedures for finding the distribution functions has been attacked by some of the most brilliant minds of physics, among them Maxwell, Boltzmann, and Gibbs. (J. Willard Gibbs was probably the most outstanding American scientist of the nineteenth century. His relative obscurity among Americans is probably due to the fact that he was a theoretician, and neither an experimentalist nor an inventor.) The basic idea introduced by these men is that the behavior of the molecules is completely random. This means that, in the absence of other information, all possibilities are equally likely. This is an assumption—an assumption that is equivalent to saying that in the face of the huge numbers of molecules in a system, we will assume total ignorance of the detailed behavior of the molecules. For example, consider standing in a room holding a helium-filled balloon and then breaking the balloon. The statistical hypothesis says that some time after breaking the balloon, when the system has reached equilibrium, it is just as likely for a molecule of helium to be in any one spot in the room as in any other, no matter how far or near these points are from the original location of the balloon. We make this assumption in spite of the fact that we know that the helium molecules all originated in the balloon.

For most systems there is some limited other information that we must use. Generally we know the total number of molecules, which stays constant. In addition, we usually know the total internal energy, which also stays constant. These pieces of information are called the *constraints* of the system.

Thus, we can begin the process of finding the equilibrium distributions by making a list of all of the possibilities that are consistent with the constraints and assume that each of these possibilities is just as likely to occur as any other. Actually, rather than make such a list explicitly, one can apply the methods of statistical analysis to the problem, but the sense of the method is conveyed by thinking about making a real list.

In order to see what is meant by the "possibilities consistent with the constraints," suppose we were to number the molecules. Then one possibility is that molecule number 1 has all the energy, and none of the others has any energy. Another similar possibility is that molecule number 17 has all the energy. Thus one *group* of possibilities has each of the molecules in turn holding all the energy while the others have no energy. Another group of possibilities is that in which two specific molecules share the energy equally. Another very large group of possibilities has the two molecules share the total energy unequally. Obviously, one can go on and on and on constructing great numbers of possibilities, each one of which is assumed to be equally likely. We call each of the possibilities in the list a *microstate* of the system.

But when we examine a gas as a system, we cannot tell one molecule from another. In any macroscopic measurement of a gas, the best we could hope to do is measure how *many* molecules have one specific amount of energy and how many have another specific amount of energy. We call one of these measurable states of the gas as a whole a

macrostate of the system. Any one macrostate could be one of many microstates. That is, if a gas is in some macrostate and the molecules are in one particular microstate, then it is possible to rearrange the molecules into other microstates without altering the macrostate.

Some macrostates correspond to relatively large numbers of microstates and some macrostates correspond to relatively small numbers of microstates. Since each microstate is equally likely, the likelihood of a particular macrostate is proportional to the number of corresponding microstates. In other words, if for a particular state of a gas, there are many ways to rearrange the individual molecules without altering the overall state of the gas, then that state is more likely to exist than a state for which only a few rearrangements are possible. Thus, even though all microstates are equally likely, different states of the gas as a whole have different probabilities.

If a gas starts out in a macrostate of low probability, then the random motions of the molecules will cause the state of the gas to change to one of higher probability. When the system is in the state of maximum probability it will tend to remain either in that state or in states whose probability is not much smaller than that one. Thus the fact that there exist equilibrium states is equivalent to the statement that there are macrostates whose probability is much higher than other nearby states. An equilibrium state is a state of maximum probability.

In this microphysical approach, the laws of thermodynamics have become statements of probability. Consider, for example, breaking a helium-filled balloon inside an evacuated box. From the thermodynamic point of view, we think of the helium gas as a system that is subject to external forces and is characterized by its temperature and pressure. Before the balloon bursts, the system is in equilibrium under the action of the external forces and it has a specific temperature. At the instant the balloon bursts, the system is not in equilibrium and the gas expands to fill the container. During the expansion we can calculate the changes in the thermodynamic variables, such as temperature and entropy, by application of the first and second laws of thermodynamics and the equation of state of the gas. From the microphysical point of view, the gas is a collection of molecules. Before the balloon bursts, this system is in a state of high probability. At the instant the walls burst, the external constraints on the system change, and the system is in a state whose probability is lower than those states through which the gas expands to fill a larger volume. Because of the random motion of the molecules and collisions with each other and the walls, the macrostate of the system changes and the gas expands. When the gas has expanded to fill the container it is in a state that has a higher probability than other possible macrostates. The gas will not spontaneously shrink back to fill only the original volume of the balloon because that state is less likely than the state in which it fills the whole container. It is not impossible that the gas will shrink, but it is very improbable.

In effect, this changes the statement that heat flows from hot objects to cold objects to the statement that heat *probably* flows from hot to cold objects. This surprising, and foolish-sounding, statement becomes less so when one recognizes that in dealing with large numbers of objects, statements about the *probable* behavior of *averages* are not very different from statements made with absolute certainty. Insurance companies with competent actuaries know only the probability of occurrence of certain events, but if they have enough policyholders they can treat these probabilities as certain knowledge. Las Vegas gambling houses know the odds, and, with enough play, they are sure to make a profit. The number of particles involved in a typical thermodynamic system is so large as to be almost beyond real comprehension. (See, for example, Problem 7.) Thus, for thermodynamic systems, probability statements are not very different from statements of absolute certainty. More important, the probability of occurrence of nonequilibrium states is exceedingly small compared to that of the equilibrium states. There is a probability that a pot of water placed on a hot stove will freeze. But that probability is so small that should it happen, most people (even physicists) would assume the existence of a trick of some sort rather than shrug it off as the occurrence of a low-probability event. The average time between recurrences of such low-probability events is much, much longer than the age of the universe. It is in this quantitative sense that we say that probability statements are not very different from statements of absolute certainty.

19-8
Entropy and probability

As we have seen, the equilibrium state of a system is a state of high relative probability. Such a state is one in which it is possible to rearrange the molecules in many different ways without altering the overall state of the system. Consider the difference between a state in which there are only a few ways to rearrange the molecules and one in which there are many ways to rearrange the molecules. One way to describe this difference is to say that the state with few rearrangements is highly *ordered*, whereas the state with many rearrangements is *disordered*. For example, a state in which only one molecule is moving and the others are standing still is very ordered, whereas one in which all of the molecules are moving randomly is very disordered. If all the molecules are moving randomly, then it is possible to make many different small changes in the location or motion of individual molecules without making distinguishable changes in the overall state of the gas. If only one molecule is moving, then it is possible to make only a few changes that do not make marked changes in the state of the gas, since, for example, having two molecules moving is very different from having only one moving. Thus an equilibrium state is a state of high disorder, or of low order.

In an earlier chapter we described the drive to equilibrium in terms of an increase in entropy. Now we see that at the same time, the drive to equilibrium is accompanied by an increase in disorder. We can then describe the entropy of a system as a measure of its disorder. In these terms, the second law of thermodynamics states that irreversible processes increase the disorder of the universe. For example, we can think of the existence of two objects, one hot and the other cold, as introducing some order in the universe, because more of the high-energy molecules are in the hot object then in the cold object. When the objects are placed in contact, heat flows and the two objects come to equilibrium. The somewhat ordered separation of the high-energy molecules disappears and the disorder of the universe has increased. In fact, we can see this natural increase of disorder in many processes. The diffusion of gases, the erosion of mountains, even the death of living things all illustrate the tendency of the disorder of the universe to increase.

The earlier statement that the direction of time is the direction in which the entropy increases can now be understood in terms of the direction in which the disorder of the universe increases.

The relation between entropy and probability was first expressed by Clausius, and can be stated in the following way. *If the total number of micro-*

states in which a system can exist is N, and the number of microstates corresponding to a particular macrostate is n, then the probability of that macrostate existing is p, where

$$p = \frac{n}{N}$$

Then the entropy S of that macrostate is given by

$$S = k \log p \qquad (19\text{-}46)$$

where k is the Boltzmann constant. With this relation we can numerically express the relation between disorder and entropy.

Questions

1. State the fundamental assumptions of the kinetic theory of ideal gases.

2. Describe several ways in which you might expect the behavior of real gas molecules to differ from the assumptions of the kinetic theory.

3. If gas in a cylinder is suddenly compressed by rapidly moving a piston, the temperature of the gas rises. Explain this on the basis of the kinetic theory.

4. If two containers, one containing gas at a high pressure and the other containing gas at low pressure but the same temperature, are connected by a tube, gas will flow from the region of high pressure to the region of low pressure. Discuss this process on the basis of the kinetic theory.

5. Explain why evaporation is a cooling process by using an extension of the kinetic theory.

6. Why cannot a temperature be assigned to a single molecule?

7. The molecules of the air in the upper atmosphere move with speeds corresponding to temperatures of approximately 1000 K, yet it is cold in the upper atmosphere. Explain this.

8. How many degrees of freedom does a particle sliding on a plane have?

9. How many degrees of freedom does a billiard ball on a billiard table have?

10. Explain why the molar specific heat of a gas is constant over a range of temperature and then rises to a new constant value for higher temperatures.

11. In order to separate gases by diffusion, the process cannot be allowed to go on for too long a time. Account for this limitation.

12. Explain the absence of helium in the earth's atmosphere.

13. Provide a reasonable explanation of the fact that the discoverer of Brownian motion was a botanist.

14. Give several examples of natural processes that increase disorder.

15. It has been argued that, since living things are ordered and since they propagate their own kind, life violates the second law. Consider living things as part of an ecological system and argue that this statement is not necessarily true.

16. Which of the following are impossible and which are unlikely? (a) A pot of water on a lighted stove freezes. (b) All the children born in Turkey in the next 10 years will be girls. (c) No one will vote in the next presidential election. (d) Your fingerprints are identical to those of Julius Caesar.

Problems

1. A hydrogen molecule has a diameter of 10 Å and moves with a speed of 1840 m/sec. (a) Find the volume swept out by this molecule in 1 sec. (b) How many molecules of hydrogen would there be in a volume this size at 1 atm at 0°C?

2. A nitrogen molecule at 0°C travels with an average speed of 493 m/sec. Such a nitrogen molecule is traveling between the walls of a room 4 m apart. How many collisions does it make with one of the walls in 10 sec?

3. Determine the average kinetic energy of the molecules of an ideal gas at 20°C.

4. Assuming the mass of an oxygen molecule to be 16 times the mass of a hydrogen molecule, determine the average speed of an oxygen molecule when the temperature of the oxygen is 20°C.

5. Determine the internal energy of a mole of an ideal gas at 200 K.

6. An ideal gas has a pressure of 1 atm, a volume of 1000 cm³, and a temperature of 100 K. How many molecules are there in this sample of gas?

7. The current estimate of the age of the universe is about 10^{10} yr. If you started counting molecules at the beginning and continued counting at the rate of 1 molecule per second, without rest, how many moles would you have counted?

8. If the molecules in 1 gm of water were distributed uniformly over the surface of the earth, how many molecules would be on 1 cm² of the surface?

9. The lowest excited energy state of internal motion of a hydrogen atom is 10.2 eV above the ground state. At approximately what temperature will a large fraction of the hydrogen molecules be in this excited state?

10. The molar heat capacity at constant volume of SO_2 is 7.5 cal/mole K. (a) Explain this value. (b) What would you expect for the molar heat capacity at constant pressure?

11. Find the separation factor for a mixture of oxygen and nitrogen.

12. Ordinary hydrogen contains about 1 part in 7000 of heavy hydrogen, or deuterium, whose atomic weight is 2. Determine the separation factor for ordinary hydrogen.

13. Air contains 78 percent nitrogen (molecular weight 28), 21 percent oxygen (molecular weight 32) and 1 percent argon (molecular weight 40). (a) Calculate the average molecular weight of air. (b) Calculate the density of air at 1 atm and 0°C. (c) Compare your result with the observed average value of 1.29×10^{-3} gm/cm³.

14. Consider an ideal gas at 20°C and 1 atm. The gas consists of N molecules. Assume that the molecules are uniformly spaced and imagine dividing the whole volume V into N cubes so that each cube contains one molecule. (a) Find the numerical value of the volume of one cube. (b) From this, find the average distance between molecules. (c) Find the ratio of this distance to the size of one molecule, which may be taken to be 10^{-8} cm.

15. Use the known value of the Avogadro number to verify that the molar volume at standard conditions is 22.4 liters.

16. Show that the average kinetic energy of a molecule of an ideal gas at room temperature and 1 atm is about 0.04 eV.

17. A rocket is launched with a speed of 24,000 mi/hr. What would the interior temperature of the rocket be if the air molecules in the rocket were moving with this speed as an average random speed?

18. A small dust particle in the air has a mass 10^8 times that of a nitrogen molecule. What is its average speed at standard conditions? Why would you expect the average displacement of this particle to be much less than the product of this speed and the time of observation?

19. Find the temperature at which the average speed of a hydrogen molecule is equal to the escape velocity from the surface of the earth. The temperature high in the upper atmosphere is about 1000 K. Would you expect to find much hydrogen there? (The fact that hydrogen is chemically active explains the fact that it has not all leaked away. Contrast with helium.)

20. Consider a system that consists of three molecules; the energy of each molecule is quantized so that a molecule may have energy $\mathcal{E}_0$, $2\mathcal{E}_0$, $3\mathcal{E}_0$, and so on. The system has total energy $5\mathcal{E}_0$. List all the microstates and find the equilibrium energy distribution.

21. List all the microstates that result when two dice are thrown. A macrostate is defined by the total number shown on the two dice. What macrostate has the maximum probability?

22. Molecules do collide with each other. One can show that if there are N molecules of a gas in a volume V, the molecules will travel a distance λ between collisions, where λ is called the *mean free path* and is given by

$$\lambda = \frac{3}{4\pi^2(N/V)D^2}$$

where D is the diameter of the molecule. Assume that an ideal gas is in a cubical container at 20°C and that the molecules have a diameter of 10^{-8} cm. At what pressure will the mean free path be the same as the size of the container? For pressures lower than this we can ignore the collisions, while for significantly larger pressures the collisions may play an important role.

23. If the probability of a particular kind of event is p_1 and the probability of an unrelated event is p_2, the probability of both occurring is the product of the probabilities $p_1 p_2$. The entropy of a system that consists of two parts is the sum of the entropies of the two parts. Show that these two statements are consistent.

24. (a) Assuming that the probability s that a molecule is in a certain volume of space, V, is proportional to the volume, or

$$s = cV$$

show that the probability P that N molecules will be found in the volume is

$$P = (cV)^N$$

(b) Using this result, show that the change in entropy in the isothermal expansion of an ideal gas from V_1 to V_2 is ΔS, where

$$\Delta S = kN \log \left(\frac{V_2}{V_1} \right)$$

transfer of heat

chapter 20

20-1
Methods of transmitting heat

We have already observed that heat will flow whenever there is a difference in temperature between two bodies or between two parts of the same body. The methods by which heat is transmitted can be classified into three distinct types, known as *conduction*, *convection*, and *radiation*. In any actual case of heat transmission a combination of any two of these methods, and in some cases all three methods, may be operating simultaneously. The principal problem in the transfer of heat is to determine the rate at which heat flows from the source at the higher temperature to the source at the lower temperature.

20-2
Conduction

The method of transferring heat by conduction can be illustrated by means of a long cylindrical copper rod that has one end placed in a gas flame, while the other end is placed in a mixture of ice and water, as shown in Figure 20-1. Heat is conducted through the copper rod from the flame to the ice-water mixture. The amount of heat that is conducted through it in any time interval, assuming that the loss of heat to the surrounding atmosphere may be neglected, can be measured by the amount of ice melted in this time. Any two points along the rod such as B and C a distance L apart are at different temperatures; let us call these temperatures T_1 and T_2, respectively. Heat flows from B to C through the copper rod because of this difference in temperature. The ratio

$$\frac{T_1 - T_2}{L}$$

is called the *temperature gradient* in this region of the conductor. The greater the temperature gradient, the greater the amount of heat that flows through this portion of the rod in any given time interval. The process of conduction may be thought of as the transfer of heat from any one point in the rod to a neighboring point because of the difference in temperature existing between these two points.

The rate at which heat is transferred by conduction is found to depend not only upon the temperature gradient but also upon the cross-sectional area of the rod. Or, stated mathematically,

$$Q \propto A \frac{T_1 - T_2}{L} t$$

or, in the form of an equation,

$$Q = KA \frac{T_1 - T_2}{L} t \qquad (20\text{-}1)$$

in which Q represents the quantity of heat that is transmitted in time t through a rod of cross-sectional area A when the temperature gradient along the length of the rod is $(T_1 - T_2)/L$. The factor K depends upon the nature of the material of the rod and the units used in expressing the other quantities; K is called the *thermal conductivity* of the substance.

The thermal conductivities of metals are generally greater than those of other solids, and silver is the best conductor of all. It is interesting to note that those substances that are good conductors of heat are also good conductors of electricity. Although long rods are suitable for the measurement

of the thermal conductivity of a metal, the measurement of the thermal conductivity of a poor conductor is best made with a very thin slab of material of large cross-sectional area. A knowledge of the thermal conductivity of poor conductors is of great practical importance, since such substances are widely used as heat insulators. The thermal conductivities of the more common substances are given in Table 20-1. Two sets of units are in use for expressing these conductivities. In scientific work K is expressed in terms of the number of calories per second transmitted through an area of 1 cm² when the temperature gradient is 1°C per cm. In engineering work K is expressed in terms of the number of Btu/hr transmitted through an area of 1 ft² when the temperature gradient is 1°F/ft.

Although conduction does take place through liquids and gases, their conductivities are very small, gases being among the poorest conductors. Many insulating materials are constructed so that they trap small quantities of air in small closed spaces and thus make use of the poor conductivity of the air for insulation and at the same time avoid the transfer of heat through the air by convection (see Section 20-3).

Illustrative example

A silver rod of circular cross section has one end immersed in a steam bath and the other end immersed in a mixture of ice and water. The distance between these ends is 6 cm, and the diameter of the rod is 0.3 cm. Calculate the amount of heat that is conducted through the rod in 2 min.

The thermal conductivity K is 0.99 cal/cm sec °C. The cross-sectional area

$$A = \frac{\pi \times (0.3)^2}{4} \text{ cm}^2 = 0.071 \text{ cm}^2$$

$$t = 120 \text{ sec}$$
$$L = 6 \text{ cm}$$

and

$$T_1 - T_2 = 100°C$$

Using Equation 20-1, we get

$$Q = 0.99 \times 0.071 \frac{100}{6} \times 120 \text{ cal}$$

$$Q = 140 \text{ cal}$$

Illustrative example

Calculate the rate at which heat flows through an insulating kaolin brick wall 6 in. thick and 16 ft² in

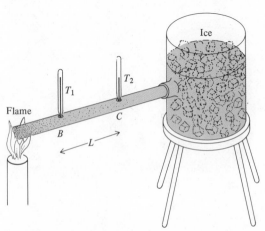

FIGURE 20-1 Method of measuring the temperature gradient along a conductor of heat to determine its conductivity.

Table 20-1 Thermal conductivities

substance	$K \ in \ \dfrac{cal}{cm \ sec \ °C}$	$K \ in \ \dfrac{Btu}{ft \ hr \ °F}$
Metals		
Aluminum	0.49	118
Brass	0.26	63
Copper	0.91	225
Gold	0.71	169
Iron	0.16	39
Lead	0.084	20
Nickel	0.14	34
Platinum	0.17	41
Silver	0.99	242
Tin	0.15	37
Tungsten	0.38	92
Insulators		
Aluminum foil,	$\left(\frac{3}{8}\text{-in. air spaces}\right)$	
crumpled	0.0001	0.025
Asbestos, sheets	0.0004	0.097
Insulating brick,		
kaolin	0.0006	0.15
Glass, window	0.0012–0.0024	0.3–0.6
Snow	0.0011	0.27
Fluids		
Air	0.000054	0.017
Water	0.0015	0.37

area if the two faces are at temperatures of 450°F and 150°F.

Assuming the conductivity of kaolin brick to be 0.15 Btu/ft hr °F for this range of temperature, we get, from Equation 20-1,

$$\frac{Q}{t} = 0.15 \times 16 \ \frac{300}{0.5} \ \frac{Btu}{hr}$$

$$= 1440 \ \frac{Btu}{hr}$$

20-3
Convection

Convection is the transfer of heat from one part of a fluid to another by the mixing of the warmer particles of the fluid with the cooler parts. As an example, consider the case of a jar of water that is heated by applying a flame at one side A, as shown in Figure 20-2. Heat is conducted through the glass to the water. As the water in contact with the glass is heated by conduction, its density decreases, and it floats to the top. Colder water moves down to replace it. This colder water in turn is heated; once

hot, it rises because of its smaller density, thus setting up a *circulation* of the liquid. During this circulation the warmer particles of the liquid mix with the cooler parts, and in a very short time a fairly uniform temperature is established through-

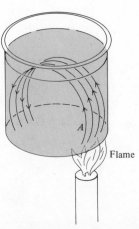

FIGURE 20-2 Convection. A circulation is set up in the fluid by heating it near one end.

out the liquid. This type of heat transfer is called *natural convection*, because the motion of the fluid is due to differences in the density of the fluid. The heat transfer can also be produced by *forced convection* by the use of a fan or pump or other mechanical device for stirring and mixing the warmer and cooler parts of the fluid.

In almost all cases of the transfer of heat by means of fluids, both convection and conduction must be considered. The heating of a room presents several interesting illustrations of convection and conduction and, to some extent, of radiation. If the room is heated by means of a radiator, heat is conducted through the walls of the radiator to the air in contact with it. This warmer air rises and displaces the cooler air, thus establishing a circulation of the air in the room. The warmer air, striking the cooler walls and windows, loses heat to the outside by conduction through the walls and windows. Fortunately, there is always a film of stagnant air close to the walls and windows so that the heat that is conducted to the outside must be *conducted* through this film of air as well as through the walls and windows. Since air is such a very poor conductor, a very thin layer of it is sufficient to form a good insulator.

20-4
Radiation

The transfer of heat by conduction and convection involves the use of material media. The transfer of heat by the process of *radiation* need not involve the use of material media. An outstanding example is the radiation of energy from the sun to the earth; by far the greatest part of the space between these two bodies is a very good vacuum. This radiant energy consists of *electromagnetic waves*, which travel with the speed of light, about 186,000 mi/sec. We shall analyze these waves and study them in greater detail in the section on light. It can be shown experimentally that all bodies radiate energy and that the rate at which energy is radiated depends upon the temperature of the body and the nature of its surface.

Not only do bodies radiate energy, but whenever radiant energy falls on the surface of a body, some of this energy is absorbed and the remainder is either reflected or transmitted. A *blackbody* is defined as one that absorbs all the radiant energy that falls upon it. Although there is no actual body that satisfies this condition, a body whose surface is coated with lampblack is very nearly a perfect

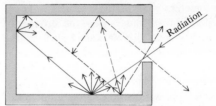

FIGURE 20-3 The hole at O acts as a blackbody.

blackbody. For laboratory work, an insulated hollow box whose interior walls can be maintained at any desired temperature and which has a small hole in one of its sides forms a very close approximation to a blackbody (see Figure 20-3). Any radiation that enters this hole will be almost completely absorbed. As shown in Figure 20-3, the radiant energy striking a wall is partly reflected and the remainder absorbed. The particular part of the wall that absorbs radiation then acts as a source, emitting radiant energy in all directions, to be absorbed and reemitted many times. Only a minute fraction of the radiation that entered the hole will come out again.

The *absorptivity* of a surface for radiant energy is defined as the fraction of the total incident radiation that is absorbed by the surface. The absorptivity of a blackbody is 1. The absorptivity of all other bodies is less than 1. Polished metallic surfaces reflect most of the radiation falling on them; their absorptivity is very small. Furthermore, bodies that are good absorbers of radiation are also good emitters of radiation, and those that are poor absorbers are poor emitters. Thus a blackbody is a good radiator, whereas a polished surface is a poor radiator. A simple experiment for demonstrating this is shown in Figure 20-4. One side of a tin can is coated with lampblack and the other side is polished. The tin can is filled with hot water.

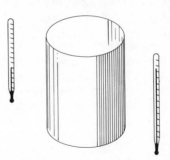

FIGURE 20-4 Thermometer opposite blackened side of can records a higher temperature.

Two thermometers with their bulbs blackened are then placed at equal distances from the surfaces of the tin can: one near the blackened surface, the other near the polished surface. The thermometer near the blackened surface will show a higher temperature reading. This is because the blackened surface, even though it is at the same temperature as the polished surface, radiates more energy in a given time.

The rate R_B at which energy is radiated from a unit area of a blackbody at absolute temperature T in unit time has been measured and found to be given by the equation

$$R_B = \sigma T^4 \qquad (20\text{-}2)$$

in which σ (Greek lower case sigma) is a constant that depends upon the units used. In the cgs system,

$$\sigma = 5.670 \times 10^{-5} \frac{\text{erg}}{\text{sec}} \text{ cm}^2 \text{ deg}^4$$

Equation 20-2 is known as the *Stefan-Boltzmann law of radiation*. The rate at which any other body at temperature T radiates energy from a unit area in unit time is given by

$$R = e\sigma T^4 \qquad (20\text{-}3)$$

in which e, known as the *emissivity* of the body, is a fraction always less than 1 and represents the ratio of the rate at which energy is emitted from the body to that emitted by a blackbody at the same temperature. Furthermore, it can be shown that the absorptivity and the emissivity of a body are the same at the same temperature. This is in agreement with the fact that good radiators are also good absorbers of radiation.

Illustrative example

Calculate the rate at which energy is radiated from a ribbon tungsten filament 1 cm long and 0.2 cm wide that is maintained at a temperature of 2727°C; the emissivity of tungsten at this temperature is 0.35.

The absolute temperature $T = 2727 + 273 = 3000$ K; the surface area emitting radiation is $2 \times 1 \times 0.2$ cm$^2 = 0.4$ cm^2. The rate at which energy is emitted from a unit area is given by Equation 20-3. Hence the rate at which energy $\mathscr{E}$ is emitted from the filament is

$$\mathscr{E} = AR = Ae\sigma T^4$$

Substituting numerical values, we get

$$\mathscr{E} = 0.4 \text{ cm}^2 \times 0.35 \times 5.670$$
$$\times 10^{-5} \frac{\text{ergs}}{\text{cm}^2 \text{ sec deg}^4} \times (3000 \text{ K})^4$$

from which

$$\mathscr{E} = 63.2 \times 10^7 \frac{\text{ergs}}{\text{sec}} = 63.2 \text{ watts}$$

Note that only a small fraction of this energy is in the form of visible light.

20-5
Heat and radiant energy

Every object, no matter what its temperature, radiates energy in the form of electromagnetic waves that travel with the speed of light. In addition, each object receives radiation from neighboring objects; some of the radiation incident upon it is absorbed by the object, and the remainder is reflected. *The difference between the amount of radiant energy the object absorbs and that which it radiates is the heat that is either added to or given out by the object.* If it absorbs more energy than it radiates, heat is added to the object; if it radiates more energy than it absorbs, heat is given out by the object.

Consider the case of an object at a high temperature T placed in a box or an enclosure whose walls are maintained at a constant temperature T_C. The rate at which energy is radiated from each unit area of the object is

$$R = e\sigma T^4 \qquad (20\text{-}3)$$

At the same time, it receives energy from the walls, and, since the absorptivity of the object is the same as its emissivity, it absorbs energy at the rate of

$$R_c = e\sigma T_C^4$$

The nature of the walls of the enclosure does not enter into the discussion since the radiation inside may be considered to be blackbody radiation. The difference between R and R_C is the amount of heat given out by unit area of the body in unit time. If A is the surface area of the body, then the amount of heat Q given out in time t is

$$Q = Ae\sigma(T^4 - T_C^4)t \qquad (20\text{-}4)$$

This shows that the rate at which heat is emitted from a body is proportional to the difference in the fourth powers of the temperatures of the body and the surroundings.

20-6
Heat insulation

A thorough understanding of the subject of the transmission of heat will enable one to solve the very important problem of heat insulation. This involves the use of proper materials for a given job as well as the development of new insulating materials. For example, gasoline storage tanks are frequently coated with aluminum or other reflecting material to reduce the absorption of radiation from the sun. Insulating materials are constructed so that they contain many small pockets of air to make use of the very low conductivity of the air; there is practically no convection, since the air is trapped in these pockets. Crumpled aluminum containing small air pockets is a very good insulator: there is practically no transfer of heat by convection; the transfer by conduction is very slight, since the crumpling of the aluminum makes the conducting path very long, while the cross-sectional area is very small; and very little heat is transferred by radiation.

The Dewar flask, sketched in Figure 20-5, is an excellent illustration of heat insulation. The Dewar flask consists of two cylindrical glass flasks sealed together at the top. The inside surface of the outer cylinder and the outside surface of the inner cylinder are silvered. Then the air between the two walls is pumped out and the space is sealed off. If hot food is placed inside the bottle and the bottle is

Cork

Silvered surfaces

Sealed tip

Vacuum

FIGURE 20-5 Section of Dewar flask.

then corked, the food will remain hot for a long time. Very little heat will be conducted along the glass or through the cork to the outside; there is practically no convection, since there is a good vacuum between the walls; and radiation is reduced considerably by the silver coatings.

20-7
Heat transfer from the human body

The food eaten by a typical well-fed person produces several thousand kilocalories per day of energy, after digestion and metabolism. Although some of this energy is transferred to other objects in the form of work, most of this energy must be transferred in the form of heat in order to keep the body temperature constant. The average temperature of the interior of a human body is approximately 37°C and varies by less than a degree for healthy people.

The metabolic energy is produced in the tissues, transferred to the blood, and then transferred from the blood to the skin and the lungs. Normally, most of this energy is transferred from the skin, although in cold winter weather about 20 percent of the energy is used to warm the air that is breathed in and to evaporate enough water to saturate the air in the lungs with water vapor. (In the most extreme circumstances, this need to warm and humidify inspired air limits the lowest temperature at which human beings can survive, no matter how well dressed they are, to a temperature of about −50°C.)

The skin loses heat by each of the methods described above; radiation, conduction, and convection. In addition, evaporation of perspiration serves to transfer heat away from the skin. Under ordinary conditions, perspiration forms a thin layer of fluid on the surface of the skin, sometimes called *insensible perspiration*. Evaporation of this fluid accounts for about 20 percent of the heat loss. In extreme conditions, when the atmospheric temperature rises or intense activity raises the rate of metabolic energy production, the other heat loss mechanisms cannot take care of the heat and the body produces large amounts of perspiration. Since evaporative losses are then the major effect, the ability of the body to maintain constant temperature is determined by the humidity as well as the temperature of the surrounding air.

Heat losses by conduction and convection are difficult to measure. Usually, however, the heat loss by conduction is negligible and the heat loss by convection is small, because of the low conductivity and low specific heat of air. When the body is immersed in water, convection losses become important.

Under ordinary conditions, approximately 60 percent of the heat lost from the body is lost by radiation from the skin. The rate of heat loss by this mechanism is then given by Equation 20-4 where the emissivity, e, is about 0.97. Equation 20-4 can be put in convenient form by writing

$$T^4 - T_c{}^4 = (T^2 - T_c{}^2)(T^2 + T_c{}^2)$$
$$= (T - T_c)(T + T_c)(T^2 + T_c{}^2)$$

Then if T and T_c are very nearly the same in magnitude,

$$T^4 - T_c{}^4 \approx (T - T_c)(2T)(2T^2)$$
$$\approx 4T^3(T - T_c)$$

Equation 20-4 can then be replaced by the approximation

$$Q \approx 4Ae\sigma T^3 (T - T_c)t \qquad (20\text{-}5)$$

or, if T remains essentially constant,

$$Q \approx K'A (T - T_c)t \qquad (20\text{-}6)$$

Equation 20-6 states that the rate of heat loss due to radiation is approximately proportional to the product of the area and the temperature difference between the body and the surrounding container. The rate of heat loss due to convection is also described by such a proportionality for small temperature differences; Equation 20-6 is sometimes called *Newton's law of cooling.*

Since the radiation takes place from the skin, the temperature T in Equation 20-6 is that of the skin. Under ordinary circumstances the body adjusts the rate of heat loss by varying the skin temperature, raising the skin temperature to increase the rate of heat loss, and lowering the skin temperature in order to decrease the heat loss. This skin temperature change is caused by a change in the blood flow to the skin.

The temperature T_c is the temperature of the walls of the container to which the heat is radiated.

Usually this is the same as the temperature of the air. However, it is fairly common for the walls, and particularly windows, of a house to have a temperature that is close to that of the outside air, even though the inside air has a significantly higher temperature. This happens because of the insulating qualities of a thin layer of static air near the walls and windows; it is common experience that the windows feel cold in winter, even in a warm room. In such a room, the temperature T_c is that of the walls and windows and is therefore somewhat lower than that of the ambient air. The exposed skin of a person in such a room will then feel cold, even if the air temperature is relatively high.

The ability of the body to regulate the heat loss by radiation is limited to a range of skin temperatures, from about 19°C to about 31°C. Above this range the blood supply cannot be increased by the body (and so evaporative cooling becomes important), whereas below this range the skin blood vessels have been constricted maximally. In this low-temperature region the body cools as if it were an inanimate object.

One interesting consequence of Equation 20-6 is that the radiative heat loss per gram of body mass is larger for small creatures than for large creatures. This is because the surface area is proportional to the square of some dimension L, which gives the size of the animal, while the mass of the creature is proportional to the cube of the dimension L. Thus the radiative heat loss per gram is inversely proportional to the dimension L.

$$\frac{Q}{m} \propto \frac{L^2}{L^3}$$
$$\frac{Q}{m} \propto \frac{1}{L}$$

Thus small warm-blooded creatures radiate more in proportion to body weight than large warm-blooded creatures, and consequently, small creatures must eat a larger amount, compared to body weight, than large creatures. Shrews and mice must eat almost continuously, whereas elephants can go for extended periods without eating.

Questions

1. Discuss the method used in heating a classroom in wintertime, describing the method of heat transfer.

2. A copper cylinder and a wooden cylinder are on a table in the room. If you pick up these cylinders, the copper cylinder will feel colder to the touch. Explain.

3. Why are stove-lid handles, coal tongs, and

other fireplace accessories made of iron rather than of copper?

4. A room is kept at a constant temperature of 70°F by means of a steam heating system. Trace what happens to the heat liberated by the condensation of the steam in the radiator of this room.

5. Gasoline storage tanks are frequently coated with aluminum foil or aluminum paint for insulation. Explain this process of insulation.

6. Steam radiators are frequently painted with aluminum paint. Discuss this in terms of the efficiency of the radiator.

7. One method of getting very high temperatures is to use a concave mirror and reflect sunlight toward a point known as the focus of the mirror. Assume the temperature of the surface of the sun to be 6000 K. Will it be possible to melt a piece of tungsten placed at the focus of the mirror?

8. A searchlight uses a mirror for reflecting the radiation coming from a hot carbon arc placed at its focus. Will the mirror get hot?

9. Discuss the convection currents in the earth's atmosphere; the principal currents are caused by heating near the equator and cooling near the poles. Describe the general wind patterns you would expect. Note that the rotation of the earth causes major deviations from this and that many small effects determine the local wind patterns.

10. Skiers usually wear many layers of clothing rather than a single heavy coat. Explain.

11. Why do snorklers wear rubber-insulated suits, despite the fact that most of them enjoy getting wet when swimming conventionally?

12. Many homes in the northern part of this country have storm windows, that is, additional windows that create an air space between them and the regular window. Show how this helps to insulate the house, and discuss the effect on the comfort of the inhabitants.

13. In many parts of the world, roofs are covered with grass, in the form of sod. Explain how this can keep the house cool in summer.

Problems

1. The water in an aluminum pot is 100°C colder than the flame that is in contact with the bottom of the pot. The bottom of the pot is 5 mm thick and 13 cm in radius. How much heat is conducted through the bottom of the pot, per second?

2. A window made of glass is 4 ft tall and 3 ft wide. The glass is $\frac{1}{4}$ in. thick. The air at the inside surface is at a temperature of 40°F, and the outside air next to the window is at 30°F. How much energy is conducted through the window per hour?

3. A blackbody has a temperature of 300 K. At what rate is energy radiated from it?

4. A hole in the door of a steel furnace has an area of 1 cm² and behaves like a blackbody. The temperature of the molten steel is 2000 K. How much power, in watts, is radiated from the hole?

5. A metal rod 80 cm long and 4 cm² in cross-sectional area conducts 60 cal of heat per minute when the ends of the rod are maintained at a difference in temperature of 80°C. Determine the coefficient of thermal conductivity of the metal.

6. A brass rod 75 cm long and 1 cm in diameter has one end immersed in a steam and the other end immersed in a mixture of ice and water. Determine the amount of heat that will be conducted through the rod in 15 min.

7. A box is made of glass 5 mm thick. How thick would the walls have to be in order to have the same insulating effect if they were made of (a) silver, (b) sheet asbestos, and (c) air (assume that you may neglect the material that holds the air in place)?

8. Water in a glass beaker is boiling away at the rate of 25 gm/min. The bottom of the beaker has an area of 350 cm² and is 0.2 cm thick. Calculate the temperature of the underside of the bottom of the beaker if K has the value 0.002 cal/cm sec °C.

9. A certain refrigerator is made of walls that may be treated as if they were 0.75-in.-thick asbestos. The refrigerated space is 4 ft high, 2.5 ft wide, and 2 ft deep. The inside of the refrigerator is maintained at 35°F and the outside air is at 72°F. (a) At what rate must heat be removed from the inside in order to maintain the interior temperature? (b) Twenty

pounds of food at room temperature are placed in the refrigerator and cooled down in 1 hr. How much heat must be removed during that hour, assuming that the food can be treated as if it were all water?

10. A tungsten lamp filament is maintained at 2000 K, has a surface area of 0.6 cm², and an emissivity of 0.3. Find the rate at which energy is radiated away.

11. A 25-watt light bulb is placed inside a closed metal can and the temperature of the can rises to 40°C. To what temperature would the can rise if the bulb were replaced with a 50-watt bulb? Assume that Newton's law of cooling applies.

12. (a) From measurements of solar radiation received at the surface of the earth it is determined that the sun is radiating energy at the rate of 6250 watts per cm² of the surface of the sun. Determine the temperature of the surface of the sun, assuming that the sun can be treated as a blackbody. (b) Find the total energy radiated from the sun in 1 sec, taking the radius of the sun to be 8×10^{10} cm.

13. Assume that a human being loses 2400 kilocal per day and that 20 percent of this is used to evaporate perspiration. (a) At what rate, in grams per hour, does the body lose water, taking the heat of vaporization at body temperature to be 580 cal/gm? (b) Treat a human

body as a cylinder 2 m tall and 15 cm in radius. Find the surface area. (c) How thick a layer would the amount of water evaporated in 10 min form on the surface?

14. A house wall consists of $\frac{1}{2}$ in. of asbestos on the inner surface and 3 in. of insulating brick, with a negligible air space between them. The inside air temperature is 68°F and the outside air temperature is 0°F. What is the temperature at the interface between the asbestos and brick?

15. The layer of ice on the top of a lake is 2 cm thick. The air temperature is 10°F and the temperature of the water just below the ice is 32°F. Find the rate at which the thickness of the ice increases, taking the conductivity of ice to be 0.004 cal/cm sec °C.

16. (a) From Newton's law of cooling show that the rate at which the temperature of a body changes is proportional to the temperature difference between the body and its surroundings. (b) If a body cools at the rate of 1 °C/min when it is 10° warmer than its surroundings, how fast will it cool when it is 5° warmer than its surroundings? (c) How fast will it cool when it is 1° warmer than its surroundings? (d) Extend this to argue that it will take an infinite time to have exactly the same temperature as its surroundings.

wave motion & sound

part 3

wave motion

chapter 21

21-1
Types of wave motion

A wave is any disturbance or shape that moves.
This general definition includes (a) *periodic waves*,
such as water waves, and (b) *pulses*, such as the
crack of a whip and those produced by applauding.

The physical interest in waves arises because
they are an important method by which distant
objects interact with each other without the transfer of matter and, in many cases, without the presence of matter between the objects. In all cases of
physical interest, the motion of waves is accompanied by the propagation of energy. Waves are a
common method of transferring energy. Examples
of this propagation of energy are electromagnetic
waves, sound waves, water waves, and elastic
waves in solids.

The most obvious direction associated with a
wave is the *direction of propagation*. For example,
the direction of propagation of typical surf waves
is *in* toward the beach; the light waves in a flashlight beam move along the beam, *away* from the
flashlight.

Consider, for example, a long rope that has one
end attached to the wall, and the other end held in
the hand, as in Figure 21-1. If the end *B* is moved
up and down, a series of elevations and depressions
are produced, which travel along the length of the
rope. The direction of propagation of the wave is
along the rope, but the individual particles of the
rope vibrate up and down, essentially at right
angles to the rope. Such a wave, in which the oscillatory motion is transverse to the direction of
propagation, is called a *transverse wave*.

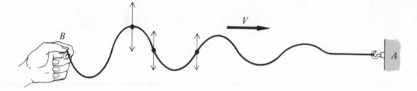

FIGURE 21-1 Transverse wave traveling along cord. Particles vibrate at right angles to direction of motion of wave along cord.

Consider a long helical spring mounted vertically, with its ends fastened to rigid supports, as in Figure 21-2. If a few turns of the coil near the lower end are set into motion up and down with a small amplitude, this vibratory motion will be transmitted to the turns above it, and a wave motion will be propagated upward. In this case the direction of vibration is parallel to the direction of propagation. Such a wave is called a *longitudinal wave*.

Ripples can be observed readily on the surface of water in a tank or in a pond. These waves are combinations of longitudinal and transverse waves; that is, the particles of water near the surface move in small elliptic paths, while the waves travel along the surface. Figure 21-3 is a photograph of ripples traveling out from a small vibrating source on the surface of the water. A single wave consists of a crest (bright region) and a trough (dark region). These waves travel out radially from the source, so lines drawn through all adjacent points that are on a crest form circles.

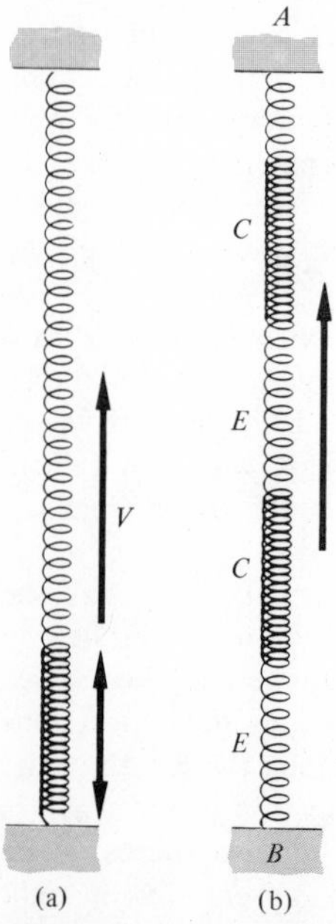

In describing the production of a transverse wave in a rope we described the motion of the hand as up and down. If the motion of the hand is restricted to the vertical direction all the vibrations at all points on the rope will also be in the vertical direction. Clearly, the motion of the hand could have been from side to side, in which case the vibrations of all the points on the rope would have been from side to side. A transverse wave in which the vibrations are confined to one direction is called a *linearly polarized wave*, or sometimes simply a *polarized wave*. If the motion of the hand randomly changed direction, then the vibrations would not be confined to any one direction, and the wave would be described as *unpolarized*. Since it makes no sense to think of a polarized longitudinal wave, the presence of polarization indicates that a wave is entirely or partially transverse.

21-2
Sine waves

Suppose that a long wire is attached at one end to a rigid support and at the other end to one prong of a tuning fork. If the tuning fork is set into vibration, the prong to which the wire is attached will oscillate with simple harmonic motion. This vibration will propagate along the wire, establishing a transverse wave in the wire. Once the wave reaches the wall, reflection will occur, and the subsequent behavior is complicated. We will initially discuss the behavior before reflection and deal with the reflection process later.

At any instant of time the shape of the wire between the leading edge of the wave and the tuning fork is a sine curve, as shown in Figure 21-4. The distance between any two adjacent crests such as C and D is called the *wavelength*, λ. The distance between two adjacent troughs, such as E and F, is also λ, as is the distance between points A and G.

FIGURE 21-2 (a) Individual turns of the spring vibrate up and down; wave travels along spring. (b) Compressions C and extensions E moving up the spring.

FIGURE 21-3 Photograph of ripples on the surface of water produced by a vibrating source. The bright circles are crests of the wave and the dark circles are troughs. (Photograph courtesy of Sylvania Electronics Defense Laboratories.)

Each point on the wave executes simple harmonic motion. Therefore if a graph were made of the vertical position of any point as a function of time, the result would be a sine curve, as in Figure 12-5. It must be emphasized that Figure 21-4 gives the displacement of each point in the wave at one *instant*, whereas Figure 12-5 gives the displacement at one *position* as a function of time.

Thus, if we focus our attention on one point on the wire, such as point F of Figure 21-4, we will see that it oscillates up and down with a period of T, or frequency f. If instead of focusing our attention on one point, we looked at the whole wire, we would see that the sinusoidal shape of Figure 21-4 is moving from left to right with velocity V. Thus the time it takes for the crest at point E to reach point F is the same as the time it takes for point F to move up to a positive maximum and then down again to a negative maximum. In other words, the time it takes for the wave to travel from E to F is equal to the period T of the simple harmonic motion. Since the distance EF is equal to the wavelength λ, the velocity V of the wave is given by

$$V = \frac{\lambda}{T} \qquad (21\text{-}1a)$$

or

$$V = \lambda f \qquad (21\text{-}1b)$$

Equation 21-1 is applicable to both transverse and longitudinal waves, and it is the fundamental equation for any wave in any medium.

The frequency of a wave may be expressed in units of vibrations per second or cycles per second; it is increasingly common to use the unit hertz (abbreviated Hz), particularly for acoustic and electromagnetic waves, where

$$1 \text{ Hz} = 1 \frac{\text{vib}}{\text{sec}} = 1 \frac{\text{cycle}}{\text{sec}}$$

All of these units have the dimension of 1/sec.

21-3
Fourier analysis

The traveling sine waves we have considered above are particularly simple. At first sight it appears, however, that very few real phenomena should be described by such waves. Most real waves start and stop rather abruptly. More important, most waves are more like a pulse or flash than a simple sine wave. Even ocean waves, which

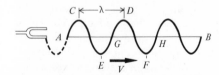

FIGURE 21-4 Transverse wave in a long wire set up by a vibrating tuning fork.

roughly look like the waves we have described above, have more irregular profiles than simple sine waves.

A great French mathematician, J. B. Fourier, demonstrated a remarkable mathematical theorem. He showed that any wave, including brief irregular pulses, could be treated as the sum of an infinite number of simple sine waves. Each of these sine waves has a different frequency and different amplidtude. Thus, in a sense, if we understand sine waves, we understand all waves.

Fourier's theorem requires that we deal with an *infinite* number of sine waves. For almost all real physical problems, the amplitudes of all but a few of these waves are very small. Thus, most real waves can be treated as the sum of a few sine waves of differing frequencies and amplitudes.

A direct illustration of this phenomenon is in the production of sound waves by musical instruments. When a clarinet sounds a particular note, the resulting sound wave is the sum of several frequencies. The lowest frequency, which has the largest amplitude, is called the *fundamental*. The pitch of the note sounded is determined by the frequency of the fundamental. The other frequencies present are called *overtones* and usually have smaller amplitudes. The number, frequencies, and amplitudes of the overtones determine the quality of the sound. The same note sounded on a violin sounds different from that sounded on a clarinet because the overtones are different. In some musical instruments the frequencies of the overtones are whole-number multiples of the fundamental frequency; such overtones are called *harmonics*.

21-4
The velocity of waves

The velocity of a wave is a property of the medium in which the wave travels. In the most general case the velocity of a wave also depends on the wavelength or frequency of the wave. A medium in which the velocity of waves is different for different wavelengths is called a *dispersive* medium, whereas a medium in which the velocity is the same for all wavelengths is called a *nondispersive* medium. Suppose, then, that we have two waves of different wavelengths in a dispersive medium, and that the two waves are superimposed at one instant. Since the velocity of the two waves is different, the waves will separate and at some time later will no longer be superimposed. This separa-

tion of waves of different wavelength is called *dispersion*. Since a pulse consists of many wavelengths, this implies that the shape of a pulse will change as the pulse propagates through a dispersive medium. Water is a dispersive medium for mechanical waves, which accounts for the fact that ocean waves change their shape as they propagate.

Many waves, such as the vibrations of wires and springs discussed above, are essentially nondispersive, as are sound waves in gases. It is this happy coincidence that accounts for the fact that the character of sound is nearly unchanged by distance; music sounds the same to the audience as to the conductor.

It is possible to write relatively simple formulas for the velocity of waves in nondispersive media. For example, the velocity of a transverse wave in a stretched wire is given by

$$V = \sqrt{\frac{S}{M/L}} \qquad (21\text{-}2)$$

where S is the tension in the wire, M is the mass of the wire, and L is its length.

Similarly, it is possible to show that the speed of sound in a gas is given by

$$V = \sqrt{\frac{\gamma P}{\rho}} \qquad (21\text{-}3)$$

where P is the pressure of the gas, ρ is its density, and $\gamma = c_p/c_v$, the ratio of the specific heats at constant pressure and constant volume. The value of γ for air is 1.4. Table 21-1 lists the speed of sound in some substances. In general, the speed of a wave depends on the inertia of the medium and on its elasticity, where the word "elasticity" roughly describes the ability of a medium to resist deformation.

Most waves are *absorbed* by the media in which they travel. That is, the amplitude of the wave decreases continuously as the wave propagates. In the most general case, the amount of absorption depends on the frequency of the wave. It is this

Table 21-1 Speed of sound (longitudinal waves)

substance	temperature in °C	speed in meters per second
Air	0	331.45
Hydrogen	0	1269.5
Carbon dioxide	0	258.0
Water	15	1447
Sea water	13	1492
Glass, Pyrex		5170
Steel		4700–5200

process that accounts for the color of deep bodies of water. Air absorbs light waves, so almost none of the ultraviolet radiation in sunlight reaches the surface of the earth; the absorption of light by air is such that if the earth were flat, one could see only about 150 miles; beyond that the light would be absorbed by the intervening air.

21-5
Huygens' principle

Visualizing and describing the progress of a wave through a medium are simple in the case of a transverse wave on a wire or a longitudinal wave in a helical spring, because these waves are one-dimensional in that the waves propagate along a single line. In the more general case represented by the motion of water waves or sound waves or light waves, the waves are two- and three-dimensional, and the description is more complex. Further, the progress of such waves is affected by the presence of barriers and walls and apertures, as well as by changes in the media in which propagation takes place.

Consider a stone dropped into a pool of water. Water waves will radiate out from the point of impact. Figure 21-5 represents these waves by a set of concentric circles. These circles could represent the crests of successive waves. In fact, recognizing that each point on the surface of the water executes simple harmonic motion, the circles could represent points on successive waves that have the same *phase* (see Chapter 12). Such lines of constant phase are called *wavefronts*. In three-dimensional waves, the wavefronts are two-dimensional *surfaces* of constant phase. The problem of describing the progress of waves is equivalent then, to describing how the wavefronts move.

The progress of a wave under a variety of conditions can be predicted with the aid of a principle first enunciated by the Dutch physicist, Christian Huygens (1629–1695). According to *Huygens' principle*, each point in a wavefront can be considered a source of waves, and the new position the wavefront will occupy after the lapse of a small time interval t can be found by drawing the envelope of all the small waves that were sent out by all the individual points on the first wavefront at the beginning of the time interval. The small waves sent out are sometimes called *Huygens' wavelets*. Several important comments must be made: first, that Huygens' principle is merely a way of *describing* the progress of waves and not a grand causal law of nature; second, that the Huygens wavelets are propagated only in the forward direction, and third, that the Huygens wavelets propagate in a way determined by the medium in which they find themselves. Thus, if some, but not all, of the Huygens wavelets enter a medium in which the velocity of propagation changes, then those wavelets will move with a velocity different from that of the remaining wavelets. The result will be a change in the shape of the wavefront.

To illustrate the method of using Huygens' principle in the simplest cases, let us consider the progress of a spherical wavefront from a point source, as in Figure 21-6. The arc of the circle AB represents the position of a section of the wavefront at a certain time. The wave is progressing through the medium with a speed V, which we assume is uniform throughout the medium. To find the new position of the wavefront at the end of a short time interval t, consider each point on the arc AB as a source of waves, and draw circles of radii equal to Vt, with each of these points as a center. The new wavefront $A'B'$ is the envelope of all these small circles; that is, $A'B'$ is tangent to each one of these circles.

The same type of construction can be used to find the subsequent position of any type of wavefront. Figure 21-7 shows a plane wavefront CD progressing with uniform speed V. Its new position after a short time t is $C'D'$, the envelope of the small waves, each of radius Vt, sent out by each of the points on the wavefront CD.

Figure 21-8 is a photograph of plane surface waves traveling in a ripple tank as described in

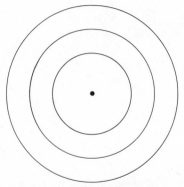

FIGURE 21-5 Water waves radiating out from the point of impact of a stone dropped into a pool.

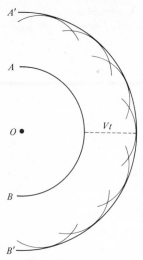

FIGURE 21-6 Huygens' construction for determining the position of wavefront AB after a time interval t. $A'B'$ is the new position of the wavefront.

Section 21-8. The crests grow less distinct as the ripples travel away from the source because such waves lose energy and their amplitudes become smaller as they move forward.

21-6
Reflection of a wave

When a wave that is traveling in one medium reaches the surface of a second medium, part of the wave is *reflected* back into the first medium and the rest penetrates the second medium; this latter part is said to be *refracted* into the second medium. Figure 21-9 shows a plane wave striking a surface S separating media I and II. The incident wavefronts are all parallel to AB and travel with speed V_1. The angle that each incident wavefront makes with the surface is called the angle of incidence i. Experiments have shown that the angle r between the reflected wavefronts and the surface is equal to the angle i; that is, *the angle of incidence is equal to the angle of reflection*. A line drawn in the direction of motion of the wave is called a *ray*; a ray is perpendicular to the wavefront. The angle of incidence i is then also the angle between the ray BC and a line perpendicular to the surface at C; such a line is called a *normal*. Similarly, the angle of reflection is the angle between the reflected ray and the normal.

The position of the reflected wave and the direction of its motion relative to that of the incident

wave can be found with the aid of Huygens' principle. Figure 21-10 shows the plane wave advancing toward the surface SS at an angle of incidence i. Let us consider the plane wavefront represented by the line AB such that point A has just reached the surface while B is still moving toward it; the angle between AB and the surface SS is the angle i. If the surface had not been there, the wavefront AB would have advanced to the position $A'C$ in a time t. But as the different parts of the wavefront reach the surface, each point on the surface, such as $A, e, f, g, \ldots, C$, is set into vibration and becomes a source of waves. To locate the new position of the wavefront, draw a circle of radius AA' with A as a center in the region above SS, another circle with e as a center and of radius ee'', and proceed in a similar manner for each point on the surface that is struck by the incident wave. Then draw the envelope $De'f'g'C$ tangent to these circles. This is the wavefront reflected by the surface. The angle r between CD and SS is the angle of reflection of the wave and is equal to the angle of incidence. The direction of motion of the reflected wave is indicated by the arrows on the rays AD, ee', and so forth.

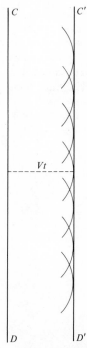

FIGURE 21-7 Huygens' construction for determining $C'D'$, the new position of the plane wavefront which started as CD at a time t earlier.

FIGURE 21-8 Photograph of plane surface waves traveling in a ripple tank; intensity diminishes with distance from source. (Courtesy Sylvania Electronics Defense Laboratories.)

The surface SS is a plane mirror reflecting the incident waves. If there is a point source P in front of a plane mirror SS, waves from P that strike the mirror will be reflected in such a way as to appear to come from a point P' behind the mirror. P' is called the *image* of P; its location can be determined by tracing the paths of a few rays from P to the mirror and then drawing the reflected rays at appropriate angles of reflection as shown in

Figure 21-11. It can readily be seen that the image P' is just as far behind the mirror as P is in front of it. Figure 21-12 is a photograph of a set of plane waves striking a concave circular mirror and being reflected from it. It will be noted that the reflected waves converge towards a small region called the *focus* of the mirror, and then diverge from the focus as they travel outward. The walls of a room usually

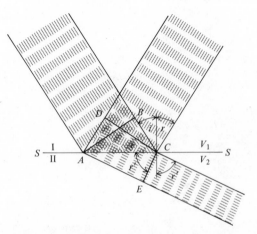

FIGURE 21-9 Reflection and refraction of a plane wave at a surface separating two media.

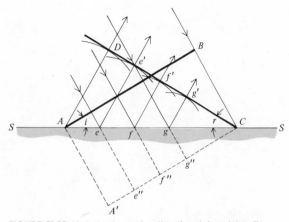

FIGURE 21-10 Huygens' construction for determining the position and direction of motion of the wavefront reflected by surface SS. AB is the incident wavefront, and CD is the reflected wavefront.

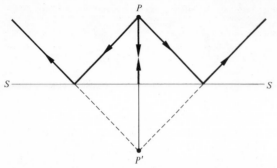

FIGURE 21-11 Reflection from a plane surface gives rise to an image P' of a point source at P.

act as mirrors for the sounds produced in the room. In rooms of average size, these reflections add to the intensity of the sounds and in this sense the reflections are valuable. In a large auditorium, however, the reflected sound may reach the hearer a considerable time after he has received the directly transmitted sound. In the case of speech this may be very objectionable; in the case of music, the overlapping of different sounds is usually pleasing to the ear.

21-7
Refraction of a wave

When a wave reaches a surface SS that separates two media, one in which its velocity is V_1 and the

second in which its velocity is V_2, part of the wave enters the second medium and travels through it at an angle r', which differs from the angle of incidence i. If V_2 is greater than V_1, then the angle r', called the *angle of refraction*, is greater than the angle i, as shown in Figure 21-9. The method for determining the position and the direction of motion of the refracted wave with the aid of Huygens' principle is illustrated in Figure 21-13. Suppose that the incident wave is a plane wave; a portion of a wavefront is represented by the line AB. A has just reached the surface SS while B is traveling toward it with the speed V_1. The position that the wavefront would have occupied if it had continued in the same medium is shown by the dotted line $A'C$. But, since part of the wavefront has entered the second medium, the new wavefront is at a different position and is traveling in a different direction. To find this new wavefront, we note that while B was traveling in the first medium with speed V_1 toward C, A was traveling in the second medium with speed V_2 for the same length of time t. Hence, using A as a center, draw a circle of radius V_2t for the wave that was emitted by point A. If we take a point at the center of the incident wavefront AB, it will reach point f on the surface a time $t/2$ after the wave reached A. For the remaining time $t/2$, the wave will progress in the second medium with speed V_2. Hence, with f as a center and with a radius $V_2t/2$, draw an arc of a circle to determine the position of the wave emitted by

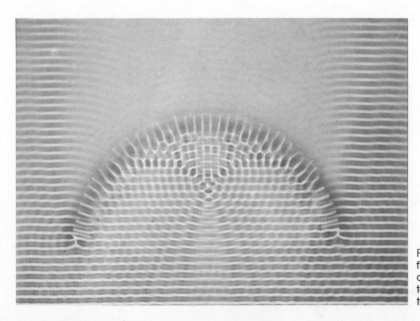

FIGURE 21-12 Photograph of reflection of plane waves from a concave circular mirror. (Courtesy of Sylvania Defense Electronics Laboratories.)

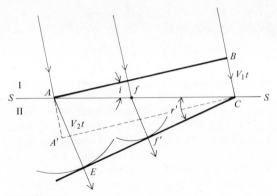

FIGURE 21-13

point f. The waves emitted by all other points on SS between A and C can be found in a similar manner. The new wavefront EC is the envelope of all of the waves emitted by the points on SS that were set into vibration by the incident wave AB.

If the speed of the wave V_2 in the second medium is greater than the speed of the wave V_1 in the first medium, then the transmitted part of the wave will be bent away from the normal. In Figure 21-13 the wavefront AB is so chosen that point A travels to point E in medium II in the same time that point B moves in medium I to the surface at C. Thus

$$\frac{BC}{AE} = \frac{V_1}{V_2}$$

Now

$$\sin i = \frac{BC}{AC}$$

and

$$\sin r' = \frac{AE}{AC}$$

therefore

$$\frac{\sin i}{\sin r'} = \frac{(BC)/(AC)}{(AE)/(AC)} = \frac{BC}{AE}$$

from which

$$\frac{\sin i}{\sin r'} = \frac{V_1}{V_2} \qquad (21\text{-}4)$$

Thus the ratio of the speeds of the waves in the two media is equal to the ratio of the sine of the angle of incidence to the sine of the angle of refraction. This ratio is called the *relative index of refraction* of the two media and will be designated

by n_r. Thus

$$n_r = \frac{\sin i}{\sin r'} = \frac{V_1}{V_2} \qquad (21\text{-}5)$$

For example, the speed of sound in air is 331 m/sec and that in water is 1447 m/sec. The relative index of refraction of water with respect to air for a longitudinal wave is

$$n_r = \frac{331}{1447} = 0.228$$

A longitudinal wave going from air to water and striking the surface at an angle of incidence of $10°$ will be refracted into the water and be transmitted at an angle r', which is given by

$$n_r = \frac{\sin i}{\sin r'}$$

and thus

$$\sin r' = \frac{\sin 10°}{0.228} = \frac{0.1737}{0.228}$$

from which

$$\sin r' = 0.7619$$

or

$$r' = 49°38'$$

21-8
Diffraction of waves

Many common phenomena, such as the bending of a sound wave around obstacles, the spreading of a sound wave in different directions after passing through a small aperture, and the directional properties of megaphones, are examples of the *diffraction of waves*. Diffraction phenomena occur not only with sound waves but with all other types of waves. The ability to produce diffraction effects is one of the criteria used to decide whether we are dealing with a wave phenomenon. A simple method of demonstrating the diffraction of a wave through a small aperture is illustrated in Figure 21-14a. $ABCD$ is a shallow rectangular tank with a glass bottom. Water is poured into this tank to a depth of about 1 in.; see Figure 21-14b. Light from a source S below the glass is transmitted through the water onto the ceiling. If a wave is started in the water by disturbing it with a stick, the form and progress of the wave can be followed by viewing its image on the ceiling. If a metal partition with a small aperture O is placed across the tank and if a

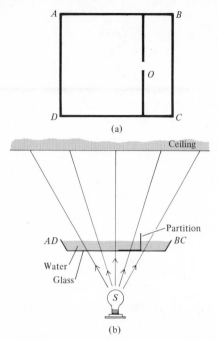

FIGURE 21-14 Shallow rectangular tank for demonstrating wave phenomena. (a) Top view showing partition in tank; narrow aperture O in partition. (b) Side view showing source of light, S, under the glass bottom of the tank, for projection of wave phenomena.

plane wave is started near AD by dipping a stick into the water so that the length of the stick is parallel to AD, the plane wave will progress toward BC. When it comes to the partition, most of the wave will be reflected from it, but a small part will go through O (see Figure 21-15). The part that goes through O will not be plane, however, but will spread out from O in a circular wave with O as the center. Thus points such as P and Q behind the partition will receive energy from the wave owing to its diffraction through the small aperture O.

Figure 21-16 is a photograph showing the diffraction of plane waves by a series of narrow apertures placed in the ripple tank. Such a series of apertures is called a *diffraction grating*. The effect produced at any point at a distance from the grating is determined by adding the ordinates of the waves reaching this point from each aperture. A diffraction grating is a very useful device for analyzing the waves incident on it. It will be discussed in greater detail in the sections on light and X rays.

21-9
Standing waves

Consider the situation in which two waves simultaneously exist in the same region of a medium. In general, the presence of one wave affects the other if only because the wave changes the medium. For example, if a transverse wave with a large amplitude is traveling through a wire, the wire will be stretched. This usually will change the properties of the wire enough to change the propagation of a second simultaneous wave. However, if the amplitude of each wave is small, the waves obey the *principle of superposition*, according to which the amplitude of the resulting wave at any point is the *algebraic sum* of the amplitudes of the separate waves. In other words, we can follow the progress of each of the waves as if the other waves were not present, and then, at any instant of time, add the effects of all of the waves.

To illustrate one consequence of this principle, consider two identical waves traveling in opposite directions. A simple way to produce this is to oscillate one end of a long wire, the other end of which is rigidly attached to a wall. When the wave reaches the wall, it will be reflected and, if we ignore absorption, it will propagate back along the wire with identical frequency and amplitude. This situation is shown in Figure 21-17, where the dark lines shows the wave propagating from B to A, and the broken line shows the reflected wave propagating from A to B. At the instant shown, the points marked with N have zero amplitude. The striking thing about this situation is that as the two waves propagate, the same points continue to have zero amplitude. That is, the points marked N (for *node*)

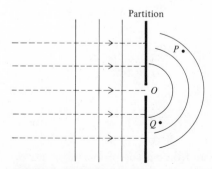

FIGURE 21-15 Diffraction of a plane wave through a very small aperture O. Points P and Q behind partition will receive waves from O.

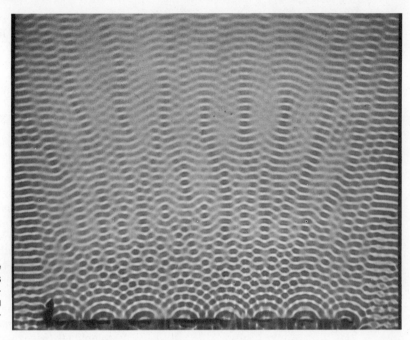

FIGURE 21-16 Photograph of the diffraction of plane waves through a series of narrow apertures. (Courtesy of Sylvania Electronics Defense Laboratories.)

do not move at all, whereas all the points intermediate between the nodes oscillate. This resulting motion is called a *standing wave*.

The method for adding the effects produced by two waves of the same wavelength and amplitude but traveling in opposite directions is illustrated in Figure 21-18. The two waves traveling in opposite directions are shown as dotted lines, with arrows to indicate the direction of motion. The resultant displacement is shown as a dark line. In the uppermost figure, the two waves produce equal and opposite displacements, so the resultant displacement of every particle is zero; this is indicated by the heavy horizontal line. In the second line each wave is shown after it has advanced one eighth of a wavelength, that is, one eighth of a period later. The resultant displacements form the heavy curves shown. After an additional eighth of a period the two waves coincide, producing a resultant wave of

twice the amplitude of either one. The other figures each show the resultant wave after an additional eighth of a period. In each figure, the nodes, marked N, remain on the horizontal axis; they experience no displacement. The segments of the wire between the nodes oscillate up and down, the displacements on either side of the node being in opposite directions. The segment between two nodes is called a *loop*, and its midpoint is called an *antinode*. From the figure one can see that the distance between adjacent nodes is half a wavelength of either of the traveling waves that combine to form the standing wave.

If both ends of a string are fixed and a traveling wave is established in the string, then the wave will propagate to one end of the string, be reflected, propagate to the other end, and propagate back again. If a standing wave is set up in the string, then the ends of the string must be nodes, because the ends of the string are fixed. Thus, standing waves can be established only if the distance between the ends of the string is a whole number of half-wavelengths:

$$L = n \frac{\lambda}{2} \qquad (21\text{-}6)$$

where L is the length of the string, n is any whole number (larger than zero), and λ is the wavelength.

FIGURE 21-17 A string vibrating in segments shows a stationary wave pattern.

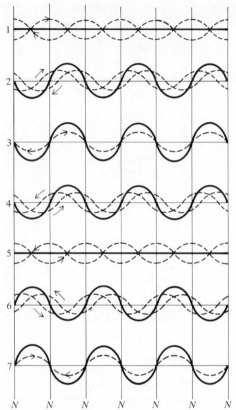

FIGURE 21-18 Method of adding two waves of equal lengths and equal amplitudes traveling in opposite directions to produce a stationary wave.

For wavelengths that do not satisfy Equation 21-6, the amplitude of oscillation will decrease rapidly after a few reflections and re-reflections.

Many vibrations can be analyzed in terms of stationary waves; these include the vibrations of strings or air columns in musical instruments, the vibrations of bars and plates, and the electromagnetic vibrations in antennas. Consider, for example, the motion of a plucked guitar string. When the string is released, its motion can be described in terms of many simple harmonic waves, according to Fourier's theorem. Of these waves, some will have wavelengths that satisfy Equation 21-6. These will set up stationary waves, while the amplitude of all the other wavelengths will rapidly diminish. Thus the string will oscillate with standing waves corresponding to only a relatively small number of wavelengths—the fundamental and

overtones of the string. Note that plucking or striking the string in different places can change the amplitudes of the Fourier components and consequently change the distribution of overtones. Piano hammers strike strings in such a position as to make the amplitude of the seventh harmonic very small.

21-10
Interference

Consider two identical waves that travel along different paths and meet at some point P. From the principle of superposition, the amplitude at P is the sum of the amplitudes due to each wave.

If the path taken by one wave is one wavelength longer than the path taken by the other, then the two amplitudes will be identical, as shown in Figure 21-19. Similarly, if the path length of one wave is longer than the other by a whole number of wavelengths, the two amplitudes will be identical. In this case the vibration at point P will have an amplitude twice that due to either wave alone. This situation is said to be one of *constructive* interference.

However, consider the situation where one wave travels a path that is half a wavelength longer than the path traveled by the other. In that case, as shown in Figure 21-20, the amplitude due to one wave is equal and opposite to the amplitude due to the other wave. Thus the sum of these two will be an oscillation with zero amplitude. Similarly, if the path difference is equal to one half-wavelength plus a whole number of wavelengths, the resulting

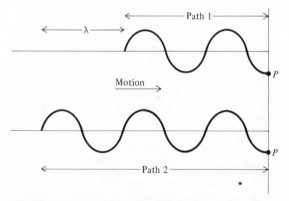

FIGURE 21-19 If the path difference between two identical waves is equal to one wavelength, then the amplitudes at point P will be identical.

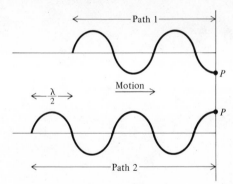

FIGURE 21-20 If the path difference between two identical waves is equal to one-half wavelength, then the amplitudes at point P will be equal and opposite.

$$\Delta P = n\lambda \quad \text{(constructive)} \quad (21\text{-}7a)$$
$$\Delta P = \left(n + \tfrac{1}{2}\right)\lambda \quad \text{(destructive)} \quad (21\text{-}7b)$$

where ΔP is the path difference between the two paths and n is any whole number. If the path difference is intermediate between those producing constructive interference and those producing destructive interference, the resulting amplitude will be intermediate betwen zero and twice the amplitude of one wave.

In general, interference is produced when two waves have a constant phase difference. This phase difference may be produced by a path difference as shown above, by a time delay, or by a combination of both. Such interference phenomena are a striking characteristic of waves, and are very important in investigations of electromagnetic waves. It was the discovery of interference phenomena that permitted the experimental verification of the wave properties of elementary particles.

amplitude will be equal to zero; this is described as *destructive* interference. In symbols:

Questions

1. Give several examples of periodic waves.

2. Give several examples of the propagation of energy by pulses.

3. Surface water waves are a mixture of transverse and longitudinal vibrations. Describe the motion of a cork floating on the surface of the water as a wave moves by the cork.

4. The water that comes out of a garden hose can be made to trace a sine wave by moving the nozzle. (a) Describe the necessary motion of the nozzle. (b) Describe the change in shape of the stream if the speed of the water increases.

5. Good audio amplifiers are designed to amplify all the frequencies that can be heard by human beings in spite of the fact that most of the notes played by an orchestra have frequencies that are significantly below the upper limit of this band. Can you account for this?

6. The typical telephone and the transmission system to which it is connected are designed to transmit only frequencies below about 2000 Hz. Explain the effect on the quality of the human voice.

7. Thunder is the sound produced by lightning. Why is it that a flash of lightning, which takes a very short time to occur, produces a long, rumbling sound?

8. Most glass is a dispersive medium for light waves. Describe a phenomenon that illustrates this property.

9. What properties of a medium determine the speed of a wave in that medium?

10. What happens to the energy of a wave that is traveling in an absorbing medium?

11. The first note of a concert is sounded in a closed room. A short time later the note is inaudible. Trace the sound and the energy associated with it from the first instant until they become inaudible.

12. Using Equation 21-3, calculate the speed of sound in air and compare with the known speed of sound in air.

13. Echoes are merely reflected sound waves. Why is the effect so startling?

14. A person standing under a window can hear sounds that originate inside the room but cannot see the source. What does this tell you about the relative diffraction of light waves and sound waves through the window? (The effect is a consequence of the fact that light waves have wavelengths that are very much smaller than the size of the aperture, whereas the wavelengths of sound are comparable in size to the aperture.)

15. What conditions are essential for the pro-

duction of standing waves in any medium?

16. Is a string in which standing waves exist ever straight?

17. A sensitive listener discovers that at certain locations in his living room he cannot hear specific notes emitted by his stereo. Give a possible physical explanation.

18. In a common demonstration of inertia a cord is tied to the bottom of a weight, which is in turn suspended from another cord as shown in Figure 21-21. Pulling slowly on the bottom cord breaks the top cord, whereas jerking the bottom cord breaks the bottom cord. Discuss this in terms of the propagation and reflection of pulses and waves.

19. If a force is applied to one end of an object, that force is communicated to the rest of the body by means of a wave, usually a compressional wave. From this argue that you would expect real objects to be deformed somewhat

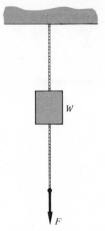

FIGURE 21-21

when starting or stopping. Under what ideal conditions would you expect that an object would not be deformed?

Problems

1. A tuning fork emits a sound wave of frequency 312 Hz. What is the wavelength of that wave in air at 0°C?

2. The rods of roof-mounted TV antennas are usually made to be half the wavelength of the signal they receive. Estimate the frequency of the TV signal waves, knowing that the waves travel with a speed of 3×10^{10} cm/sec.

3. When a 60,000 vib/sec signal is impressed on a certain crystal, waves propagate with a wavelength of 2 cm. Find the speed of the signals in the crystal.

4. The tension in a steel wire is 1 million dynes and the wire has a mass of 0.2 gm/cm. What is the speed of the transverse waves in this wire?

5. The tension in a certain wire is doubled. What happens to the speed of transverse waves?

6. How does the speed of sound in an ideal gas vary with temperature?

7. Show that the units of the right side of Equation 21-3 are those of speed.

8. Find the relative index of refraction of air and carbon dioxide for sound at 0°C.

9. A 1000-Hz oscillator is used in determining the depth of a freshwater channel. Find the length of the waves in the water.

10. Sound waves traveling at 400 m/sec enter a

medium where the speed of sound is 600 m/sec. The waves strike the boundary at an angle of incidence of 30°. In what direction will they move in the second medium?

11. Standing waves are established in a string that is 3 m long. Both ends of the string are fixed. (a) What are the three longest wavelengths? (b) Locate the nodes for each of the waves described in (a).

12. A string 90 cm long, and fixed at both ends, is plucked, and a point 30 cm from one end is touched with a finger. What is the longest wavelength of the standing wave established in the string?

13. Sound waves travel from one person to another 120 ft away in a large room. Other sound waves travel from one to the other by being reflected from the 25-ft-high ceiling at a point halfway between them. For which wavelength will these two sound waves interfere constructively, and for which will they interfere destructively? (A similar phenomenon occurs in some dome-ceilinged rooms, called "whispering galleries.")

14. The depth of water can be measured by measuring the time between emission of a sound wave and the detection of the sound wave after it has been reflected from the bottom. (a) Find the depth of the bottom if the time so measured

is 0.03 sec, assuming that the seawater is at 13°C. **(b)** Most commercial depth finders are calibrated for one specific speed of sound in water. In terms of safety is it wiser to assume a speed slightly greater than the actual speed or one slightly lower than the actual speed?

15. A stone is dropped into a well and the sound of the splash is heard 2.5 sec later. How deep is the well?

16. A sine wave is traveling in a string with a speed of 400 cm/sec and a wavelength of 1.5 m. One point of the string has a certain position at a specific instant. How much time elapses before the point is in the same position?

17. The rope marking the swimming area of a lake has floats spaced 3 ft apart. A motorboat moving parallel to the rope establishes a sine wave in the surface of the water which causes each of the floats to bob up and down 15 times per minute. When one of the floats is at its maximum vertical position, the second one from it is at its lowest position. Find the speed of the wave.

18. One end of a horizontal string is attached to a prong of a tuning fork, which is vibrating with a frequency of 400 Hz, and the other end passes over a pulley and has a 2-kg mass hanging from it. The density of the string is 0.15 gm/cm. **(a)** Determine the speed of the transverse waves in the string. **(b)** Determine the distance between the nodes when stationary waves are set up in the string.

19. A 2-m-long copper wire whose mass is 19 gm has one end attached to a prong of a tuning fork, which vibrates with a frequency of 1000 Hz. A stationary wave is set up in the wire and the distance between nodes is 8.0 cm. Determine **(a)** the wavelength of the transverse wave in the wire, **(b)** the speed of the wave, and **(c)** the tension in the wire.

20. A steel pipe 150 m long is struck at one end. A person at the other end hears the sound that travels through the metal and also the sound that travels through the air. Determine the time interval between the two sounds heard by this person.

21. **(a)** Show that the speed of sound in an ideal gas is inversely proportional to the square root of the molecular weight of the gas. **(b)** From this, explain the fact that when a person breathes helium the pitch of the voice rises.

22. Draw the Huygens construction for a set of plane waves striking a convex mirror (whose radius of curvature is large compared with the wavelength). Show that the waves diverge from a focus.

23. Determine **(a)** the index of refraction of hydrogen with respect to air for a sound wave and **(b)** index of refraction of glass with respect to air for a sound wave. A glass partition 0.6 cm thick separates a volume of air from a volume of hydrogen, where each of the gases is at atmospheric pressure. A sound wave from the air strikes the glass surface at an angle of 2° with respect to the normal. Determine the angle at which the sound wave is refracted **(c)** into the glass and **(d)** into the hydrogen.

24. A vibrator is placed 4 ft below the surface of water in a lake. The refracted waves appear to diverge from a different point. What is the apparent depth of the vibrator as determined from the sound waves refracted into the air?

25. A long horizontal tube contains air at 20°C and some fine cork dust. One end of the tube contains a movable piston, and a diaphragm attached to a prong of a tuning fork fits into the other end of the tube. The tuning fork vibrates with a frequency of 1000 Hz. The piston is adjusted until resonance occurs, at which point the cork dust piles up at the nodes of the resulting standing waves. Determine the distance between the nodes and the length of the traveling waves.

26. The air in the tube of Problem 25 is replaced with carbon dioxide, and at resonance the distance between the nodes is found to be 13.5 cm. Determine **(a)** the wavelength of the sound in carbon dioxide and **(b)** the speed of sound in carbon dioxide.

27. In order to get the maximum energy transfer to a vibrating string it should be struck near, but not exactly at, a node. If a string is struck at precisely the node of a particular overtone, then that overtone will not be sounded. Where should a string be struck in order to suppress the seventh harmonic, which is the overtone with a frequency seven times that of the fundamental? Note that this is the usual location of the hammers of a piano string.

28. A steel wire 100 cm long is tied down at both ends. The mass of the wire is 1.9 gm. When plucked, it emits a tone whose fundamental frequency is 340 Hz. **(a)** Determine the tension in the wire. **(b)** Find the frequencies of the first and second overtones of this string.

29. Two stereo speakers are 6 ft apart and both are emitting an identical 12,000-Hz note. A listener is located at a point that is 10 ft from each of the speakers. The listener slowly moves parallel to the line joining the two speakers. At what point will the sounds have zero intensity? At what point beyond that will the sounds have maximum intensity?

30. Sound of frequency 1024 Hz is transmitted into a tube that splits into two parts. Part of the sound travels in one tube, A, a distance of 50 cm, where it mixes with sound that has traveled in the other tube B. At that mixing point the two sound waves interfere. (a) For what lengths of path B will intensity maxima occur and (b) for what path lengths of B will intensity minima occur? (Assume that the tubes are filled with air at 0°C.)

31. In an experiment such as that described in Problem 30, both tubes A and B are 2 m long. One tube is filled with air and the other with hydrogen at 0°C. What is the lowest frequency that will produce an intensity maximum?

vibrations & sound

chapter 22

22-1
Introduction

Strictly speaking, the term "sound" should be reserved for those phenomena detectable by the human ear. These effects are limited to the range of frequencies between about 20 and about 20,000 Hz. The term *acoustics* is often used to describe all mechanical vibrations, whether audible or not. Further, there are two aspects of sound: one is the physical aspect, which involves the production, propagation, reception, and detection of sound; the other, which is the sensation of sound as perceived by the individual, depends on physiological and psychological effects. It is not desirable to separate the two aspects of sound completely, but the main emphasis in this book must necessarily be on the physical aspect.

22-2
Intensity and loudness

The *intensity* of a wave at any point in space is defined as *the amount of energy passing perpendicularly through a unit area at this point in unit time*. The intensity can be expressed in ergs per cm² per sec, in watts/cm², or in any other appropriate units. The intensity of the sound received from any source depends upon the rate at which the source emits energy, upon the distance of the observer from the source, and upon the reflections that the waves undergo from the walls, ceiling, floor, and objects in the room. If the size of the

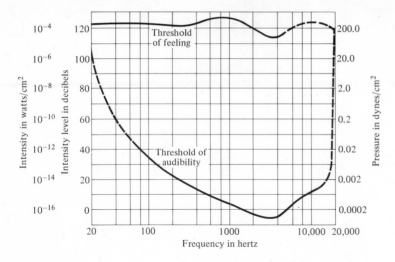

FIGURE 22-1 Range of frequencies and their intensities which are perceived by the human ear. (After H. Fletcher, *Reviews of Modern Physics*, January 1940.)

source is small in comparison with its distance from the observer and if no reflection or absorption takes place, the intensity of the sound at any place will vary inversely as the square of its distance from the source; but this is rarely the case with sound waves. In terms of the sound wave that reaches the observer, it can be shown that *the intensity depends upon the square of the amplitude of vibration of the particles in the wave.*

The *loudness* of a sound is a sensation experienced by the observer, and, although loudness is related to the intensity of the sound, the relationship between the two is not a simple one. Waves in air may be detected by the normal human ear if their frequencies lie between about 20 and 20,000 Hz and if their intensities are within a certain range; the range of intensities audible to the ear also depends on the frequency of the wave. Figure 22-1 shows the range of frequencies and their intensities that are perceived as sound by the normal human ear; the intensity of the wave is plotted along the *y* axis, while the frequency of the wave is plotted along the *x* axis. Because of the wide range of intensities, these are plotted not on a uniform scale but on a logarithmic scale. One scale shows the intensities in watts/cm². Another scale shows the intensities in terms of the pressure changes in the wave that strikes the eardrum; since the pressure in a wave varies sinusoidally, the *effective* or *root-mean-square* values of the pressure changes are used. (Compare this with the effective value of an alternating current in Chapter 27.) The lower curve represents the *threshold of audibility*. A point on this curve represents the smallest

intensity of a sound of given frequency that is just audible to the average ear. The ear is most sensitive to sounds of about 3000 Hz. Points on the upper curve represent intensities that are so great as to be painful. The region between the two curves represents the range of hearing. Above a certain intensity, known as the *threshold of feeling*, the sound is not heard but is felt by the ear as a painful sensation. The range of intensities to which the ear is sensitive is about a millionfold. Because of this large range of intensities, a *logarithmic scale* has been adopted for expressing the *level of intensities of sound*, taking the zero level at about the limit of audibility of sound. The intensity level *B* of a sound is defined as

$$B = 10 \log \frac{I}{I_0} \qquad (22\text{-}1)$$

where I is the intensity of the sound and I_0 is the zero level of intensity that is taken arbitrarily to be equal to 10^{-16} watt/cm² or 10^{-12} watt/m². The intensity level B is expressed in *decibels* (db). Thus, if a sound has an intensity $I = 10^{-14}$ watt/cm², its intensity level is

$$B = 10 \log \frac{10^{-14}}{10^{-16}} \text{ db}$$

or

$$B = 10 \log 100 \text{ db}$$

from which

$$B = 20 \text{ db}$$

Sound levels have been measured at various

places under a variety of conditions. For example, inside some noisy subway cars the sound level is about 100 db, while the threshold of feeling (or pain) is about 120 db; the sound level of a whisper in a quiet room is about 15 db.

22-3
Frequency of a musical tone

A musical tone is regarded as a pleasing sound, while a noise is usually thought of as disagreeable; there are some sounds that are difficult to classify. A musical sound, for example, can be produced by a series of regular blasts of air, while a noise results when these blasts occur at irregular intervals. This can be demonstrated by means of a disk containing five concentric rings of circular holes, as shown in Figure 22-2. In the innermost ring, these holes are irregularly spaced; the next four rings have circular holes that are regularly spaced. There are 40, 50, 60, and 80 holes in these rings, respectively. When this disk is rotated at uniform angular speed and a stream of air is directed at the innermost ring of holes, an unpleasant noise will be heard. But when the stream of air is directed against any of the other rings, a pleasant musical tone will be heard. When the stream of air is directed against the four outer rings from the second to the fifth in sequence, the *pitch* of the sound coming from the third ring will be higher than that from the second ring of holes; the pitch of the sound from the fifth ring will be heard as an octave higher

than that from the second ring. The physicists' method of describing these tones is in terms of the *frequency* of the sound produced. For example, if the disk is rotating at the rate of 10 rev/sec, the frequency of the sound produced by the ring with 40 holes in it is 400 vibrations/sec; the next ring produces 500 vibrations/sec; the one after that, 600 vibrations/sec; and the last one, 800 vibrations/sec. To a first approximation we can say that the pitch of a tone depends upon its frequency, the tone with the higher pitch having the higher frequency. Two tones an octave apart have a frequency ratio of 2:1, for example. A musician will recognize the tones that have frequencies in the ratios 4:5:6:8 as the tones comprising a major chord.

22-4
Resonance

An interesting phenomenon occurs when a body that is capable of vibrating at a definite frequency receives small impulses of the same frequency. These impulses set the body into vibration, with each succeeding impulse building up the amplitude of the vibration. This phenomenon is known as *resonance*. A simple way of demonstrating resonance is to take two tuning forks having the same natural frequency and place them a short distance apart. One tuning fork is set vibrating by a hammer blow. After a short time interval, it will be found that the other tuning fork is vibrating and emits sound. The compressions and rarefactions produced in the air by the first tuning fork set the second tuning fork vibrating. Since the sound wave and the tuning fork have the same frequency, the impulses on the tuning fork are properly timed to build up its amplitude of vibration. A steady state is reached when the energy radiated by the second tuning fork is equal to the energy it receives from the first one. Resonance of sound is sometimes called *sympathetic vibration*.

Resonance can occur between any two bodies which can vibrate with the same natural frequency. An interesting example is shown in Figure 22-3, which illustrates resonance between a tuning fork and an air column. A hollow cylindrical glass tube is inserted in a jar of water. The vibrating system is the air in the hollow tube; the length of the air column can be varied by moving the tube up or down in the water. The air column ends at the surface of the water. If a tuning fork vibrating at a known frequency is held over the open end of the hollow tube, there probably will be no appreciable

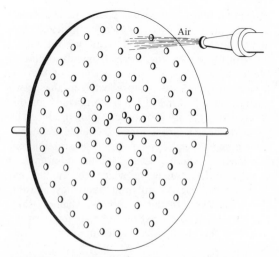

FIGURE 22-2 Construction of the wheel for showing the difference between musical tones and a noise.

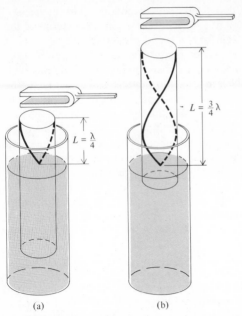

FIGURE 22-3 Resonance between an air column and a tuning fork. (a) Length of air column is λ/4. (b) Length of air column is $\frac{3}{4}$λ.

change in the loudness of the sound. But if the hollow tube is raised, at some position there will be a marked increase in loudness of the sound. At this position the air column in the tube is set into vibration with the same frequency as the tuning fork; the two are in resonance. We may think of the process of changing the length of the air column as "tuning" it to the frequency of the wave incident upon it. This may be compared to the tuning of a radio circuit to the same frequency as the incident electromagnetic wave.

A tuning fork that vibrates with a frequency f emits a wave of length λ, given by the usual equation

$$V = f\lambda$$

where V is the speed of sound in air. When an air column that is closed at one end is set into vibration, standing waves are produced in the air with a node at the closed end and an antinode or point of great motion at the other end. Since the distance between two successive nodes is half a wavelength, the distance between a node and the adjacent antinode is a quarter of a wavelength. Thus the shortest length of tube L in which the air can be in resonance with a wavelength λ is

$$L = \frac{\lambda}{4} \qquad (22\text{-}2)$$

In the above experiment, if the tube is long enough it will be found that resonance will occur again when the length of the tube is three quarters of a wavelength, for this length of air column will also have a node at the closed end and an antinode at the open end. Figure 22-3a shows the position of the tube when the length of the air column is one quarter of the wavelength of the sound. A graph of the stationary wave in the air with a node at the water surface and an antinode at the open end is shown in the air column. Figure 22-3b shows the same tube when the length of the air column is three quarters of a wavelength, together with a graph showing this mode of vibration of the stationary wave.

Tuning forks are often mounted on hollow boxes whose air columns are in resonance with the sounds emitted by these forks. The rate at which energy is radiated from such a system is greater than from the tuning fork alone.

Resonance occurs when electromagnetic waves interact with matter. When the frequency of the electromagnetic waves is the same as the frequency of vibration of one of the internal structures in the matter, then resonance can occur. In general, this will lead to absorption of energy from the electromagnetic wave and the transfer of that energy to the matter.

22-5
Quality of a musical sound

When two tones of the same pitch and same loudness are produced by two different musical instruments, such as a violin and a clarinet, the sensations produced by them are decidedly different. We recognize this difference because of the difference in *quality* or *timbre* between these two musical sounds. One of the main reasons for this difference in quality is that each sound produced by an instrument is not a tone of a single frequency but a complex sound consisting of several different frequencies. Another reason for the difference in quality is the manner in which the human ear responds to tones of different frequency and different loudness. In this section we shall consider only the effect produced by the complexity of the sound on the quality of a tone emitted by vibrating bodies.

As shown previously, the vibrations in a body

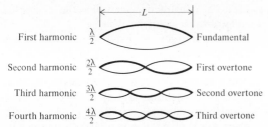

First harmonic $\frac{\lambda}{2}$ Fundamental

Second harmonic $\frac{2\lambda}{2}$ First overtone

Third harmonic $\frac{3\lambda}{2}$ Second overtone

Fourth harmonic $\frac{4\lambda}{2}$ Third overtone

FIGURE 22-4 Modes of vibration of a string that is fastened at both ends.

can be analyzed in terms of stationary waves that are set up in it by the interference of two waves traveling in opposite directions. The *fundamental mode of vibration* of a body corresponds to the longest wave, or wave of lowest frequency, that can be set up in the body. For example, in the case of a string that is fastened at both ends, the fundamental mode of vibration is one in which the string vibrates as a whole, as shown in Figure 22-4. Since the two ends are fastened, the standing wave must have nodes at these points, while the antinode is midway between them. The length L of the string is thus half the wavelength $\lambda/2$ of the transverse wave traveling in the string. The pitch of the sound emitted by the string when vibrating as a whole is called the *fundamental tone* of the string. The string can also vibrate in two parts, with a node in the center; the frequency of this sound is twice that of the fundamental tone. When vibrating in this manner, the pitch of the tone will be an octave higher than the fundamental. Other modes of vibration of the string are shown in the figure. When a string is set into vibration by plucking it or bowing it, several of these modes of vibration will be set up simultaneously; in addition to the fundamental tone, some of the higher-pitched tones or *overtones* will be emitted by the string. *The quality of the tone will depend upon the number of overtones produced and their relative intensities.*

The same thing is true of other vibrating bodies: the quality of the sound depends upon the number and relative intensities of the overtones produced. Although in the case of string instruments and wind instruments the frequencies of the overtones are whole multiples of the frequency of the fundamental tone, this is not generally true of other musical instruments, such as bells, chimes, and drums.

22-6
Vibrating air columns

Wind instruments, such as the clarinet, the trumpet, and the pipe organ, all have vibrating air columns to reinforce some of the sounds produced by the source of sound. In this discussion we shall consider only cylindrical pipes, such as those commonly used in pipe organs. There are two general classes of these pipes: the *open* pipe (that is, a pipe open at both ends) and the *closed* pipe (that is, a pipe closed at one end); the end containing the source of vibrations is always considered an open end. The vibrations can be produced in one of several ways, such as blowing air against a reed and setting it vibrating, or blowing a thin sheet of air against a thin lip at one end and setting the air into vibration. Whatever the method of setting up the vibrations, the column of air in the pipe will reinforce those modes of vibration corresponding to the standing waves that can be set up in this column.

Figure 22-5a shows several modes of vibration that can be set up in a closed organ pipe. The method for determining these modes of vibration depends upon the fact that only those vibrations can exist in the air column that have a node at the closed end and an antinode at the open end. It can be seen that the fundamental mode of vibration corresponds to a wavelength λ that is four times the length L of the air column. The first overtone possible in this case is one whose wavelength λ is four

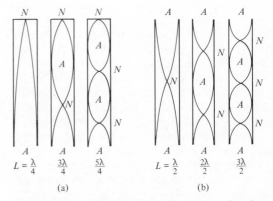

FIGURE 22-5 (a) Modes of vibration of an air column closed at one end. (b) Modes of vibration of an air column open at both ends. The nodes are at the positions marked *N*, and the antinodes are at the positions marked *A*.

thirds of the length L of the air column. The frequency of the first overtone is therefore three times the frequency of the fundamental tone. The frequency and the wavelength are related by the usual equation

$$V = f\lambda$$

where V is the speed of sound in air.

An examination of all other possible modes of vibration of an air column in a closed pipe shows that all the overtones are odd harmonics; their frequencies are in the ratios of $1:3:5:7:\cdots$.

A pipe open at both ends must have antinodes at these ends. The fundamental mode of vibration of the air in an open pipe is shown in Figure 22-5b; its wavelength λ is twice the length of the pipe. The first overtone has a wavelength λ that is equal to the length L of the pipe; its frequency is therefore twice that of the fundamental. An examination of the other possible modes of vibration shows that all the harmonics may be set up in this air column.

22-7
Musical instruments and the voice

Most musical instruments consist of two major parts: a generator and a resonator. The generator oscillates and provides the fundamental and, usually, many overtones. These vibrations are transmitted by resonance to the resonator. The resonator selects some of these frequencies (sometimes adding some of its own) and radiates relatively high-intensity sound.

In the violin, probably the most complex musical instrument, the generator is the set of strings. When bowed, these strings are plucked very rapidly because the hair of the bow alternately sticks and slips. Thus the string is excited by an irregular, rather than smooth, motion; this motion can be modified significantly, if subtly, by the skill of the violinist. The resulting vibration of the string is then rich in harmonics; the harmonic content is also varied by the location of the bowing as well as the method of bowing. The resonator is the body of the violin and the string is connected to the body by the bridge. The shape of the violin body, as well as its structure, has been developed by an empirical process so as to maximize the quality of the radiated sound. This involves the shape, materials, and construction techniques—all of which are beyond the analytic ability of physics to predict. The construction, as well as the playing, of a violin remains an art.

The resonator of the wind instruments is the body of the instrument, and the generators are the reeds or the lips of the musician or, in the flute, the air mass at the mouthpiece. Here again, the skill of the musician determines the harmonic content of the original vibrations.

When the human voice is used as a musical instrument, the generator is the vocal cords, which also serve as a valve, initiating and halting the flow of air. The resonator is the set of cavities in the head, including the throat, mouth, and to some extent the nasal cavity. The resonant frequencies of this cavity are changed by changing the size and shape, principally by moving the lips, tongue, and cheek muscles. It should be noted that when a person sings, the melody is expressed by sounding the vowels rather than the consonants. This is because the vowels are *voiced* sounds, sounds produced by the unobstructed breath. (Each of the vowels has a relatively characteristic distribution of frequencies.) The consonant sounds are produced by partial or complete obstruction of the breath.

22-8
Beats

When two sources of sound having slightly different natural frequencies, f_1 and f_2, send out waves of approximately the same amplitude into the same region of space, the effect produced at any point in this region will be a periodic variation of the intensity of the sound, known as *beats*. For simplicity, the two waves of frequencies f_1 and f_2 are drawn with the same amplitude in Figure 22-6a and b, respectively. The ordinate y is the variation in pressure of the air as a function of time, with the value $y = 0$ taken as atmospheric pressure. The effect produced at a point through which these two waves pass is obtained by adding the ordinates of the two waves, as shown in Figure 22-6c. It will be observed that at some instant of time the two waves will produce a wave of increased amplitude; this will occur when the two waves are in phase. A short time later the amplitude of the resultant wave will decrease to practically zero; this will occur when the two waves are out of phase by half a period; that is, when a compression from one wave occurs simultaneously with a rarefaction from the second wave. At this time the intensity of the sound will drop to practically zero.

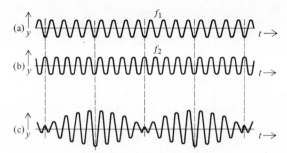

FIGURE 22-6 Beats produced by the interference of two waves of slightly different frequencies.

It can be shown that the number of times these two waves will get out of phase per unit time is equal to the difference in the frequencies of the waves. Each time these waves get out of phase a *beat* will occur, hence, the number of beats per second is $f_1 - f_2$. For example, if one tuning fork is vibrating at the rate of 256 Hz and a neighboring tuning fork is vibrating at the rate of 261 Hz, 5 beats will be heard per second. The production of beats is an interference effect of the two waves.

The phenomenon of beats is frequently used in tuning two sources of sound to the same pitch. This is a very accurate method of tuning, since the ear can perceive beats that occur only once in about 10 sec. When there are only a few beats per second, the sound produces an unpleasant effect. When the difference in frequencies is large, no beats can be distinguished in the sound produced.

22-9
The Doppler effect

When a source of sound is moving with respect to an observer, the frequency measured by the observer is different from the frequency emitted by the source. This effect is called the Doppler effect. A similar effect occurs for electromagnetic waves, such as light and radio waves, although the details are somewhat different from the case of sound waves. In analyzing the Doppler effect for sound, we must consider two distinct cases: that where the observer is standing still (with respect to the medium) while the source is moving, and that where the source is standing still while the observer is moving.

Suppose that an observer is stationary at O, as shown in Figure 22-7, and that the source of sound is stationary at S. If the source emits f vibra-

tions/sec, the length of the wave λ emitted by it will be

$$\lambda = \frac{V}{f}$$

where V is the velocity of the sound. For simplicity, let us choose S to be at a distance from O equal to the distance traveled by sound in 1 sec; that is $SO = V \times 1$ sec, where V is the speed of sound. Then, when the source and the observer are both stationary, there will be f waves in the distance SO, each of length λ. Let us now suppose that the source is moving with speed v toward O. At the end of 1 sec the source will have moved to S', where $SS' = v \times 1$ sec. During this time, the source has emitted f vibrations; the first one has already reached the observer at O, and the last one has just left the source at S'. These f vibrations are therefore located in the region $S'O$, whose length is

$$S'O = (V - v) \times 1 \text{ sec}$$

Since f waves have been emitted in this second, the length of these waves is

$$\lambda' = \frac{V - v}{f} \qquad (22\text{-}3)$$

These waves travel with the velocity of sound, V, and the frequency f' with which they reach the ear is therefore

$$f'\lambda' = V \qquad (22\text{-}4)$$

Eliminating λ' from Equations 22-3 and 22-4 yields

$$f' = f\frac{V}{V - v} \qquad (22\text{-}5)$$

In other words, more waves will now reach the ear per second than reach it when the source is stationary. This will be interpreted as a sound of higher pitch. The change in pitch produced by the

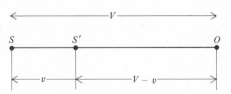

FIGURE 22-7 The sound waves that are emitted in unit time by the source as it moves from S to S' with speed v toward the observer at O are contained in the distance $S'O$.

relative motion of source and observer is known as the *Doppler effect*.

The same reasoning can be applied to show that when the source is moving away from the observer with a velocity v, the frequency f' of the sound reaching the observer is given by

$$f' = \frac{V}{V+v}f \qquad (22\text{-}6)$$

The pitch of the sound in this case is lower than the pitch of the sound when the source is stationary.

It is instructive to analyze the Doppler effect in terms of the waves emitted by the moving source. Let us assume that the source emits spherical waves, which in Figure 22-8 are drawn as circles with successive positions of the source as centers. These successive positions are shown at time intervals equal to T, the period of the vibrations emitted by the source. In the figure, S is the present position of the source, S_1 is the position of the source at a time T earlier than S, S_2 the position at a time $2T$ earlier, and S_3 the position at a time $3T$ earlier. The wave emitted when the source was at S_1 has traveled a distance VT, where V is the speed of sound; hence this wave is represented by a circle of radius VT. Similarly the wave emitted when the source was at S_2 is drawn as a circle of radius $2VT$; that emitted from S_3 is drawn as a circle of radius $3VT$. The source of sound is moving toward the right with a speed v which is less than V.

An observer in front of the moving source will receive more waves per second than if the source had been at rest. Conversely an observer behind the moving object will receive fewer waves per second than if the source had been at rest. The observer in front of the moving source will hear a higher pitched sound than the observer behind the source. When the moving source passes the observer, he will always note a drop in the pitch of the sound.

The frequency of the sound f' received by the observer in front of the source is given by Equation 22-5 and can be derived very simply by referring to Figure 22-8. The distance between successive wavefronts that reach this observer is $(V-v)T$ and is therefore the length of the wave λ' perceived by the observer; that is,

$$\lambda' = (V-v)T$$

but

$$T = \frac{1}{f}$$

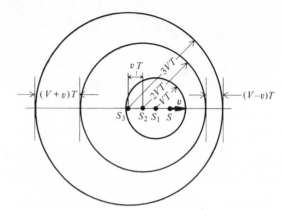

FIGURE 22-8 Waves emitted by a source moving to the right with speed v. The points S_3, S_2, S_1, S, show the positions of the source at time intervals equal to T, the period of the vibration. Therefore, these points are separated by equal distances vT.

and thus

$$\lambda' = \frac{V-v}{f}$$

and, since these waves travel with speed V in the air, the frequency f' of these waves is given by

$$f'\lambda' = V \qquad (22\text{-}4)$$

so that

$$f' = f\frac{V}{V-v} \qquad (22\text{-}5)$$

follows immediately.

Equation 22-6 can be derived in a similar manner.

Figure 22-8 is instructive in that it can be used to determine the frequency of the sound heard by a stationary observer who is not in the line of motion of the source. It will be noted that the distance between wavefronts increases from the smallest value $(V-v)T$ for an observer in front of the source to its largest value $(V+v)T$ for an observer behind the source. To an observer at right angles to the line of motion, the distance between wavefronts is simply VT or λ; that is, it is exactly the same as if the source were stationary.

When the source is stationary and the *observer* is *moving* toward the source with a velocity v, the pitch is higher than that heard when the observer is stationary, but the actual value of the new frequency is slightly different from that given by Equation 22-5. The wavelength of the sound in air remains unchanged, but as the observer moves toward the source, he receives more waves per second than he receives when standing still. If he moves toward the source with a velocity v, he will receive v/λ additional waves per second, or a total of

$$f' = \frac{V}{\lambda} + \frac{v}{\lambda}$$

or

$$f' = \frac{V + v}{\lambda}$$

Now, since

$$f\lambda = V$$

we get

$$f' = \frac{V + v}{V} \qquad (22\text{-}7)$$

which gives the new frequency of the sound heard by the observer.

In a similar manner, if the observer is moving away from the source of sound, it can be shown that the frequency f' of the sound heard by the observer is given by

$$f' = \frac{V - v}{V} f \qquad (22\text{-}8)$$

which is a sound of lower pitch than that heard by the observer when stationary.

It must be emphasized that in any one of these cases, the observer hears only one tone; he does not hear a change in pitch. Only when the motion is changed can he hear a change in pitch. Such a change in pitch can be noted when a train sounding its whistle passes an observer; the observer will then hear a drop in pitch as the train passes him.

An interesting combination of the Doppler effect and the phenomenon of beats can be produced by moving a tuning fork rapidly toward a wall. The observer will receive two sounds: one directly from the tuning fork, and one reflected from the wall. The apparent source of sound of the reflected wave is the image of the tuning fork formed by the wall acting as a plane mirror. While the tuning fork is moving away from the observer, its image is moving toward him. The direct wave from the tuning fork will have a lower pitch than the wave coming from its image, and the observer will hear beats.

Illustrative example

The siren of a fire truck is emitting a tone whose frequency is 1200 Hz. The fire truck is traveling with a speed of 60 mi/hr. A man in the street notices a drop in pitch as the truck passes him. Determine the change in frequency of the tone heard by this observer.

While the fire truck was moving toward the observer at a speed of 88 ft/sec, he heard a tone whose frequency was higher than 1200 Hz. This frequency f_1' can be determined from Equation 22-5

$$f_1' = \frac{V}{V - v}$$

yielding

$$f_1' = 1200 \times \frac{1100}{100 - 88} \text{ Hz} = 1304 \text{ Hz}$$

As the fire truck passed the observer, it moved away from him with a speed of 88 ft/sec, and the tone he heard had a frequency lower than 1200 Hz. This frequency can be determined from Equation 22-6

$$f_2' = f \frac{V}{V + v}$$

yielding

$$f_2' = 1200 \frac{1100}{1100 + 88} \text{ Hz} = 1111 \text{ Hz}$$

Hence the drop in pitch of the tone heard by the observer was due to a change in frequency of

$$f_1' - f_2' = 193 \text{ Hz}$$

22-10
Velocity of source greater than velocity of sound

When a body such as a projectile, a jet plane, or a rocket moves with a velocity v greater than the velocity of sound V in the medium, it sets up a compressional wave, as shown in Figure 22-9. The wavefront, sometimes called a *shock wave*, is a cone, with the moving body at its apex S. The cone inside which the sound waves travel can be constructed by drawing spherical waves that originated at various positions of the source during its motion. In Figure 22-9, S is the present position of the source, S_1 its position at a time t earlier, S_2 its position at a time $2t$ earlier, and S_3 its position at a time $3t$ earlier, where t is an arbitrary time interval. With S_1 as a center, we draw a circle of radius Vt; with S_2 as a center, we draw a circle of

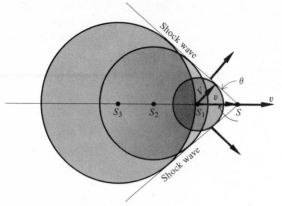

FIGURE 22-9 Waves emitted by source moving with speed $v = 1.5T$.

ratius $2Vt$; and so forth. These circles represent the present positions of the compressions which started from S_1, S_2, and S_3. The wavefront is the tangent to these circles. In a three-dimensional diagram, this wavefront would be a cone whose elements are tangent to spheres of radii Vt, $2Vt$, and $3Vt$, respectively. In the unit of time that the wave progresses a distance V, the source moves a distance v, as shown in the figure. The angle θ that an element of the cone makes with the direction of motion of the source is

$$\sin \theta = \frac{V}{v} \qquad (22\text{-}9)$$

since V is at right angles to the element of the cone and v is the hypotenuse of the right triangle.

Questions

1. What is the ratio of the frequencies of two musical tones that are an octave apart?

2. What are the three characteristics of a musical tone?

3. Describe how a loud, high trumpet tone differs from a low, soft viola tone.

4. Explain what is meant by the statement, "The sound level in the room is 25 db."

5. An open organ pipe and a closed organ pipe are tuned to the same pitch. What is the relationship between the lengths of these air columns?

6. A bugle is an open air column of fixed length. How are the different tones produced?

7. What are the generator and resonator of the kettledrum?

8. Which speech sounds are affected by a cold in the nose?

9. By what path in addition to the air does the sound of your own voice reach your ears? Describe how this might explain why most people are surprised by the sound of their own recorded voices the first time they hear them.

10. You can get water sloshing back and forth in a bathtub at only certain frequencies. Why? Extend this to explain why some bays and in- lets have large tides, while other bays have small tides.

11. Assume that you have a 256-Hz tuning fork and an out-of-tune guitar string, which is supposed to vibrate at 256 Hz. Carefully describe how you would use the beat phenomenon to tune the string, even if you do not know which way to turn the key to raise the pitch of the string and you do not know if the string starts out sharp or flat.

12. How does the apparent pitch of a sound emitted by a moving source change as the source passes a stationary observer?

13. The speed of a jet plane is often described in terms of the Mach number, the ratio of its speed to that of sound. Explain why this is a useful way of describing the speed of super-sonic planes.

14. What is the "sonic boom" produced by jet planes?

15. There is an old bridge in England with a sign on it, ordering soldiers to break step before marching across the bridge. Why this order?

16. Large suspension bridges, such as the George Washington Bridge, emit sound with a characteristic frequency of a few hertz. Discuss the possible origins of this inaudible sound.

Problems

1. Show by direct substitution that if a sound has an intensity of 10^{-12} watt/m^2, the intensity level is 0 db.

2. At what intensity will a sound have an intensity level of 2 db?

3. A siren wheel has 30 uniformly spaced holes

near its rim and is rotated by means of a stream of air. **(a)** What is the frequency of the sound wave emitted when the speed of the wheel is 35 rev/sec? **(b)** What is the wavelength of the sound in air?

4. A boy runs by a picket fence at a speed of 1 ft/sec. The spacing of the pickets is 3 in. The boy holds a stick in his hand, which rubs against the pickets, producing a sound. Find the fundamental frequency of this sound.

5. What is the lowest frequency note that can be produced by an organ pipe open at both ends and 5 ft long?

6. An organ pipe open at one end is filled with hydrogen. Find the fundamental frequency if the pipe is 6 ft long.

7. An open organ pipe sounds the note A, which has a frequency of 440 Hz. What are the first and second overtones?

8. A clarinet behaves like an organ pipe closed at the reed and open at the valve. If the distance from the reed to an open valve is 24 cm, find the frequency of the fundamental and the first two overtones.

9. An old physics book describes a trumpet as sounding notes like an open pipe. The principal notes are described as having frequencies of 256, 320, 384, 448, 512, 576, and 640 Hz. (There are many higher notes possible.) The book also says that on account of the difficulty in striking it, the lowest note is not used. What is the frequency of this lowest note?

10. A train traveling at 60 mi/hr emits a 212-Hz whistle. **(a)** What frequency does a stationary observer hear as the train approaches? **(b)** What frequency does the observer hear as the train recedes?

11. A 500-Hz tone is sounded at the same time that a 508-Hz tone is sounded. How many beats per second are heard?

12. An airplane traveling at 1800 ft/sec in air at 0°C produces a shock wave. Find the angle of the cone of the shock wave.

13. A jet plane flying at 40,000 ft altitude and at a speed of 950 mi/hr produces a shock wave. How far away is the plane when an observer on the ground detects the shock wave? (Assume that the air is homogeneous and has a temperature of 0°C.)

14. A popular band is generating sound at the threshold of pain. How much energy in watts is absorbed by the face of a listener, taking the face as a disk of radius 12 cm?

15. The intensity of sound due to independent sources is the sum of the intensities of each of the sources. The intensity level of the average whisper is 20 db. What is the intensity level due to a class of 35 whispering students, assuming they are all whispering independently and that they are all average whisperers? How many whisperers would make a sound of 65 db, the level of ordinary conversation?

16. Two identical organ pipes, 36 cm long, are sounded when they are next to each other. **(a)** What is the fundamental frequency of the sound? **(b)** A sliding collar on one of the pipes is moved up, making the pipe longer and producing 3 beats per second. How much longer is the pipe? **(c)** What would be the beat frequency between the first overtones if one somehow suppressed the fundamentals?

17. A man walks at a speed of 4 m/sec along the line between two fire stations. Both alarms sound simultaneously with a frequency of 500 Hz. Find the two apparent frequencies and the beat frequency heard by the man.

18. Two tuning forks, A and B, are observed to produce beats at the rate of 5 per second, When fork A is loaded with a bit of putty, the rate of beats increases to 7 per second. Fork A has a frequency of 440 Hz. What is the frequency of fork B when it is not loaded?

19. As a model of how a radar speed detector operates, consider a transmitter of high frequency f_0 emitting sound that strikes an automobile moving away from the transmitter with a speed v. The sound is reflected back toward the transmitter and is Doppler-shifted as if it were emitted from an object moving with twice the speed of the automobile. This reflected sound is mixed with the original transmitted sound and the beat frequency measured. Find an equation giving the beat frequency b in terms of the original frequency f_0, the speed of sound c, and the speed of the automobile, v.

electricity
& magnetism

part 4

electrostatics

chapter 23

23-1
Introduction

One simple phenomenon of electricity was known to the ancients: that when a piece of amber was rubbed, it acquired the property of attracting bits of straw, leaves, or feathers. It is believed that Thales of Miletus (c. 600 B.C.) knew of this property of amber. The Greek word for amber is elektron, hence the name *electricity* for this subject. There was practically no further development of this subject until about the seventeenth century. William Gilbert (1544–1603) and Robert Boyle (1627–1691) did some experiments and wrote on the subject, and in about 1663, Otto von Guericke (1602–1686) made an electrical machine that gave off sparks. It was not until the eighteenth century that real progress was made, however. It was then found that there were two kinds of electricity: one similar to the kind acquired by amber when rubbed with wool and the other similar to that acquired by glass when rubbed with silk. They were called *resinous* and *vitreous* electricity, respectively. These are now known as *negative* and *positive* electricity—names first introduced by Benjamin Franklin (1706–1790). Franklin made many contributions to the subject, both experimentally and philosophically. He showed the electrical character of lightning, and designed lightning rods for the protection of buildings. The subject of electricity was put on a firm mathematical foundation as a result of the experiments of Coulomb, in 1785, on the law of force between electrically charged bodies. After that, progress was very rapid.

We first introduced the concept of electric

charges and Coulomb's law of force between charges in Chapter 5. In this chapter we shall discuss this subject in greater detail, limiting the discussion to those aspects in which the charges assume fixed positions as a result of the forces that had been acting on them; hence the term *electrostatics*.

23-2
Electrical theory of matter

Since the turn of this century we have learned a great deal about the structure of matter and its behavior under the action of various external forces. Some of the important ideas have been discussed, and others will be developed as we progress. It is worthwhile at this point to outline briefly the electrical theory of matter to aid us in understanding the phenomena to be discussed later.

The basic constituents of matter are particles, most of which have the property we call *electric charge*, although some of the particles are *neutral* and have no net charge. An atom that is neutral consists of a small, positively charged, massive nucleus surrounded by small, negatively charged particles called *electrons*. The nucleus consists of positively charged particles called *protons*, and neutral particles called *neutrons*. The number of electrons surrounding the nucleus is equal to the number of protons in the nucleus of the atom. Since the charge of a proton is equal to the charge of an electron, but opposite in sign, the net charge of the atom is equal to zero.

It is possible, by the application of external forces, to remove one or more electrons from an atom. The remaining positively charged structure is called an *ion*. It is also possible to add excess electrons to an atom, yielding a negatively charged ion. Forming an ion from an atom always involves energy changes, but the amount of energy involved varies markedly from one kind of atom to another. For example, comparatively large amounts of energy are required for the removal of electrons from inert elements, whereas the atoms of chemically active elements require relatively little energy to form ions.

When atoms are brought very close together, as they are in liquids and solids, the electric charges exert forces on each other, which leads to some rearrangement of the charges. In liquid solutions, this often causes the formation of ions of both signs of charge. In metals, electrons are loosely associated with the atoms and become capable of moving freely through the material. In most solids the charges in the interior of the solid are more or less rigidly fixed, although external forces can cause some further rearrangement of the charges. In a few materials, relatively small external forces can cause the appearance of free charges capable of motion.

Thus all objects contain very large numbers of charges. Generally the number of negative charges is exactly equal to the number of positive charges, and the object is neutral. If the number of positive charges exceeds the number of negative charges, (or vice versa), then the object is charged, and has a total charge given by the *excess* charge. Generally, when we refer to *the charge* of an object we mean this total of excess charge.

For most solids the only charge that can move is that carried by electrons. Thus when an object is charged positively, this usually means that electrons have been *removed* from the object, while negatively charged objects have had electrons added to them. However, one usually introduces no serious error by treating both positive and negative charges as mobile. That is, one can usually treat a positively charged object as one which has had positive charge transferred to it *or* as one which has had negative charge removed from it.

23-3
Elementary electrical phenomena

We can use the idea that like charges repel and unlike charges attract, along with the electrical theory of matter, to understand some simple phenomena.

Charging by rubbing: The fact that amber when rubbed became charged was, as we have seen, the first phenomenon of electricity. We now know that *any* object can be charged by rubbing. Basically this happens because when two objects are rubbed together, the charges on their surfaces are brought very close together. The forces created are very large and can move electrons from one object to the other. Thus the important part of this phenomenon is not so much the rubbing as the close contact between surfaces. In fact, striking amber with wool is more effective at charging the amber than rubbing. Similarly, pressing cellophane tape on a piece of metal and then stripping the tape off is an effective way of charging the metal.

Charging by contact: If a charged object touches an uncharged object, some of the charge will be

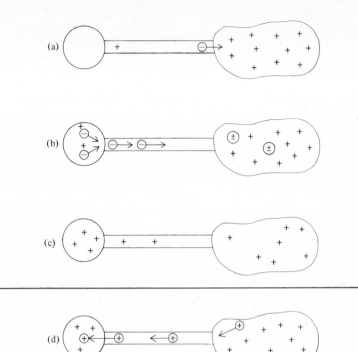

FIGURE 23-1 (a) The charged object attracts free electrons from the wire, leaving the wire charged. (b) The charged wire attracts free electrons from the neutral object. (c) The result is to leave the originally neutral object with charge and to reduce the amount of charge on the charged object. (d) A simpler description of the same process results from assuming that the positive charges can move.

transferred because the excess charges exert forces of repulsion on each other. Hence some of the excess charges can be pushed onto the neutral object.

Conduction: If a charged object is connected to an uncharged object by a piece of metal, then charge will be transferred to the neutral object through the metal. This occurs because the metal contains free charges. Initially, charge is transferred from the charged object to the metal. These transferred charges exert forces on the free charges, causing the free charges to move through the metal. When the free charges reach the uncharged object they are transferred to that object.

In the preceding paragraph we described the transfer of charge through a metal in terms of the motion of the free charge without specifying the sign of the charge. We know that in metals the charge that can move is negative, so this description is acceptable when the charged object is charged negatively. The description of the transfer of charge from a positively charged object is somewhat more involved. Figure 23-1a shows the initial situation, with a positively charged object in contact with a metal wire, which is in turn connected to a neutral object. The positive charges attract the free electrons in the wire, which move into

the charged object. This leaves the end of the wire charged positively; it then attracts free negative charges from the rest of the wire. Thus free charges move from the end of the wire through the wire into the charged object, where they cancel some of the positive charges. This motion of charges from the near end of the wire leaves the farther end of the wire charged positively, and then free negative charges are attracted from the neutral object, as shown in Figure 23-1b. The net result is that negative charges have moved out of the neutral object (leaving it positively charged), through the wire, and into the initially charged object, neutralizing some of its positive charge. The positively charged object will then have less positive charge than it started with, as shown in Figure 23-1c.

If we had assumed that there exist free positive charges, then the description would have been simpler. As shown in Figure 23-1d, we could then describe the process in terms of positive charge in the wire being repelled by positive charge in the object and moving onto the neutral object. Now we know that only free negative charges exist, and that the positive charge is fixed in most solids. However, both descriptions end up describing the same final situation, so it is simpler to talk about the motion of the positive charges. In fact, it is ex-

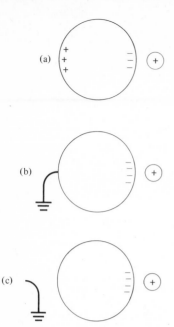

FIGURE 23-2 Charging by induction.

tremely difficult experimentally to tell which kind of charge is moving. Furthermore, there are some materials in which positive charge does move. Therefore it is common practice to describe these processes in terms of the motion of positive charge when to do so simplifies the description.

Charges can be transferred through all materials, but some materials conduct electricity better than others. Metals conduct very well; that is, almost any amount of charge placed near a metal will cause the motion of free charge. Other materials, such as ceramics or rubber, are very poor conductors; it requires large nearby charges to create free charge in these materials. Poor conductors are called insulators. As one might guess, this division into conductors and insulators is arbitrary. The ability to conduct electricity is measured numerically by a property called *conductivity*, and there exist materials with conductivities of almost every value intermediate between that of the best conductors and the best insulators. Materials with conductivities intermediate between conductors and insulators are called *semiconductors*. The conductivity of a particular material depends on its chemical composition (and purity), and external conditions, such as pressure, atmospheric humidity, temperature, and on the age of the sample. Furthermore, there are circumstances when *rela-*

tive conductivity is more important than absolute conductivity. For example, under ordinary circumstances air is an insulator. When enough charge is present, air can conduct well enough to leak charge, carry sparks, or even conduct lightning. Nevertheless it is convenient to classify materials as conductors, semiconductors, and insulators.

Charging by induction: Consider bringing a positive charge near a neutral metal object, without touching. As shown in Figure 23-2, this will cause free negative charges in the metal to move to the surface, near the adjacent positive charge. This will leave some distant portion of the metal object charged positively. Now assume that the metal object is connected by a wire to the ground. Since the earth is a comparatively good conductor, free charges will move up through the wire from the earth, neutralizing the positive charge. If the wire is now removed, the metal object will be left with excess negative charge. Thus momentarily connecting the metal object to the earth (and then disconnecting the object from the earth) while a positive charge is nearby leaves the metal object charged negatively, or charged oppositely to that of the nearby charge. Similarly if a negative charge were placed nearby and the object momentarily connected to the earth, it would be left positively charged. Instead of using a wire, this process is often done by having the experimenter touch the metal object. In that case the charge flows through the experimenter's body or over the surface of his skin.

23-4
Coulomb's law

As we saw in Chapter 5, the force between charged bodies can be described in terms of the force between point charges by Coulomb's law: the force between two point charges, q_1 and q_2, which are a distance r apart, is directed along the line joining the two points and has a magnitude given by

$$F = k \frac{q_1 q_2}{r^2} \qquad . \quad (23\text{-}1)$$

where k is given by

$$k = 9 \times 10^9 \frac{\text{nt m}^2}{\text{coul}^2}$$

in the mks system of units, where charge is measured in coulombs. In the development of the theory of electricity, the quantity 4π appears in many of the equations. Many physicists and engineers prefer to minimize the appearance of this factor in the equations that are dealt with most frequently. In many fields of science Coulomb's law is used infrequently. Hence when used in electricity and magnetism, the mks system has been *rationalized* by setting

$$k = \frac{1}{4\pi\epsilon_0}$$

in Coulomb's law. The constant ϵ_0 (epsilon zero) is called the *permittivity of free space*, and has the value

$$\epsilon_0 = \frac{1}{4\pi k} \tag{23-2}$$

$$= 8.854 \times 10^{-12} \, \frac{\text{coul}^2}{\text{nt m}^2}$$

With the aid of Coulomb's law we can understand how a charged amber rod can attract light objects. In Figure 23-3, a negatively charged amber rod is shown near a piece of paper which is neutral. The charge on the amber rod causes some rearrangement of the charges on the paper, leaving the region of the paper that is close to the rod charged positively, and the part that is farther away is left charged negatively. The positive charges on the paper are attracted to the rod, while the negative charges are repelled. But since the positive charges are closer to the rod than the negative charges, the force of attraction is larger than the force of repulsion. Thus the paper as a whole is attracted to the rod.

Once the paper touches the rod, it will acquire a net negative charge from the rod and then be repelled.

23-5
Electric field intensity

Coulomb's law of force has the same mathematical form as Newton's law of gravitation and can be treated in an analogous manner, except for the fact that the gravitational force is always a force of attraction, whereas the electric force may be either a force of attraction or a force of repulsion. As we saw in Chapter 6, we can think of the gravitational force as arising from the existence of a gravitational field. In the case of the electric force we can think of one charge as creating an electric field and the field as exerting the force on the other charge; we can focus our attention on one charge

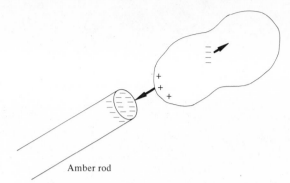

FIGURE 23-3 A charged rod attracts a neutral piece of paper by rearranging the charges in the paper. The forces exerted on the charges close to the rod are larger than the forces exerted on the more distant charges.

and think of the force exerted on that charge as an interaction between the charge and a field, and treat the origin of the field as a completely separate question. If we adopt this point of view, then we can think of every point in space as having associated with it an electric field. When a charge q is placed at a particular point in space, the electric field at that point will exert a force $\mathbf{F}$ on the charge. The force is given by

$$\mathbf{F} = q\mathbf{E} \tag{23-3}$$

where $\mathbf{E}$ is the *electric field intensity* at the location of the charge q. The mks unit of intensity is newtons per coulomb.

The electric intensity may vary from point to point in space and may be equal to zero in many regions. One can think of Equation 23-3 as a prescription for defining $\mathbf{E}$. Imagine placing a small positive charge q, called the *test charge*, at some point in space. This charge will have a force $\mathbf{F}$ exerted on it by the field. From Equation 23-3 the magnitude of $\mathbf{E}$ at the location of the test charge is

$$E = \frac{F}{q}$$

and the direction of $\mathbf{E}$ is the same as the direction of $\mathbf{F}$. If a negative charge were placed at the same location, the force would have a direction opposite to that of $\mathbf{E}$. With this definition we assume that the test charge does not alter the field.

It is often useful to visualize the electric field in a region of space in terms of *lines of force*. These lines are drawn so that the tangent to a line at any point will give the *direction* of the electric field intensity at that point; the *magnitude* of the elec-

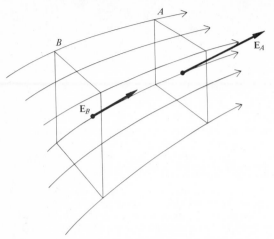

FIGURE 23-4 Representation of the electric field by lines of force. Since the lines crowd closer together at A, the field at A is larger than the field at B.

tric field intensity can be represented by the number of lines of force per unit area passing perpendicularly through a small area at that point. Figure 23-4 is such a representation of electric field intensity. Since the lines crowd closer together at A, and are farther apart at B, the electric field intensity is larger at A than it is at B. A suitable scale may be used for this representation; a commonly used scale is one in which *one line per unit area represents unit intensity*.

23-6
The electric field due to point charges

As mentioned above, the electric field is equal to the force per unit charge exerted on a test charge at a particular location. It must be emphasized that this is true only if the test charge is at rest. If the test charge is moving, then a magnetic force is also exerted. However, as we have seen, the magnetic force is proportional to the speed of the charge; thus the electric field intensity is equal to that part of the force per unit charge which is independent of the velocity of the test charge.

Since Coulomb's law states that charges exert forces on each other, it is clear that charges create electric fields. As we will see later, it is possible for electric fields to be created in other ways, but at this point we will restrict our attention to fields that are created by charges. For example, suppose that we wish to determine the electric field surrounding a point charge Q, which is at rest. Let us

imagine a small test charge q at a point distant r from Q; this test charge will experience a force of repulsion given by

$$F = \frac{Qq}{4\pi\epsilon_0 r^2}$$

The magnitude of the electric intensity at the position of the test charge is

$$E = \frac{F}{q}$$

hence

$$E = \frac{Q}{4\pi\epsilon_0 r^2} \tag{23-4}$$

If the charge Q is moving with a velocity v that is small compared to the speed of light, then the electric intensity at a point distant r from Q will remain approximately as given by Equation 23-4. If Q moves with a velocity comparable to the speed of light, then the electric field intensity will differ from that given by Equation 23-4.

The electric field intensity due to a point charge Q varies inversely as r^2. At any fixed distance r from Q, the intensity will have the same value. If we draw a sphere of radius r with the charge Q at the center, the electric field can be represented by a series of lines of force radiating from the charge and passing through the surface of the sphere, as shown in Figure 23-5. If we adopt the convention that 1 line/m² represents an intensity of 1 nt/coul, then the number of lines of force N passing through the surface of the sphere is given by the product of the electric field intensity and the area A of the sphere; thus

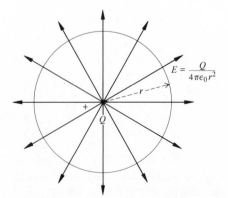

FIGURE 23-5 Q/ϵ_0 lines of force pass through a sphere of radius r enclosing a charge $+Q$ at its center.

$$N = EA$$
$$= \frac{Q}{4\pi\epsilon_0 r^2} 4\pi r^2$$

or

$$N = \frac{Q}{\epsilon_0} \qquad (23\text{-}5)$$

Thus the total number of the lines passing through a sphere of radius r is independent of the value of r. In other words, all the lines that start on the charge Q radiate out to infinity; no lines stop and no new lines are created in empty space. In addition, since the field is uniform on any one sphere of radius r, the lines spread out uniformly. If the charge were negative, the direction of the lines of force would be toward the charge. *The lines of force start on positive electric charges and end on negative charges.* Figure 23-6a shows the electric field in the neighborhood of two equal and opposite charges, and Figure 23-6b shows the field in the neighborhood of two equal charges of the same sign.

Looking at Figure 23-6, and remembering that like charges repel and unlike charges attract, one is naturally led to treat the lines of force as if they had the following two properties: (a) lines of force repel each other and (b) lines of force tend to contract along their length. Although it is difficult to put these properties in numerical terms, they turn out to be useful in the analysis of complex charge distributions, as we shall see later. It must be emphasized that lines of force are a convenient mathematical fiction and cannot be observed directly. They are useful because they can be pictured as starting and stopping only on charges, and this property is a consequence of the inverse-square character of Coulomb's law.

Since the principle of superposition applies to Coulomb's law, the electric field intensity at any point P, due to many charges, is the vector sum of the electric fields that would have been created at P by each of the charges separately.

Illustrative example

Two small charges, one of 8 μcoul and the other of -16 μcoul are spaced 0.20 m apart in air. Determine the intensity of the electric field at a point P, 0.20 m from each charge.

Referring to Figure 23-7, we find that the point P, at which the electric intensity is to be determined, is at the vertex of an equilateral triangle with the $+8$ μcoul charge at vertex A and the -16 μcoul charge at vertex B.

The electric intensity at any point due to a charge q at a distance r from it is given by

$$E = \frac{q}{4\pi\epsilon_0 r^2} = k\frac{q}{r^2}$$

(a)

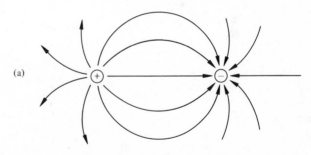

(b)

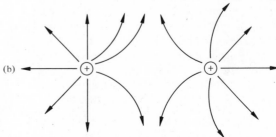

FIGURE 23-6 (a) The electric field due to equal and opposite charges. (b) The electric field due to two equal charges.

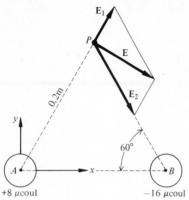

FIGURE 23-7 Calculating the electric field intensity at a point.

Putting in the numerical value for the constant in mks units, we have

$$E = 9 \times 10^9 \frac{\text{nt m}^2}{\text{coul}^2} \times \frac{q}{r^2}$$

The electric intensity E_1 at P due to the charge at A is thus

$$E_1 = 9 \times 10^9 \times \frac{8 \times 10^{-6}}{0.04} \frac{\text{nt}}{\text{coul}}$$

from which

$$E_1 = 18 \times 10^5 \frac{\text{nt}}{\text{coul}}$$

directed along the line AP away from A. Similarly the electric intensity E_2 at P due to the charge at B is

$$E_2 = 9 \times 10^9 \times \frac{16 \times 10^{-6}}{0.04} \frac{\text{nt}}{\text{coul}}$$

from which

$$E_2 = 36 \times 10^5 \frac{\text{nt}}{\text{coul}}$$

directed from P toward B along PB.

The vector sum of $\mathbf{E}_1$ and $\mathbf{E}_2$ is the electric field intensity $\mathbf{E}$ at P. Its value can be found by any of the well-known methods for adding vectors such as the parallelogram method or the analytical method. Using the latter and choosing the x direction parallel to AB as shown in Figure 23-8, we get, for the x component, E_x, of the electric field intensity

$$E_x = E_1 \cos 60° + E_2 \cos 60°$$

Substituting the appropriate numerical values for E_1 and E_2, we find that

$$E_x = 27 \times 10^5 \frac{\text{nt}}{\text{coul}}$$

Similarly, for the y component, E_y, we get

$$E_y = E_1 \sin 60° - E_2 \sin 60°$$

and, upon substitution of the appropriate numerical values, we get

$$E_y = -9\sqrt{3} \times 10^5 \frac{\text{nt}}{\text{coul}}$$

Since

$$E = \sqrt{E_x^2 + E_y^2}$$

we find that

$$E = 31.2 \times 10^5 \frac{\text{nt}}{\text{coul}}$$

The angle θ that the electric field intensity $\mathbf{E}$ makes with the x axis is given by

$$\tan \theta = \frac{E_y}{E_x} = \frac{-9\sqrt{3}}{27} = -\frac{\sqrt{3}}{3}$$

or

$$\theta = -30°$$

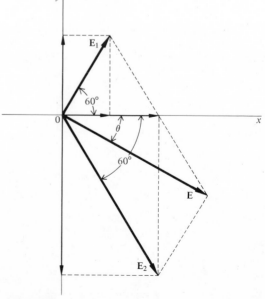

FIGURE 23-8

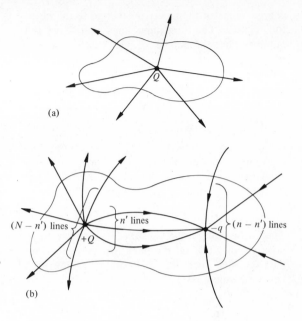

(a)

(b)

$(N - n')$ lines n' lines $+Q$ $-q$ $(n - n')$ lines

FIGURE 23-9 (a) All the lines that leave an isolated charge Q pass through an arbitrary surface surrounding Q. (b) A surface containing one positive and one negative charge. The number of lines leaving the enclosed volume is equal to $(N - n')$, where N is the number of lines that leave the positive charge and n' is the number of lines that go from the positive charge to the negative charge. The number of lines that enter the volume is $n - n'$, where n is the number of lines that go to the negative charge.

23-7
The electric field due to charge distributions

In general it is possible to calculate the electric field due to many charges by taking the vector sum of the fields due to each charge separately. For many charge distributions, however, particularly those having symmetry, it is often simpler to evaluate the field from some of the properties of the lines of force. One of the most useful methods of doing this is by the use of *Gauss' law*. This law relates the total number of lines of force leaving any volume of space to the total charge in that volume.

In order to understand the origin of the law, consider some volume of space inside of which there is a charge Q, as shown in Figure 23-9a. As we saw in the preceding section, N lines of force radiate out from that charge to infinity. Therefore, all of those lines will leave this volume of space. If a second positive charge, Q', is placed inside the volume (not necessarily at the same place as the first), then N' lines will radiate out from that charge. Although the second charge may move the lines from the first charge and thus change their distribution (as shown in Figure 23-6), the number of lines is unaffected. More important, none of the lines from the first charge can end on the second charge, since both are positive. All the lines from

both charges have to leave this volume. Thus for every charge Q_i placed inside the volume, N_i lines radiate out from the volume, and if all the charges inside the volume are positive, with total charge ΣQ_i, then the total number of lines leaving the volume is ΣN_i, where

$$\Sigma N_i = \frac{\Sigma Q_i}{\epsilon_0} \qquad (23\text{-}6)$$

The summation is made over all values of i. If the charges are all negative, then the number of lines entering the volume is given by a similar formula. In fact, if we call the number of lines *leaving* the volume a *positive* number, and the number of lines *entering* the volume a *negative* number, then Equation 23-6 describes both situations. Consider now placing a positive charge Q inside the volume and also placing a negative charge q inside the volume, as in Figure 23-9b. Then N lines will leave the positive charge, where

$$N = \frac{Q}{\epsilon_0} \qquad (23\text{-}7)$$

and n lines will go to the negative charge, where

$$n = \frac{q}{\epsilon_0} \qquad (23\text{-}8)$$

Now, of the N lines which leave the positive

charge, some of them, say n', will go to the negative charge. The remaining lines will leave the volume. Thus the number of lines leaving the volume which started on the positive charge is n_L, where

$$n_L = N - n'$$

Similarly, of the n lines which go to the negative charge, n' have come from the positive charge. The rest must come into the negative charge from outside the volume. Thus, the total number of lines that enter the volume is n_E, where

$$n_E = n - n'$$

Then, if we call all the lines that leave the volume positive lines and all the lines that enter the volume negative lines, the total number of lines that leave the volume is ΣN, where

$$\Sigma N = n_L - n_E$$

or

$$\Sigma N = (N - n') - (n - n')$$
$$= N - n$$

and using Equations 23-7 and 23-8, we get

$$\Sigma N = \frac{Q}{\epsilon_0} - \frac{q}{\epsilon_0}$$

or, if Q represents either a positive or negative charge, taken with appropriate sign,

$$\Sigma N = \frac{\Sigma Q}{\epsilon_0}$$

One can extend this argument to the situation where many charges are in the volume, and to take into account volumes of complex shapes. But Equation 23-6 is generally true: *The total number of lines of force leaving any volume of space* ΣN *is related to the total charge inside that volume of space* ΣQ, *by*

$$\Sigma N = \frac{\Sigma Q}{\epsilon_0}$$

as long as lines leaving the volume are counted as positive and lines entering are counted as negative. This relation is Gauss' law. It should be emphasized that Gauss' law applies to any arbitrary volume. In using the law, we are free to choose any volume that is convenient; typically, we choose a volume whose shape reflects the symmetry of the problem.

As an illustration of the use of the properties of lines of force, consider the following problem: *Positive charge is uniformly distributed on an infinite plane. Find the electric field created by this charge distribution.*

Figure 23-10a shows part of the plane, which is infinitely thin but extends out to infinity. In Figure 23-10b a few of the charges are shown on the edge of the section of the plane. From each of the charges, N lines radiate out to infinity. Since both sides of the plane are identical, it must be true that half of the lines radiate out in one direction and half in the other. Furthermore, since the lines repel each other, they tend to be directed perpendicular to the plane. In fact, since the charges are uniformly distributed and the plane stretches out to infinity, any one line is repelled equally by the lines above it and below it, by lines to its left and to its right. Thus, the lines must be perpendicular to the plane. From each charge on the plane, $N/2$ lines radiate out in one direction, perpendicular to the plane, and the same number radiate out perpendicular to the plane in the other direction.

In order to apply Gauss' law to this problem, we first choose a volume. In this case we choose a cylinder whose axis is perpendicular to the plane, as shown in Figure 23-10c. The cylinder has a length $2l$ and the area of its end is A. Since all of the lines are perpendicular to the plane, all of them are parallel to the curved sides of the cylinder; only the lines that go out of the ends of the cylinder leave the volume. If the number of lines leaving through one end is P, and the same number leave through the other end, the number that leave the cylindrical volume is $2P$. From Equation 23-6,

$$2P = \frac{Q}{\epsilon_0} \tag{23-9}$$

where Q is the total charge inside the cylinder. If the charge is uniformly distributed on the plane so that the charge per unit area is σ, then the total charge in the cylinder, Q, is given by

$$Q = \sigma A \tag{23-10}$$

Therefore,

$$2P = \frac{\sigma A}{\epsilon_0} \tag{23-11}$$

The electric field intensity is equal to the number of lines per unit area, or

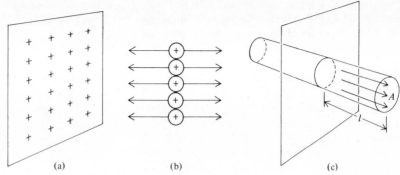

FIGURE 23-10 Application of Gauss' law to a uniform distribution of positive charge on an infinite plane.

(a)   (b)   (c)

$$E = \frac{P}{A} \qquad (23\text{-}12)$$

or

$$E = \frac{\sigma}{2\epsilon_0} \qquad (23\text{-}13)$$

It is interesting to note that this electric field intensity is independent of l, the height of the cylinder. In other words, the electric field intensity is the same at all points, no matter how far they are from the plane.

23-8
Electric potential; potential difference

A positive charge q situated in an electric field at a point A where the intensity is $\mathbf{E}$ will experience a force $\mathbf{F}$, where

$$\mathbf{F} = q\mathbf{E} \qquad (23\text{-}14)$$

In general then, if the charge is moved to some other location B, work will be done. This process can be described in the same way we described gravitational potential energy in Chapter 6; we can imagine that the charge is moved by some external agent that applies a force opposite to the force applied by the electric field. If this external force is equal to the force applied by the field during essentially all of the motion, the charge will be in equilibrium, and its kinetic energy will remain constant. Then the work done by the external agent serves to increase the electric potential energy of that system, which consists of the charge q and the charges that established the electric field. As we saw in Chapter 6, this potential energy can be thought of as shared by the charges, or alternatively as residing in the electric field.

In Chapter 6 we argued that the gravitational field is conservative; the work done by an external agent in moving an object from one place to another in a gravitational field is independent of the path over which the motion takes place. The properties of the gravitational field are a mathematical consequence of Newton's law of gravitation,

$$F = G\,\frac{m_1 m_2}{r^2}$$

whereas the properties of the electric field are a mathematical consequence of Coulomb's law,

$$F = \frac{1}{4\pi\epsilon_0}\,\frac{q_1 q_2}{r^2}$$

Since these two laws are identical in mathematical *form*, the properties of the two fields should be the same if the symbols representing one field are properly translated into symbols representing the other field. In particular, since we know that the gravitational field is conservative, we also know that the electric field is conservative. The work done by an external agent in moving a charge that is in an electric field is independent of the path. When a charge q is moved from A to B in an electric field, the change in electric potential energy depends on the location of A and B, but not on the path over which the charge moves from A to B. To calculate the change in potential energy we are at liberty to use any convenient path.

Suppose that we have some electric field. Then imagine placing a test charge q at some location A. Assume q to be small enough that it has no significant effect on the charges that created the field, and therefore the original field is not altered by the presence of the test charge (although the test charge does create a field in addition to the original field). Then if the test charge is moved from A to B, we can calculate the change in potential energy by choosing some arbitrary, convenient path that connects A and B. If the electric field is constant along this path, then the work W done by the external agent would be

$$W = Fs$$
$$= Eqs \qquad (23\text{-}15)$$

where s is the length of the path. If the electric field is not constant, then we can divide the path into short pieces in each of which we can treat the field as a constant, and we can find the work done in each of these pieces of path. The total work done by the external agent is the sum of these terms. In any case, we conclude from Equation 23-15 that the work done by the external agent is proportional to the magnitude of the test charge, q. In symbols,

$$W = qV_{AB} \qquad (23\text{-}16)$$

where V_{AB} depends on the field intensity and the location of the points A and B. The quantity V_{AB} is called the potential difference between points A and B, and is the work per unit charge that an external agent must supply in order to move any charge from A to B if the charge moves with constant kinetic energy. Equation 23-16 can be written as

$$V_{AB} = \frac{W}{q} \qquad (23\text{-}16a)$$

and is the defining equation of potential difference.

The potential difference V_{AB} is a scalar property of the electric field. It can be written in the form

$$V_{AB} = V_B - V_A \qquad (23\text{-}17)$$

where V_A is the potential of point A and V_B is the potential of point B. The potential at any point is a property of the location of that point and of the electric field. It must be emphasized that the potential at any point depends on the whole electric field, not just on the intensity at the point A. Moreover, just as in the gravitational case, an arbitrary potential that is constant everywhere in space may be added. This arbitrary reference level is chosen for convenience in any particular problem, although for a large number of problems it is convenient to choose the reference level so that the potential at a point infinitely far from the charges is equal to zero.

There are several units in use for electric potential difference. The mks unit is the *volt*, and is the same as the practical unit. In this system, work is

expressed in joules and charge in coulombs; hence, from Equation 23-16a,

$$1 \text{ volt} = 1 \, \frac{\text{joule}}{\text{coul}}$$

From Equations 23-15 and 23-16 it is clear that the dimensions of potential difference are the same as the product of the dimensions of electric field intensity and distance; hence a common unit for electric field intensity is volts/meter.

23-9
Potential due to a point charge

In Chapter 6 we asserted that if an object of mass m, which is above the surface of the earth and is initially at a distance r_i from the center of the earth (which has a mass M_e), is then moved to a distance r_f from the center of the earth, the change in gravitational potential energy is

$$\Delta \mathcal{E}_p = -GmM_e \left(\frac{1}{r_f} - \frac{1}{r_i} \right) \qquad (6\text{-}19)$$

In order to write down the electrical analog of Equation 6-19, we must remember that the gravitational force is always a force of attraction, whereas the electric force may be a force of attraction or a force of repulsion. Furthermore, mass is always a positive quantity, while charge may be positive or negative. Since the electric force between two positive charges is a force of repulsion, opposite in sign to the gravitational force between two (positive) masses, the sign of the electric analog of Equation 6-19 is reversed. Thus, if a charge q, initially at a distance r_i from a charge Q, is moved to a location which is a distance r_f from Q, the change in electric potential energy $\Delta \mathcal{E}_E$, is given by

$$\Delta \mathcal{E}_E = \frac{1}{4\pi\epsilon_0} qQ \left(\frac{1}{r_f} - \frac{1}{r_i} \right) \qquad (23\text{-}18)$$

and, using Equation 23-16, we have the potential difference between point i and point f, V_{if}, where

$$V_{if} = \frac{1}{4\pi\epsilon_0} Q \left(\frac{1}{r_f} - \frac{1}{r_i} \right) \qquad (23\text{-}19)$$

Comparing this with Equation 23-17 we see that the potential V at any point a distance r from the charge Q can be chosen to be

$$V = \frac{Q}{4\pi\epsilon_0} \left(\frac{1}{r} \right) \qquad (23\text{-}20)$$

This particular choice is equivalent to choosing the point at infinity to be the point of zero potential. Then the potential at any point is equal to the work per unit charge necessary to bring a charge of the same sign from infinity to that point. (Note that this discussion assumed that the charge Q is a point charge.) In the most general case, where we have a collection of charges distributed in space, we can arbitrarily choose one point to have zero potential and then calculate the potential difference between that point and all other points. In the particular case of a point charge, the mathematical form of Equation 23-19 makes it convenient to choose the point infinitely far from Q to be the point of zero potential; that is, setting $r_i = \infty$, so that $1/r_i = 0$. In the situation where a finite number of charges is confined to a finite volume of space, this same choice of the zero potential is similarly convenient.

23-10
Equipotential surfaces

As we have seen, an electric field may be represented by lines of force. An electric field may also be represented by *equipotential surfaces*, which can be described in the following way:

Consider any point A in an electric field. In general, there will be some surface, not necessarily of simple shape, passing through A such that all the points on the surface have the same potential. That is, the potential difference between A and *any* point on the surface is equal to zero. This surface is called an equipotential surface. *Every* point in an electric field lies on some equipotential surface (although there are special cases where an equipotential surface consists of only one point).

Consider a charge that is placed at some point in an electric field. If the charge is moved along any infinitesimally short path that is entirely in an equipotential surface, the potential difference between the initial point and end point is equal to zero, and therefore the work done is equal to zero. This could arise from either of two circumstances: either the field intensity is equal to zero everywhere along the path (and therefore the force is equal to zero) or the electric field intensity is everywhere perpendicular to the path (and therefore the work done is equal to zero). The most general statement is that the *electric field intensity is perpendicular to the equipotential surface*. In some special cases this intensity may be equal to zero.

For example, in the electric field due to a point charge Q, the equipotential surfaces are a set of concentric spheres with Q at the center, as shown in Figure 23-11a. The potential of a sphere of radius r surrounding this charge is thus

$$V = \frac{1}{4\pi\epsilon_0} \frac{Q}{r}$$

The lines of force are radial lines, which are perpendicular to the spheres. Thus by constructing perpendiculars to the equipotentials we can find the directions of the lines of force. Conversely, by drawing surfaces that are perpendicular to the lines of force we can construct the shapes of the equipotential surfaces. Thus if charges are distributed uniformly on the surface of a sphere, the lines of force will radiate out from the sphere, as in Figure 23-11b. The equipotential surfaces will be spheres concentric with the original sphere. Thus the field looks exactly the same, in the region outside the sphere, as the field created by a point charge located at the center of the sphere.

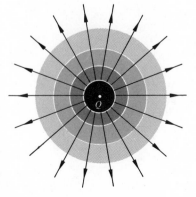 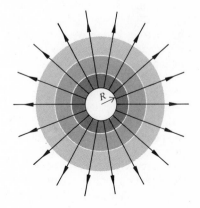

FIGURE 23-11 (a) The equipotential surfaces around a point charge are concentric spherical surfaces with the point charge at the center. (b) The equipotential surfaces outside a charged metallic sphere are spherical surfaces concentric with the charged sphere.

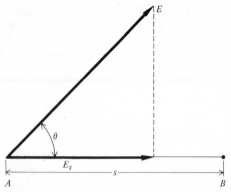

FIGURE 23-12 If an external agent moves a test charge from A to B in the presence of the constant field **E**, the work per unit charge done is $-Es \cos \theta$.

Given the equipotential surfaces we can find the magnitude of the electric field intensity, as can be seen from the following:

Consider two points A and B, which are so close together that the field intensity **E** in this region can be treated as a constant in the space between them (see Figure 23-12). The angle θ is the angle between **E** and the line joining A and B. If an external agent moves a test charge the small distance s from A to B, the work done by this external agent is the negative of the work done by the field intensity. Thus the work per unit charge (the potential difference) is given by

$$V_{AB} = -Es \cos \theta \qquad (23\text{-}21)$$

But the quantity $E \cos \theta$ is the component of the electric field intensity E_s in the direction of the line joining A and B. Thus, the *component of the electric field intensity in a direction AB is given by the potential difference between A and B, divided by the distance between A and B*.

Consider now two equipotential surfaces, as in Figure 23-13. We can start at point A on one of the surfaces and move to the other surface by any one of many paths, represented by s', s'', and s''' on the figure. Each of these paths has the same potential difference. Thus the component of the electric field intensity in each of these directions differs only because the path lengths differ. The path with the shortest path length has the largest component of electric intensity. But the direction along which a vector has the largest component is the direction of the vector itself. In other words, the direction of the electric field at point A is along the shortest path to the surface B. If this

shortest distance between A and the surface B is $s_{\min}$, then the magnitude of **E** is

$$E = \frac{V_{AB}}{s_{\min}} \qquad (23\text{-}22)$$

The equipotential surfaces can be drawn in equal increments of potential, so that the potential difference between any equipotential surface and its nearest neighbor is a constant ΔV. This forms a kind of contour map of potential. Then in the regions of space where the equipotentials crowd close together, the electric field is large. Where the equipotentials are far apart, the field is small.

Illustrative example

The proton has a charge of 1.60×10^{-19} coul. Determine (a) the electric field intensity at a point 5.3×10^{-11} m distant from it and (b) the electric potential at this point.

(a) The electric intensity in the neighborhood of a point charge is

$$E = \frac{Q}{4\pi\epsilon_0 r^2}$$

Substituting appropriate numerical values, we have

$$E = 9 \times 10^9 \times \frac{1.60 \times 10^{-19}}{(5.3 \times 10^{-11})^2} \frac{\text{nt}}{\text{coul}}$$

yielding

$$E = 51.3 \times 10^{10} \frac{\text{nt}}{\text{coul}}$$

(b) The potential at a point in the neighborhood of a point charge is given by

$$V = \frac{Q}{4\pi\epsilon_0 r}$$

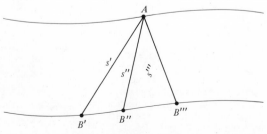

FIGURE 23-13 Each of the paths from one equipotential surface to the other differs in length.

when

$$V_{r \to \infty} = 0$$

Substituting appropriate numerical values, we have

$$V = 9 \times 10^9 \times \frac{1.60 \times 10^{-19}}{5.3 \times 10^{-11}} \text{ volts}$$

or

$$V = 27.2 \text{ volts}$$

23-11
Conductors and charge transfer

As we have seen, a conductor contains many free charges that are capable of motion. Let us assume that in some way an electric field is established inside a conductor. The field will exert forces on the free charges and they will move. The charges will move until they reach the surface of the metal, and under ordinary circumstances will remain on the surface because the air surrounding the metal is a good insulator and because the atoms of the metal exert forces on the free charges. These free charges on the surface will create an electric field intensity inside as well as outside the metal. Inside the metal the field intensity created by the surface charges is opposite in direction to the field that caused the free charges to move to the surface; the surface charges reduce the net field inside the metal. Since it is the net field inside the metal that drives the charges to the surface, this process will continue until the net field inside the metal is reduced to exactly zero. This process usually takes place very rapidly. Thus if an electric field is created inside a metal, the free charges will rearrange themselves very quickly, so charge appears on the surface of the metal and the field inside the metal is reduced to zero. Similarly, if a charge is placed inside a conductor, the free charges will rearrange themselves so as to leave the electric field intensity inside the metal equal to zero. The result will be a charge distribution on the surface of the metal and no net charge inside the metal.

If we know that all charges are at rest, then we know that the field intensity inside the conductors is equal to zero, and that all the net charge on the conductors is distributed over the surface. Since the field intensity inside a conductor is equal to zero, we know from Equation 23-21 that all points inside the conductor have the same potential; the interior of a conductor (when all the charges are at rest) is an equipotential volume. When the charges on the surface of the conductor are at rest, then the component of the electric field intensity in any direction parallel or tangent to the surface must be equal to zero; otherwise, the charges would move along the surface. Thus, the surface of the conductor is an equipotential surface. Therefore, the electric field intensity in the region of space just outside the conductor is perpendicular to the surface of the conductor.

Consider a charged conductor that is brought into contact with an uncharged conductor. The latter will acquire charge from the charged conductor. Charge will flow for a short time and redistribute itself over the surfaces of the two conductors. The electric field inside the two conductors becomes equal to zero, and the potential difference between them also becomes equal to zero; that is, the two conductors have the same potential. Another way of viewing this is to note that contact between conductors, as far as electrostatics is concerned, is equivalent to making a single conductor of the two conductors in contact.

One interesting aspect of the above concerns the distribution of charge over the surface of a conductor. We already know that the potential V of a sphere of radius R having a charge Q on its surface is given by

$$V = \frac{Q}{4\pi\epsilon_0 R}$$

The charge is distributed uniformly over the surface, and the surface density of charge S—that is, the charge per unit area—is

$$S = \frac{Q}{4\pi R^2}$$

Now if this sphere is connected to an uncharged sphere of smaller radius r by a very long, thin wire, charge will flow from the large sphere to the small sphere until both have the same potential. If they are far enough apart, the charges on one sphere will not influence the distribution of charges on the other, and we can continue to treat them as spherical distributions of charges. After the charges have ceased flowing, the larger sphere will have total charge Q', and the smaller sphere will have charge q, where

$$Q = Q' + q$$

The potential of the larger sphere will be V_R, where

$$V_R = \frac{Q'}{4\pi\epsilon_0 R}$$

and the potential of the smaller sphere will be V_r, where

$$V_r = \frac{q}{4\pi\epsilon_0 r}$$

Since the two spheres have the same potential,

$$\frac{Q'}{4\pi\epsilon_0 R} = \frac{q}{4\pi\epsilon_0 r}$$

or

$$Q' = q\left(\frac{R}{r}\right) \qquad (23\text{-}23)$$

Now the charge per unit area of the larger sphere is S', where

$$S' = \frac{Q'}{4\pi R^2}$$

and the charge per unit area of the smaller sphere is s, where

$$s = \frac{q}{4\pi r^2}$$

Therefore

$$\frac{s}{S'} = \frac{q}{Q'}\frac{R^2}{r^2} \qquad (23\text{-}24)$$

and, from Equation 23-23,

$$\frac{q}{Q'} = \frac{r}{R}$$

Inserting this in Equation 23-24, we get

$$\frac{s}{S'} = \frac{r}{R}\frac{R^2}{r^2}$$

or

$$\frac{s}{S'} = \frac{R}{r} \qquad (23\text{-}25)$$

Since

$$R > r$$

then

$$s > S'$$

or the charge per unit area is greater on the smaller sphere.

An interesting consequence of the above result is that the electric field intensity near the surface of the smaller sphere is greater than that near the surface of the larger sphere, even though the two surfaces are at the same potential.

The above results can be extended to conducting surfaces of unusual configuration—for example, metallic surfaces with sharp points. The latter may be considered as spheres of extremely small radii. When the surface is charged, the surface density of charge will be very great at the points and the electric field intensity near the points will also be very great. If there is air around the conductor, some of its molecules may be ionized by the intense electric field around the points and electric charge will flow from them, in the form of sparks or corona discharge.

23-12
Capacitance

Two conductors placed near each other form a system called a *capacitor*. Capacitors play important practical roles in electric circuits and in electronics; most often they are used to store charge or electrical energy. Ordinarily, when they are used, the total charge on one of the conductors is equal and opposite to the total charge on the other conductor. That is the total charge on the system is equal to zero; one conductor has total charge Q and the other conductor has total charge $-Q$. The magnitude of the charge on either conductor, Q, is called the charge on the capacitor. It must be emphasized that this is true even if the two conductors have different shapes or sizes.

Under the circumstances described above, the charge on the capacitor, Q, is proportional to the potential difference between the two conductors, V; that is,

$$Q = CV \qquad (23\text{-}26)$$

The constant of proportionality, C, in Equation 23-26 is called the *capacitance* of the capacitor, and depends on the geometry (the size and shape of the conductors and their relative separation) and on the nature of the material that fills the space between the two conductors. From Equation 23-26, the units of capacitance are coulombs per volt. One coulomb per volt is called a *farad*. Practical capacitances are rarely larger than a few millionths of a farad (microfarads, abbreviated μf).

Perhaps the simplest capacitor consists of two large, parallel, metal plates. See Figure 23-14. (Practical capacitors are often made in this form and then rolled or folded in order to fit into small spaces.) If the plates are large compared to the distance s between them, then we can find the potential difference between the plates by assuming that the plates are infinite in extent. Two infinite parallel plates have a uniform electric field between them. It is left to the reader to show that this field has a magnitude E, given by

$$E = \frac{\sigma}{\epsilon_0} \qquad (23\text{-}27)$$

where σ is the charge per unit area on either plate, remembering that both plates have the same magnitude of charge. Since the electric field is uniform, the potential difference V between the plates is given by

$$V = Es$$
$$= \frac{\sigma s}{\epsilon_0} \qquad (23\text{-}28)$$

The charge Q on the capacitor is the total charge on either plate. Since the plates have charge per unit area equal to σ, the charge Q is given by

$$Q = A\sigma \qquad (23\text{-}29)$$

where A is the area of one of the plates. Substituting Equations 23-29 and 23-28 in Equation 23-26, the capacitance C of this system is found to be

$$C = \frac{\epsilon_0 A}{s} \qquad (23\text{-}30)$$

23-13
The energy of a charged capacitor

Work has to be performed in charging a capacitor with a charge $+Q$ on one plate and a charge $-Q$ on the other plate. This work W can be calculated by considering the process as consisting of the transfer of very small quantities of charge from one plate to the other, starting initially with uncharged plates. As more and more charge is transferred from one plate to the other, the difference of potential between the plates increases from its zero value initially to its final value V after a charge Q has been transferred. Since V increases as Q increases, according to Equation 23-26,

$$Q = VC \qquad (23\text{-}26)$$

we can say that the average value of the potential difference between the plates during the charging process is $\frac{1}{2}V$. From the definition of difference of potential, the work done in transferring Q units of charge under a difference of potential $\frac{1}{2}V$ is

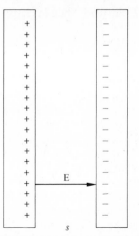

FIGURE 23-14 A parallel-plate capacitor.

$$W = \tfrac{1}{2}QV \qquad (23\text{-}31)$$

Equation 23-31 represents the work done in charging a capacitor to a difference of potential V with a charge $+Q$ on one plate and $-Q$ on the other plate.

If we combine Equations 23-26 and 23-31, the work done in charging a capacitor may also be written as

$$W = \frac{1}{2}\frac{Q^2}{C} \qquad (23\text{-}32)$$

and

$$W = \tfrac{1}{2}CV^2 \qquad (23\text{-}33)$$

Equation 23-31 represents the energy of a charged capacitor because of the work done in charging it. If the capacitor is discharged by connecting the two plates by means of a conductor, charge Q will flow through the conductor until the plates are neutral once more. While these charges are moving, the energy originally stored in the capacitor is transformed into heat and possibly other forms of energy.

Illustrative example

A capacitor having a capacitance of 4 μf is charged until the difference of potential between the plates is 150 volts. Determine (a) the charge on the capacitor and (b) the energy of this charged capacitor.

Using Equation 23-26 for determining the charge Q, we get

$$Q = 150 \times 4 \times 10^{-6} \text{ coul} = 6 \times 10^{-4} \text{ coul}$$

and, from Equation 23-31, the work done in charging this capacitor is

$$W = \tfrac{1}{2} \times 6 \times 10^{-4} \times 150 \text{ joule}$$
$$= 450 \times 10^{-4} \text{ joule}$$

23-14
Energy in an electric field

We have shown that work is done in charging the plate of a capacitor; the result is that the energy of the system is changed by an amount equal to the work done. We may think of this energy as a type of potential energy and attribute it to the new distribution of charges, or we may think of it as residing in the electric field in the space between the charges. We can calculate the energy in terms of the electric field intensity by equating it to the expression for the work done in charging a capacitor

$$\mathscr{E} = W = \tfrac{1}{2}CV^2 \qquad (23\text{-}34)$$

where $\mathscr{E}$ is the energy of the capacitor, C its capacitance, and V the potential difference between the plates. For simplicity, let us consider the case of a parallel-plate capacitor of area A and spacing s, with a vacuum between the plates. Its capacitance is given by

$$C = \frac{\epsilon_0 A}{s}$$

We have also shown that

$$V = Es$$

where E is the intensity of the electric field between the plates.

Substituting these values for C and V in the expression for the energy, we get

$$\mathscr{E} = \tfrac{1}{2}\epsilon_0 A s E^2 \qquad (23\text{-}35)$$

as the energy in the electric field between the plates of the capacitor.

The volume occupied by the electric field is As; hence the energy per unit volume, $\mathscr{E}_V$, in this field is

$$\mathscr{E}_V = \tfrac{1}{2}\epsilon_0 E^2 \qquad (23\text{-}36)$$

Although the above expression was derived for the special case of a uniform electric field, it can be shown to be generally valid. If a field is not uniform, we can consider a very small volume throughout which the electric field intensity E is uniform and derive the above result.

The concept of energy in an electric field is a particularly valuable one, theoretically as well as practically. For example, it is possible to calculate the energy in a given region of space in which an electric field exists even though the origin of the field cannot readily be traced to a distribution of charges. We shall have occasion to refer to this in our discussion of the transmission of energy by electric and magnetic fields.

23-15
Effect of dielectrics

If the space between the plates of a capacitor is filled with an insulator, or *dielectric*, the capacitance will, in general, increase. The ratio of the capacitance C of a parallel-plate capacitor that is filled with a homogeneous dielectric to the capacitance C_0 of the same capacitor with a vacuum between the plates is a constant K, which is characteristic of the dielectric; that is,

$$K = \frac{C}{C_0} \qquad (23\text{-}37)$$

The constant K is called the dielectric constant of the material. Table 23-1 gives the value of K for some materials.

The effect of a dielectric on the capacitance can be described in terms of the changes produced in the electric field in the space between the plates. Suppose that there is a vacuum between the two plates R and S of a parallel-plate capacitor, and that one has a charge $+Q$ and the other a charge $-Q$. Since each metal plate is necessarily an equipotential surface and the electric field is perpendicular to the equipotential surface, the lines of electric force will be parallel and uniformly spaced; the electric field is uniform throughout the space between the plates, except at the edges. We shall

Table 23-1

substance	dielectric constant K
Vacuum	1
Glass	5–10
Mica	3–6
Rubber	2.5–35
Water	81
Rutile (titanium dioxide)	86 along one axis 170 along perpendicular axis
Barium titanate	$\approx 10,000$

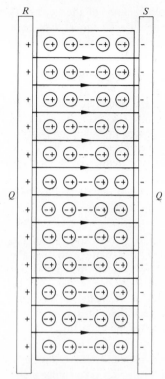

FIGURE 23-15 Charges within molecules of a dielectric are displaced by the action of the electric field between the capacitor plates R and S.

neglect the edge effect in this discussion. There are Q/ϵ_0 lines of force going from the positively charged plate to the negatively charged plate. If the area of each plate is A, the number of lines of force per unit area is $Q/\epsilon_0 A$, and this is equal to the electric field intensity E in the space between the plates.

If we now place a dielectric—that is, a nonconductor—in the space between the plates, the electric field E will produce a reorientation of the charges within the molecules of the dielectric, as shown in Figure 23-15. The negative charges of the molecules will be attracted toward the positively charged plate R, and the positive charges toward the negatively charged plate S. Since the dielectric is an insulator, these charges will move only through short distances, of the order of the molecular diameters. Some of the electric lines of force of the original electric field will end on the charges on the surface of the dielectric, so the field inside the dielectric will be smaller than the original electric field by an amount designated by K, the dielectric constant of the material. Thus the field in the medium will now be

$$E = \frac{Q}{K\epsilon_0 A} \qquad (23\text{-}38)$$

The product $K\epsilon_0$ is called the *permittivity* of the medium and is usually designated by ϵ; thus

$$\epsilon = K\epsilon_0 \qquad (23\text{-}39)$$

It is then a straightforward exercise to show that the capacitance of the filled capacitor is given by

$$C = K\,\frac{\epsilon_0 A}{s} \qquad (23\text{-}40)$$

When the space between the plates is a vacuum, $K = 1$, and Equation 23-40 becomes identical with Equation 23-30.

Questions

1. In gases that are conducting electricity, it is possible for ions of both signs of charge to move. In what common devices might you expect to find moving positive charges?

2. Describe the process by which a hard rubber comb that has been rubbed picks up bits of paper.

3. If a charged object is brought near a stream of water flowing from a faucet, the stream of water will be deflected. Try this experiment using a rubbed comb. Explain it, assuming that the water is a moderately good conductor and that the metal of the faucet is connected to the ground through the plumbing.

4. Just before entering a toll booth, automobiles are made to pass over a vertical flexible wire one end of which brushes against the car and the other end of which is embedded in the ground. What is the function of this wire?

5. The apparatus supplying oxygen to patients in hospitals is usually connected to the earth through a grounding wire. Why?

6. Compare the concept of electric field with that of gravitational field. Describe the differences if any and describe the quantities that play analogous roles.

7. When do electric field lines cross each other?

8. Sketch the field lines that arise from a long, charged cylinder.

9. Sketch the field lines due to four equal charges located at the corners of a square.

10. Give an example of a surface near a point charge through which no lines of force pass but which has an electric field at every point. (The surface need not be a closed surface.)

11. Distinguish between potential and potential difference.

12. Describe the equipotential surfaces around a very long, charged cylinder.

13. Compare the contour lines of a topographic map with the equipotential surfaces near a charge distribution.

14. In the eighteenth century, expert opinion differed over whether lightning rods should have round or pointed ends. Who was right?

15. Lightning has been known to strike more than once in the same place. Can you suggest any unusual features of such a place?

16. A charged capacitor consists of two groups of charges, Q, one positive and the other negative, separated by a potential difference V. Some students argue that since there are two groups of charge, the total energy of the system is $2QV$. Others argue that charge Q has been transferred through the potential difference V and so the total energy is equal to QV. Discuss the errors in both of these arguments.

17. A parallel-plate capacitor with air between the plates is charged to a potential difference V. Oil is now poured into the space between the plates, filling the space completely. Discuss the changes that take place in the value of the capacitance and the potential difference between the plates.

18. Why is water not used as a dielectric in capacitors, in view of its high dielectric constant and low cost?

19. A parallel-plate capacitor with air between the plates is connected to a battery so that the potential difference between the plates remains constant. The plates are then moved while still connected to the battery, increasing the separation between the plates. Discuss the change in the charge of the capacitor.

20. Examine the size of a typical laboratory capacitor. From this, estimate the volume of a 1-farad capacitor formed by connecting many such capacitors together.

Problems

1. A small sphere carries a charge of 300 μcoul. Determine the force exerted on this sphere by a charge of 120 μcoul, which is on a small sphere located 30 cm away from the first sphere.

2. Two small metallic spheres each of mass 0.2 gm are hung from a common point by means of silk threads 25 cm long. When they are charged with equal positive charges they are forced apart until the angle between the threads is 30°. Calculate the charge on each sphere.

3. Determine the electric field intensity at a point 30 cm from a small positive charge of 800 μcoul.

4. A small, positively charged body containing a charge of 25 μcoul experiences a force of 9.2 nt. Determine the electric field intensity at this point.

5. The field intensity at a certain place is 4×10^4 nt/coul. What force is exerted on a charge of 15 μcoul?

6. A charge of 0.3 μcoul experiences a force, which is measured to be 1.3 nt in an upward direction. (a) Find the magnitude and direction of the electric field intensity. (b) Determine the force that would have been exerted on a negative charge of magnitude 3.4 μcoul at this same location.

7. (a) How many lines of force terminate on one electron, using the convention stated in the text? (b) How would you change the convention, in order to have a more reasonable number, say 100 lines, terminate on the electron?

8. Find the electric field intensity outside a large plane sheet of charge, with a surface charge density of 2 μcoul/cm².

9. A large plane sheet has 10^{10} electrons per square centimeter uniformly distributed on its surface. Find the electric field intensity near the sheet.

10. The electric field intensity near the surface of a sample of metal is 3×10^4 nt/coul. Find the

charge density on the surface of the metal, assuming that the surface can be treated as a large plane.

11. It requires 3 μjoule of work to move a charge of 20 μcoul from one point to another. What is the potential difference between these points?

12. How much work does it require to move a charge 0.02 μcoul from one point to another if the potential difference between the points is 6.3 volts?

13. The potential difference between point A and ground is 107 volts, whereas the potential difference between point B and ground is 115 volts. (a) How much work does it take to move an electron from A to B if both A and B have potentials above that of the ground? (b) How much work does it take to move one electron from A to B if the potential of A is above ground potential and that of B is below ground potential?

14. Find the potential of a point that is 10^{-8} cm from an electron.

15. The equipotential surfaces in a certain region of space are plane sheets. The surfaces, which are 1 cm apart, have a potential difference of 3.5 volts. (a) Find the electric field intensity in this region of space. (b) Sketch the equipotentials and the electric field intensity vectors.

16. The equipotential surfaces in a certain place are concentric cylinders. The radius of the 50-volt surface is 1.5 cm, and the radius of the 60-volt surface is 1.8 cm. Find the electric field intensity between the cylinders.

17. A metal sphere of radius 4 cm has a charge of 50 μcoul uniformly distributed on its surface. One end of a long wire is connected to the sphere and its other end is connected to an uncharged metal sphere of radius 2 cm. How much charge flows through the wire?

18. A 1-μf capacitor is charged by a 45-volt battery. How much charge is on the capacitor?

19. When a certain capacitor is charged by a 6-volt battery, 24 μcoul flow from the battery's positive terminal. What is the capacitance of the capacitor.

20. Find the separation between the plates of an air-filled capacitor whose capacitance is 2 pf (2×10^{-12} farad) and whose plates consist of square sheets of metal, 1 m on a side.

21. A certain capacitor has the space between its plates filled with air. The plates have an area of 6 m² and are separated by 1 mm. Find the capacitance of the capacitor.

22. A 30-pf capacitor is charged by a 12-volt battery. How much energy is stored in the capacitor?

23. A charge of 450 μcoul flows from a battery charging a capacitor. The capacitor stores 3×10^{-2} joule. (a) What is the capacitance of the capacitor and (b) what is the potential difference of the capacitor when it is charged?

24. A parallel-plate air-filled capacitor with a plate separation of 1 mm has a capacitance of 20 pf. The air space is now filled with glass. What is the new capacitance of the capacitor?

25. When the empty space between the plates of a capacitor is filled with an oil, the capacitance changes from 12 μf to 38 μf. What is the dielectric constant of the oil?

26. What is the permittivity of (a) vacuum, (b) glass, and (c) water?

27. The electric field intensity in an air-filled capacitor is 10^5 volts/m. The capacitor is then filled with oil of dielectric constant $K = 6$, changing the electric field intensity. What is the new electric field intensity?

28. Two positive charges of 250 μcoul each are 20 cm apart. (a) Find the electric field intensity halfway between them. (b) Determine the potential halfway between them.

29. A 400-μcoul charge is 26 cm away from a -300-μcoul charge. Determine the electric field intensity and the potential at a point that is 8 cm away from the negative charge and on the line joining the two charges.

30. Two equal charges, each of 450 μcoul are 25 cm apart. Find the electric field intensity and the potential at a point which is 13 cm away from each charge.

31. It takes 0.25 joule to move a charge of 3 μcoul from one metal sphere to another through a wire that is 50 cm long. If the electric field in the wire is assumed to be constant, how large is the field intensity in the wire?

32. Starting from rest, an electron is moved by an external agent from a point that has a potential of 150 volts to another point which has a potential of 85 volts. When the electron arrives at the second point it has a speed of 2×10^5

cm/sec. How much work does the external agent do?

33. An electron is accelerated through 12 volts. How fast will it be moving if it started from rest?

34. Two very large sheets of charge are parallel to each other. They have equal charge densities σ, but one sheet is positively charged and the other is negatively charged. Find the electric field intensity in the space between the two sheets and the regions on either side.

35. Repeat Problem 34 for the case in which both sheets have the same charge.

36. A sheet of metal has thickness t, but is infinitely large. If charge is uniformly distributed on one surface, find the electric field intensity immediately outside that surface using Gauss' theorem. Since the field intensity inside the metal must be equal to zero, the answer must differ from that for the isolated sheet of charge given in the text. The difference arises from the fact that in order for the interior of the metal to be field-free, the other infinite surface of the metal must also have charge, and then the result of Problem 34 is applicable.

37. At a point immediately outside a charged metal sphere, the surface of the sphere looks like a large, charged plane. Use Equation 23-13 for the intensity at such a point and show that the intensity is the same as that due to a point charge located at the center of the sphere, if the sphere is uniformly charged.

38. Two spheres are very far apart; one has a radius of 2 cm and the other a radius of 3 cm. Each sphere has net positive charge of 5 μcoul. (a) Find the potential of each sphere. The spheres are then connected by a long wire. Find (b) the final potential of each sphere, (c) the charge on each sphere, and (d) the charge that has moved through the wire.

39. When the electric field intensity in air becomes larger than about 3×10^6 volts/m, the air breaks down and becomes a conductor. What is the maximum charge that can be placed on a metal sphere of radius 20 cm without causing this breakdown of the air surrounding it?

40. A 100-μf capacitor is charged with a 45-volt battery. The capacitor is then discharged through a conducting solution in a plating bath. The process is repeated 50 times. Each electron that flows through the solution causes

one atom of a metal to be deposited. How many atoms in total are deposited by this process?

41. A parallel-plate capacitor is connected to a battery and charged to 6 volts. The capacitor is disconnected from the battery and the air space is filled with rubber. (a) What is the new potential difference between the plates? (b) What is the change in energy of the capacitor? (c) What is the source of this energy?

42. Two equal and opposite charges of magnitude Q are placed a distance a apart. (a) Determine the electric field intensity at a point that is along the line joining the two charges and is a distance r from the point midway between the two charges. (b) Show that when r is very much larger than a, the intensity of the electric field E is, to a good approximation,

$$E = \frac{Q}{4\pi\epsilon_0} \frac{2a}{r^3}$$

43. Three equal charges of like sign, each of charge Q, are placed on the corners of a square of side a. Find the electric field intensity (a) at the center of the square and (b) at the fourth corner.

44. Charge is uniformly distributed on the surface of a right circular cylinder of radius R and infinite length. After convincing yourself that the lines of force radiate out uniformly from the surface, perpendicular to the axis of the cylinder, choose an appropriate surface for Gauss' law. Then show that the magnitude of the electric field intensity E at a point P, a distance y from the axis, is given by

$$E = \frac{\lambda}{2\pi\epsilon_0 y}$$

where λ is the charge per unit length on the cylinder.

45. (a) Find the potential of the surface of an electron, assuming that it is a charged sphere of radius r. (b) Assuming that this potential is the total relativistic mass energy of the electron, mc^2, divided by the charge e of the electron, determine the radius r. This number is called the "classical radius of the electron."

46. What is the maximum charge that can be stored in a parallel-plate capacitor whose plates are 1 m square? (See Problem 39 above.)

47. What is the capacitance of a capacitor, one of whose plates is a metal sphere of radius r and

the other of which is infinitely far away?

48. The plates of a parallel-plate capacitor are arranged so that the distance between them can be varied. When the distance between them is s, the capacitor is charged until the difference of potential is V. The plates are then separated until the distance between them is $2s$. Assuming that the charge Q on the plates is unchanged, determine **(a)** the difference of potential on the capacitor, **(b)** the change in energy of the capacitor, and **(c)** the work done in separating the plates.

49. Assume that the separation of the plates of a parallel-plate capacitor is changed by an external force that is nearly equal to the force of attraction of the plates. Equate the work done by this external force to the work done by the force of attraction and then equate this work to the change in energy of the capacitor when the separation changes by an amount Δs, which is small compared to the separation s. Show that the force of attraction is given by

$$F = \frac{q^2}{2\epsilon_0 A} = \tfrac{1}{2} \frac{\sigma}{\epsilon_0} q = \tfrac{1}{2} Eq$$

50. A metal doorknob is charged. Sketch the equipotentials and field lines in the space around the doorknob.

51. An electron is at rest initially in a region of space where the electric field lines curve. Show that the electron will not move along the field lines. (*Hint*: Consider the force necessary to have the electron move along an arc of a circle.)

52. An evacuated tube contains a parallel-plate capacitor whose plates are separated a distance of 1 cm. A potential difference of 3500 volts is maintained between them. **(a)** Determine the electric field intensity between the plates. **(b)** A stream of electrons directed at right angles to the electric field is sent through the space between the plates. Calculate the acceleration of these electrons. **(c)** What is the path of these electrons in the electric field?

53. Use the fact that the electric field is conservative to prove that the electric field cannot be confined to the region between the plates of a parallel-plate capacitor; there must be a fringing field. Does this change the argument of Section 23-14?

current electricity

chapter 24

24-1
Introduction

Electricity is the basis of much of our modern civilization. We have come to depend on electric motors, electric lights, electronic devices, and complicated electric circuits.

The uses and designs of such devices are limited only by the almost unbounded ingenuity of man, but by ignoring most of the details we can discuss many of these devices in terms of the transformation of energy. Lights, motors, and television sets can be thought of as *output devices*, devices whose output is energy delivered in some useful, or desired, form. The input to these devices is electrical energy. This electrical energy is itself the output of a *generator*, such as a dry cell, storage battery, dynamo, or fuel cell. The input to the generator is energy in some convenient form, such as chemical energy, heat, or the kinetic energy of moving water. We can represent this process schematically by Figure 24-1. From this point of view, electricity is a means of transporting energy from one device to another. This transportation may occur over distances of hundreds of miles, as in the case in which energy from hydroelectric stations is transported to distant homes, or it may occur over a few inches, as in the case of a flashlight. At each end of this transportation process, energy is transformed into or from electrical energy.

As an illustration of this, consider a light bulb that is connected to a dry cell, as in Figure 24-2. The cell transforms chemical energy into elec-

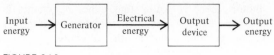

FIGURE 24-1

trical energy, and the bulb transforms the electrical energy mostly into heat and into a little light. In this process, charges move through the cell, the wires, and the filament of the bulb.

The cell has two *terminals*, to which are attached the wires leading to the bulb. The cell maintains a steady flow of charge between these terminals: charges flow out from one terminal, through one wire, through the filament, and then back to the cell through the other wire. By means of chemical processes within the cell, charges are transferred back to the first terminal.

In addition, the cell establishes a static charge distribution. The internal chemical structure of the cell separates the positive and negative charges of neutral atoms and deposits each of them in different places. Before the wires are connected to the cell, one of the terminals has a static positive charge on it, and the other terminal has static negative charge on it. Immediately after the wires are connected, charges flow; but in a very short time a new static charge distribution is established on the terminals, wires and the filament. Thus we have, simultaneously, a steady flow of charges around the circuit and a stationary distribution of charges on various parts of the circuit.

We can analyze the energy transfer in several ways, with each one somewhat more abstract than the preceding.

The most direct analysis recognizes that the static charges exert forces on the moving charges, transferring energy to them. While this is happening, the static charge distribution tends to change because energy is being extracted from the distribution. However, the cell transforms chemical energy into electrical energy and maintains the distribution. In order to carry out this kind of analysis in detail for a specific circuit, we would have to describe both the static and moving charges.

Because the charge distribution is static, the electric field established by it is constant. Therefore, we can, in a sense, replace the static charge distribution with a constant electric field, and describe the behavior of the circuit in terms of the interaction between the moving charges and this field. It turns out that for most simple circuits we

can usually know enough about the field to discuss the transfer of energy in some detail, even though we do not know all the details of the static charge distribution.

An even more abstract procedure would be to recognize that the moving charges also establish fields, magnetic and electric. Then one can replace all the charges by the fields they establish and discuss the energy transfer in terms of the interactions among these fields. This procedure becomes essential for dealing with circuits in which important parameters vary rapidly in time, but we can avoid it for simple circuits.

In this chapter we will deal with direct-current (DC) circuits only. We can adequately represent the situation by assuming that the cell or generator has established a static electric field throughout the general region of space occupied by the circuit and that a steady flow of charges is circulating through the circuit. If then we focus our attention on one of the circulating charges, we see that as it moves through the cell or the wires or the filament, it is moving from one point in the electric field to another. As it does so, it is moving from a point that has one potential to a point that has another potential. Therefore as the charge moves, the potential energy of the system changes, where this system consists of the moving charges and all of the charges that establish the field. However, we know that the field is constant in time, so although it is the total potential energy of the system that is changing, it is common to think of the changes in energy as associated with that part of the system which is moving. Therefore it is convenient to think of the potential energy of the moving charges as changing. Thus we are led to the following picture of the circuit: As the charges circulate through

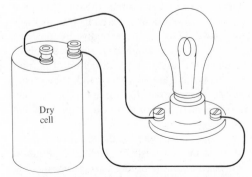

FIGURE 24-2 The dry cell transforms chemical energy into electrical energy, and the bulb transforms electrical energy into heat and light.

the wires and the bulb, they are losing electrical potential energy. When they return to the cell, electrical potential energy is transferred to them from the cell. The cell transforms some of its internal chemical energy into electrical potential energy. In other words, we can think of the potential energy as a property of the charges that are moving, just as we were able to think of the gravitational potential energy as a property of an object near the surface of the earth. As an electric charge moves around a circuit, its potential energy may undergo a series of changes. Its potential energy decreases as it goes through the circuit external to the cell, and is increased by the action of the cell.

24-2
Electric current

An electric current is a flow of charge. When currents exist in conductors they have external effects; currents create magnetic fields, generate heat, and may cause some chemical processes to occur. Generally, the detection and measurement of currents involves one or more of these external effects rather than the detailed motion of the flowing charges. It is this concern for external effects that is reflected in the definition of electric current given below.

In terms of external effects, currents have direction. A flow of positive charges in one direction in a conductor has different external effects from a flow of positive charges in the other direction. However, a flow of positive charges in one direction is essentially indistinguishable from a flow of negative charges in the opposite direction. That is, a given set of external effects may be due to positive charges flowing in one direction, or negative charges flowing in the opposite direction. Physics has adopted the *convention* that the direction of a current is the direction in which the *positive* charges move. Thus, if there really are positive charges moving in the conductor, the direction of the current is the same as the direction of motion of these charges. If, on the other hand, the moving charges are all negative, the direction of the current is opposite to the direction of motion of the charges.

In terms of external effects, the absence of flowing charge is indistinguishable from the situation in which equal amounts of like charges are flowing in opposite directions. That is, if we know from external effects that there is zero current in a conductor, then we know that either no charge is moving, or that as much charge of one sign moves in one direction as in the other direction.

These experimental observations lead to the following definition of electric current. Consider a cylindrical conductor as shown in Figure 24-3, where A is any cross section of the conductor. Then the current I is related to the *net charge Q*, which crosses the area A in time t, by the equation

$$I = \frac{Q}{t} \qquad (24\text{-}1)$$

In principle, the net charge Q is determined by counting the charges as they pass through A. In this counting process all the positive charges moving in the direction of the current and all the negative charges moving opposite to the current are counted as positive numbers. All the positive charges moving opposite to the current and all the negative charges moving in the same direction as the current are counted as negative numbers. Thus, if in any time interval, as many positive charges move in one direction as in the other, the net charge is equal to zero, and the current is equal to zero.

From Equation 24-1, the unit of current is the coulomb per second. One coulomb per second is called an *ampere*.

The relationship between the direction of the current and the direction of the motion of the charges is shown in Figure 24-4. The nature of the charges that are set in motion depends upon that nature of the conducting substance. If the conductor is a metal, the current consists of the motion of the free electrons in the metal. If the conductor is a gas, the charges that are set in motion are positive and negative ions, and under conditions of low pressure, there also may be electrons. In nonmetallic liquid conductors the current consists of positive and negative ions.

24-3
The electron gas

Metals, as we have seen, consist of relatively fixed, massive, positively charged ions, and relatively

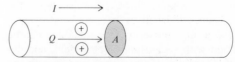

FIGURE 24-3 A current in a circuit consists of the flow of charges through any cross-sectional area A.

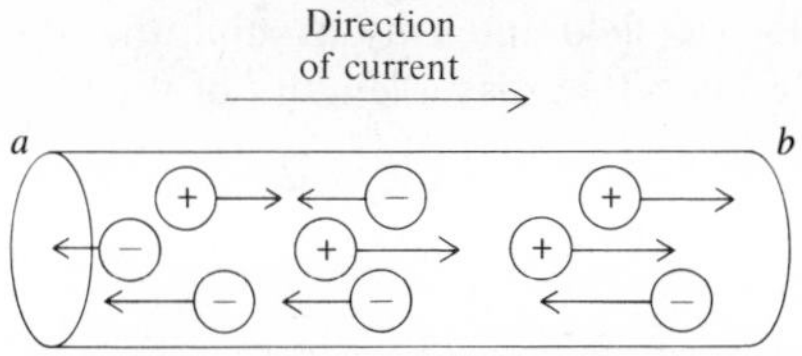

Direction
of current

a

b

FIGURE 24-4 The direction of the current is the direction of motion of the positive charges. Negative charges move in a direction opposite to that of the current.

free, light, negatively charged electrons. Ordinarily these free electrons move about in the metal, constrained only by large forces at the surfaces of the metal that serve to keep the electrons from leaving the metal. As a way of understanding many of the properties of metals it is useful to think of the electrons as if they were the molecules of a gas, the *electron gas*. It must be emphasized that quantum-mechanical effects and forces exerted by the ions play a very important role and that therefore one cannot calculate numerical values of physical properties of metals using the classical model of a gas. Nevertheless, the picture of a metal as containing a gas of electrons is adequate for a qualitative understanding of some of its physical properties.

The most important property of the electron gas is that the electrons move about randomly with large velocities; the free electrons in copper have average speeds of about 1.5×10^8 cm/sec. With these high speeds, the electrons collide with the ions very frequently; the time between collisions in copper, for example, is about 2×10^{-14} sec.

If an electric field is created in a metal by some external charge distribution, each of these electrons will experience a force. In addition to the random motion with frequent collisions, each electron will be accelerated in the direction opposite to that of the field, because the electrons have negative charge. But before an electron can acquire any significant change in velocity due to the acceleration, it collides and slows down. It then is accelerated again. Thus, this model leads to the idea that the electrons acquire a small average *drift velocity* in the direction opposite to the field. This drift velocity is superimposed on the large random velocities. In copper, the electron drift velocity is about 30 cm/sec in the presence of a field of 1 volt/cm.

The random motion of the electrons does not contribute to the current because, on the average, as many electrons will go through any cross sec-

tion in one direction as go through in the other direction. Thus, in the absence of an electric field the *net* charge passing through a cross section of the metal in any time interval is equal to zero. On the other hand, if an electric field imposes a drift velocity on the electrons, more charge will pass through in one direction than in the other. Therefore, in describing the current, we can ignore the random motion and consider only the drift velocity as significant.

Consider then a conductor that has n free electrons per unit volume, each having charge e. The average drift velocity of an electron is v_d. During a time t, each electron will move a distance d in the direction of the drift velocity (opposite to the electric field). Then,

$$d = v_d t \qquad (24\text{-}2)$$

Figure 24-5 shows such a conductor. We can calculate the current in this conductor by counting the charge that passes through surface C in time t. Let us consider a surface D, which is a distance d (as given by Equation 24-2) upstream from surface C. Then in the time t all the electrons that are at D will reach C and all the charges that are between C and D will go through C. Therefore the total charge that goes through C in the time t is the total charge contained in the region between C and D. Since the volume in this region is equal to the product of d and the area A of the surface C, the number of electrons in this region, N, is

$$N = ndA$$
$$= nv_d tA$$

and since each electron has a charge e, the total charge Q is

$$Q = Ne$$
$$= nv_d tAe$$

and, from Equation 24-1,

$$I = nv_d Ae \qquad (24\text{-}3)$$

It is sometimes useful to find the current density

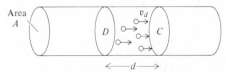

FIGURE 24-5 All the charges between surface D and surface C move through C during the time t if $d = v_d t$.

j, which is the current per unit cross-sectional area. That is,

$$j = \frac{I}{A} \tag{24-4}$$

From Equation 24-3, the current density j is given by

$$j = nv_d e \tag{24-5}$$

It is common to define a vector current density $\mathbf{j}$, which has a magnitude equal to the current per unit cross-sectional area and has the direction of the drift velocity of positive charges. In metals, the direction of the vector current density is opposite to the direction of the drift velocity of the electrons.

24-4
Ohm's law

For a large number of metals, over a wide range of conditions, the current density at any point is proportional to the electric field intensity at that point. We can write this as

$$\mathbf{j} = \sigma \mathbf{E} \tag{24-6}$$

in which σ is a constant of proportionality and is called the *conductivity* of the material. Equation 24-6 is called *Ohm's law*, and materials for which σ is a constant are called *ohmic conductors*. It should be noted that Equation 24-6 is a *definition* of the conductivity (as the ratio of the current density to the intensity). There are materials for which the conductivity is not a constant, and there are materials for which the conductivity is not a scalar; that is, its value depends on the direction of the current through the material. However, many materials are ohmic, and those materials turn out to be very useful.

Suppose a uniform electric field intensity $\mathbf{E}$ is maintained by an external generator in the piece of conductor shown in Figure 24-3. Then the current density will also be uniform, and the total current I will be

$$I = jA \tag{24-7}$$

where A is the cross-sectional area of the conductor. Substituting Equation 24-7 in Equation 24-6 we have, for the magnitudes only,

$$\frac{I}{A} = \sigma E \tag{24-8}$$

Since the electric field intensity is uniform, the potential difference V across a length l of the conductor, is

$$V = El \tag{24-9}$$

and so

$$\frac{I}{A} = \sigma \frac{V}{l} \tag{24-10}$$

or

$$V = I\left(\frac{l}{A\sigma}\right) \tag{24-11}$$

We can then define a new quantity, the *resistance*, R, of a piece of conductor of length l, cross-sectional area A, and conductivity σ, by

$$R = \frac{l}{A\sigma} \tag{24-12}$$

For those materials for which the conductivity is a constant, we see from Equation 24-12 that the resistance of a particular piece of conductor is also a constant. It is common to measure the inverse of the conductivity, which is called the *resistivity*, ρ. That is,

$$\rho = \frac{1}{\sigma} \tag{24-13}$$

Table 24-1 gives the resistivities of some common materials. And in terms of the resistivity, the resistance is given by

$$R = \frac{l}{A}\rho \tag{24-14}$$

Thus, Equation 24-11 may be written

TABLE 24-1   Resistivity of some materials

substance	resistivity in ohm meters at 0°C
Aluminum	2.83×10^{-8}
Carbon	3500×10^{-5}
Copper	1.70×10^{-8}
Copper oxide	1×10^{3}
Germanium	5×10^{-1}
Iron	10×10^{-8}
Manganin	48×10^{-8}
Nichrome	108×10^{-8}
Platinum	11×10^{-8}
Silicon	6×10^{2}
Silver	1.63×10^{-8}
Tungsten	5.5×10^{-8}

$$V = IR \qquad (24\text{-}15)$$

This equation is also often called Ohm's law. The resistance R of an element of a circuit is, from Equation 24-15,

$$R = \frac{V}{I} \qquad (24\text{-}16)$$

From Equation 24-15, we see that the units of resistance are volts per ampere. One volt/amp is called one *ohm*, often abbreviated by the symbol Ω (capital *omega*).

24-5
Energy dissipation in conductors

Consider a piece of ohmic conductor of length l in which a current I exists, as shown in Figure 24-4. Assume that the current consists of positive charges flowing from a to b. As we have seen, these charges have some average drift velocity, which we know experimentally is constant. Therefore, on the average, the charges are in equilibrium; the vector sum of the forces acting on any of the moving charges must be equal to zero. We know that an electric field intensity $\mathbf{E}$ must exist in the conductor, and that this field exerts a force $q\mathbf{E}$. Therefore some other force, $\mathbf{f}$, must be exerted on the charges; this force must be equal and opposite to the force exerted by the electric field intensity. From the condition for equilibrium,

$$\mathbf{f} = q\mathbf{E} \qquad (24\text{-}17)$$

The electric field $\mathbf{E}$ is created by some charge distribution established and maintained by the generator. The force $\mathbf{f}$ is exerted by the metal. It is the average force exerted by the positive ions and the electrons in the metal. It is very tempting to say that this force is the average force exerted in the collisions between the free electrons and the fixed ions. Unfortunately, detailed quantum-mechanical calculations indicate that this is too simplified a view. We will have to leave the question of the detailed origin of this force unanswered and simply assert that in ohmic conductors we know from experiment that the free charges move with constant average drift velocity. Therefore we conclude that the metal must exert an average force, $\mathbf{f}$, on the free charges that is exactly equal and opposite to the force exerted by the field intensity.

When one of the free charges moves through a distance l in the conductor, the electric field intensity does some work W on the charge, where

$$W = Fl$$
$$= qEl$$

and since the potential difference V is given by

$$V = El$$

we have

$$W = qV \qquad (24\text{-}18)$$

where V is the potential difference across the length l of the conductor. While the charge is moving, the metal is exerting the force $\mathbf{f}$, in the direction *opposite* to the motion. Thus, the metal does a negative amount of work on the charge, or equivalently the charge does positive work on the metal, and since $\mathbf{f}$ is equal to $q\mathbf{E}$, the negative work done by the metal is exactly equal to the work done by the field. In other words, all the work done by the field on the charge is extracted from the charge by the metal and converted into some other form of energy.

We can now look at the flow of energy as a whole in the conductor. The generator, which establishes the electric field intensity, does work on each of the moving charges that constitute the current. In time t, the total charge that flows through the conductor is

$$Q = It$$

and so the total work done by the generator on the charges is, from Equation 24-18,

$$W = ItV \qquad (24\text{-}19)$$

The moving charges act as the intermediary for transferring the work W from the generator to the conductor. This work usually appears in the form of heat, as evidenced by a rise in temperature of the metal and possibly some of the material surrounding it. If the temperature of the metal is raised to a sufficiently high value, the metal will glow; that is, a small part of the energy will be converted into light. Other changes in the properties of the metal may be produced, such as expansion, melting, and change in conductivity.

The property of the metal that is responsible for the conversion of the energy into heat is its resistance R. By the use of Ohm's law,

$$V = IR \qquad (24\text{-}16)$$

the energy supplied to the conductor

$$W = ItV \qquad (24\text{-}19)$$

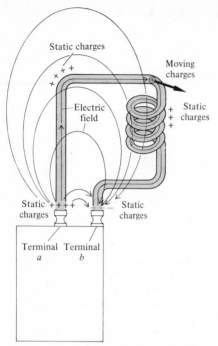

Static charges

Moving charges

Electric field

Static charges

Static charges

Static charges

Terminal *a*   Terminal *b*

FIGURE 24-6 The charge and field distributions in the neighborhood of an ohmic conductor connected to a dry cell.

can be written as

$$W = I^2Rt \qquad (24\text{-}20)$$

Thus the energy supplied to a conductor that is converted into heat is proportional to the square of the current in the conductor.

The power P that is supplied to a circuit in which there is a current I at a difference of potential V is, from the definition of power,

$$P = \frac{W}{t}$$

$$P = VI \qquad (24\text{-}21)$$

If the power is supplied to a conductor of resistance R, the rate at which energy is converted to heat is

$$P = I^2R \qquad (24\text{-}22)$$

and, equivalently,

$$P = \frac{V^2}{R} \qquad (24\text{-}23)$$

Illustrative example

An electric heating element is rated at 750 watts and 120 volts. Determine (a) the resistance of the heating element and (b) the current in the heater when in operation.

We can use Equation 24-21

$$P = VI$$

and first determine I as

$$I = \frac{P}{V} = \frac{750}{120} \text{ amp} = 6.25 \text{ amp}$$

Then solve Equation 24-16 for R, obtaining

$$R = \frac{V}{I}$$

from which

$$R = \frac{120}{6.25} = 19.2 \text{ ohms}$$

Or we can solve from Equation 24-23 and obtain

$$R = \frac{V^2}{P}$$

which gives R directly.

24-6
EMF

Consider an ohmic conductor which is connected across the terminals of a generator, as in Figure 24-6. The generator will cause an electric field to be established in the conductor. Basically the generator does this by putting charge on the terminals. The initial field thus created will cause charge to move through the conductor. If the conductor has a constant cross-sectional area, then ultimately, when equilibrium is reached, the electric field in the conductor is constant in magnitude and follows the shape of the conductor. This equilibrium field is created by static charges on the terminals of the generator and on the surface of the wires. Furthermore, this field will not be confined to the conductor. As shown in Figure 24-6, the field will exist outside the wire and inside the generator. The charges are confined to the metal because the forces exerted by the ions on the free charges are very large at the surfaces and usually prevent the charges from leaving the metal.

When a positive charge leaves terminal a and moves through the conductor, it moves under the influence of this electric field. When it reaches terminal *b*, and enters the generator, the field created by the charges opposes the motion. If no other phenomena existed, the charges would not enter the generator because the field would push them back. They would pile up on terminal *b*, and would in turn reduce the field intensity in the conductor. Ultimately, the field would disappear and the current would cease. In order to maintain the field and the current an additional force must be exerted on the charges. This force must be produced inside the generator and must be opposite to the electric field created by the charges. No matter what kind of generator we are dealing with (chemical cell, dynamo, solar cell, nuclear cell), the internal mechanism of the cell must exert such a force, which we will call a *nonelectrostatic* force, *F*. In all generators this force is exerted on charges only when they are inside the generator.

When a charge is inside the cell it has forces exerted on it by the electric field and by this nonelectrostatic force. In all real generators, the ions in the material of which the generator is made also exert a force *f*, which is of essentially the same kind as the resistive force *f* exerted by metals on the free charges. The direction of this force *f* is opposite to the motion of the charge. The direction of the electric field *E* is determined by the charge distribution and the direction of the nonelectrostatic force *F* is determined by the structure of the generator. Thus the free-body diagram for one of the free charges *e* inside the generator of Figure 24-6 is shown in Figure 24-7. In that figure the nonelectrostatic force *F* is shown as the product of the charge *e* and a nonelectrostatic field E_n. The force exerted by the static charge distribution is shown as *eE*.

The work per unit charge done by the generator on the charges is called the EMF, $\mathcal{E}$, of the generator. (The letters are an abbreviation of electromotive force, but since EMF is a work per unit charge rather than a force, it is less confusing to use the abbreviation than the full phrase.) The work per unit charge that is transferred to the field is the potential difference *V* between the terminals

of the generator. The work per unit charge extracted by the internal resistive forces, *f*, is usually proportional to the current, and may be written in the form *Ir*, where the quantity *r* is called the internal resistance of the generator. Thus we have, from conservation of energy,

$$\mathcal{E} = V + Ir \qquad (24\text{-}24)$$

In the operation of a generator, the energy per unit charge represented by $\mathcal{E}$ is derived from some other form of energy. In a chemical cell, it is obtained from the chemical energy of the constituents; in a dynamo, it is obtained from the mechanical energy of a spinning turbine. In all generators, this energy transformation is reversible. That is, it is possible to extract energy from charges that are forced to move through the generator (by some other generator) and to transform this energy into chemical energy or mechanical energy. By forcing current through a cell we can reverse the chemical reactions, and by forcing current through a dynamo we can make the dynamo act as a motor. Thus, the EMF of a generator is the *nonelectrostatic, reversible, work transferred per unit charge that moves through the generator.*

When the current is driven by the internal processes of the generator (that is, when the current moves inside the generator from the negative terminal to the positive terminal), Equation 24-24 applies. If the current is driven in the *opposite* direction, some external generator does work per unit charge equal to *V*. This work per unit charge is partly dissipated in the form of heat (*Ir*) and partly converted by the driven generator into chemical energy or mechanical energy ($\mathcal{E}$). Thus, when current is flowing in the reverse direction,

$$V = Ir + \mathcal{E} \qquad (24\text{-}25)$$

Equation 24-24 can be used to describe experimental definitions of EMF and internal resistance. From Equation 24-24 the EMF is equal to the potential difference between the terminals of the generator when no current is drawn. In the absence of a conductor between the terminals, the current will be equal to zero. Thus, the EMF of a generator is the "open-circuit" or zero-current potential difference between the terminals. Similarly, if a graph is made of the current and the potential difference between the terminals, Equation 24-24 predicts that the graph will be a straight line whose slope is equal to the internal

FIGURE 24-7 The forces exerted on a positive charge inside a generator; *E* is the electric field due to the static charge distribution, E_n is the nonelectrostatic field, and *f* is the resistive force.

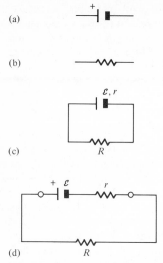

FIGURE 24-8 (a) Schematic representation of a chemical cell. (b) Schematic representation of a resistor. (c) A circuit consisting of a cell and a resistor. (d) The circuit of part (c) with the internal resistance of the cell explicitly shown.

resistance. The potential difference V is called the *terminal voltage* of the generator.

In most chemical cells, the EMF (when the cell is first manufactured) is determined by the nature of the chemicals from which the cell is made. The EMF varies very slowly as energy is transferred from the cell. The internal resistance, on the other hand, changes more rapidly as the cell transfers energy. At the point where most chemical cells stop being useful, the internal resistance is large, but the EMF is essentially the same as that of a new cell.

Generators are often loosely called *sources of EMF*, although, strictly, they are devices that transform energy from one form to another. (Unfortunately for the clarity of the language, the term "transformer" is applied to another electrical device that transforms alternating current from one voltage to another.) Generators are also sometimes called *seats of EMF*.

24-7
Simple circuits

The simplest circuits consist of generators connected to ohmic conductors. With the ideas developed in the preceding sections we can analyze such circuits and calculate such quantities as the current, the potential difference, and the power in various parts of such circuits.

Before analyzing a circuit it is usually helpful to draw a circuit diagram, where each element in the circuit is represented by a conventional symbol, as illustrated in Figure 24-8. Figure 24-8a represents a chemical cell and Figure 24-8b represents a resistor. Figure 24-8c shows a circuit diagram for a resistor connected to a chemical cell. For many purposes it is convenient to redraw such a circuit diagram in order to explicitly show the internal resistance of the cell, as in Figure 24-8d. These circuit elements are said to be connected in series. The current in the circuit is directed from the positive to the negative terminal outside the cell.

Assume that we are given the circuit of Figure 24-8d, where we are given the values of the EMF ($\mathcal{E}$), the internal resistance of the cell (r) and the resistance (R). We can proceed to calculate the current I in the circuit in the following manner: From the definition of resistance, Equation 24-15, we know that the potential difference across the resistor is V, where

$$V = IR \qquad (24\text{-}15)$$

Since the current is being driven by the cell (and not driven by some other generator) we know that the potential difference between the terminals of the cell is given by

$$V = \mathcal{E} - Ir \qquad (24\text{-}22)$$

From an examination of the figure, we see that the potential difference across the resistor R is the same as the potential difference across the cell, so

$$\mathcal{E} - Ir = IR \qquad (24\text{-}26)$$

Consequently, we have

$$I = \frac{\mathcal{E}}{R + r} \qquad (24\text{-}27)$$

Illustrative example

A chemical battery consisting of a combination of cells shown in Figure 24-9 has a total EMF of 24 volts and an internal resistance of 4 ohms. The resistor R has a resistance of 9 ohms. Determine (a) the current in the circuit, (b) the potential difference across the resistor R, and (c) the power supplied to the resistor.

Figure 24-9 shows the method of connecting a voltmeter (V) across the terminals of the resistor R to measure the potential difference across it,

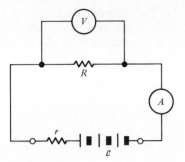

FIGURE 24-9

and also shows an ammeter (A) connected in the circuit for measuring the current I.

In this example $\mathscr{E} = 24$ volts, $R = 9$ ohms, and $r = 4$ ohms.

(a) Using Equation 24-27, we find that the current is

$$I = \frac{24}{13} \text{ amp} = 1.85 \text{ amp}$$

(b) The difference of potential V, sometimes called the voltage, across the resistor R is

$$V = IR$$
$$= 1.85 \times 9 \text{ volts}$$
$$= 16.6 \text{ volts}$$

This is also the difference of potential across the terminals of the battery. A voltmeter across the battery will not read the EMF of 24 volts, but will read 16.6 volts. The difference between the two is the drop in voltage because of the internal resistance of the battery; it is usually referred to as the Ir drop inside the cell.

(c) The power P supplied to the resistor is given by

$$P = I^2 R$$

or

$$P = (1.85)^2 \times 9 \text{ watts} = 30.7 \text{ watts}$$

24-8
Combinations of resistors

Consider now the circuit shown in Figure 24-10, where the cell is connected to two resistors of resistances R_1 and R_2. These resistors are said to be connected in series. The current I is the same in all elements of this circuit. Again, the potential

difference across the cell is given by

$$V = \mathscr{E} - Ir$$

The potential difference between points a and b is given by

$$V_{ab} = IR_1$$

and the potential difference between points b and c is given by

$$V_{bc} = IR_2$$

From the figure it is easy to see that the potential difference between points a and c is the sum of these:

$$V = IR_1 + IR_2$$
$$= I(R_1 + R_2)$$

Then the potential difference between points a and c is the same as the potential difference across the cell, so

$$\mathscr{E} - Ir = I(R_1 + R_2)$$

or

$$I = \frac{\mathscr{E}}{R_1 + R_2 + r} \qquad (24\text{-}28)$$

A comparison of Equations 24-27 and 24-28 shows that we can regard the two resistors R_1 and R_2 of Figure 24-10 as equivalent to the single resistor R of Figure 24-8. That is, we can replace the two series resistors R_1 and R_2 by a single resistor R, where

$$R = R_1 + R_2 \qquad (24\text{-}29)$$

without altering the current in the battery. This can be extended to any number of resistors connected in series. The total resistance R of a set of resistances $R_1, R_2, R_3, \ldots, R_n$ connected in series is the sum of the individual resistances, or

$$R = R_1 + R_2 + R_3 + \cdots + R_n \qquad (24\text{-}30)$$

Consider the circuit of Figure 24-11. The two

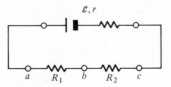

FIGURE 24-10 Two resistors in series connected to a cell.

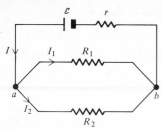

FIGURE 24-11 Two resistors in parallel connected in parallel.

resistors R_1 and R_2 are said to be connected in parallel. Here the current I through the cell divides at points a and b. Part of the current, I_1, flows through R_1 and the rest, I_2, flows through R_2. Thus we have

$$I = I_1 + I_2 \qquad (24\text{-}31)$$

As above, we know that

$$V = \mathscr{E} - Ir \qquad (24\text{-}32)$$

that the potential difference across the resistor R_1 is given by

$$V = I_1 R_1 \qquad (24\text{-}32a)$$

and that the potential difference across R_2 is given by

$$V = I_2 R_2 \qquad (24\text{-}32b)$$

It will be noted that the potential difference across each resistor is the same since they are connected across the same points a and b. Since the total current I supplied to the resistors is

$$I = I_1 + I_2 \qquad (24\text{-}31)$$

we can define the equivalent resistance R of the two resistors in parallel by the equation

$$R = \frac{V}{I} \qquad (24\text{-}33)$$

from which

$$I = \frac{V}{R}$$

From Equations 24-32 and 24-31 we can write

$$I = \frac{V}{R} = \frac{V}{R_1} + \frac{V}{R_2}$$

from which

$$\frac{1}{R} = \frac{1}{R_1} + \frac{1}{R_2} \qquad (24\text{-}34)$$

thus the reciprocal of the equivalent resistance of two resistors in parallel is the sum of the reciprocals of their individual resistances. It can be seen by solving Equation 24-34 for the equivalent resistance R

$$R = \frac{R_1 R_2}{R_1 + R_2} \qquad (24\text{-}35)$$

that the value of R is less than that of the smaller of the two resistances in parallel, and thus is less than either of the two.

In terms of the EMF $\mathscr{E}$ of the generator of internal resistance r, we can use Equation 24-26

$$\mathscr{E} - Ir = IR$$

and write

$$\mathscr{E} = I(R + r)$$

where we use the equivalent resistance for the value of R for a circuit with parallel resistors and obtain

$$\mathscr{E} = I\left(r + \frac{R_1 R_2}{R_1 + R_2}\right) \qquad (24\text{-}36)$$

Illustrative example

In the circuit sketched in Figure 24-12, the voltmeter reads 36 volts. Determine the current in each of the three resistors.

Let us first determine the equivalent resistance of the three resistors. The equivalent resistance R of the two parallel resistors of 6 ohms and 12 ohms is, from Equation 24-35

$$R = \frac{6 \times 12}{6 + 12} = 4 \text{ ohms}$$

We can consider this equivalent resistance to be in

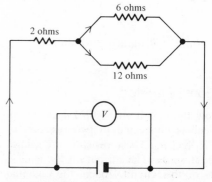

FIGURE 24-12

series with the 2-ohm resistance, so the equivalent resistance of all three resistors is 6 ohms. Since the potential difference across these resistors is 36 volts, the current I supplied by the battery is

$$I = \frac{36}{6} = 6 \text{ amp}$$

This is also the current in the 2-ohm resistor. To determine the currents in the two resistors in parallel, we make use of the fact that the potential difference is the same across each one, so

$$V = I_1 R_1 = I_2 R_2$$

from which

$$\frac{I_1}{I_2} = \frac{R_2}{R_1}$$

That is, the currents in the resistors are in the inverse ratio of the resistances. Letting $R_1 = 6$ ohms, and $R_2 = 12$ ohms, we get

$$\frac{I_1}{I_2} = \frac{12}{6} = 2$$

or

$$I_1 = 2I_2$$

and, since

$$I = I_1 + I_2 = 6 \text{ amp}$$

we get

$$3I_2 = 6 \text{ amp}$$
$$I_2 = 2 \text{ amp}$$
$$I_1 = 4 \text{ amp}$$

Questions

1. List some of the electrical devices in your home. Describe the useful energy output of these devices.

2. Figure 24-6 shows electric fields outside the wires of a circuit. Show that these fields must exist, using the fact that the electric field is conservative in your argument.

3. Electric current is not a vector, yet it has a sense described by an arrow in a diagram. List other quantities that have this property.

4. The current in a wire is driven by the electric field created by the seat of EMF. What is the effect of the field created by the electron gas?

5. A potential difference V is applied to a wire establishing a current. What happens to the drift velocity of the electrons if the potential difference is doubled?

6. What would a graph of potential difference vs current look like for an ohmic conductor? For a nonohmic conductor?

7. Equation 24-22 seems to imply that increasing resistance will increase the rate of heating, whereas Equation 24-23 seems to suggest that increasing the resistance will decrease the rate of heating. Explain.

8. Which has the larger resistance, a 25-watt light bulb or a 100-watt light bulb?

9. A 100-watt light bulb and a 25-watt light bulb are connected in series across a 110-volt line. Which will glow brighter? (In fact, one of them will not glow at all. Removing that light bulb

from its socket will turn off the one that is glowing.)

10. Electrons leave the negative terminal of a cell and flow through an external circuit toward the positive terminal; inside the cell the electrons move from the positive terminal to the negative terminal. Explain how the electrons can move this way.

11. What conditions determine how hot a resistor will become when electric current flows through it?

12. A man decides to do his own wiring at home. After wiring two lamps, he finds that they will light only when both are turned on. If one is turned off they both go out. What mistake did he make in the wiring?

13. Account for the fact that when an electric heater is turned on, the lights in the room get dimmer; when the heater is turned off, the lights get brighter again.

14. Under what circumstances is the terminal voltage of a cell less than the EMF? greater than the EMF?

15. Is it possible for the terminal voltage of a cell to be equal to zero?

16. In times of heavy demand, electric power companies sometimes reduce the voltage supplied to the mains. Why? What do they save?

17. In Edison's time it was argued that centrally generated electric power was impractical because series connection of electrical devices

required that all the devices be operating in order for any one to operate, while parallel connection required such large current-carrying capacity in the distribution lines that it would be impractical to build large enough main distribution wires. Discuss this. It is interesting to consider photographs of the central cities at the turn of the century; these often show many poles carrying dozens of wires for the distribution of power, telephone, and telegraph services. Extrapolation to present demand for electric power makes the argument impressive.

Problems

1. What is the current in a wire if a total charge of 9 coul enters one end of the wire in 0.3 sec?

2. A current of 30×10^{-3} amp flows through a wire into a plating bath. **(a)** How many coulombs pass through the bath in 30 min? **(b)** How many electrons pass through the bath in 30 min?

3. The current in a gas discharge tube consists of electrons moving in one direction and mercury ions moving in the other direction. What is the current if 2×10^{18} electrons and 1.4×10^{18} mercury ions pass through the cross-sectional area of the tube in 1 sec? Assume that all the ions have a charge whose magnitude is twice that of the electron.

4. The current density in a copper wire is 500 amp/cm². Assuming that the number of drift electrons per unit volume is 8.5×10^{22} electrons/cm³, find the drift velocity of the electrons.

5. What is the current density in a wire of radius 3 mm that is carrying 5 amp?

6. Calculate the resistance of a copper wire 4 m long and 3 mm in diameter.

7. The resistance of no. 18 copper wire is tabulated at 6.385 ohms per 1000 ft of length. What is the diameter of no. 18 copper wire?

8. A certain wire has a resistance of 20 ohms. What is the resistance of another wire, made of the same material, that is twice as long and has four times the diameter?

9. What is the resistance of a resistor through which 5 milliamp is flowing and across which the potential difference is 25 volts?

18. The engineer who designs an electrical device may design it for a voltage that is precisely the value at which the device is operated, slightly below that value, or slightly above that value. Consider the effect on the brightness and the lifetime of a bulb of operating a light bulb at a voltage somewhat below the design voltage and compare with the effect of operating somewhat above design voltage. In view of this, why are light bulbs intended for distribution in the United States marked "120 volts"? Similar considerations apply to the design voltage of most electric devices.

10. What potential difference is needed to have 5 amp flow through a 100-ohm resistor?

11. What is the resistance of a 600-watt electric heater that is designed to operate at 120 volts?

12. A 1000-watt toaster draws 10 amp. Find the resistance.

13. A 100-watt light bulb burns for 1 hr. Find the energy required.

14. A 12-volt battery of internal resistance 0.01 ohm is delivering 3.5 amp. Find the terminal voltage.

15. The voltage across the terminals of a certain cell is measured to be 20 volts while the cell is delivering 8 amp. The EMF of the cell is 24 volts. Find the internal resistance of the cell.

16. A 32-volt power supply, of negligible internal resistance, is used to charge a 12-volt cell of internal resistance 0.5 ohm. Find the current delivered by the power supply.

17. A 100-volt battery with an internal resistance of 4 ohms is supplying current to a 23-ohm load. Find the current delivered by the battery.

18. A 36-volt battery with an internal resistance of 0.4 ohm is delivering current to a resistance of 150 ohms. **(a)** Find the current. **(b)** Find the terminal voltage of the battery.

19. Given a 10-ohm resistor and a 5-ohm resistor, find the equivalent resistance of these **(a)** in series and **(b)** in parallel.

20. Five 10-ohm resistors are connected **(a)** in series and **(b)** in parallel. Find the equivalent resistance in each case.

21. A string of Christmas-tree lights consists of 8 bulbs in series. What is the potential differ-

ence across one of the bulbs when the string is connected to a 120-volt line?

22. In the Bohr model of the atom the electron of a hydrogen atom makes about 6×10^{15} rev/sec. What is the average current at a point in the orbit of the electron?

23. In a Van de Graaff generator a belt made of insulating material and 1 m wide moves with a speed of 25 m/sec. The belt moves under a line of fine points that spray charge onto the face of the belt. The charges remain on the belt and result in a current of 10^{-4} amp. (a) At what rate in coul/sec must charge be sprayed onto the belt? (b) What is the charge density on the belt?

24. A copper wire has a resistance of 5 ohms. If the same amount of copper were made into a thinner wire, twice as long, what would its resistance be?

25. Consider three pieces of wire made of aluminum, copper, and silver. Find the ratio of the resistances of these wires if they all have the same length and (a) the same diameter, (b) the same weight.

26. If copper costs $2.00/lb, at what price per pound for aluminum will it be cheaper to make conducting wire out of aluminum?

27. A 1-m piece of copper wire is connected across 6 volts. (a) What is the electric field in the wire? (b) What is the current density in the wire?

28. Show that when a current is flowing in an ohmic conductor, the rate at which heat is developed per unit volume is equal to the square of the current density, divided by the conductivity of the metal.

29. At 6 cents/kilowatt-hour, what does it cost for electric power to leave a 40-watt lamp burning for 8 hr?

30. At 6.5 cents/kilowatt-hour what does it cost to heat a 40-liter tropical fish tank from 60°F to 75°F?

31. A 12-volt automobile battery is advertised as being capable of delivering 100 amp for 2 min. (a) How many coulombs is this? (b) How much energy is stored in the battery? (c) If the battery is connected to the electrical system of the automobile by 1 m of copper wire, what is the minimum diameter of the wire if less than 1 percent of this energy is to be wasted in heating the wire, when the battery is delivering

a starting current of 100 amp? Assume that the internal resistance of the battery is negligible.

32. An electromagnet is designed to operate from 220-volt lines and draws 350 amp. The heat generated is carried off by cooling water that is circulated through the coil. The water is drawn from an artesian well and as it enters the coil it has a temperature of 10°C. As it leaves the coil it has been heated to 90°C. At what rate in liters per minute is the cooling water supplied to the magnet?

33. In a certain X-ray tube electrons are accelerated through 100,000 volts and allowed to strike a metal target in which the X rays are produced. The electron beam is equivalent to a current of 10 milliamp. (a) How many electrons per second strike the target? (b) At what rate is energy being delivered to the target by the beam? (Most of this energy is dissipated as heat.)

34. An electrician wires a house with no. 12 copper wire, which has a resistance of 1.6 ohm per 1000 ft. The circuit breakers are set to turn the current in the wires off if the current exceeds 20 amp and the line voltage input to the house is 115 volts. What is the maximum length of wire the electrician may use if the voltage at the outlet is to be within 1 percent of the input voltage?

35. When a certain cell is delivering 3 amp its terminal voltage is equal to 8.5 volts. When the same cell is being charged with 2 amp, its terminal voltage is 11 volts. Find (a) the internal resistance of the cell and (b) the EMF of the cell.

36. The terminal voltage of a certain cell when it is delivering a negligible current is 1.5 volts. When a very low-resistance wire is connected across the terminals, the battery delivers 30 amp initially. (a) What is the internal resistance of the cell? (b) What is the EMF of the cell? (c) How much current would the cell deliver to a 5-ohm resistor? (d) How large a resistor should be connected across the cell in order to dissipate $\frac{1}{2}$ watt in the resistor?

37. A 40-ohm and a 60-ohm resistor are connected in series across a 120-volt, 0.1-ohm battery. (a) Find the current in each resistor. (b) Find the potential difference across each resistor.

38. The resistors of Problem 37 are connected in

parallel and the combination connected across a 110-volt, 0.3-ohm battery. (a) Find the current in each resistor. (b) Find the terminal voltage of the cell and the rate at which the cell is delivering power to the combination of resistors.

39. Two 25-watt light bulbs are connected in series with each other and in series with a 100-watt light bulb. The bulbs are designed for 120-volt operation. The series combination is connected to a 110-volt line of negligible internal resistance. Assuming that the resistances of the bulbs are constant, find the current and power dissipation in each bulb.

40. Two resistors, R_1 and R_2, are connected in parallel and the combination is connected in parallel with a third resistor R_3. The combination of three resistors is connected to a battery. Show that the equivalent resistance, R, is given by

$$R = \frac{R_1 R_2 R_3}{R_1 R_2 + R_2 R_3 + R_3 R_1}$$

41. The three resistors of Problem 40 consist of a 700-watt electric heater, a 1000-watt electric iron, and a 150-watt lamp. (All three are designed to operate at 110 volts.) The supply EMF is 110 volts with an internal resistance of 2 ohms. Find (a) the current drawn by each appliance, (b) the current through the seat of EMF, and (c) the power drawn from the seat of EMF.

42. Assume that the free electrons that constitute a current in a wire always start from rest and, under the influence of the constant electric field, accelerate with constant acceleration for a time T. At the end of this time T, the electrons collide and come to rest. Then the average drift velocity v_d is given by

$$v_d = \tfrac{1}{2} a T$$

(a) Show that the resistance R of a length of wire of length L and cross-sectional area A is given by

$$R = \frac{2Lm}{e^2 n T A}$$

where e is the charge of the free electron, m is the mass of the electron, and n is the number of free electrons per unit volume. (b) Find the average time between collisions for a copper wire 2 m long and 4 mm in diameter. Assume that copper has 8×10^{22} electrons/cm³. (c) Find the average drift velocity of the electrons if this wire carries 2 amp. (d) What is the average distance traveled by an electron between collisions? (e) How many collisions does one electron make in traveling from one end of the wire to the other? (f) Assuming that the collisions take no time, how long does it take an electron to travel from one end of the wire to the other?

43. A wire of length L is used as a heater, dissipating power P when connected to a power supply of EMF V and negligible internal resistance. The wire is then cut in half and each half is connected to the same source of EMF. (a) What is the total power dissipated by the two half-pieces of wire? (b) Why not "improve" electric heaters by doing this?

44. A battery with an EMF of 6 volts and an internal resistance of 0.3 ohm is connected to a resistor R. The resistance of R is varied from zero to 1 ohm in steps of 0.1 ohm. Plot the power dissipated in the resistor R as a function of the value of R. Note where the maximum occurs. The result is an illustration of a general theorem called the *maximum power transfer* theorem.

45. There are four different ways of connecting three equal resistors together. (a) Draw a circuit diagram for each of these. (b) Find the equivalent resistance of the combination if each of the individual resistors has a resistance of 4 ohms.

46. When an electric motor is operating, it converts some of the input energy into heat because of the resistance of the wires in its windings. The rest of the energy is converted into mechanical energy. Thus the motor acts electrically like a battery that is being charged; the EMF of this equivalent battery measures the energy per unit charge that is converted into mechanical energy and is called the *back* EMF of the motor. (a) A certain motor has an internal resistance of 0.2 ohm and draws 20 amp when the motor is operating from 120-volt lines. Find the back EMF. (b) How much power is supplied to the motor by the lines? (c) How much power is converted into heat in the wires of the motor? (d) How much power is converted into mechanical energy? (e) What happens to

the back EMF and the input current of such a motor if the output shaft is held fixed by some external agent?

47. In demonstrating electrostatic phenomena it is common to discharge objects by touching them, providing a path to ground. Yet the human body is a poor conductor of electric current. This arises from the fact that the requirements for good insulation are very stringent for electrostatic phenomena as compared with the requirements for insulation in current phenomena. Consider, for example, a metal sphere of 4-cm radius that is charged to a potential V. (a) Find the charge on the sphere in terms of V. (b) If the sphere is in-sulated it will retain its charge. As a practical matter, the charge will leak off through the high resistance of the air and the support; if the sphere is insulated, this leakage will be slow. Find this leakage current in terms of V if the charge decreases by 0.1 percent in 1 sec. (c) Find the total resistance of the leakage path. (d) If the sphere is supported on a rod 10 cm long and 1 cm² in area, in an evacuated chamber, find the minimum resistivity of the material of which the rod is made. (Note that the resistivity of glass ranges from 10^{10} to 10^{14} ohm m, that of wood ranges from 10^8 to 10^{11} ohm m, that of quartz is 7.5×10^{17} ohm m, and that of animal tissue ranges from 0.5 to 25 ohm m.)

magnetism

chapter 25

25-1
Introduction

Man's early observation of magnetism was as a force between permanent magnets. The ability of one piece of iron to exert a force on another piece of iron, even though they were separated, was so impressive that magical properties were ascribed to magnets in the Middle Ages. The beginnings of a scientific understanding of magnetism are attributed to William Gilbert, who published a book in 1600 that summarized the experimental facts then known, and demolished most of the myths. It became clear that permanent magnets could be thought of as having two different *poles*, located at the ends of the magnet. The poles exert forces on one another, in a manner analogous to the forces exerted by electric charges. In the late eighteenth century, Coulomb made careful measurements of the forces exerted by magnetic poles. These measurements were complicated by the fact that the poles always came in pairs; it was impossible to make or find an isolated magnetic pole.

In the early nineteenth century, Faraday introduced the idea of *field*. Instead of describing the force between magnets directly, Faraday said that one magnet created a magnetic field, and that this field in turn exerted a force on the other magnet.

In 1820, Hans Christian Oersted discovered that an electric current could exert a force on a magnet —that an electric current could establish a magnetic field.

From this initial observation of the connection between electricity and magnetism grew a large body of observation and theory, culminating in the publication, in 1873, of James Clerk Maxwell's *Treatise on Electricity and Magnetism.* The theory described by Maxwell is now usually called *classical* electricity and magnetism because Einstein's theory of relativity introduced important changes to our understanding of electricity and magnetism. In this chapter we shall restrict ourselves to a classical description of magnetism. In effect, this means that we will be discussing the forces between charges whose velocities are small compared with the speed of light. Furthermore, we will in general discuss the forces as seen by one observer and ignore the question of what is seen by an observer in a second, moving frame of reference.

As we have seen in Chapter 5, moving charges exert forces on each other in addition to the electric force—forces we call the magnetic forces. In this chapter we shall treat magnetic force in two stages, first describing the magnetic field established by a moving charge and then describing the force exerted by a magnetic field. We shall then describe the magnetic effects of currents in wires.

As we shall see in the next chapter, changing magnetic fields produce electric fields and changing electric fields produce magnetic fields. Therefore, we shall further restrict the discussion of the present chapter to situations in which the fields are constant in time. This means that we shall assume that the charges that create the fields are moving with constant velocity.

25-2
The magnetic field due to a moving charge

Consider a positive charge q moving with a velocity $\mathbf{v}$, as shown in Figure 25-1. The magnetic field $\mathbf{B}$ at point P is given by

$$\mathbf{B} = k' \frac{q\mathbf{v} \times \hat{\mathbf{r}}}{r^2} \tag{25-1}$$

where r is the distance between the charge and the point P, and $\hat{\mathbf{r}}$ is a vector of *unit magnitude*, which has the direction of the line drawn from the charge to P. The quantity k' is a constant whose magnitude depends on the system of units in which the quantities are measured. In the mks system of units, the unit of magnetic field is the *weber per square meter*, or *tesla*, and the constant k' is given by

$$k' = 10^{-7} \frac{\text{weber}}{\text{amp m}} \tag{25-2}$$

Equation 25-1 is usually referred to as the law of Biot and Savart. The magnitude of the magnetic field B is, from Equation 25-1,

$$B = k' \frac{qv \sin \theta}{r^2} \tag{25-3}$$

where θ is the angle between the vector $\mathbf{v}$ and the line drawn from the charge to P.

The direction of $\mathbf{B}$ is perpendicular to the plane formed by $\mathbf{v}$ and $\hat{\mathbf{r}}$ and has a sense given by a right-hand-screw rule; that is, the rotation of a right-hand screw in the sense of rotation of $\mathbf{v}$ into $\hat{\mathbf{r}}$ advances the screw in the direction of $\mathbf{B}$, as shown in Figure 25-1b. Another way of remembering this rule is shown in Figure 25-1b; if the fingers of the right hand curl from the direction of $\mathbf{v}$ to the direction of $\mathbf{r}$, the thumb of the right hand will point in the direction of $\mathbf{B}$.

A convenient way to visualize the direction of the field is obtained by imagining the point P rotating about the direction of $\mathbf{v}$ as shown in Figure 25-1c. Every point on the circle produced by P is the same distance r from the charge and the angle θ is the same for all the points, and so the magnitude of $\mathbf{B}$ will be the same at all the points on the circle. Furthermore, the field is everywhere tangent to the circle and has the direction shown. In a sense, the field forms circles around the direction of the velocity $\mathbf{v}$. The direction of the circulation can be described in several ways. One way is to say that when an observer looks along the direction of motion of the positive charge and sees the charge moving away from him, the field circulates clockwise. A second way of saying this is in terms of another right-hand rule: when the thumb of the right hand points in the direction of $\mathbf{v}$, then for positive charge the fingers of the right hand curl naturally in the direction of circulation of the field, as shown in Figure 25-1c.

When negative charges move, they too establish magnetic fields, in a direction opposite to that which would be established by a positive charge moving in that direction. An alternative way of saying this is that the magnetic field established by a moving negative charge is the same as that of a positive charge moving in the opposite direction.

At this point it should be reiterated that the moving charge also establishes an electric field at point P. According to classical electromagnetic theory, the electric field at P is the same as the electric field that would have been established if the charge were not moving. The theory of rela-

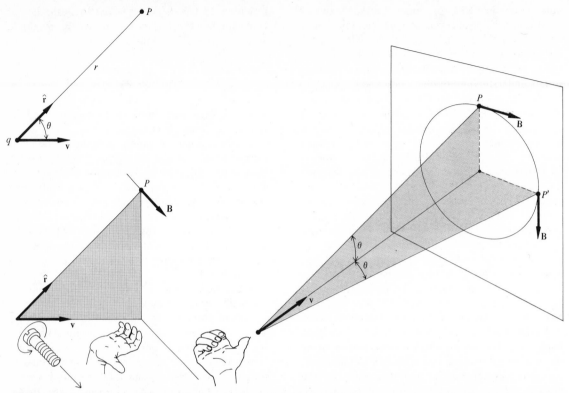

FIGURE 25-1 The direction of the magnetic field at point *P* due to a positive charge *q* moving with velocity **v**. (a) The vector **r** joining the charge and the point *p* makes an angle θ with the velocity. (b) The field is perpendicular to the plane formed by **r** and **v** and has a direction given by a right-hand rule or the direction of advance of a right-hand screw that rotates in the sense of rotation of **v** into **r**. (c) The field forms circles around the velocity.

tivity asserts that the electric field (and the magnetic field given in Equation 25-1) are changed by factors that are approximately $1 - v^2/c^2$. Thus, for particles moving with speeds that are small compared with the speed of light, we can ignore the relativistic changes.

Just as it was convenient to represent the electric field on a diagram by drawing lines of force, it is often convenient to represent the magnetic field by drawing lines of magnetic flux. If the number of lines per unit area is made proportional to the field **B**, then the density of lines indicates the magnitude of the field and the direction of the lines gives the direction of the field. There are two major differences between the lines representing the electric field and those representing the magnetic field:

1. Electric lines of force start and stop on charges, whereas magnetic lines of flux form closed curves. (As we shall see, this difference is true

only when we restrict ourselves to fields that do not vary with time.)

2. The force on a test charge is *parallel* to the electric field and (in the nonrelativistic domain) does not depend on the velocity of the charge; the force on a test charge is *perpendicular* to the magnetic field, perpendicular to the velocity of the test charge, and equal to zero if the velocity of the test charge is equal to zero.

25-3
Magnetic force on a moving charged particle

Consider a particle of positive charge q moving with a velocity **v** in a region of space in which a magnetic field **B** exists. The field will exert a force **F** on the charge given by

$$\mathbf{F} = q\mathbf{v} \times \mathbf{B} \qquad (25\text{-}4)$$

The magnitude of the force **F** is given by

$$F = qvB \sin \theta \qquad (25\text{-}5)$$

where θ is the angle between the direction of **B** and the direction of **v**. The direction of **F** is perpendicular to the plane formed by **v** and **B** and has the sense given by a right-hand rule: a right-hand screw will advance in the direction of **F** if rotated in the sense of rotation of the vector **v** into the vector **B**; the thumb of the right hand will point in the direction of **F** if the fingers curl from **v** to **B**, as shown in Figure 25-2.

One way of interpreting Equation 25-5 is that the force on a moving charge is exerted only by the component of **B** that is perpendicular to the velocity ($B \sin \theta$) whereas the force exerted by the component of **B** that is parallel to the velocity is equal to zero.

The force exerted on a moving negative particle is the same as the force on a positive particle moving in the opposite direction.

The fact that the force exerted by a magnetic field is perpendicular to the velocity has an interesting consequence: the magnetic field does no work on the charged particle. In order to see this, we recognize that during each infinitesimal interval of time, the displacement of a particle is parallel to the velocity of the particle at the beginning of that interval of time. Thus, the magnetic force will be perpendicular to the displacement and so the work done by the magnetic field will be equal to zero.

Since the magnetic field can do no work on a moving charged particle, the kinetic energy of the particle must be unaffected by the field. This means that the speed of the particle must remain constant. But the magnetic field does exert a force, which means that the particle must be accelerated. This is not a contradiction when we remember that acceleration may be restricted to a change in the *direction* of a velocity. Thus, we come to the somewhat unexpected conclusion that magnetic fields can change the direction of motion of particles, but not the speed; magnetic fields can steer charged particles, but neither speed them up nor slow them down. It must be pointed out that this conclusion is valid only for magnetic fields that do not vary in time. Changing magnetic fields create electric fields, and these in turn can change the energy of the particles.

Consider then a charged particle, of charge q, that enters a region of space in which there exists a constant magnetic field **B**. Suppose the particle has a velocity **v**, which is initially perpendicular to the field **B**. Then, as we have seen, during all of its subsequent motion it will have the same *speed*, v, but the direction of motion will change. At the first instant that it is in the field, the directions of the force, velocity, and field are given by Figure 25-2. Since the force is perpendicular to the field, the acceleration is perpendicular to the field, and therefore the change in velocity is also perpendicular to the field. Furthermore, since the initial velocity and the change in velocity are both perpendicular to the field, the new velocity must also be perpendicular to the field. Thus, the particle will continue to move in the plane that is perpendicular to the magnetic field. The motion of the

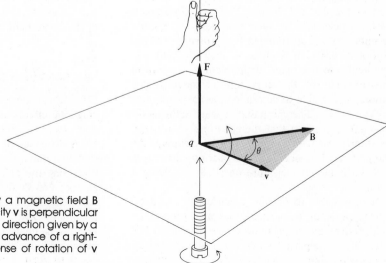

FIGURE 25-2 The force exerted by a magnetic field **B** on a charge q moving with a velocity **v** is perpendicular to the plane of **v** and **B**, and has a direction given by a right-hand rule or the direction of advance of a right-hand screw that rotates in the sense of rotation of **v** into **B**.

particle can then be described as in a plane, with constant speed and with an acceleration that is always perpendicular to the velocity. From Equation 25-5 we see that the magnitude of the acceleration is also constant. It then is almost obvious that the particle must move in a circle.

When a particle moves with speed v in a circle of radius R, it has a centripetal acceleration whose magnitude a is given by

$$a = \frac{v^2}{R}$$

This centripetal acceleration is caused by a centripetal force, **F**, whose magnitude is given by

$$F = ma$$
$$= \frac{mv^2}{R}$$

In the case of a charged particle moving in a magnetic field this centripetal force is the force exerted by the field, and has the magnitude given by Equation 25-5. Therefore, we have

$$qvB = \frac{mv^2}{R}$$

or

$$R = \frac{mv}{qB} \qquad (25\text{-}6)$$

for the radius of the circle.

Equation 25-6 shows that the radius of the circular path of a charged particle moving at right angles to the magnetic field is proportional to the momentum of the particle. Figure 25-3 is a photograph of the tracks left by such particles in a bubble chamber that is surrounded by a current-carrying coil of wire, which produces a magnetic field perpendicular to the plane of the picture. By careful measurement of the curvature of such tracks, physicists are able to determine the momentum of these particles. The spiral that appears in the figure arises from the fact that the electrons lose energy by interacting with the liquid in the chamber. As the particles lose energy the momentum decreases and the radius of curvature decreases, yielding a spiral.

It is possible to arrange electric and magnetic fields so that all the particles that pass through the fields have known velocities (see Problem 29). Thus, if all the particles that leave such a *velocity selector* then enter a magnetic field perpendicular

FIGURE 25-3 Bubble chamber tracks of charged particles in a magnetic field. (Courtesy of Brookhaven National Laboratory.)

to the field, they will move in circles whose radii are proportional to the masses of the particles. Such a system can be used to measure or select the mass of particles of specific mass or masses and is the basis of the instrument called the *mass spectrometer*.

25-4
Magnetic field of an electric current

Electric currents in metal wires produce the same magnetic effects as the flow of free charges, a fact that was first experimentally demonstrated by H. A. Rowland in 1876. Since the number of protons in a metal is equal to the number of electrons (including the moving electrons) and because the charges are almost smoothly distributed, the *electric* field outside the wire due to the current is essentially zero. In addition, the motion of the protons is small and nearly random, so the magnetic field due to the protons is approximately equal to zero. Furthermore, the drift velocity of the electrons is usually very small compared with the speed of light, so the magnetic field due to an electric current in a wire can be treated nonrelativistically, and the electric field can be ignored.

Figure 25-4 shows a tube in which a current I consists of the motion of positive charges with uniform velocity $\mathbf{v}$. From the definition of current,

$$I = \frac{\Delta q}{\Delta t} \qquad (25\text{-}7)$$

where Δq is the quantity of charge that passes through a complete cross-sectional area of the tube in time Δt. In this time interval the charge Δq will traverse a length Δs of the tube, where

$$\Delta s = v\, \Delta t \qquad (25\text{-}8)$$

Eliminating Δt from these two equations yields

$$I\, \Delta s = v\, \Delta q \qquad (25\text{-}9)$$

In other words, a piece of conductor of length Δs, carrying a current I, is equivalent to a positive charge Δq moving with a velocity v, where the velocity is in the same direction as the current.

Combining Equations 25-9 and 25-2, we obtain

$$B = k' \frac{I\, \Delta s \sin\, \theta}{r^2} \qquad (25\text{-}10)$$

for the magnitude of magnetic flux density due to a small piece of wire, of length Δs, carrying a current I. Again, Figure 25-1 can be used to describe the direction of the field. Equation 25-10 was first developed from experimental results by Biot and Savart.

Currents never occur in little straight pieces of wire; they always occur as parts of circuits. Thus, the field calculated from Equation 25-10 is only part of the field at any particular point, the part due to one piece of a circuit. In order to find the total magnetic field at any one point, one must calculate the contribution from each part of the circuit, using Equation 25-10 many times, and then add all these contributions, remembering that the field is a vector. This process is, for almost all circuits, beyond the mathematical scope of this book. Figure 25-5 shows the field for one such geometry,

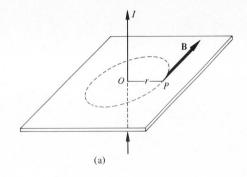

(a)

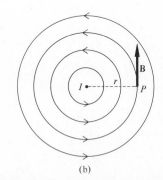

(b)

FIGURE 25-5 (a) Magnetic flux density **B** at point P at a distance r from a straight wire carrying current is circular in a plane at right angles to the current. (b) The magnetic field viewed in a plane perpendicular to I. The current is coming out of the paper toward the reader.

a very long, straight wire, carrying a current I. The result of adding all the contributions from small segments of the long wire is that the field at a distance r from the wire has the value

$$B = 2k' \frac{I}{r} \qquad (25\text{-}11)$$

Illustrative example

A long, straight wire has a current of 25 amp in it. Determine the magnetic flux density at a point 15 cm from the wire.

Using Equation 25-11 and substituting the values $I = 25$ amp, $r = 0.15$ m, and $k' = 10^{-7}$ weber/amp m we get

$$B = 2 \times 10^{-7} \frac{25}{0.15} \frac{\text{weber}}{\text{m}^2}$$

from which $B = 3.33 \times 10^{-5}$ weber/m².

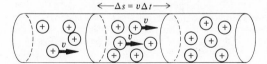

FIGURE 25-4

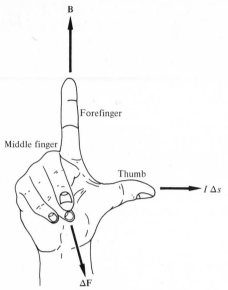

FIGURE 25-6 Three fingers of right hand can be used to show relative directions of I, F, $I \Delta s$.

25-5
The force on an electric current

Again consider a small piece of wire. Combining Equations 25-9 and 25-5, we have for the force F on a piece of wire of length Δs, carrying a current I in a field $\mathbf{B}$,

$$F = BI \Delta s \sin \theta \qquad (25\text{-}12)$$

where θ is the angle between the field $\mathbf{B}$ and the direction of the current I. The direction of the force $\mathbf{F}$ is shown in Figure 25-2, where the current is in the direction of $\mathbf{v}$.

To determine the force on a real wire, we must find the force on each segment of it, using Equation 25-12 and then sum these forces. In the special case of a long, straight wire in a uniform field, Equation 25-12 becomes

$$F = BIL \sin \theta \qquad (25\text{-}13)$$

where L is the length of the wire.

In the special case where B and L are at right angles to each other, as in Figure 25-6, $\theta = 90$ degrees and $\sin \theta = 1$, so Equation 25-12 becomes

$$F = ILB \qquad (25\text{-}14)$$

Figure 25-7a shows the direction of the force

on the wire when the current is directed upward and the magnetic field is toward the right. The relationship among these vectors is shown in Figure 25-7b. $\mathbf{F}$ is at right angles to both $I\mathbf{L}$ and $\mathbf{B}$ and directed into the paper. This can readily be verified by applying the right-hand rule.

A simple method for determining the direction of the force on a wire carrying current when placed in a magnetic field is to examine the relative directions of this magnetic field of flux density B and the magnetic field produced by the current in the wire. Figure 25-7c shows these two magnetic fields in a plane at right angles to the wire when viewed from A toward C. Since the current in the wire is coming out of the paper, its magnetic field is circular around the wire and counterclockwise. Below the wire, the two magnetic fields are in the same direction, thus producing a more intense

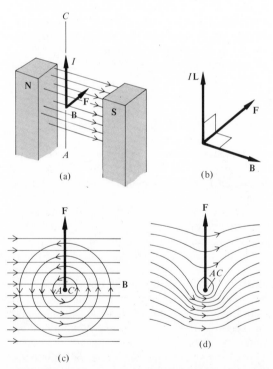

FIGURE 25-7 Wire AC between the poles of a magnet carries current I in a magnetic field of flux density B. (a) Direction of force $\mathbf{F}$ on wire is at right angles to the wire and the direction of the magnetic field. (b) Relative directions of I, B, and F; these vectors are mutually perpendicular. (c) The two magnetic fields: that due to I and the field B of the magnet; (d) The resultant of these two fields is stronger below AC and weaker above AC. The wire is forced from the stronger to the weaker field.

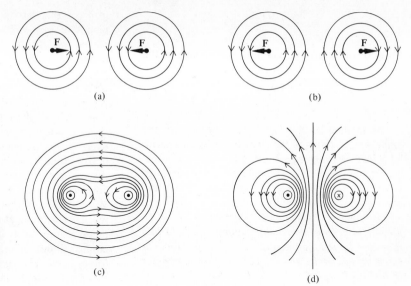

FIGURE 25-8 Magnetic field of two parallel wires in a plane at right angles to the wires: (a) when currents are in the same direction out of the paper; (b) the resultant field; (c) currents are in opposite directions; and (d) the resultant field.

(a)

(b)

(c)

(d)

magnetic field; above the wire, the two fields are in opposite directions, thus producing a weak field, as shown in Figure 25-7d. Experiment shows that the force on AC is upward; that is, the wire is forced from the stronger magnetic field toward the weaker magnetic field. The method outlined above is perfectly general; *whenever a wire carrying current is placed in a magnetic field, the force on the wire will always be directed from the stronger toward the weaker part of the* resultant *field.*

For example, if a straight wire has a length of 0.5 m in a magnetic field of 1.2 weber/m² and is at right angles to this field, then, when the current in it is 15 amp, it will experience a force

$$F = 15 \times 0.5 \times 1.2 = 9.0 \text{ nt}$$

25-6
The force between two parallel wires carrying currents

Two parallel wires carrying currents will exert forces on each other owing to the interaction of their respective magnetic fields. If the two currents are in the same direction, the force between them will be one of attraction; if the two currents are in opposite directions, the force will be one of repulsion. Figure 25-8 shows the magnetic fields in a plane at right angles to the wires. In Figure 25-8a, the currents are assumed to be coming out of the paper toward the reader. The magnetic field is circular in a counterclockwise direction around

each wire. The two fields are in opposite directions in the space between the wires. The resultant magnetic field is thus weakened in this region, as shown in Figure 25-8b, and the wires are forced toward each other, each wire going from the stronger to the weaker part of the field. Figure 25-8c shows the two magnetic fields between two parallel wires carrying currents in opposite directions. These fields, again, are drawn in a plane perpendicular to the two wires. It will be seen that the fields between the wires are in the same direction, and hence reinforce each other in this region, as shown in Figure 25-8d. The wires are therefore forced away from each other, the force on each wire being directed from the stronger to the weaker field.

We have already shown that the force on a wire carrying current I in a magnetic field of flux density B is given by

$$F = ILB \qquad (25\text{-}14)$$

when I and B are at right angles to each other. We can apply this equation to determine the force between two parallel wires. Let us consider a length L of one wire carrying current I to be in the magnetic field B of the second wire carrying current I' at a distance r from it. The value of B is given by

$$B = k' \frac{2I'}{r} \qquad (25\text{-}11)$$

Substituting this value of B into Equation 25-14 yields

$$F = k' \frac{2II'L}{r}$$

From this equation we can obtain the value of the force per unit length of wire as

$$\frac{F}{L} = k' \frac{2II'}{r} \qquad (25\text{-}15)$$

The method of defining the unit of current, the *ampere*, is in terms of the force between two parallel wires carrying equal currents. Thus the measurement of current can be reduced to the measurement of the mechanical quantities, force and distance:

An ampere is that constant current which, when flowing in each of two infinitely long parallel wires that are 1 meter apart in a vacuum, produces a force on each wire of 2×10^{-7} newton per meter of length.

If in Equation 25-15 we set

$$\frac{F}{L} = 2 \times 10^{-7} \frac{\text{nt}}{\text{m}}$$

$$I = I' = 1 \text{ amp}$$

and

$$r = 1 \text{ m}$$

we have

$$2 \times 10^{-7} \frac{\text{nt}}{\text{m}} = k' \frac{2 \times 1 \text{ amp}^2}{1 \text{ m}}$$

from which

$$k' = 10^{-7} \frac{\text{nt}}{\text{amp}^2}$$

which is the value we introduced arbitrarily in the previous chapter. The units nt/amp² are equivalent to the other units used previously, such as weber²/nt m², and nt/amp m.

Equation 25-15 is at the basis of the design of instruments for the most accurate measurements of current; such instruments, called *current balances*, have been in use since the latter part of the nineteenth century. For convenience, circular coils of wire, rather than long straight wires, are used in current balances.

25-7
Magnets

The earliest experience of human beings with magnetism was with *lodestones*, a naturally occurring iron ore that has the property of attracting small pieces of iron and steel (and, to a lesser extent, nickel and cobalt). Iron or steel needles (or bars) can be magnetized, or made into magnets, by being placed close to lodestones, and these needles can be used to make compasses. The north pole of a compass needle is that end of the compass which points toward the north pole of the earth.

An elementary fact about magnets is that they exert forces on each other as well as on unmagnetized samples of iron. We know that the origin of these forces is the motion of charges, but it is convenient to describe the forces in terms of *poles*. We can think of the simple bar magnet as containing a north pole at one end (the end that would point to the north if the magnet were used as a compass) and, at the other end, a south pole. The forces between magnets can be described, roughly, by asserting that like poles repel and unlike poles attract. By making very long, thin magnets we can examine the force between poles (assuming that the forces between the poles at the other ends of the long magnets can be ignored). Such experiments lead to the assertion that the force between poles varies inversely as the square of the distance between poles. One can even define a pole strength p, analogous to the charge q, and write an equation similar to Coulomb's law for the force between poles. In these terms, the force on a pole is proportional to the product of the pole strength p and the magnetic field B,

$$F = pB \qquad (25\text{-}16)$$

The magnetic fields established by permanent magnets can be described in terms of field lines radiating out from north poles and into south poles (analogous to the electric field lines radiating out from positive charges and into negative charges).

A major limitation of this pole picture is that it is impossible to isolate a single pole; poles always come in pairs. Further, the poles in a magnet are spread out over a region of space near the ends of the magnet, rather than being situated at a point. It is interesting to note that there are modern theories that hypothesize particles that are magnetic poles (or monopoles), but no experiment has ever observed such a monopole; isolated poles are not logically impossible but it appears that they do not exist.

The magnetic field of real magnets can be described by field lines that leave the north pole of the magnet and enter the south pole. For long, thin magnets most of the lines go to or come from infinity, while for C-shaped magnets, most of the lines go from the north pole to the south pole.

If a bar magnet is placed in a uniform magnetic field, the force on the north pole will be parallel to the field, whereas the force on the south pole will be opposite to the field. This pair of equal and opposite forces will, in general exert a torque on the magnet. The torque τ on the magnet will be proportional to the product of the field B, the pole strength p, the distance L between the poles, and the sine of the angle θ between the direction of the bar magnet and the direction of the field,

$$\tau \propto BpL \sin \theta \qquad (25\text{-}17)$$

The product pL is called the *magnetic moment M* of the magnet, and is an important property of the magnet. The magnetic field of the magnet as a whole (arising from both poles) is proportional to the magnetic moment. Further, the total force exerted on a magnet by another magnet or by a nonuniform field can be written in terms of the magnetic moment. Although the pole picture is useful, the important property of a magnet is its magnetic moment, a quantity that involves both poles and the distance between them.

A coil of current-carrying wire behaves like a bar magnet. It will exert forces on bits of iron and on magnets. It will have forces exerted on it by other magnets and by nonuniform fields and will have a torque exerted on it by a uniform field. In fact, the equations describing these effects are essentially the same as those for the bar magnet if we write for the magnetic moment of the coil,

$$M = IA \qquad (25\text{-}18)$$

where I is the current in the coil and A is its cross-sectional area. A coil of current-carrying wire is equivalent to a magnet, and every magnet can be replaced by an equivalent coil of wire. In principle, we can describe the magnetic behavior of a magnet by describing the behavior of a set of imaginary currents on the surface of the magnet. Such currents are sometimes called Amperian currents. The actual currents that give rise to the magnetic behavior of magnets are atomic in scale.

25-8
Magnetic materials

Since all materials consist of atoms that contain moving charges, it should not be surprising to dis-

cover that all materials are affected by magnetic fields. Generally these effects are subtle changes in the motion of the electrons and can be observed only by measuring small changes in the properties of the materials. For some materials, the effects are macroscopic, particularly in crystals. For example, some crystals show relatively large changes in size or shape in the presence of moderate magnetic fields; such crystals are said to be *magnetostrictive.*

When a magnet exerts a force on a bit of iron it does so because the magnetic field due to the magnet magnetizes the piece of iron, giving the iron a magnetic moment. This magnetic moment is located in the *nonuniform* field of the magnet, which exerts a force on the iron. This force attracts the iron; in other words, the iron has a force on it that tends to move it from the region where the field is weak to the region where the field is strong. A few other materials, such as steel, nickel, cobalt, and some alloys will behave in the same way; small samples will have relatively large forces exerted on them by nonuniform fields, and these forces tend to make the samples move toward the region of large field. Such materials are called *ferromagnetic* materials. All other materials will have forces exerted on them by nonuniform fields, but the forces will be significantly smaller than the forces exerted on ferromagnetic materials. These forces are usually masked by gravity and friction, but careful experiments can demonstrate their existence.

For one class of materials, called *paramagnetic* materials, the force exerted by nonuniform fields is also directed from the region of weak fields to the region of strong fields, but it is much smaller than the force on ferromagnetic materials. Paramagnetic materials are attracted weakly to permanent magnets. Examples of paramagnetic materials are oxygen, aluminum, and sodium. All the remaining materials are *diamagnetic*—that is, materials that are weakly repelled by magnets. The force exerted by a nonuniform field is toward the region of weak field. Examples of diamagnetic materials are mercury, silver, and rock salt.

Diamagnetism can be explained in terms of the motions of the electrons in orbits about the nucleus. These moving electrons are equivalent to a current in a loop of wire; it is possible to calculate the magnetic behavior of these loops and get reasonably good agreement with the behavior of diamagnetic materials.

Since all materials contain electrons that move

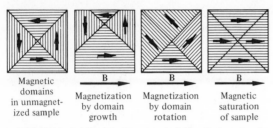

Magnetic domains in unmagnetized sample

Magnetization by domain growth

Magnetization by domain rotation

Magnetic saturation of sample

FIGURE 25-9 Schematic representation of changes in magnetic domains in a ferromagnetic substance during magnetization.

in orbits, we might expect all materials to be diamagnetic. But since there are paramagnetic materials, there must be some other source of magnetism in the atom. This is the intrinsic magnetic moment of the electron, which is usually called the *spin* of the electron. The electron behaves as if it were a spinning charged sphere; it has a magnetic moment. In most materials the electron spins are paired off; for every electron spin that points in one direction there exists another nearby electron that points in the opposite direction. As far as external magnetic effects are concerned, these paired spins cancel each other and the magnetic effects that do exist are due to the orbital motion; the material is diamagnetic. For some materials an unpaired spin does have external effects and the material is paramagnetic. Calculations based on this unpaired-spin model of paramagnetism are in reasonable agreement with the behavior of paramagnetic materials.

Ferromagnetism arises from a cooperative phenomenon; large numbers of neighboring atoms affect each other so that within a region of the material, called a *magnetic domain*, all of the atomic spins are parallel to each other. Each domain has a large magnetic moment. In an ordinary, unmagnetized sample of a ferromagnetic material there are many domains. Since the spin directions of the domains are randomly oriented, as far as external magnetic effects are concerned the net magnetic moment is equal to zero. When an external magnetic field is applied, two things may occur. (a) At low values of the field, some of the domains whose magnetic moments are favorably directed may grow in size at the expense of the other domains. (b) As the external field is increased, some of the domains are rotated so that their magnetic moments have larger components in the direction of the external field. *Magnetic saturation* is reached when all the magnetic mo-

ments are parallel to the external field. At this point the magnetic field due to the magnetic moments of the iron is a maximum. (This process is schematically shown in Figure 25-9.)

When the external magnetic field is removed from a ferromagnetic material, the domains rotate back toward their original orientation, but they do not reach their original random orientation. The cooperative effects tend to keep them lined up so that they are left with a net magnetic moment, which is usually smaller than the saturation value.

25-9
Hysteresis

A current-carrying coil of wire establishes a magnetic field in space. If a piece of ferromagnetic material is placed inside the coil, the magnetic field will be increased; the field due to the current will align the domains in the ferromagnetic material, giving the material a large net magnetic moment. This magnetic moment establishes a magnetic field, which, roughly, is in the same direction as the field due to the current.

Consider then the field **B** that exists in the ferromagnetic material. This field is the sum of two parts: the field **B**$_0$, which is caused by the current and would have been there in the absence of the ferromagnetic material; and the field **M**, which is due to the magnetic moments of the domains. The size and relative orientation of these fields depends on the geometry as well as the nature of the material, but it is possible to build simple systems in which the fields are all parallel. In such simple cases we have

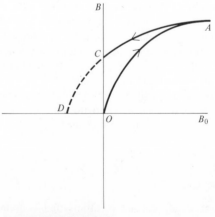

FIGURE 25-10 Hysteresis.

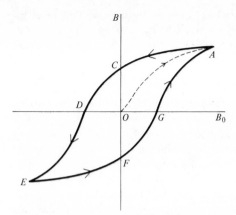

FIGURE 25-11 Hysteresis loop in cycle of magnetization of iron.

$$B = B_0 + M \qquad (25\text{-}19)$$

If we start with an unmagnetized piece of iron and zero current in the wires, all three fields will be equal to zero. If the current is now increased, the field B will increase because both B_0 and M increase. As the domains grow and rotate, the field M gets large and dominates the field in the material. A plot of B against B_0 looks like Figure 25-10; the field rises fairly sharply at first and then, as the domains saturate, it flattens out. If the current is then reduced, the curve follows the path AC; the domains tend to stay aligned and the field M does not reduce to zero. The iron has become a permanent magnet. The field represented by OC is called the retentivity. To reduce the flux density to zero—that is, to demagnetize the iron—it is necessary to reverse the current and increase its value until B_0 reaches the value at point D. This phenomenon in which the field lags behind the driving field B_0, is called *hysteresis*.

All ferromagnetic materials exhibit the phenomenon of hysteresis; one of the effects of hysteresis is that the magnetic field in (and near) a ferromagnetic material is not always the same for a given value of the current in nearby circuits, but depends on the previous treatment that the ferromagnetic material received. If an alternating current (one that varies sinuosidally with time) is sent through a coil that is wrapped around a ferromagnetic material, the curve of B against B_0 will look like Figure 25-11 for each cycle of the current. While the current in the coil goes from its maximum value to zero the corresponding value of B goes from its saturation value at A to its value at point C. As the current in the coil reverses, the field decreases to zero at point D and then reverses and reaches a saturation value at point E. Then the current decreases, and the field decreases to point F when the current is equal to zero. The current reverses and the value of the field drops to zero at point G. As the current increases to its maximum value, the field increases to A. The cycle then repeats and during the cycle the curve never goes through the origin.

It can be shown that the area of this curve is proportional to the work per unit volume that is done on the ferromagnetic material during one cycle. The work is converted into internal energy of the material and usually produces a rise in temperature of the material. Care must be taken in the design of machinery (such as motors, generators, and transformers) in which alternating currents are used, to keep the rise in temperature to a reasonable value by conducting heat away from the machine.

A simple way of demagnetizing a substance is to place it inside a coil and send alternating current through the coil. If the amplitude of the alternating current is decreased slowly, then the hysteresis loop will get smaller and smaller during successive cycles until finally, when the current in the coil is zero, the value of B will also be zero. An equivalent procedure is to remove the object slowly from the coil, which decreases B_0. This procedure of slowly removing a part from a coil carrying alternating current is used by watch repairmen to demagnetize watch parts.

Questions

1. On what factors does the magnetic field due to a moving charged particle depend?

2. How would you determine the direction of a magnetic field, using a beam of moving charges? How would you determine the direction using a compass needle?

3. It is asserted that a certain region of space contains a magnetic field but no electric field. Describe an experiment, or series of experiments, that could verify this.

4. Describe the motion of a charged particle which enters a region of space which contains a

magnetic field, if its initial velocity **(a)** is parallel to the field, **(b)** is perpendicular to the magnetic field, or **(c)** makes some angle θ with the magnetic field.

5. If the tip of the sweep-second hand of a clock were charged with an excess of electrons, what would be the direction of the magnetic field produced at the center of the clock?

6. The earth's magnetic field points approximately northward, and has a vertical component. If the electric current is flowing northward in a horizontal power line, what is the direction of the force due to the earth's magnetic field.

7. Cosmic rays are charged particles that enter the earth's atmosphere from space. The flux of cosmic rays is much higher for the region near the poles of the earth (such as northern Canada) than it is for the region near the equator. Explain, using the fact that the earth's magnetic field acts as if it arose from a large bar magnet passing through the center of the earth (with its axis slightly tilted from that of the earth).

8. A wire is set up so that it is parallel to the direction of the earth's magnetic field as indicated by a nearby compass needle. A current is sent through the wire from north to south. Which way will the compass needle be deflected if it is placed **(a)** above the wire, and **(b)** below the wire?

9. By considering the magnetic field produced by one wire and the force exerted by that field on the other, show that parallel wires carrying currents in the same direction will attract each other and that the same wires carrying currents in opposite directions will repel each other.

10. Show that a helical coil of wire will, when carrying current, tend to contract along the axis and tend to expand radially.

11. Describe a procedure by which you could make a magnetic compass without using iron or other ferromagnetic materials.

12. A student consistently uses his left hand instead of his right hand in applying the right-hand rules. Does he get the correct answers for **(a)** the magnetic field due to moving charges, **(b)** the force exerted by a magnetic field on a moving charge, and **(c)** the force exerted by a moving charge on another moving charge?

13. Account for the fact that if one end of a long iron rod is hammered while it is held parallel to the earth's magnetic field it becomes magnetized. Which end becomes the north pole?

14. Two apparently identical steel rods are found to attract each other no matter which ends face each other. Describe a procedure by which you could decide which is a magnet and which is unmagnetized.

15. The positions of the earth's magnetic poles change with time. Describe an experiment for locating the new positions of the earth's magnetic poles.

16. A small compass needle is suspended by means of a string. What will happen when it is placed **(a)** in a nonuniform magnetic field and then **(b)** in a uniform magnetic field?

17. The neutron, which is a neutral particle, has a magnetic moment. Construct a model that could account for this.

18. A common application of magnetism is in the design and construction of *relays* and *solenoids*. A relay is an electric switch or collection of switches that is operated by a lever, which is actuated by the attraction of a plate of iron to an electromagnet. A *solenoid* consists of a coil that attracts an iron core; the core operates electric circuits or mechanical systems, such as water valves. Such devices are commonly used in automobiles and electric appliances. Attempt to find some and describe their construction and operation. Discuss the advantages and disadvantages of such devices. See also the Exercises at the end of Chapter 28.

Problems

1. An electron at the origin of a coordinate system is moving parallel to the x axis with a speed of 5×10^3 cm/sec. Find the magnetic field at a point **(a)** on the x axis, 5 cm in front of the electron, **(b)** on the y axis, where $y = 5$ cm, and **(c)** $x = 5$ cm, $y = 5$ cm, $z = 0$.

2. Ten point charges, each of charge $+2 \times 10^{-8}$ coul, are uniformly distributed on the rim of a phonograph record, of radius 6 in. The record turns at $33\frac{1}{3}$ rev/min. Find the magnetic field at the center of the record.

3. A negative charge of magnitude 4×10^{-10} coul

is moving vertically with a velocity of 3×10^7 cm/sec. Find the magnetic field at a point due east of the charge, 1 mm away from the charge.

4. A charge of -2.5×10^{-8} coul is moving horizontally due south with a velocity of 3×10^4 m/sec in a horizontal magnetic field that is directed due east and has a magnitude of 3×10^{-3} weber/m². Find the force on the charge.

5. An electron is moving through the earth's magnetic field at a point where the earth's magnetic field has a magnitude of 6.5×10^{-5} weber/m². The speed of the electron is 3×10^9 cm/sec. (a) What is the maximum force that the magnetic field could exert on the electron? (b) What is the relative orientation of the velocity of the electron and the field, when the maximum force is exerted?

6. Two electrons are 10^{-6} cm apart and moving parallel to each other with identical velocities of 2×10^7 cm/sec; both velocities are perpendicular to the line joining the electrons. Find (a) the magnetic force between them and (b) the electric force between them.

7. An electron moves in a circular path of 3 cm radius in a uniform magnetic field whose magnitude is 8.5×10^{-3} weber/m². Find the speed of the electron.

8. Protons moving with a speed of 10^8 m/sec enter a uniform magnetic field and move in a circle of radius 10 cm. Find the magnitude of the magnetic field, assuming that relativistic effects can be neglected.

9. What is the magnetic field 10 cm away from a long, straight wire carrying 8 amp?

10. A long, straight wire has a resistance of 0.2 ohm and is connected to a 45-volt battery of negligible internal resistance. What is the magnitude of the magnetic field 4 mm away from the wire?

11. The magnetic field 5 mm away from a long, straight wire is 4×10^{-3} weber/m². How large a current exists in the wire?

12. A 10-cm segment of vertical wire carrying 5 amp is in a horizontal magnetic field of 1.3 webers/m². Find the force exerted on the wire.

13. A straight, horizontal wire, 25 cm long and carrying a current of 4 amp, is at an angle of 30° to a horizontal magnetic field of magnitude 0.6 weber/m². Find the magnitude and direction of the force exerted on the wire.

14. A horizontal wire 15 cm long carries a current of 6 amp, and is in a magnetic field. The force exerted by the magnetic field just supports the weight of the wire, whose mass is 1.5 gm. Find the magnitude and direction of the magnetic field.

15. Two long, straight wires are parallel and 4 cm apart. One carries a current of 2 amp and the other carries a current of 3.5 amp in the same direction. Find the magnitude and direction of the force per unit length on each wire.

16. A bar magnet 0.1 m long has a magnetic moment of 3.1×10^{-5} weber m. The magnet is placed in a uniform magnetic field with the axis of the magnet perpendicular to the direction of the magnetic field. The intensity of the field is 2.4×10^{-4} weber/m². Find the torque on the magnet.

17. A circular coil of radius 3 cm carrying 5 amp and having 50 turns is placed in a magnetic field of intensity 2.4×10^{-2} weber/m². What is the maximum torque exerted on the coil?

18. A small bar magnet is suspended from its center by means of a fine wire and placed in a uniform magnetic field with its axis initially at right angles to the direction of the magnetic field. The torque on the magnet as measured by the twist of the wire is 2.5×10^{-8} nt m. The magnitude of the magnetic field is 5.7×10^{-2} weber/m². Find the magnetic moment of the magnet.

19. A vertical magnetic field of magnitude 1.5 weber/m² is created by a large C-shaped electromagnet, as shown in Figure 25-12a. A proton moving with a speed 2.4×10^8 cm/sec enters the magnetic field in the middle of the gap, moving perpendicular to the field. Assume that the field is uniform and confined to the region between the poles, a square 50 cm on a side, as shown in Figure 25-12b. (a) Find the radius of curvature of the path and (b) find the place where the proton leaves the magnetic field.

20. The electrons in a television tube are accelerated through 17,000 volts and then enter a magnetic field of 2.1×10^{-2} weber/m². (a) Find the magnetic force on the electrons and (b) find the radius of curvature of the path of the electrons in the magnetic field.

21. Show that the magnetic field at the center of a circular coil of radius R, carrying a current I,

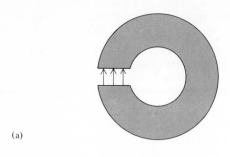

(a)

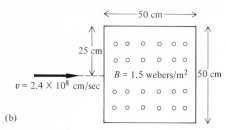

(b)

FIGURE 25-12

is given by

$$B = \frac{2k'I}{R}$$

22. Using the result of Problem 21, find the magnetic field at the center of a circular coil of 75 turns and radius 8 cm, which carries a current of 3 amp.

23. In the Bohr model of the atom, an electron moves at a speed of 2.2×10^6 m/sec in a circular orbit of radius 0.53×10^{-10} m. Find the magnetic field at the center of the orbit.

24. A long, straight wire carries a current of 1.5 amp. If the diameter of the wire is 2 mm, find the maximum magnetic field outside the wire.

25. Two long, straight wires carry equal currents of 2.4 amp. The wires are 8 cm apart. (a) Find the magnetic field halfway between the wires and (b) find the magnetic field at a point that is 5 cm away from each of the wires.

26. A straight wire carries a current of 24 amp. An electron moves parallel to the wire with a velocity of 4.5×10^7 cm/sec at a distance of 8 cm from the wire. Find the force exerted on the electron.

27. A long, straight wire carrying a current of 3 amp is parallel to the x axis, in the x-z plane

and 5 cm away from the x axis. Another long wire carrying 4 amp is parallel to the y axis, in the y-z plane and 5 cm away from the y axis. Find the magnetic field (magnitude and direction) at the origin of coordinates.

28. A rectangular coil of 200 turns of fine wire is suspended in a uniform magnetic field with its plane parallel to the direction of the field. The dimensions of the coil are 2 by 4 cm. When a current of 0.25 amp is sent through the coil, a torque of 1.2×10^{-3} nt m acts on it. Determine the magnitude of the magnetic field.

29. A group of charged particles enter a region of space in which both an electric field **E** and a magnetic field **B** exist. The initial velocities of the particles, the electric field, and the magnetic field are mutually perpendicular, just as the x, y, and z axes are mutually perpendicular. (a) Show that, with an appropriate orientation of the fields, it is possible for particles with a speed v to continue moving in a straight line and that all other particles will be deflected; the particular velocity v is given by $v = E/B$. (b) With the aid of a diagram, show the relative orientation of the two fields and the velocity of the particles that are selected by this *velocity selector*.

30. A rectangular coil of dimensions 2 by 20 cm lies on a horizontal table; the coil contains 50 turns and carries a current of 0.25 amp. A long, straight wire carrying 5 amp lies on the table and is parallel to one of the long sides of the coil, 3 cm away from the nearest point on the coil. Find the resultant force on the coil.

31. Consider a charged particle that enters a magnetic field such as the one shown in Figure 25-12. The particle may be injected into the field at any angle with the side of the square containing the field. Prove that it is impossible for the particle to remain in the region of space in which the field is contained, no matter what direction it enters and no matter with what speed it enters. (*Hint*: If the field is uniform, the orbit must be a circle, one point of which is the entry point.) This result is equivalent to the statement that one cannot trap particles in a static uniform magnetic field, a result of some significance in fusion research.

32. A charged particle of charge q and mass m is moving in a circular orbit of radius r with

speed v. Find the ratio of the magnetic moment to the angular momentum of the particle.

33. A compass needle is placed in a uniform magnetic field **B** and allowed to come to equilibrium. The needle is then given a small angular displacement and released. Show that it will execute angular harmonic motion with period T, given by

$$T = 2\pi \sqrt{\frac{I}{MB}}$$

in which I is the moment of inertia of the needle about the suspension axis in its center, M is the magnetic moment of the needle, and B is the magnitude of the field.

electromagnetism

chapter 26

26-1
Motional EMF

Consider the situation illustrated in Figure 26-1, in which a wire AC is moving through a magnetic field $\mathbf{B}$, with constant velocity $\mathbf{v}$, at right angles to the field. The velocity is also at right angles to the length L of the wire. This motion of the wire in a magnetic field will cause a separation of the charges in the wire, and consequently will establish an electric field in the wire. We can see this from the following argument: All of the protons and electrons in the wire are moving with the wire and therefore move with the velocity $\mathbf{v}$ of the wire. Each charge in the wire therefore has a force F exerted on it by the magnetic field, where

$$F = qvB \qquad (26\text{-}1)$$

Thus, in Figure 26-1, all the positive charges will have downward forces (toward the bottom of the page) exerted on them, and all the negative charges will have upward forces exerted on them. Since the protons are rigidly bound in the wire, only the electrons will move. Thus the end A of the wire will become negatively charged, while the other end of the wire will have been relatively depleted of negative charges, and will be left positively charged, as shown in Figure 26-2.

Ultimately the charges will come to rest because the separation of charge will produce an electric field in the wire. This electric field will exert forces on the charges in the wire, which will be exactly equal and opposite to the forces exerted by the magnetic field. That is, the separated charges

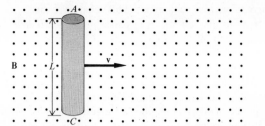

FIGURE 26-1 Wire *AC* moving with constant velocity **v** with its length *L* at right angles to lines of magnetic induction. Flux density **B** directed out of the paper.

will produce an electric field **E**, which will exert a force $q\mathbf{E}$ on any charge in the wire, where

$$qE = qvB \qquad (26\text{-}2)$$

or

$$E = Bv \qquad (26\text{-}3)$$

Therefore, when the charges have come to rest, there will be a potential difference V between the ends of the wire, where

$$V = EL \qquad (26\text{-}4)$$

or

$$V = BvL \qquad (26\text{-}5)$$

Now let us assume that the ends of the moving wire are connected to some stationary circuit. For example, we could slide the wire over a rectangular frame, as shown in Figure 26-3. In that case the separated charges would not stop at the ends of the wire, but would circulate through the circuit. We can think about this circuit in the same way we described circuits in Chapter 24, if we treat the moving wire as the source of EMF. The nonelectrostatic force, which was described in Chapter 24, is in this case the force exerted by the magnetic field, as given by Equation 26-1. The total work per unit charge done by this nonelectrostatic force is by definition the EMF, $\mathcal{E}$, and is given by Equation 26-5; therefore

$$\mathcal{E} = BvL \qquad (26\text{-}6)$$

This EMF produced by a moving conductor through a magnetic field is called *motional EMF*.

The careful reader will recognize that we have previously asserted that static magnetic fields do no work, yet we seem to be describing the EMF in Equation 26-6 as the work per unit charge done by the magnetic field. The resolution of this apparent conflict comes from recognizing that v in Equation

26-6 is only one *component* of the velocity of one of the circulating charges. When a current has been established, the circulating charges have a velocity parallel to the wire as well as the velocity v. Thus the force exerted by the magnetic field on the moving charges can be resolved into two components, one arising from the motion of the wire with velocity **v** perpendicular to the wire, and the other arising from the motion of the charges within the wire and parallel to the wire. The first component is described by Equation 26-1. The second component is perpendicular to the wire and directed opposite to v. In other words, this second component tends to stop the motion of the wire. In order to continue the motion of the wire, some external agent does work on the wire, and the magnetic field transfers this work to the circulating charges without doing any work itself.

26-2
Electromagnetic induction

We have been able to describe and understand motional EMF without introducing any new concepts. However, it is possible to think of motional EMF as one aspect of one of the most important discoveries in electricity and magnetism, *electromagnetic induction*. Electromagnetic induction was discovered independently by Michael Faraday (1791–1867) in England and Joseph Henry (1797–1870) in the United States, in 1831. Faraday's apparatus consisted essentially of two circuits placed close to each other, as represented schematically in Figure 26-4. One circuit, which we shall call the primary circuit, consisted of a battery B, a coil P of many turns of wire, and a key for opening and closing the circuit. The other circuit,

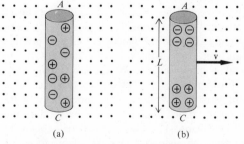

FIGURE 26-2 Motion of a wire through a magnetic field causes a separarion of the positive and negative charges, so that one end becomes positively charged and the other end negatively charged.

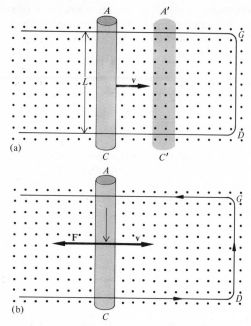

(a)

(b)

FIGURE 26-3 Determining direction of the current induced by the motion of the wire AC through a uniform magnetic field.

the secondary circuit, consisted of a coil S, and a galvanometer G, which is a device for the detection of the presence and direction of current in the secondary circuit. Faraday observed that when the key in the *primary* circuit was closed, the galvanometer in the *secondary* circuit gave a momentary deflection and then returned to its zero position and remained at the zero position as long as the key was closed. When the key was opened, there was another momentary deflection of the galvanometer, opposite in direction to the original deflection, and then the galvanometer needle returned to its zero position.

The current in the secondary circuit is not caused by a battery in the secondary circuit. The magnetic field created by a steady current in the primary circuit can exert forces on the charges in the secondary circuit only if charges are already moving. Furthermore, the current in the secondary circuit exists for only a short time, roughly the time during which the current in the primary circuit is changing. Thus, we are led to explain the current in the secondary circuit as produced by the *change* in the primary current.

While the current in the primary circuit is changing, the magnetic field created by the primary

current is also changing. This change causes a current in the secondary circuit; that is, an *electric field* is established in the secondary circuit that exerts a force on the electrons in it. A somewhat more fruitful way to describe this is as follows: The moving charges in the primary circuit create a magnetic field in the space around it. When the magnetic field is changing, an electric field is established in the secondary circuit by this changing magnetic field. We can think of the changing magnetic field as creating an electric field in a manner analogous to the creation of a magnetic field by a current—that is, by the motion of charges.

26-3
Faraday's law

Although changing magnetic fields produce electric fields even in empty space, it is instructive first to consider electromagnetic induction in circuits and then to generalize this phenomenon.

Consider then a loop of wire of area A that is in a magnetic field. If the field is uniform (constant in space but not necessarily in time) then the flux Φ through the loop is given by the product of the component of the magnetic flux density B_n, which is perpendicular to the area of the loop, and the area of the loop:

$$\Phi = B_n A \qquad (26\text{-}7)$$

If the field varies from point to point over the area within the loop, we can divide the area into small segments, within each of which the field is essentially constant, find the flux Φ_i through each of these segments, and then add all these contributions to find the total flux Φ through the loop:

$$\Phi = \Sigma\Phi_i$$
$$= \Sigma(B_n A)_i \qquad (26\text{-}8)$$

If, then, the flux through the loop changes from Φ_1 to Φ_2 during a time interval Δt, an EMF $\mathscr{E}$ will

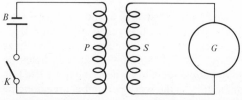

FIGURE 26-4 Faraday's experiment on electromagnetic induction.

be produced in the loop, where

$$\mathcal{E} = \frac{\Phi_2 - \Phi_1}{\Delta t} \qquad (26\text{-}9)$$

This is called *Faraday's* law. The most important thing to note about Equation 26-9 is that the EMF is generated in the loop if the flux changes for any reason; in particular, the EMF may be generated by a change in the magnetic field, a change in the area, or both. In Equation 26-9, a change of magnetic flux at the rate of 1 weber/sec will induce an EMF of 1 volt in the coil.

If a coil of wire has n turns, an EMF will be generated in each turn by the changing magnetic flux. The EMF generated in the coil will be the sum of the EMF's in the separate turns. If the change in magnetic flux is the same in each turn, then the induced EMF will be given by

$$\mathcal{E} = n \frac{\Phi_2 - \Phi_1}{\Delta t} \qquad (26\text{-}10)$$

The simplest application of Faraday's law is to the case of a circular loop of wire, of radius r, in a uniform magnetic field of flux density $\mathbf{B}$, which is perpendicular to the loop and whose magnitude changes from B_1 to B_2 in time t. Then Faraday's law gives the induced EMF as

$$\begin{aligned} \mathcal{E} &= \frac{\Phi_2 - \Phi_1}{t} \\ &= \frac{B_2 A - B_1 A}{t} \\ &= \frac{(B_2 - B_1)A}{t} \\ &= \frac{B_2 - B_1}{t} \pi r^2 \qquad (26\text{-}11) \end{aligned}$$

The induced *electric field* that gives rise to this EMF is described by lines of force that form circles tangent to the circular loop of wire, and has the same magnitude at each point of the loop. Thus, the EMF $\mathcal{E}$ is related to this induced electric field E by

$$\mathcal{E} = 2\pi r E \qquad (26\text{-}12)$$

or, from Equation 26-11,

$$2\pi r E = \left(\frac{B_2 - B_1}{t}\right) \pi r^2 \qquad (26\text{-}13)$$

and so

$$E = \left(\frac{B_2 - B_1}{t}\right) \frac{r}{2} \qquad (26\text{-}14)$$

This allows us to visualize the situation in terms of a group of magnetic field lines changing in magnitude and surrounded by circular electric-field lines.

Illustrative example

A circular coil of wire having a radius of 0.50 m is in a magnetic field directed at right angles to the plane of the coil. The magnetic flux density is changing at the rate of 0.06 weber/m² (or 0.06 tesla) per second. Determine (a) the EMF induced in the coil and (b) the electric field intensity in the coil.

(a) Using Equation 26-11,

$$\mathcal{E} = \frac{B_1 - B_2}{t} \pi r^2$$

we can write

$$\frac{B_2 - B_1}{t} = 0.06 \frac{\text{tesla}}{\text{sec}}$$

from which

$$\begin{aligned} \mathcal{E} &= 0.06 \times \pi \times (0.5)^2 \text{ volt} \\ &= 0.0150\pi \text{ volt} \\ &= 0.047 \text{ volt} \end{aligned}$$

(b) We can use Equation 26-14 to determine the electric field intensity in the coil:

$$\begin{aligned} E &= \frac{B_2 - B_1}{t} \times \frac{r}{2} \\ &= 0.06 \times \frac{0.5}{2} \frac{\text{volt}}{\text{m}} \\ &= 0.015 \frac{\text{volt}}{\text{m}} \end{aligned}$$

This electric field is tangent to the coil.

26-4
Lenz's law

The preceding discussion of electromagnetic induction has not described the direction of the induced electric field or its equivalent, the sense of the induced EMF. It is possible to establish a set of mathematical conventions that relate direction in the circular loop of wire to direction along the perpendicular magnetic field. When this is done, it is convenient to write Faraday's law in the form

$$\mathcal{E} = -\frac{(\Phi_2 - \Phi_1)}{\Delta t} \qquad (26\text{-}15)$$

where the minus sign in Equation 26-15 describes the sense of the induced EMF. For almost all purposes it is simpler to state a rule called Lenz's law: *The induced current is in such a direction as to oppose, by its magnetic action, whatever change produces the induced current.*

For example, if the magnetic field through a circular loop is increasing, the current in the loop will have such a direction that it will produce a magnetic field opposite to the original field. If the magnetic field through the loop is decreasing, the magnetic field due to the induced current will be in the same direction as the original field. Thus, if a loop is in the plane of the page and the field is into the page and increasing, the induced current will be counterclockwise.

26-5
Electromagnetic induction with a constant magnetic field

As was said earlier, motional EMF can be thought of as one aspect of electromagnetic induction. In particular we can use Faraday's law to analyze the circuit of Figure 26-3. In that circuit, the flux changes because the area of the circuit changes. At any instant the flux is the product of the field and the area of the circuit. At a time, t later, the flux will have changed because the area will have changed by an amount $(Lv\ \Delta t)$, or

$$\Phi_2 - \Phi_1 = BLv\ \Delta t \qquad (26\text{-}16)$$

and from Faraday's law

$$\mathcal{E} = \frac{\Phi_2 - \Phi_1}{\Delta t}$$

$$= BLv \qquad (26\text{-}17)$$

which is identical to Equation 26-6.

One of the most common applications of this idea is illustrated in Figure 26-5, in which a rectangular loop of wire is rotated in a constant magnetic field. As the loop is rotated, the flux through the loop changes because the magnitude of the perpendicular component of the magnetic field changes, and an EMF is produced in the loop. The simple loop shown produces an alternating EMF (as we shall see), which can be used directly or with appropriate mechanical or electrical devices to deliver direct current to an external circuit. An important thing about this system is that the output current from the rotating loop can be used to

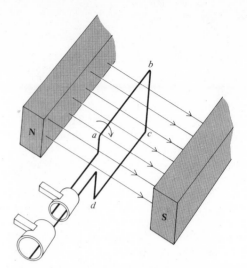

FIGURE 26-5 Essentials of an electric generator.

increase the magnetic field in which the loop turns. This in turn increases the output current. Thus it is possible to build huge generators in which both the internal magnetic field and the output current are produced in the generator. Electricity can then be produced without reliance on either permanent magnets or chemical cells. It remains true that some external agent, such as a turbine, must supply the energy, but it is this mechanism that permits the efficient transformation on a large scale of the energy in waterfalls or in steam plants into electrical energy.

The EMF produced in the rotating loop of Figure 26-5 can be calculated with the aid of Faraday's law; however, it is mathematically simpler to treat this as a case of motional EMF. Figure 26-6 is another view of the wires of Figure 26-5. Assume that the coil is rotating with a constant angular speed ω and let θ be the angle through which the coil has rotated from the vertical position in a time t. Therefore

$$\theta = \omega t$$

Since the wire ab moves in a circular path of radius r, where $2r$ is the width ad or cb of the coil, its linear velocity v is

$$v = \omega r$$

The angle between $\mathbf{v}$ and the field $\mathbf{B}$ is also θ; hence the component of $\mathbf{v}$ perpendicular to the field is $v \sin \theta$. Therefore the instantaneous EMF induced

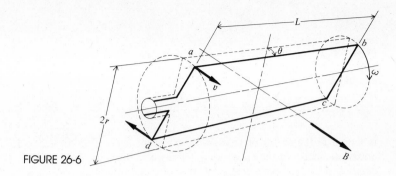

FIGURE 26-6

in the wire *ab* is

$$\mathcal{E}_1 = BLv \sin \theta$$

where L is the length of the wire. An equal EMF is induced in the wire *dc*. The magnetic field exerts forces parallel to the axis of rotation, so the forces in the sides *bc* and *ad* do not contribute to the EMF in the loop. Thus the total EMF in the loop is

$$\mathcal{E} = 2BLv \sin \theta$$

or

$$\mathcal{E} = 2BLr\omega \sin \theta \qquad (26\text{-}18)$$

and since $2r$ is the width of the loop and L its length, its area A is

$$A = 2rL$$

Hence

$$\mathcal{E} = BA\omega \sin \theta \qquad (26\text{-}19)$$

If there are n turns in this coil, the instantaneous EMF will be

$$\mathcal{E} = nBA\omega \sin \theta \qquad (26\text{-}20)$$

or

$$\mathcal{E} = \mathcal{E}_{max} \sin (\omega t) \qquad (26\text{-}21)$$

where

$$\mathcal{E}_{max} = nBA\omega \qquad (26\text{-}22)$$

is the maximum induced EMF.

A more common method of expressing the alternating EMF is in terms of the frequency f of the alternations. The angular velocity ω and the frequency f are related by

$$\omega = 2\pi f \qquad (26\text{-}23)$$

and since

$$\theta = \omega t$$

we can write the alternating EMF as

$$\mathcal{E} = \mathcal{E}_{max} \sin 2\pi f t \qquad (26\text{-}24)$$

Figure 26-7 shows this alternating EMF graphically.

26-6
Self-induction

Consider any circuit in which current is flowing. The current in that circuit will produce a magnetic field and, in general, the field will have a nonzero flux through the circuit. Thus, if the current through the circuit changes, the flux will change, thus generating an EMF in the circuit. Since this EMF will have been produced by the change in current, it will have such a direction as to oppose the change in the current.

This *self-induced EMF* is proportional to the rate of change of magnetic flux and, for simple circuits that do not contain magnetic materials, this in turn is proportional to the rate of change of current; that is,

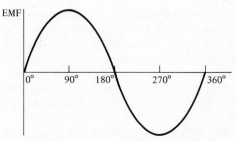

FIGURE 26-7 Graph showing the alternating EMF induced in the coil of the generator during one revolution.

$$\mathcal{E} \propto \frac{\Delta I}{\Delta t}$$

in which ΔI is the change in current in a small time Δt. This can be written in the form of an equation,

$$\mathcal{E} = L\frac{\Delta I}{\Delta t} \qquad (26\text{-}25)$$

in which L, the factor of proportionality, depends on the geometry of the circuit and is known as the *coefficient of self-inductance* of the circuit. In the mks system of units, L is expressed in henrys, and a self-inductance of 1 henry whose current changes at a rate of 1 amp/sec will give rise to an EMF of 1 volt.

Imagine now closing the switch in some DC circuit in which there is no current initially. Immediately after the switch is closed the current changes from zero to some value. This produces in the circuit a large opposing EMF, which in turn causes the current to increase less rapidly in the next instant. We conclude that the current rises gradually rather than instantly.

In a circuit that consists of a battery of EMF $\mathcal{E}$ connected to a resistor of resistance R and a self-inductance of inductance L, the current increases with time according to the equation

$$I = \frac{\mathcal{E}}{R}\left[1 - \exp\left(\frac{-t}{L/R}\right)\right] \qquad (26\text{-}26)$$

Figure 26-8 is a sketch of the graph of this equation. Although, strictly speaking, the current takes an infinite time to reach its final value ($\mathcal{E}/R$), the current reaches a value very close to this final value in a few *rise times*; a rise time is a time characteristic of the circuit and is defined by the equation,

$$T = \frac{L}{R} \qquad (26\text{-}27)$$

In most common circuits, the self-inductance is a few millihenrys at most and the resistance is at least a few ohms and so the rise time is a few thousandths of a second; the current is very close to its final value in very small fractions of a second. Sometimes it is useful to construct circuits with long rise times. Such a circuit can be used to turn on one part of an electrical system well after another part had been turned on. For such uses, circuits with large values of L are used.

Similar analyses lead to the conclusion that the current in a circuit does not drop to zero at the

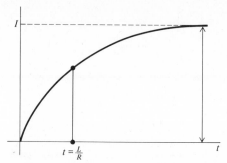

FIGURE 26-8 Growth of the current in a circuit containing inductance.

instant that the switch is opened; the self-inductance of the circuit always causes the current to decay to zero in a nonzero time interval. Very often this effect causes sparks to jump across the switch as it is opened.

If a circuit consists of a single loop of wire, its self-inductance is small. If the same length of wire is wound in the form of a coil of many turns, its self-inductance is larger. This is due to the fact that the magnetic flux through a coil of many turns is greater than that through a single large turn; furthermore the EMF induced in the coil is the sum of the EMF's induced in each of the turns. The inductance of the coil can be increased still further by putting an iron core inside the coil, since the effect of the iron is to increase the magnetic flux through the coil.

As the current i in a coil increases from zero to its steady-state value I, the magnetic field around the coil increases to its maximum or steady-state value. In this process the source of EMF that supplies the current does work. The coil has an EMF, given by Equation 26-25, that can be thought of as absorbing this work. This energy absorbed by the coil is stored in the magnetic field associated with the coil. The stored energy can be recovered when the current is decreased; usually the energy that appears in the form of sparks of switches, or in automobile spark plugs, is obtained from energy that was stored in magnetic fields.

If at some instant of time the current in a coil is i and the self-induced EMF is $\mathcal{E}$, the energy stored in the magnetic field ΔW during the next interval of time Δt is

$$\Delta W = \mathcal{E}i\,\Delta t \qquad (26\text{-}28)$$

and since

$$\mathcal{E} = L\,\frac{\Delta i}{\Delta t}$$

$$\Delta W = Li\,\Delta i \qquad (26\text{-}29)$$

If the current changes from zero to I, the average value of the product Li is $\frac{1}{2}LI$ since Li is proportional to i. Hence the energy W supplied to the inductor as the current increases from 0 to I is

$$W = \tfrac{1}{2}LI^2 \qquad (26\text{-}30)$$

26-7
Displacement current

The phenomena of electricity and magnetism that have been discussed thus far can be summarized in a series of statements or equations given by Coulomb's law for the force between electric charges, the law of Biot and Savart for the relationship between magnetic fields and currents or moving charges, and Faraday's law of electromagnetic induction for the relationship between changing magnetic fields and induced electric fields. These laws are all based on experimental evidence and were formulated in the first half of the nineteenth century.

In the middle nineteenth century, James Clerk Maxwell made the startling discovery that the basic equations of electricity and magnetism were inconsistent with each other. The primary emphasis in this analysis was on the fields, electric and magnetic. The fundamental equations were rewritten in a form that described the fields and gave the relationships among the fields and the sources of the fields (charges and currents). In this form, Maxwell showed that these equations could not all be true simultaneously and still be consistent with the idea that charge is neither created nor destroyed. It must be emphasized that this was a mathematical statement and not the result of an experiment. The way out of the dilemma was to assume that the equations were *incomplete* and to search for a mathematical way to make the equations consistent with each other while retaining their agreement with experimental evidence. The result was the assertion that *changing electric fields produce magnetic fields*. This mathematical discovery was soon verified by experiment.

One important consequence of this work is that it introduces a remarkable symmetry between electric and magnetic fields, and a very strong connection between them; that is, the sources of electric fields are charges and changing magnetic fields, and the sources of magnetic fields are moving charges and changing electric fields. In fact, the equations describing the magnetic fields created by changing electric fields are, except for units, exactly the same in form as those describing the electric fields established by changing magnetic fields.

In the equations of electromagnetism developed by Maxwell, the sources of the magnetic field are represented as the sum of two quantities. One of them is proportional to the current density due to moving charges and the other is proportional to the rate of change of electric field intensity. This latter, the rate of change of electric field, is called the *displacement current*. Using this term we can describe the source of the magnetic field as the *total current*, where the total current is the sum of the current due to moving charges and the displacement current.

The concept of displacement current permits us to assert that the *total current is continuous* or that the total current forms closed loops. For example, consider the circuit of Figure 26-9, which consists of a parallel-plate capacitor C, a resistor R, a battery B, and a switch S. When the switch S is closed, a charging current I, which consists of moving charges in the wires, exists in the metallic part of the circuit. Because the charge on the capacitor plates is changing, the electric field in the space between the plates is changing, and therefore a displacement current exists in this space. The constants in Maxwell's equation are chosen so that this displacement current is precisely equal to the charging current in the wires. Thus the total current in the circuit (moving charges in the wires and displacement current in the capacitor gap) forms a closed loop. This continuous total current

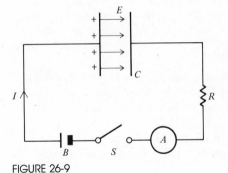

FIGURE 26-9

establishes a magnetic field in the space surrounding the circuit. When the capacitor is fully charged, the charging current is reduced to zero and, at the same time, the electric field in the capacitor stops changing and so the displacement current is reduced to zero.

26-8
Electromagnetic waves

Perhaps the most striking consequence of Maxwell's work was his prediction of the existence of electromagnetic waves and his identification of those waves with light. The mathematical development of these ideas is well beyond the scope of this book, but some comments can be made.

Let us assume that we have some distribution of moving charges, which establish both electric and magnetic fields. If the charges then accelerate, the magnetic field will change, thereby producing an electric field. This implies that the electric field changes. In turn, this implies a change in the mag-

netic field. It is clear that this cycle of changes implying changes can go on and on without reference any longer to the original charge distribution. It implies that the fields could create each other if they change at the proper rate and in the proper places. Maxwell was able to show that in fact there are circumstances in which the electric and magnetic fields can create each other. In empty space, the symmetry between the two fields discussed above requires that the two fields be mutually perpendicular. Further, in order for this mutual creation process to continue, the fields must change and move at a particular rate. It turns out that this process requires that the fields move at a rate determined by the constants in Coulomb's law and the Biot-Savart law. Solution of the equations shows that the fields must move at a velocity that is determined by the constants and turns out to be numerically identical to the velocity of light. It is a small leap to assert that light consists, in fact, of moving electric and magnetic fields. A more detailed description of some simple electromagnetic waves will be given in later chapters.

Questions

1. Upon what factors does the current in the circuit of Figure 26-3 depend?

2. Describe the direction of the current in the secondary circuit of Figure 26-4 when the switch is closed.

3. A bar magnet is inserted into a coil. With the aid of a diagram describe the direction of the induced current.

4. Describe the direction of the induced current due to removing a bar magnet from a coil. Assume that the north pole of the magnet is initially in the middle of the coil.

5. Suppose the wire of the secondary in Figure 26-4 were cut, but the cut ends remained very close to each other. What would you expect to happen if the switch in the primary were closed?

6. Devise a mechanical procedure for moving the rod of Figure 26-3 at constant speed. What is the source of the energy supplied to the rod in your procedure?

7. An airplane is flying due north in level flight, through the earth's magnetic field. Which of

the wing tips is charged with positive charge?

8. Two identical circular loops of insulated wire are placed one on top of the other on a tabletop. A switch is closed, causing current to flow clockwise in the top loop. In what direction does current flow in the bottom loop?

9. Electricians sometimes measure alternating current in a wire by using a device that consists of a coil wrapped around the wire; the coil is connected to a meter, which measures the current in the *coil*. Explain how such a device can measure the current in the *wire*.

10. A generator, such as that shown in Figure 26-5, is sometimes used as a *tachometer*, a device for measuring the angular speed of an engine. Explain how such a use can be made of a generator.

11. An electric generator is running at constant speed but supplying no current. A load is then put across the line so that the generator supplies a current I. What happens to the speed of the generator at the instant it supplies power? Where does this power come from?

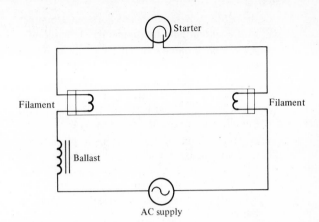

FIGURE 26-10 Fluorescent lamp circuit.

12. If the magnetic field of Figure 26-1 were not uniform, current would circulate in the rod. Explain. (Such currents are called *eddy currents*.)

13. Why is there frequently a spark across the terminals of a switch when it is opened?

14. The self-inductance of a circuit is frequently compared to inertia, and is spoken of as "electrical inertia." Use this analogy in discussing the rise of current in a DC circuit.

15. A long piece of wire in the form of a single loop is connected to a 6-volt battery. An identical piece of wire, wound in the form of a small coil of several hundred turns, is connected in parallel to the same battery. Compare the time required for the currents to reach, or approach, their steady-state values in the two cases.

16. Since mammalian blood contains charged particles, or ions, placing a large magnet near a blood vessel containing moving blood will cause the establishment of a motional EMF. Devise a procedure by which the speed of the blood can be determined by measuring the potentials of appropriate points on the walls of blood vessels.

Problems

1. A metal rod, 6 cm long, moves through a magnetic field of 0.03 weber/m^2 with a speed of 15 cm/sec, as shown in Figure 26-1. Find the EMF generated in the rod.

2. An automobile is traveling due north on a level

17. The basic circuit of a common fluorescent lamp is shown in Figure 26-10. The starter consists of a curved bimetallic strip, which makes contact with a metal pin when the strip is at room temperature, acting as a closed switch. As current flows through the bimetallic strip, it warms up and bends away from the pin, opening the switch. Because the ballast is a multiturn coil, this causes a large spark to jump through the tube, from one filament to the other. **(a)** Explain the origin of the spark. **(b)** Justify the statement that the potential difference across the tube, just before the spark, can be significantly larger than the line voltage.

18. When a conductor moves through a nonuniform magnetic field, eddy currents are established (see Question 12 above). The magnetic field exerts forces on these currents. **(a)** Show that Lenz's law implies that these forces tend to slow the moving conductor. **(b)** What happens to the kinetic energy of the conductor? (This effect can be used to devise *magnetic brakes* for moving machinery, even when the moving parts are made of nonferromagnetic materials.)

road at 50 mi/hr. The earth's magnetic field makes an angle of 45° with the horizontal (the horizontal component is due north and the vertical component is directed down). The magnitude of the field is 6×10^{-5} weber/m^2. Find the EMF generated in the horizontal bumper of the automobile; the bumper is 1.8 m long.

3. A horizontal metal rod, 3 m long, falls from rest through a height of 1 m in a region of space where the magnetic field is perpendicular to the rod and is horizontal, with a magnitude of 0.4 weber/m². What is the EMF in the rod (a) initially and (b) at the end of the fall?

4. A 50-turn coil of radius 2 cm is placed in a magnetic field of intensity 0.3 weber/m²; the field is perpendicular to the plane of the coil. The field is reduced to zero, uniformly, in a time of 0.1 sec. Find the EMF generated in the coil.

5. A 300-turn circular coil of radius 1 cm is in a magnetic field that is perpendicular to the plane of the coil. The coil is then pulled rapidly out of the field in 0.08 sec. A voltage of 0.03 volt is generated in the coil. Assuming that the change in flux is uniform, find the magnitude of the magnetic field.

6. A generator such as that shown in Figure 26-5 is rotating at 1800 rev/min in a magnetic field of 0.5 weber/m². The coil is 8 by 10 cm and contains 40 turns. What is the maximum EMF generated in the coil?

7. A bicycle generator generates 2 volts when the bicycle is traveling at 4 mi/hr. What EMF will be generated when the bicycle is traveling at 6.5 mi/hr?

8. A certain coil is spun by a motor and, when the coil is placed in a magnetic field of 0.3 weber/m², a peak voltage of 24×10^{-3} volt is generated in the coil. When the same coil is placed in a different field and spun by the same motor, a peak voltage of 53×10^{-3} volt is generated. What is the magnitude of this second magnetic field?

9. When the current in a certain coil changes from 2 amp to 1 amp in 0.2 sec, an EMF of $\frac{1}{2}$ volt is generated in the coil. What is the self-inductance of the coil?

10. The current in a 10-millihenry coil is changed at the rate of 100 amp/sec. How large an EMF will be generated in the coil?

11. How large a resistance must be connected in series with a 20-millihenry coil in order that the rise time be equal to 0.1 sec?

12. A certain coil of wire has a self-inductance of 100 millihenrys and a resistance of 80 ohms. (a) What is the rise time of the coil? The coil is connected to a 12-volt battery. Find the current in the coil (b) at the end of one rise time and (c) at the end of two rise times.

13. An 80-millihenry coil is connected to a 6-volt battery. The total resistance of the circuit is 25 ohms. Ultimately how much energy is stored in the magnetic field surrounding the coil?

14. A 150-millihenry coil is carrying a steady-state current of 3 amp when the switch in the circuit is suddenly opened. Estimate the amount of energy dissipated in the resulting spark.

15. In Figure 26-3, the distance AC is 10 cm, the speed v is 0.15 cm/sec, and the magnetic field B is 0.8 weber/m². The resistance of the circuit is 10^{-2} ohm. (a) Find the current in the circuit. (b) Find the force exerted by the magnetic field on the rod AC. (c) Find the power delivered by this force. Compare this with the energy dissipated in the circuit.

16. A square coil 8 cm on a side, having 750 turns, is rotated from a position perpendicular to the magnetic field to a position parallel to the field in 0.03 sec. If the value of B is 0.8 weber/m², what is the average value of the induced EMF, in volts?

17. A straight wire, 7 cm long, is rotated around one of its ends through an angle of 90° in 0.02 sec. If during this motion it always lies perpendicular to a uniform magnetic field of magnitude 0.4 weber/m², what is the EMF induced between its ends and (b) what is the average nonelectrostatic field induced in the rod?

18. A square metal loop, 10 cm on a side, rotates at a rate of 5 rad/sec about the y axis. A magnetic field of magnitude 0.5 weber/m² is directed along the x axis. The loop has a total resistance of 2 ohms. Find (a) the maximum EMF generated in the loop and (b) the maximum current in the loop. Plot a graph of (c) the flux through the loop as a function of time, (d) the EMF in the loop as a function of time, (e) the current in the loop as a function of time, and (f) the torque needed to keep the loop rotating at constant speed as a function of time.

19. A student arranges a piece of wire, 4 m long, in the form of a circle on the floor and connects the two free ends to a sensitive meter. The vertical component of the earth's magnetic field at that place is 6.5×10^{-5} weber/m². The student holds the wire at a point opposite to

the meter and quickly pulls the wire radially out, collapsing the circle into a flat, narrow loop of approximately zero area. The collapse of the circle takes place in 0.07 sec. What EMF is induced in the loop?

20. A long, straight wire carrying 5 amp is lying on a tabletop. A small circular coil, of 200 turns and cross-sectional area 1 cm², lies on the tabletop 10 cm away from the wire. The current in the wire is decreased to zero in 0.1 sec. (a) What is the average EMF induced in the coil? (b) With the aid of a diagram indicate the sense of the induced current.

21. From Equation 26-25, the units of self-inductance are volt-seconds per ampere. Show that these units are identical to webers per ampere.

22. Show that the units of L/R are those of time, as required by Equation 26-27.

23. The current in a certain coil is made to increase from zero at a constant rate for a time t_1 and then to decrease at the same rate until the current is equal to zero. Sketch a graph of the induced voltage in the coil as a function of time.

24. A coil with an inductance of 0.02 henry and a resistance of 4 ohms is connected to a 6-volt battery of negligible internal resistance. Find (a) the initial rate of increase of current, (b) the rate of increase of the current when the current is 1 amp, and (c) the steady-state current. (d) How much energy is stored in the magnetic field of the coil when the current has reached its steady-state value?

25. A 120-volt potential difference is applied to a coil having an inductance of 75 millihenrys and a resistance of 40 ohms. (a) What is the current in the coil after 0.01 sec? (b) At what rate is the current increasing at that time?

26. Consider a coil whose total resistance is R. Prove that if the flux through the coil changes from Φ_2 to Φ_1 in time t, the total charge Q that circulates through the coil during the time t is given by

$$Q = \frac{\Phi_2 - \Phi_1}{R}$$

27. Show that the rise time of an LR circuit is the time that the current would take to reach the steady-state value if the current continued to increase at the initial rate.

28. (a) Consider a coil of inductance L that is connected to a resistor of resistance R and a battery of EMF $\mathcal{E}$ and negligible internal resistance. Show that the rate of change of current $\Delta I/\Delta t$ is given by

$$\frac{\Delta I}{\Delta t} = \frac{\mathcal{E}}{L} - \frac{R}{L} I$$

where I is the current.

(b) Consider, now, a capacitor of capacitance C connected in series with the resistor and battery of part (a). Show that the rate of change of charge $\Delta Q/\Delta t$ on the capacitor is given by

$$\frac{\Delta Q}{\Delta t} = \frac{\mathcal{E}}{R} - \frac{Q}{RC}$$

(c) From the similarity of these two equations, show that the charge Q on the capacitor is given by

$$Q = C\mathcal{E}(I - e^{-t/RC})$$

where RC is the rise time for this circuit.

29. A long, straight wire carrying a current I lies on a tabletop. A rectangular loop of wire, a cm by b cm, lies on the table with the side of length a parallel to the wire. The resistance of the loop is R. The loop is moved with a velocity v in a direction perpendicular to the wire. Find the current induced in the loop if the nearest side is a distance c from the wire.

30. Consider the circuit of Figure 26-3 arranged in a vertical plane so that the rod falls, under the action of gravity, in the presence of a horizontal magnetic field. Take the rod to have mass m, length l, and take the resistance of the circuit to be R. Show that if friction may be neglected the rod will ultimately move with a constant terminal speed v, given by

$$v = \frac{mgR}{l^2B^2}$$

in which B is the magnitude of the magnetic field.

alternating currents

chapter 27

27-1
Measurement of alternating current

Alternating currents are used so extensively that it is worthwhile considering a few of the simpler phenomena associated with them. For example, when the terminals of a resistor are connected to a source of alternating EMF, the current in the resistor will vary continuously. The instantaneous value of the current can be determined with the aid of a device known as an oscilloscope, but for many practical purposes it is sufficient to know the *effective value* of the alternating current. This value can be determined by having the alternating current produce the same effect in a circuit that a known direct current would produce. The heating effect produced by the flow of current through a resistor is used for defining the effective value of an alternating current. *An alternating current has an effective value of one ampere when it produces the same rate of heating in a resistor as that produced by a direct current of one ampere.*

One form of an AC ammeter uses the change in length of a wire produced by the heating effect of the current that passes through it. As this wire is heated, its temperature rises and its length changes. A pointer attached to this wire moves over a scale that can be calibrated by sending a known direct current through the instrument. Since the heating effect depends upon the square of the current, the scale will not be uniform (see Figure 27-1).

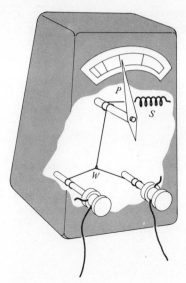

FIGURE 27-1 Essentials of a hot wire ammeter; W is the wire, S is a spring, and P is a pointer.

27-2
Instantaneous and effective values

The simplest type of alternating current is one that varies sinusoidally with a constant frequency f. Let us consider the current in a resistor that is connected to an AC generator. The instantaneous value of the voltage v across the resistor at any instant t is given by

$$v = V_m \sin 2\pi ft \qquad (27\text{-}1)$$

where V_m is its maximum value. The graph of this equation is shown in Figure 27-2. If R is its resistance, the instantaneous value i of the current in the resistor will be given by Ohm's law:

$$i = \frac{v}{R} \qquad (27\text{-}2)$$

Combining this with Equation 27-1 yields

$$i = \frac{V_m}{R} \sin 2\pi ft \qquad (27\text{-}3)$$

or

$$i = I_m \sin 2\pi ft \qquad (27\text{-}4)$$

where

$$I_m = \frac{V_m}{R} \qquad (27\text{-}5)$$

The instantaneous value of the current is plotted in Figure 27-3. If the source of supply is a 60-Hz source, one complete cycle takes $\frac{1}{60}$ sec. The current in a resistor is *in phase* with the alternating EMF; that is, the current and the voltage reach their maximum values at the same time, and also reverse direction at the same time.

The instantaneous value of the power p supplied to the resistor is

$$p = vi \qquad (27\text{-}6)$$

which, with the aid of Equation 27-2, becomes

$$p = i^2R \qquad (27\text{-}7)$$

that is, the rate at which heat is developed in the resistor depends upon the square of the current. Figure 27-4 is a graph of the square of the current in the resistor versus time. It will be noted that the square of the current is always positive, and the curve representing it has twice the frequency of the alternating current. Since the effective value of an alternating current is defined in terms of its heating effect, this effective value can be determined by averaging the squares of the current and then taking the square root of this average value. This is sometimes called the *root mean square* (or *rms*) value of the current, as well as the effective value. It can be shown that the average value of the square of the current is $I_m^2/2$; hence the effective value I of the current is

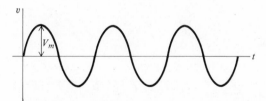

FIGURE 27-2 Instantaneous value of an alternating voltage.

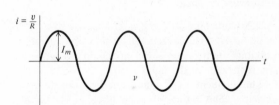

FIGURE 27-3 Instantaneous values of the alternating current in a resistor.

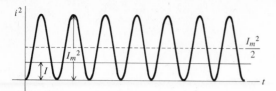

FIGURE 27-4 Instantaneous squared values of an alternating current, the average of the squared values of this current, $I_m^2/2$, and its effective value, I.

$$I = \frac{I_m}{\sqrt{2}} = 0.707 I_m \qquad (27\text{-}8)$$

where I_m is the maximum value of the alternating current. Similarly the effective value V of a sinusoidally varying EMF is

$$V = \frac{V_m}{\sqrt{2}} = 0.707 V_m \qquad (27\text{-}9)$$

where V_m is the maximum value of EMF applied to the terminals of the resistor.

From the definition of the effective value of the current, the average power supplied to the resistor can be written as

$$P = I^2 R \qquad (27\text{-}10)$$

Ohm's law holds not only for the instantaneous values of the current and voltage but also for their effective values; that is,

$$I = \frac{V}{R} \qquad (27\text{-}11)$$

so that the average power supplied to a resistor can be written as

$$P = IV \qquad (27\text{-}12)$$

In the above equations, capital letters such as I, V, and P are used to represent effective values of quantities that vary with time, while lowercase letters i, v, and p represent their instantaneous values.

Illustrative example

A 120-volt, 60-watt lamp is connected to a 120-volt, 60-Hz AC source. Determine the effective and maximum values of the current in the lamp and the resistance of the lamp.

The filament of a lamp can be considered a resistor. The rating of the source as 120 volts

gives its effective value. From Equation 27-12, the effective value of the current in the lamp is

$$I = \frac{P}{V} = \frac{60}{120} = 0.5 \text{ amp}$$

and, from Equation 27-10, the resistance R of the filament is

$$R = \frac{P}{I^2} = \frac{60}{0.25} = 240 \text{ ohms}$$

The same value could have been obtained from Equation 27-11.

The maximum value of the current is, from Equation 27-8,

$$I_m = \sqrt{2}I = 1.414 \times 0.5 = 0.707 \text{ amp}$$

27-3
Inductance in an AC circuit

If the terminals of a coil of wire are connected to a source of alternating EMF, an alternating current will flow in this coil. This alternating current will be accompanied by an alternating magnetic field around the coil. As shown previously, this alternating magnetic field will induce an EMF in this coil known as the self-induced EMF. Because of the self-induction of the coil, the current in it *lags* behind the EMF impressed on the coil; that is, the current reaches its maximum value after the impressed EMF has reached its maximum value. If the resistance of the coil is negligible, the current i in the coil is found to lag behind the impressed EMF v by a quarter of a period; this is illustrated in Figure 27-5. If the coil has appreciable resistance, then the lag between the current and the

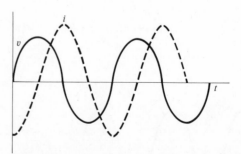

FIGURE 27-5 The current lags behind the voltage by 90° in an inductance.

impressed EMF is less than a quarter of a period. In the special case in which the inductance of the coil is negligible in comparison with its resistance, the current and the impressed voltage are in phase with each other.

The effective value I of the current in a coil which has a resistance R and a coefficient of self-inductance L can be shown to be given by

$$I = \frac{V}{\sqrt{R^2 + (2\pi fL)^2}} \qquad (27\text{-}13)$$

where V is the effective value of the impressed EMF and f is the frequency of the alternating current. The quantity $2\pi fL$ is called the *inductive reactance* of the coil and is usually designated by the symbol X_L; thus,

$$X_L = 2\pi fL \qquad (27\text{-}14)$$

The inductive reactance X_L is expressed in ohms, if the inductance L is expressed in henrys and the frequency f is expressed in hertz. Thus the simplified form of Ohm's law, $I = V/R$, is not generally applicable to an AC circuit when effective values of the current and potential difference are used. Instead, a more generalized form of Ohm's law, such as that given by Equation 27-13, must be used for AC circuits. Equation 27-13 may be written in a simplified form, as follows:

$$I = \frac{V}{Z} \qquad (27\text{-}15)$$

in which Z is called the *impedance* of the AC circuit and is also measured in ohms. In the above circuit,

$$Z = \sqrt{R^2 + (2\pi fL)^2} \qquad (27\text{-}16)$$

or

$$Z = \sqrt{R^2 + X_L^2} \qquad (27\text{-}17)$$

The effect of an inductance in an AC circuit is shown in Figure 27-6, in which a coil of many turns of wire is placed in series with a 110-volt lamp, and 120 volts from a 60-Hz source are impressed across the circuit. The lamp will glow with fair brightness. If now a soft iron core is inserted slowly into the coil, the lamp will grow dimmer, showing that the current in the circuit has been decreased. Placing an iron core inside the coil increases the changing magnetic flux in the coil and

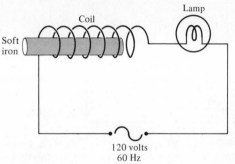

FIGURE 27-6 An iron core inside a coil increases its impedance.

hence increases its inductance. The impedance of the circuit is thus increased, and the current in it is decreased. Changing the inductance in an AC circuit is a convenient way of controlling the current.

27-4
Vector diagram for an AC circuit

For convenience in making calculations, a coil may be considered as having an inductance L in series with a resistance R, as shown in Figure 27-7. The effective voltage across R is then IR, and the effective voltage across L is IX_L. Since the voltage across the resistor R is in phase with the current I and since the voltage across the inductance L leads the current I by one quarter of a period, the voltages IR and IX_L are also out of phase by a quarter of a period and hence cannot be added algebraically. A convenient method for adding them is illustrated in Figure 27-8, in which a vector along the x axis, drawn to some convenient scale, rep-

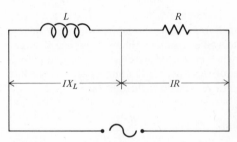

FIGURE 27-7 A circuit containing inductance and resistance.

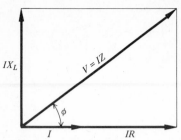

FIGURE 27-8 Vector diagram for a circuit containing inductance and resistance; the current lags behind the impressed voltage V.

resents the effective value I of the current in the circuit. The voltage drop $V_R = IR$ is in phase with this current and is represented by a vector drawn from the same origin along the x axis. The potential difference V_L across the inductance differs in phase from the current I by 90°; hence it is also out of phase by 90° with IR, the voltage drop across R. The voltage $V_L = IX_L$ is drawn along the positive y axis. The total voltage V across the terminals of the circuit is then given by the vector sum of the two voltages; thus

$$V = \sqrt{(IR)^2 + (IX_L)^2}$$
$$= I\sqrt{R^2 + X_L^2}$$

or

$$V = IZ$$

in agreement with Equations 27-13 and 27-15.

In this type of vector diagram two quantities that are out of phase by a quarter of a period are drawn at right angles to each other; they are said to be 90° out of phase. The angle ϕ between I and V is called the *phase angle* and is given by the equation

$$\tan \phi = \frac{X_L}{R} \tag{27-18}$$

This angle represents the lag between the current and the voltage, or the phase difference between them, when measured in a clockwise direction from the vector representing the impressed voltage V to the vector representing the current I.

Illustrative example

A coil having an inductance of 200 millihenrys (abbreviated mh) is placed in series with a resistor having a resistance of 30 ohms. (a) Determine the current in the circuit when 120 volts from a 60-Hz

source are impressed across the combination. (b) Determine the voltage across the coil and also across the resistor.

The inductive reactance X_L of the coil is

$$X_L = 2\pi fL = 2\pi \times 60 \times 0.2 \text{ ohms}$$
$$= 377 \times 0.2 = 75.4 \text{ ohms}$$

The impedance Z of the circuit is

$$Z = \sqrt{R^2 + X_L^2}$$
$$= \sqrt{(30)^2 + (75.4)^2} \text{ ohms}$$
$$= 81 \text{ ohms}$$

(a) The current in the circuit is

$$I = \frac{V}{Z} = \frac{120}{81} = 1.48 \text{ amp}$$

(b) The voltage across the coil is

$$V_L = IX_L = 1.48 \times 75.4$$
$$= 111.5 \text{ volts}$$

and the voltage across the resistor is

$$V_R = IR = 1.48 \times 30$$
$$= 44.4 \text{ volts}$$

These voltages are drawn to scale in the vector diagram of Figure 27-9b. The phase angle between the current and voltage is determined from the equation

$$\tan \phi = \frac{X_L}{R} = \frac{75.4}{30} = 2.51$$

yielding

$$\phi = 68°18'$$

The current lags behind the impressed voltage by 68°18'.

Voltmeters connected across the resistor, the coil, and the terminals of the circuit, as shown in Figure 27-9a, would give readings of 44.4 volts, 111.5 volts, and 120 volts, respectively.

27-5
Power in an AC circuit

The power delivered to an AC circuit by an AC generator is, at any instant, given by

$$p = vi \tag{27-6}$$

where v is the instantaneous value of the EMF and i is the instantaneous value of the current. When the circuit consists of a resistance only, the current and voltage are in phase at all times. This means that the product of the voltage and the current is always positive, and power is actually transformed from the electrical form into heat and leaves the circuit. As shown previously, the average power delivered to the resistor can be written in terms of the effective values of the current and voltage,

$$P = IV \qquad (27\text{-}12)$$

If the circuit consists of a pure inductance, the current and voltage are always out of phase by one quarter of a period, or 90°. This means that the power, as given by Equation 27-6, is sometimes positive and sometimes negative. The interpretation of this is that in that part of the cycle during which the current is increasing in one direction, energy is being supplied to build up the magnetic field around the circuit. During the next quarter of a cycle, while the current is decreasing, the magnetic field is decreasing and returning energy to the circuit by inducing an EMF in the circuit to oppose the decrease. The net result is that no energy is actually removed from the circuit; it is

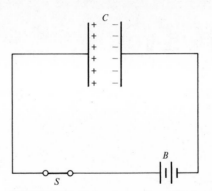

FIGURE 27-10 Capacitance in a DC circuit.

merely being stored in the magnetic field while the magnetic field is being increased and then returned to the circuit when the magnetic field is decreasing.

If the circuit contains both inductance and resistance, power is removed from the circuit only by that portion of the current that remains in phase with the voltage across the circuit. In the vector diagram of Figure 27-8, the component of V in the direction of I is $V \cos \phi$, so the power delivered to an AC circuit is, on the average,

$$P = VI \cos \phi \qquad (27\text{-}19)$$

The factor $\cos \phi$ is called the *power factor* of the circuit. When ϕ is 90° the power factor is zero, and no power is delivered to the circuit. When ϕ is 0°, the power factor is 1, and the power delivered to the circuit is VI.

27-6
Capacitance in an AC circuit

Capacitors are widely used in AC circuits. To understand the behavior of a capacitor in a circuit, let us first consider a capacitor C connected in series with a battery B and a switch S, as shown in Figure 27-10. When the switch is closed, the plates of the capacitor become charged positively and negatively, as shown. Since the plates initially were not charged, this means that there was a momentary current in the circuit in which electrons were transferred from one plate to the other until the net charge Q on each plate was

$$Q = CV \qquad (27\text{-}20)$$

where C is the capacitance of the capacitor and V is the electromotive force of the cell. If the switch S is now opened, the capacitor remains charged.

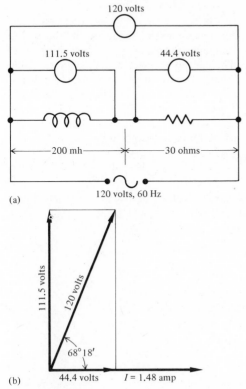

(a)

(b)

FIGURE 27-9

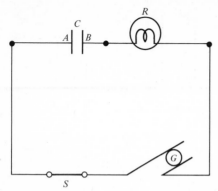

FIGURE 27-11 Capitance and resistance in an AC circuit.

It can be used as a source of electrical energy. For example, if a resistor is connected across the terminals of the capacitor, the charge Q will flow through the resistor. A current will thus have been established in this circuit, and the electrical energy that was stored in the capacitor while it was being charged has now been transformed into heat.

The behavior of a capacitor in an AC circuit can be shown with the aid of the circuit illustrated in Figure 27-11, in which the capacitor C is connected in series with a lamp R, a switch S, and an AC generator G. When the switch is closed, the lamp will glow, showing that there is a current in the circuit. Since the current is alternating, the plate A of the capacitor becomes charged positively during one half of the cycle and negatively during the opposite half of the cycle. Conversely, the plate B of the capacitor acquires a charge opposite to that of A at every instant. In the connecting wires leading to the plates of the capacitor, the current consists of the motion of electrons to and from the plates. In the space between the plates, which may be a vacuum or may contain air or some other dielectric, there is an electric field that is also alternating in direction as well as changing in magnitude. The instant the switch is closed, charges begin to build up on the two plates, positive on one, negative on the other. The rate of charging is greatest initially, and, as charges accumulate on the plates, a counter EMF equal to q/C, where q is the instantaneous value of the charge, is set up, and thus the rate of charging decreases. Therefore, the current is a maximum initially when the impressed voltage is zero and decreases in value while the voltage is increasing, the current becoming zero when the voltage reaches its maxi-

mum value (see Figure 27-12). The current reverses direction when the impressed voltage across the capacitor starts decreasing; that is, the charges on the plates are being fed back to the source. When the impressed voltage reverses direction, charges opposite in sign to that in the first half of the cycle build up on the plates, rapidly at first and at a diminishing rate as the voltage builds up to a maximum in the negative direction. It will thus be seen that the current reaches its maximum value before the voltage across the capacitor reaches its maximum value; one way of saying this is that the current *leads* the voltage. Neglecting resistance, the current leads the voltage across a capacitor by 90°, or a quarter of a period.

The effective value I of the current in a circuit containing only a capacitance C is

$$I = \frac{V}{X_C} \qquad (27\text{-}21)$$

in which V is the voltage across the terminals of the capacitor and X_C is the *capacitive reactance*, given by the equation

$$X_C = \frac{1}{2\pi f C} \qquad (27\text{-}22)$$

in which f is the frequency of the alternating current. Thus the higher the frequency and the greater its capacitance, the smaller the capacitive reactance. The capacitive reactance X_C is expressed in ohms when C is expressed in farads and f in hertz.

If an AC voltage is impressed across the terminals of a circuit consisting of a capacitor and a resistor in series, as shown in Figure 27-13, the current I in the circuit is given by

$$I = \frac{V}{\sqrt{R^2 + X_C^2}} = \frac{V}{Z} \qquad (27\text{-}23)$$

where V is the voltage across the terminals of the circuit and Z is the impedance of this circuit, given by

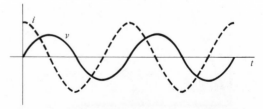

FIGURE 27-12 The current in a capacitor leads the voltage across the capacitor by 90°.

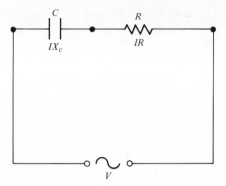

FIGURE 27-13 Capacitor in series with a resistor in an AC circuit.

$$Z = \sqrt{R^2 + X_C^2} \qquad (27\text{-}24)$$

Figure 27-14 is the vector diagram showing the relationship between the current I in the circuit, the voltage IR across the terminals of the resistor, and the voltage IX_C across the terminals of the capacitor. The vector representing IX_C is drawn in the negative y direction to show that the current leads the voltage across the capacitor by 90° or a quarter of a period. The phase angle ϕ by which the current leads the impressed voltage V is, from the figure,

$$\tan \phi = \frac{X_C}{R} \qquad (27\text{-}25)$$

Illustrative example

A circuit containing a 4.0-μf capacitor and a 250-ohm resistor is connected in series with a 60-Hz, 120-volt source. Determine (a) the current in the circuit, (b) the voltage across the terminals of the resistor, (c) the voltage across the terminals of the capacitor, (d) the phase angle, and (e) the power supplied to the circuit.

The capacitive reactance of the capacitor is

$$X_C = \frac{1}{2\pi fC} = \frac{1}{2\pi \times 60 \times 4 \times 10^{-6}} \text{ ohms}$$

$$= 663 \text{ ohms}$$

The impedance of the circuit is

$$Z = \sqrt{(250)^2 + (663)^2} \text{ ohms}$$
$$= 709 \text{ ohms}$$

(a) The current I in the circuit is

$$I = \frac{120}{709} \text{ amp} = 0.17 \text{ amp}$$

(b) The voltage V_R across the resistor is

$$V_R = IR = 0.17 \times 250 \text{ volts}$$
$$= 42.4 \text{ volts}$$

(c) The voltage V_C across the capacitor is

$$V_C = IX_C = 0.17 \times 663 \text{ volts}$$
$$= 113 \text{ volts}$$

(d) The phase angle can be found from the equation

$$\tan \phi = \frac{X_C}{R} = \frac{663}{250} = 2.65$$

from which

$$\phi = 69°20'$$

(e) The power delivered to the circuit is

$$P = VI \cos \phi = 120 \times 0.17 \times \cos 69°20'$$
$$= 7.2 \text{ watts}$$

27-7
Capacitance, inductance, and resistance in series

The most general type of series AC circuit will contain capacitance, inductance, and resistance elements. Figure 27-15 shows an AC circuit containing a capacitance C, an inductance L, and a resistance R in series with an AC generator. If V is the voltage at the terminals of the generator, the current I supplied to the circuit is given by the generalized form of Ohm's law

$$I = \frac{V}{Z} \qquad (27\text{-}26)$$

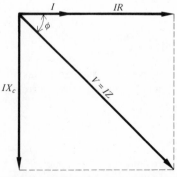

FIGURE 27-14 Vector diagram for a circuit containing a capacitance and resistance in series; the current leads the voltage.

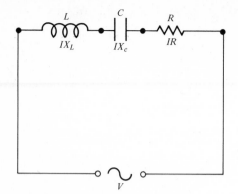

FIGURE 27-15 Circuit containing resistance, inductance, and capacitance in series.

in which the impedance Z is given by

$$Z = \sqrt{R^2 + (X_L - X_C)^2} \qquad (27\text{-}27)$$

and

$$X_L = 2\pi f L$$

$$X_C = \frac{1}{2\pi f C}$$

The reason for the negative sign with the capacitive reactance X_C is that the current in a capacitor leads the voltage by a quarter of a period, whereas the current in the inductance lags the voltage by a quarter of a period, so the voltage IX_C is 180° out of phase with the voltage IX_L. This is illustrated in the vector diagram shown in Figure 27-16, in which IX_L is drawn in the positive y direction and IX_C is drawn in the negative y direction. The difference between these two voltages is then added to the voltage IR across the resistance R to obtain the total voltage V impressed on the circuit. The phase angle between V and I is obtained from the equation

$$\tan \phi = \frac{X_L - X_C}{R} \qquad (27\text{-}28)$$

if this value is positive, the current I lags the voltage V; if this value is negative, the current I leads the voltage V.

Illustrative example

A series AC circuit consists of a capacitor of 8 μf, an inductance of 600 millihenrys, and a resistor of 48 ohms resistance. Determine (a) the current in the circuit, (b) the voltage across each element in

the circuit, (c) the phase angle, and (d) the power supplied to the circuit when the terminal voltage of the 60-Hz generator is 220 volts.

The capacitive reactance is

$$X_C = \frac{1}{2\pi \times 60 \times 8 \times 10^{-6}} \text{ ohms}$$
$$= 332 \text{ ohms}$$

The inductive reactance is

$$X_L = 2\pi \times 60 \times 0.60 \text{ ohms}$$
$$= 226 \text{ ohms}$$

The total impedance is

$$Z = \sqrt{(48)^2 + (226 - 332)^2} \text{ ohms}$$
$$= 116 \text{ ohms}$$

(a) The current in the circuit is

$$I = \frac{220}{116} = 1.90 \text{ amp}$$

(b) Constructing the vector diagram as shown in Figure 27-17, with the current I along the x axis, we find that the voltage across the resistor is

$$V_R = IR = 1.90 \times 48 \text{ volts} = 91.2 \text{ volts}$$

in phase with the current and therefore drawn along the x axis; the voltage across the capacitor is

$$V_C = IX_C = 1.9 \times 332 \text{ volts} = 630 \text{ volts}$$

in the negative y direction, since the current leads

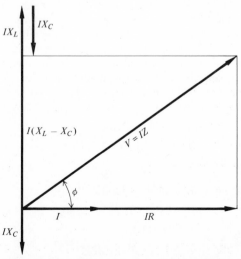

FIGURE 27-16 Vector diagram of the voltages and currents in a series circuit containing resistance, inductance, and capacitance.

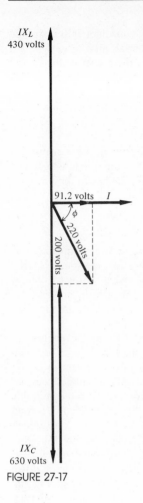

IX_L
430 volts

91.2 volts I

ϕ

220 volts

200 volts

IX_C
630 volts

FIGURE 27-17

the voltage by 90°; the voltage across the inductor is

$$V_L = IX_L = 1.9 \times 226 \text{ volts} = 430 \text{ volts}$$

in the positive y direction, since the current lags behind the voltage by 90°.

The impressed voltage V is the vector sum of these individual voltages and is given by

$$V = \sqrt{(91.2)^2 + (430 - 630)^2} = 220 \text{ volts}$$

This is a convenient way of checking the calculations.

(c) The phase angle is determined from the equation

$$\tan \phi = \frac{X_L - X_C}{R} = \frac{226 - 332}{48}$$

or

$$\tan \phi = \frac{-106}{48} = -2.21$$

from which

$$\phi = -65°40'$$

hence the current leads the voltage by 65°40'.

It will be noted that the voltage across a capacitor or an inductor in an AC series circuit may be much greater than the voltage across the terminals of the circuit.

(d) The power supplied by the generator is

$$P = VI \cos \phi$$
$$= 220 \times 1.9 \times \cos 65°40'$$
$$= 173 \text{ watts}$$

27-8
Resonance in an AC circuit

A case of very great interest is one in which the current and the impressed voltage are in phase in an AC circuit containing resistance, inductance, and capacitance. This can be established by adjusting the values of C and L so that

$$X_L = X_C \tag{27-29}$$

in which case the phase angle ϕ will be zero, since

$$\tan \phi = \frac{X_L - X_C}{R} \tag{27-28}$$

Putting in the values for X_L and X_C in Equation 27-29, we get

$$2\pi f L = \frac{1}{2\pi f C}$$

from which

$$f = \frac{1}{2\pi \sqrt{LC}} \tag{27-30}$$

When this condition is fulfilled, the current in the circuit will be a maximum and will be given simply by

$$I = \frac{V}{R}$$

since for this case

$$Z = R$$

A circuit for which Equation 27-30 holds is said to be in *resonance* at the frequency f. When the frequency of the AC supply is that given by Equation 27-30, there is a maximum transfer of energy from the generator to the circuit, since the phase angle ϕ is zero and the power factor is 1. For example, if, in the illustrative example of the previous section, sufficient inductance is added to the circuit either by inserting more iron in the inductance coils or by adding additional inductance coils so that the circuit is in resonance at 60 Hz, then the current in the circuit would be increased to its maximum value, given by

$$I = \frac{V}{R} = \frac{220}{48} \text{ amp} = 4.58 \text{ amp}$$

To determine the new value of the inductance, we can solve Equation 27-30 for L and get

$$L = \frac{1}{4\pi^2 f^2 C} = \frac{1}{4\pi^2 \times 3600 \times 8 \times 10^{-6}} \text{ henry}$$
$$= 0.88 \text{ henry} = 880 \text{ millihenrys}$$

Illustrative example

A capacitance of 0.2 μf, an inductance of 32 millihenrys, and a resistance of 300 ohms are connected in series. (a) Determine the resonance frequency of the circuit, (b) What will be the current in this circuit when a generator supplies power to it at a terminal voltage of 6.0 volts at the resonance frequency?

(a) The resonance frequency is given by the equation:

$$f = \frac{1}{2\pi\sqrt{LC}}$$

and, substituting 32×10^{-3} henry for L and 2×10^{-7} farad for C, we get

$$f = \frac{1}{2\pi\sqrt{32 \times 2 \times 10^{-10}}} \text{ Hz}$$

$$= \frac{10^5}{2\pi \times 8} = 2000 \text{ Hz}$$

(b) The current in this circuit will be a maximum when the frequency of the source is 2000 Hz, and in this case will be

$$I = \frac{V}{R} = \frac{6}{300} \text{ amp} = 0.02 \text{ amp}$$

27-9
The transformer

The electric energy that is transmitted from the generating station to the consumer is transmitted during a certain time interval, and, in designing a transmission system, it is the power or the rate at which the energy is transmitted that is of importance. If the terminal voltage of a DC generator is V, the power delivered to the transmission line is

$$P = VI \qquad (27\text{-}12)$$

where I is the current in the line. If the transmission line has a resistance R, then the rate at which heat is developed in the line is $I^2 R$; hence the power P delivered to the consumer is

$$P = VI - I^2 R \qquad (27\text{-}31)$$

A greater amount of power can be transmitted to the consumer by reducing the resistance of the power line—that is, by using wires of larger diameters—or else by transmitting the power using smaller currents. The latter method means stepping up the voltage at the generating station.

It has been found difficult to build DC generators that will develop EMF's greater than about 3000 volts. Hence, to transmit direct current at higher voltages it would be necessary to connect several generators in series. This practice is not commonly followed in this country. Another difficulty is that, for safe handling, the voltage at the consumer's end of the line must be comparatively low—not more than a few hundred volts—and no efficient methods have been developed for stepping down the voltage of a DC line. For AC generating stations, however, the problem is entirely different. With the aid of a device known as a *transformer*, it is possible to step up the voltage at the transmission line to any desired value and then to use another transformer at the consumer's end of the line to step down the voltage to a safe, usable value, the power meanwhile having been transmitted at a high voltage and low current. Modern transmission lines are operated at voltages as high as 700,000 volts.

A transformer consists of two coils near each other. In most transformers these coils are wound on closed iron cores, such as that shown in Figure 27-18a. The conventional diagram of an iron-core transformer is shown in Figure 27-18b. For special uses, particularly in some radio circuits, transformers are made without iron cores; these are usually called air-core transformers.

Suppose that the primary coil P of an iron-core transformer is connected to an AC source and that the effective voltage across its terminals is $\mathcal{E}$. Let

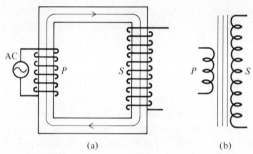

(a)   (b)

FIGURE 27-18 (a) Iron-core transformer and (b) schematic diagram; lines between the coils indicate an iron core.

us suppose initially that the terminals of the secondary coil are left open. The current that now flows through the primary coil sets up a magnetic field in the iron core. Because of the high permeability of the iron, practically the entire magnetic flux is inside the iron core. Since the current is alternating, the magnetic flux in the iron core is also alternating. This alternating magnetic flux induces an EMF in each turn of the secondary coil, and hence the induced EMF $\mathcal{E}_S$ in the secondary coil is proportional to the number of turns of wire n_S in the secondary coil. Since the same magnetic flux goes through each turn of the primary coil, an EMF will be induced in each turn of the primary coil so that the total self-induced EMF in the primary coil $\mathcal{E}_P$ will be proportional to the number of turns n_P in the primary coil. Since the magnetic flux is changing at the same rate inside each coil, we can write that

$$\frac{\mathcal{E}_S}{\mathcal{E}_P} = \frac{n_S}{n_P} \qquad (27\text{-}32)$$

Equation 27-32 holds for both the instantaneous and effective values of the EMF's.

In a well-designed transformer, $\mathcal{E}_P$ will differ very slightly from the impressed voltage $\mathcal{E}$. Thus to a very close approximation, Equation 27-32 may be written as

$$\frac{\mathcal{E}_S}{\mathcal{E}} = \frac{n_S}{n_P} \qquad (27\text{-}33)$$

If the number of turns n_S in the secondary coil is greater than the number of turns n_P in the primary coil, the transformer is called a *step-up* transformer; if the reverse is the case, it is a *step-down* transformer. For example, if the secondary coil has 1000 times as many turns as the primary coil,

the EMF $\mathcal{E}_S$ induced in the secondary will be 1000 times the voltage impressed across the primary.

When a load is connected to the terminals of the secondary coil, a current will flow in the secondary circuit and power will be supplied by it. This power must, of course, come from the source of power connected to the primary coil. This transfer of power takes place through the interactions of the magnetic fields because of the current in the primary coil and that in the secondary coil. In well-designed transformers the efficiency is as high as 98 or 99 percent. Neglecting the slight loss of power in heating the coils and the iron core, we find that the power drawn from the secondary coil must equal the power supplied to the primary coil; that is,

$$ei_P = e_s i_S \qquad (27\text{-}34)$$

where the symbols refer to the instantaneous values of the voltage and current in the primary and secondary coils, respectively. Rearranging Equation 27-34, we obtain

$$\frac{e}{e_s} = \frac{i_S}{i_P} \qquad (27\text{-}35)$$

Since Equation 27-35 holds at any instant, it also holds for the maximum values and hence for the effective values, so we can write

$$\frac{\mathcal{E}}{\mathcal{E}_S} = \frac{I_S}{I_P} \qquad (27\text{-}36)$$

which, combined with Equation 27-33, yields

$$\frac{I_S}{I_P} = \frac{n_P}{n_S} \qquad (27\text{-}37)$$

Thus the effective values of the currents in the primary and secondary circuits are in the inverse ratio of the numbers of turns in the two coils.

If we rewrite Equation 27-37 as

$$n_S I_S = n_P I_P \qquad (27\text{-}38)$$

we note that if the current in the secondary is increased, as in the case when the load on the secondary is increased, the current in the primary is also increased.

In a step-up transformer, the EMF induced in the secondary is large, but the current I_S is small, while the voltage across the primary is small and the current I_P through it is large. Both step-up and step-down transformers are used in transmitting power. A simplified version of a transmission system is shown in Figure 27-19. At the powerhouse,

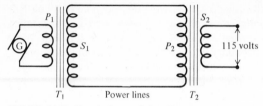

P_1 S_1 P_2 S_2 115 volts

T_1 Power lines T_2

FIGURE 27-19 Simple transmission line.

the AC generator develops electric power at, say, 120 volts; the terminals of this generator are connected to the terminals of the primary coil of a

step-up transformer T_1, which steps up the voltage to 12,000 volts at the terminals of the secondary coil. The two wires of the transmission line, which may be several miles long, connect the terminals of this secondary coil to the primary of a step-down transformer T_2 at the consumer's end of the line. Because of the voltage drop in the transmission line, the difference of potential at the primary of the step-down transformer may be, for example, only 11,550 volts. If the ratio of turns between P_2 and S_2 is 100:1, the EMF at the terminals of S_2 will be about 115 volts, suitable for use with a great many electric appliances.

Questions

1. Discuss the advantages of AC over DC.

2. Discuss the advantages of DC over AC.

3. What effect does the self-inductance of a coil have on the current (a) when connected to a DC source and (b) when connected to an AC source.

4. Discuss the effect of putting a capacitor (a) in a DC circuit and (b) in an AC circuit.

5. A lamp rated at 120 volts, 60 watts is placed in a circuit containing both inductance and capacitance. Under what conditions would it be possible for the lamp to operate at normal brightness when the circuit is connected to a 120-volt AC source?

6. Design a system that could be used to dim theater lights without using a transformer. Assume that the lights are incandescent bulbs. (The best system would dissipate as little power as possible, and would have as little inertia as possible so that the dimming could be accomplished quickly and efficiently.)

7. What properties of an electric circuit determine the resonance frequency of the circuit?

8. When a circuit is operating at its resonance frequency, what is the relationship between the voltage across the resistors and the voltage impressed on the terminals of the circuit?

9. Under what conditions can the voltage across an inductance in a series AC circuit exceed the total voltage impressed across the terminals of the circuit?

10. Under what conditions can the voltage across the terminals of a capacitor in a series AC circuit exceed the total voltage applied to the terminals of the circuit?

11. Under what conditions can current be supplied to an AC circuit without delivering power to it?

12. Alternating-current generators are usually rated in kilovolt-amperes instead of kilowatts to designate their output. Can you give a reason for this? Under what conditions would the number of kilovolt-amperes be the same as the number of kilowatts supplied by the generator?

13. Suppose that you have an automobile lamp that is designed to be operated at 12 volts and that you have available two power sources: a 110-volt DC source and a 110-volt AC source. With the aid of appropriate diagrams, show how you could operate this lamp (a) on the DC source and (b) on the AC source.

14. Describe a procedure by which you might make a variable (a) resistor, (b) inductor, and (c) capacitor.

15. Discuss the practical difficulties of constructing AC circuits having resonance frequencies that are (a) very high, say 10^{12} Hz, and (b) very low, say $\frac{1}{2}$ Hz.

16. Capacitors are often placed in parallel across power supplies to divert high-frequency signals, and coils are often placed in series to block low-frequency signals. Discuss these procedures.

17. A *parallel* combination of a capacitor C and inductor L has an equivalent impedance of infinity at the resonance frequency given by Equation 27-30. Give a physical argument to explain this.

Problems

1. Find the maximum value of an alternating EMF whose effective value is 115 volts.

2. Find the resistance of the filament of a 120-volt, 100-watt lamp.

3. Find the maximum current supplied to a 120-volt, 25-watt lamp.

4. Find the inductive reactance of a 50-millihenry coil when supplied by a 60-Hz, 120-volt line.

5. At what frequency will a 0.6-henry coil have an inductive reactance of 150 ohms?

6. Find the impedance of an inductor having an inductance of 0.5 henry and a resistance of 100 ohms, when used at 400 Hz?

7. A certain coil has a resistance of 50 ohms when used in a DC circuit and an impedance of 100 ohms when used in a 50-Hz circuit. Find the inductance of the coil, assuming that both the resistance and inductance are constant.

8. Find the phase angle for a 1.5-henry coil with a resistance of 75 ohms when the coil is used in a 1000-Hz circuit.

9. In a certain AC circuit containing a resistor in series with a coil, the phase angle is known to be 30°. The total resistance in the circuit is known to be 150 ohms and the frequency is known to be 400 Hz. Find (a) the inductive reactance and (b) the inductance.

10. A 1-millihenry coil having a resistance of 1 megohm is connected to a 150-volt source at 10^7 Hz. Find (a) the inductive reactance, (b) the impedance, and (c) the current.

11. A 10-amp current, leading by 30°, is supplied by a 110-volt source. Find (a) the power factor and (b) the power.

12. A certain motor uses 500 watts from a 110-volt source that is supplying 5 amp. What is the power factor?

13. Find the capacitive reactance of a 2-pf capacitor at 10^3 Hz.

14. At what frequency will a 25-μf capacitor have a capacitive reactance of 100 ohms?

15. A 4-pf capacitor is in series with a 1-megohm resistor at 10^4 Hz. Find the impedance of the circuit.

16. Find the impedance of a 15-μf capacitor in series with a 10-ohm resistor at 500 Hz.

17. The phase angle of a circuit containing a capacitor and a 25-ohm resistor is 30° at 60 Hz. Find the capacitance of the capacitor.

18. At what frequency will a 50-millihenry coil be in resonance with a 25-μf capacitor?

19. A 7-μf capacitor is in series with a 1-ohm resistor and a variable inductor. At what value of the inductance will the circuit be in resonance at 60 Hz?

20. A bell transformer is designed to step 110 volts down to 18 volts. The secondary has 36 turns. How many turns are there in the primary?

21. A transformer has 50 turns in the primary and 1000 in the secondary. What is the voltage on the secondary when the primary is connected to a 110-volt line?

22. A coil with an inductance of 30 millihenrys and a resistance of 15 ohms is supplied with a current of 5.0 amp from an AC source of 60 Hz. Find (a) the voltage across the terminals of the coil and (b) the power supplied to the coil.

23. An inductance of 0.4 henry is connected in series with a resistance of 75 ohms. The terminals of this circuit are connected to a 50-Hz generator with a terminal voltage of 140 volts. Determine (a) the current in the circuit, (b) the phase angle, and (c) the power supplied to the circuit.

24. A 4-μf capacitor is connected in series with a 600-ohm resistor and the circuit is supplied with power from a 230-volt, 60-Hz generator. Find (a) the current in the circuit and (b) the power supplied to the circuit.

25. A lamp rated at 60 watts, 120 volts, is connected in series with a capacitor of 12-μf capacitance to a 60-Hz, 120-volt source. (a) Determine the current in the circuit. (b) Compare the brightness of the lamp in this circuit with the normal brightness of the lamp.

26. A 60-ohm resistor is connected in series with an inductance of 1.2 henrys and a capacitance of 20 μf. (a) Determine the impedance of this circuit when it is supplied with power from a 60-Hz, 220-volt source. (b) Determine the current.

27. A 40-ohm resistor is connected in series with a capacitance of 10 μf and an inductance of 25 millihenrys. A 60-Hz current of 2.5 amp flows in the circuit. Determine (a) the voltage across each element of the circuit, (b) the total voltage

applied to the terminals of the circuit, and (c) the phase angle between the current and the applied voltage.

28. Consider two circuits: one has a 0.1-μf capacitor in series with a 50,000-ohm resistor and the other has an inductance of 5 henrys in series with a 35-ohm resistor. Compare the current in these two circuits when each is supplied with a 50-volt signal at (a) 10^3 Hz (an audio frequency) and (b) 10^6 Hz (a radio frequency).

29. A 5-millihenry coil that has a resistance of 9 ohms is in series with an 8-ohm resistor and a 40-μf capacitor. A 500-Hz source of power produces a current of 1 amp in the circuit. (a) Find the impedance of the circuit. (b) Find the effective value of the input voltage and the maximum value. (c) Find the phase angle. (d) Find the effective value of the voltage across the resistor, the capacitor, and the coil.

30. A 48-ohm resistor is connected in series with an inductance of 400 millihenrys and a capacitance of 8 μf. (a) What is the resonance frequency of this circuit? (b) What current will flow in this circuit when it is supplied from a 110-volt source at the resonance frequency? (c) Determine the voltage across the capacitor under these conditions.

31. An iron-core transformer has 100 turns in the primary and 1800 turns in the secondary, and is operated from a 120-volt, 60-Hz generator. Determine (a) the EMF induced in the secondary, (b) the current in the secondary circuit when it is taking 2.4 kw of power with a power factor of 0.75, and (c) the current in the primary.

32. A transformer is used to step down the voltage of a transmission line from 13,200 volts to 220 volts. (a) What is the ratio of the turns on the two windings? (b) If the secondary supplies 25 amp, determine the current in the primary.

33. The input current to the primary of a transformer is 1 amp and lags 10° behind the 120-volt input. The input to the secondary load is 6 volts with a power factor of 0.7. The efficiency of the transformer is 98 percent. Find the current in the secondary.

34. As the frequency of the supply voltage to an RLC circuit is increased, the impedance reaches a minimum at f_0, the resonance frequency given by Equation 27-30. For a lower frequency f_L, the impedance has a value Z that is larger than R, the resistance of the circuit. There exists a higher frequency f_H, for which the impedance is also equal to Z. Show that the resonance frequency f_0 is the geometric mean of f_H and f_L.

electronics

chapter 28

28-1
Introduction

Some of the most important applications of electricity involve electronics. Electronic devices include electron tubes, transistors, and integrated circuits. They are used extensively in radios, television sets, computers, and control systems, which have become significant, if not vital, parts of our technology and culture.

A complete description of these devices is well beyond the scope of this book. In this chapter we shall describe a few of the elements and characteristics of such devices in order to indicate some of the basic principles of electronics.

28-2
Thermionic emission

The development of the modern vacuum tube depends to a very great extent on a discovery made by Edison, in 1883, in his study of methods for improving the electric light bulb. Edison put a metal plate into a glass tube containing a filament and evacuated the tube. The filament F was heated by means of a battery A, as shown in Figure 28-1. Another battery B and a galvanometer G were connected in series between the plate P and one side of the filament F. When the positive terminal of B was connected to the plate, a current flowed through the galvanometer; but when the battery was reversed so that the negative terminal of B was connected to the plate, no current flowed through the galvanometer. Edison merely noted this effect but made no use of it. The explanation of this effect

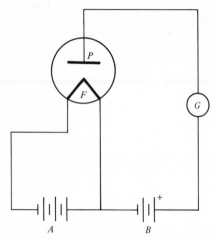

FIGURE 28-1 Circuit for showing thermionic emission.

is that the heated filament emits negative electrons; this is known as *thermionic emission* of electrons. When the plate is positive with respect to the filament, the electrons are attracted to the plate and flow through the circuit back to the filament. When the plate is negative with respect to the filament, the electrons are repelled by the plate, and there is no current in the circuit from F to P, G, and B.

When the filament is heated to a temperature T, some of the free electrons of the metal evaporate from the solid into the space around it, forming what may be called an *electron gas* around the solid. Equilibrium is reached when just as many electrons leave the filament as reenter it from the electron gas around it. When a difference of potential is established between the filament and plate, sometimes called the *plate voltage*, electrons are attracted to the plate, and other electrons evaporate from the filament to replace them. When the plate voltage is made sufficiently large, the electrons are moved to the plate as fast as they are emitted by the filament. A further increase in the plate voltage does not increase the current through the tube. Figure 28-2 shows the current through the tube— that is, the plate current I—as a function of the difference of potential V between plate and filament. At first, the current I increases as V increases, but, beginning at point B, an increase in V does not affect the current. This value of the current is called the *saturation current*, and, for any one filament, the saturation current depends upon the temperature. If the temperature of the filament

is increased from T to T', the saturation value of the current is increased. It is to be noted that Ohm's law does not apply in this case.

Although heated filaments are used as sources of electrons in many types of electron tubes, most tubes used in radio, television, and other forms of communication use a tungsten filament as an auxiliary heater H to raise the temperature of another surface, called the cathode K, so that it becomes the emitter of the electrons. One reason for this is the desire to use an equipotential surface as the source of electrons. The filament is not an equipotential surface since its two ends are connected to a source of power and maintained at a difference of potential. By using a separate cathode K connected to one end of the filament, we get an electron emitter at a single potential. Another reason is that many substances that are very good electron emitters cannot readily be made into filaments. Such substances can be put on other surfaces and used as cathodes. The conventional representation of a cathode K is shown in the diode of Figure 28-3a. Sometimes, however, the heating filament is omitted completely from the diagram, as in Figure 28-3b.

28-3
The rectifier tube

A vacuum tube containing a cathode K and a plate P is called a *diode*; it can be used to rectify alternating current. Figure 28-4 shows one method for using a diode as a rectifier. Alternating current

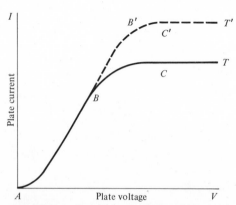

FIGURE 28-2 Curves showing the thermionic current as a function of the voltage at different filament temperatures.

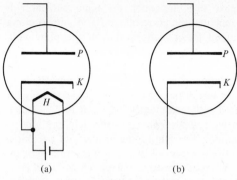

(a)   (b)

FIGURE 28-3 Representation of a diode. K is the cathode and P is the plate; H is the heating filament.

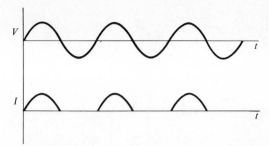

FIGURE 28-5 The action of the diode as a rectifier.

is supplied to the primary of a transformer; the secondary of this transformer consists of two separate coils, one to step down the voltage to the value required to operate the heater H, and the other to provide the proper operating voltage for the plate circuit. The external load to which the direct current is to be delivered is represented by the resistance R_L, which is connected between the cathode K and the plate P.

The cathode K is heated to the required temperature for electron emission by the heater H. Whenever the plate P becomes positive with respect to

the cathode K, electricity will flow through the circuit; no electricity will flow when the plate is negative with respect to the cathode. Thus there will be a current in the circuit only during the positive half of the cycle. Figure 28-5 shows the alternating difference of potential from the AC generator, and directly underneath it is the graph of the rectified current through the external load R_L.

When full-wave rectification is desired, either two diodes are used, or else a single tube containing one cathode and two plates is used. Figure

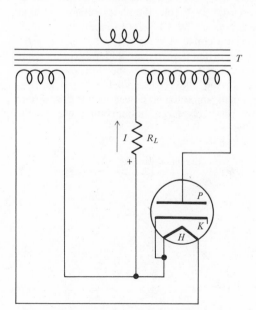

FIGURE 28-4 A rectifier circuit using a diode; direct current is supplied to the external load R_L.

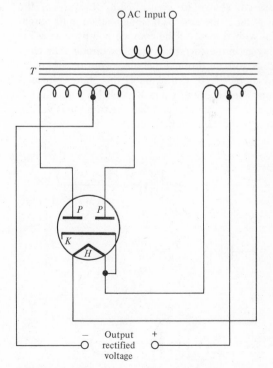

FIGURE 28-6 Full-wave rectifier using tube with two plates.

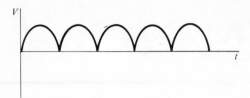

FIGURE 28-7 Output voltage of a full-wave rectifier.

28-6 shows a circuit using the latter type of tube as a rectifier, and Figure 28-7 shows the rectified output voltage of this circuit.

28-4
The triode

The three-electrode tube, called a *triode*, in addition to the cathode K and plate P, has an electrode in the form of a wire mesh, usually called a *grid*, G. The grid is placed between the cathode and the plate. The introduction of the grid into the electron tube opened up many new uses for these tubes. Only a few such uses can be sketched. To understand the operation of such a tube, consider the diagram shown in Figure 28-8. Because of the position of the grid G, a small change in its potential with respect to the cathode will produce a big change in the current in the plate circuit. The electrons go right through the spaces in the wire mesh on their way to the plate. If the plate voltage is kept fixed at some value V_1 and the potential difference between the grid and the cathode is varied, we get the characteristic curve labeled V_1, shown

in Figure 28-9, in which the plate current I is plotted against the grid voltage V_G. A similar curve is obtained for a higher value of the plate voltage V_2.

Since triodes are normally operated at less than the saturation value of the plate current, the electrons are not drawn to the plate as fast as they are emitted by the filament. They constitute what is known as a *space charge* in the neighborhood of the filament. The amplifying property of a triode is due to the position of the grid in the region of the space charge. As we have seen, a small change in the potential of the grid produces a large change in the plate current if the plate voltage is kept constant. One method of rating a tube is in terms of its voltage amplification. The *voltage amplification constant* μ of a tube is defined as the ratio of the change in plate voltage to the change in the grid voltage needed to keep the current in the plate circuit constant. This can be determined from the voltage characteristic curves of Figure 28-9. Thus if the grid voltage is changed by an amount ΔV_G, the plate voltage will have to be changed from V_1 to V_2 to keep the plate current constant; hence

$$\mu \doteq \frac{V_2 - V_1}{\Delta V_G} \qquad (28\text{-}1)$$

Suppose, for example, that the voltage amplification factor is 10. This means that if the grid voltage is decreased by 1 volt, the plate voltage will have to be increased by 10 volts to maintain the plate current at a constant value.

It is important to recognize that the grid serves to *control* the plate current but that the source of the energy to the plate circuit load is the *B battery*. Very small amounts of power in the grid circuit control relatively large amounts of power in the

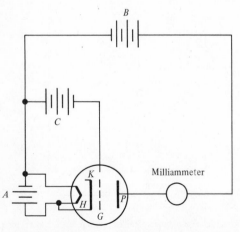

FIGURE 28-8 Triode and circuit for obtaining the characteristic curves.

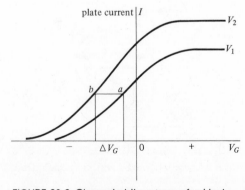

FIGURE 28-9 Characteristic curves of a triode.

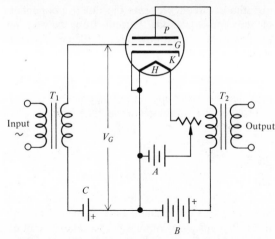

FIGURE 28-10 Single-stage, transformer-coupled triode amplifier.

plate circuit. In this sense the triode can be thought of as a power amplifier as well as a voltage amplifier. In some applications the triode is used merely as a switch, but most often it is used as an amplifier.

The use of a triode as an amplifier is shown in the circuit illustrated in Figure 28-10. The grid is kept at a negative potential with respect to the cathode by means of a battery C; this potential is usually referred to as the *grid bias*. The value of this grid bias is chosen so that the steady current in the plate circuit has a value that is in the center of the straight portion of the characteristic curve; this is the point L on the curve, shown in Figure 28-11. If an alternating potential difference is applied between grid and cathode, the current in the plate circuit will fluctuate in value; these fluctuations will follow the alternations impressed on the grid. This fluctuating current, in going through the primary coil of the transformer T_2, will produce a fluctuating magnetic field, which, in turn, will induce an alternating EMF in the secondary coil having a form similar to that impressed on the grid of the tube. In addition to the voltage amplification produced by the tube, additional amplification is obtained if T_2 is a step-up transformer. The terminals of the secondary coil of T_2 can be connected to another tube amplifier circuit, thus increasing the amplification still further. Several stages of amplification are quite common. When transformers are used between the various stages, we speak of the amplifier as a transformer-coupled amplifier. Resistors or capacitors may be used

instead of transformers to couple the different stages.

One of the most important uses of the triode is as a generator of oscillations in electric circuits. There are many different types of oscillator circuits, one of which is shown in Figure 28-12. The grid bias is obtained from the difference of potential between the terminals of the resistor R, through which a very small current is flowing. The grid circuit has a coil of inductance L and a capacitor of capacitance C. An additional coil of inductance L_1 is in the plate circuit, and its position with respect to coil L can be varied. Suppose that there is a change in the grid potential owing to a change in current in the circuit; this will produce a change in the current in the plate circuit, and, because of the magnetic coupling between the coils L_1 and L, an additional change will be produced in the grid circuit. In this way energy is fed back from the plate circuit to the grid circuit. If the amount of energy fed from the plate circuit into the grid circuit is greater than the energy lost by this circuit,

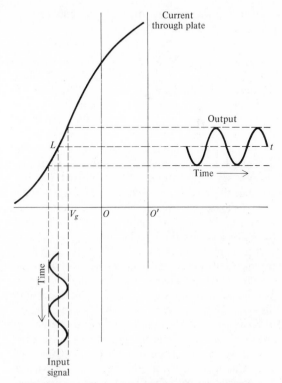

FIGURE 28-11 Action of the triode as an amplifier; an alternating voltage applied to the grid produces similar variations in the plate current.

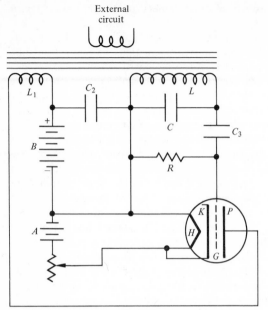

FIGURE 28-12 A triode oscillator circuit using a tickler coil.

oscillations will be set up in this circuit of frequency

$$f = \frac{1}{2\pi\sqrt{LC}} \qquad (28\text{-}2)$$

These oscillations may be transmitted to an external circuit by coupling it to the coil L.

To start the original changes in the current in the circuit in order to be able to feed energy from the plate circuit to the grid circuit through the coils L_1 and L is a simple matter. For example, there are always small variations in the current flowing through the circuit, particularly when the circuit is first closed. No matter how small these variations are, they are quickly amplified until oscillations are set up in the circuit.

It is interesting to note that the source of power for the oscillating circuit is the B battery in the plate circuit, which, of course, supplies direct current. The function of the triode, therefore, is to convert the DC form of energy into AC energy. When oscillations are set up, both direct and alternating current flow in the plate circuit and also in the grid-to-cathode circuit. The capacitor C_2 is put into the circuit as a shunt across the battery to provide a low-impedance path for the alternating current in the plate circuit; the direct current cannot flow through the capacitor. The capacitor C_3 in the grid-to-cathode circuit prevents the direct component of the grid-to-cathode current from flowing in the coil L, so it must flow through the grid resistor R. This ensures the proper bias of the grid with respect to the filament. Since a fraction of the alternating current also flows in R, there will be provided the variations in the grid potential needed to produce the variations in the plate current.

28-5
Impure semiconductors; diodes

In recent years the transistor and integrated circuit have replaced the vacuum tube in many applications. Perhaps more important, the development of these solid-state devices has permitted the development of electronic devices that at one time were impractical. The basic physics of these devices is quantum mechanical, but we can describe some of their significant characteristics.

When in the form of pure crystals, semiconductors such as germanium have a relatively high resistance because the electrons are fairly tightly bound to the crystal structure. As the temperature is increased some of these electrons are liberated and the crystal becomes conducting, but at low temperatures these pure semiconductors are poor conductors. However, it is possible to make germanium crystals into conductors by adding appropriate impurities. For example, if a very small amount of antimony (Sb) is added to the germanium crystal, the antimony atoms will become part of the germanium crystal. However, each antimony atom has five valence electrons (electrons involved in the bonding of the atom to the crystal), whereas each germanium atom has four such electrons. Thus at each site at which an antimony atom exists in the crystal, there exists an extra electron, which would not be there if a germanium atom were at that site. This extra electron is free to move and will be set in motion by the application of a small electric field. A germanium crystal with this type of impurity is called an n-type crystal, the n standing for negative. On the other hand, if an atom of indium (In), which has only three valence electrons, replaces a germanium atom in the crystal, there will be an absence of an electron, or a "hole," left in the crystal at this site. If an electric field is applied to a crystal containing such holes, electrons from other parts of the crystal will flow toward the holes, leaving holes in the other parts of the crystal. Thus the current in this type of

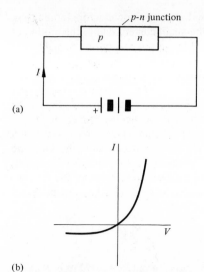

(a)

(b)

FIGURE 28-13 (a) A junction diode; (b) characteristic curve of a junction diode.

crystal can be considered as the motion of these holes. A hole in a region that normally would be occupied by an electron is equivalent to the presence of a positive charge at this place. A germanium crystal with this type of impurity is called a p-type crystal, where p stands for positive.

It is possible, by the use of very sophisticated chemical and physical techniques, to produce a single crystal of semiconductor, in which one part of the crystal is p-type while the rest is n-type. The boundary between these two is called the p-n junction. When a potential difference V is connected to the terminals of such a crystal, as shown in Figure 28-13a, the resulting current I follows the curve shown in Figure 28-13b. The important thing about this curve is that the current is large when the positive terminal of the battery is connected to the p-type semiconductor (in this configuration the crystal is said to be *forward-biased*) while the current is very small when the positive terminal is connected to the n-type semiconductor (in this configuration the crystal is said to be *reverse-biased*). Thus such a p-n junction acts as a *rectifier*, and can be used instead of a vacuum diode. In such applications, the principal advantages of the semiconductor are that it requires no heater current and that it can be made very small.

This directional characteristic of the p-n junction can be understood by recognizing that when the crystal is forward biased, the electric field in the crystal drives holes in the p region to flow into

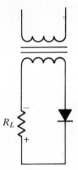

FIGURE 28-14 Half-wave rectifier using a solid-state diode.

the n region and electrons in the n region to flow into the p region. But when the crystal is reverse-biased, the field drives electrons from the p region into the n region and holes from the n region into the p region. However, at usual temperatures there are almost no free electrons in the p region and almost no holes in the n region.

Figure 28-14 shows a diode used as a half-wave rectifier; the diode is represented by a conventional symbol. Figure 28-15 shows two diodes used as a full-wave rectifier. These figures should be compared with Figures 28-4 and 28-6. The relative simplicity of these diode circuits arises from the absence of the heater elements.

Figure 28-16 shows a full-wave rectifier that can be used to rectify an AC signal without the use of a center-tapped transformer.

28-6
Transistors

A transistor is a crystal having either two p-type regions separated by an n-type region, in which

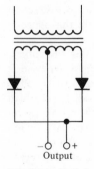

FIGURE 28-15 Full-wave rectifier using two solid-state diodes.

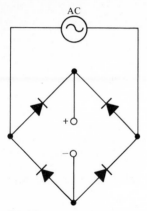

FIGURE 28-16 Full-wave bridge rectifier.

case it is known as a *p-n-p* transistor, or it may have two *n*-type regions separated by a *p*-type region, in which case it is called an *n-p-n* transistor. Figure 28-17 is a schematic diagram of a circuit containing an *n-p-n* transistor; the electrode connected to the *p* region is called the *base*, one wire or electrode connected to an *n* region is called the *emitter*, and the other wire or electrode is called the *collector*.

A transistor behaves as a power amplifier. Its action can be demonstrated by connecting one battery B_1 so that the base is positive with respect to the emitter, and a second battery B_2 puts an opposite bias on the collector. The electric field in the left half of the crystal is directed so that electrons and holes move toward the common junction and electrons move into and through the *p* region. The conductivity of the crystal is thus large for current in this direction. If the battery terminals of B_1 were reversed, the electrons and holes would move away from the common junction so that only a small current would result; the conductivity of the crystal would thus be small for current from base to emitter. Therefore, this crystal has a low resistance for current in one direction and a high resistance for current in the opposite direction.

It will be noted from the figure that the voltage bias on the collector with respect to the base is opposite to that of the emitter with respect to the base; hence the collector region will have a high resistance when the emitter region has a low resistance. If the thickness of the intermediate region is very small, many electrons from the *n* region on the left will go through the *p* region to the *n* region

on the right and toward the collector. In addition, there will be the small current due to the motion of electrons and holes of the region on the right.

If we call the current in the emitter branch I_e and the resistance of this branch R_e, the power developed in it is $I_e^2 R_e$. Similarly, the power developed in the collector branch is $I_c^2 R_c$, where I_c is the current in the collector branch and R_c is its resistance. The value of R_c is usually much greater than R_e, mainly because of the way the electric fields produced by the batteries B_1 and B_2 are biased in the different sections of the crystal. The ratio of the power delivered to the collector to that in the emitter is

$$\frac{I_c^2 R_c}{I_e^2 R_e} \qquad (28\text{-}3)$$

Even though I_c may be only slightly larger than I_e, there will still be a large gain in power in the ratio R_c/R_e.

The amplifier of Figure 28-17 is called a *common-base* amplifier. Figure 28-18 shows a *common-emitter* amplifier. (The figure uses the conventional symbol for a transistor, rather than the schematic representation of Figure 28-17.) If the battery E_2 is chosen to have an EMF appropriately larger than the EMF of battery E_1, the emitter-base bias and the collector-base bias can be approximately the same as the amplifier of Figure 28-17. In that case

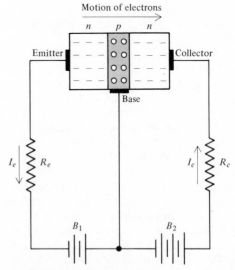

FIGURE 28-17 Schematic diagram of a circuit containing an *n-p-n* transistor.

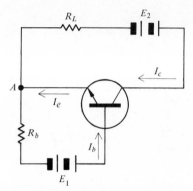

FIGURE 28-18 Schematic diagram of a common-emitter amplifier.

the emitter current I_e and the collector current I_c are approximately the same. Since the emitter current divides at point A, the base current I_b is much smaller than the collector current; that is,

$$I_e = I_b + I_c$$
$$I_c = I_e - I_b$$

and

$$I_e \approx I_c$$
$$I_b << I_c$$

For a typical transistor (type 2N592) the base current will be about 100 microamp and the collector current will be a few milliamperes. Thus, there will be a current amplification. If the load resistor R_L is larger than R_b, there will be a voltage and power amplification as well.

In many modern applications of transistors, single crystals are built, with many junctions. Parts of this crystal are then etched away, leaving many different transistors. Thin layers of metal can be added, forming conductors, resistors, and capacitors. The result is a complex *integrated circuit*, containing dozens of components, in a container often less than an inch or so across.

28-7
The cathode-ray tube

The modern cathode-ray tube, widely used in physics laboratories, consists of a highly evacuated glass tube, as shown schematically in Figure 28-19, that contains a cathode K plus its heater, a grid G, an anode A with a small aperture in it, and two sets of deflecting plates V_1V_2 and H_1H_2. The tube widens out at the end opposite the cathode. This face is coated with a phosphor that fluoresces when electrons strike it, converting the energy of the electrons into visible radiation.

By means of appropriate circuits that supply the necessary currents and potential differences, electrons from the cathode are accelerated through the grid to the anode, and most of them pass through the small aperture in the anode and then travel to the screen S, producing a small spot of light at the point of impact. On the way to the screen, the electrons can be deflected vertically when they pass through the space between the plates V_1V_2 by an electric field applied to them. They may also be deflected horizontally when passing through the space between the plates H_1H_2 by a suitable electric field. The electrons can also be deflected by means of magnetic fields produced by magnets or electromagnets outside the tube. The latter method is used in cathode-ray tubes used as picture tubes in television sets. The intensity of the electron beam is varied by the changes in the potential of the grid G produced by the variations in the electromagnetic waves received from the transmitting antenna. The variations in intensity of the electron beam produce variations in the intensity of light emitted by the phosphor producing the pattern or picture seen by the viewer.

In the cathode-ray tubes used in television sets, the electron beams are deflected by magnetic fields rather than by electric fields. These deflection fields are produced by coils of wire wrapped around the outside of the neck of the tube. These coils form the *yoke* around the neck of the tube.

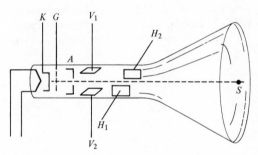

FIGURE 28-19 Schematic diagram of a cathode-ray tube.

Exercises

1. Examine a light bulb that has been used for some time. What evidence is there that the metal of the filament has been evaporated?

2. Thermionic emission occurs in an incandescent lamp. Discuss the behavior of the electrons after they leave the filament.

3. (a) What is the direction of the current in a vacuum-tube diode? (b) What is the nature of the charge flowing through the tube?

4. Explain what is meant by the saturation current of a vacuum diode.

5. The output voltage of a full-wave rectifier is not a constant voltage. The wave shown in Figure 28-7 can be thought of as a signal that is primarily DC with several high-frequency Fourier components. To eliminate these high-frequency components and produce a smooth DC signal, a *filter* is needed. Describe how (a) a capacitor or (b) an inductor might be used as a filter.

6. Explain why the straight-line portion of the characteristic curve is used when a triode is used as an amplifier.

7. What is the difference between *n*-type germanium and *p*-type germanium?

8. Compare the resistance of a forward-biased and reverse-biased *p-n* junction.

9. One of the most common uses of the transistor is as an electrically controlled switch. Show how the collector current in Figure 28-14 can be turned off by a change in the emitter bias.

10. Show how the plate current in a triode can be turned off by a change in the grid bias.

11. Figure 28-20 shows the usual way of representing a common-emitter amplifier, particularly when the amplifier is used in computer or logic circuits. In that figure the input and output signals are the voltages of the points shown.

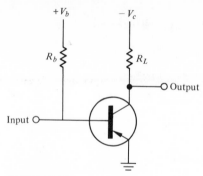

FIGURE 28-20

(a) Show that when the input is a negative voltage, the *p-n-p* transistor is turned on and current will flow in the collector circuit. (b) Show that when the input is a positive voltage, the transistor will be turned off and essentially no current will flow in the collector circuit. (c) Show that if the input voltage switches from negative to positive, the output voltage will switch from a high value to a low value. The amplifier then *inverts* and amplifies.

12. (a) What happens to the output voltage of a two-diode full-wave rectifier if one of the diodes burns out and becomes an open circuit? (b) What happens to the output voltage of the two-diode full-wave rectifier if one of the diodes burns out and becomes a short circuit?

13. Transistors have been used in many applications where relays had been used (see Question 18 in Chapter 25). What are the advantages of the transistor over the relay? (Note, however, that relays are still in common use for switching large currents or high power, particularly in applications where low speed is not an important disadvantage. It is possible to build multiple-switch relays that handle large currents at fairly low cost.)

light

part 5

geometric optics

chapter 29

29-1
Introduction

In a very large number of circumstances, the wave nature of light maybe neglected. When the objects, distances, and apertures involved are large compared to the wavelengths of light, diffraction effects are usually negligible and the light may be treated as if it were a collection of rays. Geometric optics is an approximate method of treating the behavior of light, based on the assumption that light rays travel in straight lines in a homogeneous medium and change direction only at the boundary between homogeneous media.

When a light ray traveling in a homogeneous medium strikes the boundary between that medium and a second medium, the ray is in general both *reflected* and *refracted*; the *incident* ray gives rise to a *reflected* ray (traveling in the original medium) and a *refracted* ray (traveling in the second medium). For simplicity we can assume that the boundary is a plane and then the geometry is governed by a set of simple rules.

1. The reflected and refracted rays both lie in the plane formed by the incident ray and the normal to the boundary, as shown in Figure 29-1a.
2. The angle between the incident ray and the normal, the *angle* of *incidence i*, is equal to the angle between the reflected ray and the normal, the *angle of reflection r*; that is,

$$i = r \qquad (29\text{-}1)$$

This is known as the law of reflection.

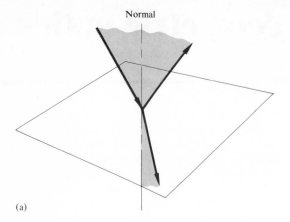

(a)

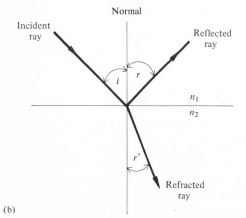

(b)

FIGURE 29-1 Light reflected and refracted at a surface separating two media.

3. The angle of incidence, i, is related to the angle of refraction, r' (the angle between the refracted ray and the normal), by

$$n_1 \sin i = n_2 \sin r \qquad (29\text{-}2)$$

where n_1 is the index of refraction of the medium in which the incident ray is traveling and n_2 is the index of refraction in the medium in which the refracted ray is traveling. This is known as Snell's law of refraction.

Light rays are also governed by the *principle of reversibility*, which states that if a reflected or refracted beam is reversed in direction it will retrace its original path.

All of these laws can be thought of as empirical laws, although they can be shown to be consequences of Maxwell's equations. In one form or another they have been known for centuries; a Dutch astronomer, Willebrod Snell, discovered

the law of refraction in 1621, although he wrote it in a slightly different form. Equation 29-2 is commonly known as Snell's law, although the form of Equation 29-2 was discovered by Descartes.

The intensity of refracted and reflected *beams* of light depends on the size of the beams, the absorption of the media, the nature of the surface, and to some extent the diffraction effects we are neglecting. Therefore we shall, in general, neglect the intensity. However, we can note empirically that very often the intensity of the refracted beam is essentially equal to zero. When light strikes a metallic surface, for example, almost all of the refracted beam is absorbed. However, if the metal is very thin, then light can be transmitted through the metal. Similarly, very little light is reflected from insulators, such as glass, for most angles of incidence. However, almost all the light is reflected from the surface of insulators if the angle of incidence is very close to 90°. The principle of conservation of energy requires that the sum of the energy in the reflected beam and the energy in the refracted beam must equal the energy in the incident beam. However, detailed utilization of this principle requires a knowledge of the amount of energy absorbed.

Geometric optics can be thought of as a logical system (like geometry) in which the rules given above play the role of the axioms and postulates. From these rules we can derive a great many theorems describing the behavior of systems, such as lenses and mirrors, and for the behavior of combinations of these systems, such as microscopes. In the following we shall describe a few of these results, particularly for simple but commonly used systems. In this process we shall derive formulas to describe the optical behavior of these systems, usually by describing the behavior of a few representative rays. This process, called *ray tracing*, is the fundamental procedure for geometric optics. In order to describe an optical system fully, we should trace the behavior of many rays, so that we can understand the behavior of rays that take complex as well as simple paths through an optical system. In the past, this meant that the analysis of optical systems involved laborious, repeated applications of the laws of reflection and refraction or the use of mathematical approximations. The development of large digital computers has allowed optical designers to do exact ray tracing relatively quickly. In a sense the computer has permitted the optical designer to return to the simple rules

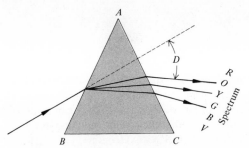

FIGURE 29-2 Deviation and dispersion of a beam of white light by a triangular prism.

stated above and use those as the only physics needed to design lenses, mirrors, and prisms.

29-2
Refraction and dispersion

The *absolute index of refraction* of a medium is the ratio of the speed of light in a vacuum to its speed in the medium. Now the speed of light in a vacuum is the same for all colors, but this is not true in a material medium. The speed of light depends not only on the nature of the medium but also on the particular color or wavelength of light. This can easily be demonstrated by sending the light through some transparent substance arranged so that the surface through which the light enters the medium is inclined at an angle A to the surface through which the light leaves the medium (see Figure 29-2). A triangular prism of the material is a very convenient form for this experiment. If a very narrow beam of white light is incident on side AB, it will be observed that the emergent beam is not white but consists of an array of colors extending from red through orange, yellow, green, blue, and violet. The prism is said to *disperse* the incoming radiation into its *spectrum*.

This dispersion is due to the fact that each particular color or wavelength travels with a different speed through the glass. The white light incident upon the surface AB at some angle i is refracted and dispersed. The different wavelengths or colors that constitute the white light travel through the material at different speeds and are refracted again at the surface AC. The ray entering the prism is bent toward the normal, and when it leaves the prism it is bent away from the normal, thus deviating it still further from its original direction. The angle of deviation D for any one color depends upon the material of the prism, the angle A of the prism, and the angle of incidence of the light.

It will be observed that red light is deviated least and violet light shows the greatest deviation. Thus red light travels fastest through the glass. The index of refraction of glass for red light is smallest. In specifying the index of refraction of a substance, it is also necessary to specify the color or the wavelength of the light used. Table 29-1 lists the absolute indices of refraction for a few substances for several wavelengths; the latter are expressed in angstroms (Å).

If a very narrow beam of *monochromatic light*—that is, light of a single wavelength such as the yellow light from a sodium flame or a sodium arc—is sent through a triangular prism ABC that is made of some transparent substance, the beam will be deviated from its original direction through an angle D, the angle of deviation. This angle may be found experimentally, or it may be computed with the aid of Snell's law, provided that the index of refraction of the substance is known. If the angle of incidence is changed by rotating the prism with respect to the incident beam, for example, the angle of deviation will also change. It may increase or decrease in value, depending upon the direction of rotation of the prism. Suppose that the prism is rotated so that the angle of deviation decreases; it

Table 29-1 Absolute indices of refraction

wavelength *substance*	7682 Å	6563 Å	5893 Å	4861 Å	4047 Å
Borosilicate crown glass	1.5191	1.5219	1.5243	1.5301	1.5382
Dense flint glass	1.6441	1.6501	1.6555	1.6691	1.6901
Water 20°C	1.3289	1.3311	1.3330	1.3371	
Carbon disulfide 18°C		1.6198	1.6255	1.6541	
Diamond			2.417		

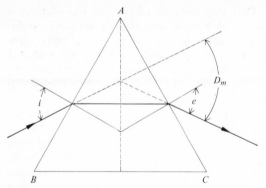

FIGURE 29-3 Angle of minimum deviation.

will be found that the angle of deviation will reach a minimum value D_m, and any further rotation of the prism will produce an increase in the angle of deviation (see Figure 29-3). When the angle of deviation is a minimum, the angle of incidence i and the angle of emergence e are equal.

There is a simple relationship connecting the angle of minimum deviation D_m, the angle A between the two refracting faces of the prism, and the index of refraction n of the material of the prism for the particular wavelength of light used. This relationship can be derived from simple geometrical considerations. The necessary angles are shown in Figure 29-4. Since D_m is an exterior angle to the isosceles triangle whose base angles are each $i - r'$, we get

$$D_m = i - r' + i - r' = 2i - 2r'$$

The angle opposite A in the quadrilateral formed by the sides of the prism and the normals is equal to $\pi - A$. Since the sum of the angles of a triangle is equal to π, we can write

$$r' + r' + \pi - A = \pi$$

or

$$2r' = A$$

Hence

$$D_m = 2i - A$$

or

$$i = \frac{D_m + A}{2}$$

Using Snell's law

$$n = \frac{\sin i}{\sin r'}$$

we get

$$n = \frac{\sin \frac{1}{2}(A + D_m)}{\sin \frac{1}{2}A} \tag{29-3}$$

If a transparent substance is fashioned in the form of a prism of refracting angle A, its index of refraction for any desired wavelength can be determined by measuring the angle of minimum deviation for this particular wavelength.

Illustrative example

When a narrow beam of sodium light was sent through a particular glass prism whose refracting angle was 60°, the angle of minimum deviation was found to be 51°. Determine the index of refraction of this glass for yellow sodium light.

In this case, $A = 60°$ and $D_m = 51°$; substituting these values in Equation 29-3, we obtain

$$n = \frac{\sin 55°30'}{\sin 30°}$$

from which

$$n = 1.65$$

29-3
Refraction effects

When light goes through a transparent medium with parallel surfaces, as shown in Figures 29-5 and 29-6, the light is *displaced* but not *deviated*; when the light emerges, it travels parallel to its original direction. When white light is used, rays of

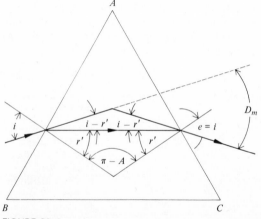

FIGURE 29-4

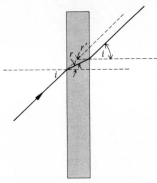

F IGURE 29-5 Displacement of a ray of light by a transparent plate with parallel surfaces.

different colors are refracted by different amounts upon entering the glass, but they become parallel again after emerging from it. When viewed by the eye, these rays are brought to a focus in a single point, and the effect produced is the same as that of a single beam of white light.

When a divergent beam of light passes through a plate of glass with parallel surfaces, the rays, on emerging, are displaced by different amounts, as shown in Figure 29-7. If the divergence is not too large, the rays seem to come from a point I displaced from the original point O by an amount

depending on the thickness of the glass. By reversing the directions of the rays in this diagram, we get the effect produced by the plate of glass on a bundle of rays that are converging toward a point or focus. The focal point or point of convergence of the rays is shifted away from the plate of glass. This method has been used to change the focal length of camera lenses.

Another interesting effect produced by the refraction of a narrow bundle of rays is illustrated in Figure 29-8a. Rays from some point O at the bottom of a pool of water are viewed by an observer whose eye is vertically above this point in air. The narrow bundle of rays entering the eye appears to come from a point I above point O. The pool of water seems shallower; its apparent depth is equal to the depth of the water divided by its index of refraction, which is 1.33.

The shallowing effect produced by refraction can be expressed in terms of the absolute index of refraction n of the transparent medium. Referring to Figure 29-8b we can write

$$n = \frac{\sin i}{\sin r'}$$

Now

$$\sin i = \frac{AB}{IB}$$

FIGURE 29-6 Photograph of a beam of light incident upon rectangular plate of glass. Note the reflected rays at both surfaces. The transmitted rays are parallel to the incident rays but displaced slightly. (Courtesy of Bausch and Lomb.)

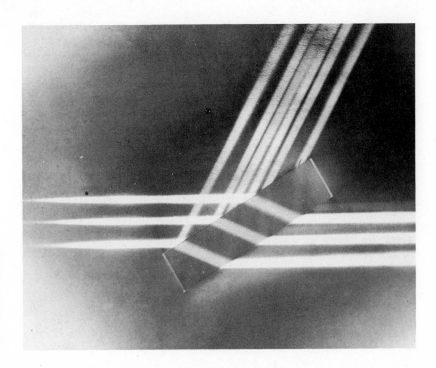

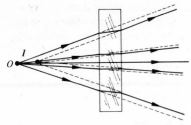

FIGURE 29-7 The point of divergence of a beam of light appears displaced after passing through a plate of glass with parallel faces.

and

$$\sin r' = \frac{AB}{OB}$$

hence

$$n = \frac{OB}{IB}$$

When angles i and r' are both small, IB is practically equal to the distance IA, and $OB = OA$. Now OA is the depth d of the object below the surface and IA is the distance of the image below the surface—that is, the apparent depth d_a of the object. Hence

$$n = \frac{d}{d_a} \qquad (29\text{-}4)$$

29-4
Critical angle; total reflection

A ray of light traveling in a region of high index of refraction toward one of smaller index of refraction —from water toward air, for example—may penetrate through the surface of separation, provided that the angle of incidence r' in the water is less than a certain value called the *critical angle* of incidence and denoted by c. If the angle of incidence is greater than this critical angle, the light will not be able to penetrate the surface but will be *totally reflected* back into the water.

A few typical rays from a point source O in water are shown in Figure 29-9. The ray OA strikes the surface normally and is not deviated. The ray OB strikes the surface at an angle r' and enters the air at a larger angle i. From Snell's law,

$$\frac{\sin r'}{\sin i} = \frac{1}{n}$$

where n is the absolute index of refraction of the

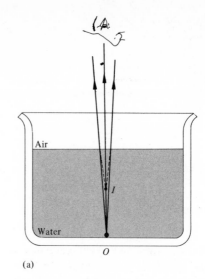

(a)

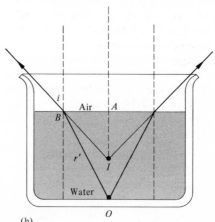

(b)

FIGURE 29-8 Shallowing effect produced by refraction.

medium. The limiting case is one for which the angle $i = 90°$; the sine of i for this case is 1 and is its maximum value. The ray OC strikes the surface

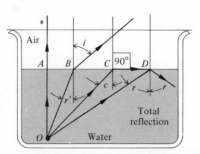

FIGURE 29-9 Critical angle of refraction; total reflection.

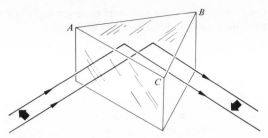

FIGURE 29-10 A totally reflecting prism.

at such an angle that the refracted ray in air makes an angle of 90° with the normal; that is, the refracted ray travels along the surface. The angle that this ray OC makes with the normal is called the *critical angle* of incidence and is denoted by c. Applying Snell's law to this case, we get

$$\frac{\sin c}{\sin 90°} = \frac{1}{n}$$

or

$$\sin c = \frac{1}{n} \qquad (29\text{-}5)$$

The value of the sine of the critical angle for any substance is the reciprocal of its index of refraction. Taking $n = 1.33$ for water, the critical angle is 48°46'. If the angle of incidence is greater than the critical angle, the ray is totally reflected back into the medium. The ray OD illustrates the case of total reflection. For rays that are incident at angles greater than the critical angle, the surface acts as a perfectly reflecting surface. It is for this reason that, whenever possible, glass prisms arranged for total reflection are used instead of silvered surfaces as reflectors in optical instruments.

29-5
Totally reflecting prisms

Totally reflecting glass prisms have found wide applications in such optical instruments as binoculars, periscopes, and range finders. The index of refraction of the glass used in making these prisms ranges from 1.50 to 1.65. When $n = 1.50$, the critical angle is 41°50'. If a ray of light going through the glass strikes the surface at an angle greater than this critical angle, the ray of light will be totally reflected. Figure 29-10 shows total reflection in a 45° prism. The rays enter one face normally, strike the hypotenuse at an angle of 45°, and are reflected through an equal angle, leaving

the other face of the prism normally. Since the rays enter and leave normally, no dispersion takes place.

Another way of using this prism is shown in Figure 29-11. The rays enter the hypotenuse face normally, undergo two total reflections, and leave the same face normally. Not only is the path of the rays reversed, but the rays themselves are reversed from top to bottom. Prism binoculars use two such prisms at right angles to each other. One produces a reversal from right to left and the other a reversal or, better, an inversion from top to bottom. Other varieties of shapes of prisms are used in optical instruments to displace the path of light around obstacles and to invert and reverse images.

29-6
Light pipes; fiber optics

There are many interesting applications of the phenomenon of total internal reflection, among which is the transmission of light along a predetermined path in a transparent medium. Let us consider the simple case of a cylinder of glass whose length L is large in comparison with its radius r, as shown in Figure 29-12; its ends A and B are polished and plane. Light rays from some source are shown entering face A near the axis of the cylinder O, forming a cone of light. Many of these rays will strike the cylindrical surface at angles of incidence greater than the critical angle and will be totally internally reflected. These rays will undergo several more total reflections and emerge from face B. Other rays from the source that enter face A and strike the cylindrical surface at angles of incidence smaller than the critical angle will be only partially reflected; moreover, they will

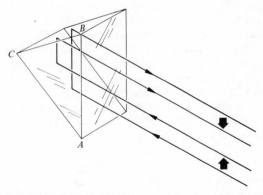

FIGURE 29-11 Two reflections and an inversion by a totally reflecting prism.

FIGURE 29-12 Light from source O undergoing total internal reflections in a cylinder. The cylinder acts as a light pipe.

be further reduced in intensity upon each additional reflection. If the glass is of good quality and the cylindrical surface is very clean, as much as 60 percent of the light entering face *A* will emerge from face *B*. If the cylinder is curved so that the radius *R* of the curved section is much greater than the radius *r* of the cylinder, the light will still be guided along this pipe by multiple internal reflections without serious loss of intensity.

The foregoing principles have been known and applied for many years. But the fact that a transparent substance, even if bent, is still a good light pipe, provided that *R* is much greater than *r*, suggested the idea of making fibers of transparent materials. Most materials are very flexible when drawn into fine fibers; this includes glass fibers. Optical fibers have been made with diameters of 0.05 mm or less. Many such fibers are placed together in very careful alignment either with their cylindrical axes accurately parallel, or else arranged in a predetermined pattern, depending upon the particular optical use for which it is intended. These fibers are held in some container or coating and the ends of the bundles of fibers are made accurately plane and polished. Each fiber transmits light from a small area of an object and forms an image at the other face.

One advantage of a bundle of optical fibers is that one end of it, which we shall call the objective, can be brought very close to the object to be imaged or explored, and thus transmit a greater quantity of light to a distant viewing point or detecting instrument. The objective may be moved along the object by scanning, while the other end of the light pipe remains fixed to some recording device. Because of its flexibility, the light pipe may be bent around opaque obstacles to obtain images of otherwise hidden or inaccessible objects. The light pipe may be used as a two-way path, as sketched in Figure 29-13, sending light from a source down one set of fibers to illuminate the object, and have the light scattered by the object conducted back to the viewing device along the other fibers. The field of fiber optics is comparatively new and has already found innumerable and

ingenious applications. No doubt, newer and even more ingenious applications will be found.

29-7
Image formation

An ideal lens can form an image of an object. All of the rays of light that leave one point of the object and go through the lens are brought together, or focused, at a single point of the image, and the images of two points that are close together in the object are also close together. The image may be magnified (or reduced), turned upside down, or turned left to right, but there is still a one-to-one correspondence between points in the object and points in the image.

If we start out knowing that an optical system does form images in this ideal way, or if we assume that it does, then we can locate the image in a fairly simple way. Since all of the rays from a given point are focused at a single point, we need trace only two rays from one point on the object and locate

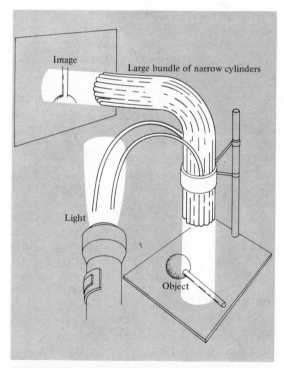

FIGURE 29-13 Light pipe. Light from a source illuminates objects; light from object transmitted by light pipe forming an image of the object.

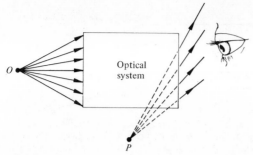

FIGURE 29-14 An optical system forms a virtual image when rays leaving the system diverge, apparently from a point *P*.

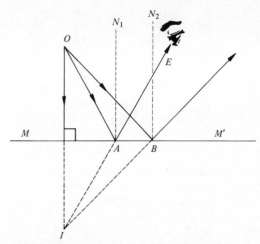

FIGURE 29-15 Method of determining the position of the image formed by a plane mirror.

the point where these are focused. (If we are unlucky enough to have chosen two rays that happen to leave the optical system so that they are parallel to each other, we shall have to trace a third ray.) Since we have assumed that the image is not distorted, we can describe the image completely by locating the image of three or four properly chosen points.

An image in which the light from an object point is actually brought to a focus is called a *real* image. A real image will affect photographic film, and if a real image is formed on a screen or ground glass, a human being will see the image located at the screen. Some optical systems form another kind of image, called a *virtual* image. Suppose that all the light rays that leave a point *O* of an object and go through an optical system spread out so that they are diverging, apparently from some point *P*, as shown in Figure 29-14. Then if the diverging rays of light enter the eye of an observer, they will be indistinguishable from rays of light that actually left point *P*; the observer will see an image at point *P*. Such a virtual image will not affect photographic film placed at *P* and it cannot be focused on a screen, but it can be seen.

Perhaps the most common virtual image is that formed by a plane mirror. The method for determining the position of the image of any point of an object placed in front of a plane mirror is illustrated in Figure 29-15. *MM'* is a plane mirror and *O* is any point of an object. Since a point is determined by the intersection of two lines, it will be sufficient to take any two rays from *O* that strike the mirror and are reflected from it. Normals are drawn at the points of incidence *A* and *B*, and the angles of reflection are made equal to the respective angles of incidence. The two reflected rays do not meet in

front of the mirror, but when produced backward they meet in a point *I*. *I* is the *image* of point *O*. It is easy to prove, by simple geometry, that *I* lies just as far behind the mirror as *O* does in front of it. Further, the line *OI* is perpendicular to the plane of the mirror.

This method of image construction can be extended to an object of finite size. Figure 29-16 shows the image *A'B'* of an object *AB*. It will be

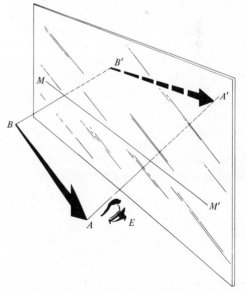

FIGURE 29-16 The image formed by a plane mirror appears reversed to an observer at *E*.

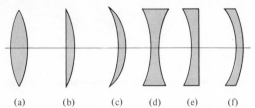

FIGURE 29-17 Spherical lenses. (a), (b), and (c) are converging lenses; (d), (e), and (f) are diverging lenses.

noted that to an observer at E, the head of the arrow AB appears on the left, but when viewed in the mirror, the head of the arrow $A'B'$ appears on the right. The image is said to be *reversed* from right to left. The size of the image formed by a plane mirror is the same as the size of the object.

It should be noted that real optical systems do not form images in this ideal way. In general, the images of points are small volumes rather than points, and the images have distortions of several kinds. Further, the imaging properties of real optical systems depend on the wavelength, so color-related distortions exist. For many purposes these aberrations are negligible and for many applications it is possible to design rather complex systems that minimize these effects.

29-8
Lenses

Lenses, either singly or in combinations, are used for the formation of images of objects that send light to them. Lenses are most frequently made of transparent material with spherical surfaces. Spherical surfaces are used because they are easiest to make. For certain special cases, other surfaces may be made to eliminate certain defects of spherical lenses or to provide certain needed effects. A plane surface is a special case of a spherical surface; a plane can be considered a sphere of infinite radius. Spherical lenses are classified as either *converging lenses* or *diverging*

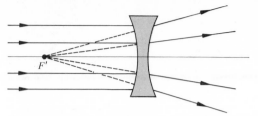

FIGURE 29-19 Rays parallel to the principal axis are diverged by the diverging lens. After passing through the lens, the rays appear to come from the virtual focus at F'.

lenses. Three common forms of converging lenses are shown in Figure 29-17a, b, and c, and three common forms of diverging lenses are shown in Figure 29-17d, e, and f. It will be observed that a converging lens is thicker at the center than at the edges, whereas a diverging lens is thinner at the center than at the edges.

A beam of parallel light incident on a converging lens will be converged toward a point F, known as the *focus* of the lens (see Figure 29-18). A beam of parallel light incident upon a diverging lens will diverge after passing through the lens as though it came from a point F', its focus, as shown in Figure 29-19. F' is called a *virtual focus*, since the rays do not actually pass through it.

The action of the lens is due to the refraction of the light as it enters and leaves the spherical surfaces bounding the lens. Figure 29-20 shows the effect of a converging lens on the wavefronts of the parallel beam. The wavefronts incident on the first surface of the lens are plane and parallel. The part of the wavefront that passes through the center of the lens is retarded more than the part that goes through the outer part of the lens. The emerging wavefront is not plane; it can be shown that it is spherical, with its center at the focus. Since the wavefronts travel perpendicular to their surfaces,

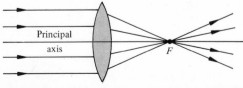

FIGURE 29-18 Rays parallel to the principal axis are converged toward the principal forces by the converging lens.

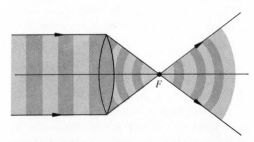

FIGURE 29-20 Change in wavefront produced by a converging lens on a parallel beam of light incident upon the lens.

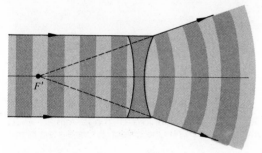

FIGURE 29-21 Change in wavefront produced by a diverging lens on a parallel beam of light incident upon the lens.

they converge toward the center F. Parallel wavefronts incident on a diverging lens, as shown in Figure 29-21, are retarded more at the thicker edge of the lens that at the center. The emerging wavefront is spherical, with its center at F'. These wavefronts travel perpendicularly to their surfaces, diverging from the center F'.

29-9
Thin lenses

A thin lens is one whose thickness is small in comparison with its diameter. Only thin lenses will be considered in this book. Figure 29-22 shows a section of a thin, converging lens whose two surfaces are parts of spheres, one of radius R_1 and the other of radius R_2, with the centers of these spheres at C_1 and C_2, respectively. The line passing through these centers of curvature is called the *principal axis*. A *principal focus* of a lens is a point of convergence of a beam of light that is parallel to the principal axis. F and F' in Figure 29-23 are principal foci. There are two principal foci, one on each side of the lens. The focal length f of a thin lens is the distance from the lens to either principal focus, measured along the principal axis. The focal length of a lens depends upon the kind of glass used and upon the radii of the spherical surfaces

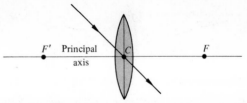

FIGURE 29-23 Principal axis, principal foci, and optical center of a thin lens.

of the lens. We shall assume that the lens is so thin that all distances can be measured from a line through the center of the lens perpendicular to the principal axis. The *optical center C* of a lens is a point such that any ray passing through it will not be deviated. In the case of a thin lens whose surfaces have the same curvature, the optical center coincides with the geometrical center.

A plane drawn through a principal focus perpendicular to the principal axis is called a *focal plane* of the lens. A beam of parallel light that is inclined at a small angle to the principal axis, as shown in Figure 29-24, after passing through a converging lens will be focused at a point in the focal plane. To determine the position of the focal point F_2, it is necessary merely to trace that ray in the beam that passes through the optical center of the lens. Since this ray goes through the lens without deviation, the point F_2 at which it intersects the focal plane is the point of convergence of all the other rays in the parallel beam.

29-10
Image formation by converging lenses

With the aid of a few well-chosen rays, it is a comparatively simple matter to construct graphi-

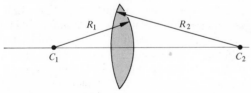

FIGURE 29-22 The center of curvature of the surfaces of a lens.

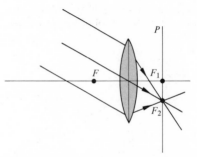

FIGURE 29-24 A parallel beam of light inclined to the principal axis is focused at a point F_2 in the focal plane P.

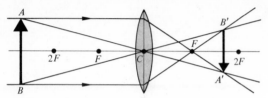

FIGURE 29-25 Image formed by a converging lens when the object distance is greater than 2f.

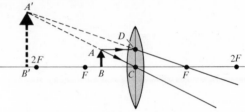

FIGURE 29-27 Virtual enlarged image formed by a converging lens when the object distance is less than f.

cally the image formed by a converging lens. Figure 29-25 shows an object AB at a distance from the converging lens greater than 2f, where f is its focal length. Two rays from any one point on the object, traced through the lens, are sufficient to determine the image of this point. In the figure, one ray from point A is chosen parallel to the principal axis; it is deviated by the lens so that it passes through the principal focus F. A second ray from point A is taken so that it passes through the optical center and travels on without deviation. These two rays intersect at a point A', which is the image of A. A similar construction yields B', the image of B. We could do this for other points on the object and obtain their images as well. These points would lie along A'B', which is the image of AB. It will be observed that this image is *smaller* than the object, is *inverted*, and is *real*; that is, light rays actually pass through the image. This image can be seen by looking along the axis from a point beyond the image, or the image may be formed on a screen placed where the image is, and then the image may be viewed from any convenient position.

Figure 29-26 shows another type of image formation by a converging lens. In this case the object lies between the focus F and a point distant 2f from the lens. The image is *real*, *inverted*, and *larger* than the object. Point B is chosen on the axis in order to simplify the diagram.

A special case intermediate between the two just discussed occurs when the object is at a dis-

tance from the lens equal to twice the focal length. The image in this case is the same distance from the lens on the other side. It is real, inverted, and the same size as the object. It can be shown that this is the minimum distance between object and real image for any position of the object from the converging lens. This forms a convenient method for determining the focal length of a converging lens: when the object and image are the same size, the focal length of the lens is one fourth the distance between them.

Figure 29-27 shows a third type of image formed by a converging lens. The object distance is less than the focal length of the lens. It will be observed that the rays from A do *not* meet after passing through the lens, but, when projected back, they meet in point A' on the same side as the object. The image in this case is *virtual*, *erect*, and *larger* than the object. The virtual image can be seen only by looking through the lens; it cannot be formed on a screen.

29-11
The lens equation

In the preceding section we showed how to determine by graphical means the relative positions of the image produced by a thin converging lens for different positions of the object. These same diagrams enabled us to determine the relative sizes of image and object, and also the nature of the image—that is, whether it was real or virtual. The graphical method is applicable to all cases; the same results can also be obtained by means of a simple equation called the *lens equation*, which can be derived most readily from the graphical representation. In Figure 29-28, AB is an object situated at a distance s to the left of the lens of focal length f, and A'B' is the image formed by this lens at a distance s' to the right of the lens. All distances are measured from a line through the

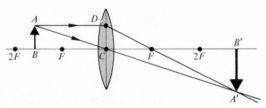

FIGURE 29-26 Real enlarged image formed by a converging lens when the object distance is greater than f and less than 2f.

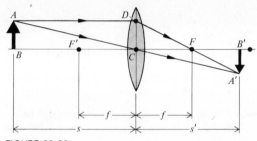

FIGURE 29-28

center C of the lens perpendicular to the principal axis; the latter passes through the two principal foci F and F'. Triangles ABC and $A'B'C$ are right triangles by construction; further, angle ACB = angle $A'CB'$ since they are vertex angles. Hence

$$\frac{A'B'}{AB} = \frac{CB'}{CB} = \frac{s'}{s} \qquad (29\text{-}6)$$

Triangles CDF and $A'B'F$ are similar since they are both right triangles with vertex angles (DFC and $B'FA'$) that are equal. Therefore

$$\frac{A'B'}{CD} = \frac{B'F}{CF} = \frac{s'-f}{f}$$

Now

$$CD = AB$$

by construction, and therefore

$$\frac{A'B'}{AB} = \frac{s'-f}{f} \qquad (29\text{-}7)$$

From Equations 29-6 and 29-7 we get

$$\frac{s'}{s} = \frac{s'-f}{f}$$

and thus

$$s'f = ss' - sf$$

Dividing by $ss'f$, this becomes

$$\frac{1}{s} = \frac{1}{f} - \frac{1}{s'}$$

or

$$\frac{1}{s} + \frac{1}{s'} = \frac{1}{f} \qquad (29\text{-}8)$$

Equation 29-8 is the *lens equation*. In using this equation, we start with light coming toward the lens from the left; s is positive when the object is to the left of the lens; s' is positive when the image is to the right of the lens; f will then be positive for a converging lens and negative for a diverging lens.

Equation 29-6 gives the relative sizes of image and object when they are at right angles to the principal axis. We now define a term called the *magnification*, m, as follows:

$$m = -\frac{s'}{s} \qquad (29\text{-}9)$$

where s' is the image distance and s is the object distance. The negative sign is used to indicate that when s' and s are both positive, the image is inverted relative to the object. *The positive and negative signs are to be used only when numerical values are substituted for s, s', and f.*

We shall illustrate the method of using Equations 29-8 and 29-9 in a few typical cases.

Illustrative example

An object 6 cm tall is situated 45 cm from a converging lens of 15 cm focal length. Determine the position, size, and nature of the image.

This case is illustrated in Figure 29-25. Let us solve Equation 29-8 for s':

$$\frac{1}{s} + \frac{1}{s'} = \frac{1}{f} \qquad (29\text{-}8)$$

from which

$$\frac{1}{s'} = \frac{1}{f} - \frac{1}{s}$$

$$= \frac{s-f}{sf}$$

or

$$s' = \frac{sf}{s-f} \qquad (29\text{-}10)$$

In this case,

$$s = 45 \text{ cm} \quad \text{and} \quad f = 15 \text{ cm}$$

hence

$$s' = \frac{45 \times 15}{45 - 15}$$

or

$$s' = 22.5 \text{ cm}$$

Since s' is positive, the image is real and on the right side of the lens.

Using Equation 29-9 for the magnification produced by the lens, we obtain

$$m = -\frac{s'}{s}$$

$$m = -\frac{22.5}{45} = -\frac{1}{2} \qquad (29\text{-}9)$$

that is, the height of the image is half that of the object, or 3 cm high. The negative sign indicates that the image is inverted. Thus the image is real, inverted, and smaller than the object.

Illustrative example

Suppose that the object of the above example is moved to a position 20 cm from the converging lens; determine the position, size, and nature of the image.

This case is illustrated in Figure 29-26. Using $s = 20$ cm and $f = 15$ cm, Equation 29-10 yields, for the position of the image,

$$s' = \frac{20 \times 15}{20 - 15} = 60 \text{ cm}$$

The magnification produced by this lens is now

$$m = -\frac{60}{20} = -3$$

Thus the image is real, three times larger than the object, and inverted.

Illustrative example

Suppose that the above object is now moved to a position 10 cm from the lens; determine the position, size, and nature of the image.

This case is illustrated in Figure 29-27. Using $s = 10$ cm and $f = 15$ cm, Equation 29-10 yields, for the position of the image,

$$s' = \frac{10 \times 15}{10 - 15} = -30 \text{ cm}$$

The magnification produced by this lens is now

$$m = -\frac{-30}{10} = +3$$

The fact that s' is negative means that the image is to the left of the lens—that is, on the same side as the object; hence the lens now forms a virtual image. The magnification is now $+3$; that is, the image is upright with respect to the object, and its height is three times that of the object.

29-12
Image formation by diverging lenses

Figure 29-29 shows the image $A'B'$, formed by the diverging lens of the object AB. The rays from any point such as A, after passing through the lens, diverge and do not meet; when projected back, they meet in point A', which is the virtual image of A. A diverging lens cannot form a real image of a real object. This statement can be verified by experiment, or by graphical construction using a variety of positions for the object, or by an analysis of Equation 29-10. The experiment consists of attempting to focus the image formed by the diverging lens on a screen. A lamp or an illuminated arrow can be used as an object. No position of the diverging lens will be found that will give an image on the screen. One might inquire how such images can be seen; the answer, of course, is they are seen with the aid of the eye, which can be considered an optical instrument containing a converging lens. The eye forms a real image with the diverging light coming from the lens. Optical instruments constructed for visual use generally form virtual images.

In Equation 29-10, using our sign convention, f

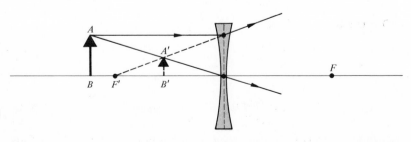

FIGURE 29-29 Virtual image formed by a diverging lens.

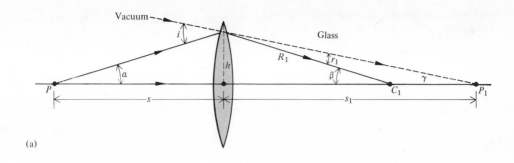

(a)

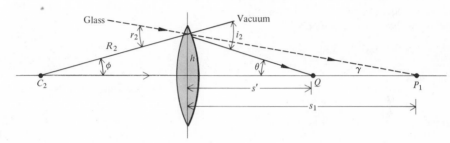

FIGURE 29-30   (b)

for a diverging lens will always be a negative number. For a real object, s is positive. The numerator of the fraction in Equation 29-10 will therefore always be negative and the denominator will always be positive, so s' will always be negative; that is, the image will be virtual. The graphical proof, if needed, is left to the student.

Suppose that, in the case illustrated in Figure 29-29, s is 25 cm and $f = -15$ cm; then, from Equation 29-10,

$$s' = \frac{25(-15)}{25 - (-15)} = -\frac{25 \times 15}{40} = -9.4 \text{ cm}$$

Since s' is negative, the image is virtual.

The magnification produced by this lens is

$$m = -\frac{-9.4}{25} = +0.38$$

The image is thus smaller than the object and is upright with respect to the object.

29-13
The lensmaker's equation

The focal length of a lens depends upon the index of refraction n of the glass used and the radii of curvature, R_1 and R_2, of its spherical surfaces. The relationship between these quantities and the focal length f of the lens can be derived for the simple case of a very thin lens, using some reasonable approximations. One of these approximations is that the rays from the object through the lens to the image make small angles with the principal axis; these are called *paraxial* rays. Another is that the lens is so thin that no appreciable error is introduced by measuring object and image distances from the center of the lens instead of from the surfaces. For simplicity let us trace two rays from the point P of the object through the first glass surface of radius R_1 as shown in Figure 29-30a.

The point P is situated on the principal axis; one ray along the principal axis goes through the spherical surface at the vertex V; a second ray inclined at a small angle α to the principal axis is incident on the first surface at an angle i, is refracted towards the normal, and makes an angle r_1 with it. The two rays meet at point P_1. Using Snell's law, we can write

$$\frac{\sin i}{\sin r_1} = n$$

and since we are considering only paraxial rays, for which

$$\frac{\sin i}{\sin r_1} = \frac{i}{r_1} = \frac{\tan i}{\tan r_1}$$

we can write

$$\frac{i}{r_1} = n$$

or

$$i = nr_1$$

Using the theorem that an angle exterior to a triangle is equal to the sum of the opposite interior angles, we have from the figure

$$i = \alpha + \beta \qquad \text{and} \qquad \beta = \gamma + r_1$$

from which

$$r_1 = \beta - \gamma$$

hence

$$\alpha + \beta = n(\beta - \gamma)$$

Further

$$\alpha = \frac{h}{s} \qquad \beta = \frac{h}{R_1} \qquad \gamma = \frac{h}{s_1}$$

so

$$\frac{h}{s} + \frac{h}{R_1} = n\left(\frac{h}{R_1} - \frac{h}{s_1}\right)$$

which can be put in the form

$$\frac{1}{s} = (n - 1)\frac{1}{R_1} - \frac{n}{s_1}$$

or

$$\frac{1}{s} + \frac{n}{s_1} = (n - 1)\frac{1}{R_1} \qquad (29\text{-}11a)$$

The rays that travel through the glass meet the second spherical surface and are refracted into the air (or vacuum) and meet at point Q on the axis as shown in Figure 29-30b. Using the notation shown in the figure, and following the steps used above, we have

$$n = \frac{i_2}{r_2}$$

or

$$i_2 = nr_2$$

$$i_2 = \theta + \phi \qquad \text{and} \qquad r_2 = \gamma + \varphi$$

hence

$$\theta + \varphi = n_2(\gamma + \varphi)$$

Further

$$\theta = \frac{h}{s'} \qquad \varphi = \frac{h}{R_2} \qquad \gamma = \frac{h}{s_1}$$

and thus

$$\frac{1}{s'} + \frac{1}{R_2} = n\left(\frac{1}{s'} + \frac{1}{R_2}\right)$$

or

$$\frac{1}{s'} - \frac{n}{s_1} = (n - 1)\frac{1}{R_2} \qquad (29\text{-}11b)$$

Adding Equations 29-11a and 29-11b, we get

$$\frac{1}{s} + \frac{1}{s'} = (n - 1)\left(\frac{1}{R_1} + \frac{1}{R_2}\right) \qquad (29\text{-}12)$$

Comparing this equation with the lens equation, we see that

$$\frac{1}{f} = (n - 1)\left(\frac{1}{R_1} + \frac{1}{R_2}\right) \qquad (29\text{-}13)$$

The sign convention used is that R_1 and R_2 are positive when the surfaces are convex and negative for concave surfaces.

Illustrative example

Glass, whose index of refraction $n = 1.65$, is used for making lenses. (a) Determine the focal length of a lens if one surface has a radius of curvature of +12 cm and the other has a radius of curvature of +18 cm. (b) Determine the focal length of a plano-concave lens made with a spherical surface having a radius of curvature of −25 cm.

(a) For this double convex lens we can write

$$\frac{1}{f} = (1.65 - 1)\left(\frac{1}{12} + \frac{1}{18}\right)$$

from which

$$\frac{1}{f} = \frac{3.25}{36}$$

or

$$f = 11.1 \text{ cm}$$

(b) The radius of curvature R_1 of a plane surface is infinite, so

$$\frac{1}{R_1} = 0$$

hence for this plano-concave lens, we can write

$$\frac{1}{f} = (1.65 - 1)\left(\frac{1}{-25}\right) = -\frac{0.65}{25}$$

or

$$f = -38.5 \text{ cm}$$

29-14
Combinations of lenses

Two or more lenses are frequently used in combination to produce a desired result. Although a general discussion of this topic is beyond the scope of this book, a few simple combinations of two lenses will be discussed here because of their importance in the design of many optical instruments.

If two thin lenses are placed in contact so that their principal axes coincide, and if the thickness of the combined lens system is still sufficiently small that it may be treated as a single thin lens, then the focal length f of this system is given by

$$\frac{1}{f} = \frac{1}{f_1} + \frac{1}{f_2} \tag{29-14}$$

where f_1 and f_2 are the focal lengths of the individual lenses. For example, if a converging lens of 15 cm focal length is placed in contact with another converging lens of 10 cm focal length with their principal axes coinciding, then the focal length f of this system is given by

$$\frac{1}{f} = \frac{1}{15} + \frac{1}{10} = \frac{2 + 3}{30}$$

or

$$f = 6 \text{ cm}$$

If a converging lens of focal length 10 cm is placed in contact with a diverging lens of -15 cm focal length with their principal axes coinciding, then the focal length of the combination is given by

$$\frac{1}{f} = \frac{1}{10} - \frac{1}{15} = \frac{3 - 2}{30}$$

or

$$f = 30 \text{ cm}$$

If the two lenses are separated, the simplest method is to treat each lens separately, first determining the image formed by the lens nearer the object and then using the image thus formed as the object for the second lens. Either the graphical method or the lens equation can be used in the solution of this problem.

Illustrative example

An object is placed 14 cm in front of a converging lens of 10 cm focal length. Another converging lens of 7 cm focal length is placed at a distance of 40 cm to the right of the first lens. Determine the position and character of the final image.

The graphical solution is shown in Figure 29-31. $A'B'$ is the image formed by the first lens and is real, inverted, and larger than the object AB. $A'B'$ is used as the object for the second lens, which forms the final image $A''B''$. The latter is virtual, larger than $A'B'$, and upright with respect to it; hence the final image is inverted with respect to the original object.

We can solve this problem by two successive applications of the lens equation. For the first lens $L_1, f_1 = 10$ cm and $s_1 = 14$ cm; hence

$$s_1' = \frac{14 \times 10}{14 - 10} = 35 \text{ cm}$$

The subscripts 1 refer to the first lens. Since the two lenses are 40 cm apart, the image $A'B'$ is 5 cm from the second lens L_2. Using $A'B'$ as the object for the second lens, then $s_2 = 5$ cm and $f_2 = 7$ cm and therefore

$$s_2' = \frac{5 \times 7}{5 - 7} = -17.5 \text{ cm}$$

that is, the final image is virtual and to the left of the second lens.

Since the second lens magnifies the image produced by the first lens, the total magnification m is the product of the magnification m_1 produced by the first lens and the magnification m_2 produced by the second lens; that is,

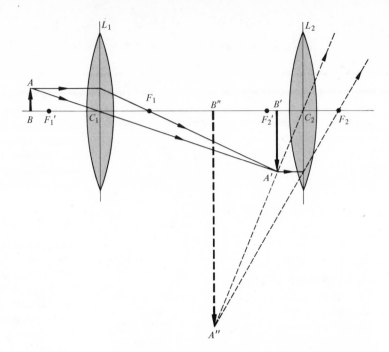

FIGURE 29-31

$$m = m_1 m_2$$

or

$$m = \frac{s_1'}{s_1} \cdot \frac{s_2'}{s_2}$$

which in this case becomes

$$m = -\frac{35}{14} \cdot \frac{17.5}{5} = -8.75$$

The negative sign shows that the final image is inverted with respect to the original; the magnification produced by this lens system is 8.75.

29-15
Spherical mirrors

A spherical mirror consists of a small section of the surface of a sphere with one side of the surface covered with a polished reflecting material, usually silver or aluminum. If the outside or convex surface is silvered we have a *convex mirror*, shown in Figure 29-32; if the inside or concave surface is silvered we have a *concave mirror*, shown in Figure 29-33. Most mirrors used commercially are made of glass, with the rear surface silvered and then covered with a coating of lacquer or paint for protection. Mirrors for astronomical telescopes and for other accurate scientific work are silvered on the front surface. In a mirror silvered on the back surface, light is reflected not only from the silvered surface but also from the front glass surface, giving rise to two images, one fainter than the other. There is also absorption of some of the light that enters the glass on its way to and from the silvered surface. With front-surface mirrors, these difficulties are avoided. In the following discussion, only front-surface mirrors will be considered.

The principal axis of the mirror is a line through the center of curvature C of the mirror and the

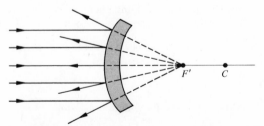

FIGURE 29-32 A beam of light parallel to the principal axis is reflected from a convex spherical mirror. The reflected beam appears to diverge from the virtual focus F'.

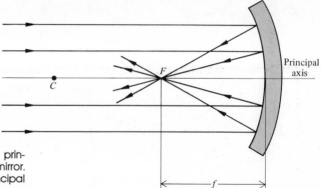

FIGURE 29-33 A beam of light parallel to the principal axis is reflected from a concave spherical mirror. The reflected beam converges toward the principal focus, F.

vertex or center of the portion of the spherical surface used. A bundle of rays parallel to the principal axis, when reflected from a concave mirror, will go toward the principal focus F. The distance from F to the vertex or center of the mirror is its focal length f. A bundle of rays parallel to the principal axis, when reflected from a convex mirror, will diverge as though it came from the principal focus F'; F' is a virtual focus.

It can readily be shown that the principal focus F of a spherical mirror is halfway between the center of curvature C and the vertex V when the angles of incidence are small. In Figure 29-34, a ray AN is drawn parallel to the principal axis CFV of a concave mirror of radius R. The line CN is a radius of the sphere and hence is normal to the spherical mirror at N. The angle ANC is the angle of incidence i, and the angle CNF is the angle of

reflection r. The angle r is thus equal to the angle i. Further, the angle NCF is also equal to i, since CN is a transversal to two parallel lines. Hence the triangle CFN is an isosceles triangle. Thus CF = FN. When the angle i is small, FN does not differ appreciably from FV, so for small angles of incidence F is halfway between the center C of the mirror and the vertex V. Since CV = R, the radius of the sphere, FV = CF = R/2. Rays parallel to the principal axis are reflected through point F, the principal focus; hence the focal length of the mirror is

$$f = R/2 \qquad (29\text{-}15)$$

29-16
Image formation by concave mirrors

Just as in the case of a converging lens, so in the case of a concave mirror, three positions of the object are chosen to illustrate the three types of images formed from real objects. In Figure 29-35 the object distance s is greater than the radius R of the sphere or greater than 2f. One ray from point A of the object is taken parallel to the principal axis; this ray is reflected from the mirror through the principal focus. A second ray from point A is drawn through the center of curvature C; this ray strikes the mirror normally and is reflected back on itself. The two reflected rays meet in A', which is the image of A. The image A'B' is real, inverted, and smaller than the object.

Figure 29-36 shows an enlarged, real, inverted image formed by a concave mirror when the object is placed between the principal focus and the center

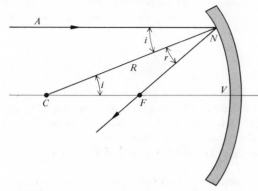

FIGURE 29-34 The principal focus of a concave mirror is halfway between the center of curvature C and the vertex V.

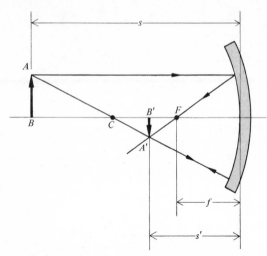

FIGURE 29-35 A real, inverted, smaller image is formed by a concave mirror when the object distance is greater than 2f.

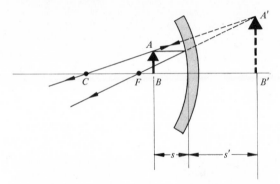

FIGURE 29-37 A virtual, erect, enlarged image is formed by a concave mirror when the object distance is less than f.

of curvature. One ray from A is taken parallel to the principal axis and is reflected back through the principal focus. The second ray from A is taken through the principal focus and is reflected back parallel to the principal axis. They meet in A', the image of A. The image $A'B'$ is real, inverted, and larger than the object.

When the object distance is less than the focal length, as illustrated in Figure 29-37, the rays from A, when reflected by the mirror, do not meet; when projected on the other side of the mirror, they

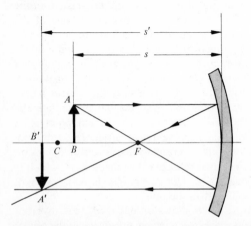

FIGURE 29-36 A real inverted enlarged image is formed by a concave mirror when the object distance is between C and F.

meet in A', which is the virtual image of A. The image in this case is virtual, erect, and enlarged.

29-17
The mirror equation

An equation relating the distance s of the object from the mirror to the distance s' of the image from the mirror, in terms of its focal length f or its radius of curvature R, can be derived in a manner analogous to that used in deriving the lens equation. Referring to Figure 29-38, since triangles ABC and $A'B'C$ are similar, we can write

$$\frac{AB}{A'B'} = \frac{BC}{B'C} = \frac{s-R}{R-s'}$$

If we drop a perpendicular from D to E onto the principal axis, then, since the triangles DEF and $A'B'F$ are similar,

$$\frac{DE}{A'B'} = \frac{EF}{B'F}$$

Now $DE = AB$, and, if we restrict this discussion to rays that make very small angles with the principal axis and if the section of the sphere used for a mirror is very small, then E may be considered to coincide with the vertex V, so $EF = f$ and $B'F = s' - f$.

Substituting these values in the above equation, we get

$$\frac{AB}{A'B'} = \frac{f}{s'-f}$$

Equating the two expressions for the ratio $AB/A'B'$, we have

$$\frac{s-R}{R-s'} = \frac{f}{s'-f}$$

Now, since

$$R = 2f$$

we can write

$$\frac{s-2f}{2f-s'} = \frac{f}{s'-f}$$

Clearing fractions and rearranging terms, we obtain the equation

$$s'f + sf = ss'$$

and, dividing each term by $ss'f$, we get

$$\frac{1}{s} + \frac{1}{s'} = \frac{1}{f} \qquad (29\text{-}16)$$

for the mirror equation. It will be noticed that this is identical with the lens equation; however, one must be careful about the sign convention. In the mirror equation as in the lens equation, the light is assumed to come from the left; the distance s is

positive when the object is to the left of the mirror. Since a mirror reflects light, the image distance s' will also be considered positive when the image is on the left. If s' is negative, the image is virtual and on the right, that is, behind the mirror. The focal length f is positive for a concave mirror and negative for a convex mirror. *The positive and negative signs are to be used only when numerical values are substituted for s, s', and f.*

The mirror equation may also be written in terms of the radius R of the spherical mirror. Since $f = R/2$, the mirror equation becomes

$$\frac{1}{s} + \frac{1}{s'} = \frac{2}{R} \qquad (29\text{-}16a)$$

The magnification m produced by a mirror is defined in the same manner as that produced by a lens and is given by

$$m = -\frac{s'}{s} \qquad (29\text{-}9)$$

Illustrative example

An object is placed 50 cm from a concave mirror of 30 cm radius. Determine the position and character of the image.

The graphical solution of this problem is shown in Figure 29-35. The focal length f of this mirror is $R/2$ or 15 cm. If we solve Equation 29-16 for the distance s', we get

$$s' = \frac{sf}{s-f}$$

Since $s = 50$ cm and $f = 15$ cm,

$$s' = \frac{50 \times 15}{50 - 15} = 21.4 \text{ cm}$$

The magnification produced by this mirror is

$$m = -\frac{s'}{s} = -\frac{21.4}{50} = -0.43$$

The image is 21.4 cm from the mirror; it is real, inverted and smaller than the object.

29-18
Image formation by convex mirrors

No matter where the object is placed in front of a convex mirror, its image will always be virtual,

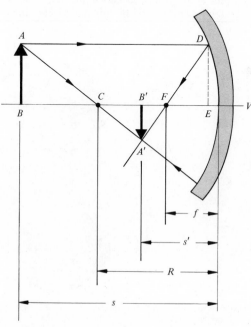

FIGURE 29-38

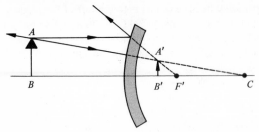

FIGURE 29-39 A virtual, erect, smaller image is formed by a convex mirror for any position of a real object.

erect, and smaller than the object, as shown in Figure 29-39. A ray from A parallel to the axis is reflected back as though it came from the virtual focus F'. A second ray, taken normal to the mirror, is reflected back on itself. These two rays do not meet; when projected behind the mirror they meet in point A', which is the virtual image of A. The image $A'B'$ is virtual, erect, and smaller than the object.

Suppose that the focal length of the convex mirror is -15 cm and that the object is placed 45 cm in front of the mirror. The distance of the image from the mirror is

$$s' = \frac{sf}{s-f} = \frac{45(-15)}{45-(-15)} = -11.25 \text{ cm}$$

The magnification produced by this mirror is

$$m = -\frac{s'}{s} = -\frac{-11.25}{45} = +0.25$$

The image is virtual and is situated 11.25 cm behind the mirror. Since the magnification is positive, the image is upright and, in this case, is one fourth the size of the object.

29-19
The magnifying glass

When an object is to be examined minutely, it is usually brought as close as possible to the eye. The closer it is brought, the larger is the *visual angle*, a, that it subtends at the eye, and the larger is its image on the retina (see Figure 29-40).

In the human eye, the light is focused on the retina R in Figure 29-41, by the combination of the cornea C, the aqueous fluid A, and the lens L. This compound optical instrument has a variable focal length because the eye muscles can distort

the radii of curvature of the lens. This permits people to see objects within a certain *range of accommodation*, the distance between the *near point* and the *far point*. The far point is that point which is sharply imaged on the retina when the eye muscles are in their most relaxed state, whereas the near point is that point which is sharply imaged when the eye muscles exert their greatest effort. For the normal (or *emmetropic*) eye the far point is at infinity and the near point is about 25 cm from the eye. We shall, in the subsequent discussion, assume that objects or images can be examined at this distance, but no closer, and that this imposes a limit on the size of the retinal image and the smallness of the detail that can be seen. (The nearsighted or *myopic* eye has a far point that is at a finite distance from the eye, usually because the eye itself is too long. The farsighted, or *hyperopic*, eye is usually too short. With age, most people lose some of the ability to focus on nearby objects. This normal farsightedness is called *presbyopia*.)

The visual angle can be increased and a magnified retinal image obtained with the aid of a converging lens used as a *magnifying glass*. The most common way of using a magnifying glass is to place the object at or within the principal focus F of the lens so that parallel light enters the eye from each point of the object, as shown in Figure 29-42; that is, the image formed by this lens is at infinity.

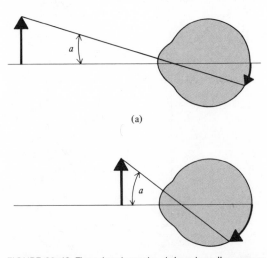

(a)

FIGURE 29-40 The visual angle determines the apparent size of any object.

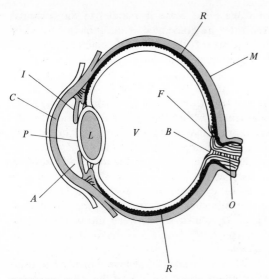

FIGURE 29-41 Longitudinal section of a human eye.

Either a hyperopic or emmetropic eye can use such an image.

The *angular magnification M* of a converging lens used in this manner is the ratio of the two visual angles *a* and *A*, or

$$M = \frac{A}{a} \qquad (29\text{-}17)$$

For the emmetropic eye, to a close approximation,

$$a = \frac{S}{25 \text{ cm}} \qquad (29\text{-}18)$$

where *S* is the size of the object in centimeters and the near point is expressed in centimeters, and

$$A = \frac{S}{f} \qquad (29\text{-}19)$$

where *f* is the focal length of the magnifying lens. Therefore,

$$M = \frac{A}{a} = \frac{25 \text{ cm}}{f} \qquad (29\text{-}20)$$

if *f* is expressed in centimeters, or

$$M = \frac{10 \text{ in.}}{f} \qquad (29\text{-}20a)$$

if *f* is expressed in inches.

The *power P* of a lens is expressed in *diopters*, where

$$P = \frac{1}{f}$$

when *f*, the focal length, is measured in meters. A converging lens has positive power and a diverging lens has negative power.

Another way of using a converging lens as a magnifier is to place the object between the principal focus and the lens, as shown in Figure 29-43. In this case the image will be virtual and will be in focus when it is at the near point—that is, 25 cm from the eye. If *S'* is the size of the image, the visual angle it forms at the eye is *S'*/25 cm. If *S* is the size of the object, then when it is viewed from the near point by the unaided eye, the visual angle formed by it is *S*/25 cm. The angular magnification *M*, which is the ratio of these two visual angles, is thus

$$M = \frac{S'}{S}$$

Hence, in this particular case, the angular magnification is the same as the linear magnification *m*, so we can write

$$M = -\frac{s'}{s}$$

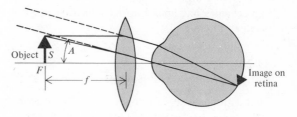

(a)

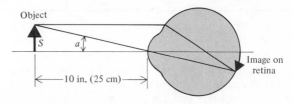

(b)

FIGURE 29-42 Use of a converging lens as a simple magnifying glass.

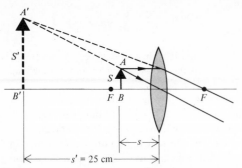

FIGURE 29-43 Use of a magnifying glass.

The value of s can be found from the lens equation, yielding

$$s = \frac{s'f}{s' - f}$$

Hence

$$M = -\frac{s' - f}{f} = \frac{-s'}{f} + 1$$

Now

$$s' = -25 \text{ cm}$$

and thus

$$M = \frac{25 \text{ cm}}{f} + 1 \qquad (29\text{-}21)$$

A simple converging lens used as a magnifier will produce a magnification of 25 cm/f when the object is placed at the principal focus and will produce a magnification of (25 cm/f) + 1 when the object is placed inside the focus to produce an image at the near point. For intermediate cases the magnification will lie between these two values. For example, if the focal length of a thin lens used as a magnifier is 5 cm, its magnification will be 5, or 6, or some intermediate value. For lenses of very short focal length and therefore of high power, there is very little difference between the two extreme values and thus the magnification may be taken as 25 cm/f.

29-20
The astronomical telescope

The astronomical telescope, used for viewing distant objects, consists of two converging lenses: an objective lens L_1, and an eye lens L_2. Points from a distant object can be considered as sending sets of parallel rays to the objective. If the rays

from one point come in parallel to the principal axis, they are converged at the principal focus of the objective. If the rays from some other point of the object come in inclined at some small angle to the principal axis, they are focused in a point which is in the *focal plane* of the objective. When the eye lens L_2 is placed at a distance equal to its focal length f from the principal focus of the objective, rays that come from the image in the focal plane of the objective will leave the eye lens as parallel rays, as shown in Figure 29-44. In other words, parallel rays entering the objective lens leave the eye lens as parallel rays. This is the normal adjustment of the telescope for distant objects. The distance between the two lenses is, of course, $F + f$, where F is the focal length of the objective.

It will be observed that rays which come from object points above the axis of the telescope enter the eye as though they came from below the axis. This type of telescope therefore gives an inverted and reversed image. To see this image, the eye must be focused for infinity. If the distant object subtends a visual angle A as seen with the unaided eye, it will subtend a visual angle B when viewed with the aid of the telescope. The ratio of these angles determines the ratio of the retinal images, and is therefore the angular *magnification M* produced by the telescope. Thus

$$M = \frac{B}{A} \qquad (29\text{-}22)$$

In Figure 29-44 the limiting rays from the object enter the center of the objective lens, making an angle A with each other; upon emerging from the objective lens, they still make an angle A with each other. These two rays go to the end points of the image I, a distance F from the objective. The rays from the end points of the image I make an angle B with each other at the center of the eye lens and when they enter the eye. This image is at a distance f from the eye lens. From Figure 29-44 it can be seen that

$$M = \frac{B}{A} = \frac{F}{f} \qquad (29\text{-}23)$$

that is, the magnification of a telescope is the ratio of the focal lengths of the objective lens and the eye lens. By using eyepieces of different focal lengths, different magnifications can be obtained with the same objective.

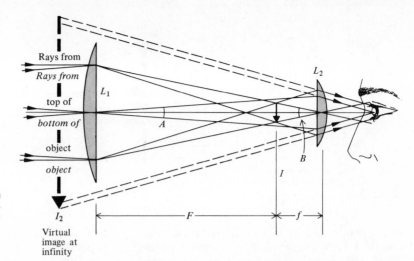

Rays from

Rays from

top of

bottom of

object

object

L_1

A

L_2

B

I

F

f

I_2

Virtual
image at
infinity

FIGURE 29-44 Optical diagram of a simple astronomical telescope.

29-21
The terrestrial telescope

Since the astronomical telescope produces inverted and reversed images, it is unsuitable for most terrestrial uses. The astronomical telescope can be modified to produce an erect image by inserting a converging lens between the focal plane of the objective and the eye lens. If f_1 is the focal length of this converging lens, it should be placed at a distance $2f_1$ from the focal plane of the objective to produce an erect image, without magnification, at a distance $4f_1$ from the focal plane. The eyepiece is then placed at a distance f from this image. The length of the terrestrial telescope formed this way has been increased by four times the focal length of the erecting lens. Usually, terrestrial telescopes use an erecting system consisting of two converging lenses with a diaphragm or stop between them to correct for spherical aberration.

A better way to make a terrestrial telescope is

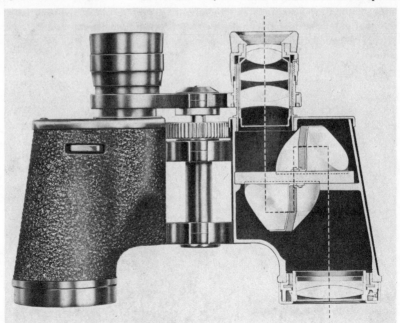

FIGURE 29-45 Prism binoculars. Two 90° prisms are used for reversing and inverting the image. (Courtesy of Bausch and Lomb.)

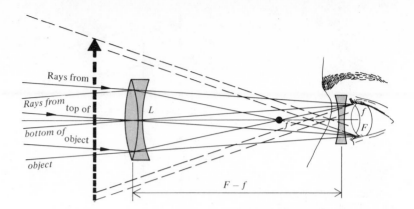

FIGURE 29-46 Optical diagram of a Galilean telescope or opera glass.

to use two 90° prisms, one for inverting the image and the other for reversing it, as shown in Figure 29-45. This method is used for getting erect images in prism binoculars. A pair of prism binoculars consists of two telescopes joined together. Each telescope contains a pair of right-angled prisms arranged to invert and reverse the image formed by the telescope.

A third method for producing erect images is to make a Galilean type of telescope. This uses a diverging lens for the eyepiece, as shown in Figure 29-46. The distance between the objective and the eyepiece is $F - f$ where F is the focal length of the objective and f is the numerical value of the focal length of the eyepiece. Parallel rays that enter the objective are converged toward its focal plane; however, the diverging lens inter-

cepts them and deviates them so that they emerge as parallel rays. From the figure, it will be noted that rays that come from object points above the axis appear to come from above the axis after passing through the telescope; the image as seen by the eye, therefore, is erect. Galilean telescopes are used extensively in opera glasses.

29-22
The compound microscope

The compound microscope consists of two systems of converging lenses; in the diagram sketched in Figure 29-47 they are shown as single converging lenses. The objective lens L_1 has a very short focal length and the object is placed very close to, but just outside, the principal focus of this lens. A real,

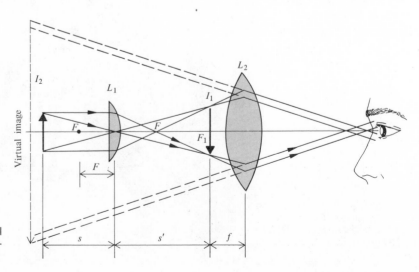

FIGURE 29-47 Simplified optical diagram of a compound microscope.

inverted image I_1 is formed at a distance s' from the objective lens. The eye lens L_2 is then used as a simple magnifying glass for viewing this image. If the eye is focused for parallel light, the eye lens L_2 is moved until the image I_1 is at its principal focus F_2, so parallel rays enter the eye as though coming from a virtual image I_2 at infinity.

The magnification m_1 produced by the objective lens is

$$m_1 = -\frac{s'}{s} \qquad (29\text{-}24)$$

where s is the distance of the object from the lens. To a very close approximation, the object distance s can be taken as the focal length F of the object lens, and therefore

$$m_1 = -\frac{s'}{F} \qquad (29\text{-}25)$$

The magnification m_2 produced by the eye lens is

$$m_2 = \frac{25 \text{ cm}}{f} \qquad (29\text{-}26)$$

where f is the focal length of the eye lens in centimeters. Hence the total magnification M of the compound microscope is

$$M = m_1 m_2 = -\frac{s'}{F} \times \frac{25 \text{ cm}}{f} \qquad (29\text{-}27)$$

Magnifications of several hundred diameters are common with compound microscopes. Thus, if the magnification is 400 diameters, the area of the object is magnified 160,000 times. The object must therefore be strongly lighted to provide sufficient illumination so that the image can be seen. This is usually accomplished by focusing onto the object light from a source by means of a mirror and a condensing lens (see Figure 29-48).

Questions

1. Explain why white light, after passing through a plate of glass with parallel faces is not dispersed into its component colors.

2. A beam of light is converging toward a screen. A plane, parallel plate of glass is put in the path of this converging beam. In which direction will the point of convergence be shifted? Justify your answer with the aid of a diagram.

3. A fish in a lake looks at a fisherman standing on the shore. Does the fisherman appear to the fish to be taller than, shorter than, or equal to his full height?

4. Why does a swimming pool appear shallower than it really is?

5. Totally reflecting prisms are usually designed so that the light enters and leaves the prism at an angle of 90° to a surface. Can you account for this?

6. Why are totally reflecting prisms preferred to silvered mirrors in optical instruments?

7. A front-surface mirror is a mirror formed by placing a metal surface, usually aluminum, on the front of a flat plate of glass or plastic. What advantage does such a mirror have over the common back-surface mirror?

8. A half-silvered mirror is one in which the metal coating is thin enough to transmit some light as well as reflecting some. If such a mirror is placed in the wall separating a well-lighted room from a dark room, explain how such a mirror can be used as a window into one room and a mirror in that same room.

9. What is the smallest length of mirror you can use to see your whole figure imaged in it?

10. Using a plane mirror near a sheet of paper, print in capital, block letters the words CHOICE QUALITY so that they appear upright when viewed in the mirror. Examine the appearance of these words on the sheet of paper; account for the difference in appearance of these two words.

11. Consider an object that is on the line midway between two mirrors that make a right angle between them. Then consider the image of this object as formed by one of the mirrors as the virtual object for the other mirror. (a) Locate this image of the virtual object. The three images and the object should form the corners of a square. (b) Try this with two small mirrors. Also use different angles and see the formation of multiple images. (c) From this, attempt to explain the construction of a kaleidoscope.

12. Consider two vertical mirrors that are at right angles to each other and consider a horizontal

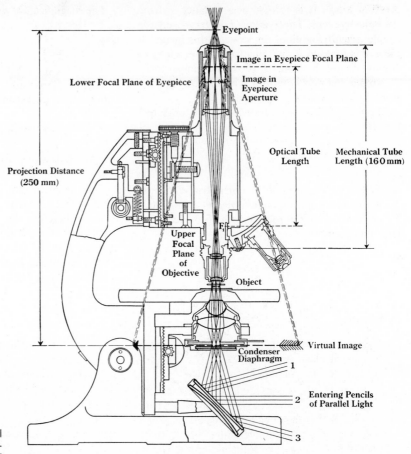

Eyepoint

Image in Eyepiece Focal Plane

Lower Focal Plane of Eyepiece

Image in Eyepiece Aperture

Optical Tube Length

Mechanical Tube Length (160 mm)

Projection Distance (250 mm)

F

Upper Focal Plane of Objective

Object

Virtual Image

Condenser Diaphragm

1

2   Entering Pencils of Parallel Light

3

FIGURE 29-48 Complete optical system of a compound microscope. (Courtesy Bausch and Lomb.)

Optical System of a B&L Laboratory Type Microscope.

ray that strikes one mirror and is reflected to the second mirror. Show that the ray will emerge from the second mirror parallel to its initial direction, no matter what the initial direction.

13. Consider three mirrors that form the corner of a cube. Extend the argument of Question 12 to show that any ray that strikes this system will emerge parallel to its initial direction. This configuration of reflecting surfaces is used to make reflectors such as those used on bicycles and reflecting materials such as those used to make highway signs.

14. From Figure 29-15, show that the image formed in a plane mirror is **(a)** just as far behind the mirror as the object is in front of the mirror and **(b)** the same size as the object.

15. Most truck drivers use both plane mirrors and convex mirrors as side-view mirrors, for looking to the rear. Discuss the advantages of each.

16. As you approach a concave mirror, such as a shaving or makeup mirror, your inverted image becomes blurred and then becomes a magnified erect image. At what point does the blurring occur?

17. A photographer has a lens of focal length 150 mm, a lens of 50 mm, and a lens of 30 mm. Which is the telephoto lens, which the wide-angle lens, and which the normal lens?

18. The f number of a photographic lens is the focal length divided by the diameter of the aperture. Usually the f numbers of a camera, the f stops, change by a factor of $\sqrt{2}$. Explain why the photographer must double the exposure time when he increases from one f stop to the next larger one.

19. For a real object, can (a) a converging lens form a virtual image that is smaller than the object and (b) a diverging lens form a virtual image that is smaller than the object?

20. Watchmakers often use dual-magnification *loupes*; the high-magnification lens is used for inspection and the low-magnification lens is used while actually working on the watch. Why not use just one or the other?

21. One of the important characteristics of an optical system is the depth of field, which is the thickness of the object that is in acceptable focus for a given position of the object, lens, and film or observer's eye. (a) How would you expect the depth of field to change as the size of the aperture changes? (b) How would you expect the depth of field to change as the magnification changes?

22. Show that myopia might be corrected with a diverging lens.

23. Show that hyperopia might be corrected with a converging lens.

24. Why do middle-aged (and older) people often require bifocals or trifocals?

25. Eyestrain usually is caused by requiring the eye muscles to remain tensed for an extended period of time. Discuss the significance of this for the user of (a) a simple magnifier, (b) a telescope, and (c) a microscope.

26. Discuss the similarities and differences between a telescope and a microscope.

27. Very large astronomical telescopes use mirrors instead of lenses for their objectives. Can you suggest some of the advantages of a mirror over a lens as a telescope objective?

28. Suppose that you are looking at yourself in a concave mirror and your head is located between the focal point and the center of curvature. As was shown in Figure 29-36, the mirror forms a real, inverted image. However, your eyes intercept the light before it can converge to form an image, so what you are looking at is a virtual object. Show that this object is erect; that is, show that in order to see your chin you have to look down and to see the top of your head you have to look up. The image you see when your head is beyond the center of curvature is erect because then you can look at the rays diverging from the real image.

Problems

Note: Unless otherwise specified, take the index of refraction of water to be 1.33 and the index of refraction of glass to be 1.5 in the following problems.

1. A beam of light strikes the surface of water at an angle of 30° with the normal. Find (a) the direction of the beam in the water, (b) the direction of the reflected beam, and (c) the speed of light in the water.

2. A beam of light strikes the surface of a block of glass at an angle of 15° with the surface of the glass. Find (a) the direction of the reflected beam and (b) the direction of the refracted beam.

3. What can you conclude about the index of refraction from the order of colors given in Figure 29-2?

4. Using Table 29-1, find (a) the velocity in water of light of wavelength 5893 Å, and (b) the velocity in flint glass of light of wavelength 4047 Å.

5. The actual depth of a swimming pool is 15 ft. How deep does it appear to a person standing on the edge of the pool?

6. A certain rough diamond in the shape of a cube 2 cm on a side has a very small black imperfection in its geometric center. How far from the surface will the imperfection appear to be if viewed under yellow light of wavelength 5893 Å?

7. What is the critical angle for yellow sodium light, wavelength 5893 Å, in diamond?

8. What is the smallest possible value of the index of refraction for the glass used in making a totally reflecting prism of the type shown in Figure 29-10?

9. A 5-ft 2-in. person and a 6-ft person are roommates, (a) What is the shortest mirror in which

both of them can see their entire figures and (b) how should the mirror be placed on the wall?

10. Repeat the ray-tracing analysis of Figures 29-25 through 29-27 for a diverging lens.

11. Find the magnification of a converging lens in terms of the object distance s and the focal length f of the lens.

12. A point source is 25 cm from a converging lens of focal length 10 cm. Where is the image?

13. A converging lens of focal length 18 cm forms a real image 24 cm from the lens. Where is the object?

14. A student focuses the image of the laboratory windows on a piece of paper with a lens and estimates the distance between the lens and the paper to be the focal length of the lens. The distance between the lens and the windows is 6 m. What distance will the student estimate for the focal length of the lens if the true focal length is (a) 5 cm and (b) 1 m?

15. An observer views an object through a diverging lens of focal length 28 cm. The object is 15 cm from the lens. Locate the image.

16. Glass of index 1.55 is used to make a lens. Both surfaces are identical convex surfaces. Find the radius of curvature if the focal length of the lens is 30 cm.

17. A symmetric double convex lens is made of glass of index 1.48 and has a focal length of 18 cm. Find the radii of curvature.

18. A plano-convex lens has a focal length of 30 cm. The radius of curvature of the convex side is 17 cm. Find the index of refraction of the glass.

19. A converging lens of focal length 15 cm is placed in contact with a diverging lens. The combined focal length is 34 cm. Find the focal length of the diverging lens.

20. An object is placed 25 cm in front of a concave mirror, the radius of curvature of which is 55 cm. Locate the image.

21. An object is placed 25 cm in front of a convex mirror, the radius of curvature of which is 55 cm. Locate the image.

22. Water is placed in a flat-bottomed glass jar. Trace the path of a ray of light that strikes the surface of the water at 30° with respect to

the normal and passes through the water and into the glass bottom. Find (a) the angle of refraction in the water, (b) the relative index of the water and glass, and (c) the angle of refraction in the glass.

23. The angle between the two refracting surfaces of a crown-glass prism is 60°. Trace the path of a ray of yellow sodium light through the prism, if the angle of incidence at one side is 50°. Find (a) the angle of refraction at the first surface and (b) the angle of deviation.

24. A hollow glass prism is filled with water; the cross section of the prism is an equilateral triangle. Trace a ray of light through the prism when the angle of refraction at the first surface is 25°. (Ignore the presence of the glass.)

25. A low-power microscope is used to measure the index of refraction of a plate of glass that is 1.2 cm thick. The microscope is first focused on the bottom of the glass; the microscope tube is then raised 0.8 cm and is found to be in focus on the top surface of the glass. Determine the index of refraction of the glass.

26. When the refracting angle A of a prism is small, the angle of minimum deviation D_m is also small. Show that for a prism of small angle, the angle of deviation is given by

$$D_m = A(n - 1)$$

Use the fact that the sine of the angle is equal to the angle for small angles measured in radians.

27. An object 5 mm long is placed 40 cm in front of a converging lens whose focal length is 25 cm. Find (a) the position and (b) the size of the image. (c) Is the image erect or inverted? (d) Is the image real or virtual?

28. The lens of Problem 27 is replaced by a diverging lens of the same magnitude of focal length. Repeat Problem 27 for this lens (with the same object and object distance).

29. An object 2 cm long is placed 10 cm in front of a converging lens of focal length 30 cm. Find (a) the position, (b) the size, and (c) the character of the image.

30. Repeat Problem 30 with the lens replaced by a diverging lens of focal length 30 cm.

31. A projector lens is placed 3 cm from a slide and an image is focused on a screen 7 m from

the lens. Find (a) the focal length of the lens and (b) the magnification of the image.

32. An illuminated object and a screen are 6 m apart. A converging lens placed between them forms an image on the screen that is 20 times as large as the object. Determine (a) the distance between the object and lens and (b) the focal length of the lens.

33. An object is placed in front of a converging lens of focal length 25 cm. The real image is four times as large as the object. How far from the lens is the object?

34. The lens of Problem 33 forms a virtual image of an object, where the image is four times as large as the object. Where is the object?

35. A thin-walled spherical flask with a radius of 8 cm contains a transparent liquid. A narrow beam of parallel light falls normally on the surface of the sphere and is observed to be focused on a point on the surface of the flask opposite the entrance surface. Neglecting the refraction through the thin-walled surface, determine (a) the index of refraction of the liquid in the flask and (b) trace several rays through the liquid.

36. An object is placed 50 cm from a convex mirror whose radius is 30 cm. Determine (a) the location of the image and (b) the magnification. (c) Repeat parts (a) and (b) for an object placed 50 cm from a concave mirror having a radius of 30 cm.

37. A person 5 ft 7 in. tall stands 8 ft away from a garden ornament, which is a reflecting sphere, 1 ft in diameter. (a) Locate the image of the person. (b) Find the size of the image. (c) Describe the image.

38. A dentist uses a concave mirror whose radius of curvature is 1 cm. The mirror is held 20 mm from a pinhole cavity. Describe the image.

39. A convex mirror whose radius of curvature is 30 cm is placed on the wall of a room. The wall opposite the mirror is 12 ft away and a person stands 4 ft in front of the mirror. Locate the position of the image of (a) the person and (b) the opposite wall. (c) From this, attempt to explain why such mirrors were popular decorations in eighteenth century homes.

40. A concave spherical mirror has a radius of curvature of 20 cm. Make a diagram of the mirror to scale and show rays incident on the mirror that are parallel to the axis and at distances of 1, 2, 3, 4, 5, and 10 cm from the axis. Use a protractor to construct the reflected rays.

41. Repeat Problem 40 for a convex mirror of the same curvature.

42. The distance between the lens and retina for the normal eye is about 1.7 cm. (a) Find the range of focal lengths through which the optical system of the eye can be adjusted. (b) Give the equivalent powers of the optical system.

43. A person can adjust the power of the lens of his eye from 58 to 64 diopters. With the eye muscles relaxed he can see a very distant object, such as a star, clearly. (a) Find the distance between lens and retina and (b) locate the near point.

44. A middle-aged man has a near point that is 65 cm, so he must hold a book at arm's length to read it. (a) Assuming that the retina is 1.7 cm distant from the lens, find the power of the eye lens. (b) What power eyeglass lens will reduce the near point to a normal, comfortable 25 cm, assuming that the two lenses, eye and eyeglass, can be treated as if they were in contact?

45. A small laboratory telescope has an objective whose focal length is 22 cm and an eye lens whose focal length is 2.5 cm. (a) What is the magnification of the telescope? (b) How far apart are the lenses?

46. An astronomical telescope has an objective lens whose focal length is 80 cm. The telescope is 82 cm long when adjusted for parallel light. (a) What is the magnification of the telescope? (b) What is the focal length of the eye lens?

47. A compound microscope has an objective of 4.5 mm that forms an image of an object placed 4.8 mm from the lens. The eye lens has a focal length of 2 cm. (a) Locate the image formed by the objective lens. (b) What is the magnification of the microscope when it is adjusted so that parallel light enters the eye?

48. An object 5 mm long is placed 10 cm from a converging lens of focal length 25 cm. A second lens of focal length 12 cm is placed 20 cm beyond the first lens. Find the position and size of the image.

49. An object is placed 20 cm in front of a converging lens of focal length 10 cm. A diverging lens, also of focal length 10 cm, is placed 5 cm

behind the first lens. Find the position and size of the image if the object is 1 cm long.

50. Two lenses, of focal lengths 25 cm and −15 cm, respectively, are 10 cm apart. Locate the image of an object 100 cm away from the closer lens **(a)** if the closer lens is the converging lens and **(b)** if the closer lens is the diverging lens.

51. A converging lens forms an image on a screen 60 cm from it. A thin, diverging lens is interposed between them at a distance of 40 cm from the converging lens. It is now found necessary to move the screen 15 cm away from the lens in order to produce a sharp image.

Determine the focal length of the diverging lens.

52. A cylindrical glass rod 50 cm long has its ends rounded so that they are convex spherical surfaces, each of radius 12 cm. The index of refraction of the glass is 1.60. A parallel beam of light enters one end of the rod, parallel to the axis of the cylinder. **(a)** Find the focal point of the first surface and **(b)** trace a few rays through the rod.

53. Prove that the distance between an object and the real image formed by a thin, converging lens is always larger than four times the focal length of the lens.

electromagnetic waves

chapter 30

30-1
Introduction

The study of electromagnetic waves has been one of the most important parts of physics throughout its long history. Until the middle of the nineteenth century, physicists studied optics without knowing that light waves are electromagnetic waves. Still, a great deal of experimental information was obtained about the behavior of light waves, from which detailed and accurate models were constructed (see Figure 30-1). Simultaneously, physicists were also studying electricity and magnetism. This culminated in Maxwell's brilliant theoretical work, which showed that electromagnetic waves were identical to the known properties of light. Maxwell showed, then, that light waves were electromagnetic waves of a particular range of frequencies.

Subsequently, the study of electromagnetic waves followed three main routes. First, efforts were made to find and, if possible, produce and control electromagnetic waves of frequencies other than those of light waves. From these attempts came the development of radio, radar, and television, and the discovery of X rays, gamma rays, and cosmic rays. The result is that almost the entire conceivable spectrum of electromagnetic waves has been observed, as shown in Figure 30-2. The second major route of study has been an investigation of the interaction of electromagnetic waves with matter. From this has come our present understanding of the structure of matter; in a sense, solid-state physics, atomic and nuclear physics, and quantum theory have developed from an attempt to understand electromagnetic waves. The third major route of study, that of attempting to

FIGURE 30-1 James Clerk Maxwell (1831–1879). Mathematician and physicist. He put the laws of electricity in mathematical form. These equations form the foundations of electromagnetic theory. He predicted the existence of electromagnetic waves and originated the electromagnetic theory of light, and also made outstanding contributions to the molecular theory of heat. (Courtesy of *Scripta Mathematica*.)

understand the propagation of electromagnetic waves, has led to the development of the theory of relativity, which we have described earlier.

A related idea is that all our knowledge of astronomy is based on an analysis of the electro-magnetic radiation received at the earth. Thus, our understanding of the structure of the extraterrestrial universe is based, in large measure, on our understanding of the theory of electromagnetic radiation.

As is shown in Figure 30-2, the electromagnetic spectrum is divided into several major regions (although the boundaries between the regions are generally imprecise): radio, microwaves (or radar waves), infrared, visible light, ultraviolet, X rays, and high-energy photons. In principle, all of these waves demonstrate all of the properties of waves that were described in Chapter 21. However, there are practical limitations. For example, the wavelength of some cosmic-ray photons is smaller than the size of atoms, so it is difficult to imagine constructing solid walls or apertures for such waves. Similarly, there are limitations on the experiments that can be performed on electromagnetic waves whose wavelengths are as large as or larger than the earth. Nevertheless, we will in this chapter describe the general properties of electromagnetic waves without restricting the discussion to one portion of the spectrum. The interaction of electromagnetic waves with matter will be left for the section on modern physics.

Maxwell predicted the existence of electromagnetic waves in 1864. In 1887 Heinrich Hertz (1857–1894) succeeded in producing these waves by setting up electrical oscillations in a circuit containing a capacitor and an inductor. The circuit consisted of a capacitor of capacitance C, a coil of inductance L, and a spark gap S, as shown in Figure 30-3. When a sufficiently high voltage was applied to the terminals of the spark gap to produce a spark in the air between the terminals, electrical oscillations were set up in this circuit of frequency

$$f = \frac{1}{2\pi\sqrt{LC}} \qquad (30\text{-}1)$$

and electromagnetic waves of this frequency were radiated by these oscillations. The spark-gap

FIGURE 30-2 The complete electromagnetic spectrum. Because of the wide range in wavelengths, the latter have been drawn to a logarithmic scale.

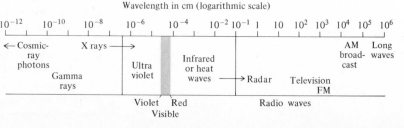

Wavelength in cm (logarithmic scale)

10^{-12} 10^{-10} 10^{-8} 10^{-6} 10^{-4} 10^{-2} 10^{-1} 1   10 10^2 10^3 10^4 10^5 10^6

← Cosmic-ray photons   X rays →

Gamma rays

Ultra violet

Infrared or heat waves

→ Radar

Television FM

AM broadcast   Long waves

Violet   Red
Visible

Radio waves

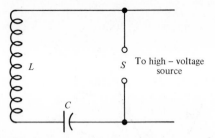

FIGURE 30-3 A spark-gap oscillator circuit.

method for producing oscillations in a circuit was in practical use for many years but has since been replaced by vacuum-tube and transistor methods.

Hertz (for whom the unit of frequency is named) detected these waves by using an antenna that was simply a piece of wire with small knobs on the ends and bent into a circle, forming a spark gap between the knobs. The electromagnetic waves caused a spark in the gap. Hertz was able to show that the waves were reflected by metal, that they could be focused by a large concave metallic reflector, and that they demonstrated many of the other properties of light. He was able to measure the wavelength of the waves and, with the frequency given by Equation 30-1, found that the speed of the waves was that of light, as prediced by Maxwell.

In our study of sound it was shown that resonance can occur between any two elastic bodies that can vibrate with the same natural frequency. It was suggested by Sir Oliver Lodge in 1890 that the same type of phenomenon could be produced with electric circuits that had the same natural frequency. He succeeded in setting up oscillations in one circuit containing a spark gap and in receiving the electromagnetic waves radiated from it in a neighboring circuit by *tuning* it to the same frequency. He detected this wave by observing a spark produced across a small spark gap in the same circuit. If the first circuit has an inductance L_1 and capacitance C_1, and the second circuit has an inductance L_2 and capacitance C_2, then the condition for electric resonance is

$$\frac{1}{2\pi\sqrt{L_1 C_1}} = \frac{1}{2\pi\sqrt{L_2 C_2}}$$

from which

$$L_1 C_1 = L_2 C_2 \qquad (30\text{-}2)$$

When the condition represented by Equation 30-2 is satisfied, the two circuits are said to be in resonance. This idea is at the basis of the method

for *tuning* a radio receiver or television receiver. When an electromagnetic wave of frequency f strikes a conductor, it sets up an EMF that produces a current in the conductor. If the conductor is part of a circuit containing inductance L and a variable capacitor, then the circuit can be tuned by varying the capacitance until the resonance frequency of the circuit is equal to the frequency of the incident electromagnetic waves.

30-2
The nature of electromagnetic waves

When charges are accelerated, they produce electric and magnetic fields, which propagate in the form of electromagnetic waves. These waves transport energy and momentum that has been transferred to them from the accelerating charges.

For example, if a beam of high-energy electrons is permitted to strike a metal plate, the electrons will be brought to rest. This deceleration of the electrons will cause the emission of electromagnetic waves, usually in the form of X rays. Similarly, the production of radio waves, microwaves, ultraviolet light, visible light, infrared light, and gamma rays can be traced back to the acceleration of charged particles. Generally, the electromagnetic waves are produced by the acceleration of a large number of charged particles in a macroscopic body. Usually, the radiation contains a large number of frequencies, and the electric and magnetic fields are essentially random in direction.

It is possible and useful to construct simple electromagnetic waves by restricting the acceleration of the charges. For example, in a straight wire, or antenna, electrons can be forced to oscillate with simple harmonic motion. This antenna will then radiate electromagnetic waves of the same frequency as that of the simple harmonic motion of the electrons. Further, if the wire is surrounded by a uniform medium (such as empty space or air), the fields will be confined to directions determined by the direction of the antenna.

Consider the antenna in Figure 30-4, which is shown as parallel to the z axis. Such an antenna will radiate in almost all directions, but the figure shows only the fields that constitute the wave propagated along the y axis. The fields are shown as they are at one instant of time. At each point along the direction of propagation, there exist both electric and magnetic fields. The electric field at any point is parallel to the z axis, and its magnitude and direction vary along the direction of propagation in the

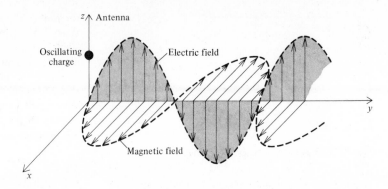

FIGURE 30-4 Propagation of plane electromagnetic waves along the y axis. The waves arise from the oscillation of charges in an antenna on the z axis. The electric and magnetic fields are at right angles to each other; the electric field is wholly in the yz plane and the magnetic field is wholly in the xy plane.

form of a sine curve. Similarly, the magnetic field at any point is parallel to the x axis, and its magnitude and direction vary sinusoidally along the direction of propagation. Further, the fields are in phase; that is, both fields have maximum values at the same point, and both are equal to zero at the same point. At subsequent instants of time, the pattern of fields moves along the y axis away from the antenna. Thus, the electric field at any point varies sinusoidally with time between a maximum in the positive z direction and a maximum in the negative z direction. Similarly the magnetic field at any point varies from a maximum in the positive x direction to a maximum in the negative x direction.

The waves propagated in other directions have similar geometric relationships. In general,

1. Both fields are perpendicular to each other and perpendicular to the direction of propagation, where the relative directions of the two fields and the direction of propagation are the same as those in Figure 30-4.
2. The electric field is confined to the plane formed by the direction of propagation and the antenna; the magnetic field is perpendicular to that plane.
3. The fields are in phase with each other and vary sinusoidally in space and time. At any point the magnitudes of the fields are related to each other by

$$E = cB \qquad (30\text{-}3)$$

where c is numerically equal to the speed of light.

All of the waves produced by this simple antenna will therefore be *transverse* and *polarized* and will have the same frequency as that of the oscillating charge. The waves will propagate with a velocity determined by the properties of the medium. If the medium surrounding the antenna is a vacuum, the waves will propagate with the speed c, where c is equal to the speed of light in vacuum. The relationship among the frequency f of the waves, the wavelength λ, and the speed of propagation c is

$$c = f\lambda \qquad (30\text{-}4)$$

In general, electromagnetic waves are a mixture of waves of varying frequencies and directions of polarization. It should also be noted that when the waves propagate through nonisotropic media or conducting media, the simple relationships given above no longer hold.

30-3
Energy and momentum of electromagnetic waves

In general, electromagnetic waves are useful because they transport energy. In the case of the simple polarized waves described above, one can write down a simple equation that describes the rate at which energy is transported. The energy is moving in the direction of propagation of the wave, so we may fix our attention on a plane that is perpendicular to that direction. Then, the amount of energy S transferred through a unit area of that plane in a unit time is given by

$$S = \frac{1}{\mu_0} EB \qquad (30\text{-}5)$$

Since

$$E = Bc$$

we have

$$S = \frac{c}{\mu_0} E^2 \qquad (30\text{-}6)$$

It should be noted that Equation 30-6 gives the *instantaneous* rate of energy transmission. The fields of electromagnetic waves vary rapidly in time, so one usually detects the average value of the energy transmission.

The idea that electromagnetic waves carry energy is not surprising in view of such well-known phenomena as the warming effect of sunlight, the ability of X rays to alter biological molecules, and the use of microwaves in cooking. The idea that electromagnetic waves carry momentum is, on the other hand, somewhat unexpected, although it is well founded in both theory and experiment.

If an electromagnetic wave strikes an object and is then absorbed by it, momentum will be transferred to the object. This transfer of momentum produces an effect called the *radiation pressure*. According to Maxwell, if an object absorbs an amount of electromagnetic energy $\mathcal{E}$, in time t, then the momentum p transferred to the object is

$$p = \frac{\mathcal{E}}{c} \qquad (30\text{-}7)$$

If the same amount of energy is reflected, then the amount of momentum transferred is

$$p = \frac{2\mathcal{E}}{c}$$

Because c, the speed of light, is so large, the momentum transferred in most processes is very small; therefore, radiation pressure is difficult to detect. However, radiation pressure is a significant effect in the interactions of light with elementary particles and in the intense electromagnetic fields in the neighborhood of stars.

30-4
Polarization

As we have seen, the electromagnetic waves produced by the simple antenna of Figure 30-4 are transverse, linearly polarized waves. That is, along a given direction of propagation, say the y direction, the electric field vectors are always parallel to a given line at right angles to this direction, say the z axis, and the magnetic field vectors are parallel to the x axis.

The simplest way to detect such waves, in principle, is to allow the waves to strike another metal wire, called the receiving antenna. The fields will then exert forces on the free charges in the metal of the receiving antenna and currents will flow. These currents in turn can be measured. We can simplify the subsequent discussion by recognizing that we can usually ignore the magnetic field because, from Equation 30-3, the magnetic field is significantly smaller than the electric field, and so the forces exerted by the magnetic field will usually be negligible compared with those exerted by the electric field. In other words, the principal effect of the waves will be the currents in the antenna caused by the electric field of the wave. It is then obvious that the maximum currents will be generated in the receiving antenna when that antenna is parallel to the emitting antenna. As the receiving antenna is rotated, the currents will become smaller and, in general, will approach zero when the two antennas are perpendicular to each other.

Consider then a beam of electromagnetic radiation emitted by a collection of antennas that are randomly oriented. In such a beam the electric field vectors are randomly oriented, although they must be perpendicular to the direction of propagation. That is, the receiving antenna will have large currents independent of the orientation of the antenna, as long as the antenna is perpendicular to the direction of propagation. Such a beam is said to be *unpolarized*.

If an unpolarized beam of electromagnetic radiation strikes an antenna that is oriented perpendicular to the direction of propagation, large currents will be induced in the antenna. These currents will consist of oscillatory motions of the charges in the antenna and, consequently, the receiving antenna will emit electromagnetic radiation. This radiation will be emitted in almost all directions, including that perpendicular to the original direction of propagation, and will be polarized (see Figure 30-5). We can describe this by saying that the original unpolarized beam is scattered by the antenna and that the scattered radiation is polarized, or that the scattering process polarizes the radiation. The existence of polarization by scattering is one of the most sensitive tests of the transverse character of electromagnetic waves. A common example of this phenomenon is the polarization of skylight.

Light from the sun is unpolarized. However, the blue light from the sky is not directly propagated from the sun, but is scattered from particles in the sky. In the scattering process the particles behave in much the same way as the antenna just discussed, and the scattered light is polarized.

When light is transmitted through matter, its electric and magnetic fields exert forces on the charged particles that constitute the atoms. The atoms in turn can radiate electromagnetic waves.

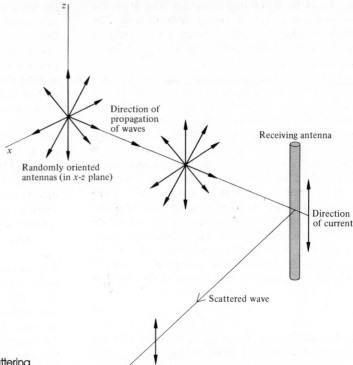

FIGURE 30-5 Polarization produced by scattering.

The atoms in *crystals* are bound together in an ordered structure, and therefore the atoms can collectively respond to electromagnetic waves in a way that can depend on the structure of the crystal and on the direction of polarization and propagation of light. In other words, the transmission of light through crystals can depend in part on the polarization of the light.

Some crystals show the property of *dichroism*; that is, they absorb most or all of the light having electric fields in a specific direction, whereas they absorb much less of the light in which the electric field vectors are perpendicular to that direction. Thus, if an unpolarized beam of light is transmitted through a dichroic crystal, almost all of the light after transmission will be polarized in a specific direction.

Some crystals show the property of *double refraction*. When a beam of unpolarized light is incident on such a crystal, the beam is split inside the crystal into two beams. Each of these beams is polarized, and each travels with a different speed and generally in a different direction. By appropriate choice of double-refracting crystal (and usually with some cutting and shaping), it is possible to select one of the two beams.

Thus, there exist devices, called polarizers, that from an initially unpolarized beam select and transmit only light polarized in a specific direction.

If polarized light enters a polarizer so that the direction of polarization of the light is perpendicular to the direction of polarization of the polarizer, then no light will be transmitted by the polarizer; all of it will be absorbed. Consider then a beam of unpolarized light that strikes a polarizer. The emerging light will be polarized in a specific direction. If this light strikes a second polarizer that is parallel to the first then all of the polarized light will be transmitted. If, however, the second polarizer is rotated through 90°, all of the light will be absorbed. Thus, by its rotation, the second polarizer can be used to analyze the direction of polarization of the light, and it is therefore called an

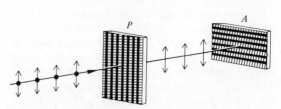

FIGURE 30-6 Crossed polarizers.

analyzer (see Figure 30-6). Note that in this process the polarizer P absorbed some of the initially unpolarized light, and that the analyzer A can absorb the rest.

Light can also be polarized by the process of reflection. Generally, when an unpolarized beam of light is reflected, the reflected beam is also unpolarized. However, the average intensity of the electric field in some directions is larger than in others. At a certain angle of incidence i, all of the reflected light is polarized, with the electric field vectors parallel to the reflecting surface. This specific direction is determined by the condition that the reflected and refracted beams are at right angles to each other. Using this condition, it can be shown (see Problem 15) that

$$\tan i = n \qquad (30\text{-}8)$$

where n is the index of refraction of the substance. This process of polarization by reflection can be used as the polarizer and then a second reflection can be used as an analyzer. Much of the annoying glare we are subject to is due to reflection from horizontal surfaces, such as roadways and automobile hoods and this light is therefore partly polarized. Consequently, polarizing sunglasses with their axes of polarization vertical are useful in reducing this glare.

30-5
Discovery of X rays

A very interesting portion of the electromagnetic spectrum is the short-wavelength region, usually termed the X-ray region. The wavelengths here are of the order of 10^{-8} or 10^{-9} cm. The discovery of these radiations by W. C. Roentgen in 1895 is usually said to mark the beginning of modern—that is, twentieth-century—physics. Roentgen made this discovery while studying the electric discharges in gases contained in glass tubes. While operating such a tube at very low pressure, he observed that a platinum-barium cyanide screen at some distance from the tube fluoresced. He shielded the tube so that no visible radiation could reach the screen although the fluorescence could still be observed. He also interposed various absorbing materials between the tube and the screen and found that, although the intensity of the fluorescence was reduced, it was not completely obliterated. He interpreted the phenomena as being due to radiations, which, coming from the walls

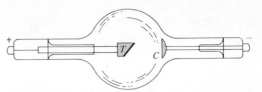

FIGURE 30-7 A gas discharge X-ray tube; C is the cathode and T is the target.

of the tube, penetrated these otherwise opaque materials and, upon reaching the screen, caused it to fluoresce. He called these radiations X rays. The X rays were produced when the cathode rays struck the glass walls of the electric discharge tube.

Ever since their discovery X rays have played an important part in the investigations of atomic physics and have added immeasurably to our knowledge of the structure of the atom. The uses of X rays were not confined merely to the physics laboratory. Almost immediately after their discovery they were used by physicians as aids in diagnosis and later in therapeutics. Industry turned to the use of X rays for the study of the properties and internal structure of materials, and for the examination of castings to determine the presence of flaws so that they could be eliminated before the castings were incorporated into the final product.

X rays are produced whenever a stream of electrons strikes some substance. There are three types of tubes in general use for the production of X rays. An early type, shown in Figure 30-7, utilizes the electrons liberated in a low-pressure gas-discharge tube by the bombardment of the cathode by the positive ions moving under the influence of the difference of potential between the target and cathode. Since the electrons come from the cathode, they are sometimes referred to as cathode rays. These cathode rays move perpendicular to the surface of the cathode when they leave it. The cathode rays can be focused on any desired region by properly curving the cathode. If the difference of potential between the target and cathode is V, the electrons reach the target with an amount of energy Ve, where e is the charge of an electron. The target then becomes the source of X rays. These gas X-ray tubes are usually operated at about 30,000 to 50,000 volts.

The second type of tube differs from the first one in that the source of electrons is a heated filament and the vacuum in the tube is made as high as possible (see Figure 30-8). The filament may be heated

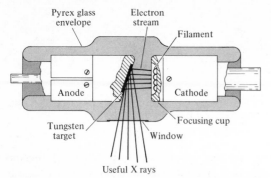

FIGURE 30-8 A modern Coolidge-type X-ray tube. (Courtesy of General Electric Company.)

by a battery or a step-down transformer. The filament is usually surrounded by a metallic cup shaped to produce the desired focusing of the electron beam. One of the chief advantages of the heated-filament type of X-ray tube is the greater ease in controlling the current and voltage of the tube. A high voltage applied to the terminals of the tube accelerates the electrons to the target; the target then becomes the source of X rays. This type of tube is sometimes referred to as a Coolidge X-ray tube. Tubes of this type have been operated at voltages from a few hundred volts to about a

million volts. The higher the voltage across the tube, the greater the penetrating power of the X rays produced.

A device called a *betatron*, developed by D. W. Kerst in 1941, is the third type of X-ray tube now coming into general use wherever X rays of very great penetrating power are desired. A betatron consists of a doughnut-shaped vacuum tube placed between the poles of a large electromagnet (see Figure 30-9). The mode of operation of a betatron can be seen from the diagram of Figure 30-10, which shows the doughnut-shaped tube in cross section. Electrons from a heated filament F are accelerated by a small difference of potential through a grid G. An alternating magnetic field is applied perpendicular to the path of the electrons. This produces two effects: the electron is made to travel in a circular path of radius R perpendicular to the magnetic field and, since the field is changing, an induced EMF is produced, which is tangent to the circular path and thus accelerates the electron, increasing its speed and giving it additional kinetic energy as it circulates in this path. The pole pieces of the electromagnet have to be carefully shaped so that the magnetic field at every instant will be of the right form in order to keep the electrons moving in the same circular orbit. The alternating magnetic field is produced

FIGURE 30-9 The 100,000,000 electron volt betatron. The betatron tube is in the center between the poles of the electromagnet. (Courtesy of General Electric Company.)

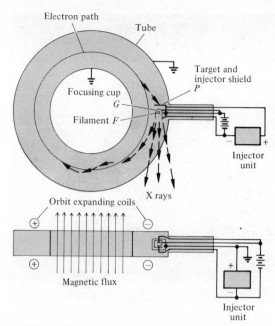

FIGURE 30-10 Path of an electron in a betatron tube.

by supplying 60-Hz alternating current to the field coils of the electromagnet. The electrons are injected into the tube for a very short time at the beginning of a cycle of the alternating current, and then continue traveling around the circular orbit until the magnetic field reaches its maximum value in 1/240 sec. Each electron makes several hundred thousand revolutions in this quarter of a cycle. During each revolution the electron gains additional energy. When the electron has acquired its maximum energy, current is sent through an auxiliary set of coils; this changes the magnetic field so that the electron now moves in a larger orbit and strikes the back of the plate P, which acts as the target and is the source of the X rays. Betatrons are now being operated at energies up to 300 MeV, and modified forms of the betatron are being designed to operate at still higher energies.

30-6
Some properties of X rays

X rays are invisible to the eye, but they can be detected by their blackening of a photographic plate or by the ionization they produce in their passage through a gas or vapor. The intensity of the X rays can be measured by the ionization they produce in a specially designed ionization chamber, one type of which is illustrated in Figure 30-11. The X rays enter the chamber through a thin window made of mica or aluminum and ionize the gas in the chamber. A difference of potential between the rod R and cylinder C causes these ions to move; the motion of these ions constitutes the current in the ionization chamber. This current, though very small, can be measured with an electrometer E, or it may first be amplified with the aid of an amplifying circuit, and this amplified current can then be measured with a galvanometer.

Besides being able to blacken a photographic plate and to ionize gases, X rays can penetrate various thicknesses of substances, including those that are opaque to visible radiation. In their passage through matter, some of the X rays are absorbed, their energy being converted into other forms, while some pass through and can be detected and measured. The fraction of the energy absorbed depends upon the atomic number of the substance, upon its density and thickness, and upon the wavelength of the incident X rays.

If a beam of X rays is sent through a composite substance made of different types of materials, the photograph can be used to reveal the nature of the materials and their locations inside the substance (see Figures 30-12 and 30-13).

We shall consider some other properties of X rays in later chapters of this book.

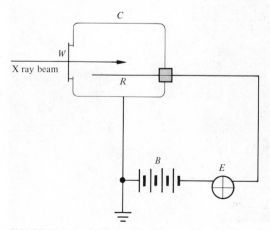

FIGURE 30-11 Ionization chamber and electrometer for measuring the intensity of X rays.

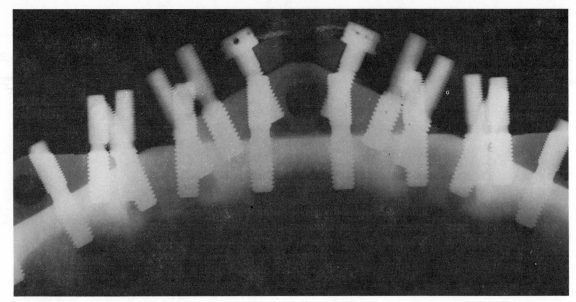

FIGURE 30-12 Radiograph of an airplane motor's crankcase showing the exact position and depth of studs. (Courtesy of General Electric Company.)

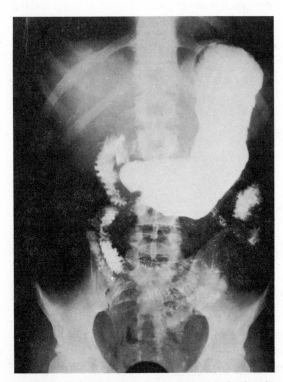

FIGURE 30-13 Radiograph showing opaque barium in a stomach. (Courtesy of General Electric Company.)

Questions

1. Arrange the following in order of increasing wavelength; **(a)** red light, **(b)** X rays, **(c)** short waves as used in radio broadcasting, **(d)** ultraviolet light.

2. Arrange the radiations listed in Question 1 in order of increasing frequency.

3. What kind of electromagnetic radiation has a wavelength of about the size of **(a)** a house, **(b)** a human being, **(c)** a fly, and **(d)** a bacterium.

4. The wavelength of an electromagnetic radiation is about the same size as that of the antenna from which it is radiated. From this, estimate the size of the antenna from which the following are radiated: **(a)** visible light, **(b)** gamma rays, **(c)** television signals.

5. American TV receiving antennas are horizontal, whereas those in England are vertical. What can you conclude from this?

6. What is the wavelength radiated by a 60-Hz power transmission line?

7. State the relative directions of the electric field, magnetic field, and velocity of a polarized electromagnetic wave **(a)** in terms of a

right-hand rule and **(b)** in terms of a cross product.

8. A window screen is essentially transparent to sunlight, yet a fire screen absorbs a relatively large amount of the heat radiated from a fireplace. How could you account for this in terms of the wavelengths of the radiations?

9. The sun loses energy by radiating electromagnetic waves. Does the sun lose momentum in this process?

10. It has been suggested that the radiation pressure from solar radiation could be used as a "wind" to "sail" an interplanetary vehicle. How could such a vehicle return to its starting point?

11. Using an appropriate crystal, a piece of Polaroid, or a pair of sunglasses with polarizing lenses, investigate the polarization of light scattered from the sky. If possible, verify the relative orientations shown in Figure 30-5. (Note that this experiment should be done when the sky is blue. When the sky is gray, the light has usually been scattered several times and so it is difficult to determine the relative orientation.)

12. Why are sunglasses made with polarizing lenses usually better than those made with colored glass?

13. Describe the use of a sheet of polarizing material as **(a)** a polarizer and **(b)** an analyzer.

14. Television tubes do produce some X rays of relatively low penetrating power. **(a)** Where are they produced? **(b)** Why is the danger from these X rays more serious for color TV than for black-and-white TV? **(c)** How can the danger be eliminated or minimized?

15. List some of the uses of X rays.

16. In recent years archaeologists and art historians have made extensive use of X rays. Discuss the advantages of X-ray analysis of **(a)** Egyptian mummies and **(b)** the internal structure of metal castings, such as statues and jewelry.

17. It is now possible to make X-ray motion pictures. Discuss some of the uses that might be made of such a development, particularly for medical diagnosis, evaluation of therapy, and research.

Problems

1. The AM broadcast band covers the range from 550 to 1600 kilohertz (kHz). Find the corresponding wavelengths.

2. The FM broadcast band covers the range from 88.1 to 108 megahertz (MHz). Find the corresponding wavelengths.

3. The electric field in a plane-polarized electromagnetic wave has a magnitude of 60 volts/m. Find the magnitude of the magnetic field.

4. A very intense laser beam contains electric fields of magnitude 2×10^6 volts/m. **(a)** Find the magnitude of the magnetic field and **(b)** find the force exerted by this radiation on a stationary electron.

5. A certain lamp radiates 5 watts/m². Find the average magnitude of the electric field in the electromagnetic radiation, assuming that it can be treated as a polarized wave.

6. The electric field in a certain radiation has a magnitude of 150 volts/m. Find the rate at which energy is being transmitted per unit area by this radiation, in watts per square meter.

7. A light beam of intensity 3 watts/m² falls normally on a plane, square mirror, 1 cm on a side. **(a)** How much energy falls on the mirror in 1 min? **(b)** How much momentum is transferred to the mirror in that time? **(c)** What is the average force exerted on the mirror?

8. A certain flashlight emits 5 watts of light, which can be treated as if it were uniformly spread over a circular cross section, of radius 2 cm. **(a)** How much momentum is transferred to the flashlight per minute? **(b)** What is the average force exerted on the flashlight?

9. At what angle of incidence will the light reflected from glass of index 1.65 be polarized?

10. When a beam of light is incident on a glass surface at an angle of 58°, the reflected beam is polarized. What is the index of refraction of the glass?

11. Assume that a light bulb radiates uniformly in all directions. What is the average force exerted by the light on an object that has an area of 1 cm² and is located 1 m from a 100-watt

light bulb? Assume that the object is perfectly absorbing.

12. The intensity of the sunlight reaching the earth is about 1400 watts/m². (a) Find the average magnitude of the electric field in this radiation, assuming it were polarized. (b) Find the average force exerted by this radiation on the earth, assuming the earth to be a flat disk of radius 4000 mi.

13. A polarized electromagnetic wave has a vertical electric field, whose maximum magnitude is 5 volts/m. The wave passes through an analyzer. Find the maximum value of the electric field in the beam that emerges from the analyzer if the polarization axis of the analyzer is (a) vertical, (b) horizontal, and (c) at an angle of 45° with the vertical.

14. An initially unpolarized beam passes through two polarizers whose axes make an angle of 45°. (a) What is the ratio of the magnitude of the electric field in the emerging beam to the magnitude of the electric field in the beam transmitted by the first polarizer? (b) What is the ratio of the intensity of the emerging beam to the intensity of the beam transmitted by the second polarizer? (c) Assuming that the first polarizer reduces the intensity by one half, what is the ratio of the emerging beam intensity to the incoming beam intensity?

15. Assuming that polarization by reflection occurs when the reflected and refracted beams are at right angles to each other, prove that this requires that (see Equation 30-8)

$$\tan i = n$$

where the incident beam makes an angle of incidence i and is traveling in a vacuum and n is the index of refraction of the reflecting material.

16. The critical angle for a certain substance is 45°. What is the polarizing angle for that substance?

interference & diffraction of eletromagnetic waves

chapter 31

31-1
Interference of light from two sources

As we saw in Chapter 21, interference between two identical waves occurs when the two waves have a constant phase difference. Suppose that we have two sources of electromagnetic waves of the same frequency at points S_1 and S_2 of Figure 31-1 and suppose that they send out waves simultaneously. The figure shows the crests of the waves as solid lines and the troughs as dashed lines. The points where the crests from one source meet the crests from the other source are points where the two waves are in phase, and therefore are points where constructive interference occurs. The points where the crests from one source meet the troughs from the other are points where the waves are 180° out of phase, and where destructive interference occurs. In terms of light, the points of constructive interference will be bright, whereas the points of destructive interference will be dark. If the electromagnetic waves fall on a screen, P, the points B and O will be bright and the points D will be dark.

In order for these interference effects to be seen, the sources S_1 and S_2 must emit electromagnetic radiation in such a way that the phase difference between the sources is constant. If this condition is satisfied, the sources are said to be *coherent*. It is fairly easy to build oscillators in the radio and microwave range of frequencies that produce radiation with constant phase, and therefore it is not difficult to build two coherent sources. Until relatively recently it was essentially impossible to produce light with constant phase; light is produced by discrete atomic processes and therefore the phase of the radiation from ordinary

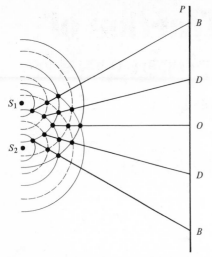

FIGURE 31-1 Interference of light from two sources.

light sources changes randomly from instant to instant, and so it was impossible to have two light sources with constant phase difference. The relatively recent development of lasers has made it possible to produce light of constant phase, and so it is possible to do the simple experiment described above with visible light.

However, it is possible to use ordinary light sources to show interference by allowing the light from a *single* source to pass through a slit and then permitting the light diffracted by that slit to fall on two slits, as shown in Figure 31-2. Here, even though the phase of the light at S varies randomly, the phase difference between S_1 and S_2 depends on the difference in length of the two paths SS_1 and SS_2.

The positions of the light and dark regions on the screen can be determined with the aid of Huygens' principle. Figure 31-2 shows light from a

source S going through two narrow slits S_1 and S_2 toward a screen a distance L from them. Let us assume that the light is monochromatic with a wavelength λ. For a point P on the screen to be bright, light reaching it from the two slits must be in phase; that is, the light path from S_2 to P must exceed that from S_1 to P by an integral number of wavelengths, or

$$S_2P - S_1P = S_2A = n\lambda$$

where n is an integer ($n = 0, 1, 2, 3, \ldots$) and S_2A is the path difference.

The angle at A can be considered to be a right angle when the distance to the screen is very large. For this case,

$$S_2A = d \sin \theta$$

where d is the distance between the slits and θ is the angle between S_1S_2 and S_1A. From the figure, it can be seen that the angle PCO is equal to θ; hence

$$\sin \theta = \frac{OP}{CP} = \frac{x}{L}$$

where x is the distance from the central image to P and, for the small angles involved, $CP = L$. Hence

$$\frac{x}{L} = \frac{n\lambda}{d}$$

or

$$x = n \frac{\lambda L}{d} \tag{31-1}$$

To get an idea of the order of magnitude of the quantities involved in this type of experiment, suppose that the source is a sodium lamp emitting an intense yellow line of wavelength 5893 Å, that the distance between the slits is 0.5 mm or 0.05 cm, and that the screen is 200 cm away. The first

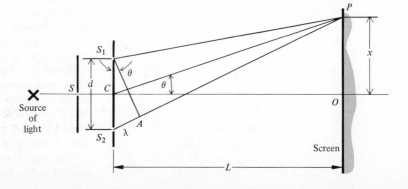

FIGURE 31-2

FIGURE 31-3 Photograph of the interference pattern produced by the passage of light through two narrow slits that are close together. (Reproduced by permission from *University Physics*, 4th ed., Sears and Zemansky, 1970. Addison-Wesley, Reading, Mass.)

bright region, for which $n = 1$, will be displaced a distance x from the central image, for which $n = 0$, by an amount

$$x = \frac{5893 \times 10^{-8} \times 200}{0.05} \text{ cm}$$

or

$$x = 0.24 \text{ cm} = 2.4 \text{ mm}$$

This will also be the spacing between successive bright spots.

If white light is used instead of sodium light, each color will produce its own set of interference bands, and these will overlap. The central image at O will be very bright, and there will be a series of colored bands on either side of O. These interference bands can be viewed very easily by the reader if he looks at the light from the bulb of an automobile lamp (such as is used in a high intensity or Tensor lamp) through two pinholes punched close together in a piece of cardboard. Figure 31-3 is a reproduction of an interference pattern produced by two narrow slits.

31-2
Interference from thin films

Interference phenomena are very common; for example, the colors that are observed on soap bubbles are due to interference of light reflected from the two film surfaces. The colors observed on wet pavements are due to the interference effects produced by thin films of oil which reflect light from their two surfaces.

Consider monochromatic light incident on a thin film of oil at almost normal incidence, as shown in Figure 31-4. Ray I strikes the first surface at A; it is partly reflected into the air and partly refracted into the oil toward B, at which point part of the light is again reflected to C and some of it is re-

fracted out into the air. Ray II, which strikes point C of the top surface, undergoes similar reflections and refractions. Now at C two rays meet and travel on into the air along CF: part of ray II, which is reflected at C, and part of ray I, which traveled through the oil and was reflected at B. The difference in path between these two rays is $AB + BC$, or approximately twice the thickness t of the oil film. If these two waves meet in phase at C, it will be a bright spot; if they meet out of phase by 180°, C will be a dark spot.

If the thickness of the film varies from point to point, there will be some points in which the two waves will reinforce each other, producing bright spots, and other points in which the waves meet out of phase, producing dark spots. If yellow sodium light is the source of illumination, the film will show yellow regions separated by dark spaces. If white light is used, each component wavelength will produce its own interference pattern. For example, where the thickness of the film is just right to produce a dark spot for yellow light, the other wavelengths will be present because the thickness of the film is not sufficient to cause the two reflected portions to be completely out of phase. What is observed at this point is a mixture of all the colors in varying intensities except yellow. This spot may appear blue. At a different point in which interference completely removes

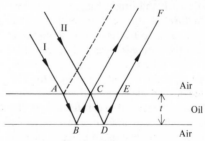

FIGURE 31-4 Interference from thin film.

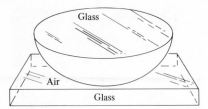

FIGURE 31-5 Formation of Newton's rings.

the red light, a composite of all the other colors will be seen; this may appear green. In this manner a thin film of oil on the pavement, when reflecting white light from the sky, shows a variety of colors.

Newton observed and studied the interference effects produced by the thin film of air between the convex surface of a lens and the flat surface of a plane piece of glass. Figure 31-5 shows the experimental arrangement. The two pieces of glass make contact only at the center, so the thin film of air has a wedge-shaped section. Suppose that monochromatic light is incident on the convex surface of the lens; part of it is reflected and part refracted into the air and reflected from the flat glass plate. Wherever the two waves meet in phase, they will reinforce each other, and those points will be bright spots; wherever the two waves meet out of phase by half a wavelength, they will annul each other, and those points will be dark spots. Because of the circular symmetry of the arrangement, a series of dark and bright rings will be seen; these are called *Newton's rings*. If a source of white light is used, the rings will appear colored.

If we try to determine the positions of the bright regions from the difference in path of two waves, we find the surprising result that where the thickness of the film is such that two waves meeting at a point would be expected to produce a bright spot, this point is found to be a dark spot. Conversely, where the thickness of the film is such that two waves meeting would be expected to produce a dark spot, that point is found to be a bright spot. The reason for this interchange of the expected positions of the dark and bright rings is the difference in the nature of the reflection of the two waves. One wave is reflected from a glass-air surface and the other wave is reflected from an air-glass surface. This difference in reflections introduces a change of phase of half a wavelength. The waves reflected at the glass-air surface suffer no change in phase, whereas those reflected at the air-glass surface suffer a change in phase of half a wavelength. This conclusion receives further

verification by noting that the central spot in Newton's rings is black. In this region the thickness of the air film is negligible; hence the only change in phase is that produced on reflection from the second surface. Thus a dark spot will be observed in the center regardless of the wavelength of the light. This can be seen clearly in Figure 31-6. Figure 31-6a is a photograph of Newton's rings made with sodium light, and Figure 31-6b is a photograph of Newton's rings made with white light.

Thomas Young, the first person to give the above explanation of Newton's rings on the wave theory of light, proceeded to show the correctness of the interpretation of the central black spot. He reproduced the Newton-ring experiment by using an oil film between a crown-glass lens and a flint-glass plate. The oil used has an index of refraction intermediate between that of crown glass and that of flint glass; hence the light underwent the same change of phase at each surface, and thus these phase changes canceled each other. In this case the center spot was white, and the other bright and dark regions were shifted in a corresponding manner.

Interference bands can be easily produced by making a wedge-shaped air film between two plane pieces of glass, as shown in Figure 31-7. If monochromatic light is used, a series of parallel bright and dark lines will be observed if the two glass surfaces *A* and *B* are perfectly plane. If these surfaces are not perfectly plane, the shape of these bands will vary from place to place. If one surface is known to be perfectly plane from previous tests, it can be used to test the flatness of other surfaces by observation of the interference pattern produced either when a wedge-shaped film of air is set up between them or when they are placed in contact.

A beautiful demonstration of interference produced by thin films, suitable for viewing by a large audience, is one, originally suggested by Pohl, in which light from a small, intense mercury arc is allowed to fall on a thin sheet of mica placed about 5 cm away. Light reflected from these two surfaces is then allowed to fall on a large screen a meter or so away. A circular interference pattern, such as that shown in Figure 31-8 will be clearly visible. In this experiment the sheet of mica had a thickness of about 0.05 mm. The size of the pattern can be judged from the meter stick placed near the right edge of the screen.

The interference produced by thin films has been

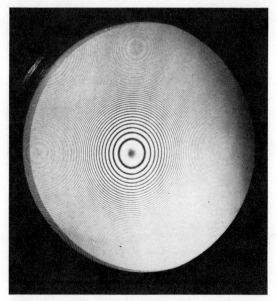
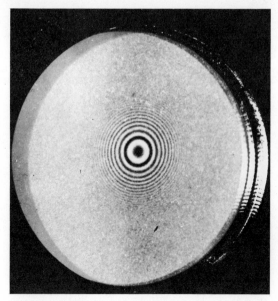

FIGURE 31-6 (a) Photograph of Newton's rings made with sodium light. (b) Photograph of Newton's rings made with white light. (Courtesy of Bausch and Lomb.)

used to reduce the loss of light that takes place at the glass surfaces of lenses used in optical instruments. Under the best conditions, about 5 percent of the incident light is reflected at each surface, so that, if there are six such surfaces, only about 75 percent of the light is transmitted through the instrument. By coating each surface with a thin film of a suitable material, this reflection loss can be diminished considerably. The film must satisfy two conditions for best results: (a) its thickness must be a quarter of a wavelength of light in the film, and (b) its index of refraction should be equal to the square root of the index of refraction of the glass to which it is applied.

Since the index of refraction varies with the wavelength, and white light contains many different wavelengths, it is not possible to reduce the reflection completely to zero, but good results can never-

theless be obtained. For example, the index of refraction of one common type of flint glass is 1.66. The square root of this is 1.29. It is difficult to find a substance of such low index of refraction. Lithium fluoride has an index of refraction of 1.39, and calcium fluoride has an index of refraction of 1.43. If the thickness of the film is made one quarter the wavelength of sodium light, the reflection will be reduced to about 1 percent. One method of putting

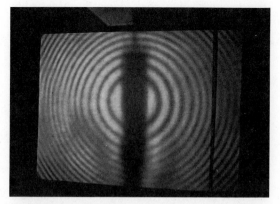

FIGURE 31-8 Interference rings produced by mercury light reflected from a thin sheet of mica. Meter stick on right side of screen for comparison. Center shadow is that of the mercury arc. Light photographed through green filter. Thickness of mica is 0.05 mm. (Photograph courtesy of M. W. Zemansky, H. Semat, and I. Antman.)

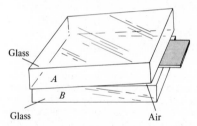

Glass

A

B

Glass   Air

FIGURE 31-7 A wedge-shaped air film.

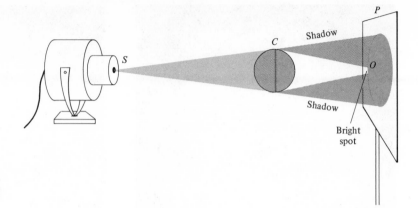

FIGURE 31-9 Diffraction produced by the passage of light
past a spherical obstacle.

the thin film on the glass is to heat the calcium
fluoride in a good vacuum and allow the calcium
fluoride to vaporize and deposit on the glass in the
vacuum. This or a similar process is used to produce the so-called coated lenses of modern cameras
and other optical instruments.

31-3
Diffraction of light

We have already noted that in the case of sound,
the waves bend around corners and also spread
out in going through a narrow aperture; that is,
sound waves are *diffracted*. Since the interference
experiments show definitely that light is propagated
as a wave motion, it should be possible to show
diffraction phenomena with light. One very interesting diffraction effect, originally suggested by
Fresnel, is outlined in Figure 31-9. *S* is a point
source of light and *C* is the circular cross section
of an obstacle such as a sphere. If it were not for
diffraction effects, there would be simply a sharp
circular shadow, the so-called *geometrical shadow*,
cast on the screen *P*. But because light is propagated as a wave motion, the light will bend around
the rim of the sphere; according to Huygens'
principle, each point on the rim can be considered
as sending waves of light into the region of shadow.
Now point *O* in the center of the shadow is equally
distant from each point on the rim of the sphere;
therefore the waves reaching point *O* will all be in
the same phase and *O* will be a bright spot. This
spot can be seen in a photograph of the shadow or
it may be located with the aid of a magnifying glass
or a small telescope (see Figure 31-10).

Another diffraction effect can be observed when
light from a point source passes through a narrow

aperture, such as a pinhole in a card. If the light
strikes a screen several feet from the pinhole, the
pattern on the screen will consist not of a single
spot of light but of a series of dark and bright rings
surrounding the central bright spot. The light
spreads out into the region of the geometrical
shadow, producing this circular diffraction pattern.
This diffraction pattern may be observed readily
by looking through a single pinhole at a small
source of light such as the bright filament of an

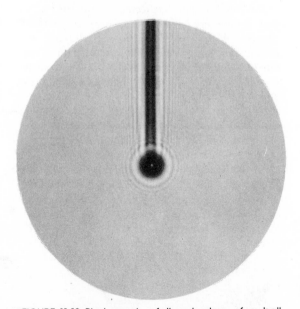

FIGURE 31-10 Photograph of the shadow of a ball
bearing supported on a rod. Note the diffraction pattern around the rod and ball bearing and the bright
spot in the center of the shadow of the ball bearing.
(Reproduced by permission from *College Physics*, 4th
ed., Sears, Zemansky, and Young, 1974, Addison-
Wesley, Reading, Mass.)

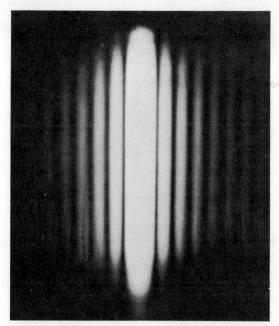

FIGURE 31-11 Photograph of a diffraction pattern produced by the passage of light through a narrow slit. (Reproduced by permission from *College Physics*, 3rd ed., Sears and Zemansky, 1960. Addison-Wesley, Reading, Mass.)

automobile lamp bulb placed 10 to 15 ft from the observer.

31-4
Diffraction through a narrow slit

When a beam of light passes through a narrow slit and falls on a screen at some distance from it, the pattern on the screen will consist both of a bright image of the slit and a series of light and dark lines or fringes on either side of the central bright spot, as shown in Figure 31-11. Only a very small portion of the incident wavefront passes through the narrow slit to produce this diffraction pattern. The appearance of the bright and dark regions on the screen can be explained by assuming that each point of this section of the wavefront acts as a source of light sending out wavelets in all directions. The effect produced at any point on the screen depends upon the phase relationships of those wavelets which reach this point.

Imagine AB of Figure 31-12a to be the edges of the slit, greatly magnified, and the wavefront approaching it to be a plane monochromatic wave;

every point in this wavefront sends out waves in the same phase. Point C is in the center of the slit, and CO is a perpendicular line from the slit to the screen. For every point in the wavefront in AC that sends light to O, there is a symmetrically placed point in CB that sends light to O in the same phase; hence O will be a bright spot.

Let us now consider a point D on the screen above O such that the distance BD exceeds the distance AD by one wavelength λ. To a very good approximation, CD exceeds AD by $\lambda/2$; similarly, BD exceeds CD by $\lambda/2$. Now consider a typical point e at a distance x below A sending waves to D; there is a corresponding point f an equal distance below C sending waves to D; the waves from these two points will be out of phase by $\lambda/2$ on reaching D. Similarly, for every point between A and C sending waves to D, there is a corresponding point between C and B sending waves to D, such that the waves from these two points reach D out of phase by $\lambda/2$; hence D will be a dark region. There will be a symmetrical point D' below O that will also be dark.

As we proceed along the screen away from D, the above conditions will no longer hold, so these points will increase in brightness until a point E is reached, such that the distance BE exceeds AE by $3\lambda/2$. If we divide the aperture AB into three equal parts, then, using the same type of argument as

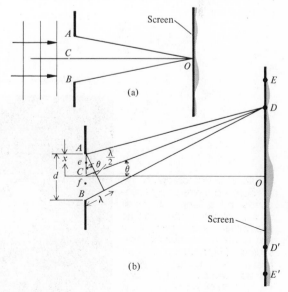

FIGURE 31-12

above, we find that waves from corresponding points in two adjacent sections will be out of phase by λ/2 upon reaching E, thus contributing nothing to the light at E. Only those points in the third remaining section will be sending waves to E that will not be out of phase by λ/2, even though not entirely in phase; hence point E will be bright, but not as bright as the central region. There will be a a symmetrical point E' below D' that will be as bright as E and for the same reason.

The above type of argument can be used to explain the appearance of other bright and dark regions in the diffraction pattern. Wherever the distances from the edges of the slit to a point on the screen differ by a whole number of wavelengths, that point will be dark. In such cases the slit may be considered as consisting of an even number of equal regions; waves from pairs of adjacent regions will reach this point out of phase by λ/2, and thus this point will be dark. The positions of these dark regions can readily be found. Referring to Figure 31-12, we note that

$$\sin \theta = \frac{\lambda}{d} \qquad (31\text{-}2)$$

for the first dark spot, where d is the width of the slit and θ is the angle between CO and CD. Thus the central image extends over a comparatively large region, with maximum brightness at O and minimum at D and D'. The screen will become bright again beyond D; the second dark region will occur at a point such that the angle θ will satisfy the equation

$$\sin \theta = \frac{2\lambda}{d} \qquad (31\text{-}3)$$

31-5
Diffraction and resolving power

Diffraction phenomena can also be observed when light passes through large apertures such as the lenses of telescopes and microscopes. The effect of such phenomena is to *limit the resolving power* of the instrument—that is, the ability of the instrument to show increasingly greater detail. If light from a point source is sent through a converging lens, for example, its image will not be a sharp point even if the lens has been corrected for spherical and chromatic aberrations. To determine the exact shape of the image and the distribution of light in it, one must consider the contributions made to it by the wavelets coming from every point of the wavefront that comes through the lens. The detailed analysis of this problem is beyond the scope of this book; but, using arguments similar to those of the preceding section, it can be shown that the image will consist of a central bright spot surrounded by dark and bright rings, as shown in Figure 31-13.

The size of the central disk of the diffraction pattern can be shown to depend upon three factors: (a) the focal length of the lens, (b) the wavelength of the light used, and (c) inversely, the diameter of the lens.

If we consider two points sending light through an optical system, the image of each point will be a diffraction pattern. If the points are close together, these patterns may overlap and it will not be possible to distinguish them as two separate points. Two images are said to be resolved if the dark ring of one pattern passes through the center of the disk of the other pattern or if the two central disks are

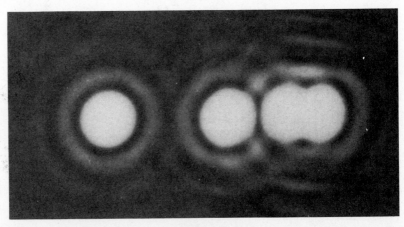

FIGURE 31-13 Photograph of the diffraction patterns of light produced by a lens. Four point sources of light were used. The two patterns on the right can just be resolved as due to two sources. (Reproduced by permission from *College Physics*, 4th ed., Sears, Zemansky, and Young, 1974. Addison-Wesley, Reading, Mass.)

separated a distance equal to the radius of one of them. If two points cannot be resolved by an instrument, merely increasing its magnification without increasing the resolving power serves no useful purpose.

In the case of a telescope, one method of getting higher resolving power is to increase the diameter of the objective. If the optical instrument is used for visual work, the resolution is limited by the wavelength of light. If a photographic plate is used to record the image, then light of shorter wavelength—ultraviolet light—may be used to increase the resolution. Microscopes have been built with quartz lenses so that ultraviolet light can be used for increasing their resolving power.

An entirely different type of microscope, known as an *electron microscope*, is capable of resolving points that are as close together as 5 Å. This is about 800 times the theoretical limit of the best optical microscopes.

31-6
The diffraction grating

The *diffraction grating* is widely used for the measurement of the wavelength of light and for spectrum analysis. Diffraction gratings are used in one of two ways, either as *reflection* gratings or as *transmission* gratings. A reflection grating consists of a series of parallel rulings or scratches made on a polished reflecting surface. The number of rulings varies from about 400 per centimeter in some grat-

ings to about 6000 per centimeter in others. A transmission grating has a series of parallel rulings made on a flat glass surface. The light is transmitted through the spacings between the scratches. Good diffraction gratings are difficult to make. For many ordinary purposes, replicas are used. These replicas can be made by pouring a solution of collodion in ether over a ruled grating. After the ether has evaporated, the thin collodion layer is stripped off and placed between two flat glass plates. The collodion retains the impression and acts as a fairly good diffraction grating.

To understand the action of a grating, let us consider a set of plane, parallel waves incident on a transmission grating, as shown in Figure 31-14a. The spaces between rulings can be considered a series of equally spaced narrow slits, a few of which are shown in the figure. The light that passes through the grating can be considered as coming from these slits, and, according to Huygens' principle, the slits can be considered sources of waves. These waves will be circular in a plane perpendicular to the rulings. For the sake of simplicity, let us assume that the incident light is monochromatic and of wavelength λ. Since the incident wavefront is plane and parallel to the plane of the grating, the light emerging from each of the slits at any one time is all in the same phase and spreads out in concentric circles from each slit as a center. The circular wavefronts from a few of these slits are shown in the figure; the distance between successive wavefronts is λ, the wavelength of the incident

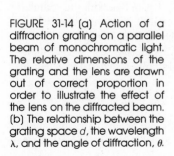

FIGURE 31-14 (a) Action of a diffraction grating on a parallel beam of monochromatic light. The relative dimensions of the grating and the lens are drawn out of correct proportion in order to illustrate the effect of the lens on the diffracted beam. (b) The relationship between the grating space d, the wavelength λ, and the angle of diffraction, θ.

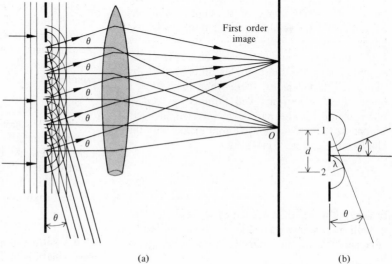

(a)

(b)

light. The resultant wavefront produced by these individual waves from all the slits can be found by drawing a line tangent to the circles. All points on this line will be in the same phase; therefore this line is a wavefront. Several such resultant wavefronts are shown in the figure. One set is parallel to the incident wavefronts and proceeds in the same direction. These parallel wavefronts can be brought to a point focus by a converging lens and will produce what is known as the *central image* at O.

Another set of wavefronts is shown traveling at an angle θ to the original direction of the beam. It will be observed that each such wavefront is tangent to circular wavefronts from adjacent slits that differ in phase by *one whole period*. These parallel wavefronts can also be brought to a point focus by a converging lens. This point is called the *first-order image*. Let us consider waves emitted from adjacent slits 1 and 2 that differ in phase by one whole period and whose path difference is one wavelength λ. From Figure 31-14b it can be seen that the resultant wavefront makes an angle θ with the plane of the grating such that

$$\sin \theta = \frac{\lambda}{d} \qquad (31\text{-}4)$$

where d is the distance between two slits, from which

$$\lambda = d \sin \theta \qquad (31\text{-}5)$$

The longer the wavelength, the greater the angle θ. Thus, if violet light is used, its first-order image will be deviated through a smaller angle than that for the first-order image of red light. Equations 31-4 and 31-5 are valid only for light all of which is incident normally on the surface of the grating.

If a plane is drawn tangent to circles from adjacent slits so that these circles differ in path by two whole wavelengths, then, when focused at a point, these wavefronts will produce a *second-order image* for the particular wavelength used. The angle θ_2 that such a wavefront makes with the plane of the grating is given by

$$\sin \theta_2 = \frac{2\lambda}{d} \qquad (31\text{-}6)$$

In general, wavefronts formed by waves that differ in path by n whole wavelengths, when emitted by adjacent slits, will be inclined at an angle θ_n, given by

$$\sin \theta_n = \frac{n\lambda}{d}$$

or

$$n\lambda = d \sin \theta_n \qquad (31\text{-}7)$$

There will be first-order, second-order, and so on, images on the other side of the central image. If white light is incident on the diffraction grating, a series of spectra will be obtained on each side of the central image. The central image itself will be white; θ is zero for the central image and so is n; that is, the waves of all wavelengths are in the same phase. For $n = 1$, we get the first-order spectrum with violet light deviated least—that is, closest to the central image—and red light deviated by the greatest amount. Then farther on there will be a second-order spectrum, again extending from violet to red, but fainter than the first-order spectrum. The highest order n_{max} in which a given wavelength can appear may be determined by setting $\sin \theta_n = 1$, its maximum value; Equation 31-7 then yields

$$n_{max} = \frac{d}{\lambda}$$

31-7
The diffraction-grating spectrometer

A spectrometer, with a diffraction grating mounted on its table, forms one of the most convenient instruments for measuring the wavelength of light. Light from some source illuminates the slit S of the collimator C, as shown in Figure 31-15. The parallel light coming out of the collimator goes through the grating G, which sends out waves in various directions; images of the slit will be formed in directions making angles θ_n with the direct beam, when the wavelengths satisfy the grating equation

$$n\lambda = d \sin \theta_n$$

When the telescope T is in line with the collimator, θ is zero and the central image is observed in its field of view. As the telescope is rotated about the spectrometer axis, the shortest wavelength is observed in the first order, when the angle θ_1 is reached such that

$$\lambda = d \sin \theta_1$$

If the source sends out white light, the shortest wavelength is in the violet region. The other wave-

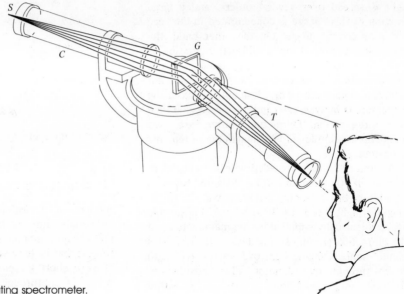

FIGURE 31-15 The diffraction-grating spectrometer.

lengths then appear at larger angles up to the longest visible wavelength, which appears red. It will be noted that the order in which the colors appear in the grating spectrum is the reverse of that in the spectrum produced by a prism; in the spectrum formed by a diffraction grating the violet is deviated least, whereas in the prism spectrum the red is deviated least.

There is a break after the red in the first-order spectrum, and then the violet appears once more (see Figure 31-16). This is the beginning of the second-order spectrum, and the angle of the telescope θ_2 satisfies the condition

$$2\lambda = d \sin \theta_2$$

The third-order spectrum, however, overlaps the second-order spectrum. This can be seen from the value of $d \sin \theta$ for violet light in the third order, which is approximately

$$d \sin \theta_3 = 3 \times 3800 \text{ Å} = 11,400 \text{ Å}$$

and its value for red light in the second order, which is approximately

$$d \sin \theta_2 = 2 \times 7500 \text{ Å} = 15,000 \text{ Å}$$

Thus, irrespective of the number of lines per centimeter in the grating, the third order will always overlap the second order. This is illustrated in Figure 31-16.

The prism spectroscope has one advantage over the diffraction-grating spectroscope in that all of the energy that goes through the prism is concentrated in a single spectrum. In a diffraction-grating spectroscope the energy from the source of

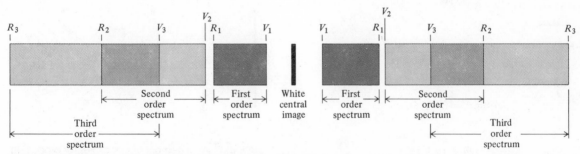

FIGURE 31-16 Relative positions of the first three orders of spectra produced by a diffraction grating on either side of the central white image. Notice that the second and third orders overlap considerably.

light is spread over several orders, and a large fraction of this energy is concentrated in the zero order or central image. On the other hand, the dispersion produced by a diffraction grating is given by Equation 31-7, which is simple, whereas the dispersion produced by a prism does not follow a simple law. For most optical glass, the dispersion is much greater in the violet region than in the red region. Diffraction gratings are used where great accuracy in wavelength measurements is desired.

A circular scale is provided with the spectrometer so that the angle θ can be measured for each position of the telescope; if d is known, the wavelength of light is then easily measured. The grating space d is usually supplied by the manufacturer of the diffraction grating. If it is desired to check this value, monochromatic light of known wavelength is sent through the collimator and the positions of the first- and second-order images are measured on both sides of the central image. The grating space d can then be calculated from these results.

Illustrative example

Light from a mercury arc is observed through a diffraction grating spectrometer. The spectrum consists of a series of 7 lines: 2 yellow, 1 green, 1 blue-green, 1 blue, and 2 violet. The wavelength of the green line is 5461 Å, where 1 Å is 10^{-8} cm. Determine the angular deviation of this green line in the first and second orders, if the grating has 6000 lines per centimeter ruled on it.

The distance between rulings is

$$d = \frac{1}{6000} \text{ cm}$$

The angle θ_1 in the first order satisfies the equation

$$\sin \theta_1 = \frac{\lambda}{d}$$

hence

$$\sin \theta_1 = \frac{5461 \times 10^{-8}}{1/6000}$$

from which

$$\sin \theta_1 = 0.32766$$

or

$$\theta_1 = 19°7'36''$$

In the second order,

$$\sin \theta_2 = \frac{2\lambda}{d}$$

from which

$$\sin \theta_2 = 0.65532$$

or

$$\theta_2 = 40°56'38''$$

Notice that angle θ_2 is not twice the angle θ_1.

31-8
Diffraction and interference of X rays

As a result of the fundamental work of Laue and his collaborators in 1912, it was definitely shown that X rays are of the same nature as light and radio waves; that is, they are electromagnetic waves but of very short wavelength. If a narrow pencil of X rays is sent through a small thin crystal, such as a piece of rock salt, and received on a photographic plate a short distance away, as shown in Figure 31-17, the photograph will show a series of small spots arranged in a definite pattern. This pattern, as indicated in Figure 31-18, is a diffraction pattern formed by the action of the ions of the crystal because of their regular arrangement in the crystal. The crystal in this case acts as a three-dimensional diffraction grating, and the pattern obtained on the photographic plate is called a Laue pattern. From the distribution and intensities of the points on the photographic plate, the arrangement of the ions in the crystal can be deduced.

A slightly different arrangement of the X-ray beam and crystal used by Bragg gives a simpler pattern that is more easily interpreted. This is shown in Figure 31-19, in which X rays coming

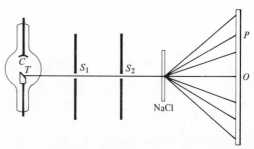

FIGURE 31-17 Arrangement of apparatus for producing a Laue diffraction pattern, using a rock-salt crystal. S_1 and S_2 are pinholes, and P is the photographic plate.

FIGURE 31-18 Photograph of Laue diffraction pattern of rock-salt. (Photograph courtesy of J. G. Dash.)

from the target T of the X-ray tube pass through two narrow slits and then are incident upon the face of a crystal, which is mounted on a spectrometer table. The crystal scatters the X rays in all directions, but the photographic plate is set so as to receive only that part which comes from the face of the crystal. The angle θ between the incident beam and the face of the crystal is changed slowly by rotating the crystal. In general, the photograph will show a series of sharp lines against a continuous background. An ionization chamber can be used instead of a photographic plate. In this case the ionization chamber measures the intensity of the X-ray beam entering it. With a very narrow slit in front of the window of the ionization chamber, it is found that the intensity of the X-ray beam coming from the crystal face is a maximum when this beam makes an angle θ with the crystal. It is for this reason that the beam is sometimes said to be "reflected" from the crystal.

A simple explanation of the action of the crystal in this case can be given with the aid of Figure 31-20, in which the ions of the crystal are arranged in layers parallel to the surface of the crystal. The distance d between atomic layers is shown greatly enlarged. The X-ray beam incident upon the crystal at an angle θ to its face penetrates the crystal and is scattered by the ions in all directions. Consider two rays I and II very close together, and consider only that part of the scattered beam which makes an angle θ with the surface of the crystal. Ray I strikes the upper surface at A and is reflected; ray II strikes the next layer at B and is reflected. These two rays, which are so close together that they give a single impression on the photographic plate, have traveled different distances. From the figure it is evident that ray II has traveled a longer distance than ray I. If originally they started out in phase, they will now differ in phase because of the difference in their paths. This difference in path is $CB + BD$, which is related to the distance d between atomic planes and the angle θ by the relationships

$$CB = d \sin \theta$$
$$BD = d \sin \theta$$

or

$$CB + BD = 2d \sin \theta$$

If this difference in path is a whole wavelength λ or a whole number n of wavelengths, then the two rays will reinforce each other after scattering and will produce an intense spot on the photographic plate or be registered as an intense beam by the ionization chamber. Thus, whenever

$$n\lambda = 2d \sin \theta \qquad (31\text{-}8)$$

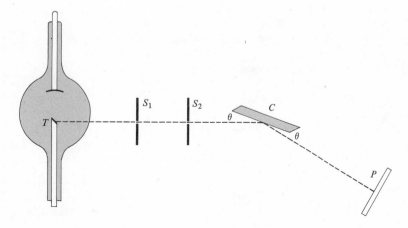

FIGURE 31-19 The single crystal X-ray spectrometer with photographic plate. S_1 and S_2 are narrow slits, C is the crystal, and P is the photographic plate.

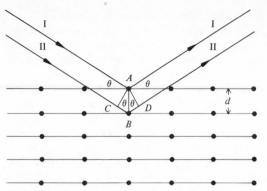

FIGURE 31-20 Reflection of X rays from atomic planes.

there will be reinforcement of the waves "reflected" by the crystal.

Equation 31-8 is called the *Bragg equation*; it is of fundamental importance in determining X-ray wavelengths and in the analysis of the structure of crystals. If the distance d between atomic or ionic planes is known, then the wavelength can readily be determined. The integer n is called the order of the spectrum; when n is 1, the path difference $CB + BD$ is 1 wavelength and is in the first-order spectrum; when n is 2, the path difference is 2 wavelengths, and the wavelength is said to appear in the second order. It will be noted that a crystal acts as a three-dimensional diffraction grating for X rays; the distance d is frequently referred to as the *crystal grating space*.

Illustrative example

An X-ray line of wavelength 1.541 Å is reflected from a quartz crystal in which the distance d between atomic planes is 4.255 Å. Determine the angle between the X-ray beam and the atomic planes (a) in the first order and (b) in the second order.

Solving the Bragg equation for sin θ, we get

$$\sin \theta = \frac{n\lambda}{2d}$$

(a) Now $\lambda = 1.541 \times 10^{-8}$ cm, $d = 4.255 \times 10^{-8}$ cm, and, in the first order, $n = 1$; calling θ_1 the angle between the X-ray beam and the atomic planes for the first-order reflection, we have

$$\sin \theta_1 = \frac{1.541}{2 \times 4.255}$$

from which

$$\sin \theta_1 = 0.1811$$

and

$$\theta_1 = 10°26'$$

(b) In the second order, $n = 2$; calling the angle between the X-ray beam and the atomic planes θ_2 for the second-order reflection, we have

$$\sin \theta_2 = \frac{2 \times 1.541}{2 \times 4.255}$$

from which

$$\sin \theta_2 = 0.3622$$

and

$$\theta_2 = 20°45'$$

31-9
X rays and crystal structure

In using the Bragg equation it is essential that we know independently either the X-ray wavelength λ or the spacing d between atomic or ionic planes. At the time of the discovery of X-ray diffraction and interference the crystal grating space of rock salt was determined from other data, and the value d so obtained was used to measure X-ray wavelengths. Figure 31-21 shows the arrangement of the ions in a rock-salt crystal. The crystal is known to be a cube from crystallographic studies; the centers of the ions are at the corners of the cubes.

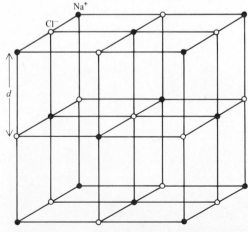

FIGURE 31-21 The arrangement of sodium (Na) ions and chlorine (Cl) ions in a crystal of salt.

It will be observed that each sodium ion (Na^+) is surrounded by 6 chlorine ions (Cl^-), and that each chlorine ion is surrounded by 6 sodium ions. If d is the length of the cube, then the volume of each cube is $V = d^3$. This is the volume associated with each ion. If M is the gram-molecular weight of sodium chloride and if ρ (rho) is its density, then the volume v of 1 mole of sodium chloride is

$$v = \frac{M}{\rho}$$

Now, there are $2N_0$ ions in each mole of sodium chloride, where N_0 is the Avogadro number; hence the volume associated with each ion is

$$V = \frac{v}{2N_0} = \frac{M}{2\rho N_0}$$

The distance d between ions is therefore

$$d = \sqrt[3]{\frac{M}{2\rho N_0}} \qquad (31\text{-}9)$$

All the quantities on the right-hand side of Equation 31-9 are known; hence d can be calculated as follows:

The gram-molecular weight M is the sum of the gram-atomic weights of sodium and chlorine and is

$$M = 22.990 + 35.453 = 58.453 \text{ gm}$$

The density is

$$\rho = 2.164 \, \frac{\text{gm}}{\text{cm}^3} \quad \text{at } 18°C$$

and

$$N_0 = 6.0222 \times 10^{23}$$

hence

$$d = \sqrt{\frac{58.453}{2 \times 2.164 \times 6.0222 \times 10^{23}}} \text{ cm}$$

from which

$$d = 2.820 \times 10^{-8} \text{ cm}$$

or

$$d = 2.820 \text{ Å} \quad \text{at } 18°C$$

With the value of the grating space of rock salt known, it is now possible to measure X-ray wavelengths with a single-crystal spectrometer and study X-ray spectra, and, using known wavelengths, to determine the grating spaces of other crystals. However, in order to use a crystal with an X-ray spectrometer, the crystal must be of a reasonable size, say 1 or 2 cm in length and width. Such large crystals, sometimes called single crystals, may be found in nature, or may sometimes be grown from saturated solutions in the case of various types of salt crystals or by the slow cooling of molten material in the case of metals.

31-10
Diffraction of X rays by powdered crystals

The ordinary solid is not a large crystal, but is made up of many very small crystals called *microcrystals*. Even when the solids are in powder form, they consist of many microcrystals. A very powerful method of X-ray analysis of these microcrystals was developed by A. W. Hull, and independently by P. Debye and P. Scherrer. This method consists in sending a very narrow pencil of X rays of a

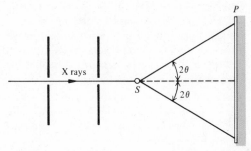

FIGURE 31-22 Method of obtaining X-ray diffraction patterns using a powder.

FIGURE 31-23 X-ray powder diffraction pattern of aluminum. (Reproduced by permission of A. W. Hull.)

FIGURE 31-24 X-ray powder diffraction of tungsten (From a photograph by L. L. Wyman, supplied by A. W. Hull.)

single wavelength or of a few known wavelengths through a very small sample of the powder or solid, as shown in Figure 31-22. Since the powder or solid consists of a great many microcrystals oriented at random, there is some probability that one of these microcrystals will be oriented so that its atomic planes make an angle θ with the incident radiation, which will satisfy the Bragg equation for the particular wavelength λ of the incident X rays. Because of the random orientation of these microcrystals, there undoubtedly will be other microcrystals whose atomic planes make the same angle θ with the incident X rays but in different planes of incidence. In each plane of incidence, the angle between the original beam and the reflected beam will be 2θ. A photographic plate, placed at right angles to the incident beam at a convenient distance from the sample, will record the diffraction pattern. It will be found that this

pattern consists of a series of circles, as shown in Figure 31-23. Each circle is the intersection of a cone of central angle 4θ and the plane of the photographic plate. The radius of such a circle depends upon the distance of the plate from the sample, the wavelength of the incident X rays, the order n of the diffraction, and the grating space d of the crystal.

In a variation of the above method, the photographic plate is replaced by a photographic film bent in the form of a cylinder with the sample at its center. Holes are cut in the film so that the direct pencil of X rays can enter and leave this camera without blackening the film. When this film is unrolled and developed, we obtain the type of pattern shown in Figure 31-24.

The X-ray powder diffraction method is capable of very high precision and is being used extensively for the analyses of crystal structures.

Questions

1. In order for two rays of light to interfere constructively, what condition must be met by (a) the path differences and (b) the phase differences?

2. What happens to the energy in two rays which interfere destructively?

3. What happens to the spacings between two bright lines in the interference pattern produced by light from two slits when the slits are placed closer together?

4. Explain why you cannot see interference between the two headlights of a distant oncoming automobile.

5. Why are the colors of thin oil films on pavements most frequently seen when the pavement is wet?

6. Suppose that in the experiment on Newton's rings, we first used red light and then blue light. Which set of rings would have the larger diameter and therefore the greater spacing between rings?

7. Could Newton's rings be observed with trans-

mitted light? If so, would the center spot be dark?

8. As a soap bubble evaporates, its surface appears dark just before the bubble bursts. Explain.

9. When white light is incident on a coated nonreflecting lens, light reflected from it is purple. Account for this color.

10. When a high-index film of quarter-wave thickness is deposited on a piece of glass, the reflectivity is increased. Show how several different pieces of glass coated in this way could be used to separate a white light image into several component colors. Such a system is used in color television cameras to split the beam with minimum loss of light.

11. When a distant line source of light (such as a showcase lamp) is examined through a slit, the image broadens as the slit is made narrower. Explain this.

12. Distinguish carefully between the resolving power of a telescope and its magnifying power.

Can the magnifying power of a telescope be increased without increasing its resolving power?

13. Discuss the meaning of the resolving power of a microscope. What factors limit the resolving power? Two microscopes have equal magnification. Do they necessarily have the same resolving power? Can you suggest a way to increase the resolving power of a microscope so that it can distinguish large molecules of about 1000 Å in length?

14. Holding your hand over your eye, look through the narrow slits formed between your fingers at a long, fluorescent tube that is parallel to the slits. Account for the dark lines that appear between your fingers.

15. What difference is there, if any, between the interference pattern formed by two slits and that formed by a diffraction grating with the same spacing between the slits?

16. Which would give more orders of spectra, a diffraction grating with 500 lines per centimeter or one with 5000 lines per centimeter?

17. Look at a street lamp through a piece of window screening. Attempt to account for the image.

18. Discuss the influence of X-ray crystallography on modern biology. (Read, for example, *The Double Helix* by J. D. Watson, published by Atheneum Press.)

Problems

1. Yellow sodium light whose wavelength is 5893 Å comes from a single source and passes through two narrow slits 0.8 mm apart. The interference pattern is observed on a screen 200 cm away. How far apart are the adjacent bright bands?

2. Light from a mercury arc, after passing through a green filter, is then incident on two narrow slits 0.05 cm apart. The interference pattern is formed on a screen 220 cm away. The distance between two adjacent green lines is found to be 2.40 mm. Determine the wavelength of the light.

3. Light of 4300 Å falls on two narrow slits and the third-order bright line occurs at an angle of 30° with the forward direction. How far apart are the slits?

4. Light of wavelength 5000 Å falls on a slit that is 3×10^{-4} cm wide. Find the angular deviation of the first dark band.

5. When light of wavelength 5600 Å falls on a slit, the first two dark fringes on either side of the central maximum are separated by 60°. How wide is the slit?

6. Light of wavelength 5200 Å falls on a diffraction grating having 3000 lines per centimeter. Find the angular position of the first-order image.

7. Light of wavelength 4400 Å falls on a grating and the second-order image is seen at an angle of 30° from the central image. Determine the number of lines per centimeter in the grating.

8. An X-ray beam of wavelength 1.1 Å is reflected from a crystal and the first-order beam is found to make an angle of 12.6° with the crystal planes. Find the crystal grating space.

9. An X-ray beam of wavelength 1 Å is allowed to fall on an optical diffraction grating having 6000 lines per cm. Find the angular position of the first-order line, assuming that the material of which the grating is made can be treated as if it were opaque to X rays.

10. The grating space of sodium chloride crystals is 2.82 Å. Find the position of the second-order Bragg scattering for X rays of wavelength 0.5 Å.

11. Five percent of the incident light that strikes an air-glass surface is reflected back. What percentage of the incident light is transmitted after passage through an optical system containing five surfaces?

12. Calculate the thickness, in centimeters, of a nonreflecting film for coating a lens of index of refraction 1.62 assuming sodium light, of wavelength 5893 Å in vacuum, is incident on it. The wavelength of light in a medium of index of refraction n is λ/n.

13. (a) Referring to Figure 31-7, show that if the thickness of the object placed between the two glass plates to produce the wedge-shaped air film is T, and if monochromatic light of wavelength λ is incident normally on the top surface, then the total number N of dark lines produced is given by

$$N = \frac{2T}{\lambda}$$

(b) Calculate the total number of dark lines that will be produced when sodium yellow light is incident normally on a wedge-shaped air film produced by inserting a piece of steel of thickness 0.02 cm between the two glass plates.

14. A human hair is placed between two glass plates at one end, forming a wedge of air. When illuminated with mercury green light of wavelength 5461 Å, 85 dark lines are observed. How thick is the hair?

15. A soap bubble has a thickness of 2500 Å and an index of refraction of 1.33. It is illuminated with white light. Show that it will appear to be blue.

16. Two glass plates 15 cm long are in contact at one end and separated at the other end by a piece of metal foil, 10^{-3} in. thick. The plates are illuminated normally by light of wavelength 6600 Å. (a) How far apart are the resulting dark lines? (b) How many dark lines would there be if the plates were illuminated with light of wavelength 4800 Å?

17. A diffraction grating has 1000 lines per centimeter. Light of wavelength 5800 Å falls on the grating. How many images can be seen?

18. A diffraction grating having 4500 lines per centimeter has light from hydrogen incident on it. At what angle will blue light from hydrogen, 4861 Å, fall in (a) the first order and (b) the second order?

19. A grating having 6000 lines per centimeter is illuminated with red light of wavelength 7281 Å. Find the angular position of this line for all the orders in which it appears.

20. Light of wavelength 4500 Å falls on a slit of width 8000 Å. Estimate the width of the central image of the slit on a screen that is 3 m from the slit.

21. A diffraction pattern is formed by sending green light of wavelength 5500 Å through a circular aperture 0.8 mm in diameter and allowing it to fall on a screen 1.5 m away. Determine the diameter of the first dark ring of the diffraction pattern.

22. The yellow lines in the spectrum of sodium, sometimes called the D line, consists of two lines very close together when viewed with a spectroscope of moderate resolving power. The wavelengths of these lines are $D_1 = 5896$ Å and $D_2 = 5890$ Å. Determine the angular separation of these lines when examined with a diffraction grating having 5500 lines per centimeter and viewed in (a) the first order and (b) the second order.

23. Derive a formula for the Nth Newton's ring in the following way: Consider a lens of radius of curvature R in contact with a plane surface. At a distance r from the point of contact, the thickness of the air wedge is t. (a) From the Pythagorean theorem, show that

$$R^2 = r^2 + (R - t)^2$$

(b) Show that if t is very small compared to R, then t is given, to good approximation, by

$$t = \frac{r^2}{2R}$$

(c) From this, show that the radius of the Nth dark ring r_N is given by

$$r_N = \sqrt{N\lambda R}$$

24. The resolving power of a telescope may be defined quantitatively by $1/\theta$, where θ is the angular separation of two objects that can just be resolved or separated by the instrument. For a telescope with circular lenses, it can be shown that the resolving power is given approximately by

$$\frac{1}{\theta} = \frac{a}{1.22\lambda}$$

where a is the diameter of the objective lens. Treat the human eye as a simple telescope and assume that the resolving power is limited by diffraction. (a) What is the maximum distance at which the headlights of an oncoming car can be resolved? (Assume that the headlights are 120 cm apart, the wavelength of the light is 5500 Å, and that the diameter of the eye pupil is 7 mm at night.) (b) At what maximum distance could the headlights be resolved if viewed through a telescope with an objective lens 3 cm in diameter?

modern physics

part 6

atomic physics & quantum mechanics

chapter 32

32-1
Introduction

Until late in the nineteenth century, physics was solidly based on the work of Newton and Maxwell. Classical physics had achieved extensive quantitative success in describing a broad range of phenomena. Even thermodynamics and statistical mechanics, which were to some extent distinct from the rest of classical physics, seemed to be consistent with classical mechanics and electromagnetism. As we have seen, the theory of relativity introduced revolutionary changes to the structure of physics at the turn of the century. At the same time the discovery of quantum mechanics introduced a second set of revolutionary ideas.

The concept of atoms was put on firmer ground with the development of statistical mechanics. In particular, the success of the theory in describing the behavior of gases led to the general acceptance of the idea that matter is composed of atoms and molecules. This led to an attempt to understand the interaction of electromagnetic radiation and matter—the absorption and emission of light. As we shall see in the rest of this chapter, it became necessary to alter some of the ideas of classical physics. First it became necessary to assume that electromagnetic waves sometimes behaved as though they consisted of particles, called photons. The fact that electromagnetic radiation had a dual nature, sometimes behaving like waves and sometimes like particles, was mirrored later with the discovery that particles, such as electrons, sometimes behaved like waves in that they exhibited interference and diffraction. Perhaps most important of all, it was discovered that the laws of

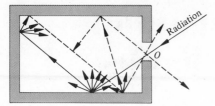

FIGURE 32-1 The hole at *O* acts as a blackbody.

physics, as then known, had to be altered. The behavior of atoms and their constituents could not be described in terms of Newton's laws (even when modified to take special relativity into account). It became necessary to invent new equations to replace Newton's laws in order to describe the behavior of atomic and subatomic particles. These new equations, which form the basis of the theory called quantum mechanics, described such objects on the atomic scale in a radically new way, while simultaneously retaining the classical description of macroscopic objects.

32-2
Blackbody radiation

Every object, no matter what its temperature, radiates energy in the form of electromagnetic waves that travel with the speed of light. In general, the intensity, frequency distribution, and temperature-dependence of this emitted radiation depends on the material. However, it is possible to describe the behavior of an idealized heated solid, called a *blackbody*, as described in Chapter 20. A blackbody is defined as a body that absorbs all of the radiation incident on it; that is, none of the radiation is reflected. The light-emitting properties of such a body turn out to be independent of its particular material; thus the blackbody plays a role similar to that of the ideal gas.

The blackbody used in *experimental* investigations usually consists of a hole in one side of a hollow box with well-insulated walls, as shown in Figure 32-1. A tungsten filament inside the hollow box can be heated to any desired temperature. The radiation inside the box will consist of that emitted by the filament, by the walls, and that entering from the outside through the hole in the box. The radiation interacts continually with the atoms and molecules of the walls of the box and with the tungsten filament. There is absorption and reemission of radiation from these atoms and molecules, and ultimately equilibrium is established between the

radiation and matter. A small amount of this radiation emerges from the blackbody and is typical of the radiation inside the box. This radiation, also called *thermal radiation*, is characteristic of the temperature of the filament and walls of the box with which it is in equilibrium.

The radiation from a blackbody is sent through a spectrometer or spectrograph, which disperses the radiation into its component wavelengths; this is usually called the *spectrum* of the radiation. The intensities—that is, the energies within very narrow ranges of wavelengths—are then measured throughout the spectrum. Figure 32-2 is a graph of the relative intensity as a function of the wavelength of the radiation. The lower curve shows the distribution of intensities in blackbody radiation at a temperature T_1. It will be observed that there is one particular wavelength λ_1 at which the intensity is a maximum. This is typical of blackbody radiation. The upper curve shows the distribution of intensities in blackbody radiation at a higher temperature T_2. This has the same general features as the lower curve except that the intensities are greater at all wavelengths and the particular wavelength λ_2 at which the intensity is a maximum is shifted toward shorter wavelengths. A very simple relationship is found to exist between the wavelength at which the intensity is a maximum and the temperature of the blackbody. This relationship, known as *Wien's displacement law*, is

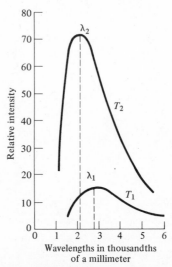

FIGURE 32-2 Blackbody radiation curve showing the distribution of intensities among the different wavelengths at two different temperatures.

$$\lambda_1 T_1 = \lambda_2 T_2 = b = \text{constant} \qquad (32\text{-}1)$$

The constant b, as evaluated experimentally, is

$$b = 0.2897 \text{ cm K} \qquad (32\text{-}2)$$

This suggests a convenient method for determining the temperature of a blackbody or a body that closely approximates one. The light from such a body is examined spectroscopically, and the wavelength that has the greatest intensity is determined. Its temperature can then be calculated from Equation 32-1. For example, the sun approximates a blackbody; its most intense radiation has a wavelength of 5×10^{-5} cm; this corresponds to a temperature of about 6000 K. Temperatures of some of the bright stars have been determined by this method.

One of the major reasons physicists have been interested in blackbodies is that it is possible to invoke fairly general thermodynamic arguments to predict the characteristics of the radiation, even though we may not have a detailed knowledge of the structure of solids or a detailed understanding of the interaction between the radiation and the solids.

From such thermodynamic arguments, coupled with electromagnetic theory, it is possible to show that

1. The spectrum of a blackbody should depend on the temperature but not on the material of which the body is made.
2. The temperature T (measured on the Kelvin scale) of the blackbody is related to the wavelength, λ_m, at which maximum emission occurs, by Wien's law:

$$\lambda_m T = \text{constant} = b \qquad (32\text{-}1)$$

3. The total energy, R, radiated by the blackbody per unit area per unit time is related to the temperature by the Stefan-Boltzmann law, which was given in Chapter 20 as

$$R = \sigma T^4 \qquad (20\text{-}2)$$

Thus classical physics was able to predict the maximum of the spectrum using Wien's displacement law, and the area under the curve using the Stefan-Boltzmann law. However, in order to predict the *shape* of the curve, detailed calculations had to be made based on models of the interaction between the radiation and the atoms of which the body was made. All such attempts failed. In fact, the best of these calculations predicted that the

rate of emission should become *infinite* at very short wavelengths—a result described as the *ultraviolet catastrophe*.

In 1900 Max Planck proposed an empirical formula to describe the shape of the blackbody spectrum. This formula gives the dependence of the intensity R_λ at a particular wavelength λ on the temperature T and the wavelength, as follows:

$$R_\lambda = \frac{c_1}{\lambda^5} \frac{1}{e^{-c_2/\lambda T} - 1} \qquad (32\text{-}3)$$

The constants c_1 and c_2 are usually written in terms of other constants h and k as

$$c_1 = 2\pi c^2 h \qquad (32\text{-}4)$$

and

$$c_2 = \frac{hc}{k} \qquad (32\text{-}5)$$

where c is the speed of light; h, Planck's constant, is given by

$$h = 6.624 \times 10^{-27} \text{ erg sec} \qquad (32\text{-}6)$$

and k, Boltzmann's constant, is given by

$$k = 1.381 \times 10^{-16} \frac{\text{erg}}{\text{K}} \qquad (32\text{-}7)$$

Equation 32-3 was an empirical formula; it described the curve but was not based on any theory. In fact it was not possible to derive the equation from classical physics. Planck was able to derive the equation only by introducing a radical alteration of the ideas of classical physics. In describing the interaction between the radiation and matter, Planck postulated that radiation is emitted or absorbed in discrete units or quanta; each such unit is called a *quantum of radiation*. A quantum of radiation possesses a certain amount of energy $\mathscr{E}$ that is proportional to the frequency f of the radiation; that is,

$$\mathscr{E} \propto f$$

or

$$\mathscr{E} = hf \qquad (32\text{-}8)$$

where f is the frequency of the radiation and h is Planck's constant, which is given by Equation 32-6.

There was no precedent in classical physics for this postulate of quantization. Planck proposed the postulate solely as a way of deriving the equation for the spectral distribution of blackbody radiation. Planck himself described the postulate as heuristic and for many years was loathe to accept the idea as true. Within a few years, however, the work of Einstein, Bohr, and others confirmed the existence of light quanta, as we shall see in following sections.

32-3
Further evidence for the photon

In 1905, the same year in which he proposed the special theory of relativity, Albert Einstein extended the work of Planck and invoked the idea of the quantization of light to explain the photoelectric effect. This theory (for which he won the Nobel Prize) provided solid evidence for the existence of the photon. With the acceptance of this explanation of the photoelectric effect, it became difficult to consider the quantization of light as merely a convenient way to derive the blackbody spectrum.

In essence, the photoelectric effect is the ejection of electrons from metals by incident electromagnetic radiation. Since electromagnetic radiation consists of electric and magnetic fields, it is not surprising that radiation can exert forces on the electrons in metals that could push electrons out of the material. However, the details of the process turned out to be surprising. Using classical ideas, it is possible to make some predictions about the photoelectric effect that turn out to disagree with experimental evidence.

Let us consider first some of these reasonable, but erroneous predictions. Classically, the energy of a beam of electromagnetic radiation, or the intensity of the beam, depends on the magnitude of the electric and magnetic fields, and not on the wavelength or frequency. That is, for any frequency it is possible to construct a beam of arbitrary intensity by varying the size of the electric field. In the photoelectric effect, the beam of radiation strikes the metal and transfers some of its energy to the electron, which is ejected. It is reasonable, therefore, to assume that the energy of the ejected electron should increase with increasing intensity of the incident beam of light. Since the electron has mass, it is possible that its inertia will prevent it from responding to the oscillation of very-high-frequency waves, and so it might happen that ultraviolet radiation would be less effective at ejecting electrons than infrared radiation. In the absence of incident light, the electrons are bound to the metal; the electrons cannot be ejected unless some minimum amount of energy is added to them.

Thus if a very-low-intensity beam of light is incident on the metal, we might expect that no electron could be ejected until after this minimum amount of energy had been added to the metal. That is, for low-intensity light beams, we might predict that there would be a time delay between the ejection of one electron and the next, and this time delay would increase as the intensity of the light beam was decreased.

The experimental data show marked disagreement with these reasonable predictions. For any particular wavelength and intensity of incident light, the emitted electrons have a distribution of energies up to some maximum kinetic energy $\mathcal{E}_m$. This maximum kinetic energy of the electrons turns out to be *independent* of the intensity of the light beam and to vary linearly with the frequency of the light beam. Thus electrons ejected by ultraviolet light have much more energy than electrons ejected by infrared light. Further, for each material there exists a characteristic minimum cutoff frequency f_c. If light of frequency less than f_c is incident on a material, then no electrons will be ejected. Finally, it turns out that the time interval between electron emissions is *random* and sometimes so short that according to the classical theory the beam could not have transmitted enough energy to eject the second electron.

In order to explain the photoelectric effect, Einstein asserted that when the light interacted with the electrons it did so in the form of photons colliding with the electrons. An electron was emitted when, and only when, a photon struck the electron and transferred enough energy to the electron. In Einstein's model, the electron absorbed the entire photon, whose energy was hf, and the electron was emitted with kinetic energy $\frac{1}{2}mv^2$, where

$$hf = \tfrac{1}{2}mv^2 + \mathcal{E}_s \qquad (32\text{-}9)$$

where $\mathcal{E}_s$ is called the *work function* of the metal and represents the energy necessary to get the electron through the surface of the metal. This model fit the data extremely well, and this explanation of the photoelectric effect was probably the most important step in firmly establishing the particle character of radiation. Equation 32-9 is usually called Einstein's photoelectric equation.

Additional verification of the photon idea was provided by the Compton effect, which will be discussed in the next chapter.

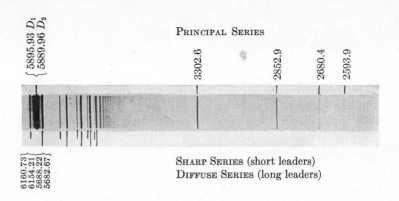

PRINCIPAL SERIES

FIGURE 32-3 Photograph of the emission spectrum of sodium. The lines are classified into different series. The numbers represent the wavelengths of the lines in Å. (Reprinted by permission from *Atomic Spectra and Atomic Structure*, G. Hertzberg. Dover Publications, New York.)

SHARP SERIES (short leaders)
DIFFUSE SERIES (long leaders)

32-4
Spectra

As we have seen, Planck was able to explain the continuous spectra emitted by blackbodies. Heated gases at relatively low pressures emit distinctly different types of spectra; these are called *line* or *band* spectra. A typical line spectrum, that of sodium vapor, is shown in Figure 32-3. These line and band spectra are characteristic of the atoms, molecules, and ions that emit the light and are used extensively as tools for identification of particular materials. Similarly, gases absorb light in a characteristic way. Figure 32-4 shows the absorption spectrum of sodium vapor.

At the beginning of the twentieth century a very large number of these characteristic line spectra had been measured and tabulated. Although some regularities had been noted, there was no theoretical model that predicted the observed frequencies or the distribution of intensities. Furthermore, there appeared few simple connections among the spectra, and no way to summarize the data short of simply tabulating the numbers.

Hydrogen, the simplest element, has a relatively simple spectrum, with four prominent lines in the visible region of the spectrum. In 1885 Johann Balmer discovered a formula that gave the wavelengths of these lines to very good accuracy. Balmer's formula can be put in the form

$$\frac{1}{\lambda} = R\left(\frac{1}{2^2} - \frac{1}{n^2}\right) \qquad (32\text{-}10)$$

in which λ is the wavelength of a spectral line for a given value of n, and n is a whole number greater than 2. For example, when $n = 3$, the wavelength is that of the red line of the hydrogen spectrum, known as the H_α line. Similarly, when $n = 4$, the wavelength is that of the blue line, H_β. The constant R is now called the Rydberg constant and had the value

$$R = 109{,}677.76 \text{ cm}^{-1}$$

Balmer had predicted the existence of lines with values of n equal to 7, 8, . . . , and at least 30 of these lines have been found. As may be seen in Figure 32-5, these lines crowd closer and closer together and approach a limit, which is called the *series limit*. This remarkable achievement of Balmer remained empirical until Bohr provided a theoretical basis for the formula. The work of Bohr has subsequently been modified and extended, but it played such an important role in the development of atomic physics that it is worthwhile presenting the original simplified version (see Section 32-5).

When X rays from any target are analyzed with a crystal spectrometer, the spectrum is found to consist of a series of sharp lines superposed on a continuous background of radiation (see Figure 32-6). The energy in the continuous radiation, sometimes referred to as the *continuous spectrum*,

FIGURE 32-4 Photograph of the absorption spectrum of sodium, showing some of the lines in the ultraviolet region. (Reproduced by permission from *Atomic Spectra and Atomic Structure*, G. Hertzberg. Dover Publications, New York.)

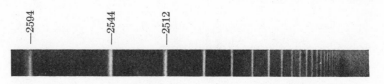

FIGURE 32-5 Photograph of the emission spectrum of hydrogen, showing the Balmer series lines in the visible and the near ultraviolet regions. The numbers represent the wavelengths of the corresponding lines in Å. (Reprinted by permission from *Atomic Spectra and Atomic Structure*, G. Hertzberg. Dover Publications, New York.)

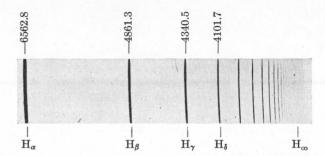

is found to depend upon the voltage across the tube, the current in it, and the atomic number of the element that constitutes the target; the higher the atomic number, the greater the energy, if the other quantities remain constant. The sharp lines that are superposed on the continuous radiation are found to be characteristic of the element of the target.

The first systematic study of the characteristic X-ray spectra of the elements was made by Moseley in 1913. He used a modification of the Bragg method; the crystal spectrometer and the photographic plate were placed in an evacuated chamber to avoid absorption in air of the X rays of long wavelength. Each element investigated was used as the target of an X-ray tube. He found that all the elements gave similar types of spectra; the lines emitted by each element were classified into two groups or series: a group of short wavelength, called the *K* series, and a group comparatively long wavelength, called the *L* series. These two series are widely separated from one another in wavelength, as illustrated in Figure 32-7 for the case of silver. Other investigators have found two other series of still longer wavelengths in the heavier elements, classified as *M* series and *N* series. For example, the frequency of the most intense line of the *K* series can be written in the form

$$f = cR(Z - 1)^2 \left(\frac{1}{1^2} - \frac{1}{2^2} \right)$$ (32-11)

where R is the Rydberg constant, c is the velocity of light, and Z is the atomic number of the element emitting this line of frequency f.

32-5
Bohr's theory of the hydrogen atom

During 1910 and 1911 Ernest Rutherford and his co-workers performed a series of experiments in which they measured the scattering of α particles (particles emitted by certain radioactive materials and having the same properties as positively charged helium ions) by thin metal foils. From these experiments it became clear that an atom consists of a small, massive, positively charged nucleus, and light, negative electrons located very far from the nucleus.

This picture of the atom immediately raised the difficulty that the Coulomb force of attraction between the nucleus and the electrons should cause the atom to collapse, unless the electrons were moving in orbits similar to the orbits of the planets about the sun. Such a planetary atom is particularly simple to describe in the case of hydrogen, which consists simply of a single proton and a single electron. In that case the force on the electron is the Coulomb force,

$$F = \frac{1}{4\pi\epsilon_0} \frac{e^2}{r^2}$$ (32-12)

This force provides the centripetal force, so

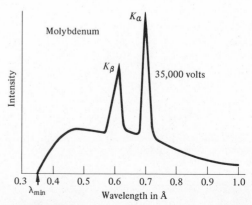

FIGURE 32-6 Characteristic K_α and K_β lines superposed on the continuous X-ray spectrum of molybdenum. Note the sharp cutoff at the wavelength minimum.

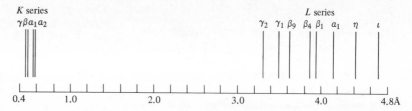

FIGURE 32-7 Relative positions of the K and L X-ray series lines of silver.

$$F = \frac{mv^2}{r} \qquad (32\text{-}13)$$

The total energy of this system is equal to the sum of the potential energy of the system and the kinetic energy of the electron, since the nucleus may be assumed to be at rest because it is so massive compared to the electron. Thus, the total energy $\mathcal{E}$ is given by

$$\mathcal{E} = -\frac{1}{4\pi\epsilon_0} \frac{e^2}{r} + \tfrac{1}{2}mv^2 \qquad (32\text{-}14)$$

From Equations 32-12 and 32-13 we have

$$mv^2 = \frac{1}{4\pi\epsilon_0} \frac{e^2}{r} \qquad (32\text{-}15a)$$

or

$$\tfrac{1}{2}mv^2 = \tfrac{1}{2}\left(\frac{1}{4\pi\epsilon_0}\right)\frac{e^2}{r} \qquad (32\text{-}15b)$$

and, substituting Equation 32-15b into Equation 32-14, we obtain

$$\begin{aligned}
\mathcal{E} &= -\frac{1}{4\pi\epsilon_0}\frac{e^2}{r} + \tfrac{1}{2}\left(\frac{1}{4\pi\epsilon_0}\right)\frac{e^2}{r} \\
&= -\tfrac{1}{2}\left(\frac{1}{4\pi\epsilon_0}\right)\frac{e^2}{r} \\
&= -\frac{1}{8\pi\epsilon_0}\frac{e^2}{r} \qquad (32\text{-}16)
\end{aligned}$$

The minus sign for the total energy indicates that it is assigned the value zero when the electron is far removed from the nucleus, or when r is very great, theoretically when $r = \infty$. As the electron gets closer to the nucleus, the atom loses energy, presumably in the form of radiation. From kinetic theory it is possible to make estimates of the size of the atom, r, and therefore to estimate the total energy. These estimates are in reasonable agreement with the amount of energy necessary to remove the electron to infinity, the ionization energy.

There is nothing in this model that predicts that the atom should emit light in the form of sharp

spectral lines. Further, if the atom emits light, then the total energy of the atom should decrease. This means that the energy as given by Equation 32-16 should go toward larger negative values, and therefore that the radius of the atom should decrease. Thus, in this model, the atom should collapse. Reasonable calculations of the time it takes for the atom to collapse predict that the atom can exist for only very small fractions of a second.

In 1913, only two years after the concept of the nuclear atom had been demonstrated, Niels Bohr published the first of several papers that described his theory of the structure of atoms. The theory assumes the planetary atom described above, and introduces two hypotheses to overcome the difficulties encountered. The first postulate restricts the value of the angular momentum an electron can have in a circular orbit to those that satisfy a certain restrictive condition, which may be put in the form of the equation

$$mvr = n\frac{h}{2\pi} \qquad (32\text{-}17)$$

where mv is the momentum of the electron at any point in its circular orbit of radius r, and mvr is the angular momentum of the electron. The quantity n is an integer of any of the values 1, 2, 3, . . . , and h is the Planck constant. Bohr's first postulate can then be stated as follows: only those electron orbits are permissible for which the angular momentum of the electron is a whole multiple of $h/2\pi$. The orbits that satisfy this restrictive condition are usually called *stationary orbits*. The integer n is called a *quantum number*. In subsequent modifications of the theory it became necessary to introduce other quantum numbers to describe the orbit, and call the integer n the *principal quantum number*.

The second postulate states that whenever the energy of the atom is decreased from its initial value $\mathcal{E}_i$ to some final value $\mathcal{E}_f$, the atom emits radiation of frequency f in the form of whole quanta or photons of energy hf, given by the equation

$$\mathscr{E}_i - \mathscr{E}_f = hf \qquad (32\text{-}18)$$

It is interesting and simple to calculate the radii and the energies of the permissible orbits and then to compare the frequencies of the radiations predicted by Bohr's second postulate with those actually observed experimentally.

The radii of the permissible orbits can be obtained by eliminating v from Equations 32-15 and 32-17 and solving for r, yielding

$$r = n^2 \, \frac{\epsilon_0 h^2}{\pi m e^2} \qquad (32\text{-}19)$$

The radii of the orbits are then given in terms of constants and the integers $n = 1, 2, 3, \ldots$ characterizing the particular orbits. By substituting the known values of h, m, and e, and setting n equal to 1, we find that the radius of the first orbit is

$$r_1 = 0.529 \times 10^{-10} \text{ m}$$

This value is in good agreement with the size of the hydrogen atom as determined by kinetic theory.

The energy of any orbit characterized by quantum number n can be determined by eliminating r from Equations 32-19 and 32-16, yielding

$$\mathscr{E} = -\frac{1}{n^2} \frac{m e^4}{8 \epsilon_0^2 h^2} \qquad (32\text{-}20)$$

Applying Bohr's second postulate, we find that the frequency f of the radiation emitted when an electron goes from its initial orbit n_1 to another orbit n_2 is

$$f = \frac{\mathscr{E}_i - \mathscr{E}_f}{h}$$

and therefore

$$f = \frac{m e^4}{8 \epsilon_0^2 h^3} \left(\frac{1}{n_f^2} - \frac{1}{n_i^2} \right) \qquad (32\text{-}21)$$

It is convenient to rewrite Equation 32-21 in terms of the wavelength of the emitted radiation, since that is the measured quantity. Since

$$f\lambda = c$$

we have

$$\frac{1}{\lambda} = \frac{f}{c}$$

and so

$$\frac{1}{\lambda} = \frac{m e^4}{8 \epsilon_0^2 h^3 c} \left(\frac{1}{n_f^2} - \frac{1}{n_i^2} \right) \qquad (32\text{-}22)$$

This equation is identical in form to the Balmer formula (Equation 32-10). Inserting the known values of the constants, we obtain

$$\frac{m e^4}{8 \epsilon_0^2 h^3 c} = 10{,}974{,}000 \text{ m}^{-1}$$
$$= R$$

which is in good agreement with the experimental value of the Rydberg constant; so Equation 32-22 can be written

$$\frac{1}{\lambda} = R \left(\frac{1}{n_f^2} - \frac{1}{n_i^2} \right) \qquad (32\text{-}23)$$

Bohr's theory provides an explanation of the emission of radiation by means of a mechanical model of the atom. The hydrogen atom was pictured as a series of concentric rings surrounding the nucleus, as shown in Figure 32-8, with the electron in one of them at any instant. These rings represent the stationary orbits of the electron, and their radii are known from Equation 32-19. Each ring is characterized by a quantum number n and a definite energy, $\mathscr{E}_n$, given by Equation 32-20. If the electron were initially in the orbit for which $n = 3$ and then "jumped" to the orbit for which $n = 2$, radiation would be emitted of such wavelength as to yield the red line of the visible spectrum. If it jumped from orbit $n = 4$ to orbit $n = 2$, the wavelength of the radiation corresponded to the blue line. In other words, an electron starting from any orbit for which n was greater than 2 and then jumping to the orbit for which $n = 2$, would emit radiation of wavelength equal to that of one of the lines of the Balmer series. Since any small quantity of hydrogen contains an enormous number of atoms, when hydrogen is excited to emit radiation, all the lines of the Balmer series are emitted simultaneously.

In addition to the Balmer series other spectral lines of hydrogen were known; some in the far-ultraviolet region had been discovered by Lyman in 1906, while others in the infrared region had been discovered by Paschen in 1908. The lines of the Lyman series are accounted for by jumps of electrons from outer orbits to the innermost orbit, for which $n_f = 1$. In a similar manner the lines of the Paschen series are accounted for by jumps of electrons from outer orbits to the orbits for which $n_f = 3$. Lines of two other series in the far-infrared region for which $n_f = 4$ and for which $n_f = 5$ were later found by Brackett and Pfund. The relative

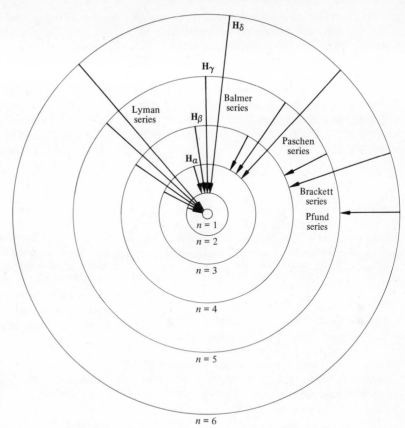

positions of the lines of the known series of hydrogen are shown in Figure 32-9.

The theory accounts for the absorption spectra by assuming that an atom may absorb a photon *only* if the energy hf so absorbed brings the electron precisely to one of the permissible orbits. If the energy of the photon is too small or too large then the atom will not absorb the photon, and the electron will remain in its initial orbit. Normally,

most of the atoms of hydrogen gas have their electrons in the innermost orbit. Thus one would expect that most of the absorption would be from $n = 1$ to orbits with higher n; thus the absorption spectrum of hydrogen should be at the same frequency as the Lyman series of emission lines. When the hydrogen is heated, enough energy is added to the atoms that an appreciable number may be found with electrons in orbits with $n = 2$ and

FIGURE 32-9 Relative positions of the lines of the different spectral series of hydrogen. The upper scale is in Å, whereas the lower scale is the reciprocal of the wavelength in centimeters.

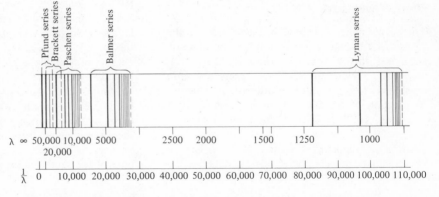

$n = 3$. Therefore absorption lines corresponding to the Balmer and Paschen series appear only in hydrogen that is heated or otherwise excited.

Consider then an atom that is in the state where $n = 1$. If a photon of energy corresponding to any of the lines in the Lyman series strikes the atom, the atom will absorb the photon and the electron will end up in one of the larger orbits. If the photon has energy $\mathcal{E}_L$, the limiting energy of the Lyman series, where

$$\mathcal{E}_L = \frac{me^4}{8\epsilon_0{}^2 h^2} \qquad (32\text{-}24)$$

then the final state of the atom will be that where $n = \infty$, $r = \infty$, or the electron will be free of the atom; that is, the atom will be ionized. If the photon has any energy hf larger than $\mathcal{E}_L$, the electron will have some kinetic energy, $\frac{1}{2}mv^2$, where

$$\tfrac{1}{2}mv^2 = hf - \mathcal{E}_L \qquad (32\text{-}25)$$

The energy of the lowest state of the hydrogen atom is that for which the principal quantum number n is equal to 1. This is sometimes called the *ground state*. The minimum energy necessary to remove the electron from the hydrogen atom is thus equal to $\mathcal{E}_L$. By substituting the known values of the constants into Equation 32-24, we obtain

$$\mathcal{E}_L = 21.74 \times 10^{-19} \text{ joule} = -13.58 \text{ eV}$$

This is the minimum energy that must be supplied to the hydrogen atom to ionize it, and is called the *ionization energy*. The atom may be ionized by photons of energy greater than or equal to $\mathcal{E}_L$ or by a beam of electrons having at least this energy. If the incident energy is greater than $\mathcal{E}_L$, the emitted electron will have kinetic energy given by Equation 32-25.

32-6
Energy-level diagrams

The very simplicity of the Bohr theory leads to the danger that the theory may be taken too literally. In present theory we cannot think of the electron as a point moving in an orbit. Therefore we are most likely to be led astray in attempting to picture the motion of the electron in its orbits and in "jumping" or "falling" from one orbit to another. It is safer, and somewhat simpler, to ignore the electron and think of the atom as a whole. That is, we can think of permissible energy states of the

atom characterized by a quantum number n, with the energies given by Equation 32-20. Rather than considering the electron as jumping from one orbit to another, we can think of the *atom* making a transition from a *state* n_i to a *state* n_f, and in that process the energy of the atom changes and a photon is either absorbed or emitted. The energy of the photon is equal to the difference in energies of the initial and final states of the atom. In addition, the ionized atom has a continuous range of positive energies.

This description of the hydrogen atom can be summarized in a simple and elegant manner by means of an energy-level diagram, shown in Figure 32-10. In this diagram the energies of the atom in the states corresponding to the different quantum numbers n are plotted on a vertical scale. Two different sets of numbers are shown on the diagram. The numbers on the right are the reciprocals of wavelengths in centimeters. It will be recalled that a quantity of energy, $\mathcal{E}$, may be expressed in terms of a frequency f by

$$\mathcal{E} = hf$$

and since for electromagnetic radiation

$$f\lambda = c$$

we have

$$\mathcal{E} = hc\left(\frac{1}{\lambda}\right) \qquad (32\text{-}26)$$

Thus the reciprocal of the wavelength, $1/\lambda$, when multiplied by hc will yield the energy of the state. The scale for $1/\lambda$ in reciprocal centimeters is chosen with the zero level as that of the ionized state of the atom. The difference between the numbers for any two energy levels gives the reciprocal of the wavelength of the radiation emitted when the atom makes a transition from the upper level to the lower one.

The numbers on the left give the energies of the levels in electron volts, but with the zero level as the lowest level and the ionization energy as the energy of the highest level corresponding to a bound atom. Above this region the electron is free. The numbers on the vertical lines between any two levels represent the wavelengths, in angstrom units, of the radiation emitted when the energy of the atom changes from one level to the other. All lines ending at the same level represent spectral lines of the same series; an attempt has been made

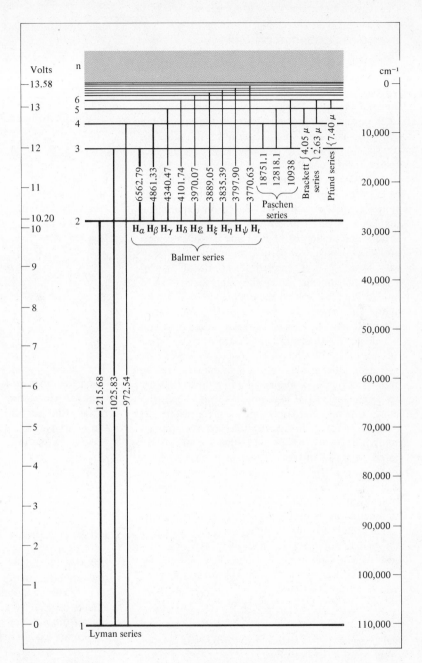

FIGURE 32-10 The energy-level diagram for hydrogen.

to represent the relative intensities of the spectral lines by the thickness of the lines in the diagram.

The advantage of the energy-level diagram is that it is independent of any particular model of the atom; only experimental data are used in its construction. The location of the levels is determined from the fact that when an atom emits or absorbs light of frequency f, the energy of the atom changes from one level, $\mathcal{E}_i$, to another, $\mathcal{E}_f$, and

$$hf = \mathcal{E}_i - \mathcal{E}_f$$

Using this condition, the energy-level diagrams of a very large number of atoms, in addition to hydrogen, have been constructed.

32-7
The correspondence principle

The existence of sharp spectral lines, the stability of atoms, the details of the emission of light by hot solids, the photoelectric effect—all of these were unexplainable by classical physics. It became clear to Bohr and the other physicists of the turn of the century that some modification of the laws of physics would have to be made. However, in their search for the correct form of these modifications, the physicists were guided by an idea that Bohr called the *correspondence principle*, which may be stated generally as follows. Based on experience, the laws of classical physics have correctly described the motion of objects and particles over a broad range of size of particle and distance. Classical physics gave correct descriptions of macroscopic phenomena and, through successes of kinetic theory, was able to describe, correctly, many phenomena on the scale of atomic size. Classical physics fails to describe phenomena on a scale smaller than atoms, and so must be modified for this range of sizes and masses. However, the two descriptions, classical and quantum, overlap in the description of atoms. In other words, it must be demanded of the new physics that it give the same (or approximately the same) results as classical physics in descriptions of phenomena involving masses and distances that are large compared to those of atoms but small compared to macroscopic phenomena. The correspondence principle suggested the quantization of angular momentum (Equation 32-17) to Bohr. Part of this argument can be used to illustrate the correspondence principle in the following.

Consider an electron in a planetary orbit, as described in Section 32-5, where the total energy of the orbit was given by classical physics as

$$\mathcal{E} = -\frac{1}{8\pi\epsilon_0}\frac{e^2}{r} \qquad (32\text{-}16)$$

and classically, the speed in the orbit was related to the size of the orbit by

$$mv^2 = \frac{1}{4\pi\epsilon_0}\frac{e^2}{r} \qquad (32\text{-}15a)$$

If we describe the orbit in terms of the frequency of rotation, f, then we have

$$v = 2\pi f r \qquad (32\text{-}27)$$

Eliminating v between Equations 32-27 and 32-15a,

and substituting in Equation 32-16, we get

$$\mathcal{E}^3 = \frac{me^4}{32\epsilon_0{}^2}f^2 \qquad (32\text{-}28)$$

Consider now a quantum description of this same system. From the Balmer formula alone we can assume that the energy levels of the atom have the form

$$\mathcal{E} = -\frac{hRc}{n^2} \qquad (32\text{-}29)$$

and therefore when the atom makes a transition from a state with quantum number n to one with quantum number $n - 1$, it emits a photon of frequency

$$f = Rc\left[\frac{1}{(n-1)^2} - \frac{1}{n^2}\right]$$

This equation may then be rewritten as

$$f = Rc\left[\frac{2n-1}{n^2(n-1)^2}\right] \qquad (32\text{-}30)$$

The classical and quantum descriptions overlap when the electron is in a large orbit, or when n is large. In Equation 32-30, when n is large we can replace the quantity $(2n - 1)$ by $2n$, and replace the quantity $n^2(n - 1)^2$ by n^4. Therefore, when n is large, Equation 32-30 becomes approximately

$$f \approx Rc\,\frac{2n}{n^4}$$
$$\approx \frac{2Rc}{n^3} \qquad (32\text{-}31)$$

Further, according to classical physics, when an electron is moving in a circular orbit with frequency f, it should radiate electromagnetic radiation with that same frequency. That is, for large n, the frequency of the photon, given by Equation 32-31, should be exactly equal to the frequency of rotation of the electron. Thus, for orbits with large n, combining Equations 32-31 and 32-29 gives

$$\mathcal{E}^3 = \frac{h^3Rc}{4}f^2 \qquad (32\text{-}32)$$

Now, the correspondence principle requires that Equation 32-32 agree exactly with 32-28, or

$$\frac{h^3Rc}{4} = \frac{me^4}{32\epsilon_0{}^2}$$

or

$$R = \frac{me^4}{8\epsilon_0^2 h^2 c}$$

which is precisely the result arrived at earlier.

The correspondence principle takes many forms. Here we have illustrated one form, which may be restated as: *quantum results become identical to classical results in the limiting case of large quantum numbers.* Another useful way of invoking the correspondence principle arises from recognizing that many quantum phenomena depend on the photon character of light. However, if the energy of one photon, hf, is small compared to other energies in the situation being considered, then the photon character of the light may be ignored. In other words if hf is vanishingly small, then a particular problem may be treated classically. Therefore, if h can be treated as vanishingly small, then quantum-mechanical results should be identical to classical results. In a sense then, we can treat quantum mechanics as the correct description of nature and recognize that classical physics describes macroscopic phenomena correctly because h is very small compared to macroscopic quantities.

32-8
De Broglie waves

The next major step in the development of the new mechanics came in the form of a hypothesis proposed by L. de Broglie in 1924. De Broglie noted that the energy $\mathcal{E}$ and momentum p of the photon were related by

$$\mathcal{E} = pc$$

and that

$$\mathcal{E} = hf$$
$$= \frac{hc}{\lambda}$$

Therefore for a photon the momentum is related to the wavelength by

$$p = \frac{h}{\lambda} \qquad (32\text{-}33)$$

This equation relates a particle property of light, the momentum of the photon, to a wave property of light, the wavelength. De Broglie assumed a certain symmetry in nature and hypothesized that *particles*, such as electrons and protons, had wave properties and that the wavelength of these waves

are related to the momentum by Equation 32-33.

Within two years after de Broglie had put forth this hypotehsis, two experiments on the diffraction and interference of waves associated with electrons were performed successfully, one by G. P. Thomson and the other by C. Davisson and L. H. Germer.

One of the postulates of the Bohr theory was that an electron moves in a stationary orbit with an angular momentum equal to $nh/2\pi$. On the basis of the de Broglie hypothesis, this electron has associated with it a wave of length

$$\lambda = \frac{h}{mv} \qquad (32\text{-}34)$$

Suppose that an electron is moving in a circular orbit of radius r; the waves move with the electron along this path. In the course of a few revolutions, different parts of the wave will overlap and if they are not to interfere destructively, they must produce a standing wave. The condition for a standing wave is that the length of the path, $2\pi r$, should be a whole multiple of the wavelength λ, or

$$2\pi r = n\lambda \qquad (32\text{-}35)$$

Substituting Equation 32-34 into 32-35, we get

$$2\pi r = \frac{nh}{mv}$$

or

$$mvr = n\frac{h}{2\pi}$$

which is the Bohr quantization condition (Equation 32-17). Thus we can replace the picture of electrons moving as points in an orbit with one in which the electron is in some way associated with a wave. If, in circulating about the nucleus, the wave establishes standing waves, the electron can exist in that orbit. If the wave interferes with itself destructively, the electron may not remain in that orbit.

No method has yet been found for measuring the velocity w of a de Broglie wave; however, we can find its relationship to the velocity v of the particle associated with it. To do this, let us start with the relativistic expression for the total energy $\mathcal{E}$ of a particle of mass m,

$$\mathcal{E} = mc^2$$

where m is the relativistic mass of the particle whose rest mass is m_0. The momentum of this

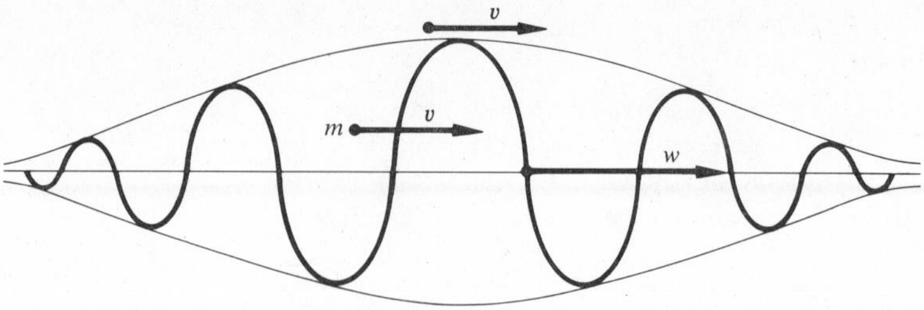

FIGURE 32-11 A group or packet of de Broglie waves associated with a particle whose velocity is v.

particle is

$$p = mv$$

so

$$\frac{\mathcal{E}}{p} = \frac{c^2}{v} \qquad (32\text{-}36)$$

If we make the further assumption that the relationship

$$\mathcal{E} = hf$$

holds for the de Broglie waves, then, since

$$\lambda f = w$$

we have

$$f = \frac{w}{\lambda}$$

and thus

$$\mathcal{E} = \frac{h}{\lambda} w$$

But

$$p = \frac{h}{\lambda}$$

so

$$\frac{\mathcal{E}}{p} = w \qquad (32\text{-}37)$$

Hence, from Equations 32-36 and 32-37, we see that

$$w = \frac{c^2}{v} \qquad (32\text{-}38)$$

is the relationship between the velocity v of the particle and the velocity w of the de Broglie waves associated with it.

The velocity v of a particle is always less than the velocity of light c; hence the velocity of the

de Broglie wave is always greater than the velocity of light. There is no contradiction between the above statement and the fundamental postulate of the special theory of relativity that no signal or energy can be transmitted with a speed greater than c. The energy is transmitted with the velocity v of the particle. The de Broglie waves can be thought of as consisting of a group or packet of waves occupying an extended region of space with the particle somewhere inside this region, as shown in Figure 32-11. The exact position of the particle inside the packet cannot be described with certainty. If we focus our attention on some particular aspect of the group of waves, say the position of maximum amplitude, we can show that this aspect of the group of waves moves with the velocity v of the particle. The group of waves can be shown to consist of individual waves containing a wavelength $\lambda = h/mv$ of comparatively large amplitude and others of slightly different wavelengths and smaller amplitudes.

32-9
The concept of probability

When we discussed the nature of light we indicated that demonstrating the phenomenon of diffraction was critical in showing that light is a wave. If a beam of light is allowed to pass through a slit, as in Figure 32-12, and then is detected on a screen S, the existence of light at points such as C or D demonstrates that diffraction has occurred. As we have seen, the intensity of light at various points on the screen depends on the size of the slit, the distance between the slit and the screen, the original intensity of the beam, and the wavelength of the light. From electromagnetic theory, we know that the intensity at any point is proportional to the square of the electric field E in the wave at that point. From the photon point of view, the intensity of the light is proportional to the number of photons per second at that point. Thus the number of photons per second is proportional to the square of the electric field.

Consider now what would happen if we were to send a very low-intensity beam of photons through the slit, say a beam in which one photon per minute was directed at the slit. If we were to place a photographic plate at the screen and expose it to the beam for several weeks, we would obtain a photograph that is indistinguishable from one exposed for a brief time to a high-intensity beam. That is, during the several weeks of exposure, the greatest number of photons would have traveled straight through the slit, a smaller number would have been deflected by a small angle, and an even smaller number would have been deflected by large angles. However, since all the photons are identical, there is no way to know beforehand in what direction a particular photon will travel after it goes through the slit. The shape of the diffraction pattern gives the *probability* that a photon will move in any particular direction. The photon intensity, and therefore the probability, is given by the square of the electric field. Thus, if we examine the behavior of a

beam of light particles, then the probability distribution is determined by the wavelength and the square of the intensity of the associated wave.

In a similar way we can examine the behavior of a beam of particles such as electrons. In principle at least, we could perform exactly the same experiment using a beam of electrons and observe a diffraction pattern. In order to explain the observed diffraction pattern, we assume that the electrons are accompanied by a de Broglie wave, whose wavelength is determined by the momentum of the electron beam, as given by Equation 32-33. The amplitude of this wave, ψ, determines the intensity of the beam in that the number of electrons per unit area per second is proportional to the square of the amplitude of the wave. That is, the square of the amplitude of the de Broglie wave gives the probability distribution for the electrons.

These probability waves accompanying particles exhibit interference. Imagine sending a beam of electrons through two slits and detecting them on a screen S, as shown in Figure 32-13. If we counted electrons at various places on the screen we would get an intensity distribution P, which is similar to the intensity distribution of Young's double-slit experiment. If we were to cover up one of the holes we would have repeated the diffraction experiment above, and obtained an intensity distribution such as P_1. If we were to close the other hole instead, we would obtain a similar pattern, P_2, which is shifted over.

The striking thing is that the intensity distribution obtained with both slits open is not equal to the sum of the two single-slit patterns. Opening a second slit increases the number of electrons arriving at some points, but it also decreases the number of electrons arriving at other points. It is impossible to explain this phenomenon and think of the electrons as point particles traveling along linear paths. It is however fairly simple to calculate the intensity distribution by treating the de Broglie waves of the electrons as interfering with each other, and then finding the probability of detecting an electron as proportional to the square of the amplitude of the resulting wave.

32-10
The Heisenberg uncertainty principle

One of the more striking consequences of the wave nature of particles is described by a general principle first enunciated by Heisenberg, and called the

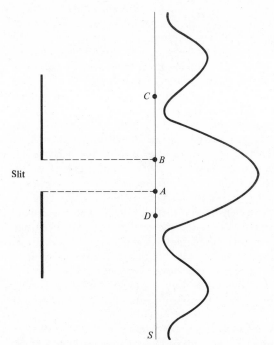

FIGURE 32-12 The diffraction pattern due to a single slit. The curve on the right gives the intensity on the screen S. The existence of light in the shadow, such as at point D, demonstrates that diffraction has occurred.

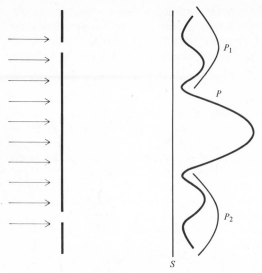

FIGURE 32-13 Diffraction due to two slits. The curve P_1 gives the intensity on the screen S when one slit alone is open, whereas the curve P_2 gives the intensity when the other slit alone is open. The curve P gives the intensity when both slits are open.

uncertainty principle, or indeterminancy principle. The principle may be stated in several forms, one of which is that in any *simultaneous determination of the position and momentum of a particle there is an uncertainty Δx in its position as measured by its coordinate x and an uncertainty Δp_x in its momentum p_x in the x direction, such that*

$$\Delta x \, \Delta p_x \geq h \qquad (32\text{-}39)$$

where h is the Planck constant. These uncertainties are not due to errors in construction of apparatus, but are an intrinsic part of nature. They express the wave nature of particles. The principle asserts that if we somehow contrive to make the uncertainty in momentum very small, then of necessity the uncertainty in position will become very large. Similarly, if the uncertainty in position is made very small, then the uncertainty in momentum will be very large.

One way to understand the origin of the uncertainty principle is to recognize that the momentum of the particle is related to the wavelength of the de Broglie wave by Equation 32-33. However, if the wave is a pure sine wave, it extends out to infinity in both directions. Therefore, the probability of finding the particle at any one place is about

the same as the probability of finding the particle at any other place. In terms of the uncertainties, if a particle moving in the x direction is described by a de Broglie wave that has a single known wavelength, then the uncertainty of momentum, Δp, is zero, but then it may be located anywhere along the x axis, and so the uncertainty of position, Δx, is infinite. If, however, we know that the particle is located in some small region of space, then we know that the amplitude of the de Broglie wave is zero outside that region of space. But as we saw, when describing the Fourier theorem, in order for a wave to be confined to a pulse, or to a small region of space, it must contain many frequencies or wavelengths. The more closely confined in space the pulse is, the larger the range of wavelengths must be. That is, a particle confined to a region of space consists of a group of waves of many wavelengths, or a *wave packet*, as shown in Figure 32-11. But if this wave packet contains many wavelengths, then the momentum of the associated particle may be any one of a range of values; in other words, if the uncertainty in the location of the particle is reduced, the uncertainty in momentum of the particle is increased.

The uncertainty principle can be connected to the single-slit diffraction experiment. The width of the slit gives an uncertainty in position, since we know that any electron that went through the slit could not have been outside the slit, but we have no way of knowing precisely where in the slit it was. Accompanying this uncertainty in position is an uncertainty in the component of momentum *transverse* to the beam direction. That is, any one electron may be moving sideways as well as forward. Therefore, some of the electrons will be displaced sideways after passing through the slit; it is these electrons that are diffracted. If we attempt to decrease the uncertainty in position, we must make the slit narrower. But narrowing the slit widens the diffraction pattern, which is equivalent to increasing the uncertainty in momentum. Similarly, widening the slit narrows the pattern, thereby decreasing the uncertainty in momentum but increasing the uncertainty in position.

Still another way to look at the uncertainty principle is to recognize that the photon character of electromagnetic radiation places limits on the kinds of measurements that may be made, even in principle. In order to make a measurement of the position of a particle we must have a measuring instrument interact with the particle. This means that at least one photon must interact with the particle. The less energy that photon has, the less it will disturb the particle. But low-energy photons have large wavelengths, and the larger the wavelength of the photon, the less information the photon can transmit about the *position* of the particle. In order to define the position of the particle more effectively, we must use a short-wavelength photon, which will necessarily have high energy. After the photon interacts with the particle, some of the momentum of the photon will have been transferred to the particle and the uncertainty in the momentum of the photon will be large.

Another form of the uncertainty principle is that in any simultaneous measurement of energy $\mathscr{E}$ and time t there are uncertainties $\Delta\mathscr{E}$ and Δt such that

$$\Delta\mathscr{E}\,\Delta t \geq h$$

That is, if an event occurs at a precisely known time, Δt is small and therefore there is a large uncertainty in the energy associated with the event. For example, many energy levels of atoms have very short lifetimes; an atom in such a state very rapidly emits a photon and makes a transition to another level. Since the uncertainty in the time when the emission takes place cannot be longer than the lifetime, such states have small uncertainties in time and therefore large uncertainties in energy. Thus the energy of such a short-lifetime state is necessarily uncertain by a large amount; all we can know is that the energy lies somewhere in a wide range, and therefore such levels are sometimes called wide or broad energy levels. On the other hand, a level that exists for a long time, or a stable level, has a large time uncertainty and therefore the energy may be measured with precision. Such levels are sharp.

32-11
Quantum mechanics

As we have seen, the behavior of particles is determined by the de Broglie waves, at least to the extent that these waves determine the probability of making specific measurements. In principle, then, solving a physics problem requires the calculation of the amplitude, phase, and wavelength of the de Broglie waves for all the particles in the problem. This is analogous to the statement in classical physics that solving a physics problem requires the calculation of the positions and velocities of all the particles in the problem. The classical physics problem is solved by writing down the

fundamental equation of motion, Newton's second law, and then solving the resulting set of equations. In order to solve the quantum-mechanical problem in physics, it was first necessary to discover the fundamental equation—the equation that described the de Broglie waves. In 1925 Heisenberg discovered a mathematical form of this fundamental equation, and in 1926 Schrödinger discovered a different form, which, it was possible to show, was identical in significance to that of Heisenberg. In 1928 Dirac discovered the fundamental equations of relativistic quantum mechanics.

The Schrödinger formulation of quantum mechanics provides a set of rules for writing down an equation, called the Schrödinger equation, for any particular problem. The solution to this equation is the wave function ψ, which is essentially the same as the de Broglie wave. The mathematical operations associated with this equation are formidable, and for most problems the only solutions available are approximations.

An integral part of the theory is a set of rules for calculating the results of measurements. Since uncertainty and probability are such important parts of quantum theory, these calculations usually make statistical rather than absolute statements. That is, we imagine that many identical experiments can be made to measure one quantity, say the velocity of an electron at a particular place in the apparatus at a particular time in the experiment. In general, quantum theory predicts that such measurements, no matter how carefully made, will produce a range of values. The theory provides a set of rules for calculating the distribution of these measured values. If the experiment were done once, rather than many times, the calculation would provide the probability that any specific value would be measured in that experiment. Quantum mechanics, like relativity, places heavy emphasis on the distinction between concepts that are constructs, and cannot be measured, and concepts that are capable of being measured. For example, the orbit of a particle cannot be measured; the uncertainty principle limits the precision with which the position and velocity of the particle may be measured.

For closed systems, such as atoms, the theory shows that the system can exist in one of many discrete states, called *stationary states*. If the system is completely isolated from all interactions it will remain in that state. Real systems, however, are never completely isolated and, through interactions with each other and with the electromagnetic fields in space, they make transitions from one stationary state to another. The theory provides a way of calculating the rate of these transitions, which gives the *intensity* of the spectral lines. If a transition occurs between two stationary states, each of which is characterized by a wave function of specific frequency, then these transitions can occur only if the emitted radiation has a frequency equal to the beat frequency of the two states. This result is essentially the Bohr condition for emission of radiation.

32-12
The particle in a box

Although we cannot here solve a quantum-mechanical problem, we can illustrate the nature of some of the results by considering the behavior of de Broglie waves in a simple situation. Consider a particle that is confined to the interior of a cubical box having inpenetrable, perfectly elastic walls. Classically, the particle would bounce back and forth with constant energy, and any energy is acceptable. Quantum-mechanically, however, the particle is restricted to states where the de Broglie waves do not interfere destructively with themselves. If a de Broglie wave starts at one wall with a particular phase, travels to the opposite wall, and returns to the first wall, it must reach the first wall with that same phase. Otherwise, after many collisions with the walls the de Broglie waves will average to zero, and the probability of finding the particle in the box will become zero. Thus, in order for the particle to exist, a whole number of waves must fit into the round trip from one wall to the other and back again. Thus,

$$n\lambda = 2d$$

where d is the distance between the walls, λ is the de Broglie wavelength, and n is an integer. Then, substituting Equation 32-33, we have

$$\frac{nh}{p} = 2d$$

or

$$p = \frac{nh}{2d} \qquad (32\text{-}40)$$

That is, only particles of the specific momenta given above may exist in the box. The energy of the particle is similarly quantized; we may write

$$\mathcal{E} = \frac{p^2}{2m} \qquad (32\text{-}41)$$

for the kinetic energy $\mathcal{E}$ of the particle. Therefore,

$$\mathcal{E} = n^2 \frac{h^2}{8md^2} \qquad (32\text{-}42)$$

The particle can exist only with specific energies. It is interesting to note that since n must be at least equal to 1, there is a minimum energy. The particle cannot exist in the box with zero energy.

The wave function for the particle has the form of a sine curve. The square of this function gives the probability of finding the particle at each point in the box. It is interesting that this probability varies from point to point in the box; the particle is more likely to be at some places than at others.

32-13
Modern description of the hydrogen atom

The application of wave mechanics to the hydrogen atom has led to a completely new picture or description of atomic structure. It is no longer possible to assign a very definite orbit, such as a circle, to the path of the electron around the nucleus. What is determined, instead, is the probability of finding the electron at any given distance r from the nucleus. Further, its motion is no longer confined to a plane; the electron may be found anywhere around the nucleus. The problem thus becomes a three-dimensional one. In the Schrödinger solution of the hydrogen problem, it is assumed that there are de Broglie waves associated with the electron, and solutions are obtained for the amplitudes of these waves when the total energy of the electron is $\mathcal{E}$ and its potential energy is

$$V = -\frac{1}{4\pi\epsilon_0} \frac{e^2}{r} \qquad (32\text{-}43)$$

It is found that solutions exist for negative values of the total energy $\mathcal{E}$ given by

$$\mathcal{E} = -\frac{1}{n^2} \frac{me^4}{8\epsilon_0{}^2h^2} \qquad (32\text{-}44)$$

where n is an integer called the *principal quantum number* and may have the values

$$n = 1, 2, 3, 4, \ldots \qquad (32\text{-}45)$$

This is exactly the same result as that obtained by Bohr.

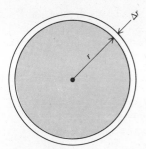

FIGURE 32-14 Spherical shell of inner radius r and thickness Δr.

Another result is that the angular momentum of the electron is given by

$$\sqrt{l(l + 1)}\ \frac{h}{2\pi} \qquad (32\text{-}46)$$

where l is an integer, called the *angular momentum quantum number*, whose value depends upon n; it may have any value from zero to $(n - 1)$. This may be stated as

$$l = 0, 1, 2, \ldots, (n - 1) \qquad (32\text{-}47)$$

This result differs considerably from that of the Bohr theory except for very large values of l. The solution of the Schrödinger equation yields the values of the amplitude ψ as a function of the coordinates of the electron for various values of n and l. Since ψ^2 is a measure of the probability of finding the electron at any point in space, $\psi^2\,\Delta v$ is

the probability of finding the electron in a volume Δv. If we consider the volume Δv to be a thin spherical shell of inner radius r and outer radius $r + \Delta r$, as shown in Figure 32-14, its volume is $4\pi r^2 \Delta r$, where $4\pi r^2$ is the surface area. Thus $\psi^2 r^2$ will be a measure of the probability of finding the electron at a distance r from the nucleus within a spherical shell of thickness Δr. Figure 32-15 is a graph in which $\psi^2 r^2$ is plotted as ordinate and the distance r of the electron from the hydrogen nucleus is plotted as abscissas. Curves are drawn for a few selected values of n and l. The curve for $n = 1$ and $l = 0$ shows a maximum at a distance of $r = r_1 = 0.529$ Å, equal to that of the radius of the first Bohr orbit. It may also be found at any other point within a range of about three such radii with the same value of the total energy $\mathcal{E}_1$. The curve for $n = 2$, $l = 1$, shows a maximum at $r = 4r_1$, the radius of the second Bohr orbit. Thus an electron with a total energy $\mathcal{E}_2$ may be found almost anywhere outside the nucleus, but the probability is a maximum at the position of the Bohr orbit. Thus the simple Bohr picture of the hydrogen atom must be replaced by some cloud of points such as that shown in Figure 32-16, where the density of points or clouds represents the probability of finding the electron in that region. Since the electron carries a charge $-e$, some think of this as an *electronic cloud* around the nucleus. Values of n and l other than those shown in Figure 32-16 lead to more complex probability distributions.

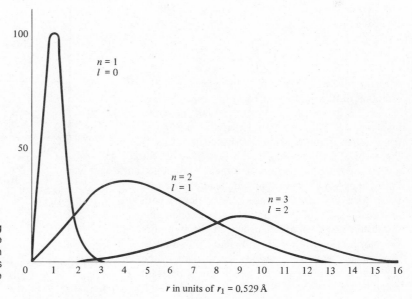

FIGURE 32-15 Curves showing the probability of finding the electron in the hydrogen atom within a spherical shell of radius r and thickness Δr for some values of n and l.

r in units of $r_1 = 0.529$ Å

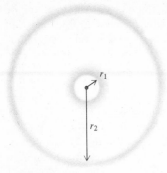

FIGURE 32-16 Electron cloud around the nucleus of a hydrogen atom for values of $n = 1$, $l = 0$, and $n = 2$, $l = 1$.

The complete solution of the Schrödinger equation for hydrogen also yields a third quantum number, which we shall designate as m_l. The value of this *magnetic orbital quantum number* is determined by the particular value of the angular momentum quantum number l of the electron; it may have any integral value from $-l$ through zero to $+l$. Thus if $l = 2$, m_l may have any of the values -2, -1, 0, $+1$, $+2$. It will be recalled that angular momentum is a vector quantity. It is possible to prescribe a direction to the angular momentum only with respect to some preferred direction such as that provided by an external magnetic field of flux density **B**. If we represent the angular momentum of the electron in its orbit by a vector **l** perpendicular to the plane of the orbit, its direction when in a magnetic field of flux density **B** cannot have any arbitrary value but is confined to certain angles θ such that

$$\cos \theta = \frac{m_l}{l} \qquad (32\text{-}48)$$

where m_l is the projection of the vector **l** in the direction of **B**, as shown in Figure 32-17.

Very early in the development of quantum mechanics it was discovered that the spectra of some other atoms could be described in terms of three quantum numbers; n, l, and m_l. However, for a very large number of atoms the spectra contained many lines which were doubled; it appeared that there were two closely spaced energy levels with identical values of n, l, and m_l. To explain this Goudsmit and Uhlenbeck proposed the hypothesis that the electron could be treated as if it had an *intrinsic angular momentum*; even though the electron generally behaves like a point mass, it could be treated here as if it were a spinning sphere.

The doubling of the energy levels arose from the fact that, in the presence of an external magnetic field, of flux density **B**, the vector **s** representing the *spin angular momentum* can take one of two possible orientations, either parallel and in the same direction as **B**, or parallel and oppositely directed to **B**, as shown in Figure 32-18. We sometimes refer to these as the components of **s** along the direction of **B** and designate them by m_s. Thus $m_s = \pm \frac{1}{2}$ in units of $h/2\pi$.

The *total angular momentum of the electron*, designated by the letter **j**, again in units of $h/2\pi$, will thus be the vector sum of the orbital angular momentum **l** and the spin angular momentum **s** or

$$\mathbf{j} = \mathbf{l} + \mathbf{s} \qquad (32\text{-}49)$$

with the restriction that j must be an odd half integer. Thus, for the first Bohr orbit, $l = 0$ and $s = \frac{1}{2}$, so $j = \frac{1}{2}$. For the second Bohr orbit, $l = 1$, j can have either of two values

$$j = l + \frac{1}{2} = \frac{3}{2}$$

or

$$j = l - \frac{1}{2} = \frac{1}{2}$$

Similarly for all other orbits, there will be two possible values for the total angular momentum j.

The energy-level diagram for hydrogen now

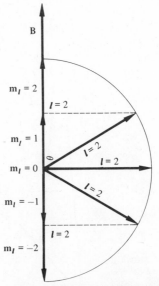

FIGURE 32-17 The projection of the orbital angular momentum vector **l** on the direction of the external magnetic field **B** yields m_l.

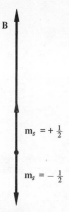

FIGURE 32-18 The projection of the spin angular momentum **s** in the direction of the external magnetic field **B** yields $m_s = +\frac{1}{2}$ or $-\frac{1}{2}$.

becomes much more complex than that shown in Figure 32-10. For example, for $n = 3$, instead of a single level there will be five different levels, as shown in Figure 32-19. When $n = 3$, l may have the values, 0, 1, or 2, and the corresponding values for j will be

$$l = 0, \quad j = \tfrac{1}{2}$$
$$l = 1, \quad j = \tfrac{1}{2} \text{ or } \tfrac{3}{2}$$
$$l = 2, \quad j = \tfrac{3}{2} \text{ or } \tfrac{5}{2}$$

Except for the state $l = 0$ (sometimes called an S state) all the other states are doublets. It will be recalled that transitions between two states give rise to spectral lines. The theory predicts that the most probable transitions will be those for which l changes by ± 1, and j either remains unchanged or changes by ± 1. These *selection rules*, as they are called, may be stated as

$$\Delta l = \pm 1 \tag{32-50}$$
$$\Delta j = 0 \quad \text{or} \quad \pm 1$$

The possible transitions from the orbits of principal quantum $n = 3$ to $n = 2$, corresponding to the H_α line (red line) of hydrogen, are shown in Figure 32-19. There should be five components to this line. However, because of the difficulty of resolving all of the components and the great differences in their relative intensities, only two components have actually been observed and measured. The same diagram can be used for the deuterium atom with suitable modifications of the spacings of the energy levels. A third component can be seen in the corresponding H_α line of deuterium.

32-14
Assignment of electrons to atoms

We have shown that the state of an electron in an atom can be specified by four quantum numbers, n, l, m_l, and m_s; each state will have a certain amount of energy and a certain amount of angular momentum. We have thus far restricted the discussion to the single electron of the hydrogen atom. The calculation of the states of the electrons in more complex atoms is extremely difficult and most often impossible. However, some help is obtained from experimental evidence provided by spectral analyses and studies of chemical properties, electrical properties, and magnetic properties. There is one additional guiding principle, attributable to W. Pauli, called the Pauli *exclusion principle*. This principle states that *no two electrons in an atom can exist in the same state*. For example, in the case of helium ($Z = 2$) the lowest state would be $n = 1$; for this state $l = 0$, $m_l = 0$, and $m_s = \pm\frac{1}{2}$. Thus when the two electrons of helium are in their lowest state, their spins must be oppositely directed, as shown in Figure 32-20. Table 32-1 shows the possible assignment of electrons to normal states of atoms for values of $n = 1, 2,$ and 3, consistent with the Pauli exclusion principle.

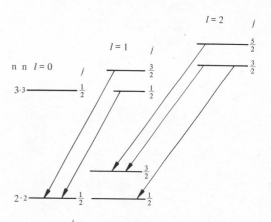

$$\Delta l = \pm 1 \qquad \Delta j = \pm 1, \text{ or } 0$$

FIGURE 32-19 Some of the energy levels of hydrogen and the possible transitions from $n = 3$ to $n = 2$ (not drawn to scale).

A glance at Table 32-1 shows that the electrons having the same value of the principle quantum number n form a *group* or a *shell*, and those with the same value of l for a given value of n form a subgroup. For example, the electrons in the group $n = 3$ can be in three subgroups corresponding to $l = 0$, 1, and 2. The maximum numbers of electrons in these subgroups are 2, 6, and 10, respectively, for a total of 18 electrons. The Appendix gives the present assignment of electrons to the atoms of the elements now known, from atomic number $Z = 1$ to atomic number $Z = 103$.

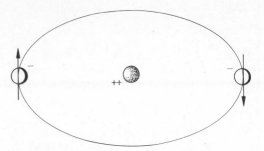

FIGURE 32-20 The two electrons of helium have their spins oppositely directed in the lowest or normal state.

Table 32-1 Possible number of electrons in a given group

n	l	m_l	m_s	number of electrons in subgroup	number of electrons in completed group
1	0	0	$+\frac{1}{2}$	2	2
1	0	0	$-\frac{1}{2}$		
2	0	0	$+\frac{1}{2}$	2	
2	0	0	$-\frac{1}{2}$		
2	1	-1	$+\frac{1}{2}$		8
2	1	-1	$-\frac{1}{2}$		
2	1	0	$+\frac{1}{2}$		
2	1	0	$-\frac{1}{2}$	6	
2	1	1	$+\frac{1}{2}$		
2	1	1	$-\frac{1}{2}$		
3	0	0	$+\frac{1}{2}$	2	
3	0	0	$-\frac{1}{2}$		
3	1	-1	$+\frac{1}{2}$		
3	1	-1	$-\frac{1}{2}$		
3	1	0	$+\frac{1}{2}$	6	
3	1	0	$-\frac{1}{2}$		
3	1	1	$+\frac{1}{2}$		
3	1	1	$-\frac{1}{2}$		
3	2	-2	$+\frac{1}{2}$		18
3	2	-2	$-\frac{1}{2}$		
3	2	-1	$+\frac{1}{2}$		
3	2	-1	$-\frac{1}{2}$		
3	2	0	$+\frac{1}{2}$		
3	2	0	$-\frac{1}{2}$	10	
3	2	1	$+\frac{1}{2}$		
3	2	1	$-\frac{1}{2}$		
3	2	2	$+\frac{1}{2}$		
3	2	2	$-\frac{1}{2}$		

32-15
Solids

The energy-level structure we have described is applicable to the electrons in isolated atoms. When atoms are combined to form molecules or solids, the outermost, or valence, electrons in any one atom are located in electric and magnetic fields caused by the other atoms. Generally, these fields split the energy levels in a complex manner. In solids the effect of the splitting is to transform a single energy level into many closely spaced levels —so many and so closely spaced that the level is usually described as an *energy band*. The energy-level structure of a solid becomes, then, a group of bands separated by gaps. The valence electrons may have any energy within one of the bands, but may not have an energy in one of the gaps.

If the temperature of the solid is low, then the average energy of the electrons will be low, and almost all the electrons will be confined to the lowest band. However, the Pauli principle applies to the electrons, and so only one electron may be in any one state. That is, we can think of the electrons as filling in the levels within the lowest band, starting at the lowest energy, and being forced by the Pauli principle to occupy higher and higher levels in that band. If the number of levels in a particular band is smaller than the number of electrons, then some of the electrons will have to fill in the lowest levels in the next higher band.

Consider a situation in which the last electrons just fill the band. In order to add energy to any electron, it must be raised to the next highest band, across an energy gap. That is, a minimum energy must be added to the electron in order to make it move. If, on the other hand, an electron is in a band that is only partially filled, it may absorb almost any small amount of energy. Thus, an insulator is a material in which essentially all the electrons are in filled bands, whereas a conductor is a material in which many electrons exist in partially filled bands.

A semiconductor is an insulator in which the energy gap above the filled band is not large compared with the average thermal energy of the electrons. Thus, in a semiconductor, an appreciable number of the higher-energy electrons will be excited into the unfilled band by thermal energy. The effect of the impurities described in Chapter 28 is to create new levels in the gap, close to the conduction band.

32-16
The optical maser or laser

The usual type of light source emits radiation in all directions with no definite phase relationships among the waves emitted by its atoms and molecules. Thus, in all previous experiments designed to show interference effects from two light sources, it was always necessary to start with a single source and divide this into two beams in order to ensure known phase relationships between them. For example, in the Young interference experiment, light from a single source was sent through two small apertures, which then acted as two sources of light; if placed symmetrically with respect to the single source, light waves from the two apertures came out in phase with each other. The light coming from these two sources is called *coherent* light because of this relationship.

The *optical maser* or *laser* is a new type of source that emits *coherent* radiation in the visible, or near-visible, region of the spectrum. The laser has been developed into a source of extremely intense, well-collimated coherent light containing a few sharp spectral lines. The idea of a laser is an outgrowth of an earlier development of the *maser*, in 1954, by C. H. Townes of the U.S. and Basov and Prokhorov of the USSR. The term *maser* comes from the first letters (an acronym) of the words: microwave amplification by the stimulated emission of radiation. (The term *laser* uses light as the first word of the acronym.) The maser is an extremely accurate frequency standard in the microwave band and has been used for the measurement of time intervals with very high precision. As mentioned in Chapter 1, the optical maser is being used as the new standard for the measurement of time. The extension of the maser to the optical region was suggested by A. L. Shawlow and C. H. Townes in 1958 and first produced experimentally by T. H. Maiman in 1960.

Since that time literally thousands of physicists and engineers have entered this new field, resulting in the production of a large variety of lasers utilizing different materials for a variety of uses. Here we will simply present the essential ideas of laser operation, confining the discussion to two kinds—the ruby laser and the helium-neon laser.

In our earlier discussion of the emission of light from an atom such as hydrogen, we found that the atom can be in any one of a large number of energy states, the energy of any one state being repre-

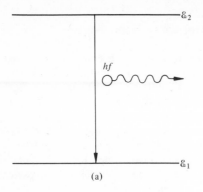

(a)

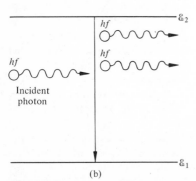

(b)

FIGURE 32-21 (a) Spontaneous emission of photon hf. (b) Stimulated emission by an incident photon hf results in an additional photon hf with the two waves in phase.

sented by an energy level. Given a large number of atoms, most atoms will be in the normal or ground state, whose energy will be designated as $\mathcal{E}_1$. A much smaller number of atoms will be in the higher-energy states or *excited* states; the energies of these states will be designated by $\mathcal{E}_2$, $\mathcal{E}_3$, and so forth. Radiation is emitted only when an atom goes from an excited state to a lower-energy state, the frequency f of the radiation being given by the Bohr equation

$$\mathcal{E}_2 - \mathcal{E}_1 = hf \qquad (32\text{-}18)$$

where h is the Planck constant. Atoms in the normal state can be raised to an excited state in a variety of ways; one is by the absorption of radiation of appropriate frequency, another is by electron bombardment, and a third is by inelastic collisions with other atoms and molecules.

There are several important conditions that must be fulfilled for the production of coherent radiation

in a laser. One is that the number of atoms N_2 in the excited state $\mathcal{E}_2$ should be greater than N_1, the number of atoms in the lower-energy state $\mathcal{E}_1$. This condition is sometimes referred to as *population inversion*. Another condition that must be fulfilled is that the atoms should be situated in a radiation field containing radiation of the same frequency as that emitted when the atom goes from state $\mathcal{E}_2$ to state $\mathcal{E}_1$. The reason for this is that the photons of this radiation field are the ones that will *stimulate* the atoms in state $\mathcal{E}_2$ to emit radiation; an important aspect of this stimulated emission is that the waves of the two photons are in phase, as illustrated schematically in Figure 32-21. Atoms in state $\mathcal{E}_2$ may emit photons of energy hf in going to state $\mathcal{E}_1$ spontaneously, in which case there are no particular phase relationships between them, or they may be stimulated to emit this radiation by the photons incident upon them from the radiation field, in which case the waves of these photons are in phase or coherent. The probability that an atom in state $\mathcal{E}_2$ will be stimulated to emit radiation depends upon the density of photons of the proper frequency, or the intensity of this radiation field. Hence, to ensure laser action it is essential to build up the intensity of the radiation field of the proper frequency in the space occupied by the excited atoms. This region is sometimes called a *resonant cavity*.

One method of obtaining the coherent radiation field is illustrated schematically in Figure 32-22, which shows a long tube with two plane, parallel mirrors or reflectors, M_1 and M_2, at its ends. Suppose that some monochromatic light of wavelength λ is produced in this tube. Some of the light will leave the tube through its walls, and some will travel to the mirrors. Those waves that strike the mirrors normally will be reflected to the opposite ones; thus the tube will contain waves traveling from M_1 to M_2 and back again. By adjusting the distance between the mirrors, standing waves can

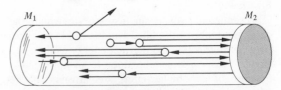

FIGURE 32-22 An optical cavity containing radiation reflected from mirrors M_1 and M_2 and back to M_1 stimulating emission from some atoms.

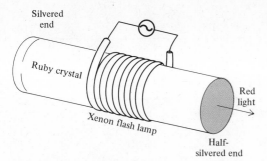

FIGURE 32-23 Schematic diagram of ruby laser.

be set up in this tube analogous to those produced in an elastic medium such as a string or an air column. We thus have a cavity containing a coherent radiation field. If the tube contains atoms that can be raised to the appropriate excited state $\mathcal{E}_2$ by some other means, these atoms can be stimulated by the cavity radiation to emit photons of the same frequency and wavelength as the cavity radiation; furthermore, the radiation will be coherent.

The ruby laser was the first one constructed; it consists of a fairly long, single crystal of ruby. The entire crystal is the optical cavity; its two ends are ground optically flat, and silvered. One of these mirrors may be only partly silvered to permit some of the light to emerge; or one end may be left clear and a mirror placed at a distance from it so that part of the light may be deflected from the beam. Ruby consists of aluminum oxide, Al_2O_3, plus a very small percentage of chromium oxide, Cr_2O_3; the chromium ions, Cr^{3+}, are the sources of the laser light. The ruby crystal is surrounded by a "flash tube," as shown in Figure 32-23; this tube contains a gas such as xenon, which, when flashed on, emits light in the yellow-green region of the spectrum. Figure 32-24 shows the three energy levels involved in this particular laser action. When the external light is flashed on, the yellow-green light from the xenon is absorbed by some of the Cr^{3+} ions in ruby, raising them to level $\mathcal{E}_3$; this level has a lifetime of about 10^{-7} sec. These excited ions give up some of their energy to the crystal lattice in a radiationless transition and drop to excited state $\mathcal{E}_2$; this is a metastable state and has a lifetime for spontaneous emission of about 3 milli-sec. If the intensity of the light from the xenon is very great, the number of ions N_2 in the energy

state $\mathcal{E}_2$ can be built up to a much higher value than the number of ions N_1 in state $\mathcal{E}_1$.

Some chromium ions will go from state $\mathcal{E}_2$ to $\mathcal{E}_1$ spontaneously, emitting photons of red light of wavelength 6943 Å. Some of this light is reflected back and forth between the reflecting ends of the ruby crystal, quickly forming the cavity radiation in it. This radiation stimulates the emission of photons by other chromium ions in state $\mathcal{E}_2$, causing them to drop to state $\mathcal{E}_1$, and resulting in the production of a large pulse of coherent, well-collimated, monochromatic radiation. The lifetime of state $\mathcal{E}_2$ is reduced considerably by the action of the cavity radiation, down to the order of microseconds. The xenon flash lamp is then activated once more, sending an intense pulse of yellow-green light through the sides of the crystal, "pumping" chromium ions to state $\mathcal{E}_3$, from which they drop by radiationless transitions to state $\mathcal{E}_2$, building up the population of such states to a value N_2 greater than N_1, the population of ions in state $\mathcal{E}_1$. Cavity radiation is quickly built up, stimulating the emission of a pulse of red laser light in a well-collimated beam. The amount of luminous energy emitted in each pulse can be very large—of the order of several joules or more. Since this burst lasts for a few microseconds, the power of the luminous laser light is of the order of millions of watts. The intensity of this beam, which is the

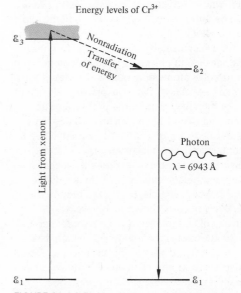

FIGURE 32-24 Energy levels and transitions involved in emission of laser light by ruby crystal.

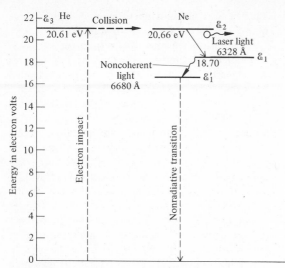

FIGURE 32-25 Helium-neon energy levels involved in the production of red laser light of 6328 Å wavelength.

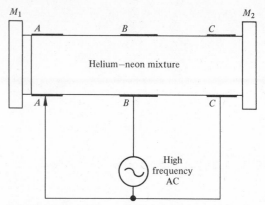

FIGURE 32-26 Helium-neon laser with external electrodes A, B, and C connected to source of high frequency alternating current. M_1 and M_2 are mirrors.

energy per unit time per unit cross-sectional area, or power per unit area, can be increased still more by focusing the beam with a lens.

The helium-neon laser differs in its operation from the ruby laser in that four energy levels are involved in its operation, one in helium and three in neon. Most of the spectral lines emitted in its laser action are in the infrared region, but there is one line, whose wavelength is 6328 Å, in the red region. The four energy levels involved in the emission of this line are shown in Figure 32-25. The light from a helium neon laser is emitted continuously rather than in pulses. A helium-neon mixture in the ratio of 5 atoms of helium to 1 atom of neon is put inside a glass tube at a pressure of about 0.6 mm of mercury; that is, the pressure of helium is 0.5 mm and that of neon is 0.1 mm. The atoms of helium and neon are raised to the excited states by electron impact. The electrons are produced by the ionization of some of the atoms by means of a high-frequency electromagnetic field produced by an outside circuit; the electrodes are mounted outside the glass tube as shown in Figure 32-26.

Electrons of suitable energy raise some atoms of helium to the metastable state $\mathcal{E}_3$, which has an energy of 20.61 eV above the ground state, and also raise some atoms of neon to state $\mathcal{E}_2$, which has an energy of 20.66 eV. The population N_2 of state $\mathcal{E}_2$ is further increased by transfer of energy from helium to neon during collisions. The difference in energy of 0.05 eV between states $\mathcal{E}_3$ and $\mathcal{E}_2$

comes from the kinetic energies of the colliding atoms. Photons of red light of wavelength 6328 Å are emitted by neon atoms that go from state $\mathcal{E}_2$ to state $\mathcal{E}_1$. Some of these photons travel between

FIGURE 32-27 Photograph of a 33-ft long laser. The extraordinary length makes possible a greater amplification of the laser beam each time it travels the length of the tube. (Courtesy Bell Telephone Laboratories.)

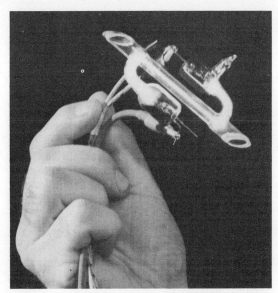

FIGURE 32-28 Miniature gas laser operates continuously at room temperatures. When placed in an appropriate optical cavity, it oscillates at only one frequency of visible red light. (Courtesy of Bell Telephone Laboratories.)

mirrors at the ends of the tube and form the cavity radiation that stimulates other excited neon atoms to emit the laser light. Atoms in excited state $\mathcal{E}_1$ go to state $\mathcal{E}_1'$ with a lifetime of the order of 10^{-8}

sec, so that the population N_1 of state $\mathcal{E}_1$ can be much smaller than N_2 in order to ensure laser action. State $\mathcal{E}_1'$ is a metastable state; atoms in this state go to the ground state of neon by a radiationless transition, mostly by collision with the walls of the tube. In the design of a helium-neon laser it is essential to have the proper ratio of diameter to length of the tube so that neon atoms in state $\mathcal{E}_1'$ can diffuse rapidly to the walls of the tube and give up their energy to it. The light emitted when atoms go from state $\mathcal{E}_1$ to state $\mathcal{E}_1'$ is not laser light. Figure 32-27 is a photograph of a very long helium-neon laser used for experimental work. Figure 32-28 is a photograph of a miniature gas laser with windows at the Brewster angle so that the light emitted is linearly polarized. The reflecting mirrors are outside the tube.

The range of applications of lasers will undoubtedly grow as we learn more about the physics involved. It has already been used in retinal surgery and is finding wide applications in biophysics. Light from the ruby laser has been reflected from the surface of the moon back to the earth. One of the expected applications is as a carrier wave for communications systems. The frequency of the light from a helium-neon laser, for example, is of the order of 10^{14} Hz, with a very high degree of accuracy. Techniques are now being developed to modulate this beam. Success in this effort will provide thousands of new communication channels.

Questions

1. Explain why a hole in the side of a hollow box with well-insulated walls can be used as an experimental blackbody.

2. Objects at room temperature radiate. Explain why we cannot see in a dark room at night.

3. In heat-treating metals, craftsmen are told to heat metals to a "cherry red." In other heat-treating processes of metals they are told to heat the metal to a "straw color." Which is hotter?

4. Describe how the photoelectric effect is used in (a) an electric eye, (b) a photographer's light meter, and (c) the sound system of a motion picture projector.

5. Why is it permissible to use a classical, rather than relativistic, expression for the kinetic energy of the electron in describing the photoelectric effect?

6. Ordinary photographic film can be handled safely in a darkroom that is illuminated with red light, yet the film would be ruined if the same intensity of yellow light were present in the darkroom. Explain.

7. Describe two experiments that can be explained only on the basis that light propagates as a wave.

8. Describe two experiments that show that light behaves as a stream of particles.

9. How could you use spectroscopic evidence to decide whether the moon shone from light reflected from the sun or from internal light?

10. Attempt to explain why iron vapor emits a line spectrum, whereas heated, solid iron emits a continuous spectrum.

11. Explain why the fifth Balmer line was not discovered along with the first four.

12. State the postulates of the Bohr theory.

13. The Balmer series is observed as an absorption spectrum for stars but is rarely observed in the absorption spectrum in the laboratory. Explain.

14. Explain the X-ray spectrum given by Equation 32-11 in terms of the energy-level diagram of Figure 32-10.

15. Discuss some of the changes that might occur if the value of Planck's constant were suddenly to become large and all the laws of physics were to remain otherwise unchanged.

16. An electron and a proton both have the same speed. Which has the larger de Broglie wavelength?

17. What is the significance of the amplitude of a de Broglie wave?

18. A beam of electrons is aimed through two small holes in a metal plate toward an array of electron counters beyond the metal plate. In one experiment each hole is left open for 1 hr while the other is kept closed. In the second experiment both holes are left open for 1 hr. Would you expect the total number of counts in any one counter to be the same or different in the two experiments? How would the total number of counts compare?

19. The velocity of a photon is always exactly c in vacuum. Why is this not inconsistent with the uncertainty principle?

20. The amplitude of the de Broglie wave for the particle in a box discussed in Section 32-12 is equal to zero for some points in the box. Discuss the fact that the particle can move through the box yet pass through points for which its probability of occurrence is equal to zero. Describe the kinds of measurements that would have to be made.

21. Compare the values of the angular momentum of an electron of hydrogen as given (a) by the Bohr theory and (b) by quantum mechanics.

22. An energy level or state of an atom is described by a set of quantum numbers. List one such set.

23. State the Pauli exclusion principle.

24. What is the significant difference between the energy-level structure of an insulator and that of a semiconductor?

Problems

1. Find the wavelength at which the maximum intensity of radiation is emitted by a healthy human being (98.6°F). Assume that the person radiates like a blackbody.

2. At what temperature will a blackbody radiate with maximum intensity at 6000 Å (yellow light)?

3. The chickadee has a body temperature of 105°F and has a surface area of about 4 cm². Assuming that the bird can be treated as a blackbody, find the rate at which it radiates energy.

4. Treat the sun as a blackbody with a temperature of 6000 K and a radius of 7×10^8 m. Find the rate at which it radiates energy.

5. Find the energy of a photon emitted as part of a beam of sodium yellow light, wavelength 5893 Å.

6. At what frequency would a photon have an energy of 1 erg?

7. The work function of a particular metal is equal to 2 eV. Find the maximum kinetic energy of the photoelectrons ejected by light of wavelength 4200 Å.

8. Photoelectrons are not emitted from a certain metal if the wavelength of the light incident on the surface is larger than 3400 Å. Find the work function of the metal.

9. Calculate the wavelengths of the four visible lines in the Balmer series directly from the Balmer formula.

10. Calculate the wavelength of the series limit of the Balmer series.

11. Calculate the frequency of the most intense line of the K series for silver, for which Z is equal to 47.

12. Calculate the radii of the second and third Bohr orbits.

13. Find the speed of the electron in the first Bohr orbit.

14. The energy of a particular state of an atom is 5.36 eV and the energy of another state is 3.45 eV. Find the wavelength of the light emitted when the atom makes a transition from one state to the other.

15. The sodium D lines (5890 Å and 5896 Å) are emitted by transitions from two closely spaced

levels to a single level. Find the separation in energy of the two closely spaced levels.

16. Find the wavelength of an electron that has a kinetic energy of 1500 eV.

17. An electron has a wavelength of 1.5 Å. Find the momentum of the electron.

18. What is the de Broglie wavelength of an aircraft that weighs 10 tons and is flying at 1500 mi/hr?

19. What is the velocity of the de Broglie wave associated with an electron in the first Bohr orbit? (See Problem 13.)

20. Find the velocity of the de Broglie wave associated with the electron of Problem 17.

21. Assume that the uncertainty of position for an electron in a Bohr orbit is given by the diameter of the orbit. Find the uncertainty of momentum for an electron in the first Bohr orbit.

22. A certain atomic state has a lifetime of about 10^{-2} sec. Find (a) the width in energy of this level and (b) the associated spread in frequencies.

23. Find the first two allowed energies of an electron confined to a box that is the size of the first Bohr orbit.

24. Find the first two allowed energies for a neutron (mass 1.67×10^{-27} kg) that is confined to a box about the size of a nucleus, 2×10^{-13} cm.

25. What is the numerical value of the angular momentum that is in a state with $l = 2$?

26. What is the maximum value of m_l, for a state for which $l = 3$?

27. An electron is in a state where $n = 5$, $l = 3$, $m_l = -2$. What is the numerical value of the component of its angular momentum along the direction of an external magnetic field?

28. What is the maximum value of the total angular momentum quantum number for an electron in a state for which the principal quantum number is equal to 4?

29. How many subgroups are there for a group characterized by $n = 4$?

30. Show that a state with angular momentum l can contain $2(2l + 1)$ electrons.

31. Before Planck discovered the formula given by Equation 32-3, Wien had guessed at a similar formula. Wien's formula is

$$R_\lambda = \frac{c_1}{\lambda^5} e^{-c_2/\lambda T}$$

where the symbols have the same meaning as they do in Equation 32-3 and the constants have the same values. Show that Wien's formula is a good approximation to Planck's if λT is small compared with c_2.

32. In the usual experiment to investigate the photoelectric effect, the electrons emitted by the photoelectric surface S are collected by another electrode C. If the potential difference between C and S is made sufficiently negative, however, the electrons will not have enough energy to reach C. The minimum potential difference V necessary to stop the most energetic electrons is called the *stopping potential*. In a certain experiment the stopping potentials measured for several wavelengths of incident light are given by the following table. (a) Find the work function of the material used to emit the photoelectrons and (b) find Planck's constant from these data by drawing a graph.

V (volts)	1.48	1.5	0.93	0.62	0.36	0.24
λ (Å)	3660	4050	4360	4920	5460	5790

33. The average human being requires about 3×10^{-10} erg of visible light incident on the cornea in order to have the sensation of vision. (a) How many photons of light of wavelength 5100 Å does this correspond to? (b) Of the energy incident on the cornea, 4 percent is reflected and 50 percent is absorbed by the material of the eye before the light reaches the retina. Of the remainder, 20 percent is absorbed by the visual purple, the material responsible for the sensation of light. How many photons are needed by the visual purple to give the sensation of light?

34. Calculate the equivalent of the Balmer formula for an atom that consists of a helium nucleus and one orbital electron. The helium nucleus has a mass four times that of the proton and has twice the charge of a proton. Assume that the Bohr theory can be applied directly.

35. Find the time it takes for an electron in the nth Bohr orbit to make one revolution. Compare this with a typical lifetime (10^{-10} sec).

36. As an illustration of the correspondence principle, calculate the frequency of revolution in the Bohr orbit and the frequency of the photon emitted in a transition to the next lowest orbit from orbits whose initial values are (a) $n = 2$, (b) $n = 100$, and (c) $n = 1000$.

37. Ultraviolet light of wavelength 1850 Å is incident on a tube containing hydrogen. Determine the minimum kinetic energy with which an electron will be ejected from hydrogen.

38. A photon and an electron each have a wavelength of 10^{-8} cm. Find (a) the energy and (b) the momentum of each.

39. A crystal with a crystal spacing of 1.5 Å is used as a diffraction grating for electrons. The first-order maximum is found at an angle of 30° to the original direction of the electron beam. Find (a) the momentum and (b) the speed of the electrons.

40. An electron is sent through a slit whose width is 0.04 cm. Determine the uncertainty in the momentum of the electron in a direction parallel to the width of the slit.

41. Assume that an electron is known to be inside a nucleus of diameter 2×10^{-13} cm. Find the uncertainty in energy of this electron. Compare this with the Coulomb potential energy of an electron and proton at this separation.

42. Assume that Planck's constant was suddenly to increase to 1 joule sec. Find the uncertainty in transverse velocity of a pitched baseball, which weighs 150 gm. Assume that the uncertainty in transverse position is the maximum width of a baseball bat, about 8 cm.

43. Imagine that you tried to take a photograph of an electron in the first Bohr orbit. (a) Estimate a reasonable size for the wavelength of the radiation you would use, recognizing that the wavelength should be small compared to the size of the orbit. (b) Calculate the energy of one photon of this radiation. (c) Assume that this photon scatters from the electron and changes its wavelength by h/mc (see the discussion of the Compton effect in the next chapter). Find the change in energy of the photon. (d) Compare this change in energy with the binding energy of the electron.

44. Consider a particle in a box of length d, in the first excited state, for which n is equal to 2. (a) Where is the probability of finding the parti-

cle equal to zero? (b) Where is the probability a maximum? (c) Repeat for the state with n equal to 3.

45. An electron in a box makes a transition from the first excited state to the ground state and emits a photon of energy 5 eV. What is the size of the box?

46. List the quantum numbers l and m_l for all states with the principal quantum number n equal to 4.

47. How are the electrons arranged in (a) lithium, $Z = 3$, (b) sodium, $Z = 11$, and (c) potassium, $Z = 19$? (d) Relate these arrangements to the chemical behavior and position in the periodic chart of these elements.

48. Consider the two-slit experiment shown in Figure 32-13. Assume that one slit is closed and that the electron beam is so arranged that the probability P_1 of finding an electron on the screen is almost uniform over a broad region and that this probability distribution is equal to P_2, the distribution obtained with the other slit open. Show that with both slits open, the maximum probability of finding an electron at a particular location is equal to $4P_1$ and that the minimum probability is equal to zero.

49. Consider a rigid diatomic molecule, which consists of two atoms, each of mass M, separated by a distance d. Assume that the system rotates with angular velocity ω about the midpoint of the line connecting the atoms, and that the angular momentum L is quantized by a Bohr condition:

$$L = \frac{lh}{2\pi}$$

(a) Show that the kinetic energy of rotation $\mathcal{E}_r$ is given by

$$\mathcal{E}_r = \frac{l^2 h^2}{4\pi^2 M d^2}$$

(b) Find the ground-state kinetic energy of rotation for a hydrogen molecule, for which $M = 1.7 \times 10^{-24}$ gm and $d = 0.74 \times 10^{-8}$ cm.

nuclear physics

chapter 33

33-1
Introduction

Almost all of the behavior of real objects can be explained on the basis of the physics we have described up to this point. The atom can be viewed as a dense, charged nucleus surrounded by a cloud of electrons. The structure and interactions of this electron cloud are determined by the electromagnetic interaction, as governed by the laws of quantum mechanics; these interactions determine the structure and chemical behavior of molecules, and therefore of liquids and solids. The gross behavior of large samples is determined by the laws of electromagnetism and gravity, as governed by quantum mechanics and relativity and, to very good accuracy, approximated by the classical laws of physics. The behavior of samples consisting of large numbers of elements is accurately described by the laws of thermodynamics and statistical physics. In a sense, physics has discovered the laws governing the domain of the practical.

From this point of view the nucleus is merely the center of the atom, providing most of the atom's mass and determining its behavior through the fact that the total number of electrons in the neutral atom is equal to the total number of positive charges in the nucleus. In order to understand most of the practical world, we need not know anything about the internal structure of the nucleus. Nevertheless, the curiosity of physicists has led them to attempt to understand the structure of nuclei. This attempt has produced a vast amount of data, much of it only dimly understood. A profusion of particles and particle-like entities have been discovered, and large amounts of energy-level data have been accumulated. Most physicists believe that in order to understand these data, new fundamental laws of physics will have to be invented. These new laws will necessarily be approximated by the laws of quantum mechanics and relativity, which are known to apply in the domain of atoms and electrons, but they may in their subtleties reveal aspects of nature of great pragmatic as well as abstract significance.

The attempt to understand the nucleus has proceeded along several tracks:

1. Since the charge, mass, and electrical and magnetic properties of nuclei are consistent with the idea that all nuclei are composed of protons and neutrons, attempts have been made to understand the forces among neutrons and protons. The hope is that if we could completely understand these forces, we could understand the processes by which complex nuclei are constructed.

2. Through measurements of the emission and absorption spectra of nuclei, a great deal of information about the internal energy of nuclei has been accumulated. Since nuclei can emit and absorb charged particles and neutral particles in addition to photons, the spectra of nuclei are very complex. Some of these emission or absorption processes involve the change of one kind of nucleus to a chemically or physically different nucleus; these processes are similar to chemical reactions, and are called *nuclear reactions*.

3. A particularly interesting class of nuclear reactions involves very-high-energy particles used as probes of the nucleus. From this class of experiment has come the discovery of many very short-lived particles. The study of these objects has become a distinct branch of physics, called particle physics.

This chapter will provide a brief description of the physics of the nucleus, and the next chapter will introduce the subject of particle physics.

33-2
General properties of nuclei

For most purposes we can regard nuclei as composed of protons and neutrons. For many purposes it is convenient to use the term *nucleon* to refer to a particle that is either a proton or a neutron.

Any particular nucleus is characterized by the number of protons, Z, the number of neutrons, N, and the total number of nucleons, A.

The number of protons, Z, determines the positive charge of the nucleus (as numerically equal to Z times the charge of an electron), which in turn determines the number of electrons forming the electron cloud of the corresponding atom. The number of electrons, which is the atomic number, determines the chemical identity of the atom. For example, a nucleus with $Z = 6$ is the nucleus of a carbon atom.

The mass of the proton is approximately the same as the mass of the neutron, and therefore the mass of a nucleus is approximately equal to A times the mass of a nucleon. (As we shall see below, there are small corrections for the binding energy as well as the difference in mass between the proton and the neutron.) Therefore, the quantity A is sometimes called the *mass number* of the nucleus. From the definition of A, N, and Z, we know that

$$A = N + Z \qquad (33\text{-}1)$$

A particular species of nucleus, characterized by a specific atomic number and specific mass number, is called a *nuclide*.

In a convenient and widely used notation, a nuclide is represented by the chemical symbol, with the mass number written as a superscript in the upper left and the atomic number written as a subscript in the lower left; in this notation, $^{13}_{6}C$ represents a nucleus containing 6 protons, and a total of 13 nucleons; this nuclide therefore contains 7 neutrons. Since the chemical symbol tells us the number of protons, the subscript is sometimes omitted; the nucleus described above is sometimes represented by ^{13}C.

Two nuclei which have the same Z but different values of A are said to be *isotopes*; the term "isotope" is also used to describe atoms formed from such nuclei. For example, naturally occurring carbon is a mixture of two isotopes, $^{12}_{6}C$ and $^{13}_{6}C$.

Two nuclides with the same value of the mass number A but different values of Z are called *isobars*. Two isobars are said to be *mirror nuclides* if the number of protons, Z, of one nucleus is equal to the number of neutrons, N, of the other.

Most nuclei are slightly deformed spheres, with a radius R (in centimeters) given approximately by

$$R = (1.2 \times 10^{-13}) A^{1/3} \qquad (33\text{-}2)$$

where A is the mass number of the nucleus. From Equation 33-2, the nuclear volume V, which is proportional to the cube of the radius, is proportional to A, the number of nucleons:

$$V = \tfrac{4}{3}\pi R^3 = \tfrac{4}{3}\pi(1.2 \times 10^{-13})^3 A \qquad (33\text{-}3)$$

The mass M of a nucleus is roughly

$$M = M_p A \qquad (33\text{-}4)$$

where M_p is the mass of the proton. Combining Equations 33-3 and 33-4 we reach the conclusion that the nuclear density is roughly constant.

The Appendix lists the atomic masses in atomic mass units (u), in which the atomic mass of the most abundant isotope of carbon, $^{12}_{6}C$ is exactly 12. The energy equivalent of 1 u is 931.5 MeV, while

$$1 \text{ u} = 1.66053 \times 10^{-24} \text{ gm}$$

Nuclei are bound systems; in order to separate a nucleus into a collection of free neutrons and protons, a specific amount of energy must be added to the nucleus. From the relativistic principle of energy conservation, the mass of a system consisting of a nucleus plus its binding energy must equal the sum of the masses of Z protons and N neutrons. Thus, the mass of a nucleus is less than the mass of Z protons and N neutrons; the difference between these two masses is called the mass defect of the nucleus. The mass defect is simply the binding energy expressed in units of mass.

The binding energy per nucleon is a measure of the stability of a nucleus and is plotted in Figure 33-1 (as a function of A). As can be seen from the figure, the binding energy per nucleon is fairly constant for A larger than 50, although there is a small decrease for large A, and there is a very sharp rise for small A.

Nuclei contain internal energy that is quantized, and thus the nucleus has energy levels between which transitions can occur. Like atoms, nuclei can emit photons, although the energies of these photons are usually millions of times larger than the energies associated with atomic transitions. Nuclei may make transitions by three other processes: *beta decay*, in which the nucleus emits an electron (or a positively charged electron, called a *positron*) and a neutrino; *alpha particle emission*, in which the nucleus emits an alpha particle; and *fission*, in which the nucleus splits into two pieces of nearly equal mass. In each of these processes the residual nucleus (or nuclei) has an atomic number different from that of the original nucleus, and the last two processes also involve changes in the mass number of the nucleus. Nuclei can also

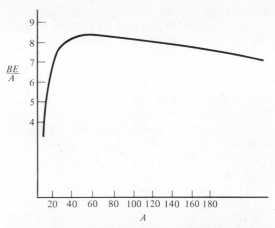

FIGURE 33-1 Graph of the binding energy per nucleon as a function of the mass number, A, of the nucleus.

transfer some of their energy directly to an atomic electron, causing that electron to be emitted from the atom. This process is called *internal conversion*. In addition, the nucleus may *capture* an atomic electron, annihilate it, and emit a photon.

A given nucleus can usually undergo several of these energy transfer processes and, when in a particular energy level, a nucleus may sometimes be able to make two or three alternative transitions. Those nuclei that emit alpha particles, beta particles, or photons (usually called gamma rays) are said to be *radioactive*.

In addition to charge, mass, size, and energy-level structure, several other gross properties of nuclei have been measured. Since both the neutron and proton have spin and magnetic moment, nuclei generally have these properties. The spin of a nucleus is equal to the vector sum of the spins of the nucleons of which it is composed. The magnetic moment associated with the nuclear spin exerts forces on the atomic electrons that produce splittings of the *atomic* spectral lines similar to the splitting caused by the electron spin. In addition to these properties, measurements have been made of the deviations from sphericity of many nuclei; most nuclei turn out to be cigar-or football-shaped.

33-3
Alpha particles

We now know that alpha particles are the nuclei of helium atoms and consist of two protons and two neutrons. After Becquerel discovered natural radioactivity in 1896, Rutherford discovered that the rays emitted by radioactive materials were of two kinds: a soft radiation, easily absorbed by matter, which he called alpha radiation; and a more penetrating radiation, which he called beta radiation. The charge-to-mass ratio of alpha rays was measured by measuring the deflection of the rays in known electric and magnetic fields. The charge was measured by counting the total number of alpha particles emitted by a sample of radioactive material in a specific time interval, collecting these particles, and measuring the total charge of these particles. In 1908 Rutherford was able to collect enough alpha particles in a container to measure the optical spectrum of the gas that was formed. This direct measurement showed that alpha particles are identical to helium nuclei.

It should be pointed out that in this process of investigating radioactivity the physicists were faced by two simultaneous problems: the phenomena themselves and the need to develop instruments that could measure these completely new processes. The development of radiation detection and measurement devices has been a major part of physics during the whole of this century.

A very important property of an alpha particle is the speed with which it travels when emitted by a radioactive element. Its speed can be measured by making it travel at right angles to a very intense magnetic field and measuring the radius of its circular orbit. The results of such experiments show that alpha particles move with speeds of the order of 1.6×10^9 cm/sec. This is a very high speed for such massive particles. But much more interesting is the fact that the speed is characteristic of the isotope that ejects it, as can be seen from Table 33-1. In some cases, such as radon (with

$A = 222$), all the alpha particles have the same speed. In other cases, such as radium (with $A = 226$), the alpha particles have one of two definite speeds. In still other cases, there are several characteristic speeds.

The fact that alpha particles have known mass, charge, and velocity made them very useful tools for investigation of the properties of nuclei. Rutherford, Geiger, and Marsden performed a historic series of experiments in which they observed the scattering of alpha particles by thin, metal foils. This series of experiments led Rutherford to propose his nuclear theory of the atom. Rutherford calculated the expected orbits of the alpha particle using classical physics, since quantum mechanics was not invented until 15 years later. Subsequently, the calculation was done quantum mechanically. The striking result is that the correct quantum mechanical calculation agrees exactly with the classical calculation. This exact agreement occurs because the deflecting force obeys an inverse-square law. If the force obeyed any other law, such as an inverse-cube law, the quantum-mechanical calculation would give results very different from the classical one. Thus Rutherford could postulate the nuclear picture of the atom only because, by an accident of nature, his classical calculations happened to give a correct description of the scattering of alpha particles.

In general, however, the behavior of alpha particles follows quantum-mechanical rather than classical laws. An interesting illustration of such behavior involves the process of emission of alpha particles from nuclei, the theory of which will be described in the next section.

When a nucleus emits an alpha particle it is transformed into a residual nucleus. Because the alpha

Table 33-1   Velocities of alpha particles from some isotopes

element	atomic number Z	mass number A	older name of isotope		velocities in 10^9 cm/sec
Bismuth	83	212	Thorium C	α_1	1.711
				α_2	1.705
				α_3	1.665
				α_4	1.645
				α_5	1.642
Polonium	84	210	Polonium		1.597
		214	Radium C′		1.922
		215	Actinium A		1.882
Radon	86	222	Radon		1.625
Radium	88	226	Radium	α_1	1.517
				α_2	1.488

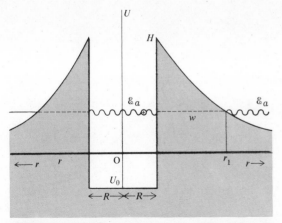

FIGURE 33-2 Potential energy U of a nucleus and an alpha particle as a function of the distance r from the center of the nucleus. H is the height of the potential barrier and w is its width.

particle carries off two units of charge and four nucleon masses, the atomic number Z of the residual nucleus is lower by two and its mass number A is lower by four than the original nucleus. This process, like all radioactive decay, occurs randomly in time. That is, if we observe one nucleus, there is no way to tell beforehand when that nucleus will emit an alpha particle. However, for a sample containing a large number of nuclei, the rate of disintegration is constant. This rate of disintegration is expressed in terms of the *half-life*; the half-life is defined as the *time required for half the nuclei in a sample to have emitted an alpha particle*. The constancy of the half-life implies that at the end of one half-life, half of the nuclei will have disintegrated, and during the next half-life, one half of those remaining will have disintegrated. Thus at the end of two half-lives, one quarter of the original nuclei will remain. The half-lives of natural alpha particles range from a fraction of a microsecond to 10^{10} years.

33-4
Theory of alpha-particle emission

The emission of an alpha particle from a radioactive nucleus is a comparatively rare event; only alpha particles of certain specific energies can escape from a given nucleus. From the experiments on the scattering of alpha particles by nuclei we know that Coulomb's law of force holds outside the nucleus, even as close as 10^{-12} cm. The poten-

tial energy of an alpha particle with respect to a nucleus is proportional to $1/r$; this is shown in Figure 33-2. Inside the nucleus the alpha particle is subject to two types of forces: (a) electric forces of repulsion, and (b) strong attractive forces due to the other particles, the specifically nuclear forces, which are independent of the charges on the particles. The exact nature of the latter type of force is not known; for simplicity we shall assume that the resultant of all the forces on the alpha particle in the nucleus produces a constant potential energy, $-U_0$, that extends out to a distance R from the center. Let us call this distance R the radius of the nucleus and join the two curves by a vertical line that intersects the Coulomb potential curve at a point H. This point H is called the *height of the potential barrier*. The height of the potential barrier can readily be calculated for any given nucleus. For the heavy nuclei such as radium and uranium, it is of the order of 30 MeV. On the basis of classical physics, the height of the potential barrier represents the minimum energy that an alpha particle must have in order to escape from the nucleus. Experimental evidence, however, shows that this does not happen; the alpha particles that do escape have energies in the range of 5–10 MeV. Furthermore, only alpha particles with very specific energies manage to escape from specific nuclei.

Alpha-particle emission can best be understood on the basis of quantum mechanics. Imagine that an alpha particle in the nucleus has acquired an amount of energy $\mathcal{E}_\alpha$ as a result of its interaction with the nucleons, and that it is moving toward the nuclear surface. Here it encounters the potential barrier of height H, greater than $\mathcal{E}_\alpha$, and is reflected from it towards the opposite side. We may imagine the alpha particle as moving back and forth in this potential well of height H at a level $\mathcal{E}_\alpha$ above the bottom. The alpha particle has a de Broglie wave associated with it of wavelength $\lambda = h/mv$; this wave moves back and forth in this potential well. Each time it strikes the wall, part of it is transmitted and part reflected. The *probability of penetrating* the barrier of width w, also called the *transparency* of the barrier, is the ratio of the square of the amplitude of the transmitted wave to the amplitude of the incident wave. The transparency T is very small and depends upon the energy of the alpha particle for a given potential-barrier height, and the width of the barrier. If the alpha particle moves with a velocity v, the frequency with which it strikes the potential barrier is $v/2R$. The product of this frequency and the transparency is equal to the probability of escape per unit time of the alpha particle; this is also the reciprocal of the average lifetime of the nucleus. The results obtained by the above method are in qualitative agreement with the results of experiment. They have been used for calculating the radius R of the nuclei that decay by alpha-particle emission by measuring the kinetic energy $\mathcal{E}_\alpha$ of the particles that escape from the nucleus. The width w of the potential barrier can then be determined with the aid of Coulomb's law. It is through such calculations that Equation 33-2 for the radius R of a nucleus was determined.

We might speculate for a moment on the reverse of alpha-particle emission—that is, on the bombardment of nuclei by alpha particles. We have already mentioned one aspect of this problem, the single scattering of alpha particles by nuclei in the Rutherford experiment. If an alpha particle having an energy $\mathcal{E}_\alpha$ and an initial velocity v is directed toward a nucleus of charge Ze along a line joining their centers, then, on the basis of classical physics, the alpha particle should come to rest at a distance r_1 and then be reflected back along the same line. On the basis of quantum mechanics, however, there is a probability that the alpha particle will penetrate the potential barrier, even though its energy is less than the height H of the potential barrier. This probability of penetration will have a maximum value for that energy that is equal to the energy of the alpha particle in the nucleus of charge $(Z + 2)e$. Such experiments were performed successfully by Rutherford and co-workers beginning in 1919, years before the quantum theory was developed.

33-5
Radioactive disintegration by beta-particle emission

When a radioactive isotope of atomic number Z, neutron number N, and mass number A disintegrates by the emission of a beta particle—that is, an electron—the atomic number of the product nucleus becomes $Z + 1$, but the mass number A remains unchanged. Hence the neutron number must decrease to $N - 1$. For example, the radioactive isotope of bismuth, $Z = 83$, $A = 210$, also sometimes known as radium E, emits a beta particle; the product nucleus is polonium, $Z = 84$, $A = 210$. The half-life of this distintegration is 5

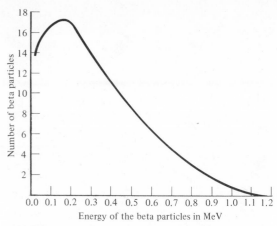

FIGURE 33-3 Distribution of energy among beta particles emitted in the beta decay of bismuth, $A = 210$, or radium E.

days. We can write this radioactive disintegration as a nuclear reaction equation as follows:

$$^{210}_{83}\text{Bi} \rightarrow {}^{210}_{84}\text{Po} + \beta + Q \qquad (33\text{-}5)$$

where Q is the disintegration energy or the energy released in this reaction and comes from the difference in masses of initial and final particles. On the basis of the above reasoning we might expect to find that the beta particles emitted by ^{210}Bi all have the same velocity. However, when these velocities are measured by a magnetic spectrograph, it is found that there is a continuous distribution of velocities up to a maximum velocity; that is, there is a continuous distribution of energies among them. Such a distribution of energies is called an *energy spectrum*. Figure 33-3 shows the distribution of energies among the beta particles emitted by ^{210}Bi. The ordinate is the number of beta particles within a given energy range, and the abscissa is the energy of these particles expressed in MeV. It will be observed that there is a continuous distribution of energy up to the maximum value or *end-point energy*.

One might hope to account for the existence of beta decay by assuming that a neutron in a nucleus breaks up into a proton and an electron, and that the electron is subsequently emitted by a process similar to that of alpha emission. However, such a picture fails to explain the continuous distribution of energy. Further, it is inconsistent with the conservation of angular momentum. It was suggested by W. Pauli that these difficulties might be overcome if it were assumed that *two* particles were emitted in beta decay—an electron and an undetected neutral particle. This neutral particle, which was named the *neutrino*, has a negligible mass, and a spin of $\frac{1}{2}$ (in units of $h/2\pi$).

In 1934 E. Fermi proposed a theory to explain beta decay. In this theory he assumed that the electron and neutrino were *created* at the same time that a neutron is transformed into a proton. That is, we do not think of a neutron as containing the proton, electron, and neutrino, but rather that some force operates to *create* the electron and neutrino. This force is the weak nuclear force referred to in Chapter 5. The Fermi theory of beta decay, sometimes now called the *theory of weak interactions*, has been somewhat modified in the intervening years, but in essence it has been successful at describing the phenomena. The theory explains the natural beta decay we have described, and in addition explains the similar emission of positrons that occurs in artificially produced isotopes. It also explains other processes, such as atomic electron capture, the decay of an isolated neutron (into proton, electron, and neutrino), and the interactions of several of the more recently discovered particles (in particular, the decay of the muon).

A major modification of the Fermi theory was made by Lee and Yang in 1956 when they showed that there are two kinds of neutrino. One has its spin oriented antiparallel to the direction of the motion of the particle. This particle is called the neutrino, ν. The other, the antineutrino, $\bar{\nu}$, has its spin parallel to the direction of motion. Thus the beta decay of the neutron occurs by the process

$$n \rightarrow p + e^- + \bar{\nu} \qquad (33\text{-}6)$$

Similarly, it is possible for a proton inside a nucleus to decay, emitting a positron by the process

$$p \rightarrow n + e^+ + \nu \qquad (33\text{-}7)$$

33-6
The nucleon-nucleon interaction

The fact that nuclei exist at all demonstrates that there exists a nuclear force that acts among nucleons (in addition to the electric and gravitational forces). Since the Coulomb force between protons in nuclei is so large and repulsive, there must exist a very strong attractive force. One of the basic objectives of nuclear physics is to arrive at a complete understanding of the force between two pro-

tons, the force between two neutrons, and the force between a proton and a neutron. At this time there is no complete theory, and much of the available knowledge is too complex and technical for discussion here. But it is possible to describe some of the procedures used to attack the problem and to describe some of the conclusions.

Some of the general properties of the nuclear force may be deduced from the general properties of nuclei. The stability or existence of nuclei indicates that the nuclear force must be attractive between two protons that are separated by distances of approximately the size of a nucleus. It is possible that for separations that are small compared with the size of the nucleus the force may be repulsive. There are several indications, including direct measurements of the scattering of neutrons or protons by protons, that the force must become vanishingly small for separations larger than about 10^{-13} cm, the size of the nucleus.

The fact that the nuclear density is constant and the fact that the binding energy per nucleon is roughly constant lead to the conclusion that the nuclear force has the property of *saturation*. The Coulomb force and the gravitational force obey the principle of superposition; the force exerted on particle A by particle B is independent of how many other particles have forces exerted on them by B. A force exerted by a particle B is said to be saturated if the force can be exerted on only a finite number, n, of particles. If the nuclear force did not have the property of saturation, then every nucleon in a nucleus would exert a force on every other nucleon. Then adding a nucleon would increase the force on *each* nucleon, and so the binding energy *per nucleon* should increase with increasing mass number. Similarly, the increase in force due to the addition of a nucleon should make the nucleus contract and become more dense. (Note that the equivalent effects *are* observed in atoms, which are governed by the electromagnetic force, a force that does not saturate.) Thus we reach the conclusion that a given nucleon can exert a force on a limited number of other nucleons —probably only the nearest neighbors—and that it exerts no force at all on the other nucleons in the nucleus. Thus the range of the force exerted by a nucleon in a nucleus cannot be very large.

One of the most striking aspects of the nuclear force is that it is *charge-independent*; the proton-proton force is the same as the neutron-neutron force, and is the same as the proton-neutron force if we subtract out the Coulomb force. This property is a justification of the use of the term "nucleon."

Strictly speaking, we know only that the nuclear force is charge-independent for low-energy systems. For these systems there are several pieces of evidence leading to the conclusion of charge-independence. It is possible to do direct experiments, scattering protons off hydrogen (whose nucleus is a proton) and scattering neutrons off hydrogen. The Pauli principle limits the kind of information that can be deduced from such experiments, but in general they are consistent with the hypothesis that the *p-p* force is the same as the *n-p* force at low energies. The force between two neutrons is difficult to measure directly, since there are no targets of pure neutrons, but one can perform experiments involving scattering neutrons from deuterium, whose nuclei are deuterons—a bound state of a proton and a neutron.

Additional indications of the validity of charge independence comes from the stability of the naturally occurring nuclides. It is known that for low A, the neutron number is approximately equal to the number of protons and that for high A the number of neutrons exceeds the number of protons. If the force between two neutrons were larger than the force between two protons, then the most abundant stable nuclei would have more neutrons than protons, which is not the case for low A. As the mass number increases, the Coulomb repulsion among the protons increases, and because the nuclear force is saturated, more neutrons are required to hold the nucleus together so the high A data are not inconsistent with charge-independence. In addition, the energy-level structure of mirror nuclei are very similar (except for the effects of the Coulomb force). Since mirror nuclei differ only in that the number of neutrons in one is the same as the number of protons in the other, this indicates the charge-independence of the nuclear force.

In order to account for the shape of the deuteron, the nucleus of $^2_1$H, and some of the details of the scattering of nucleons, it is necessary to assume that the force between two nucleons depends on the angle between the spin vectors of the two particles.

In summary, for low energies, the nuclear force between two nucleons is known to be *attractive, short range, spin-direction-dependent, charge-independent*, and *saturable*.

The fact that the nuclear force is charge-independent means, as we have seen, that the proton and neutron are indistinguishable as far as

the nuclear force is concerned. That is, there exists a single kind of particle, the nucleon, that appears in two forms, or states. One of these states of the nucleon is the proton and the other state is the neutron. This situation is analogous to the fact the atomic electron can be in either of two states: one state with its spin parallel to an external magnetic field, and the other with its spin antiparallel to the field. Physicists have therefore postulated an imaginary space called *isotopic-spin space*. In this imaginary space a nucleon has associated with it a vector quantity called the *isotopic spin, T*, which has a magnitude $\frac{1}{2}$. When the isotopic-spin vector points "up" in this space, the nucleon is observed (by real observers in real space) to be a proton, and when the isotopic spin points "down" in isotopic-spin space, the nucleon is observed to be a neutron. Thus, a proton is a nucleon whose z component of isotopic spin, T_z, is equal to $+\frac{1}{2}$, whereas the neutron is a nucleon whose z component of isotopic spin is $-\frac{1}{2}$. It must be emphasized that the z axis is a direction in a purely fictional space. This strange and awkward convention has been adopted because it turns out to be useful to describe a broad range of phenomena. In particular, it is possible to define the isotopic spin of *systems* of nucleons (nuclei). Perhaps more important, many of the particles discovered since 1950 can be classified in terms of isotopic spin. Further, the nuclear force, or strong interaction, conserves total isotopic spin, a fact that is useful in describing nuclear reactions and the interactions of the particles.

33-7
Nuclear reactions

One of the main sources of information about nuclear properties is the analysis of nuclear reactions. In the most common type of reaction, a beam of particles, x, strikes a sample target nucleus, X. After the interaction, a residual nucleus, Y, is left and emerging from the process is a particle y. The beam particle, x, may be a photon, proton, neutron, alpha particle, or even a nucleus, as may the emerging particle, y. Such a reaction is described by the relation

$$x + X \rightarrow Y + y \qquad (33\text{-}8)$$

or in the notation $X(x, y)Y$. Thus, reactions in which a *neutron* beam strikes a target nucleus and a *proton* emerges are called (n, p) reactions. Reactions in which the beam particle x and the emerging particle y are the same kind of particle are called *scattering* reactions. If a scattering process occurs without change of kinetic energy of the beam particle, it is said to be an *elastic* scattering; otherwise, it is said to be *inelastic*.

In any reaction the total charge is conserved, which can be described by saying that the total atomic number is conserved. The total mass number is also conserved, but this is not the same as saying that mass is conserved. In general, the masses of the initial particles are not the same as the masses of the final particles; the difference between the total mass of the initial particles and the total mass of the final particles is called the reaction energy, Q, or simply the Q value. Since the total relativistic energy is conserved, the Q value is equal to the change in kinetic energy in the process. If the total kinetic energy of the final particles exceeds that of the initial particles, Q is a positive number, and the sum of the masses of the final particles is less than that of the initial particles. If Q is negative, the sum of the final kinetic energies is less than the sum of the initial kinetic energies and the sum of the final masses is larger than the sum of the initial masses. Equation 33-8 is sometimes written so as to include the Q value explicitly, in the form

$$x + X \rightarrow Y + y + Q \qquad (33\text{-}9)$$

For example, when radium, $Z = 88$, $A = 226$, emits an alpha particle, the product is an atom of the element radon, $Z = 86$, $A = 222$. This disintegration may be written as a nuclear reaction equation as follows:

$$^{226}_{88}\text{Ra} \rightarrow\, ^{222}_{86}\text{Rn} + \,^{4}_{2}\text{He} + Q$$

The value of Q can be calculated from the known atomic masses of the particles involved as follows:

mass of Ra	$= 226.02536$ u
mass of Rn	$= 222.01753$ u
mass of He	$=\ \ \ 4.00260$ u
mass of final products	$= 226.02013$ u
mass difference $= Q$	$=\ \ \ 0.00523$ u

Since 1 u = 931.5 MeV, the disintegration energy Q may be expressed as

$$Q = 4.87 \text{ MeV}$$

The kinetic energy of the alpha particle emitted by radium, as determined from its velocity is 4.78 MeV. The difference between the two values can be accounted for by the fact that when an alpha particle is emitted, the product nucleus recoils. Although the momentum of the alpha particle is equal and opposite to the momentum of the recoil nucleus, the velocity of the latter is much smaller and its kinetic energy is also much smaller.

Many of the phenomena encountered in the study of nuclear reactions may be understood best in terms of the *compound-nucleus* theory, which was proposed by Niels Bohr in 1936. In this theory the beam particle x enters the target nucleus X, and the two together form an intermediate compound nucleus X^*. The energy brought into the compound nucleus is distributed randomly among the nucleons in the compound nucleus, and after some time the compound nucleus decays by one of the processes we have described earlier.

The random distribution of the energy among the constituents of the compound nucleus means that the compound nucleus "forgets" how it was formed in the sense that the beam nucleon becomes indistinguishable from one of the nucleons in the target nucleus. It is possible to form a given compound nucleus from different target nuclei, using different beam particles; in general, the subsequent decay of these compound nuclei will be indistinguishable from each other. This compound-nucleus theory requires that the compound nucleus live, or exist, for a time long enough for the random redistribution to have occurred. That means roughly that the compound nucleus must have a lifetime that is large compared with the time it takes a nucleon to travel across a nucleus. Experimentally measured lifetimes of compound nuclei are as large as 10^{-14} sec, which is consistent with this requirement.

One series of reactions that has been studied carefully is the set of (α, p) reactions, using aluminum as the target. With a fixed alpha-particle energy, the outcoming protons have several discrete energies. Using the compound nucleus theory, this implies that the compound nucleus has several discrete energy levels. Each of the discrete proton energies comes from the decay of a compound nucleus in one of its discrete energy levels.

This energy-level structure of the compound nucleus can be investigated further by varying the energy of the alpha-particle beam. When the beam energy is precisely the right value to leave the compound nucleus in one of its energy levels, the probability of formation of that energy level is higher than if the beam particle brings in energy that leaves the compound nucleus intermediate between two levels. Thus the yield of outgoing protons of a particular energy should rise to a maximum and then decrease as the energy of the alpha-particle beam is varied through this range. Such a peak in the rate of a nuclear reaction is called a *resonance*. In the (α, p) reactions on aluminum, at least 18 such resonance peaks have been observed. Similar resonance studies have been carried out with many targets and many beam particles, resulting in very detailed energy-level tables for short-lived compound nuclei.

The compound-nucleus theory is very useful, but it does not adequately describe reactions involving very light nuclei because there are too few nucleons for the assumption of randomness to be valid. In addition, it is doubtful that the theory is applicable to processes in which the energy of the beam particle is very large compared to the energy of the compound-nucleus levels.

Some reactions are particularly interesting because the radioactive nuclei produced are useful. For example, when ordinary sodium, ^{23}Na, is bombarded with deuterons, radioactive ^{24}Na is formed by the process, as follows:

$$^{23}_{11}\text{Na} + ^{2}_{1}\text{H} \rightarrow (^{25}_{12}\text{Mg}) \rightarrow ^{24}_{11}\text{Na} + ^{1}_{1}\text{H} \quad (33\text{-}10)$$

where $^{25}_{12}$Mg is the compound nucleus. The radio-active sodium decays by the process

$$^{24}_{11}\text{Na} \rightarrow ^{24}_{12}\text{Mg} + \beta^{-} + \bar{\nu} \quad (33\text{-}11)$$

with a half-life of 14.8 hr. Thus ordinary table salt, NaCl, can be made radioactive. When table salt is ingested by animals, the motion of the sodium ions can be followed by detecting the beta rays. Such *tracer* studies have become an essential tool in many sciences. Radioactive nuclei have been used to tag molecules or parts of molecules, and these have been used to gain detailed knowledge of many chemical and biochemical processes. The technique has been used to deliver therapeutic doses of radioactivity to specific organs of the body, and as an aid to medical diagnosis.

Another interesting reaction is that described by

$$^{14}_{7}\text{N} + ^{1}_{0}n \rightarrow (^{15}_{7}\text{N}) \rightarrow ^{14}_{6}\text{C} + ^{1}_{1}\text{H} \quad (33\text{-}12)$$

This process is believed to occur naturally when cosmic-ray neutrons strike atmospheric nitrogen. The resulting carbon nucleus, ^{14}C, is radioactive, and decays via

$$^{14}_{6}\text{C} \rightarrow ^{14}_{7}\text{N} + \beta^{-} + \bar{\nu} \quad (33\text{-}13)$$

with a half-life of 5580 years. The naturally occurring radioactive carbon is absorbed by living organisms. When the organism dies, it ceases to absorb ^{14}C, and the ^{14}C that is present decays. Thus the percentage of ^{14}C in wood, bone, and other organic material decreases at a known rate. Assuming that the original percentage of ^{14}C is known, it is possible to determine the age of archaeological and geological samples that contain carbon. Most such carbon dating has been made on the assumption that the production of ^{14}C has been constant, and such dates have been fairly accurate for periods extending back about 40,000 years. Recently, there has been some evidence, from dates obtained from tree rings, that the production of ^{14}C in the atmosphere has varied somewhat and therefore the ages obtained from carbon dating may have to be modified.

33-8
Nuclear models

In an attempt to describe nuclear reactions, physicists have developed several models. Each of these models is used to describe some aspect of a nuclear process or specific kinds of reactions. Since each is limited, in no sense can it be said that physics has a definitive model of nuclear structures. Their value lies in the fact that each does summarize some data, and the hope that one or more may suggest or lead to a complete description of the nucleus. In what follows we shall briefly describe (a) the statistical model, (b) the liquid-drop model, (c) the alpha-particle model, (d) the collective model, (e) the optical model, and (f) the shell model.

The statistical model assumes that, because of the very strong interactions among the nucleons, we cannot describe the motions of individual nucleons but must treat the nucleus mathematically as if it were a gas, analogous to the electron gas of solids. This model is basically the same as the compound nucleus model, and has the limitation that it describes averages, omitting the effects of significant details.

The liquid-drop model also ignores the individual nucleon, and treats the nucleus as a drop of liquid that is incompressible and has surface tension. In this model, decay by particle emission is analogous to evaporation. The model provides a picture for fission. When energy is added to the nucleus,

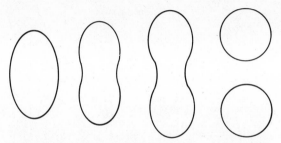

FIGURE 33-4 In the liquid-drop model of the nucleus, oscillations can distort the nucleus enough to split the nucleus into two pieces.

oscillations are set up that distort the shape of the nucleus. If the drop is distorted enough, as shown in Figure 33-4, the Coulomb repulsion between the pieces can be large enough to overcome the surface tension. The result is that the nucleus splits into two separate pieces. The model has been able to describe the qualitative features of the fission process, but it describes the quantitative features poorly.

The alpha-particle model assumes that a nucleus is a cluster of alpha particles. The model describes some characteristics of a few light nuclei, where A is equal to four times some integer, $A = 4n$. However, even for this very limited class of nuclei, the model is sometimes inaccurate.

In the liquid-drop model the nucleus is thought of as a collective entity; however, as we shall see, in the shell model each nucleon is thought of as moving independently. The collective model combines these two approaches. In this model it is assumed that the collective behavior of the nucleons may produce distortions of the nuclear shape, that the nucleus may have a permanent nonspherical shape, and, further, that oscillations of the surface may occur. Within the nucleus the individual nucleons will move independently, subject to a nonspherical potential. The model has been particularly useful in describing nuclear deformations and the energy levels that arise from these deviations from sphericity. The model has also been useful in providing detailed corrections to the predictions of the shell model.

The optical model is most useful in the description of high-energy scattering processes. The simplest way to treat such processes is to treat the nucleus as if it exerted some average force on the scattered particle. Quantum-mechanically, this is equivalent to saying that the nucleus changes the wavelength of the particle. The scattering of the particle is analogous to the scattering of light by a transparent sphere. However, it is possible for the particle to be absorbed by the nucleus and form a compound nucleus. In order to take this into account, the model treats the nucleus as if it were not quite transparent; the model is sometimes called the "cloudy crystal ball" model.

The shell model is an extension of the closed-shell picture, which is so useful in describing atoms, to the description of nuclei. The idea of closed nuclear shells was first proposed by Elsasser in 1934; in 1948 Maria G. Mayer summarized the experimental evidence to show the existence of periodicity in nuclear properties. Many nuclear properties change markedly when either the number of protons, Z, or the number of neutrons, N, approach certain values. In particular, stable nuclei occur when either of these numbers is equal to 2, 8, 20, 50, 82, or 126. These numbers are called, picturesquely, *magic numbers*. The existence of these magic numbers has been interpreted as meaning that closed shells are formed at these values. The model provides a mechanism to predict the magic numbers and to predict some properties of nuclei with closed shells or almost closed shells.

A great deal of evidence has been accumulated for the existence of magic numbers. For example, the most abundant naturally occurring nuclei are the most stable. In a plot of the relative abundance of nuclei, pronounced peaks occur for the magic-number nuclei, such as ^{16}O (N and $Z = 8$), ^{40}Ca (N and $Z = 20$), ^{118}Sn ($Z = 50$). Similarly, the binding-energy data show that nuclei with magic numbers are more stable than nuclei with slightly different values of N or Z. Nuclei with Z equal to one of the magic numbers have large numbers of stable isotopes; for example, tin ($Z = 50$) has the largest number of stable isotopes (10). The stable isotopes of nuclei with a magic number have a very low probability of absorbing a slow neutron compared with that of neighboring nuclei.

In order to explain the magic numbers, the model makes several assumptions. First, it assumes that each particle moves independently, in the field created by the other nucleons. Further, it assumes that the spin of the nucleon is strongly coupled to the orbital angular momentum; that is, it is assumed that the energy of a state where the spin angular momentum is parallel to the orbital angular momentum is very different from the energy of the state where the two angular momenta are antiparallel. With the aid of the Pauli principle, the model is then able to describe the shell structure of the nuclei. The model is particularly useful in describing the lowest energy levels of magic number and neighboring nuclei. It must be emphasized that the shell structure is a structure of energy and that it would be misleading to assume that the nucleons are arranged in layers in space. Further, the shell structure exists for the neutrons and protons independently.

33-9
Radioactive series

In the years following the discovery of the radioactivity of uranium by Becquerel, many radioactive elements and isotopes of elements were discovered. Most of the naturally radioactive isotopes were found to be genetically related and fitted into one of three *radioactive series*. These are known as the *uranium series*, the *thorium series*, and the *actinium series*. Radium, for example, is a member of the uranium series. This series starts with the isotope of uranium, $Z = 92$ and $A = 238$, which has a very long half-life of 4.51×10^9 years; the other members of the series are formed through a succession of alpha-particle emissions, or beta-particle emissions, or both, as shown in Figure 33-5. Originally these radioactive isotopes were given special names, but the modern method is simply to identify them by the appropriate values of A for a given value of Z.

As we have already seen, the emission of an alpha particle by an isotope having given values of Z and A results in the production of a new isotope of atomic number $Z - 2$ and mass number $A - 4$. On the other hand, the emission of a beta particle leaves the mass number unchanged but increases the atomic number of the product nucleus to $Z + 1$, where Z is the atomic number of the parent nucleus.

Each of the three naturally radioactive series terminates with a nonradioactive isotope of lead. In the uranium series this isotope has a mass number of 206, in the actinium series its mass number is 207, and in the thorium series its mass number is 208.

The half-lives of the radioactive isotopes of the uranium series are indicated in Figure 33-5. If a sample of ore containing uranium is analyzed chemically and the percentage of uranium and lead measured, it is possible to calculate, from the ratio

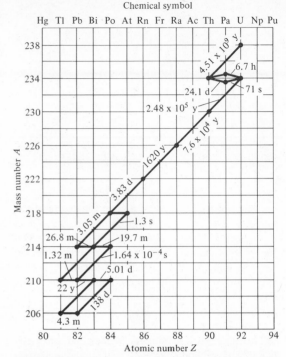

FIGURE 33-5 The naturally radioactive uranium series. The half-lives of the disintegrations are expressed either in years (y), days (d), hours (h), minutes (m), or seconds (s).

of uranium to lead, how long a time must have elapsed for the production of this amount of lead. This value can thus be used as an indication of the age of the earth.

In this discussion of radioactive decay, it was tacitly assumed that a given isotope will have only one mode of disintegration—that is, either by alpha-particle emission or by beta-particle emission. However, a glance at Figure 33-5 will show many interesting cases of *branching*. For example, the isotope of polonium with $Z = 84$ and $A = 218$, also known as radium A, will decay most often (in 99.96 percent of the cases) with the emission of an alpha particle forming the product nucleus $Z = 82$, $A = 214$, a radioactive isotope of lead. In a few cases (0.04 percent), however, the nuclei will emit beta particles forming the product nucleus $Z = 85$ and $A = 218$, an isotope of a comparatively new element called astatine.

In the course of experiments in which nuclei of some of the lighter elements were bombarded with alpha particles, M. and Mme. Joliot-Curie in 1934 observed that the bombarded substances continued to emit radiations even after the source of alpha particles had been removed. Ionization measurements and magnetic deflection experiments showed that the radiations consisted of *positrons*. Furthermore, the intensity of the radiation was found to decrease with the time, in the same way as radiations from a naturally radioactive substance. The half-life of the positron radiation was measured in each case studied. This was the first time that radioactivity was ever induced in a substance. One of the reactions, for example, was the bombardment of boron by alpha particles accompanied by the emission of neutrons according to the reaction equation

$$_{5}^{10}B + _{2}^{4}He \rightarrow (_{7}^{14}N) \rightarrow _{7}^{13}N + _{0}^{1}n \quad (33\text{-}14)$$

Now, nitrogen of mass number 13 is not a stable isotope. It disintegrates with the emission of a positron according to the reaction equation

$$_{7}^{13}N \rightarrow _{6}^{13}C + \beta^{+} + \nu \quad (33\text{-}15)$$

with a half-life $T = 10$ min. The symbol β^{+} is used to represent the positron, since it has a positive charge of the same magnitude as that of the beta particle. The carbon atom of mass number 13 is a stable isotope of carbon.

The identification of the radioactive atom as nitrogen was made certain by chemical analysis. Since the amount available for analysis is very minute, special methods have to be developed for each case. For example, in the boron reaction, the target was made of boron nitride. After bombarding it with alpha particles for several minutes, it was heated with caustic soda. One of the products of this chemical reaction was gaseous ammonia, NH_3. This was found to be radioactive with a half-life of 10 min, showing that the nitrogen of the ammonia was the radioactive element.

Since the discovery of artifically induced radioactivity, it has become possible to induce radioactivity in all elements—that is, to produce one or more radioactive isotopes of every known element. It has also been possible to produce new elements, both to fill in gaps that had existed in the periodic table of elements and to extend the number of elements to more than 100. These developments were made possible by two essentially different types of devices: one known as *particle accelerators*; the other known as *nuclear reactors*.

33-10
The curie

The activity of a radioactive substance is sometimes expressed in terms of a unit known as the *curie*. By definition, the curie is equal to 3.70×10^{10} disintegrations/sec. This number is very close to the number of alpha particles per second emitted by 1 gm of radium. The curie, as a unit of radioactive disintegration, is independent of the nature of the particle emitted in this process.

A radioactive substance whose strength is 1 curie is a very intense source of radiation. Submultiples of the curie in common use are the millicurie ($= 10^{-3}$ c $= 1$ mc) and the microcurie ($= 10^{-6}$ c $= 1$ μc). For example, a beta-ray source that is rated at 4 μc emits 14.8×10^4 beta particles/sec. This is also equal to the number of atoms of the substance that disintegrate per second.

33-11
Nuclear fission

All the nuclear disintegrations described thus far have concerned the emission of comparatively light particles, such as electrons, positrons, protons, and alpha particles. A new type of nuclear process, known as *nuclear fission*, was discovered by Hahn and Strassmann in 1939 in a series of experiments in which uranium was bombarded by neutrons. Chemical analysis of the products of disintegration showed the presence of barium, $Z = 56$, lanthanum, $Z = 57$, and other elements of medium atomic weights. The interpretation of these results is that the capture of a neutron by a uranium nucleus produces a compound nucleus of uranium that is unstable and disintegrates into two particles of intermediate masses; for example, if one of the particles is barium, $Z = 56$, the other particle is krypton, $Z = 36$. Cloud-chamber photographs (see Figure 33-6) have amply verified this hypothesis.

The particles that are produced in the fission of uranium have energies of the order of 200 MeV. The source of this energy is the difference in mass between the initial products—that is, the neutron and the uranium nucleus—and the final or fission products. There is a decrease in mass of about 0.1 percent in this process; thus, in the nuclear fission of 1 kg of uranium, there is a decrease in mass of 1 gm, which corresponds to about 25×10^6 kw hr of energy.

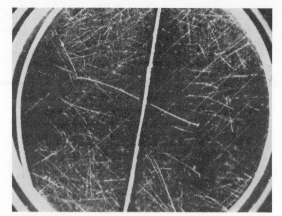

FIGURE 33-6 Cloud chamber photograph showing the fission of uranium. The foil in the center of the cloud chamber is coated with uranium and bombarded by neutrons. The tracks of the two heavy fission particles can be seen coming from the foil where a uranium atom has undergone fission as the result of the capture of a neutron. (Photograph courtesy of J. K. Boggild, K. K. Brostom, and T. Lauriten.)

The masses of the fission products are found to be those of unstable isotopes; that is, they have many more neutrons than the stable isotopes of the corresponding elements. One of the first questions investigated was the manner in which these unstable fission products disintegrated, particularly whether any of the excess neutrons were emitted in this process. Early experiments showed that between 2 and 3 neutrons were emitted per nuclear fission. The process can now be represented schematically, as shown in Figure 33-7. When a neutron is captured by a uranium nucleus of mass number 238, a new isotope of uranium of mass number 239 is formed; in the process of nuclear fission, the latter splits into two isotopes of intermediate masses, say barium and krypton, with the prompt emission of 2 neutrons. A variety of other pairs of nuclei may be produced in the fission process, all of them radioactive, most of them decaying to a stable form by the emission of beta rays; gamma rays are also emitted by many of these isotopes.

In addition to uranium, thorium, $Z = 90$, and protoactinium, $Z = 91$, have been found to be fissionable by the capture of neutrons, and a new element, plutonium, $Z = 94$, is also fissionable by the capture of neutrons. Fission may also occur spontaneously, by excitation of a nucleus with high-energy gamma rays and by bombardment of heavy nuclei with protons, deuterons, or alpha

particles. In the following sections, we shall discuss only neutron-induced nuclear fission.

33-12
A nuclear-fission chain reaction

The concept of a nuclear chain reaction is very simple; if a single nuclear-fission process involving the capture of one neutron results in the release of energy and simultaneously the release of more than one neutron, it should be possible to arrange the mass of fissionable material in such a way to ensure the capture of the newly released neutrons. Or, stated another way, the mass of fissionable material should be so arranged that at any one place the number of new neutrons produced should be equal to the number of free neutrons originally present at that place. The ratio of these two numbers of neutrons is called the *multiplication factor* K. If $K = 1$, the chain reaction will be self-sustaining; if K is less than 1, the process will ultimately come to a halt; if K is greater than 1, the neutron density will increase and may lead to an explosive reaction. A mass of fissionable material so arranged that the multiplication factor is equal to or greater than 1 constitutes a *nuclear reactor*.

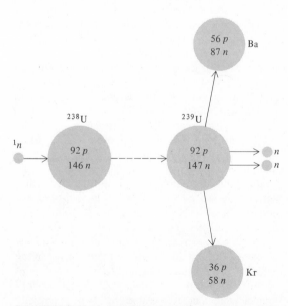

FIGURE 33-7 Nuclear fission of uranium. A fast neutron is captured by a nucleus of uranium of mass number 238, forming ^{239}U; the latter splits into two comparatively massive particles, krypton and barium, with the simultaneous emission of two fast neutrons.

In order to be able to design a nuclear reactor, it is essential to know the conditions under which neutrons are captured by nuclei and the conditions under which such capture of neutrons results in the fission of the product nuclei. We shall restrict this discussion to the fission of uranium. Ordinary uranium consists of 3 isotopes: one of mass number 238, another of mass number 235, and a third of mass number 234. The most abundant of these is ^{238}U—about 99.3 percent abundance. The amount of ^{234}U in ordinary uranium is negligible. ^{235}U constitutes about 0.7 percent of ordinary uranium. Experiments show that ^{238}U is fissionable only if it captures *fast* neutrons—that is, neutrons having energies of 1 MeV or greater. On the other hand, ^{235}U is fissionable with neutrons of any speed and particularly with *slow* neutrons—that is, neutrons having energies corresponding to the thermal energies at ordinary temperatures. These energies are much less than 1 eV.

The neutrons released in nuclear fission have a wide range of energies. In the case of the fission of ^{235}U, these energies extend up to about 17 MeV, with a maximum number having energies of about 0.75 MeV. If such neutrons are captured by other uranium nuclei, they produce nuclear fission. However, not every collision between a fast neutron and a uranium nucleus results in capture of the neutron; the collision may simply produce a decrease in the energy of the neutron. Thereafter, the probability of its capture will be very small; additional collisions will produce further reductions in the energy of the neutrons. At some particular values of energy, the neutron will be readily captured by ^{238}U, but such a capture does not result in nuclear fission. Instead, the newly formed isotope of uranium, ^{239}U, emits a gamma-ray photon and then becomes radioactive, emitting a beta ray with a half-life of 23 min. The nuclear reaction equations are

$$^{238}_{92}\text{U} + {}^{1}_{0}n \rightarrow (^{239}_{92}\text{U}) \rightarrow {}^{238}_{92}\text{U} + \text{gamma ray}$$

then

$$^{239}_{92}\text{U} \rightarrow {}^{239}_{93}\text{Np} + \beta^- \qquad T = 23.5 \text{ min} \quad (33\text{-}16)$$

The new element thus formed, called neptunium, Np, is itself radioactive, emitting a beta particle with a half-life of 2.35 days. The product nucleus formed in this reaction is plutonium, Pu, of atomic number 94. The reaction in which this is formed is

$$^{239}_{93}\text{Np} \rightarrow {}^{239}_{94}\text{Pu} + \beta^- \qquad T = 2.35 \text{ days}$$

It is followed by

$$^{239}_{94}\text{Pu} \rightarrow {}^{235}_{92}\text{U} + {}^{4}_{2}\text{He} \qquad T = 24{,}360 \text{ yr} \quad (33\text{-}17)$$

The isotope of plutonium is radioactive, emitting an alpha particle, but it has a very long half-life—24,360 yr. In this sense, it is a comparatively stable element. It will be noted that neptunium and plutonium are *transuranic* elements—that is, elements with atomic numbers greater than that of uranium. When plutonium disintegrates with the emission of an alpha particle, the resulting nucleus is ^{235}U. The plutonium isotope formed in the above process is fissionable by the capture of neutrons of any energy and is thus similar to ^{235}U as far as the fission process is concerned. Since it is chemically different from uranium, it can be separated more readily from the uranium metal than the uranium isotope of mass number 235.

If ordinary uranium is to be used in a nuclear reactor, it is essential to avoid loss of neutrons by nonfission capture. Since slow neutrons can produce fission in ^{235}U and since the probability of capture varies inversely with the speed of the neutron, one method of ensuring its fissionable capture is to slow down the neutrons very rapidly to thermal energies. This is done with the aid of a *moderator*—that is, a light element in which the probability of nuclear capture of a neutron is negligible, but in which collisions between neutrons and nuclei will cause a rapid decrease of the energy of the neutron. Deuterium and carbon are two elements suitable for use as moderators.

The first nuclear reactor, or *uranium pile* as it is sometimes called, was operated successfully in Chicago on December 2, 1942; it was built under the direction of E. Fermi and operated by groups headed by W. H. Zinn and H. L. Anderson. A schematic diagram of the construction of a graphite (carbon)-moderated uranium pile is shown in Figure 33-8. Rods of uranium metal are embedded in blocks of graphite; rods of boron metal are inserted at various places in the pile to control the flux of neutrons; boron nuclei capture neutrons very readily. No special source of neutrons is needed to start this pile; there are always neutrons present from cosmic rays, or from spontaneous fission, to start the nuclear reactor. The mode of its operation can be understood by referring to Figure 33-9. Suppose that a neutron is captured by a uranium nucleus, so that fission results and that 2 new neutrons are released, with energies of about 1 MeV each. These neutrons then make

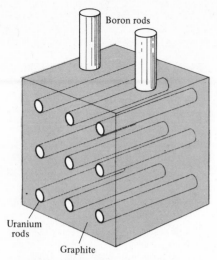

FIGURE 33-8 Schematic diagram of a uranium pile. Cylindrical rods of uranium are embedded in a large mass of graphite that acts as a moderator to slow down neutrons. Boron rods, which are inserted into the pile, control its rate of activity.

several collisions with nuclei of the moderator, graphite, until their energies are reduced to thermal energies. Whenever one of these slow neutrons is captured by ^{235}U, fission will again occur with the release of, say, 2 neutrons. Some neutrons may be lost through the surface of the reactor; one way to reduce this loss is to make the reactor very large; the increase in the surface area is proportional to the square of its linear dimension, and the volume is proportional to the cube of the linear dimension. Other neutrons may be lost through capture by impurities or through nonfissionable capture by ^{238}U. But if $K = 1$, the reaction will be self-sustaining. To prevent the multiplication factor from becoming excessive, boron rods are inserted to various depths in the pile to absorb the excess neutrons. One other control factor may be mentioned here; that is, not all the neutrons are emitted promptly in nuclear fission; a small percentage of the neutrons are delayed, some by 0.01 sec, others by as much as 1 min.

A whole new field of nuclear science and engineering has been opened as the result of the discovery of nuclear fission and following the successful construction of the first nuclear reactor. Nuclear reactors designed for many different purposes are now in operation throughout the world. Some are used as sources of energy for power plants; others are used for experimental purposes. One such reactor is shown in Figure 33-10. A nuclear reactor is one of the best sources of neutrons for use in physical, chemical, and biological experiments. It is also a source of radioactive isotopes for medical and industrial uses.

33-13
Nuclear fusion

In some of the reactions mentioned previously, nuclei of low mass numbers interact to form nuclei of higher mass numbers. In 1939 Bethe suggested that such processes probably take place in the sun and other stars where the high temperatures in the interiors are sufficient to provide the necessary energy to overcome the Coulomb repulsion of the electric charges of the nuclei. One type of reaction suggested was essentially the *fusion* of four hydrogen nuclei to form a nucleus of helium. This is not likely to occur as a single reaction but could occur in a series of reactions, depending upon the state of evolution of the star. One such series could be a proton-proton chain, such as the following:

$$_1^1\text{H} + _1^1\text{H} \rightarrow _1^2\text{H} + \beta^+$$

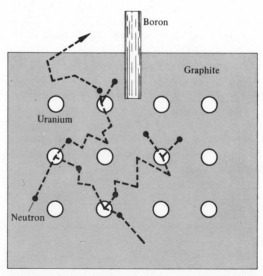

FIGURE 33-9 Schematic diagram of the action of a neutron in a uranium pile based on the assumption that each fission yields two neutrons. The circles represent rods of uranium; the small circular dots represent neutrons. Sudden changes in direction of the neutron path are due to collisions with nuclei of the moderator, graphite.

FIGURE 33-10 View of the Gulf-General Atomic TRIGA reactor core. The reactor is designed for research, training, and isotope production; it rests at the bottom of a tank 21-ft deep, which is inside a concrete-lined pit. The surrounding earth and 16 ft of demineralized water provide the required shielding. (Courtesy of Gulf Energy and Environmental Systems.)

followed by

$$\,_1^1H + \,_1^2H \rightarrow \,_2^3He \qquad (33\text{-}18)$$

The chain ends either with the reaction

$$\,_1^1H + \,_2^3He \rightarrow \,_2^4He + \beta^+ \qquad (33\text{-}19)$$

or, more probably, with the reaction

$$\,_2^3He + \,_2^3He \rightarrow \,_2^4He + \,_1^1H + \,_1^1H \qquad (33\text{-}20)$$

Another possible series of reactions in which four protons are converted into a helium nucleus could be through the intermediary of carbon and nitrogen as *catalysts*; this is usually referred to as the *carbon-nitrogen cycle*, and consists of the following reactions:

$$\,_6^{12}C + \,_1^1H \rightarrow (\,_7^{13}N) \rightarrow \,_7^{13}N + \text{gamma ray}$$
$$\,_7^{13}N \rightarrow \,_6^{13}C + \beta^+ \qquad T = 9.96 \text{ min}$$

$$\,_6^{13}C + \,_1^1H \rightarrow (\,_7^{14}N) \rightarrow \,_7^{14}N + \text{gamma ray} \qquad (33\text{-}21)$$

and

$$\,_7^{14}N + \,_1^1H \rightarrow (\,_8^{15}O) \rightarrow \,_8^{15}O + \text{gamma ray}$$
$$\,_8^{15}O \rightarrow \,_7^{15}N + \beta^+ \qquad T = 124 \text{ sec}$$
$$\,_7^{15}N + \,_1^1H \rightarrow (\,_8^{16}O) \rightarrow \,_6^{12}C + \,_2^4He \qquad (33\text{-}22)$$

It will be noted that this chain of reactions can be started with either carbon or nitrogen. No matter which process is involved, the amount of energy released is the difference in mass between four hydrogen atoms and one helium atom; since

$$\,_1^1H = 1.007825 \text{ u}$$

$$4 \times \,_1^1H = 3.03130 \text{ u}$$

and

$$\,_2^4He = 4.00260 \text{ u}$$

then

$$\Delta m = 0.02870 \text{ u}$$

which is equivalent to about 26.7 meV per helium nucleus formed. A small amount of this energy is carried away by the neutrinos formed in the radioactive parts of the cycle. Bethe estimated that if the sun were to continue radiating energy at its present rate, about 30 billion years would be needed for the sun to exhaust its supply of protons. Other nuclear processes also take place in the sun and in the other stars; the particular type of process depends upon the state of evolution of the star. There are several interesting theories of the evolution of the stars and of the elements in the stars; these are left for other studies.

The temperatures produced in the explosion of a fission bomb (sometimes called an *A-bomb*) are of the order of millions of degrees—that is, as hot as, or hotter than, the interior of the sun. It is thus not surprising that the first attempts to produce nuclear fusion would be as a military weapon using an A-bomb to produce the necessary high temperatures. The actual methods used are, of course, military secrets, but thermonuclear bombs or hydrogen bombs have been produced. For the past 20 years experiments have been in progress in many parts of the world to produce controlled thermonuclear fusion processes, thus far with very little success. Practically all the methods involve the use of the heavier isotopes of hydrogen, since these reactions proceed at a faster rate. The reactions with deuterons, together with the energy released in each one, are

$$_1^2\text{H} + _1^2\text{H} \rightarrow _1^3\text{H} + _1^1\text{H} + 4.02 \text{ MeV} \qquad (33\text{-}23a)$$

$$_1^2\text{H} + _1^2\text{H} \rightarrow _2^3\text{He} + _0^1 n + 3.25 \text{ MeV} \qquad (33\text{-}23b)$$

When deuterium and tritium are used, a much larger amount of energy is released; thus

$$_1^2\text{H} + _1^3\text{H} \rightarrow _2^4\text{He} + _0^1 n + 17.6 \text{ MeV} \qquad (33\text{-}24a)$$

and

$$_1^2\text{H} + _2^3\text{He} \rightarrow _2^4\text{He} + _1^1\text{H} + 18.3 \text{ MeV} \qquad (33\text{-}24b)$$

Both $_1^3\text{H}$ and $_2^3\text{He}$ are products of the deuteron-deuteron reactions.

Although the above fusion reactions can be and have been produced with particles that received their energy in particle accelerators, the aim in thermonuclear fusion is to trap these particles in some container to start the reaction with a small number of nuclei, and then have the energy from these reactions raise the temperature of the system so that the fusion processes will proceed spontaneously. The method most commonly used involves a gas-discharge tube containing deuterium in which the electric discharge produces a *plasma*—that is, a region in which the gas is completely ionized. The present method of confining the gas is to trap its particles in a magnetic field so shaped that the particles move within the region and are reflected at the ends. It is hoped in this way to build up the temperature of the system so that the deuterons will have sufficient energy to react with one another to start the process of thermonuclear fusion.

The term "temperature" as used in this discussion refers to the average kinetic energy of the particles, assuming it to be about kT, where k is the Boltzmann constant. At room temperature, for example, which is about 300 K, the average kinetic energy of a particle is about 1/40 eV. Hence

$$1 \text{ eV} \approx 12{,}000 \text{ K} \qquad (33\text{-}25)$$
$$1 \text{ keV} \approx 12 \times 10^6 \text{ K}$$
$$1 \text{ MeV} \approx 12 \times 10^9 \text{ K}$$

33-14
Photon emission and absorption

A photon can be thought of as a neutral particle that has energy $\mathcal{E}$ and momentum p; since a photon always travels with the speed of light c it can be thought of as having zero rest mass (particles of nonzero rest mass cannot travel with the speed of light because then they would have infinite energy and momentum). The relation between the energy and momentum of a photon is given by:

$$\mathcal{E} = pc \qquad (33\text{-}26)$$

As we have seen, nuclei have a complex internal energy structure; that is, nuclei can exist in different *states*, where each state has a distinct total energy, although all the states of a particular nucleus have the same number of neutrons and protons. The state of a particular nucleus with the lowest total energy is called the *ground state*, the state with the next largest total energy is called the *first excited state*, the one with the next largest energy is called the *second excited state*, and so on. A nucleus in a given state can be thought of as a particle (even though we know that it has internal structure) with a rest mass M, where M is related to the total energy $\mathcal{E}$ of the state by

$$\mathcal{E} = Mc^2$$

That is, each state of a particular nucleus differs from the other states in having a different mass: the ground state has mass M_0, the first excited state has mass M_1, the second excited state has mass M_2, and so on. When a nucleus is in one of the excited states it will (very rapidly) emit a photon (sometimes called a gamma ray) and end up either in one of the lower excited states or in the ground state. If it ends up in an excited state, it will emit another photon, and then perhaps another, ultimately ending up in the ground state. (At any stage the nucleus may emit an alpha particle or beta particle, but we shall here consider only photon emission.) In the emission process the energy of the photon will be exactly equal to the difference in total energy of the initial and final states of the nucleus.

Consider then a nucleus that is at rest in one of its excited states, of energy $\mathcal{E}'$ and mass M', and then emits a photon of energy U such that the nucleus ends up in its ground state with mass M_0. In this process both energy and momentum are conserved. Since we have assumed that the initial nucleus was at rest, the total momentum of the system must remain zero, and since the emitted photon must have some momentum, the ground-state nucleus must *recoil*, with momentum p. Thus we can write, considering conservation of energy,

$$\mathcal{E} = M'c^2 = \mathcal{E}_0 + U \qquad (33\text{-}27)$$

where $\mathcal{E}_0$ is the total energy of the recoiling ground-state nucleus. Now, from Equation 11-23 we have, for the energy of the recoiling nucleus,

$$\mathcal{E}_0{}^2 = p^2c^2 + M_0{}^2c^4 \qquad (33\text{-}28)$$

We can substitute for $\mathcal{E}_0$ in this equation, from Equation 33-27,

$$\mathcal{E}_0 = M'c^2 - U$$

and thus Equation 33-28 becomes

$$(M'c^2 - U)^2 = p^2c^2 + M_0{}^2c^4 \qquad (33\text{-}29)$$

From conservation of momentum for the emission process we know that the momentum p of the recoiling nucleus must equal the momentum of the photon and therefore from Equation 33-26 we know that

$$p = \frac{U}{c} \qquad (33\text{-}30)$$

Then, substituting Equation 33-30 into Equation 33-29, we get

$$(M'c^2 - U)^2 = U^2 + M_0{}^2c^4 \qquad (33\text{-}31)$$

or

$$M'^2c^4 + U^2 - 2M'c^2U = U^2 + M_0{}^2c^4$$

and so

$$M'^2c^4 - 2M'c^2U = M_0{}^2c^4 \qquad (33\text{-}32)$$

Now each state of the nucleus has a definite energy and mass, and the difference is a fixed, well-defined energy called the Q *value*. That is, we can define Q by

$$M'c^2 = M_0c^2 + Q \qquad (33\text{-}33)$$

Then from Equation 33-33 we have

$$M_0c^2 = M'c^2 - Q$$

and squaring this equation, we get

$$M_0{}^2c^4 = M'^2c^4 + Q^2 - 2M'c^2Q \qquad (33\text{-}34)$$

Combining Equations 33-32 and 33-34, we obtain

$$M'^2c^4 - 2M'c^2U = M'^2c^4 + Q^2 - 2M'c^2Q$$

or

$$U = Q - \frac{Q^2}{2M'c^2}$$

that is,

$$U = Q\left[1 - \frac{Q}{2M'c^2}\right] \qquad (33\text{-}35)$$

Thus, although the difference in energies between the initial and final states at rest is a definite number Q, the emitted photon has an energy U, which is less than Q. This difference arises from the fact that some of the initial energy must appear in the kinetic energy of the recoiling nucleus.

We can now think about the inverse process, absorption. In this process a photon strikes a nucleus that is at rest and is absorbed by the nucleus. The resulting composite nucleus is one of the excited states. As was said above, each of the excited states of the nucleus has a definite energy and mass. Thus, in order for the nucleus to make the transition from the ground state to one of the excited states, a definite energy must be added to it. If the photon can add this amount of energy to the nucleus and still conserve energy and momentum, the photon will be absorbed (and disappear). One of the interesting and unexpected results of quantum mechanics is that if the photon can transfer only some *other* amount of energy to the nucleus, then the photon will not be absorbed. That is, if the photon has either too much or too little energy then the photon is unaffected by the nucleus, whereas it is absorbed if it has just the right amount of energy.

In order for a photon to be absorbed by a nucleus it must *transfer* the Q value to the nucleus. Before the collision the photon has momentum as well as energy. Therefore, since momentum must be conserved, the composite nucleus must have momentum and must therefore be moving (assuming that the absorber was at rest initially). This implies that the composite nucleus must have kinetic energy as well as rest mass energy. This kinetic energy must have been supplied by the photon. Thus, the photon must have *more* energy than the Q value, because some of its energy must provide kinetic energy for the composite nucleus and some of its energy must provide the mass-energy difference between the ground state and the excited state.

For a particular combination of excited and ground states the *emitted* photon has *less* energy than the Q value, whereas in order for the photon to be *absorbed* it must have *more* energy than the Q value. Thus an emitted photon cannot be absorbed by the same kind of nucleus; in other words, a particular nucleus is transparent to its own emitted radiation. This effect is, as a practical matter, modified by several effects: the Q values are not perfectly sharp (that is, for a particular transition the Q value may have any value in a particular narrow *range*); the nuclei generally have small motions due to thermal energy; and the energy of the nuclei may be modified by electric and magnetic fields.

Mössbauer, who received the Nobel Prize in 1961, discovered an interesting effect. When certain nuclei are part of a crystal, they are bound so tightly to the crystal structure that they cannot recoil independently. When they recoil, the whole crystal (or a significant part of it) recoils, and so the M' of Equation 33-35 is the mass of a very large number of atoms. Thus, the energy of the emitted photon is almost exactly the Q value. Similarly the energy necessary for a photon to be absorbed is also the Q value. The nucleus *can* absorb the emitted photon. Experimentally, this can be shown by having the photons emitted by one crystal and absorbed by a second crystal. If the absorbing crystal is then given a small velocity, the absorption process would have to take place between a photon and a *moving* nucleus. This would require a different amount of photon energy and so the photon will not be absorbed. Absorber velocities as low as 1 cm/sec can destroy the absorption. If the emitter and absorber nuclei are present as impurities in different crystals, then the emitter and absorber nuclei will be in different electric and magnetic fields. In general, then, the excited states will have different Q values and the emitted photon will not be absorbed. If the absorber is then given a small velocity, it may be possible for the photons to be absorbed. Analysis of these processes can yield important information about the fields in the crystals that provides useful data about the structure of the crystals. Thus an essentially relativistic nuclear effect can be used as an important analytic chemical tool.

33-15
The Compton effect

Another very convincing type of evidence concerning the corpuscular nature of radiation is provided by the *Compton effect*. Suppose a beam of X rays of frequency f and wavelength λ is incident upon a substance that then scatters it in various directions. If we examine the scattered beam with an X-ray crystal spectrometer, we find that the beam consists of X rays not only of frequency f and wavelength λ but also of lower frequency f' and thus of higher wavelength λ'. The explanation of this effect, given by A. H. Compton, is that the X-ray photon collides with an electron and is scattered by it, and, if the electron is free to move, the photon gives up some of its energy to the electron. Using the principle of conservation of energy, we can write

$$hf = hf' + \tfrac{1}{2}mv^2 \qquad (33\text{-}36)$$

where hf is the energy of the incident photon, hf' is the energy of the scattered photon, and $\tfrac{1}{2}mv^2$ the kinetic energy of the scattered electron. Since this is a collision between two particles, we can apply the principle of conservation of momentum to this process. Figure 33-11 shows the incident photon with momentum hf/c, the photon scattered through an angle θ with momentum hf'/c, and the electron recoiling as a result of the collision with momentum mv in a direction making an angle ϕ with the original direction of the incident radiation. From the principle of conservation of momentum, we get the equations

$$\frac{hf}{c} = \frac{hf'}{c}\cos\theta + mv\cos\phi \qquad (33\text{-}37)$$

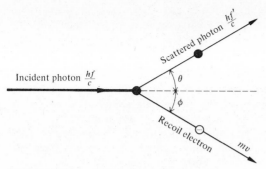

FIGURE 33-11 The Compton effect; scattering of a photon and recoil of an electron as a result of a collision.

and

$$0 = \frac{hf'}{c} \sin \theta - mv \sin \phi \qquad (33\text{-}38)$$

Equation 33-37 states that the initial momentum along the original direction of the X-ray beam is

equal to the components in this direction of the final momenta of the scattered photon and the recoil electron. Equation 33-38 is a similar expression for the components of the momenta at right angles to the original direction of motion.

If the above three equations are solved for the frequency of the scattered photon, and the result is then converted to their corresponding wavelengths, the result obtained is

$$\lambda' - \lambda = \frac{h}{mc} (1 - \cos \theta) \qquad (33\text{-}39)$$

where λ' is the wavelength of the ray scattered at an angle θ and λ is the original wavelength. The change in wavelength depends only upon the angle of scattering and not upon the substance. The results predicted by Equation 33-39 have been amply verified by many careful experiments. The energy and momentum of the recoil electron have also been measured, and the results were found to be in agreement with the predictions obtained from solutions of the above equation.

Questions

1. Which of the following properties of an atom are due primarily to the structure and properties of the nucleus and which are due primarily to the structure and properties of the atomic electrons: (a) the mass, (b) the valence, (c) the optical spectrum, (d) the X-ray spectrum, (e) the gamma-ray spectrum?

2. (a) What property has the same value for two isotopes? (b) What property has different values?

3. What nuclide is a mirror nuclide to $^{11}_{6}C$?

4. Is there an atomic-mass defect? If so, estimate its magnitude, expressed as a fraction of the atomic mass.

5. List the processes by which a nucleus in a particular state can make a transition to another state.

6. If the nuclei are plotted on a graph in which the y axis is the neutron number N and the x axis is the proton number Z, the stable nuclei fall roughly on a curve. Where would you expect to find the β^- emitters on the graph? The positron emitters?

7. Give the change in A, Z, and N of the nucleus for each of the following: (a) alpha emission, (b) beta emission, (c) positron emission, (d) gamma emission, (e) internal conversion.

8. Distinguish between soft and hard radiation.

9. As an interesting illustration of the interplay between the phenomena of nuclear physics and the instrumentation, investigate the story of the early work of Madame Curie. In particular, attempt to discover the role played by Monsieur Curie.

10. How is it that a naturally occurring isotope can have a half-life of a few seconds or minutes? Should it not have disappeared before physicists began to investigate the phenomena?

11. What is meant by: (a) half-life, (b) barrier penetration, (c) Q value, (d) saturation of a force, (e) compound nucleus, (f) magic number?

12. Describe some of the properties of the force between nucleons.

13. Illustrate the meaning of saturation by describing the chemical bonding force.

14. Distinguish between nuclear reactions that are elastic and those that are inelastic.

15. How can radioactivity be used to measure the age of the earth?

16. Discuss some of the applications of radioactivity to biological research and medical therapy.

17. Discuss the fission reactor as a source of en-

ergy. Describe the source of energy, the fuel, and the waste products. Discuss the environmental problems associated with the reactor as a source of energy and compare these with the problems associated with conventional sources of energy.

18. Discuss the fusion process as a source of energy. Compare this with the fission reactor and conventional sources of energy.

19. Radioactive isotopes are sometimes used for medical therapy or for diagnostic purposes. What factors other than the half-life of the isotope will influence the total dosage of radioactivity that results at a particular organ from a specific amount of material ingested by the body?

20. Why are high-energy electrons rather than low-energy electrons used as probes to investigate the distribution of charge in nuclei?

21. The most common element in the universe is hydrogen. Explain why helium is the next most common.

22. For many biological applications we are interested in the ionizing ability of a radioactive source, or the rate at which it transfers energy to matter, rather than in the number of disintegrations per second. For such applications the strength of radioactivity is measured in *rads*, where a 1-rad source deposits 100 ergs/gm. The rad is a large unit, since a few hundred rads in a short period of time can lead to serious illness or even death. (a) Calculate the temperature rise of water caused by 200 rads. (b) Would such a temperature rise be perceptible to the touch?

Problems

1. For the following, replace X with the appropriate number or chemical symbol: (a) $_{2}^{3}X$ (b) $_{40}^{97}X$ (c) $_{X}^{30}P$ (d) $_{X}^{15}N$.

2. For each of the nuclides in Problem 1, find the number of neutrons.

3. Find Z, N, and A for each of the following nuclides: (a) $_{6}^{13}C$, (b) $_{11}^{24}Na$, (c) $_{92}^{234}U$.

4. Consider a 12-gm sample of pure $_{6}^{12}C$. (a) How many atomic electrons are there in the sample? (b) How many nuclear protons? (c) How many nuclear neutrons?

5. Calculate the nuclear radius of the following: (a) $_{2}^{4}He$, (b) $_{6}^{12}C$, (c) $_{82}^{210}Pb$.

6. Find the volume of the following nuclides: (a) $_{5}^{12}B$, (b) $_{11}^{21}Na$, (c) $_{92}^{238}U$.

7. Find the density of nuclear matter in $_{56}^{141}Ba$.

8. A radioactive nuclide has a half-life of 3 min, and yields a stable nuclide. After 9 min, what fraction of the original sample is radioactive?

9. A radioactive sample gives 12,000 counts/min. Exactly 24 hr later it gives 3000 counts/min. What is the half-life of the sample?

10. Write the equation for each of the following processes: (a) Radon emits an alpha particle. (b) ^{24}Na emits a negative beta particle. (c) ^{22}Na emits a positron. (d) ^{10}B emits a gamma particle.

11. Calculate the binding energy of the deuteron, $_{1}^{2}H$.

12. Calculate the binding energy per nucleon for ^{20}Ne.

13. Calculate the binding energy per nucleon for ^{57}Fe.

14. Calculate the binding energy per nucleon for ^{207}Pb.

15. In a certain decay the products have a total mass that is 0.012 u less than the mass of the initial nuclide. How much kinetic energy is divided among the products?

16. The alpha emitter ^{210}Po has a mass of 209.98287 u and the alpha particle has a mass of 4.00260 u. After the decay 5.40 MeV of kinetic energy are shared by the alpha particle and the residual ^{206}Pb. Find the mass of the residual nuclide.

17. Write the equations for the following processes: (a) The formation of $_{15}^{32}P$ in an (n, p) reaction, (b) the (α, p) reaction with ^{14}N as the target, (c) the (n, α) reaction with ^{10}B as the target.

18. How much energy is required to break up a $_{6}^{12}C$ nucleus into three alpha particles?

19. How many beta particles are emitted per second by a 5-mc sample of ^{214}Pb?

20. From the known masses of the neutron and proton, find the maximum kinetic energy of the beta particle emitted in the decay of a neutron into a proton.

21. An archaeologist finds an old wooden beam, in which the ratio of radioactive to stable carbon is one fourth that of new wood. Estimate the age of the beam.

22. The alpha particle can be decomposed by the addition of energy into (a) two deuterons, (b) a ^{3}He nucleus and a neutron, (c) a ^{3}H nucleus and a proton, and (d) two protons and two neutrons. Find the amount of energy required for each process.

23. Tritium decays by beta emission. (a) Write the reaction equation and (b) calculate the maximum energy of the emitted beta particle.

24. An alpha particle is moving with a velocity v toward a heavy nucleus of atomic number Z along a line joining their centers. Derive an expression for the distance of closest approach between them, assuming that nonrelativistic equations may be used.

25. An alpha particle having a kinetic energy of 4 MeV is moving toward a gold nucleus along the line joining their centers. Use the result of Problem 24 and calculate the distance of closest approach of the alpha particle to the gold nucleus.

26. A sample of polonium is placed in a magnetic field of magnitude 2 webers/m². Calculate the radii of curvature of the paths of the alpha particles whose initial velocities are perpendicular to the magnetic field.

27. Using the data of Table 33-1, for the lowest velocity alpha particle emitted by bismuth-212 find (a) the kinetic energy of the emitted alpha particle, (b) the momentum and (c) the energy of the recoil nucleus, and (d) the Q value of the disintegration.

28. Using the data of Table 33-1, calculate (a) the momentum of the alpha particle emitted by radon and (b) the de Broglie wavelength associated with this alpha particle. (c) Compare this wavelength with the radius of the radon nucleus.

29. Assume that the speed of the alpha particle inside the nucleus is twice that of the emitted particle. (a) Calculate the radius of $^{226}_{88}$Ra. (b)

Using this result and the data of Table 33-1, estimate the frequency with which the alpha particle inside radium-226 makes collisions with the potential barrier. (c) Using this result and the known half-life of radium-226, 1620 yr, find the transparency of the potential barrier.

30. Calculate the potential energy of an alpha particle that is a distance of 2×10^{-12} cm from the center of a polonium-210 nucleus. Express the result in MeV.

31. (a) Polonium-210 emits an alpha particle; write the reaction equation. (b) Determine the kinetic energy of the alpha particle. (c) If the strength of the sample is 4 mc, determine the number of atoms of lead formed in 1 hr.

32. Fill in the missing symbols represented by the question mark: (a) $^{14}_{7}N(p,n)$?, (b) $^{107}Ag(?,n)$ ^{107}Cd, (c) $^{11}B(\alpha,?)^{14}N$, (d) $^{17}O(?,\alpha)^{15}N$.

33. Slow neutrons are difficult to detect because they create no ions. If a counter is filled with boron trifluoride, the neutrons cause an (n, α) reaction on ^{10}B. The resulting alpha particles can be detected. (a) Write the reaction equation and (b) calculate the disintegration energy for this process, assuming that the kinetic energy of the neutrons is negligible.

34. Determine the Q value of the (α, p) reaction with ^{27}Al as the target.

35. Determine the Q value of the (α,n) reaction in which the compound nucleus is $^{13}_{6}C$.

36. (a) Write the reaction equation for the process illustrated in Figure 33-7. (b) Determine the difference in mass between the initial particles and the final products of this reaction. (c) Express the result of (b) in MeV. (d) How much energy is released in this reaction?

37. If 200 MeV of energy is released per fission, (a) how many fissions take place per second in a power reactor that is operating at 10,000 kilowatts? (b) How many ^{235}U nuclei will have undergone fission in 1 yr? (c) What is the total mass of the ^{235}U fuel thus consumed in 1 yr? (d) What is the mass equivalent of the energy produced in 1 yr?

38. What mass of tritium would be consumed per day in a fusion reactor based on the process of Equation 33-24a if the reactor were operating at 10,000 kilowatts?

39. (a) Calculate the potential energy of two deuterons separated by 10^{-12} cm. **(b)** Calculate the temperature at which the deuterons would have a kinetic energy of this magnitude, assuming that the deuterons are the "molecules" of an ideal gas. In reality at this temperature all the atoms would be ionized and we would have a plasma rather than a gas.

40. In a Compton scattering experiment it is desired to select X-ray photons such that the scattered wavelength will have a maximum value equal to twice that of the incident wavelength. **(a)** Determine the angle at which the scattered wavelength is a maximum. **(b)** Determine the wavelength of the incident photons to satisfy the above conditions.

41. In calculating the Q value for a reaction we usually use the atomic masses rather than the nuclear masses, because that automatically takes the atomic electrons into account. **(a)** For a beta decay in which a negative electron is emitted, show that the Q value is equal to the difference in *atomic* mass between the initial nucleus (the parent nucleus) and the final nucleus (the daughter). Assume that the binding energies of the atomic electrons cancel out and can be ignored. **(b)** For a positron emission, show that the Q is equal to two electron masses *less* than the difference in atomic mass between the parent and daughter nucleus:

$$Q = M_a({}^A_Z X) - M_a({}^{\ \ A}_{Z-1} Y) - 2m$$

in which $M_a({}^A_Z X)$ is the atomic mass of the atom of atomic number Z and mass number A, and m is the mass of the positron or electron.

42. Estimate the height of the Coulomb barrier for alpha emission from radon, by treating the alpha particle and the residual nucleus as charged spheres that are just touching. Calculate the electrostatic potential energy of this system.

43. In a typical resonance reaction an incident particle of mass m and speed v strikes a nucleus and a compound nucleus of mass M is formed. Some of the kinetic energy $\mathscr{E}_k$ of the incident particle is transferred to internal energy of the compound nucleus, $\mathscr{E}_k'$, and the rest is transferred to kinetic energy of the compound nucleus. Assume that the problem may be treated nonrelativistically and that the target nucleus is at rest. Show that

$$\mathscr{E}_k' = \mathscr{E}_k\left(1 - \frac{m}{M}\right)$$

particle physics

chapter 34

34-1
Particles and antiparticles

In Chapter 33 we referred to the emission of positrons by certain radioactive nuclei. The positron, which has the same mass as the electron but opposite charge, was first discovered experimentally by Carl D. Anderson in 1932 in the course of an investigation of cosmic-ray phenomena. Figure 34-1 is a photograph by Anderson of the path of a positron in a cloud chamber; the path is curved because the chamber is in a magnetic field. While the particle passed through the sheet of lead, it lost energy and the curvature of the path changed. This permitted Anderson to determine the direction of motion of the particle. From the known direction of the magnetic field, the curvature, and the direction of motion, he could determine the sign of the charge of the particle to be positive. The length and density of the track verify that the mass of the particle is approximately that of the electron.

The existence of the positron had been predicted by Dirac before Anderson's experiment. Dirac's prediction was based on his reformulation of quantum mechanics, which was published in 1928. The Dirac theory, unlike the formulations of Schrödinger and Heisenberg, was consistent with special relativity. Also, the Dirac equation included the spin of the electron explicitly: Dirac was able to solve several problems that had not been solved by the Schrödinger-Heisenberg formulations. For most problems involving atomic phenomena, the Schrödinger equation remains useful because the Dirac equation reduces to the Schrödinger equation when describing the behavior of nonrelativistic electrons.

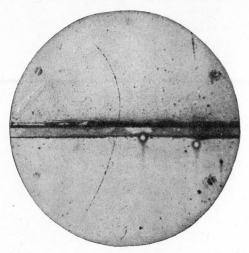

FIGURE 34-1 Cloud chamber photograph of a positron in a magnetic field. The positron originated at the bottom of the chamber and passed through a sheet of lead 6 mm thick. The magnetic field is directed into the paper. (Photograph courtesy of Carl D. Anderson.)

Positrons and electrons exhibit two striking phenomena, *pair production* and *annihilation*. Pair production occurs when a photon disappears and a pair of particles, one positron and one electron, appears. Annihilation is the inverse process, in which a positron interacts with an electron and both particles disappear and photons appear. In both processes, both energy and momentum must be conserved. Thus a photon cannot produce a positron-electron pair unless it has energy equal to or greater than $2mc^2$, in which m is the mass of the electron. Similarly, the total photon energy remaining after annihilation must be equal to the sum of the kinetic energies of the particles and the rest mass energies of the particles. In order for energy *and* momentum to be conserved in the annihilation process at least two photons must be produced. The process may be represented by the equation

$$e^+ + e^- \rightarrow 2\gamma \qquad (34\text{-}1)$$

where γ represents the photon. Similarly, a single photon cannot produce a single e^+e^- pair and both conserve energy and momentum. However, if some other particle, such as a proton, electron, or nucleus, is nearby, it can absorb the recoil momentum; pair production occurs in such processes as

$$\gamma + p \rightarrow e^- + e^+ + p \qquad (34\text{-}2)$$

$$\gamma + e^- \rightarrow e^- + e^+ + e^- \qquad (34\text{-}3)$$

$$\gamma + \text{Pb} \rightarrow e^- + e^+ + \text{Pb} \qquad (34\text{-}4)$$

The electron and positron, which differ in that they have opposite charge are a *particle-antiparticle* pair. It is now believed that for *every* kind of particle there exists an antiparticle that has opposite electromagnetic properties. *Antiprotons*, which are negative particles with the mass of a proton, have been observed and are now produced so copiously in accelerators that beams of antiprotons can be utilized in scattering experiments. Similarly, all charged particles have antiparticles, which differ from the particle in that they have opposite charge; the antiparticles of charged particles can be *distinguished* from the particles because of the difference of charge. The antiparticles of some neutral particles can also be distinguished from the particle. For example, antineutrons are neutral, like neutrons, but have a magnetic moment whose direction relative to the spin is opposite to that of the neutron. For some neutral particles, on the other hand, the antiparticle is indistinguishable from the particle; the particle is its own antiparticle.

It is convenient, then, to introduce the following terminology: particles of integral spin (0, 1, 2, . . . , in units of $h/2\pi$) are called *bosons*, and particles of half integral spin ($\frac{1}{2}$, $\frac{3}{2}$, $\frac{5}{2}$, . . . , in units of $h/2\pi$) are called *fermions*. Using these terms, particles and antiparticles are distinguishable from each other for all fermions (charged and neutral) and for charged bosons, but are indistinguishable for neutral bosons.

In the current theory, particles and antiparticles are treated symmetrically. Particles and antiparticles are produced, annihilated, interact, and decay in a manner governed by the fundamental interactions as limited by the conservation laws, but there appears no reason in the theory to treat a particle, such as an electron, as any more fundamental than an antiparticle, such as a positron.

In spite of this symmetry it is true that all the stable matter we observe is composed of electrons, protons, and neutrons, rather than the antiparticles of these. This basic fact may indicate that our understanding of the fundamental theory is incomplete, or it may be that the predominance of matter is some sort of accident, or chance occurrence. It is possible that this is a local phenomenon; the matter in other parts of the universe may be composed of antiparticles. Some physicists have speculated that of all the matter in the universe, half is ordinary matter, composed of particles, and the other half is antimatter, composed of antiparticles.

34-2
The discovery of the pion and the muon

In the early 1930s the list of known or assumed elementary particles was fairly short; it consisted of the electron, the proton and neutron, the photon and neutrino, and the antiparticles of each of these.

One of the outstanding successes of quantum mechanics is that it is able to describe the electromagnetic interaction among charged particles in terms of the transfer or *exchange* of photons between the particles. Classically, one pictures the interaction between two charged particles in terms of the fields established by the charges. Quantum-mechanically, one describes the interaction by assuming that the particles emit and absorb photons. Similarly, the binding force between atoms (chemical bonding) can be described in terms of an exchange of electrons. In 1935 Hideki Yukawa proposed a theory that explained the origin of the binding force between two nucleons in terms of the exchange of a hypothetical particle, called a *meson*. Yukawa's theory predicted that the meson would have a mass approximately 200 times the mass of an electron. The name was given to the particle because this mass was intermediate between that of the heavy elementary particles known at the time (the nucleon) and that of the electron.

Yukawa also predicted that the meson would decay into an electron and neutrino. The mechanism for this decay would be exactly the same interaction (now called the weak interaction) that Fermi had postulated to explain nuclear beta decay. In 1939 mesons were discovered by Blackett and Wilson, and independently by Neddermeyer and Anderson in cloud-chamber photographs.

After World War II, physicists returned to a study of mesons. From the work of the experimentalists Conversi, Pancini, and Piccioni, as interpreted by the theorists Fermi, Teller, and Weisskopf, it was clear that the interaction of the experimentally observed mesons with nuclear matter is significantly weaker than the Yukawa theory had predicted.

R. E. Marshak and H. A. Bethe suggested that there were two mesons: the heavier particle interacted strongly, as predicted by Yukawa, and decayed rapidly into the lighter particle, which was the weakly interacting particle observed by Blackett and the others. At about the same time, Occhialini, Powell, and Perkins observed a decay that was consistent with this interpretation.

It is now known that the heavier particle, now called a pion (π), has a mass of 270 electron masses and has a spin of zero. It is therefore a strongly interacting boson. In current usage, the term *meson* is used to describe all *strongly interacting bosons*, and so the pion is now classified as one of many mesons. The lighter particle described by Marshak and Bethe is a fermion (spin $\frac{1}{2}$), with a mass of approximately 207 electron masses, and is now called a muon (μ). As a *weakly interacting fermion, with a mass less than that of the proton*, the muon is classified as a *lepton* (rather than a meson).

34-3
Properties of pions and muons

Muons are always created in a weak interaction; they are the products of the decay of pions and other particles. They exist in two charge states (positive and negative); no neutral muons have ever been observed. They have a mass of $206.77m_e$, or 105.659 MeV; and a spin of $\frac{1}{2}$ (in units of $h/2\pi$).

A muon decays into an electron, a neutrino, and an antineutrino. In the last few years it has become clear that there are in reality *four* neutrinos: a neutrino and antineutrino, which are always associated with electrons ($\nu_e, \bar{\nu}_e$) and a neutrino and antineutrino ($\nu_\mu, \bar{\nu}_\mu$) which are always associated with muons. In terms of these particles, the decay of the muon occurs through the processes

$$\mu^- \rightarrow e^- + \bar{\nu}_e + \nu_\mu \qquad (34\text{-}5a)$$
$$\mu^+ \rightarrow e^+ + \nu_e + \bar{\nu}_\mu \qquad (34\text{-}5b)$$

These decay processes occur with a half-life of 1.5×10^{-6} sec. This half-life is measured in a frame of reference in which the muons are at rest.

Except for the fact that it decays, the muon behaves like a heavy electron. In fact, in the process of slowing down, negative muons often are bound in atoms, replacing orbital electrons, and forming atoms. Since the mass of the muon is so much larger than that of an electron, the muon orbits are much smaller than those of electrons. The muon orbits penetrate the nucleus, and therefore the radiation emitted by these mesic atoms provides information about the structure of the nucleus.

Pions are strongly interacting bosons of spin 0; they exist in three charge states (π^+, π^-, π^0), where the charged pions are the antiparticles of each other and have the same charge as the electron or positron, whereas the neutral pion is its own antiparticle. The mass of the charged pions is $273.14m_e$ or 139.58 MeV, whereas the neutral pion has a mass of $264m_e$ or 134.98 MeV.

Pions are produced in many processes. They are commonly produced in nucleon-nucleon collisions, nucleon-antinucleon annihilations, and the decay of many of the particles which will be mentioned later.

Charged pions decay principally into muons, and occasionally into electrons, via processes such as

$$\pi^+ \to \mu^+ + \nu_\mu \qquad (34\text{-}6a)$$
$$\pi^+ \to e^+ + \nu_e \qquad (34\text{-}6b)$$

The half-life for the decay of charged pions is 1.8×10^{-8} sec.

This decay of the charged pion proceeds via the weak interaction. The pion cannot decay via a strong interaction because in any decay the product particle must be less massive than the original particle, and there exists no strongly interacting particle that is lighter than the pion. The neutral pion can decay directly into photons, since it has no charge to conserve. This permits the neutral pion to decay via the electromagnetic interaction, which is stronger than the weak interaction. Thus the half-life of the neutral pion is 1.4×10^{-16} sec, which is much faster than the decay of the charged pion. The decay of the neutral pion occurs via processes such as

$$\pi^0 \to \gamma + \gamma \qquad (34\text{-}7a)$$
$$\pi^0 \to e^+ + e^- + \gamma \qquad (34\text{-}7b)$$

34-4
Some conservation laws

The problem of particle physics is simply that no one has yet discovered the fundamental laws of nature governing the particles. There exists no fundamental equation (or equations) that predicts the existence of the particles, their properties, or their interactions.

In the absence of such a theory, physicists have placed great reliance on the conservation laws. As we have seen in classical physics, conservation laws, such as the principles of conservation of momentum and energy, are very powerful, particularly for the solution of problems for which we have

limited data. Thus while searching for the fundamental laws of nature that govern particle phenomena, particle physicists have adopted the interim strategy of searching for the conservation laws that govern these phenomena. Knowing which quantities are conserved and the conditions under which they are conserved often provide an explanation of why certain processes occur and why others do not. Further, because of the statistical nature of quantum mechanics, such laws often provide an explanation of the rate at which a process occurs. Perhaps most important, it is hoped that a knowledge of all the conservation laws will lead to the discovery of the fundamental laws.

The keystones of physics are the principles of the conservation of momentum and energy (which in the special theory of relativity are combined into a single statement). Perhaps the single most important assumption that runs through all of physics is that energy and momentum are conserved. As we have seen, the apparent violation of this principle led to the assumption of the existence of the neutrino. At this time it is difficult to imagine physicists surrendering this assumption.

Similarly it is believed that *angular momentum* is always conserved. In particle interactions, the total angular momentum of the system usually consists of two parts: the orbital angular momentum, which has the form mvr for each particle, and the intrinsic angular momentum or spin of each of the particles. Up to this point, whenever a new particle has been discovered it has always been possible to assign to it a *unique* spin and then to show that angular momentum is conserved in all processes involving that particle.

Equally fundamental is the assumption that electric charge is conserved. This assumption hinges on the idea that equal quantities of positive and negative charges are together equivalent to the absence of charge. Thus, for example, the principle of conservation of charge states that a system that consists of neutral particles may, after some interaction, become a system of charged particles, but the number of positive charges must be exactly equal to the number of negative charges, in terms of the electronic charge e.

These conservation laws arose in classical physics and have been carried over to the realm of particles. Several conservation laws have been discovered in the course of studying particles. Of these, we shall mention two at this point: conservation of baryons and conservation of leptons.

In order to describe these conservation laws we must introduce a procedure for counting particles. In this procedure, when we count each kind of particle (such as electrons or protons) we count each particle as plus one particle, and count each antiparticle as minus one particle. Thus a system that consists of one proton and one antiproton is counted as having zero protons, a system that consists of two protons and one antiproton is counted as having one proton, and a system that consists of one antiproton is counted as having minus one proton.

Once we have defined this procedure for counting particles we can group particles together into classes and count the number of members of that class. One such class we have already defined is the nucleon. For any system we can count the number of nucleons by adding the total number of protons and neutrons and subtracting the total number of antiprotons and antineutrons. Until fairly recently it was believed that when counting was done in this way, the total number of nucleons was conserved. In fact, for most ordinary processes, including radioactivity, the total number of nucleons *is* conserved. In recent years, however, a number of particles have been discovered that decay, yielding nucleons. In order to describe these processes, as well as the more common processes that conserve nucleons, a new class of particle has been defined, the *baryon*. A baryon is a fermion whose mass is at least as large as that of a nucleon and which can interact via the strong interaction. Thus the nucleon is a subclass of the baryon. All experimental information supports the assumption that the *total number of baryons is always conserved*.

For example, in the beta decay of a neutron

$$n \rightarrow p + e^- + \bar{\nu}_e \qquad (34\text{-}8)$$

the neutron and proton are each counted as one baryon, while the electron and antineutrino are each counted as zero baryons. Similarly in a process for producing antiprotons, in which a proton beam strikes hydrogen nuclei,

$$p + p \rightarrow p + p + p + \bar{p} \qquad (34\text{-}9)$$

the total number of baryons is constant with the value two.

Popular expositions on fission and fusion processes and articles about the Einstein equation $\mathcal{E} = mc^2$ have often implied that it is possible to convert all of a sample of matter into photons.

However, from the principle of baryon conservation, we know that if the process starts with a given number of nucleons it must end with the same number of baryons (plus any number of baryon-antibaryon pairs). Thus complete annihilation of matter is impossible unless we start with equal numbers of nucleons and antinucleons.

Another class of particles is called *leptons*. A lepton is a fermion that does not interact via the strong interaction but does interact via the weak (or Fermi) interaction. At this point the known leptons are the electron, muon, and the neutrinos ν_e and ν_μ, along with their antiparticles. Using the particle-counting procedure described above, it is now believed that the *total number of leptons is always conserved*. Thus in the neutron beta decay, Equation 34-8, the initial number of leptons is equal to zero and the final number of leptons is also zero, because the electron is counted as one lepton and the antineutrino is counted as minus one lepton. Similarly lepton conservation is shown by the process described by Equations 34-2, 34-3, and 34-4, where the photon counts as zero leptons (as well as zero baryons). The processes described by Equations 34-5a and b and Equations 34-6a and b illustrate an even more restrictive conservation principle. If one defines the *electron* number as the number of electrons plus the number of ν_e, and defines the *muon* number as the number of muons plus the number of ν_μ, then the electron number is conserved separately and the muon number is conserved separately.

The principle of lepton conservation specifically forbids the radiative decay of the muon,

$$\mu^- \rightarrow e^- + \gamma \qquad (34\text{-}10)$$

It should be noted that there is no conservation law for mesons; mesons appear and disappear, limited only by the other conservation laws. For example, the processes described by Equations 34-6 illustrate the disappearance of a meson.

34-5
Quantum numbers

In Chapter 32 we saw that a particular level or state of an atom is specified by listing the quantum numbers of that state. Transitions between states are governed by selection rules, which describe the allowed changes in the quantum numbers in the transition. These selection rules are determined by the conservation laws. For example, the selection rule on j (see Section 32-13) states that the angular momentum of the system (including the photon) is conserved.

Similarly, it has become common to describe a particle by listing quantum numbers that specify the properties of the particle. Reactions among the particles are then governed by selection rules for these quantum numbers. Some of these quantum numbers arise from simple extensions of the discussion of the last section. For example, each particle has a *charge quantum number Q*, which gives the charge of the particle in units of the charge on the proton. Similarly, each particle has a *spin quantum number*, usually represented by J, which gives the spin of the particle in units of $h/2\pi$.

We can express the principle of baryon conservation by assigning to each particle a *baryon number B*. This is most simply done by assigning the baryon number $B = 0$ to the photon, the leptons, and the mesons, while assigning the baryon number $B = 1$ to the baryons, and assigning the baryon number $B = -1$ to all the antibaryons. Then the conservation of baryons can be stated as: *the total baryon number B is constant in all reactions.* In a similar manner we can express the principle of lepton conservation by assigning a lepton number to all particles.

In Chapter 33 we were able to describe the neutron and proton as distinct states of a single particle, the nucleon, because the strong interaction is charge-independent. This allowed us to assign a quantum number, the isotopic spin T, to the nucleon and to assign a quantum number T_z to each of the two states. The fact that the neutron and proton are states of the nucleon is expressed by asserting that the value of T_z is a projection of T on some axis; this relation between T and T_z is analogous to the relation between s and $\mathbf{m}_s$ as shown in Figure 32-18. The important thing about Figure 32-18 is that there are only two possible values ($+\frac{1}{2}$ and $-\frac{1}{2}$) for the projection of a vector of magnitude $\frac{1}{2}$. Since there are only two states of the nucleon, this suggests a two-valued quantum number; hence for the nucleon $T = \frac{1}{2}$, with $T_z = +\frac{1}{2}$ for the proton and $T_z = -\frac{1}{2}$ for the neutron.

Since the strong interaction is charge-independent for all strongly interacting particles, the concept of isotopic spin has been extended to assign a value of T and T_z to all the strongly interacting

particles. Some such particles such as the pion, exist in three charge states (π^+, π^0, π^-). Such *charge triplets* are assigned $T = 1$, with each of the charge states assigned one of the values $T_z = 1$, 0, or -1. All charge *doublets,* such as the nucleon, are assigned isotopic spin $T = \frac{1}{2}$, with one of the two charge states having $T_z = +\frac{1}{2}$, and the other having $T_z = -\frac{1}{2}$. Charge *singlets,* which have only one charge state, are assigned isotopic spin $T = 0$ and $T_z = 0$.

34-6
The strange particles

The discovery of the pion and muon lengthened the list of known particles but, on the whole, physicists found the discoveries hopeful, if not satisfying. By extending Yukawa's theory and combining it with some of the techniques successfully used to describe the interaction between photons and electrons it was hoped that the strong interaction might be completely understood. Some physicists were somewhat puzzled by the existence of the muon because they took the point of view that all of the other particles played some role in nature that could not be played by the other particles, whereas the muon seemed to be nothing more than a heavy electron. However, it seemed clear that the interaction of the muon could be understood in terms of the weak interaction.

This relatively neat situation was disturbed in the 1950s and 1960s by the discovery of a bewildering number of particles (and particle-like entities called resonances). The list of particles grew with amazing rapidity and continues to grow.

The first group of particles were discovered in cloud-chamber and bubble-chamber photographs as kinks in tracks and V-shaped tracks. Consequently, the particles were at first called V particles. It quickly became clear that some of these were mesons, now called *kaons* (from their original name of K mesons), and others were baryons.

Some of the properties of these baryons—(the lambda (Λ), sigma (Σ), and xi (Ξ)—are listed in Table 34-1. From the point of view of the historical development of the subject, the most striking thing about these properties is that the lambda and the sigma decay into nucleons and pions with a lifetime of about 10^{-10} sec. That these facts are important can be seen from the following arguement:

Table 34-1

particle	mass, MeV	half-life, sec	spin	common decay modes
Λ^0	1155.5	1.7×10^{-10}	$\frac{1}{2}$	$\Lambda^0 \rightarrow \Gamma + \pi^-$
				$\rightarrow n + \pi^0$
Σ^+	1189.5	5.6×10^{-11}	$\frac{1}{2}$	$\rightarrow p + \pi^0$
				$\rightarrow n + \pi^+$
Σ^0	1192.5	$<10^{-11}$	$\frac{1}{2}$	$\rightarrow \Lambda^0 + \gamma$
Σ^-	1197.0	1.1×10^{-10}	$\frac{1}{2}$	$n + \pi^-$
Ξ^0	1314.9	2×10^{-10}	$\frac{1}{2}$	$\Lambda + \pi^0$
Ξ^-	1321.3	1.2×10^{-10}	$\frac{1}{2}$	$\Lambda + \pi^-$

The particles were produced, typically, in an experiment in which proton or pion beams struck nuclei. Most important, they were produced copiously in such processes. All of the data are consistent with the assumption that these particles are produced as a consequence of the *strong* interaction. From what is known about the strong interaction, this process takes place in a time interval of about 10^{-23} sec. Thus the *production* process implies that the particles are strongly interacting particles. The decay of the particles ends with strongly interacting particles. Therefore, we would normally assume that the decay also occurs via a strong interaction and that the half-life of the decay process should be close to the 10^{-23} sec, which is the time scale characteristic of strong interactions. It is true that conservation laws may limit the decay process as compared with the production process, but it is difficult to imagine that the conservation laws can limit the decay enough to introduce a factor of 10^{13} and leave the lifetime 10^{-10} sec. This contrast between the production process and the decay process is difficult to understand.

A solution to this problem was proposed independently by Gell-Mann and Nishijima in a remarkable display of physical insight. Basically, the proposal was that there exists a *new* quantum number (strangeness, S) associated with particles, and that this quantum number is conserved in strong interactions but not in weak interactions. At a time when relatively little data existed, Gell-Mann and Nishijima were able to assign unique values of this quantum number to each of the particles, explain many of the difficulties, and most important, make remarkably accurate predictions. This quantum number S is connected to T_z, Q, and B by the following formula:

$$Q = T_z + \frac{B + S}{2} \qquad (34\text{-}11)$$

Then the hypothesis was made that S is conserved in strong interactions, but not in weak interactions. Note that charge conservation and baryon conservation hold for both strong and weak interactions; thus, from Equation 34-11, strangeness conservation is equivalent to the assertion that T_z is conserved in strong interactions but not in weak interactions.

Table 34-2 gives the quantum numbers of the mesons and the baryons.

In order to explain the data, Gell-Mann and Nishijima had to assume that the kaon is two charge doublets (K^+ and K^0; K^- and a second neutral kaon called the $\overline{K^0}$) rather than a triplet. The two neutral kaons (K^0 and $\overline{K^0}$) are produced in strong interactions and differ in that $\overline{K^0}$ has positive strangeness, whereas K^0 has negative strangeness. Because these two kaons differ only in strangeness, they can decay into each other in a weak interaction, which does not conserve

Table 34-2

particle	I-spin T	T_3	baryon number B	strangeness S	charge Q
π^+	1	$+1$	0	0	$+1$
π^0	1	0	0	0	0
π^-	1	-1	0	0	-1
K^+	$\frac{1}{2}$	$+\frac{1}{2}$	0	$+1$	$+1$
K^0	$\frac{1}{2}$	$-\frac{1}{2}$	0	$+1$	0
K^0	$\frac{1}{2}$	$+\frac{1}{2}$	0	-1	0
K^-	$\frac{1}{2}$	$-\frac{1}{2}$	0	-1	-1
η	0	0	0	0	0
p	$\frac{1}{2}$	$+\frac{1}{2}$	$+1$	0	$+1$
n	$\frac{1}{2}$	$-\frac{1}{2}$	$+1$	0	0
$\bar{p}$	$\frac{1}{2}$	$-\frac{1}{2}$	-1	0	-1
$\bar{n}$	$\frac{1}{2}$	$+\frac{1}{2}$	-1	0	0
Λ	0	0	$+1$	-1	0
$\overline{\Lambda}$	0	0	-1	$+1$	0
Σ^+	1	$+1$	$+1$	-1	$+1$
Σ^0	1	0	$+1$	-1	0
Σ^-	1	-1	$+1$	-1	-1
Ξ^0	$\frac{1}{2}$	$+\frac{1}{2}$	$+1$	-2	0
Ξ^-	$\frac{1}{2}$	$-\frac{1}{2}$	$+1$	-2	-1
Ω^-	0	0	$+1$	-3	-1

strangeness. This explains the experimental fact that the decay process for both the $\overline{K}^0$ and K^0 occur with a mixture of two lifetimes, 5.3×10^{-8} sec and 0.86×10^{-10} sec.

One of the most striking predictions of this theory is that of *associated production.* The nucleons and pion have zero strangeness. Thus, in a production process that starts with nucleons and pions, the final state must contain *at least two strange particles,* one with positive strangeness and the other with negative strangeness.

The Λ, Σ, and Ξ may not decay via the strong interaction (which requires that strangeness be conserved) because they do not have enough mass to decay into at least one baryon (to conserve B), and at least one kaon (to conserve S). Thus, when they do decay, they do so via the weak interaction. Similarly, the kaons must decay via the weak interaction. Thus, the half-lives of the strange particles should be comparable to the half-lives of beta decay rather than the interaction times characteristic of strong interactions.

In recent years it has become common among particle physicists to use, instead of strangeness, a different quantum number, the *hypercharge, Y,* which is defined by

$$Y = S + B \qquad (34\text{-}12)$$

and in terms of Y, Equation 34-11 becomes

$$Q = T_z + \frac{Y}{2} \qquad (34\text{-}13)$$

The principle of conservation of strangeness becomes, then, the assertion that the hypercharge Y is conserved in strong (and electromagnetic) interactions but not in weak interactions.

34-7
Resonances

As we have seen, certain nuclear reactions have a maximum yield at specific incoming beam energies. These *resonances* occur because the compound nucleus has a definite energy-level structure. Similar effects were discovered in the early 1950s by Fermi and others in pion-scattering experiments. In these experiments a beam of pions was directed at the protons in a target and the probability of scattering was measured. The probability of scattering turned out to vary with the energy of the pions, with a sharp maximum at specific energies. This is interpreted to mean that the pion beam and the target proton form an intermediate *system* during the scattering process and that this system has one or more energy levels. Since this first discovery many similar resonances have been found.

The term "resonance" is used because the mathematical description of these processes is essentially the same as that of ordinary mechanical and electrical resonance processes, when the particle interactions are described in quantum-mechanical terms. However, the same general description is used for two cases: one where the intermediate state of definite energy is a bound state of two particles (such as the hydrogen atom in a particular energy level), and the other where the intermediate state is a particle, which is created out of the initial particles and which decays into the same kinds of particles or other particles (such as a neutron, which may be created and which decays via the weak interaction). When a resonance is discovered we cannot know whether the resonance is a bound state or a particle. In fact, it is no longer clear that there is a difference between a bound state and a particle; both have definite masses, definite lifetimes, and definite quantum numbers. Experimentally, an atom and a neutron both behave as particles. We think of an atom as a bound system because we can describe its properties in terms of an assumed structure of nucleus and electrons. We assume that a neutron is a particle because we cannot describe a structure in terms of other particles. As long as the number of known particles was small, there was no difficulty in assuming that each was structureless and fundamental. However, now that the number of particles and resonances is in the hundreds, this assumption is at least uncomfortable. Currently there is no way of knowing which of the particles or resonances are fundamental and which are bound states. It is possible, for example, that the proton is a stable bound system of several particles. There even are theories that assert that *none* of the particles are fundamental, but that each can be thought of as a bound system of some of the others. In any case, it is common to refer to all of the particles and resonances as particles.

In the large array of particles a few, such as the electron and proton, are absolutely stable. These absolutely stable particles do not decay at all. Others, such as the neutron and the pion-proton resonance described above, decay with definite lifetimes into other particles. Some of these lifetimes are relatively long and others are relatively short. It is convenient to classify these lifetimes using a time scale that is characteristic of particles. We can measure the distance traveled in units of the wavelength of the particle. If the particle travels many wavelengths in one lifetime, then the lifetime is long, whereas if the particle travels few wavelengths in one lifetime, then the lifetime is short. Using this scale, particles with lifetimes of more than 10^{-18} second live for long times, and can be described as stable in some sense. In this sense, all of the particles in Table 34-1 are stable.

The most stable of the resonances is the η (eta), which was discovered by Pevsner and others in 1961. They studied the reaction

$$\pi^+ + d \rightarrow p + p + \pi^+ + \pi^- + \pi^0 \quad (34\text{-}14)$$

and found that some of the pion triplets arose from the decay of a resonance with a mass of 550 MeV, the η (550). Subsequently it was discovered that in addition to the process

$$\eta \rightarrow \pi^+ + \pi^- + \pi^0 \quad (34\text{-}15)$$

the η decayed via a purely electromagnetic process

$$\eta \rightarrow \gamma + \gamma \quad (34\text{-}16)$$

The rate of decay of the η via the electromagnetic process is significantly larger than the three-pion decay rate. The lifetime of the eta is about 10^{-10} sec and so it is relatively stable. These two facts, relatively long lifetime and electromagnetic decay, indicate that the eta should be classified as a stable meson, along with the pion and kaon, rather than with other resonances that decay rapidly.

Extensive work has been done in discovering resonances and determining their masses, lifetimes, and such quantum numbers as spin, strangeness, baryon number, and iostopic spin. This work has been so productive that the known data can no longer be summarized in a short table, and few particle physicists can rely on their memories to retain all the data. The tables of data published for particle physicists resemble the atlases common to astronomy and biology more than they resemble a table in this book.

34-8
SU(3) and quarks

Because of the huge amount of data accumulated by particle physicists, there have been extensive efforts to organize and classify the results in a way that might lead to a unifying theory.

One of the most interesting approaches involves the ideas of *symmetry*. Knowing the specific kind of symmetry displayed by a particular process permits us to deduce a good deal about it. For example, just knowing that the Coulomb interaction is spherically symmetric permits us to conclude that each of the energy levels of the hydrogen atom consists of $(2l + 1)$ levels. These levels are said to be *degenerate* in that they would have the same energy if the only interaction were the Coulomb interaction. Other interactions, such as the magnetic interaction, shift each of the degenerate levels differently, removing the degeneracy. We can turn this argument around and state that knowing the numbers of degenerate levels in each level permits us to conclude that the basic interaction must be spherically symmetric.

Attempts have been made to extend this kind of argument to the particles by assuming that particles can be grouped together into *multiplets*. The multiplet is thought of as a single level (of mass) characteristic of some basic interaction, and secondary interactions split each multiplet into several particles of different mass. The number of levels into which a multiplet can be split is determined by the symmetry characteristic of the basic interaction.

In making these arguments, physicists have invoked ideas that have generalized the ordinary concepts of symmetry. Symmetry usually is a geometric idea, implying that some shape or object is unchanged, or is *invariant,* by such a *process* as translation, rotation, or reflection. This idea can be extended from the ordinary space described by geometry to the mathematical spaces described by coordinate systems. A mathematical entity described in some coordinate system possesses symmetry if it is invariant after processes analogous to translation, rotation, or reflection. Mathematicians have developed a general theory of symmetry that has great strength and intellectual beauty. This theory has been used by physicists to understand the properties of crystals and the structure of complex molecules. Gell-Mann and Ne'eman independently applied part of this theory—a part called SU(3)—to the problem of particles.

From this SU(3) theory, Gell-Mann and Ne'eman argued that there should exist multiplets of particles consisting of ten particles (decouplet),

Table 34-3

	B	Y	T_z	Q	spin
q_1	$\frac{1}{3}$	$\frac{1}{3}$	$\frac{1}{2}$	$\frac{2}{3}$	$\frac{1}{2}$
q_2	$\frac{1}{3}$	$\frac{1}{3}$	$-\frac{1}{2}$	$-\frac{1}{3}$	$\frac{1}{2}$
q_3	$\frac{1}{3}$	$-\frac{2}{3}$	0	$-\frac{1}{3}$	$\frac{1}{2}$

eight particles (octet), and one particle (singlet). The eight spin $\frac{1}{2}$ baryons (proton, neutron, Λ^0, Σ^+, Σ^0, Σ^-, Ξ^0, Ξ^-) constitute one octet. Similarly, the eight mesons, (π^+, π^0, π^-, K^0, $\bar{K}^0$, K^+, K^-, η) constitute another. Each of these multiplets is split into eight particles with different values of hypercharge Y, and z component of isotopic spin T_z. There existed a group of nine particles which Gell-Mann and Ne'eman stated were really a decouplet (multiplet with ten members). They predicted that a tenth particle would be found and were able to predict the mass (and quantum numbers) of this tenth particle, which they called the Ω^-. An extended search for this particle resulted, in 1964, in the discovery of the particle. This particle has a mass of 1676 MeV, baryon number 1, strangeness -3, and exists in only the negative charge state. It cannot decay via the strong interaction (and conserve charge, energy, and hypercharge), so it must decay via the weak interaction. Therefore it has a long half-life (10^{-10} sec). This discovery of a predicted particle was a triumph of the theory.

This success led to further speculation. The SU(3) theory can be thought of as describing the symmetry of systems which are formed by combining three different objects in different ways. This implies that there might exist three fundamental particles which combine to form the octets and decouplets of Gell-Mann and Ne'eman. Gell-Mann proposed that three such particles exist, which he called *quarks*, and a similar suggestion was made by Zweig, in which he called the particles *aces*. The quantum numbers of the quarks (q_1, q_2, and q_3) are listed in Table 34-3.

The baryons can be thought of as combinations of three quarks (or antiquarks). For example, the nucleons can be thought of as a combination of two q_1's and one q_2, whereas the Σ can be formed from two q_1's and one q_3. Similarly, the mesons can be formed by combining two quarks.

With some simple assumptions, this quark model has been able to make predictions about the masses, lifetimes, and magnetic moments of the mesons and baryons that are in good agreement with the experimental data. However, in spite of extensive searches, no one has ever directly observed a quark. It is possible that they will be observed in the future. On the other hand, it is possible that they do not exist and that the success of the theory is an accident.

34-9
Symmetry and conservation laws

The idea of symmetry exists throughout physics. In fact, some apparently trivial statements about symmetry have fundamental consequences.

In writing the laws of nature we always refer to some coordinate system, but we make the assumption that we are at liberty to move the origin of coordinates. Among other things this implies that we can *translate* the coordinate system without changing the laws of nature. This can be described by saying that the laws of nature are *symmetric under translation*, or invariant under translation. This is not the same as saying that nature itself is invariant to translation. Moving from one place to another on the surface of the earth does change the force of gravity, the weather, the local magnetic field, and so forth. However, the laws of nature retain the same form. The surprising thing about this statement is that it can be shown that the principle of conservation of momentum is a *consequence* of this symmetry. In a sense, translational symmetry and conservation of momentum are different ways of expressing the same fact about nature.

Similarly, the principle of conservation of angular momentum is a consequence of the rotational symmetry of the laws of nature. That is, the fact that the laws of nature are unchanged by rotation through a finite angle implies that angular momentum is conserved.

The principle of conservation of energy is a consequence of the fact that the laws of nature are invariant under translations of time. That is, the fact that the laws of nature have had and will have the same form implies that energy is conserved.

This connection between conservation laws and symmetry applies to all the conservation laws. The fact that some quantity is conserved allows us to conclude that the laws of nature are symmetric and unchanged by some operation. Conversely, if the laws of nature exhibit some symmetry, this implies that some quantity will be conserved. However, the symmetry associated with some conservation laws, such as the conservation of charge, is poorly understood.

In the development of particle physics, several new kinds of symmetry have been invoked.

For any process involving particles, it is always possible to imagine a second, related process in which every particle in the first process is replaced by its antiparticle. If the laws governing the two processes have the same form, then the process is *invariant under the interchange of particle and antiparticle*. The process of interchanging particle and antiparticle (which, it must be emphasized, is a mathematical rather than experimental process) is called *charge conjugation*. The strong interaction processes are invariant under charge conjugation, but weak interactions are not. Symbolically, this is represented by the statement that a quantum number C is conserved by the strong interactions.

One simple symmetry that appears in classical physics is the symmetry of a mirror image. There are objects in nature that are different from their mirror images, the most common example of which is the human hand. The mirror image of a left hand is a right hand. Thus the human hand is *not* invariant to a mirror-image interchange. In particle physics, the interchange of a system and its mirror image is called a *parity interchange*, and a process that is invariant under this interchange is said to conserve parity.

Classical physics is symmetric under these mirror-image interchanges; classical physics conserves parity. At first thought this seems not surprising, perhaps even trivial. However, we have in several places described the laws of nature using right-hand rules, or right-hand-thumb rules. In a mirror image universe each of these would become a left-hand rule, and it is possible that the laws of nature would change. On close examination it turns out that, for the classical laws of mechanics and electromagnetism, the right-hand rules are always used twice in just such a way that they are indistinguishable from left-hand rules used twice. For example, in the description of torque and angular momentum, *both* quantities are described by a right-hand rule. In a mirror image, both would be described by a left-hand rule, leading to identical predictions. Most important, the magnetic field produced by a moving charge is described by a right-hand rule. In a mirror image the field would be reversed, because a left-hand rule would apply. However, the force produced by the field is also described by a right-hand rule, and this

too would be reversed in a mirror. This double reversal means that in the mirror image, the force (which is the observable quantity) would be unchanged.

Early in the history of particle physics it was assumed that *all* the laws of nature were mirror-image-invariant, and therefore conserved parity. For example, the atomic selection rule on l (Equation 32-50) is a mathematical consequence of conservation of parity. This became important because it was possible to assign an intrinsic parity to particles, analogous to the intrinsic spin. In 1956 Lee and Yang pointed out that the assumption of parity conservation was not universally true. Although strong interactions conserve parity, weak interactions do not. The mirror images of some naturally occurring weak interaction processes are not possible processes. The explanation for this may lie in the nature of neutrinos. As we have seen, the direction of the spin of the neutrino is opposite to that of its momentum. The spin of the antineutrino is parallel to that of its momentum. This implies that the mirror image of a neutrino is an antineutrino. Thus in the mirror image of a weak interaction (which always involves neutrinos), the neutrinos are subject to a particle-antiparticle interchange. But the other particles are not subject to a particle-antiparticle interchange. Thus the mirror image process is intrinsically different from the original process, which may explain why it cannot happen.

The last line of reasoning leads to the possibility that weak interactions are invariant to a simultaneous interchange of particle and antiparticle and mirror-image inversion. Most experimental evidence supports this propostion, although there are some contrary data.

A related symmetry is time reversal. The basic laws of classical physics are independent of the direction of time, and in general it is assumed that the laws of physics are also invariant to a reversal of time. However, there exists a very powerful mathematical theorem, called the CPT theorem, which is essentially that the laws of nature are invariant to the simultaneous operations of time reversal, mirror-image interchange, and particle-antiparticle interchange. The theorem is proved on the assumption of some very basic ideas, which are equivalent to the assumption of causality, that cause always precedes effect. The theorem is accepted by most physicists as almost axiomatic; should it turn out not to agree with nature, it would cause a revision of some of the most basic ideas of science. With respect to the present discussion, the CPT theorem states that if the laws of nature are not invariant under time reversal, then they cannot be invariant under the simultaneous mirror-image interchange and charge conjugation that we described above. Thus, the current indications are that the laws of nature are in fact invariant under time reversal, although there are some data to indicate that the weak interaction may not be invariant under time reversal.

Questions and Problems

1. State the direction of motion of the particle and the direction of the magnetic field in Figure 34-1.

2. Which of the four fundamental forces influences (a) a neutron, (b) a pion, (c) an electron, and (d) a neutrino?

3. Assume that a particle can be thought of as a spinning sphere. Find the angle between the angular momentum and magnetic moment for (a) an electron and (b) a positron.

4. Electron-positron pairs always yield photons after annihilation. Why would a pair with sufficient energy not form mesons?

5. Why is it possible to produce pion beams that travel many meters in spite of the relatively short lifetime of the pion?

6. None of the following reactions can occur. State at least one conservation law which prevents the reaction:

   (a) $\mu^+ \rightarrow e^+ + \nu + \gamma$
   (b) $n \rightarrow p + \bar{p} + \pi^0$
   (c) $\Lambda^0 \rightarrow p + K^-$
   (d) $\Sigma^+ \rightarrow \Lambda^0 + \bar{p} + \pi^0$
   (e) $\Sigma^- \rightarrow n + K^-$
   (f) $\gamma + p \rightarrow \pi^+ + \pi^- + \pi^0$
   (g) $p + p \rightarrow \Xi^0 + K^0 + \pi^+$

7. Because of energy conservation, a particle at rest cannot decay into two particles whose rest mass is larger than the rest mass of the initial particle. Can the decay occur if the initial particle has enough kinetic energy?

8. What is the simplest reaction that will produce antiprotons in a process in which protons interact with protons?

9. Explain why a Σ cannot decay into a Λ via a strong interaction but can decay into a Λ by a weak interaction.

10. What combination of quarks will produce the properties of (a) a proton and (b) a Λ?

11. Find the radius of the first Bohr orbit of a muon that replaces the electron in a hydrogen atom.

12. A charged pion at rest decays to a muon and a neutrino. Find the kinetic energy of the neutrino.

13. A kaon at rest decays into two pions. Assume that both pions have the same mass and find the kinetic energy of each pion.

14. When very high-energy photons pass through matter, one always observes some photons with energy 0.51 MeV emitted by the material. Explain the origin of these photons.

15. A beam consisting of a mixture of protons, muons, and neutrinos is directed at a thick lead wall. Which particles stop first and which stop last?

16. The proton is the baryon of lowest rest mass and is stable. The electron is the lepton of lowest rest mass and is also stable. The pion is the meson of lowest rest mass and is not absolutely stable. Explain this difference.

17. Very high-energy beams of electrons or muons are used to investigate high-energy electromagnetic processes. Why are charged pion beams not used in this way?

18. Why are the electron and the muon not assigned isotopic-spin quantum numbers?

19. The uncertainty principle connects the lifetime of particles with the uncertainties in their masses. Find the uncertainty in the mass of the neutral pion.

20. Yukawa estimated the mass of the pion from an argument similar to the following. Assume that the strong interaction occurs because a particle of mass M is exchanged between nucleons. In the exchange the particle travels with a speed c for a time t, through a total distance R. Assume that the uncertainty in energy is the mass energy of the particle and the uncertainty in time is the time of travel. (a) Show that $M = h/Rc$. (b) Assume that R is equal to 10^{-13} cm and calculate M.

the general theory of relativity

chapter 35

35-1
Introduction

As we have seen, the special theory of relativity arose from an attempt to understand some electromagnetic phenomena. Throughout the discussion of the theory we explicitly restricted the discussion to inertial frames of reference and explicitly excluded gravity. With these restrictions, the theory has been very fruitful, providing an accurate description of nature. The ideas of the special theory have been verified again and again, so the language and ideas of the special theory of relativity are now an essential part of physics. Without the theory it would be impossible to describe much of the physics of the twentieth century.

The general theory, on the other hand, developed out of an attempt by Einstein to resolve some philosophical problems. It seemed to Einstein that the fact that the inertial mass of an object is always equal to the gravitational mass of that object is more than mere accident; rather, it appeared to reveal some fundamental truth about nature. In addition, Einstein felt that it should be possible to write the laws of nature so that they would be independent of coordinate system—that the restriction to inertial frames of reference was unnecessary. These ideas led to the general theory of relativity, a theory in which gravity plays a central role. In contrast to the special theory, the general theory of relativity has had only very limited experimental verification and until very recently was of interest only to an extremely small number of physicists. In recent years, interest in the general theory has heightened, partly because of the discovery of quasars, pulsars, and the suggestion of the existence of other astronomical phenomena that may permit direct tests to be made of the theory.

The general theory of relativity is expressed in mathematical language well beyond the scope of this book, so this chapter will be descriptive in nature and attempt mainly to indicate some of the basic ideas.

35-2
The postulates of general relativity

We have, in general, treated the fact that the inertial mass and gravitational mass of any object are identical as an accident. Einstein's belief in the simplicity and symmetry of nature led him to ascribe deep significance to this fact.

We can reexamine some of the consequences of the identity of the two kinds of mass by considering some simple situations. Consider an elevator that is at rest near the surface of the earth, as in Figure 35-1a. Since the elevator is in the gravitational field of the earth, if a physicist in that elevator were to drop an object, he would measure the acceleration of that object to be 9.8 m/sec². Because the inertial mass and the gravitational mass of all objects are identical, all objects will have this same acceleration. Further, if an object is suspended from a spring balance in the elevator, the tension in the spring, and therefore the reading of the spring balance, will be equal to the weight of the object.

Consider now this same elevator isolated in space, where the gravitational field of the earth is negligible. The elevator is suddenly accelerated, as shown in Figure 35-1b, with an acceleration of 9.8 m/sec² relative to an inertial coordinate system fixed in space. An object that is dropped inside the elevator will move with constant speed relative to the inertial coordinate system. Since the elevator itself is accelerating relative to the coordinate system, the dropped object will accelerate relative to the elevator, as shown in Figure 35-1b, with an acceleration of 9.8 m/sec². Similarly, the suspended object will be accelerating, along with the elevator, relative to the inertial coordinate system. Thus the spring will exert a force on the object, and so the tension in the spring will be exactly equal to the weight of the suspended object.

Thus, to a physicist in the elevator, the two situations are indistinguishable. As far as measurements inside the elevator are concerned, the accelerating elevator might be at rest in a gravitational field, or the elevator near the surface of the

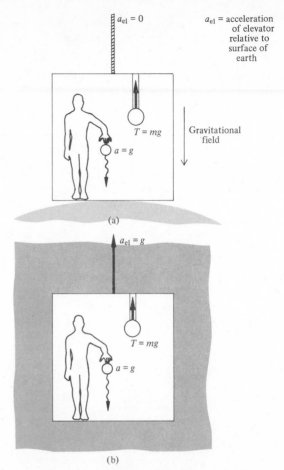

FIGURE 35-1 (a) An elevator at rest in the gravitational field near the surface of the earth. A freely falling object has an acceleration equal to g relative to an observer in the elevator. A spring balance reads the weight of the hanging object. (b) The same elevator accelerating with acceleration equal to g in a region of space with negligible gravitational field. The freely falling object will have an acceleration, relative to the observer, equal to g, and the spring balance will read the weight of the object.

earth might be in space and accelerating. (It should be noted that this is true only if the elevator is small enough so the earth's field can be treated as a constant in magnitude and direction.) In Einstein's words, either mode of interpreting the experimental facts, ". . . violates neither reason nor known mechanical laws." Einstein elevated this observation to a basic postulate, called the *principle of equivalence*. One way of stating the principle of equivalence is that *no physical measurements, confined to a region of space in which the gravita-*

tional field is constant, can distinguish between a situation in which that region of space is at rest in a gravitational field and one in which that region of space is accelerating with constant acceleration. This simple statement has several significant consequences, which we shall examine in subsequent sections.

In both classical Newtonian physics and in special relativity, the laws of nature have the same form in all *inertial* reference frames, but will have different forms in noninertial frames. The Austrian philosopher, physicist, and psychologist Ernest Mach (1838–1916) questioned the reality of inertial frames of reference. Mach argued that inertial reference systems were mental constructs and that in order to discuss them one had to imagine the absence of matter. But in reality, all the matter in the universe always exists. All of the effects of inertia could be due to the effects of the distant matter in the universe. Consider a bucket, partly full of water, that is spinning as shown in Figure 35-2. As the water spins, it tends to move out from the center, tangentially, but it is prevented from leaving by the walls of the bucket. Therefore, the water will pile up near the walls of the bucket and the surface of the water will assume a curved shape. Newton had discussed this as an illustration of the absolute nature of inertia and had argued that this demonstrated that there existed an inertial frame of reference. He suggested that this frame of reference was fixed with respect to the stars. Mach argued on the other hand that when the bucket rotated, it did so with respect to the stars; the curvature of the surface of the water could be thought of as arising from the relative motion of the water and the stars. Mach asserted that if it were possible to rotate the stars while keeping the bucket still, the water would pile up near the walls of the bucket, because the phenomenon depends on the relative motion. Newton asserted that rotation of the stars would have no effect on a bucket of water at rest in an inertial reference system.

Einstein took the position that the laws of nature should be written in a form that was independent of any coordinate system. That is, he believed that it should be possible to write the laws of nature in a form which was valid in noninertial *and* inertial frames. One form in which he stated the *general principle of relativity* was that *all frames of reference are equivalent for the formulation of the general laws of nature whatever may be their state of motion.*

This hypothesis required that the laws of nature (and in particular, the laws of motion and gravitation) be rewritten in such a way that they took the same mathematical form in all frames of reference. Einstein was able, by 1917, to publish such laws of motion and gravitation. These laws are complex mathematical statements, well beyond the scope of this book. Further, the equations are so difficult to solve that they have been applied to very few cases and consequently, there is very limited evidence for the validity of the laws.

35-3
The bending of light beams

The principle of equivalence is a simpler statement than the general principle of relativity. One of the most striking consequences of the principle of equivalence is that a beam of light will be bent by the presence of a gravitational field. One can see this by considering the path of a beam of light in the elevator laboratory discussed above. For example, as shown in Figure 35-3, we could send a beam of light by directing the light from a flashlight through two holes placed on opposite faces of a box. If the elevator, box, and flashlight were all at rest with respect to an inertial reference frame, the beam of light would travel straight across the box. That is, the light leaving hole *A* would travel straight across the box and leave through hole *B*. If however, the elevator were accelerating upward, then the light leaving hole *A* would travel straight across the box to where hole *B had been.* The light would strike the wall

FIGURE 35-2 As the bucket rotates, the surface of the water assumes a curved shape.

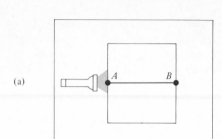

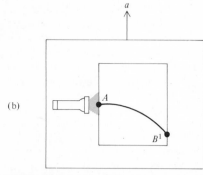

FIGURE 35-3 (a) The path of a beam of light in a stationary elevator, in the absence of gravity, is straight. (b) If the elevator is given an acceleration, a, the path of the beam of light relative to the elevator is curved.

of the box at a point *below* hole B. In other words, in the accelerating frame of reference of the accelerating elevator, the light beam would not travel in a straight line, but would instead curve downward, as in Figure 35-3b. According to the principle of equivalence, the situation in the accelerating elevator is indistinguishable from that in an elevator that is at rest in the presence of a gravitational field. That is, if the experiment were performed in the elevator while it was at rest near the surface of the earth, the path of the light beam would have the same curvature as shown in Figure 35-3b. Light beams are bent by gravity.

This prediction of the principle of equivalence has been verified by experiment. Consider the light that in traveling from a star to the earth passes near the sun. The large gravitational field of the sun will deflect the beam as shown in Figure 35-4. Thus, the light that reaches the earth will appear to come from a position that is shifted with respect to the actual position of the star. It must be noted that this effect occurs when the light passes near the sun and so occurs when the stars are observed in the *daytime*. At least until now it has been difficult to observe with accuracy the

positions of stars in the daytime under ordinary circumstances. The experiment thus requires that the apparent positions be measured during a solar eclipse. During the eclipse the positions of stars that appear to be near the sun are measured, and these positions are compared with the positions expected on the basis of nighttime observations of those same stars. These nighttime observations must be made roughly six months earlier than the eclipse measurements. Einstein predicted that the angle θ of Figure 35-4 was at most 1.7 seconds of arc. This prediction was first verified in measurements made in 1919, and has been verified subsequently in several other measurements. It must be emphasized that the effect is very small, that the measurements are very difficult, and therefore that these results cannot be considered definitive.

35-4
The effects of gravity on length and time

Another striking consequence of the principle of equivalence is the prediction that the presence of gravitational fields will alter the measurement of length and time.

Following Einstein let us consider a frame of reference S', which consists of a disk rotating about the point O with angular velocity ω, as shown

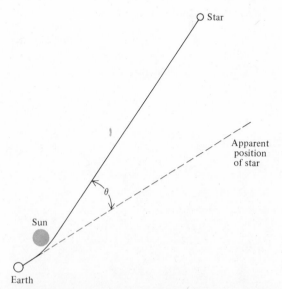

FIGURE 35-4 Light from a distant star is bent by the gravitational field of the sun. The apparent direction to the star is deflected by an angle θ from the real direction to the star.

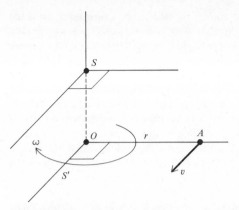

FIGURE 35-5 A frame of reference S' that is rotating with respect to an inertial frame of reference S. The point A has a velocity v relative to the inertial reference frame. This relative velocity is tangential; from the special theory of relativity tangential measurements are contracted, but radial measurements are not.

in Figure 35-5. The frame S' is rotating with respect to another (inertial) frame S. As we have seen, each point in the rotating frame of reference is accelerated radially inward. Thus, the point A has an acceleration a, directed in toward O, where

$$a = r\omega^2 \tag{35-1}$$

and the point A has a velocity v, where

$$v = r\omega \tag{35-2}$$

Consider then an observer in the inertial, non-rotating frame of reference S. According to the special theory of relativity, if that observer compares the readings of a clock that is at rest with respect to him, with the readings of a clock at point A (and therefore moving with velocity v), he will observe that the clock at point A is running slow. On the other hand, a clock at point O will run at exactly the same rate as the clock in the fixed frame of reference. Thus, if an observer at O, on the disk, compares his clock with one at A, also on the disk, he will observe that the clock at A runs slow compared to the clock at O. In fact, if the time between the ticks of the clock at O is t_o, and the time between ticks of the clock at A is t_A, then, from Chapter 10, we have

$$t_A = \frac{t_o}{\sqrt{1 - \dfrac{v^2}{c^2}}} \tag{35-3}$$

or

$$t_A = \frac{t_o}{\sqrt{1 - \dfrac{r^2\omega^2}{c^2}}} \tag{35-4}$$

We know, from the principle of equivalence, that to observers on the rotating reference frame the motion is indistinguishable from the situation where the disk is at *rest* in the presence of a gravitational field. That is, the observer at A cannot distinguish between the situation in which he is subject to a centripetal acceleration a and the situation in which he is in a gravitational field of intensity g. This equivalent gravitational field is directed radially outward and has a magnitude g given by

$$g = a = r\omega^2 \tag{35-5}$$

Thus, the principle of equivalence allows us to assert that the clock at point A, which is in a gravitational field of intensity given by Equation 35-5, runs slower than the clock at point O, which is in a gravitational field of zero intensity. In other words, gravitational fields make clocks run slower; an exact calculation shows that the rate of a clock depends on the gravitational *potential* at the location of the clock. It must be noted that this result is different from that described in Chapter 10 in that both observers at A and O agree on which clock runs slow. The argument can be extended to the assertion that a clock at a point of high gravitational potential will run slower than a clock at a point of low potential.

Consider a clock A that is a distance r_1 from an object of mass M_1, and a clock B that is a distance r_2 from an object of mass M_2. Then, in general, the two clocks will run at different rates. If the gravitational fields are not too large, it can be shown that the rates of the two clocks are connected by the following *approximation*:

$$\frac{t_A}{t_B} = 1 + \frac{\left(\dfrac{GM_1}{r_1} - \dfrac{GM_2}{r_2}\right)}{c^2} \tag{35-6}$$

A useful clock for testing Equation 35-6 is an atom or nucleus that emits electromagnetic waves of a definite frequency. The argument above asserts that an atom at a high gravitational potential will emit radiation of lower frequency than that same atom would if it were at a lower gravitational potential. That is, the characteristic frequencies emitted by an atom or nucleus will be shifted toward lower frequencies if the atom or nucleus is placed in a strong gravitational field. Atoms near the surface of most stars are in much larger gravitational fields than atoms near the surface of the earth, and therefore one might expect this shift of frequencies for the light emitted from the stars. Since visible, ultraviolet, and, X radiation is shifted toward the red end of the spectrum, this phenomenon is called the *gravitational red shift*. The frequency shift predicted by Equation 35-6 for light emitted by atoms at the surface of the sun is approximately two parts per million, a shift that is essentially too small to be detected. Large red shifts that seem to be in agreement with Equation 35-6 are observed for other astronomical bodies, but the masses and sizes of these bodies are not known accurately, and further, the gravitational red shift is difficult to separate from shifts due to other phenomena. However, in 1960, Pound and Rebka, using the Mössbauer effect, were able to measure the red shift arising from the change in gravitational field due to a change in altitude of less than 100 ft at the surface of the earth. The predicted shift was a few parts in 10^{15}, and the measurements agreed with the prediction to within a few percent. In addition to verifying the gravitational red shift, the experiment is remarkable because it measured such a tiny effect.

Two other recent experiments have verified the effect of gravitational potentials on natural clocks. Shapiro at MIT has measured the time for radar signals to travel from earth to Mars and return; he has also measured the time of radar signals reflected from Venus. When the radar signals were in such positions that they passed near the sun, Shapiro measured delays in the return of the radar pulses. Similarly radio astronomers have measured the pulse rate of pulsars (see Section 35-7). When the radiation from the pulsars passes near the sun, the pulses observed (by the astronomers on earth) are delayed. Both sets of observations agree with the numerical predictions of general relativity to within about 6 percent.

Let us return to consider the rotating frame of reference S' of Figure 35-5. To an observer on the fixed inertial frame of reference S, each point on the rotating disk (except O) is moving with a tangential velocity v. Therefore, according to the results of special relativity, tangential lengths are contracted, but radial lengths are not contracted. An observer at O, on the disk, is at rest relative to the inertial frame, and so measurements made by an observer at O agree with those made in the

inertial frame of reference. Consider then what would happen if an observer at O were to draw a circle on the disk, concentric with O, and measure the circumference and the diameter of the circle. In order to measure the circumference he would place a meter stick tangential to the circle. But this meter stick would be contracted, and he would measure a circumference larger than that measured at rest. To measure the diameter he would place the meter stick along the radius. This meter stick would not be contracted, and the radius would be identical with the rest measurement. If then, the observer at O were to calculate the ratio of the circumference of the circle to the diameter, he would obtain a number larger than π! This curious result becomes more striking when we invoke the principle of equivalence. Since the rotating disk is indistinguishable from a disk at rest in the presence of a gravitational field, this argument states that in the presence of a gravitational field the ratio of the circumference to the diameter of a circle will be different from π. In short, gravity changes geometry. This strange result will be discussed further in the next section.

35-5
The geometry of space

Euclidian geometry arose from a set of empirical rules based on measurement, but it has been abstracted to a logical system. The system consists of all of the logical deductions that are a consequence of a particular set of axioms and postulates. The axioms and postulates are simply asserted or assumed; they need not correspond to reality. For example, the fifth postulate of Euclid asserts that in a plane, one and only one line may be drawn parallel to a given line through a given point. This assumption is reasonable but is not proved in Euclidian geometry; it is merely asserted as a postulate. It is possible to construct consistent, logical (though unfamiliar) geometries with a different fifth postulate. We could postulate that no parallel lines may be drawn. Or we could postulate that an infinite number of parallel lines could be drawn. In either case we could construct a geometry that was internally consistent; that is, we could prove many theorems, and these theorems would not contradict each other. It is true that a theorem based on the "no parallel lines" postulate might contradict one based on the "one

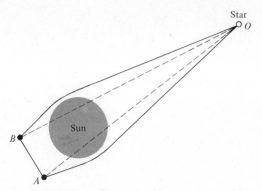

FIGURE 35-6 The sun's gravitational field causes light rays from O to bend. The dotted lines are the paths of the light in the absence of the sun.

and only one" parallel line postulate, but each system would be *internally* consistent.

Looked at in this way, geometry becomes an intellectual exercise distinct from the behavior of natural objects. In order to connect geometry with nature, one must appeal to experiment and devise a scheme that expresses the primitive indefinables of geometry in terms of measurement. For example, the concept of straight line is an undefined primitive of geometry. In order to apply geometry to nature, we must first decide on some mechanism for defining a straight line. An appealing and useful suggestion is to define a straight line as the path of a light ray. This certainly is consistent with common practice in which surveyors and carpenters use line-of-sight measurements as practical equivalents of a straight line. It is clear that general relativity or the principle of equivalence makes this definition less intuitively obvious. However, it is difficult, if not impossible, to imagine a simpler definition.

If we accept this definition of a straight line, then the bending of light rays by gravity forces us to conclude that Euclidian geometry is not applicable to nature. Consider, for example, light from a star, O, as shown in Figure 35-6, which grazes the sun. Light that reaches the observers at A and B will have been bent as shown, whereas light will travel from A to B along the straight line shown. Since we define a straight line as the path of a light ray, the solid lines connecting O, A, and B form a triangle whose sides are "straight." The dotted lines are Euclidian straight lines and are the paths of light rays in the absence of the sun. In the presence of the sun, it is clear from the figure that the sum of the interior angles of the triangle is greater than 180°. That is, the presence

of the sun, or rather of the gravitational field due to the sun, changes the geometry of space.

A useful way to visualize these effects is to consider the results of ordinary geometric measurements made on the surface of a sphere. On the surface of a sphere the shortest distance between two points is the arc of a great circle (a circle whose center is the center of the sphere) that passes through both points. On the surface of the earth, the equator and the lines of longitude are such great circles, while the parallels of latitude are not. Consider then, two points A and B on the equator, and the north pole as shown in Figure 35-7. If we define a straight line on the surface of the sphere to be the shortest distance between two points, the figure NAB is a triangle whose sides are "straight lines." Since the angles NAB and NBA are each right angles, the sum of the interior angles of this triangle is greater than 180°. Similarly, the equator is the circumference of a circle whose center is N. The circumference of this circle is the length of the equator and the radius is the length of the line NA. Thus, the ratio of the circumference to the diameter is 2, rather than π. The surface of the sphere is a two-dimensional curved space, which is non-Euclidian.

In a similar way, general relativity asserts that ordinary space, or rather four-dimensional space-time, is curved by the presence of matter. The presence of matter makes the geometry of space non-Euclidian. This curvature of space-time depends on the density and distribution of matter.

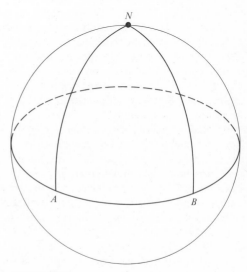

FIGURE 35-7

From this point of view we can think of the motion of objects under the influence of gravity in the following way: the presence of matter changes the geometry of space, introducing curvature, or bumps and valleys. The object moves through this space along the shortest path, moving down hill and along valleys. In other words, space can no longer be thought of as independent of matter, and the effects of matter can be thought of as alterations in the properties of space.

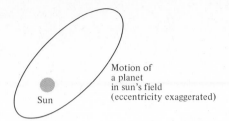

35-6
The perihelion of Mercury

The explanation of the elliptical orbits of the planets was among the principal successes of the Newtonian theory. It is therefore interesting that one of the few direct checks of the general theory of relativity, as distinct from the principle of equivalence, is an explanation of a small discrepancy in the motion of one of the planets.

According to Newtonian theory, the motion of a planet is an ellipse only if the planet moves in the gravitational field of the sun. That is, the orbits of the planets are ellipses only if one can neglect the forces exerted by the other planets and the other objects in the universe, and further, only if the sun is spherically symmetric. Generally the effect of the other planets is to introduce a slow rotation of the elliptical orbit, as shown in Figure 35-8. This rotation is called the *precession of the perihelion*. By the turn of this century it was possible to calculate the precession of the perihelion of all the planets to great precision, and compare them with very careful astronomical measurements. Excellent agreement was reached except for the planet Mercury. For Mercury the discrepancy amounted to a motion through an angle of approximately 42 *seconds* of arc per *century*.

Einstein's theory of gravitation (based on the general principle of relativity) predicted a precession of the perihelion of the planets slightly different from that predicted by the Newtonian theory. For Mercury, the prediction was for an additional precession of 43 seconds of arc per century. This almost exact agreement with the observations was essentially the only direct check of the theory until the recent astronomical measurements mentioned above. At the present time (1973) many experiments are being planned to investigate the detailed predictions of general relativity. These involve accurate radar measure-

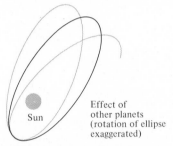

FIGURE 35-8 (a) Motion of a planet in the gravitational field of the sun (eccentricity exaggerated). (b) The effect of other planets is to rotate the ellipse (effect exaggerated).

ments and the use of satellite probes and orbiting observatories. One interesting proposal by Fairbanks is to put a very accurate gyroscope in orbit around the earth. A measurement of the precession of this gyroscope during one year should provide an accurate check on the theory.

The theory of general relativity predicts the existence of gravitational waves, analogous to electromagnetic waves. Weber has performed a complex experiment that, he claims, has observed such waves. At this time, several laboratories are making measurements in order to attempt to verify Weber's measurements.

The equations describing general relativity have no stable solutions, if matter is present in space. That is, they predict that the universe must expand. This prediction led many physicists to reject, or at least doubt, the theory when it was first published. With the discovery by Hubble in 1929 of the red shift of the galaxies, astronomers accepted the hypothesis that the universe is expanding.

Dicke has proposed an alternate formulation of the general theory, which differs somewhat from that of Einstein. At the present time the experimental data are not precise enough to distinguish between the two theories unequivocally.

35-7
Quasars, pulsars, and black holes

In the past few years a series of interesting astronomical discoveries have been made that have aroused the hope that more direct tests of general relativity may be possible. In the 1950s several astronomical objects, called *quasars* (from quasistellar), were found. Careful analysis of the spectra emitted by these objects led to the conclusion that the spectra had very large red shifts. In order to account for these very large red shifts, one had to assume that either these objects were receding from the earth with unprecedented large velocities (in some cases with v/c values as large as 0.8), or that the intensity of the gravitational field near the surface of these objects is extremely large. In the 1960s another group of objects, called *pulsars*, were discovered. These objects emitted very regular pulses of light or X rays with periods from about 1 sec to about 0.01 sec. The Crab Nebula pulsar appears to be the remnant of the explosion of a supernova, which was observed by Chinese astronomers in about 1054. In order to explain the observed properties of pulsars it is assumed that they are *neutron stars*, objects with a mass approximately equal to that of the sun, a radius of a few kilometers, and consisting mostly of free particles (mostly neutrons) rather than atoms and ions. The density of such an object would be about 10^{14} times as large as the density of the earth, and therefore the gravitational field in the neighborhood of a neutron star would be very large.

Both quasars and pulsars are poorly understood. The amount of observational data available is increasing rapidly, and the level of theoretical interest is high. The indications are strong that these objects have very intense gravitational fields associated with them, and that therefore the methods of general relativity may have to be invoked to explain their properties. This has stimulated a great deal of research in general relativity, and revived interest in some of the speculations suggested by the theory.

One such speculation is the possibility that there may exist *black holes*. A black hole is a region of space with so large a density of matter, and therefore so large a gravitational field, that the bending of light rays would prevent the escape of any electromagnetic radiations from the object. Such an object would absorb all the nearby matter and radiant energy and emit nothing; hence the name.

These things are interesting because they involve gravitational field intensities that are so large that the predictions of Einstein's theory differ markedly from those of Newton's theory of gravitation. Rather than describe the distinction in terms of the intensity of the gravitational field, it is sometimes convenient to describe a quantity called the *gravitational radius* of an astronomical body, which is defined in the following:

In Newtonian theory, the velocity of escape v of an object from the gravitational field G of a body of mass M and radius R is given by

$$v = \sqrt{\frac{2GM}{R}} \tag{35-7}$$

As we have seen, relativistic equations usually involve the ratio v/c rather than v alone. Therefore, we can rewrite Equation 35-7 as

$$\frac{v}{c} = \sqrt{\frac{2GM}{Rc^2}} \tag{35-7a}$$

As we have seen, the limiting speed of objects is c, the speed of light. Therefore, Equation 35-7a can be accurate only if the ratio v/c is small compared to unity. An equivalent statement is that the classical result must give incorrect results when it predicts that v is equal to or greater than c. That is, when the right side of Equation 35-7a is approximately equal to one, classical theory may not be used to describe the gravitational field or its effects. That is, if

$$R = \frac{2GM}{c^2} \tag{35-8}$$

then general relativity *must* be used. The quantity R given by Equation 35-8 is called the *gravitational radius* of an object of mass M. If the gravitational radius of an object is small compared with the actual radius of the object, then it may be described in terms of classical Newtonian theory. However, if the gravitational radius is comparable to, or larger than, the actual radius, then the gravitational field is large enough to require that the description be relativistic. The gravitational radius of the earth is about 1 cm, whereas the gravitational radius of the sun is about 3 km. Black holes, if they exist, are objects whose gravitational radii are equal to or larger than the actual radii of these objects.

Questions

1. Consider a very wide elevator that is at rest near the surface of the earth. Show that careful measurement of the direction of the gravitational field at several points across the width of the elevator can distinguish between this situation and the one in which the elevator is accelerating with constant acceleration.

2. Consider a very tall elevator that is at rest near the surface of the earth. Show that careful measurements of the magnitude of the gravitational field can distinguish between this situation and the one in which the elevator is accelerating with constant acceleration.

3. Discuss the limitations that are placed on the principle of equivalence by the answers to Questions 1 and 2.

4. Assume that the ratio of inertial to gravitational mass is not a constant, but depends on the atomic number of the material. Describe an experiment that could show that the principle of equivalence is not valid.

5. A spaceship is accelerating in a gravity-free region of space. A light source at the nose of the ship is emitting light of a specific frequency, and an instrument at the tail of the ship measures the frequency of the light. Is the measured frequency larger than, equal to, or less than the emitted frequency? Discuss your answer in terms of the principle of equivalence.

6. In the Pound-Rebka experiment, the frequency of a photon was shifted as the photon moved through a gravitational field. Since the energy of a photon is proportional to the frequency, this implies that the energy of the photon changes as it moves through the gravitational field. What happens to the energy lost by the photons?

7. A balloonist floats at 6000 ft. Does his watch run fast or slow compared with the watch of a member of the ground crew? Assume that both have the same velocity.

8. Does the bending of the light rays by the sun make the stars appear to move toward or away from the sun?

9. If a photon of yellow light falls near the surface of the earth, is it shifted toward the blue or the red?

Problems

1. How far does an initially horizontal light beam fall if it travels a horizontal distance of 10 m near the surface of the earth? Use the principle of equivalence. (Note that the correct relativistic answer is twice as large.)

2. Using the principle of equivalence calculate the vertical deflection of an initially horizontal beam of light that travels a horizontal distance equal to the diameter of the sun. Assume that the gravitational field is constant and has a magnitude equal to the magnitude of the gravitational field at the surface of the sun. From this, estimate the angular deflection of the beam.

3. A radioactive substance at the surface of the earth emits gamma rays with a frequency of 2.5×10^{18} Hz. What frequency would this same material emit at the top of a building 400 m high?

4. Consider a photon of energy $\mathcal{E}$ and frequency ν, with $\mathcal{E} = h\nu$. Assume that the photon can be treated as having a gravitational mass m which is related to $\mathcal{E}$ by

$$\mathcal{E} = mc^2$$

(a) Show that the gravitational potential energy $\mathcal{E}_p$ of this photon at a point a distance r from an object of mass M is given by

$$\mathcal{E}_p = -\frac{GM}{r}\frac{h\nu}{c^2}$$

(b) Assume that the change in energy $\Delta\mathcal{E}$ is equal to the negative of the change in gravitational potential energy, $\Delta\mathcal{E}_p$, when the photon moves from one point to another. Show that when the photon moves from a point that is a distance r_1 from an object of mass M_1 to a point that is a distance r_2 from an object of mass M_2, the frequency is shifted by the same amount as predicted by Equation 35-6.

$$\frac{\Delta\nu}{\nu} = \frac{GM_1}{r_1} - \frac{GM_2}{r_2}$$

5. Light is observed from a star with frequencies that are shifted by one part in 10^{14}. If the star has the same mass as the sun, how large is the star?

6. A turntable of radius 5 m rotates so that a point on the rim is moving with a speed of 0.9 c. What is the ratio of the circumference of the turntable to the diameter of the turntable, as measured by an observer on the turntable?

7. Estimate the fractional shift in the frequency of the gamma rays emitted by ^{52}Mn as the gamma rays move from the surface of the nucleus to infinity. Use Equation 33-2 to estimate the size of the nucleus and Equation 33-4 to estimate the mass.

8. Calculate the gravitational radius of the moon.

9. When two objects of identical mass m are separated by a distance d, they have a gravitational potential energy $\mathcal{E}_p$, such that

$$\mathcal{E}_p = -\frac{GM^2}{d}$$

The mass equivalent M of this energy is given by

$$-\mathcal{E}_p = Mc^2$$

If the two objects are so close to each other that M is larger than m, then general relativity must be used to describe the gravitational interaction. How close can two electrons be before general relativity must be invoked?

appendices

tables

appendix A

Table 1 Four-place logarithms of ordinary numbers

N	0	1	2	3	4	5	6	7	8	9
10	0000	0043	0086	0128	0170	0212	0253	0294	0334	0374
11	0414	0453	0492	0531	0569	0607	0645	0682	0719	0755
12	0792	0828	0864	0899	0934	0969	1004	1038	1072	1106
13	1139	1173	1206	1239	1271	1303	1335	1367	1399	1430
14	1461	1492	1523	1553	1584	1614	1644	1673	1703	1732
15	1761	1790	1818	1847	1875	1903	1931	1959	1987	2014
16	2041	2068	2095	2122	2148	2175	2201	2227	2253	2279
17	2304	2330	2355	2380	2405	2430	2455	2480	2504	2529
18	2553	2577	2601	2625	2648	2672	2695	2718	2742	2765
19	2788	2810	2833	2856	2878	2900	2923	2945	2967	2989
20	3010	3032	3054	3075	3096	3118	3139	3160	3181	3201
21	3222	3243	3263	3284	3304	3324	3345	3365	3385	3404
22	3424	3444	3464	3483	3502	3522	3541	3560	3579	3598
23	3617	3636	3655	3674	3692	3711	3729	3747	3766	3784
24	3802	3820	3838	3856	3874	3892	3909	3927	3945	3962
25	3979	3997	4014	4031	4048	4065	4082	4099	4116	4133
26	4150	4166	4183	4200	4216	4232	4249	4265	4281	4298
27	4314	4330	4346	4362	4378	4393	4409	4425	4440	4456
28	4472	4487	4502	4518	4533	4548	4564	4579	4594	4609
29	4624	4639	4654	4669	4683	4698	4713	4728	4742	4757
30	4771	4786	4800	4814	4829	4843	4857	4871	4886	4900
31	4914	4928	4942	4955	4969	4983	4997	5011	5024	5038
32	5051	5065	5079	5092	5105	5119	5132	5145	5159	5172
33	5185	5198	5211	5224	5237	5250	5263	5276	5289	5302
34	5315	5328	5340	5353	5366	5378	5391	5403	5416	5428
35	5441	5453	5465	5478	5490	5502	5514	5527	5539	5551
36	5563	5575	5587	5599	5611	5623	5635	5647	5658	5670
37	5682	5694	5705	5717	5729	5740	5752	5763	5775	5786
38	5798	5809	5821	5832	5843	5855	5866	5877	5888	5899
39	5911	5922	5933	5944	5955	5966	5977	5988	5999	6010
40	6021	6031	6042	6053	6064	6075	6085	6096	6107	6117
41	6128	6138	6149	6160	6170	6180	6191	6201	6212	6222
42	6232	6243	6253	6263	6274	6284	6294	6304	6314	6325
43	6335	6345	6355	6365	6375	6385	6395	6405	6415	6425
44	6435	6444	6454	6464	6474	6484	6493	6503	6513	6522
45	6532	6542	6551	6561	6571	6580	6590	6599	6609	6618
46	6628	6637	6646	6656	6665	6675	6684	6693	6702	6712
47	6721	6730	6739	6749	6758	6767	6776	6785	6794	6803
48	6812	6821	6830	6839	6848	6857	6866	6875	6884	6893
49	6902	6911	6920	6928	6937	6946	6955	6964	6972	6981
50	6990	6998	7007	7016	7024	7033	7042	7050	7059	7067
51	7076	7084	7093	7101	7110	7118	7126	7135	7143	7152
52	7160	7168	7177	7185	7193	7202	7210	7218	7226	7235
53	7243	7251	7259	7267	7275	7284	7292	7300	7308	7316
54	7324	7332	7340	7348	7356	7364	7372	7380	7388	7396
N	0	1	2	3	4	5	6	7	8	9

Table 1 Four-place logarithms of ordinary numbers *(cont.)*

N	0	1	2	3	4	5	6	7	8	9
55	7404	7412	7419	7427	7435	7443	7451	7459	7466	7474
56	7482	7490	7497	7505	7513	7520	7528	7536	7543	7551
57	7559	7566	7574	7582	7589	7597	7604	7612	7619	7627
58	7634	7642	7649	7657	7664	7672	7679	7686	7694	7701
59	7709	7716	7723	7731	7738	7745	7752	7760	7767	7774
60	7782	7789	7796	7803	7810	7818	7825	7832	7839	7846
61	7853	7860	7868	7875	7882	7889	7896	7903	7910	7917
62	7924	7931	7938	7945	7952	7959	7966	7973	7980	7987
63	7993	8000	8007	8014	8021	8028	8035	8041	8048	8055
64	8062	8069	8075	8082	8089	8096	8102	8109	8116	8122
65	8129	8136	8142	8149	8156	8162	8169	8176	8182	8189
66	8195	8202	8209	8215	8222	8228	8235	8241	8248	8254
67	8261	8267	8274	8280	8287	8293	8299	8306	8312	8319
68	8325	8331	8338	8344	8351	8357	8363	8370	8376	8382
69	8388	8395	8401	8407	8414	8420	8426	8432	8439	8445
70	8451	8457	8463	8470	8476	8482	8488	8494	8500	8506
71	8513	8519	8525	8531	8537	8543	8549	8555	8561	8567
72	8573	8579	8585	8591	8597	8603	8609	8615	8621	8627
73	8633	8639	8645	8651	8657	8663	8669	8675	8681	8686
74	8692	8698	8704	8710	8716	8722	8727	8733	8739	8745
75	8751	8756	8762	8768	8774	8779	8785	8791	8797	8802
76	8808	8814	8820	8825	8831	8837	8842	8848	8854	8859
77	8865	8871	8876	8882	8887	8893	8899	8904	8910	8915
78	8921	8927	8932	8938	8943	8949	8954	8960	8965	8971
79	8976	8982	8987	8993	8998	9004	9009	9015	9020	9025
80	9031	9036	9042	9047	9053	9058	9063	9069	9074	9079
81	9085	9090	9096	9101	9106	9112	9117	9122	9128	9133
82	9138	9143	9149	9154	9159	9165	9170	9175	9180	9186
83	9191	9196	9201	9206	9212	9217	9222	9227	9232	9238
84	9243	9248	9253	9258	9263	9269	9274	9279	9284	9289
85	9294	9299	9304	9309	9315	9320	9325	9330	9335	9340
86	9345	9350	9355	9360	9365	9370	9375	9380	9385	9390
87	9395	9400	9405	9410	9415	9420	9425	9430	9435	9440
88	9445	9450	9455	9460	9465	9469	9474	9479	9484	9489
89	9494	9499	9504	9509	9513	9518	9523	9528	9533	9538
90	9542	9547	9552	9557	9562	9566	9571	9576	9581	9586
91	9590	9595	9600	9605	9609	9614	9619	9624	9628	9633
92	9638	9643	9647	9652	9657	9661	9666	9671	9675	9680
93	9685	9689	9694	9699	9703	9708	9713	9717	9722	9727
94	9731	9736	9741	9745	9750	9754	9759	9763	9768	9773
95	9777	9782	9786	9791	9795	9800	9805	9809	9814	9818
96	9823	9827	9832	9836	9841	9845	9850	9854	9859	9863
97	9868	9872	9877	9881	9886	9890	9894	9899	9903	9908
98	9912	9917	9921	9926	9930	9934	9939	9943	9948	9952
99	9956	9961	9965	9969	9974	9978	9983	9987	9991	9996
N	0	1	2	3	4	5	6	7	8	9

Table 2 Four-place values of functions and radians

Degrees	Radians	Sin	Cos	Tan	Cot	Sec	Csc		
0° 00′	.0000	.0000	1.0000	.0000	——	1.000	——	1.5708	**90° 00′**
10	029	029	000	029	343.8	000	343.8	679	50
20	058	058	000	058	171.9	000	171.9	650	40
30	.0087	.0087	1.0000	.0087	114.6	1.000	114.6	1.5621	30
40	116	116	.9999	116	85.94	000	85.95	592	20
50	145	145	999	145	68.75	000	68.76	563	10
1° 00′	.0175	.0175	.9998	.0175	57.29	1.000	57.30	1.5533	**89° 00′**
10	204	204	998	204	49.10	000	49.11	504	50
20	233	233	997	233	42.96	000	42.98	475	40
30	.0262	.0262	.9997	.0262	38.19	1.000	38.20	1.5446	30
40	291	291	996	291	34.37	000	34.38	417	20
50	320	320	995	320	31.24	001	31.26	388	10
2° 00′	.0349	.0349	.9994	.0349	28.64	1.001	28.65	1.5359	**88° 00′**
10	378	378	993	378	26.43	001	26.45	330	50
20	407	407	992	407	24.54	001	24.56	301	40
30	.0436	.0436	.9990	.0437	22.90	1.001	22.93	1.5272	30
40	465	465	989	466	21.47	001	21.49	243	20
50	495	494	988	495	20.21	001	20.23	213	10
3° 00′	.0524	.0523	.9986	.0524	19.08	1.001	19.11	1.5184	**87° 00′**
10	553	552	985	553	18.07	002	18.10	155	50
20	582	581	983	582	17.17	002	17.20	126	40
30	.0611	.0610	.9981	.0612	16.35	1.002	16.38	1.5097	30
40	640	640	980	641	15.60	002	15.64	068	20
50	669	669	978	670	14.92	002	14.96	039	10
4° 00′	.0698	.0698	.9976	.0699	14.30	1.002	14.34	1.5010	**86° 00′**
10	727	727	974	729	13.73	003	13.76	981	50
20	756	756	971	758	13.20	003	13.23	952	40
30	.0785	.0785	.9969	.0787	12.71	1.003	12.75	1.4923	30
40	814	814	967	816	12.25	003	12.29	893	20
50	844	843	964	846	11.83	004	11.87	864	10
5° 00′	.0873	.0872	.9962	.0875	11.43	1.004	11.47	1.4835	**85° 00′**
10	902	901	959	904	11.06	004	11.10	806	50
20	931	929	957	934	10.71	004	10.76	777	40
30	.0960	.0958	.9954	.0963	10.39	1.005	10.43	1.4748	30
40	989	987	951	992	10.08	005	10.13	719	20
50	.1018	.1016	948	.1022	9.788	005	9.839	690	10
6° 00′	.1047	.1045	.9945	.1051	9.514	1.006	9.567	1.4661	**84° 00′**
10	076	074	942	080	9.255	006	9.309	632	50
20	105	103	939	110	9.010	006	9.065	603	40
30	.1154	.1132	.9936	.1139	8.777	1.006	8.834	1.4573	30
40	164	161	932	169	8.556	007	8.614	544	20
50	193	190	929	198	8.345	007	8.405	515	10
7° 00′	.1222	.1219	.9925	.1228	8.144	1.008	8.206	1.4486	**83° 00′**
10	251	248	922	257	7.953	008	8.016	457	50
20	280	276	918	287	7.770	008	7.834	428	40
30	.1309	.1305	.9914	.1317	7.596	1.009	7.661	1.4399	30
40	338	334	911	346	7.429	009	7.496	370	20
50	367	363	907	376	7.269	009	7.337	341	10
8° 00′	.1396	.1392	.9903	.1405	7.115	1.010	7.185	1.4312	**82° 00′**
10	425	421	899	435	6.968	010	7.040	283	50
20	454	449	894	465	6.827	010	6.900	254	40
30	.1484	.1478	.9890	.1495	6.691	1.011	6.765	1.4224	30
40	513	507	886	524	6.561ʹ	012	6.636	195	20
50	542	536	881	554	6.435	012	6.512	166	10
9° 00′	.1571	.1564	.9877	.1584	6.314	1.012	6.392	1.4137	**81° 00′**
		Cos	Sin	Cot	Tan	Csc	Sec	Radians	Degrees

Table 2 Four-place values of functions and radians *(cont.)*

Degrees	Radians	Sin	Cos	Tan	Cot	Sec	Csc		
9° 00′	.1571	.1564	.9877	.1584	6.314	1.012	6.392	1.4137	**81° 00′**
10	600	593	872	614	197	013	277	108	50
20	629	622	868	644	084	013	166	079	40
30	.1658	.1650	.9863	.1673	5.976	1.014	6.059	1.4050	30
40	687	679	858	703	871	014	5.955	1.4021	20
50	716	708	853	733	769	015	855	992	10
10° 00′	.1745	.1736	.9848	.1763	5.671	1.015	5.759	1.3963	**80° 00′**
10	774	765	843	793	576	016	665	934	50
20	804	794	838	823	485	016	575	904	40
30	.1833	.1822	.9833	.1853	5.396	1.017	5.487	1.3875	30
40	862	851	827	883	309	018	403	846	20
50	891	880	822	914	226	018	320	817	10
11° 00′	.1920	.1908	.9816	.1944	5.145	1.019	5.241	1.3788	**79° 00′**
10	949	937	811	974	066	019	164	759	50
20	978	965	805	.2004	4.989	020	089	730	40
30	.2007	.1994	.9799	.2035	4.915	1.020	5.016	1.3701	30
40	036	.2022	793	065	843	021	4.945	672	20
50	065	051	787	095	773	022	876	643	10
12° 00′	.2094	.2079	.9781	.2126	4.705	1.022	4.810	1.3614	**78° 00′**
10	123	108	775	156	638	023	745	584	50
20	153	136	769	186	574	024	682	555	40
30	.2182	.2164	.9763	.2217	4.511	1.024	4.620	1.3526	30
40	211	193	757	247	449	025	560	497	20
50	240	221	750	278	390	026	502	468	10
13° 00′	.2269	.2250	.9744	.2309	4.331	1.026	4.445	1.3439	**77° 00′**
10	298	278	737	339	275	027	390	410	50
20	327	306	730	370	219	028	336	381	40
30	.2356	.2334	.9724	.2401	4.165	1.028	4.284	1.3352	30
40	385	363	717	432	113	029	232	323	20
50	414	391	710	462	061	030	182	294	10
14° 00′	.2443	.2419	.9703	.2493	4.011	1.031	4.134	1.3265	**76° 00′**
10	473	447	696	524	3.962	031	086	235	50
20	502	476	689	555	914	032	039	206	40
30	.2531	.2504	.9681	.2586	3.867	1.033	3.994	1.3177	30
40	560	532	674	617	821	034	950	148	20
50	589	560	667	648	776	034	906	119	10
15° 00′	.2618	.2588	.9659	.2679	3.732	1.035	3.864	1.3090	**75° 00′**
10	647	616	652	711	689	036	822	061	50
20	676	644	644	742	647	037	782	032	40
30	.2705	.2672	.9636	.2773	3.606	1.038	3.742	1.3003	30
40	734	700	628	805	566	039	703	974	20
50	763	728	621	836	526	039	665	945	10
16° 00′	.2793	.2756	.9613	.2867	3.487	1.040	3.628	1.2915	**74° 00′**
10	822	784	605	899	450	041	592	886	50
20	851	812	596	931	412	042	556	857	40
30	.2880	.2840	.9588	.2962	3.376	1.043	3.521	1.2828	30
40	909	868	580	994	340	044	487	799	20
50	938	896	572	.3026	305	045	453	770	10
17° 00′	.2967	.2924	.9563	.3057	3.271	1.046	3.420	1.2741	**73° 00′**
10	996	952	555	089	237	047	388	712	50
20	.3025	979	546	121	204	048	356	683	40
30	.3054	.3007	.9537	.3153	3.172	1.049	3.326	1.2654	30
40	083	035	528	185	140	049	295	625	20
50	113	062	520	217	108	050	265	595	10
18° 00′	.3142	.3090	.9511	.3249	3.078	1.051	3.236	1.2566	**72° 00′**
		Cos	Sin	Cot	Tan	Csc	Sec	Radians	Degrees

Table 2 Four-place values of functions and radians *(cont.)*

Degrees	Radians	Sin	Cos	Tan	Cot	Sec	Csc		
18° 00′	.3142	.3090	.9511	.3249	3.078	1.051	3.236	1.2566	**72° 00′**
10	171	118	502	281	047	052	207	537	50
20	200	145	492	314	018	053	179	508	40
30	.3229	.3173	.9483	3346	2.989	1.054	3.152	1.2479	30
40	258	201	474	378	960	056	124	450	20
50	287	228	465	411	932	057	098	421	10
19° 00′	.3316	.3256	.9455	.3443	2.904	1.058	3.072	1.2392	**71° 00′**
10	345	283	446	476	877	059	046	363	50
20	374	311	436	508	850	060	021	334	40
30	.3403	.3338	.9426	.3541	2.824	1.061	2.996	1.2305	30
40	432	365	417	574	798	062	971	275	20
50	462	393	407	607	773	063	947	246	10
20° 00′	.3491	.3420	.9397	.3640	2.747	1.064	2.924	1.2217	**70° 00′**
10	520	448	387	673	723	065	901	188	50
20	549	475	377	706	699	066	878	159	40
30	.3578	.3502	.9367	.3739	2.675	1.068	2.855	1.2130	30
40	607	529	356	772	651	069	833	101	20
50	636	557	346	805	628	070	812	072	10
21° 00′	.3665	.3584	.9336	.3839	2.605	1.071	2.790	1.2043	**69° 00′**
10	694	611	325	872	583	072	769	1.2014	50
20	723	638	315	906	560	074	749	985	40
30	.3752	.3665	.9304	.3939	2.539	1.075	2.729	1.1956	30
40	782	692	293	973	517	076	709	926	20
50	811	719	283	.4006	496	077	689	897	10
22° 00′	.3840	.3746	.9272	.4040	2.475	1.079	2.669	1.1868	**68° 00′**
10	869	773	261	074	455	080	650	839	50
20	898	800	250	108	434	081	632	810	40
30	.3927	.3827	.9239	.4142	2.414	1.082	2.613	1.1781	30
40	956	854	228	176	394	084	595	752	20
50	985	881	216	210	375	085	577	723	10
23° 00′	.4014	.3907	.9205	.4245	2.356	1.086	2.559	1.1694	**67° 00′**
10	043	934	194	279	337	088	542	665	50
20	072	961	182	314	318	089	525	636	40
30	.4102	.3987	.9171	.4348	2.300	1.090	2.508	1.1606	30
40	131	.4014	159	383	282	092	491	577	20
50	160	041	147	417	264	093	475	548	10
24° 00′	.4189	.4067	.9135	.4452	2.246	1.095	2.459	1.1519	**66° 00′**
10	218	094	124	487	229	096	443	490	50
20	247	120	112	522	211	097	427	461	40
30	.4276	.4147	.9100	.4557	2.194	1.099	2.411	1.1432	30
40	305	173	088	592	177	100	396	403	20
50	334	200	075	628	161	102	381	374	10
25° 00′	.4363	.4226	.9063	.4663	2.145	1.103	2.366	1.1345	**65° 00′**
10	392	253	051	699	128	105	352	316	50
20	422	279	038	734	112	106	337	286	40
30	.4451	.4305	.9026	.4770	2.097	1.108	2.323	1.1257	30
40	480	331	013	806	081	109	309	228	20
50	509	358	001	841	066	111	295	199	10
26° 00′	.4538	.4384	.8988	.4877	2.050	1.113	2.281	1.1170	**64° 00′**
10	567	410	975	913	035	114	268	141	50
20	596	436	962	950	020	116	254	112	40
30	.4625	.4462	.8949	.4986	2.006	1.117	2.241	1.1083	30
40	654	488	936	.5022	1.991	119	228	054	20
50	683	514	923	059	977	121	215	1.1025	10
27° 00′	.4712	.4540	.8910	.5095	1.963	1.122	2.203	1.0996	**63° 00′**
		Cos	Sin	Cot	Tan	Csc	Sec	Radians	Degrees

Table 2 Four-place values of functions and radians *(cont.)*

Degrees	Radians	Sin	Cos	Tan	Cot	Sec	Csc		Degrees
27° 00′	.4712	.4540	.8910	.5095	1.963	1.122	2.203	1.0996	**63° 00′**
10	741	566	897	132	949	124	190	966	50
20	771	592	884	169	935	126	178	937	40
30	.4800	.4617	.8870	.5206	1.921	1.127	2.166	1.0908	30
40	829	643	857	243	907	129	154	879	20
50	858	669	843	280	894	131	142	850	10
28° 00′	.4887	.4695	.8829	.5317	1.881	1.133	2.130	1.0821	**62° 00′**
10	916	720	816	354	868	·134	118	792	50
20	945	746	802	392	855	136	107	763	40
30	.4974	.4772	.8788	.5430	1.842	1.138	2.096	1.0734	30
40	.5003	797	774	467	829	140	085	705	20
50	032	823	760	505	816	142	074	676	10
29° 00′	.5061	.4848	.8746	.5543	1.804	1.143	2.063	1.0647	**61° 00′**
10	091	874	732	581	792	145	052	617	50
20	120	899	718	619	780	147	041	588	40
30	.5149	.4924	.8704	.5658	1.767	1.149	2.031	1.0559	30
40	178	950	689	696	756	151	020	530	20
50	207	975	675	735	744	153	010	501	10
30° 00′	.5236	.5000	.8660	.5774	1.732	1.155	2.000	1.0472	**60° 00′**
10	265	025	646	812	720	157	1.990	443	50
20	294	050	631	851	709	159	980	414	40
30	.5323	.5075	.8616	.5890	1.698	1.161	1.970	1.0385	30
40	352	100	601	930	686	163	961	356	20
50	381	125	587	969	.675	165	951	327	10
31° 00′	.5411	.5150	.8572	.6009	1.664	1.167	1.942	1.0297	**59° 00′**
10	440	175	557	048	653	169	932	268	50
20	469	200	542	088	643	171	923	239	40
30	.5498	.5225	.8526	.6128	1.632	1.173	1.914	1.0210	30
40	527	250	511	168	621	175	905	181	20
50	556	275	496	208	611	177	896	152	10
32° 00′	.5585	.5299	.8480	.6249	1.600	1.179	1.887	1.0123	**58° 00′**
10	614	324	465	289	590	181	878	094	50
20	643	348	450	330	580	184	870	065	40
30	.5672	.5373	.8434	.6371	1.570	1.186	1.861	1.0036	30
40	701	398	418	412	560	188	853	1.0007	20
50	730	422	403	453	550	190	844	977	10
33° 00′	.5760	.5446	.8387	.6494	1.540	1.192	1.836	.9948	**57° 00′**
10	789	471	371	536	530	195	828	919	50
20	818	495	355	577	520	197	820	890	40
30	.5847	.5519	.8339	.6619	1.511	1.199	1.812	.9861	30
40	876	544	323	661	501	202	804	832	20
50	905	568	307	703	1.492	204	796	803	10
34° 00′	.5934	.5592	.8290	.6745	1.483	1.206	1.788	.9774	**56° 00′**
10	963	616	274	787	473	209	781	745	50
20	992	640	258	830	464	211	773	716	40
30	.6021	.5664	.8241	.6873	1.455	1.213	1.766	.9687	30
40	050	688	225	916	446	216	758	657	20
50	080	712	208	959	437	218	751	628	10
35° 00′	.6109	.5736	.8192	.7002	1.428	1.221	1.743	.9599	**55° 00′**
10	138	760	175	046	419	223	736	570	50
20	167	783	158	089	411	226	729	541	40
30	.6196	.5807	.8141	.7133	1.402	1.228	1.722	.9512	30
40	225	831	124	177	.393	231	715	483	20
50	254	854	107	221	385	233	708	454	10
36° 00′	.6283	.5878	.8090	.7265	1.376	1.236	1.701	.9425	**54° 00′**
		Cos	Sin	Cot	Tan	Csc	Sec	Radians	Degrees

Table 2 Four-place values of functions and radians *(cont.)*

Degrees	Radians	Sin	Cos	Tan	Cot	Sec	Csc		
36° 00'	.6283	.5878	.8090	.7265	1.376	1.236	1.701	.9425	**54° 00'**
10	312	901	073	310	368	239	695	396	50
20	341	925	056	355	360	241	688	367	40
30	.6370	.5948	.8039	.7400	1.351	1.244	1.681	.9338	30
40	400	972	021	445	343	247	675	308	20
50	429	995	004	490	335	249	668	279	10
37° 00'	.6458	.6018	.7986	.7536	1.327	1.252	1.662	.9250	**53° 00'**
10	487	041	969	581	319	255	655	221	50
20	516	065	951	627	311	258	649	192	40
30	.6545	.6088	.7934	.7673	1.303	1.260	1.643	.9163	30
40	574	111	916	720	295	263	636	134	20
50	603	134	898	766	288	266	630	105	10
38° 00'	.6632	.6157	.7880	.7813	1.280	1.269	1.624	.9076	**52° 00'**
10	661	180	862	860	272	272	618	047	50
20	690	202	844	907	265	275	612	.9018	40
30	.6720	.6225	.7826	.7954	1.257	1.278	1.606	.8988	30
40	749	248	808	.8002	250	281	601	959	20
50	778	271	790	050	242	284	595	930	10
39° 00'	.6807	.6293	.7771	.8098	1.235	1.287	1.589	.8901	**51° 00'**
10	836	316	753	146	228	290	583	872	50
20	865	338	735	195	220	293	578	843	40
30	.6894	.6361	.7716	.8243	1.213	1.296	1.572	.8814	30
40	923	383	698	292	206	299	567	785	20
50	952	406	679	342	199	302	561	756	10
40° 00'	.6981	.6428	.7660	.8391	1.192	1.305	1.556	.8727	**50° 00'**
10	.7010	450	642	441	185	309	550	698	50
20	039	472	623	491	178	312	545	668	40
30	.7069	.6494	.7604	.8541	1.171	1.315	1.540	.8639	30
40	098	517	585	591	164	318	535	610	20
50	127	539	566	642	157	322	529	581	10
41° 00'	.7156	.6561	.7547	.8693	1.150	1.325	1.524	.8552	**49° 00'**
10	185	583	528	744	144	328	519	523	50
20	214	604	509	796	137	332	514	494	40
30	.7243	.6626	.7490	.8847	1.130	1.335	1.509	.8465	30
40	272	648	470	899	124	339	504	436	20
50	301	670	451	952	117	342	499	407	10
42° 00'	.7330	.6691	.7431	.9004	1.111	1.346	1.494	.8378	**48° 00'**
10	359	713	412	057	104	349	490	348	50
20	389	734	392	110	098	353	485	319	40
30	.7418	.6756	.7373	.9163	1.091	1.356	1.480	.8290	30
40	447	777	353	217	085	360	476	261	20
50	476	799	333	271	079	364	471	232	10
43° 00'	.7505	.6820	.7314	.9325	1.072	1.367	1.466	.8203	**47° 00'**
10	534	841	294	380	066	371	462	174	50
20	563	862	274	435	060	375	457	145	40
30	.7592	.6884	.7254	.9490	1.054	1.379	1.453	.8116	30
40	621	905	234	545	048	382	448	087	20
50	650	926	214	601	042	386	444	058	10
44° 00'	.7679	.6947	.7193	.9657	1.036	1.390	1.440	.8029	**46° 00'**
10	709	967	173	713	030	394	435	999	50
20	738	988	153	770	024	398	431	970	40
30	.7767	.7009	.7133	.9827	1.018	1.402	1.427	.7941	30
40	796	030	112	884	012	406	423	912	20
50	825	050	092	942	006	410	418	883	10
45° 00'	.7854	.7071	.7071	1.000	1.000	1.414	1.414	.7854	**45° 00'**
		Cos	Sin	Cot	Tan	Csc	Sec	Radians	Degrees

Table 3 Atomic masses of some isotopes Based on $C^{12} = 12.00000^a$

Mass No. A	Atomic No. Z	Element	Atomic Mass in u	Mass No. A	Atomic No. Z	Element	Atomic Mass in u
1	0	n	1.00867	30	14	Si	29.97376
1	1	H	1.00783	35	17	Cl	34.96885
2	1	H	2.01410	37	17	Cl	36.96590
3	1	H	3.01605	38	18	A	37.96272
3	2	He	3.01603	39	19	K	38.96371
4	2	He	4.00260	40	18	A	39.96238
6	3	Li	6.01513	40	20	Ca	39.96259
7	3	Li	7.01601	40	19	K	39.96400
7	4	Be	7.01693	41	19	K	40.96184
8	3	Li	8.02249	41	20	Ca	40.96228
8	4	Be	8.00531	42	19	K	41.96241
9	4	Be	9.01219	42	20	Ca	41.95878
10	5	B	10.01294	57	26	Fe	56.93539
11	5	B	11.00931	57	27	Co	56.93629
11	6	C	11.01143	58	26	Fe	57.93327
12	6	C	12.00000	59	27	Co	58.93319
13	6	C	13.00354	60	27	Co	59.93380
13	7	N	13.00574	68	30	Zn	67.92486
14	6	C	14.00324	84	36	Kr	83.91150
14	7	N	14.00307	107	47	Ag	106.90497
15	6	C	15.01069	107	48	Cd	106.90652
15	7	N	15.00011	109	47	Ag	108.90470
15	8	O	15.00307	138	56	Ba	137.90501
16	8	O	15.99492	184	74	W	183.95089
17	8	O	16.99912	197	79	Au	196.96655
18	8	O	17.99916	205	81	Tl	204.97446
18	9	F	18.00095	207	82	Pb	206.97590
19	9	F	18.99840	209	83	Bi	208.98042
19	10	Ne	19.00189	210	84	Po	209.98287
20	9	F	19.99999	214	84	Po	213.99519
20	10	Ne	19.99244	222	86	Rn	222.01753
21	10	Ne	20.99385	224	88	Ra	224.02022
22	10	Ne	21.99138	226	88	Ra	226.02536
22	11	Na	21.99444	230	90	Th	230.03308
23	11	Na	22.98977	232	90	Th	232.03821
23	12	Mg	22.99414	235	92	U	235.04393
24	11	Na	23.99097	238	92	U	238.05076
24	12	Mg	23.98504	239	93	Np	239.05294
25	12	Mg	24.98584	239	94	Pu	239.05216
27	13	Al	26.98154	250	97	Bk	250.07849
28	13	Al	27.98191	255	53	Md	255.09057
28	14	Si	27.97851				

^{a} Data computed from 1961 Nuclidic Mass Table. L. A. Konig, J. H. E. Mattauch, and A. H. Wapstra, *Nuclear Physics*, Vol. 31, p. 18, 1962.

Table 4 Atomic weights of the elements

	symbol	atomic number	atomic weight (amu)		symbol	atomic number	atomic weight (amu)
Actinium	Ac	89		Germanium	Ge	32	72.5_9
Aluminum	Al	13	26.9815^a	Gold	Au	79	196.9665^a
Americium	Am	95		Hafnium	Hf	72	178.4_9
Antimony	Sb	51	121.7_5	Helium	He	2	4.00260^b
Argon	Ar	18	$39.94_8^{b,c}$	Holmium	Ho	67	164.9303^a
Arsenic	As	33	74.9216^a	Hydrogen	H	1	$1.008_0^{b,c}$
Astatine	At	85		Indium	In	49	114.82
Barium	Ba	56	137.3_4	Iodine	I	53	126.9045^a
Berkelium	Bk	97		Iridium	Ir	77	192.2_2
Beryllium	Be	4	9.01218^a	Iron	Fe	26	55.84_7
Bismuth	Bi	83	208.9806^a	Krypton	Kr	36	83.80
Boron	B	5	10.81^c	Lanthanum	La	57	138.905_5^b
Bromine	Br	35	79.904	Lawrencium	Lr	103	
Cadmium	Cd	48	112.40	Lead	Pb	82	207.2^c
Calcium	Ca	20	40.08	Lithium	Li	3	6.94_1^c
Californium	Cf	98		Lutetium	Lu	71	174.97
Carbon	C	6	$12.011^{b,c}$	Magnesium	Mg	12	24.305
Cerium	Ce	58	140.12	Manganese	Mn	25	54.9380^a
Cesium	Cs	55	132.9055^a	Mendelevium	Md	101	
Chlorine	Cl	17	35.453	Mercury	Hg	80	200.5_9
Chromium	Cr	24	51.996	Molybdenum	Mo	42	95.9_4
Cobalt	Co	27	58.9332^a	Neodymium	Nd	60	144.2_4
Copper	Cu	29	63.54_6^c	Neon	Ne	10	20.17_9
Curium	Cm	96		Neptunium	Np	93	237.0482^o
Dysprosium	Dy	66	162.5_0	Nickel	Ni	28	58.7_1
Einsteinium	Es	99		Niobium	Nb	41	92.9064^a
Erbium	Er	68	167.2_6	Nitrogen	N	7	14.0067^b
Europium	Eu	63	151.96	Nobelium	No	102	
Fermium	Fm	100		Osmium	Os	76	190.2
Fluorine	F	9	18.9984^a	Oxygen	O	8	$15.999_4^{b,c}$
Francium	Fr	87		Palladium	Pd	46	106.4
Gadolinium	Gd	64	157.2_5	Phosphorus	P	15	30.9738^a
Gallium	Ga	31	69.72	Platinum	Pt	78	195.0_9

Table 4 Atomic weights of the elements *(cont.)*

	symbol	atomic number	atomic weight (amu)		symbol	atomic number	atomic weight (amu)
Plutonium	Pu	94		Sulfur	S	16	32.06_c
Polonium	Po	84		Tantalum	Ta	73	$180.947_9{}^b$
Potassium	K	19	39.10_2	Technetium	Tc	43	98.9062^d
Praseodymium	Pr	59	140.0977^a	Tellurium	Te	52	127.6_0
Promethium	Pm	61		Terbium	Tb	65	158.9254^a
Protactinium	Pa	91	231.0359^a	Thallium	Tl	81	204.3_7
Radium	Ra	88	$226.0254^{a,d}$	Thorium	Th	90	232.0381^a
Radon	Rn	86		Thulium	Tm	69	168.9342^a
Rhenium	Re	75	186.2	Tin	Sn	50	118.6_9
Rhodium	Rh	45	102.9055^a	Titanium	Ti	22	47.9_0
Rubidium	Rb	37	85.467_8	Tungsten	W	74	183.8_5
Rutheium	Ru	44	101.0_7	Uranium	U	92	238.029^b
Samarium	Sm	62	150.4	Vanadium	V	23	$50.941_4{}^b$
Scandium	Sc	21	44.9559^a	Xenon	Xe	54	131.30
Selenium	Se	34	78.9_6	Ytterbium	Yb	70	173.0_4
Silicon	Si	14	$28.08_6{}^c$	Yttrium	Y	39	88.9059^a
Silver	Ag	47	107.868	Zinc	Zn	30	65.3_7
Sodium	Na	11	22.9898^a	Zirconium	Zr	40	91.22
Strontium	Sr	38	87.62				

[a] Elements with no isotopes. These elements exist as one type of atom only.
[b] Elements with one predominant isotope (around 99 to 100% one isotope).
[c] Elements for which the variation in isotopic abundance in earth samples limits the precision of the atomic weight.
[d] Most commonly available isotope (radioactive).

Table 5 Distribution of the electrons in the atoms

X-Ray Notation		K	L		M			N			
Quantum Numbers n, l		1,0	2,0	2,1	3,0	3,1	3,2	4,0	4,1	4,2	4,3
Element	Atomic Number Z										
H	1	1									
He	2	2									
Li	3	2	1								
Be	4	2	2								
B	5	2	2	1							
C	6	2	2	2							
N	7	2	2	3							
O	8	2	2	4							
F	9	2	2	5							
Ne	10	2	2	6							
Na	11		Neon Configuration 10 electron core		1						
Mg	12				2						
Al	13				2	1					
Si	14				2	2					
P	15				2	3					
S	16				2	4					
Cl	17				2	5					
A	18				2	6					
K	19		Argon Configuration 18 electron core					1			
Ca	20							2			
Sc	21						1	2			
Ti	22						2	2			
V	23						3	2			
Cr	24						5	1			
Mn	25						5	2			
Fe	26						6	2			
Co	27						7	2			
Ni	28						8	2			
Cu	29						10	1			
Zn	30						10	2			
Ga	31						10	2	1		
Ge	32						10	2	2		
As	33						10	2	3		
Se	34						10	2	4		
Br	35						10	2	5		
Kr	36						10	2	6		

Table 5 Distribution of the electrons in the atoms *(cont.)*

X-Ray Notation		K L M	N			O			P			Q	
Quantum Numbers n, l		1 2 3	4,0 4,1	4,2	4,3	5,0	5,1	5,2	6,0	6,1	6,2	7,0	7,1
Element	Atomic Number Z												
Rb	37	Krypton Configuration 36 electron core				1							
Sr	38					2							
Y	39			1		2							
Zr	40			2		2							
Nb	41			4		1							
Mo	42			5		1							
Ma	43			6		1							
Ru	44			7		1							
Rh	45			8		1							
Pd	46			10									
Ag	47	Palladium Configuration 46 electron core				1							
Cd	48					2							
In	49					2	1						
Sn	50					2	2						
Sb	51					2	3						
Te	52					2	4						
I	53					2	5						
Xe	54					2	6						
Cs	55	Xenon Configuration 54 electron core							1				
Ba	56								2				
La	57	Shells 1,0 to 4,2 contain 46 electrons				2	6	1	2				
Ce	58				1	2	6	1	2				
Pr	59				2	2	6	1	2				
Nd	60				3	2	6	1	2				
Pm	61				4	2	6	1	2				
Sm	62				5	2	6	1	2				
Eu	63				6	2	6	1	2				
Gd	64				7	2	6	1	2				
Tb	65				8	2	6	1	2				
Dy	66				9	2	6	1	2				
Ho	67				10	2	6	1	2				
Er	68				11	2	6	1	2				
Tm	69				13	2	6	0	2				
Yb	70				14	2	6	0	2				
Lu	71				14	2	6	1	2				

Table 5 Distribution of the electrons in the atoms *(cont.)*

X-Ray Notation		K L M N		O			P			Q	
Quantum Numbers n, l		1 2 3 4	5,0 5,1	5,2	5,3	6,0	6,1	6,2	7,0	7,1
Element	Atomic Number Z									
Hf	72			2		2				
Ta	73			3		2				
W	74			4		2				
Re	75			5		2				
Os	76			6		2				
Ir	77			7		2				
Pt	78	Shells		9		1				
Au	79	1,0 to 5,1		10		1				
Hg	80	contain		10		2				
Tl	81	68 electrons		10		2	1			
Pb	82			10		2	2			
Bi	83			10		2	3			
Po	84			10		2	4			
At	85			10		2	5			
Rn	86			10		2	6			
Fr	87	Radon Configuration							1	
Ra	88	86 electron core							2	
Ac	89					2	6	1	2	
Th	90				1	2	6	1	2	
Pa	91				2	2	6	1	2	
U	92				3	2	6	1	2	
Np	93				4	2	6	1	2	
Pu	94				5	2	6	1	2	
Am	95				6	2	6	1	2	
Cm	96				7	2	6	1	2	
Bk	97				8	2	6	1	2	
Cf	98				9	2	6	1	2	
E	99				10	2	6	1	2	
Fm	100				11	2	6	1	2	
Md	101				12	2	6	1	2	
No	102				13	2	6	1	2	
Lw	103				14	2	6	1	2	

Table 6   Conversion factors relating MKS and CGS units

quantity	symbol	MKS unit		CGS unit
Length	L, l	1 m	=	100 cm
Area	A, a	1 m²	=	10^4 cm²
Volume	V, v	1 m³	=	10^6 cm³
Energy	$\mathscr{E}, U$	1 joule	=	10^7 ergs
Power	P	1 watt	=	10^7 ergs/sec
Mass	m, M	1 kg	=	10^3 gm
Force	F, f	1 nt	=	10^5 dyne
Electric charge	Q, q	1 coul	=	3×10^9 statcoul $= 0.1$ abcoul
Electric intensity	E	1 nt/coul $=$ 1 volt/m	=	$\frac{1}{3} \times 10^{-4}$ statvolt/cm $= \frac{1}{3} \times 10^{-10}$ abvolt/cm
Electric current	I, i	1 amp	=	3×10^9 statamp $= 0.1$ abamp
Electric potential	$V, \mathscr{E}$	1 volt	=	3.336×10^{-3} statvolt $= 10^8$ abvolt
Resistance	R	1 ohm	=	1.113×10^{-12} statohm $= 10^9$ abohm
Capacitance	C	1 farad	=	9×10^{11} statfarad $= 10^{-9}$ abfarad
Inductance	L	1 henry	=	1.113×10^{-12} stathenry $= 10^9$ abhenry
Magnetic flux	Φ	1 weber	=	10^8 maxwells $= \frac{1}{300}$ esu
Magnetic field	H	1 amp/m	=	$4\pi \times 10^{-3}$ oersted $= 10^{-3}$ abamp/cm
Magnetic flux density	B	1 weber/m² $=$ 1 tesla	=	10^4 gausses

Table 7   Some useful mathematical formulas

arithmetic and algebra

1. $x^a x^b = x^{a+b}$

   Therefore $10^5 \, 10^7 = 10^{12}$

2. $\dfrac{x^a}{x^b} = x^{(a-b)}$

   Therefore $\dfrac{10^6}{10^5} = 10^1$

3. $x^{-a} = \dfrac{1}{x^a}$

   Therefore $10^{-2} = \dfrac{1}{10^2}$

4. $x^0 = 1$

5. $(x^a)^b = x^{ab}$

6. If
$$ax^2 + bx + c = 0$$
   then
$$x = \frac{-b \pm \sqrt{b^2 - 4ac}}{2a}$$

geometry and trigonometry

1. The sum of the interior angles of any triangle is 180°.
2. The sum of the two acute angles of a right triangle is 90°.
3. The corresponding sides of similar triangles are proportional.
4. The corresponding angles of similar triangles are equal.
5. Two angles are equal if their sides are mutually perpendicular.

For a right triangle of hypotenuse c and sides a and b and in which the angle opposite a is A and the angle opposite b is B:

6. $c^2 = a^2 + b^2$ (Pythagorean theorem)

7. $\cosine A = \dfrac{b}{c}$

8. $\sine A = \dfrac{a}{c}$

9. $\tan A = \dfrac{a}{b}$

In a triangle of sides a, b, and c, with angle A opposite side a, and angle B opposite side b, and angle C opposite side c.

10. $c^2 = a^2 + b^2 + 2ab \cos C$ (law of cosines)

11. $\dfrac{a}{\sine A} = \dfrac{b}{\sine B} = \dfrac{c}{\sine C}$ (law of sines)

12. $\sin(x + y) = \sin x \cos y + \cos x \sin y$
13. $\cos(x + y) = \cos x \cos y - \sin x \sin y$
14. $\sin 2x = 2 \sin x \cos x$
15. $\cos 2x = \cos^2 x - \sin^2 x$
16. $\sin x \pm \sin y = 2 \sin \frac{1}{2}(x \pm y) \cos \frac{1}{2}(x \mp y)$

approximations

If x is small, then:

1. $(1 \pm x)^m \approx 1 \pm mx$

   (Note that this approximation may be used for fractional and negative values of m.)

2. $\sin x \approx x$ (if x is expressed in radians).
3. $\tan x \approx x$ (if x is expressed in radians).

the greek alphabet

appendix B

Lower-case letter	Capital letter	Name of letter
α	A	alpha
β	B	beta
γ	Γ	gamma
δ	Δ	delta
ϵ	E	epsilon
ζ	Z	zeta
η	H	eta
θ	Θ	theta
ι	I	iota
κ	K	kappa
λ	Λ	lambda
μ	M	mu
ν	N	nu
ξ	Ξ	xi
o	O	omicron
π	Π	pi
ρ	P	rho
σ, ς	Σ	sigma
τ	T	tau
υ	Υ	upsilon
ϕ	Φ	phi
χ	X	chi
ψ	Ψ	psi
ω	Ω	omega

answers to odd-numbered problems

appendix C

Chapter 1

3. 1 qt $\approx$ 0.908 l
5. —
7. 360 mi, 311.7 mi
9. (a) 218.4 ft; (b) 0.36 ft
11. (a) 900 mi; (b) radius of the earth

Chapter 2

1. (a) 50 mi north; (b) 12 mi southwest; (c) 20.3 mi, 65.3° N of W; (d) 8.6 mi, 4.5° S of W
3. (a) 144.2 m, 93.7°; (b) 42.9 m; 24.8°; (c) 7.9 m, $-54.3°$
5. 50.6 mi, 98 mi
7. 5.17 mi
9. 12.5 ft/sec, 0
11. $a = 0.47$ mi/sec², 7.7 g
13. 5.4 mi/hr, $\theta = \pm 21.8°$
15. (a) 6×10^{17} cm/sec²; (b) 10^{-9} sec
19. 3.13 sec
21. 144 ft
23. 24.2 m/sec, 4.94 sec
25. 84 ft
27. (a) 6 m/sec; (b) 0.49 sec
29. $\sqrt{2}$ times the first at 135°
31. 7° 39′ W of N
33. 132 ft upstream
35. (a) 0.5 sec; (b) 52.5 ft/sec, $\theta = 17.7°$
37. 5 ft
39. 19.6 m, 60 m
41. 9.2 sec
43. (a) 600 cm/sec; (b) 747 cm/sec; (c) 284.7 cm
45. (a) —; (b) 25.5×10^3 m

Chapter 3

3. 520 lb
5. 43.75 lb
7. 2 m/sec^2
9. 17.7 lb
11. 620 cm/sec^2, 98 dynes
13. 16.7 dynes
15. 162.5 lb, 182.5 lb
17. (a) 0.352 ft/sec^2; (b) 46.8 ft/sec; (c) 363.3 ft
19. (a) 5.55 × 10^{-4} sec; (b) 1.17 × 10^8 dyne
21. (a) 4.7 sec; (b) < 143 ft/sec^2
23. 965 lb
25. (a) 7160 nt; (b) the ground; (c) 3560 nt
27. —
29. (a) 189.7 cm/sec^2; (b) 189.7 cm/sec^2; (c) 4.74 × 10^5 dynes
31. —
33. 229 lb
35. (a) 5.2 m/sec^2; (b) 55.4 nt
37. (a) 1.83 sec; (b) 1.3 sec
39. —
41. tan$^{-1}(a/g)$

Chapter 4

1. 371 nt, 642 nt
3. 210 lb
5. 40 lb
7. 5 ft-lb, down at hinges
9. 4600 lb, 4800 lb
11. 56.6 nt
13. 92.3 lb; 61.6 lb parallel and 76.9 lb perpendicular to lineman
15. (a) 210 lb; (b) 210 lb; (c) 207 lb
19. 95.6 lb on wall, 180 lb vertical, 95.6 lb horizontal on ground
21. (a) zero; (b) 30 nt; (c) 58.8 nt; (d) 78.4 nt; (e) 58.8 nt
23. 33.5 lb
25. 4 lb, 0.165 lb, 0.165 lb, 9.85 lb, 9.85 lb
27. 218 lb

Chapter 5

1. 26.7 × 10^{-11} nt
3. 40 nt
5. 1.57 m/sec^2
7. 30,000 m
9. (a) 735 nt; (b) 0.82 nt; (c) 126 nt; (d) 3.5 × 10^8 m from earth
11. (a) 10^{-47} nt; (b) 2.3 × 10^{-8} nt; (c) 2.5 × 10^{22} m/sec^2
13. $F_{near} - F_{far} = 2.3 \times 10^{-9}$ nt; 7 percent
15. (a) 810 nt; (b) 1946 nt

Chapter 6

1. 630 ft-lb
3. 1247 ft-lb
5. 5.36 × 10^{-6} erg
7. 27 × 10^3 ft-lb
9. 10 joules
11. 759.4 ft-lb
13. 5 × 10^2 m
15. 4.9 × 10^8 joules/sec, 4.9 × 10^6 bulbs
17. (a) 2.18 × 10^5 ft-lb; (b) 619 lb
19. 42 cm
21. 1.62 m/sec
23. 2.34 × 10^3 m/sec
25. (a) 12 × 10^3 ft-lb; (b) 24 × 10^3 lb
27. $\frac{4}{3}r_e$
29. 1.67 × 10^4 m/sec
31. 2.2 × 10^8 cm/sec
33. (a) 24.8 ft/sec; (b) 2.53 ft

Chapter 7

1. 26.9 ft/sec^2
3. 14.5 ft/sec^2
5. 24 × 10^3 dynes
7. 0.66 lb
9. 16° 53′
11. 7.74 × 10^{11} m
13. 1.02 × 10^4 sec
15. 0.17
17. (a) 4057 ft; (b) 1200 lb; (c) 0.0784 nt
19. 73.3 lb
21. (a) —; (b) 1.033
23. 1.96 × 10^{30} kgm; 1.36 gm/cm^3
25. (a) zero; (b) 0.42 m/sec^2; (c) 0.42 m/sec^2; (d) 2° 28′
27. (a) $T_1 = \frac{2}{\sqrt{3}}mg$; (b) $\left(2\,gl\left[1 - \frac{\sqrt{3}}{2}\right]\right)^{1/2}$, (c) 1.27 mg
29. (a) 22.1 m/sec; (b) 17.1 m/sec
31. 4.2 × 10^{11} m, 1.78 × 10^4 m/sec, 14.8 × 10^7 sec, 4.22 × 10^4 m/sec

Chapter 8

1. 420 kg m/sec
3. 9900 lb sec
5. 1100 lb
7. 20 m/sec
9. 0.27 ft/sec
11. 2 mi/hr
13. 6.2 kg m/sec
15. 4.9 × 10^{-23} m/sec
17. 65.6 mi/hr, 10.6 mi/hr
19. (a) 5.77 ft/sec; (b) 0.4 ft-lb

21. (a) 0.8 ft/sec; (b) 0.57 ft
23. 3.4 ft/sec, 17° with original direction
29. 1447 lb

Chapter 9

1. at 6
3. $1\frac{7}{8}$ rad/sec
5. (a) 7.3×10^{-5} rad/sec; (b) 3.4×10^{-2} m/sec²;
   (c) 2.4×10^{-2} rad/sec²
7. 0.14 lb sec²-ft
9. (a) 100π rad/sec²; (b) 39.6×10^3 rev
11. 7250 joules
13. 8.63×10^4 erg sec
15. 2.80×10^{34} joule sec
17. 8767 years
19. 11.2 sec
21. 134.6 sec
23. $V = MR\,\omega/m$
25. $\sqrt{3\ gL}$
27. (a) 16.3 rad/sec²; (b) 196 cm/sec²; (c) 392×10^3 dynes
29. (a) 6.19×10^7 erg sec; (b) 1.08×10^8 dyne cm
   (c) 1.74 rad/sec

Chapter 10

1. $0.8, \frac{5}{3}$
3. $\frac{4}{3} \times 10^{-10}$ sec
5. 4.8 ft
7. 13.3 years
9. (a) 4.63×10^3 m; (b) zero; (c) 0.99 c, 0
11. 0.99999999955 c
13. c
15. 6.5 cm
17. -36.5×10^5 km, 75 km, 75 km, 12.8 sec
19. no
21. 0.98 c
25. (a) $\sqrt{1 - v^2/c^2}\, L \cos\theta$, $L \sin\theta$;

   (b) $L\,(1 - (v^2/c^2) \cos^2\theta)^{1/2}$;

   (c) $\tan^{-1}\left(\dfrac{1}{\sqrt{1 - v^2/c^2}} \tan\theta\right)$

27. —
29. —

Chapter 11

1. 12×10^{10} gm cm/sec
3. 1.02×10^{-6} erg, 0.2×10^{-6} erg
5. 0.866 c
7. 0.98 c, 4.6 GeV/c
9. $(1 - 4.9 \times 10^{-4})\ c$, 30.015 GeV/c, 5.33×10^{-26} kg
11. $\$1.5 \times 10^6$

13. 0.44×10^{10} kg/sec, 0.63×10^8 kg/year
15. (a) 0.0717 amu; (b) 0.0474 c

Chapter 12

1. 6.4 in.
3. 245 cm/sec² upward
5. 99.3 cm
7. 32.2 cm/sec
9. 98.7 dynes cm/rad
11. (a) 0.49 sec; (b) 9×10^5 dynes, up; (c) 10^3 cm/sec²
13. (a) 4834 dynes/cm; (b) 53.8 cm/sec; (c) 483 cm/sec²
15. (a) 39.25 cm/sec; (b) 13 cm; (c) 0.2 sec, 1.0 sec
17. (a) 0.34 nt; (b) 1.36 m/sec²; (c) 0.44 m/sec
19. 0.71 joule
21. 70 sec
23. 2.03 sec
25. —
27. $T = 2\pi\sqrt{d/2g}$

Chapter 13

1. 2.25 lb/in.²
3. 49.2 cm³, 54.9 cm³
5. no
7. 4105 dynes
9. 1.33×10^{-3} gm/cm³
11. 2.12×10^6 gal
13. (a) 1.57×10^{-7} cm³; (b) 2.36×10^{-4} erg
15. 10.3 lb/in.²
17. (a) 14 cm³; (b) 4.56 gm/cm³
19. (a) 4.53×10^4 lb/ft²; (b) 9.7×10^6 tons
21. (a) 0.445 ft; (b) 3.07×10^4 lb; (c) zero
23. (a) 8.92 lb/in.²; (b) 567 lb
25. 1.13×10^4 lb
27. 180 lb
29. 8.1 in. from bottom
31. 15.6 lb/ft²
33. 9.4 percent
35. (a) 6.62×10^2 cm/sec; (b) 4.67 liter/sec
37. (a) 27.7 ft/sec; (b) 434 cm³/sec
39. 200 gm, 2.5 gm/cm³
41. 0.014 p/LW
43. —

Chapter 14

1. 1.67×10^{-2}
3. 2.5×10^{-2}
5. 252.4 dynes
7. 1.46×10^{-4} cm radius
9. (a) 4.68×10^9 dynes/cm²; (b) 2.34×10^{-3}; (c) 0.47 cm

11. 1.88 cm²
13. 7.78 × 10⁻⁴ in.
15. 235 m
17. 7.4 cm, 4.94 cm, 2.46 cm
19. −0.575 cm, −0.433 cm, −0.309 cm
21. (a) 208 dynes; (b) 1040 erg
23. 52 dynes/cm²
25. 1.85 cm
27. 41.7 dynes/cm²
29. YA/L

Chapter 15

1. 37° C, 39.4° C
3. −37.97° C
5. −40°
7. 0.206 ft
9. 0.022 cm³
11. 63°
13. 5727° C, 10,341° F
15. 60.02 ft
17. 3.3° C
19. 47.8 × 10⁶ dynes
21. 1.0225 liter
23. 2.69 × 10⁹ dynes/cm²
25. (a) 1.0001 sec; (b) 9.5 sec/day
27. (a) —; (b) $(\beta - 3\alpha) \, l_0 \, \Delta T$

Chapter 16

1. 10.75° C
3. 0.53 cal/gm° C
5. 24,000 cal
7. 34 ft³
9. 1.66 kg
11. 0.096 cal/gm° C
13. 0.11° C
15. 5.625 joules/cal
17. 253 atm
19. 535.8 cal
21. —
23. 1219 cal, 1219 cal
25. 22.2 gm

Chapter 17

1. 21,600 cal
3. 5250 cal
5. 4 × 10⁶ cal
7. 32.4 percent
9. 12.4 mm Hg
11. 38.9° C
13. 59 gm
15. 5.4 min, 31.4 min
17. 45.2° C
19. (a) 188.4 gm; (b) —
21. (a) 169.2 joules; (b) 2091 joules
23. 6.9 × 10⁻⁶ cm radius

Chapter 18

1. (a) 0.107; (b) 3572 cal
3. 0.421
5. 0.373
7. 5.1 cal/K
9. 0.293 cal/K
11. (a) 0.562; (b) 2810 cal; (c) 2190 cal
13. (a) 0.0137 cent; (b) 9510 cal
15. 6.42 × 10⁶ cal/kwh
17. 2.08 cal/K
19. (a) —; (b) 0.64 mole; (c) 5094 cal; (d) 4829 cal;
    (e) 265 joules; (f) 0.052
21. —

Chapter 19

1. (a) 1.44 × 10⁻⁹ cm³; (b) 4 × 10¹⁰ molecules
3. 4.04 × 10⁻¹⁴ erg
5. 3.93 × 10²⁴ erg
7. 5.24 × 10⁻⁷ mole
9. 1.2 × 10⁵ K
11. 1.07
13. (a) 28.96; (b) 1.20 × 10⁻³ gm/cm³
15. —
17. 1270 K
19. 1010 K
21. —
23. —

Chapter 20

1. 52 × 10³ cal/sec
3. 6.11 × 10⁶ erg/sec cm²
5. 0.25 cal/cm sec° C
7. (a) 206 cm; (b) 0.833 mm; (c) 1.13 × 10⁻² mm
9. (a) 2.64 × 10³ BTU/hr; (b) 740 BTU/hr
11. 60° C
13. (a) 34.5 gm/hr; (b) 1.88 × 10⁴ cm²; (c) 3.06 × 10⁻⁴ cm
15. 3.05 × 10⁻⁴ cm/sec

Chapter 21

1. 1.06 m
3. 12 × 10⁴ cm/sec
5. multiply by $\sqrt{2}$
7. —
9. 1.45 m
11. (a) 6 m, 3 m, 2 m; (b) —
13. 10 m, 5 m . . . , 20 m, $\frac{20}{3}$ m . . .
15. 29.8 m
17. 3 ft/sec
19. (a) 16 cm; (b) 16 × 10³ cm/sec; (c) 2.43 × 10⁸ dyne
21. —
23. (a) 0.261; (b) 0.064; (c) 33°; (d) 7° 40′
25. 0.17 m, 0.343 m
27. $L/8, 2L/8, \ldots$

29. 0.075 ft, 0.15 ft
31. 634.8 Hz

Chapter 22

3. (a) 1050 Hz; (b) 0.32 m
5. 108.7 Hz
7. 880 Hz, 1320 Hz
9. 64 Hz
11. 8 beats/sec
13. 3.2×10^4 ft, horizontally
15. 35.4 db, 3.2×10^4
17. 494 Hz, 506 Hz, 12 Hz
19. $b = \dfrac{1}{1 + c/2v} f_0$

Chapter 23

1. 3.6×10^3 nt
3. 8×10^7 nt/coul
5. 0.6 nt
7. (a) 1.8×10^{-8} lines; (b) —
9. 9.04×10^5 nt/coul
11. 0.15 volt
13. (a) -12.8×10^{-19} joule; (b) 3.55×10^{-17} joule
15. (a) 3.5×10^2 volts/m; (b) —
17. 16.7 μcoul
19. 4 μf
21. 53.1×10^{-9} f
23. (a) 3.38 μf; (b) 133 volts
25. 3.17
27. 1.67×10^4 volts/m
29. 5.3×10^8 nt/coul, -1.375×10^7 volts
31. 1.67×10^5 nt/coul
33. 2.05×10^6 m/sec
39. 6.67 μcoul
41. (a) $2.4 - 0.171$ volts; (b) $2.5 - 35$ times original energy; (c) —
43. (a) $Q/2\pi \epsilon_0 a^2$; (b) $2.83 \, Q/4\pi \epsilon_0 a^2$
45. (a) —; (b) 2.81×10^{-15} m
47. $4\pi \epsilon_0 r$
49. —
51. —
53. —

Chapter 24

1. 30 amp
3. 0.768 amp
5. 17.7 amp/cm²
7. 1.02×10^{-3} m
9. 5000 Ω
11. 24 Ω
13. 36×10^4 joules
15. 0.5 Ω
17. 3.7 amp
19. (a) 15 Ω; (b) 3.33 Ω
21. 15 volts
23. (a) 10^{-4} coul/sec; (b) 4×10^{-6} coul/m²

Chapter 24

25. (a) 2.83:1.70:1.63; (b) 1:2:2.24
27. (a) 6 volts/m; (b) 3.53×10^8 amp/m²
29. 1.92 cents
31. (a) 12,000 coul; (b) 1.44×10^5 joules; (c) 4.25×10^{-3} m
33. (a) 6.25×10^{16} electrons/sec; (b) 10^3 watts
35. (a) 0.5 Ω; (b) 10 volts
37. (a) 1.2 amp; (b) 72 volt, 48 volt
39. 0.0849 amp, 4.15 watts, 1.04 watts
41. (a) 4.85 amp, 6.93 amp, 1.04 amp; (b) 12.9 amp; (c) 1082 watts
43. (a) 4 P; (b) —
45. (a) —; (b) 12 Ω, 6 Ω, 2.67 Ω, 1.33 Ω
47. (a) 4.4×10^{-12} V; (b) 4.4×10^{-15} V; (c) 2.5×10^{14} ohm; (d) 2.3×10^{11} ohm m

Chapter 25

1. (a) zero; (b) 3.2×10^{-22} weber/m²; (c) 1.13×10^{-22} weber/m²
3. 12×10^{-6} weber/m²
5. (a) 3.12×10^{-14} nt
7. 2.64×10^6 m/sec
9. 1.6×10^{-5} weber/m²
11. 100 amp
13. 0.3 nt, vertical
15. 3.5×10^{-5} nt
17. 1.70×10^{-4} nt m
19. (a) 1.6 cm; (b) —
21. —
23. 3.99 weber/m²
25. (a) zero; (b) 1.15×10^{-5} weber/m²
27. $B_x = 1.6 \times 10^{-5}$ weber/m², $B_y = 1.2 \times 10^{-5}$ weber/m²
29. —
31. —
33. —

Chapter 26

1. 2.7×10^{-4} volt
3. (a) zero; (b) 5.3 volts
5. 2.55×10^{-2} weber/m²
7. 3.25 volt
9. 10^{-1} henry
11. 0.2 Ω
13. 0.23 joule
15. (a) 0.012 amp; (b) 9.6×10^{-4} nt; (c) 14.4×10^{-7} watt
17. (a) 2.20 volt; (b) 31.4 volts/m
19. 1.18×10^{-3} volt
21. —
23. —
25. (a) 2.99 amp; (b) 0.2 amp/sec
27. —
29. $2k'Iabv/Rc(c + b)$

Chapter 27

1. 162.6 volts
3. 0.147 amp
5. 39.8 Hz
7. 0.276 henry
9. (a) 86.6 Ω; (b) 34.5 millihenry
11. (a) 0.866; (b) 953 watts
13. $7.96 \times 10^7 \ \Omega$
15. $4.1 \times 10^6 \ \Omega$
17. 183 μf
19. 1.43×10^5 henry
21. 2200 volts
23. (a) 0.959 amp; (b) 59° 10′; (c) 68.8 watts
25. (a) 0.368 amp
27. (a) 100 volts, 662.5 volts, 23.55 volts; (b) 647 volts; (c) −81°
29. (a) 18.7 Ω; (b) 18.7 volts, 26.4 volts; (c) 24° 30′; (d) 8 volts, 7.96 volts, 18.1 volts
31. (a) 2160 volts; (b) 1.48 amp; (c) 26.7 amp
33. 28.7 amp

Chapter 28 (no answers)

Chapter 29

1. (a) 22° 5′; (b) 30°; (c) 2.25×10^{10} cm/sec
3. —
5. 11.3 ft
7. 24° 26′
9. (a) $\approx$41 in.; (b) top is halfway between top of head and eyes of taller
11. $-f/(s-f)$
13. 72 cm from lens
15. −9.77 cm
17. 17.3 cm
19. −26.8 cm
21. −275 cm
23. (a) 30° 10′; (b) 39° 19′
25. 1.5
27. (a) 66.7 cm; (b) 8.33 mm; (c) inverted; (d) real
29. (a) −15 cm; (b) 3 cm; (c) virtual, erect
31. (a) 2.99 cm; (b) 233
33. 31.25 cm
35. (a) 2; (b) —
37. (a) −0.242 ft; (b) 2.3 in.; (c) virtual, upright
39. (a) −13.35 cm; (b) −14.4 cm; (c) —
41. —
43. (a) 1.72 cm; (b) 16.8 cm
45. (a) 8.8; (b) 24.5 cm
47. (a) 56.8 mm; (b) 158
49. −30 cm, 2 cm long, virtual
51. −46.7 cm
53. —

Chapter 30

1. 545 m, 187.5 m
3. 2×10^{-7} weber/m²

5. 1.43×10^{-7} volt/m
7. (a) 1.8×10^{-2} joule; (b) 1.2×10^{-10} kg-m/sec; (c) 2×10^{-12} nt
9. 49° 21′
11. 2.65×10^{-12} nt
13. (a) 5 volts/m; (b) zero; (c) 3.54 volts/m
15. —

Chapter 31

1. 1.47 mm
3. 2.58×10^{-4} cm
5. 11.2×10^{-5} cm
7. 5682 lines/cm
9. 6×10^{-5} rad
11. 77.4 percent
13. (a) —; (b) 678 lines
15. —
17. 35
19. 25° 55′, 60° 54′
21. 10.3 mm
23. —

Chapter 32

1. 9.345×10^{-4} cm
3. 2.19×10^6 ergs/sec
5. 3.37×10^{-12} erg
7. 0.96 eV
9. 6565 Å, 4863 Å, 4342 Å, 4103 Å
11. 5.22×10^{18} Hz
13. 2.19×10^8 cm/sec
15. 3.43×10^{-15} erg
17. 4.416×10^{-19} gm cm/sec
19. 4.11×10^{12} cm/sec
21. 6.66×10^{-19} gm cm/sec
23. 2.15×10^{-10} erg, 8.62×10^{-10} erg
25. 2.34×10^{-27} gm cm³/sec
27. 2.11×10^{-27} gm cm²/sec
29. 4
31. —
33. (a) 77; (b) 7
35. 1.52×10^{-16} sec
37. 3.32 eV
39. (a) 8.83×10^{-19} gm cm/sec; (b) 9.69×10^8 cm/sec
41. 0.604 erg $\approx 5 \times 10^7$ times coulomb energy
43. —
45. 5.49×10^{-8} cm
47. —
49. (a) —; (b) 1.19×10^{-14} erg

Chapter 33

1. (a) He; (b) Zr; (c) 15; (d) 7
3. (a) 6, 7, 13; (b) 11, 13, 24; (c) 92, 142, 234
5. (a) 1.9×10^{-13} cm; (b) 2.75×10^{-13} cm; (c) 7.13 $\times 10^{-13}$ cm

7. 2.29×10^{14} gm/cm^3
9. 12 hr

Chapter 33

11. 2.24 MeV
13. 8.78 MeV/nucleon
15. 11.2 MeV
17. —
19. 18.5×10^7 β/sec
21. 11,160 years
23. (a) —; (b) 18.6 keV
25. 3.39×10^{-12} cm
27. (a) 9.725×10^{-6} erg; (b) 11.4×10^{-15} gm cm/sec; (c) 0.187×10^{-6} erg; (d) 9.91×10^{-6} erg
29. (a) 7.31×10^{-13} cm; (b) 2.08×10^{21} collisions/sec; (c) 9.39×10^{-33}
31. (a) —; (b) 5.3 MeV; (c) 5.33×10^{11} atoms
33. (a) —; (b) 4.46 MeV
35. 5.7 MeV
37. (a) 3.13×10^{17} sec^{-1}; (b) 9.88×10^{24} year^{-1}; (c) 3.85 kg; (d) 3.51 gm

39. (a) 2.31×10^{-7} erg; (b) 1.11×10^9 K
41. —
43. —

Chapter 34

1. up into paper
3. (a) 180°; (b) 0°
5. —
7. —
9. —
11. 2.58×10^{-13} m
13. 107.3 MeV
15. —
17. —
19. 4.65×10^{-5} MeV

Chapter 35

1. 5.44×10^{-15} m
3. $(1 - 4.36 \times 10^{-14}) \times 2.5 \times 10^{18}$ Hz
5. 2.12×10^{12} m
7. 1.43×10^{-16}
9. 6.75×10^{-58} m

index

Aberrations, 422
Absolute thermodynamic temperature scale, 257
  zero, 218
Absorption of photon by nucleus, 525
Absorption spectrum, 481
Absorptivity, 283
Abundance of nuclei, 517
Acceleration, 22
  angular, 121
  average, 23
  due to change in direction, 97
  due to change in speed, 97
  due to gravity, 28
  instantaneous, 23
  in uniform circular motion, 99
  radial, 123
  straight line motion with constant, 123
  tangential, 98, 123
  vector representation of angular, 124
Accommodation, range of, 434
Aces, 540
Acoustics, 307
Actinium series, 518
Adhesion, 203
Adiabatic process, 233
Air columns, vibrating, 311
Air speed, 188
Airplane, lift of, 186
Alpha particle, 482, 509
  emmission of, 509, 518
    theory of, 511
  model of nucleus, 517
  speed of, 510
Alternating current, measurement of, 388
Alternating EMF, 381
Amber, 321
Ampere, the, 346, 368
Amplification, voltage, 406
Amplifier, transistor, 410
  vacuum tube, 407

Amplitude of simple harmonic motion, 163
Analyzers of polarized light, 450
Anderson, C. D., 531, 533
Anderson, H. L., 521
Angle of attack, 187
    of contact, 203
    of deviation, 415
    of incidence, 296, 413
    of reflection, 296, 413
    of refraction, 414
    visual, 434
Angstrom, the, 10
Angular displacement, velocity, and acceleration, 121
Angular momentum, conservation of. 534, 540
    of electron, Bohr theory of, 483
    quantum number, 495, 496
Annihilation, 532
Anode, CRT, 411
Antenna, 447, 449
Antimatter, 533
Antineutrino, 533
    spin of, 513
Antineutrons, 532
Antinode, 301
Antiprotons, 532
Aphelion, 110
Apparent depth, 417
Archimedes' principle, 179
Associated production, 538
Atom, 477, 482
Atomic mass unit, 509
    planes in crystals, 470
    transitions, 486
    weight, 230
Atwood machine, 49
Audible range, 308
Audibility, threshold of, 308
Avogadro's hypothesis, 267
    number, 268
    number, determination of, 273

β (relativistic), 142
Back EMF, 358
Balance wheel, 168
Balmer series, 481, 484
Bands, energy, 499
Band spectrum, 481
Banking of a curved road, 101
Barometer, 182
Barrier penetration by alphas, 511
Baryons, 535
    conservation of, 535
    number, 536
Base of transistor, 410
Baseball, curving of, 188
Basov, N., 499
Beats, 312
Becquerel, 509

Bernoulli's theorem, 185
Beta decay, 509, 512, 518
    radiation, 510
Betatron, 452
Bethe, H., 522, 533
Bias, forward, 409
    grid, 407
    reverse, 409
Bimetallic strip, 219
Binding energy, 93
    nuclear, 509
Binding of molecules, 533
Binoculars, prism, 419, 437
Biot and Savart, law of, 361, 383
Blackbody, 283, 478
Black hole, 551
Blackett, 533
Bode's law, 111
Bohr, N., 481, 483, 488, 515
    postulates, 483
    theory of hydrogen atom, 482
Boiling point, 239
Boltzmann, 274, 479
    constant, 268, 479
Bomb, fission and fusion, 524
Bonding, covalent, 196
    ionic, 195
    metallic, 196
    molecular, 196
Boson, 532
Buoyant force, 180
Boyle, Robert, 321
    law of, 182, 232
Brackett, 484
Bragg, 468, 482
    equation, the, 470
Brahe, T., 106
Branching, 518
Breaking strength, 198
Bridge rectifier, 410
Broadcast band, 455
Brown, Robert, 272
Brownian motion, 272
Btu, the, 225
Bubble chamber, 244
Bubble, pressure in, 207
Bulk modulus, 200
    of gas, 202

Caloric, 224
Calorie, the, 225
Capacitance, 336
Capacitor, 336
    displacement current in, 383
    in AC circuits, 393
Capillarity, 202, 205
Carbon, radioactive, 516
Carbon dating, 516

Carbon-nitrogen cycle, 523
Carnot cycle, 254
  efficiency of, 256
  entropy change in, 257
Carnot, Sadi, 255
Cathode, 404
Cathode-ray tube, 411
Cathode rays, 451
Cell, electric, 345
Celsius temperature scale, 215
Center of gravity, 58, 114
  of mass, 114
  of momentum, 114
Centigrade temperature scale, 215
Centrifugal force, 100
Centrifuge, 190
Centripetal force, 99
CGS system, 42
Chain reaction, 520
Charge conjugation, 541
  conservation of, 534
  distribution on circuits, 345
  free, 323
  independence of nuclear force, 513
  of an object, 322
Charging by contact, 323, 335
  by induction, 324
  by rubbing, 322
Circle, reference, 161
Circuit diagram, 352
Circular motion, uniform, 98
Classical physics, 477, 488
Clausius, 257, 277
Clock, effect of gravity on, 547
Closed shells, nuclear, 517
Cloud chamber, 244
Cloud, electronic, 495
Cloudy crystal ball model, 517
Coating of lenses, antireflection, 461
Coefficient of linear expansion, 219
  self-induction, 382
  volume expansion, 220
Coherent light, 499
Coherent sources, 457
Cohesion, 202
Coil, EMF in due to changing flux, 379
Collective model of nucleus, 517
Collector of transistor, 410
Collisions, 114
  elastic, 115
  probability of in gases, 265
Combinations of lenses, 429
  of resistors, 353
Common base amplifier, 410
Common emitter amplifier, 410, 412
Components of a vector, 16
Compound nucleus, 515
Compressibility of gases, 182

Compression, adiabatic, 234
Compressive strain, 199
  stress, 199
Compton effect, the, 526
Concave mirror, 430
Concentration, equilibrium of, 258
Condensation, 237
Conduction of heat, 280
Conductivity, electrical, 323, 324, 348
  ohmic, 348
  thermal, 281
Conductors, static charge distribution on, 335
Conservation laws, 91, 534
  and selection rules, 535
Conservative field, 85, 331
Consonants, 312
Constant acceleration, equations of motion for, 25
Constant-volume gas thermometer, 216
Constructive interference, 302, 457
Contact force, 38
Continuity, equation of, 184
Control rods, 522
Convection, 282
Conventional current, 346
Converging lens, 422, 423
Conversi, 533
Conversion factors, length, 7
Convex mirror, 430
Coolidge tube, 452
Cooling, Newton's law of, 286
CPT theorem, the, 541
Copernicus, N., 106
Cornea, 434
Correspondence principle, 488
Coulomb, 321
  balance, 73
  law of, 72, 324, 383
  value of constant in law of, 74
Covalent bonding, 196
Critical angle, 418
  point, pressure, temperature, 241
Crystals, 195, 450, 470
  diffraction of X rays by, 468
  grating space, 470
  impure, 196, 408
Curie, the, 519
Current density, 348
  caused by changing magnetic fields, 378
  electric, 346
  total, 383
Curved path, motion along, 97
Cutoff frequency, photoelectric, 480

D lines of sodium, 504
Davisson, C., 489
de Broglie, L., 489
  velocity of waves of, 490
Debye, P., 471

Decibel, 308
Decouplet, 539
Deflection in CRT, 411
Degenerate levels, 539
Degrees of freedom, 268
Demagnetization, 371
Density, 174
Descartes, 414
Destructive interference, 303, 457
Deviation, angle of minimum, 416
Dew point, 247
Dewar flask, 285
Diamagnetic materials, 369
Diatomic molecule, energy of, 506
Dichroism, 450
Dicke, 550
Dielectric, 338
Dielectric constant, 338
Diffraction of light, 462
  of particles, 492
  of waves, 299
  of X rays, 468
  of X rays by powders, 472
Diffraction grating, 300, 465
Diffusion, 259
  of gases, 271
Diode, solid state, 408
  vacuum tube, 404
Diopter, the, 435
Dirac, P.A.M., 493, 531
Dispersion, 294, 415
Displacement, angular, 121
  current, 383
  of a particle, 13
  of ships, 180
Dissipative forces, 91
Distance, 13
Distribution function, 274
Diverging lens, 422, 426
Domains, magnetic, 370
Doppler effect, 313
Double refraction, 450
Drift velocity, 347
Dynamics, relativistic, 152
Dyne, the, 42

Earth, distance from sun to, 10
  mass of, 10, 78
  radius of, 78
Eddy currents, 385
Edison, 403
Effective value of AC, 388, 390
Efficiency, thermodynamic, 253
Einstein, A., 137, 138, 273, 480, 543, 545
Elastic limit, 198
Elastic scattering, nuclear, 515

Elasticity, 196
  energy of, 89
  modulus of, 197
Electric charge, 72
Electric current, force on, 366
Electric field due to charge distribution, 329
  due to changing magnetic field, 378, 379
  due to infinite plane of charge, 330
  in electromagnetic wave, 448
  intensity, 325
Electromagnetic induction, 377
Electromagnetic waves, 384, 445
Electromagnetism, 72
Electron, 322
  beam, 411
  capture, 509
  charge of, 74
  gas, 346, 404
  mass of, 10, 74
  microscope, 465
Electron volt, the, 156
Electrons, assignment to atoms, 497
Elsasser, 517
EMF, 350
  alternating, 380
  induced, sense of, 379
  motional, 377
Emission of photon by nucleus, 524
  spectrum, 481
  thermionic, 403
Emissivity, 284
Emitter of transistor, 410
Emmetropic eye, 434
End-point energy of beta spectrum, 512
Energy, 80
  average per degree of freedom, 269
  bands in solids, 499
  conservation of, 91, 534, 540
    in collisions, 115
    mechanical, 85
  dissipation of, 349
  distribution of, 274
  equipartition of, 268
  gravitational potential, 84, 86
    at large altitudes, 92
  indeterminacy of, 493
  in electric field, 338
  in magnetic field, 393
  in simple harmonic oscillators, 169
  internal, 228
  kinetic, 83
    level diagram, 486, 496
    levels of compound nucleus, 515
    of Bohr orbits, 484
    of charged capacitor, 337
    of electromagnetic waves, 448

Energy (continued)
    of liquid surface, 204
    of moving fluids, 186
    of particles, 86
    of photoelectrons, 480
    of rotation, 125
    of rotation of rigid body, 129
    of springs, 88
    of systems, 87
    relativistic, 154, 156
    rest mass, 155
    solar, 158
    stored in magnetic field of coil, 383
    total of planetary atom, 483
Energy transfer, electric, 344
    by magnetic field, 377
    maximum to strings, 305
Engine, gasoline, 251
    steam, 252
Entropy, 257
    and probability, 276
    and time, 260
    change in processes, 258
Equation of state, 229
Equilibrium, dynamic, 52
    of rigid body, 56
    state, 228
    states, 275, 276
    static, 51
    thermal, 214
    the drive to, 258
Equipartition, 268
Equipotential surface, 333
Equivalence, principle of, 72, 544
Equivalent resistance, parallel, 354
    series, 353
Erg, the, 82
Escape velocity, 93
Eta particle, 539
Ether, the, 139
Evaporation, 241
Exchange of photons, electrons, and mesons, 533
Excited states, 508
    nuclear, 524
Expansion of liquids, thermal, 220
    of solids, 218
    of the universe, 550

f stop, 440
Fahrenheit temperature scale, 215
Fairbanks, 550
Far point, 434
Farad, 336
Faraday, M., 360, 377
Faraday's law, 378, 383
Farsightedness, 434

Fermi, E., 513, 521, 533, 538
Fermi theory, 513
Fermion, 532
Ferromagnetic materials, 369
Fiber optics, 419
Field, electric, 325
    gravitational, 87
    magnetic, 361
Filament, 403
First law, Newton's, 36, 38
    of thermodynamics, 227
First order image, 466
Fission, 509, 516, 519
Fizeau, 137
Floating bodies, 180
Flow of fluids, 184
    of heat, 224
Fluorescent lamp, 385
Fluids, 173
Focal length, 423
    determination of, 424
    of mirror, 431
    of thin lens, 427
Focal plane, 423
Focus, 297
    of lens, 422
    of mirror, 431
Foot, the, 6
Force, centrifugal, 100
    centripetal, 99
    concurrent, 52
    contact, 38
    definition of, 37
    external, 121
    friction, 38
    fundamental, 68
    impulsive, 113
    internal, 121
    normal, 38
Forced convection, 283
Fourier, J. B., 294
Fourier analysis, 293
Frame of reference, 12, 140
    inertial, 39
Franklin, B., 321
Free-body diagram, 43
Free fall, 27
Frequency of simple harmonic motion, 163
    of waves, 293
Fresnel, 462
Friction, 38, 61, 62
    fluid, 189
Front-surface mirror, 430
Fundamental frequency, 294
    mode of vibration, 311
    particles, 538

Fusion, 242
  heat of, 237, 243
  nuclear, 522

γ, heat capacity ratio, 233
  relativistic, 142
G, value of, 70
g, variation with altitude, 70
Galilean telescope, 438
  transformation, 142
Galileo, G., 27, 166
Galvanometer, 378
Gas-vapor distinction, 241
Gases, kinetic theory of, 264
Gasoline engine, 251
Gauss law, 329
Geiger, 510
Gell-Mann, M., 537, 539
Generators, 380, 398
Geometry, effect of gravity on, 548
Germer, L. H., 489
Gibbs, J. W., 274
Gilbert, W., 321, 360
Goudsmit, 496
Graham's law, 272
Gravitational field, 544
  mass, 71, 543
  radius, 551
  waves, 550
Gravity, acceleration due to, 28
  Newton's law of, 69, 107
Grid, vacuum tube, 406
  CRT, 411
Ground state, 486
  nuclear, 524
Group, electron, 498
Gyroscope, 131

Hafele and Keating experiment, 146
Hahn, 519
Hairspring, 168
Half-life, 145, 511
Half lives of isotopes in uranium series, 518
Harmonic, seventh, 302
Heat, 224
  conduction, entropy change in, 258
  flow, 229
    methods of, 280
  of fusion, 237, 243
  of sublimation, 237
  of vaporization, 237, 240
  pump, 262
  rate of development in resistor, 389
Heisenberg, equation of, 493
  uncertainty principle, 491
Helium, formation in fusion, 522
Helium-neon laser, 502
Henry, J., 377

Hertz, H., 446
Hertz, the, 293
Holes, in semiconductors, 408
Hooke's law, 197
Horsepower, the, 93
Hubble, 550
Hull, A. W., 471
Human body, heat transfer from, 285
  voice, 312
Humidity, 246
Huygens, C., 166, 295
Huygens' principle, 295
Hydraulic brakes, 178
Hydraulic press, 178
Hydrogen atom, bomb, 524
  modern description of, 494
  spectrum of, 481
Hygrometer, 246
Hypercharge quantum number, 538
Hyperopia, 434
Hysteresis, 370

Ice, phases of, 246
Ice point, the, 215
Ideal gas, work done by, 231
Ideal gas law, 229
Image, 297
  real, 421
  virtual, 421
Impedance, RC, 395
  RL, 391
  RLC, 396
Impulse, 112
Indeterminancy principle, 491
Index of refraction, 299, 415
Inductance, 382
  in AC circuit, 390
Inertia, 39
  demonstration of, 304
  electrical, 385
  moment of, 124
    rigid bodies, 129
Inertial coordinate system, 40, 544, 545
  mass, 71, 543
Infrared radiation, 446
Instantaneous values in AC, 389
Insulation, heat, 285
Insulators, electrical, 324, 499
Integrated circuit, 411
Intensity of a wave, 307
Interference, 302
  of light from two sources, 457
  of particle waves, 489, 491
  of X rays, 468
Internal conversion, 509
Internal energy, 228
  of gases, 269

Internal resistance, 351
Invariance, 539, 540
Ion, 322
Ionic bonding, 195
Ionization chamber, 453
    energy, 483, 486
Iron, magnetism of, 370
Irreversibility, 254
Isobar, 508
Isothermal process, 232
Isotope, 508
    radioactive, 518
Isotopic spin, 514, 536

Joliot-Curie, M. and Mme., 519
Joule, James Prescott, 226
Joule, the, 82
Jupiter, eclipses of, 137
Jurin's law, 206

k, value of, 74
k', 361
K series, X ray, 482
Kaon, 536
Kelvin, Lord, 255
Kelvin temperature scale, 216, 218
Kepler, J., laws of, 106
Kerst, D. W., 452
Kilocalorie, 225
Kinetic theory of gases, 264

L-series, 482
Lag-lead, capacitor, 394
    inductor, 390
    RLC, 396
Lambda particle, 536
Laser, 458, 499
Laue, 468
Lee, T. D., 513, 541
Length, concept of, 5
    contraction, 144
    standard of, 5
Lens, 422
    equation, 424
    human, 434
    image formation, by converging, 423
        by diverging, 426
Lenses in contact, 429
Lensmakers' equation, 427
Lenz's law, 379
Leptons, 535
Lifetime, 493, 539
Light, 384
    bending of by gravity, 545, 546, 549
    pipes, 419
    speed of, 136, 138
    speed measurement of, 137
Line of action of a vector, 14

Line spectrum, 481
Linear expansion, coefficient of, 219
Lines of force, 325
    and equipotentials, 333
    convention for density, 326
    electric, properties of, 327
Liquid drop model of nucleus, 516
Lodge, O., 447
Longitudinal waves, 292
Loop, rotating, EMF in, 380
Lorentz contraction, 145
    transformation, 140, 143
Loudness, 308
Loupe, 441
Lyman series, 484

Mach, E., 545
Mach number, 316
Macrophysics, 213
Macrostate, 275
Magic numbers, 517
Magnetic field, 368
    due to current, 364
    due to magnets, 368
    energy stored in, 382
    in electromagnetic wave, 448
Magnetic moment, 369
Magnetic orbital quantum number, 496
Magnetism, 75
Magnetostriction, 369
Magnets, 360, 368
Magnification, 425
    angular, 435
    of microscope, 439
    of telescope, 436
Magnifying glass, 434
Maiman, T. H., 499
Marsden, 510
Marshak, R. E., 533
Maser, 499
Mass, concept of, 8
    defect, 509
    number, 508
    of nucleus, 509
    operational definition of, 40
    relativistic, 154, 156, 157
    spectrometer, 364
    standard of, 8
    unit of, 8
Maxwell, J. C., 136, 274, 361, 383, 414, 446
Mayer, M. M., 517
Measurement, limits on, 492
Mechanical equivalent of heat, 226
Medium, propagation of waves in, 291
Melting, 237, 242
Membrane, semipermeable, 259
Mendel, Gregor, 261
Mesic atom, 533

Meson, 533
Metastable state, 501
Meter, the, 5
Michelson, A. A., 137, 139
Michelson and Morley experiment, 139
Microscope, compound, 438
Microstate, 275
Microwaves, 446
Melting point, dependence on pressure, 244
Metallic bonding, 196
Microcrystal, 471
Microphysics, 213
Mirror equation, the, 432
  image formation, by convex, 433
    by concave, 431
  image interchange, 541
  nuclei, 514
  nuclide, 508
  plane, image in, 421
  spherical, 430
Mixtures, method of, 225
MKS system, 42
Moderator, 521
Molar heat capacity, 270
  volume, 267
Mole, 230
Molecules, 267
  bonding, 196
  in kinetic theory, 264
Moment of force, 53
  of inertia, 124
    measurement of, 169
Momentum, angular, conservation of, 127, 534, 540
  angular of particle, 125
  angular of system, 127
  indeterminacy of, 492
  linear, of charged particles, 364
    conservation of, 114, 534, 540
    of particle, 112
    of system, 113
  of electromagnetic waves, 449
  relativistic, 151, 153
Monopole, 368
Moon, distance from earth, 78
  mass of, 78, 95
  radius of, 78, 96
Moseley, 482
Mössbauer effect, 526, 548
Motion along a curved path, 97
  in a vertical circle, 103
  of a rigid body, 120
  relationship between linear and angular, 122
  simple harmonic, 159
Motional EMF, 376
M series, 482
Multiplets of particles, 539
Multiplication factor, 520
Muon, properties of, 148, 533

Musical instruments, 302, 312
Myopia, 434

N series, 482
n-type crystals, 409
Natural convection, 283
Near point, 434
Nearsightedness, 434
Neddermeyer, 533
Ne'eman, 539
Neutrino, 513, 533
  and parity, 541
Neutron, 322
  capture, 521
  decay, 513
  stars, 551
Newton, I., 137, 545
  first law of, 38
  law of cooling, 286
  law of universal gravitation, 69
  laws of, 36
  rings, 460, 474
  second law, 40
    relativistic, 157
  third law, 46
Newton, the, 42
Nishijima, 537
Node, 300
Nonelectrostatic force, 351, 377
Nonequilibrium states, 244
Normal, 296
Normal boiling point, 239
Normal eye, 434
Normal force, location of, 63
Nuclear forces, 77
  characteristics of, 514
  magnetic resonance, 132
  models, 516
  potential, 511, 517
  radius, 508
  reactions, 514
  spin, 517
  transitions, 509
Nucleon, 508, 514
  conservation of, 535
Nucleon-nucleon interaction, 513
Nucleus, 507
  number of particles in, 508
Nuclide, 508

Occhialini, 533
Octet, 540
Oersted, H. C., 360
Ohm, the, 349
Ohm's law, 348, 349, 390, 391
Oil film, colors on, 459
Omega minus particle, 540
Operational definition, 5

Optical center, 423
Optical model of nucleus, 517
Orbit, radius of first Bohr, 484
Orbits, Bohr, quantum mechanical, 495
  electron, 489
  permissible, 484
Order and entropy, 276
Organ pipes, 311
Orifice, velocity of fluid from, 188
Oscillator, 407
Overtones, 294, 311

p-type crystals, 409
Packet, wave, 490
Pair production, 532
Pancini, 533
Parallel, resistors in, 354
Paramagnetic materials, 369
Paraxial rays, 427
Parity, conservation of, 541
Particle in a box, 494
Particle-antiparticle pairs, 532
Pascal's principle, 178
Paschen series, 484
Path difference and interference, 303
Path, thermal, 228
Pauli, W., 512
Pauli exclusion principle, 497
Pendulum, physical, 168
  seconds, 170
  simple, 166
    period of, 162, 167, 169, 171
Perihelion, 110
  of Mercury, 550
Period, in circular motion, 105
  of planets, 106
  of simple harmonic motion, 161, 162
  of waves, 293
Perkins, 533
Permittivity, 339
  of free space, 324
Perpetual motion, 254
Perrin, 273
Perspiration, 285
Pevsner, 539
Pfund series, 484
Phase angle, 392
  RC, 395
  RLC, 396
Phase change on reflection, 460
Phase, in simple harmonic motion, 165
Phases, of a substance, 237
  of matter, 173
Photoelectric effect, 480
Photoelectric equation, Einstein's, 480
Photon, 477, 479, 483, 489
  emission and absorption, 524
  scattering, 527

Piccioni, 533
Pinhole, diffraction from, 462
Pion, 533
  properties of, 534
Pitch of sound, 309
Planck, Max, 479
Planck's constant, 479
Planetary model of atom, 482
Planets, motion of, 106
Plasma, 524
Plate of vacuum tube, 403
Plutonium, 521
Pohl, 460
Point charge, electric field due to, 326
Polarization, 449
Polarized electromagnetic waves, 448
  waves, 292
Polarizers, 450
Pole strength, 368
Poles, magnetic, 360, 368
Population inversion, 500
Position of an object, 13
  indeterminancy of, 493
Positron, 509, 579
  discovery of, 531
Potential barrier, 511
Potential difference, 331
Potential due to point charge, 332
  electric, 331, 332
Potential energy in circuits, 345
Pound, 548
Pound mass, 8
Powder diffraction, 471
Powell, 533
Power, 93
  factor, 393
  in AC circuit, 392
  of lens, 435
Precession, 130
  of the earth, 135
  of planetary perihelion, 550
Presbyopia, 434
Pressure, 175
  and curved surfaces, 206
  due to weight of liquid, 175
  equilibrium of, 258
  in confined liquid, 178
  in kinetic theory, 265
  in moving fluid, 186
  of the atmosphere, 182
  units of, 175, 177
Principal axis, 423
  of mirror, 430
Principal focus, 423
Principal quantum number, 483, 494
Prism, dispersion in, 415
  totally reflecting, 419
Probability and diffraction, 491

Probability and diffraction (continued)
 and thermodynamics, 275
Probability distribution in hydrogen atom, 495
Projectile motion, 29
 altitude of, 31
 range of, 30
Prokhorov, A. M., 499
Propagation of electromagnetic waves, 448
Propagation of waves, direction of, 291
Proton, 322
 charge of, 76
 mass of, 76
Proton-proton chain, 522
Ptolemy, C., 106
Pulsar, 548, 551
Pumping in laser action, 501

Q value, 515, 525
Quality of musical sound, 310
Quantization, 488
 of energy, 269
 of orbits, 483
Quantum jump, 484
Quantum mechanics, 477
Quantum number, 483, 495, 496, 497
 of particles, 535
Quantum of radiation, 479
Quark, 539
 model, 540
Quasars, 551

R, value of, 230
Rad, the, 528
Radar, 317
Radian, the, 121
Radiation, 283
 from accelerated charges, 447
 fields in, 447
 pressure, 449
 temperature dependence of, 478
Radiationless transition, 501
Radio, 446
Radioactive decay, 511
 series, 518
Radioactive tracers, 516
Radioactivity, 509
 beta emission, 512
 induced, 519
Radius of nucleus, 508
Radius of orbit of particle in magnetic field, 364
Range of nuclear force, 513
Range of projectile, 30
Ray, 296
Ray tracing, 414, 424
Reactance, capacitive, 394
 inductive, 391
Reaction energy, 515
Reactor, nuclear, 520

Reaumur scale of temperature, 222
Rebka, 548
Reciprocal wavelength, 486
Recoil, 115
Rectification, full wave, 405, 409
 half-wave, 405, 409
Rectifier, solid state, 409
 tube, 404
Red shift, 150
 gravitational, 146, 548
 of the galaxies, 550
Reflection, of a wave, 296
 grating, 465
 laws of, 413
 phase change in, 460
 polarization by, 451
Refraction of a wave, 296, 298
Refrigerator, 254, 262
Regelation, 244
Relative humidity, 246
Relative velocity, 20
Relativity, general theory of, 543, 545
 special theory of, 138
  postulates, 138
Relay, 372
Replica grating, 465
Resistance, 348
Resistivity, 348
Resolving power, 464
 of telescope, 474
Resonance, 309, 447
 AC, 397
 reactions, 516
Resonances, particle, 538
Resonant cavity, 500
Resultant, 15
Retentivity, 371
Retina, 434
Reversible process, 254, 255
Reversibility, principle of, 414
Right-hand rule, 361
Rigid body, as a collection of particles, 128
 definition of, 120
Ripple tank, 300
Rise time, 382, 387
Rockets, 116
 thrust of, 117
Roemer, O., 137
Roentgen, W. C., 451
Root mean square value of current, 389
Rotation, 121
 average energy of molecular, 269
Rotational invariance, 540
Rowland, H. A., 364
Ruby laser, 499
Rumford, Count, 225
Rutherford, E., 482, 510
Rydberg constant, the, 484, 487

S state, 497
Satellites, artificial, 108
Saturated air, 242
Saturated vapor pressure, 242
  of water, 246
Saturation current, 404
Saturation of forces, 70
  magnetic, 370
  of nuclear force, 513
Scalar quantities, 14
Scattering of alpha particles, 510
Scattering reactions, 515
Scherrer, P., 471
Schrödinger equation, 493
Second law of Newton, 36, 40
  of thermodynamics, 254, 258
Second-order image, 466
Selection rules, 497
  and conservation laws, 535
Self-induction, 381
Semiconductors, 324, 408, 409
Sense of a vector, 14
Separation factor in diffusion, 272
Series limit, Balmer, 481
  Lyman, 486
Series, radioactive, 518
  resistors in, 353
  spectral, 484
  X ray, 482
Shadow, geometric, 462
Shapiro, 548
Shawlow, A. L., 499
Shell, atomic, 498
  nuclear, 517
Shock wave, 315
Sigma particle, 536
Simple harmonic motion, angular, 168
  equations of motion, 164
Simultaneity, 143
Siphon, 189
Skylight, polarization of, 449
Slit, diffraction of waves through, 463
  diffraction of particles through, 491
Slug, the, 42
Snell's law, 414
Soap film, colors on, 459
Sodium, radioactive, 516
Solar radiation, 288
Solenoid, 372
Solids, 499
Sound, 307
  level of, 308
Space charge, 406
Space, curvature of, 549
  geometry of, 548
Spark gap, 446
Sparks, 382
Specific gravity, 181

Specific heat, 225
  at constant pressure, 232
  at constant volume, 232
  of a gas, 232, 270
Spectrometer, diffraction grating, 466
Spectrum, 415, 467
  beta energy, 512
  electromagnetic, 445
  of gases, 481
  of thermal radiation, 478
  X ray, 481
Speed, average, 21
  instantaneous, 22
  of body in uniform motion, 19
  of light, 136, 138
  of sound, 294
Spin angular momentum quantum number, 496
Spin dependence of nuclear force, 514
Spin of electrons, 370
  of nucleus, 509
  of neutrino, 513
  of particles, 534
Spin quantum number, 536
Spring, energy of and force exerted by, 89
  simple harmonic motion of, 159
Stable particles, 538
Standing waves, 300
  in cavity radiation, 501
Stars, bending of light from, 546
  fusion in, 522
  temperature of, 479
States of electrons in atoms, 497
Stationary orbits, 483
Stationary state, 493
Statistical physics, 274
Statistical mechanics, 213
Statistical model of nucleus, 516
Steam engine, 252
Steam point, the, 215
Stefan-Boltzmann law, 284, 479
Stimulated emission of radiation, 500
Stopping potential, 505
Straight line, definition of, 549
Strain, 196
  compressive, 199
  tensile, 197
Strange particles, 536
Strangeness, conservation of, 537
  quantum number, 537
Strassmann, 519
Streamline, 184, 186
  in air, 187
Stress, 196
  compressive, 199
  tensile, 197
String instruments, 311
SU(3), 539
Sublimation, 245

Sublimation (continued)
  heat of, 237
Sun, mass of the, 79
  radius of the, 79
Sunlight, intensity of, 456
Supercooled liquids, 244
Superposition, principle of, 69, 300, 513
Supersaturated vapor, 244
Surface tension, 203
Symmetry, 539
  and conservation laws, 540
  of electric and magnetic fields, 383
Sympathetic vibration, 309
Synchronization of clocks, 143

Tachometer, 384
Telescope, astronomical, 436
  Galilean, 438
  terrestrial, 437
Teller, 533
Temperature, 214
  determination of for blackbody, 479
  gradient, 281
  of ideal gas, 267
  regulation of human body, 286
  scale, absolute, 256
  scales, 215
Tensile strain and stress, 197
Tension, 197
Tension in ropes, 45
  of liquid surface, 204
Terminal voltage, 351
Tesla, the, 361
Test charge, 325
Test particle, 88
Thales of Miletus, 321
Thermal conductivity, 281
  expansion of liquids, 220
  expansion of solids, 218
  equilibrium, 258
  neutrons, 521
  radiation, 478
Thermocouple, 214
Thermodynamics, 213
  first law of, 214, 227
  second law of, 254, 258
Thermometer, 214
Thermometric property, 214
Thermonuclear fusion, 524
Third law, Newton's, 36, 46
Thompson, G. P., 489
Thorium series, 518
Threshold of audibility, 308
  of feeling, 308
Time and entropy, 276
  dilation, 145
  direction of, 260
  reversal, 541

Time and entropy (continued)
  standard of, 7
Torque, 53, 124
  on magnet, 369
Torricelli's theorem, 189
Torricelli vacuum, 238
Torsion constant, 168
Total internal reflection, 418
Townes, C. H., 499
Tracers, 516
Transformation, Galilean, 142
  Lorentz, 143
Transformer, 398
Transformer coupled amplifier, 407
Translation invariance, 540
Transistors, 409
Transition, atomic, 486
Transmission grating, 465
Transmission line, 398
Transparency, 511
Transuranic elements, 521
Transverse waves, 291
  electromagnetic, 448
Triode, 406
Triple point, 216, 245
Triple point cell, 216
Tuned circuit, 447
Twin paradox, 146
Two-slit interference, 458

Uhlenbeck, 496
Ultraviolet catastrophe, 479
Ultraviolet radiation, 446
Unbalanced force, 37
Uncertainty principle, 491
Units of length, 6
Units of mass, 8
Units, systems of, 41
Uranium, fission of, 521
Uranium pile, 521
Uranium series, 518

V particles, 536
Vapor-gas distinction, 241
Vapor pressure, 239
Vapor, saturated, 239
Vaporization, 237
  curve, 240
  heat of, 240
Vector, addition, 14
  analytic, 16
  parallelogram, 16
  polygon, 15
  cross product, 55
  definition, 14
  diagram AC, 391, 395, 396
  resolution of, 16
  scalar product, 82

Velocity, angular, 121
  average, 22
  instantaneous, 22
  of body in uniform motion, 19
  of gas molecules, 267, 271
  of particle in simple harmonic motion, 163
  relative, 20
  of transverse waves in wire, 294
Velocity of waves, 293
  relative, classical, 141
    relativistic, 147
  selector, 364, 374
  vector representation of angular, 124
Venturi tube, 186
Vertical circle, motion in, 103
Violin, 312
Viscosity, 189
Visible light, 446
Visual angle, 434
Vocal cords, 312
Volt, the, 332
Volume expansion, coefficient of, 220
von Guericke, O., 321
von Mayer, Julius Robert, 226
Vowels, 312

Watson, J. D., 473
Watt, the, 93
Wave, motion, 291
Wave packet, 490, 492
Wavefronts, 295
Wavelength, 292
  determination of X ray, 470
  of light, measurement of, 468
  of particles, 489
Waves and particles, 477
  de Broglie, 489
  probability, 491

Weak nuclear force, 513
Weber, 550
Weber per square meter, the, 361
Wedge shaped film, interference in, 460
Weight, 38
Weight and mass, distinction between, 42
Weight density, 174
Weisskopf, 533
Wein's blackbody formula, 505
Wien's displacement law, 479
Wilson, 533
Work, 81
Work-energy equation, 85
Work function, 480
Work, sign of in thermodynamics, 229
Working substance, 252, 255

X rays, 446
  diffraction and interference of, 468
  discovery of, 451
  properties of, 453
  spectrum of, 481
X-ray tubes, 451
Xi particle, 536

Yang, C. N., 513, 541
Yard, the, 6
Year, the, 7
  tropical, 7
Young, T., 460
Young's modulus, 197
Yukawa, H., 533

Zeroth law of thermodynamics, 214
Zinn, W. H., 521
Zweig, 540